International Conference on Applications and Design in Mechanical Engineering (ICADME 2017)

Journal of Physics: Conference Series Volume 908

Penang, Malaysia
21 – 22 August 2017

ISBN: 978-1-5108-5063-7
ISSN: 1742-6588

Printed from e-media with permission by:

Curran Associates, Inc.
57 Morehouse Lane
Red Hook, NY 12571

Some format issues inherent in the e-media version may also appear in this print version.

Printed by Curran Associates, Inc. (2017)

For permission requests, please contact the Institute of Physics
at the address below.

Institute of Physics
Dirac House, Temple Back
Bristol BS1 6BE UK

Phone: 44 1 17 929 7481
Fax: 44 1 17 920 0979

techtracking@iop.org

Additional copies of this publication are available from:

Curran Associates, Inc.
57 Morehouse Lane
Red Hook, NY 12571 USA
Phone: 845-758-0400
Fax: 845-758-2633
Email: curran@proceedings.com
Web: www.proceedings.com

Table of contents

Volume 908

International Conference on Applications and Design in Mechanical Engineering (ICADME 2017)

21–22 August 2017, Penang, Malaysia

Accepted papers received: 9 October 2017
Published online: 29 October 2017

Preface

Papers

Applied Mechanics

Design and Manufacturing

Thermo, Energy and Fluid

IOP Conf. Series: Journal of Physics: Conf. Series **908** (2017) 011001 doi:10.1088/1742-6596/908/1/011001

Preface

The International Conference on Applications and Design in Mechanical Engineering 2017 (ICADME 2017) was held on 21-22 August 2017 in Penang, one of the most popular tourist destinations in Malaysia. ICADME 2017 was organized by Mechanical Engineering Programme in School of Mechatronic Engineering, Universiti Malaysia Perlis.

The ICADME 2017 aims to provide a platform for participants from diverse field of Mechanical Engineering backgrounds to exchange ideas and findings to promote cooperation, networking as well as knowledge sharing in both engineering practice and research in higher education. The conference includes the fields of Applied Mechanics, Themofluids and Energy as well as Design and Manufacturing. The conference received more than 100 papers from Malaysia and other countries over the world. After a rigorous peer-review process, 84 papers were accepted. The accepted papers that focused on theory, applications, design and experimental studies relevant to mechanical engineering field were invited to present in this conference.

I wish to record my gratitude to all reviewers, both local and international, for their time and effort in reviewing the papers. Their feedbacks to the authors have been excellent and useful to ensure the high quality of papers for the conferences. My sincere thanks also go to all sponsors for their generosity. Last but not least, I would like to thank also the organizing committee members for their hard work, cooperation and suggestions in ensuring the success of this conference.

Ir. Dr. Khairul Salleh Basaruddin
Chairman
ICADME 2017

ICADME 2017
BAYVIEW HOTEL
Organized By:

ICADME 2017
BAYVIEW
HOTEL

IOP Conf. Series: Journal of Physics: Conf. Series **908** (2017) 011002 doi:10.1088/1742-6596/908/1/011002

ICADME 2017

ICADME 2017

IOP Conf. Series: Journal of Physics: Conf. Series 908 (2017) 011002 doi:10.1088/1742-6596/908/1/011002

IOP Conf. Series: Journal of Physics: Conf. Series **908** (2017) 011002 doi:10.1088/1742-6596/908/1/011002

IOP Conf. Series: Journal of Physics: Conf. Series **908** (2017) 011003 doi:10.1088/1742-6596/908/1/011003

ORGANIZING COMMITTEE FOR ICADME 2017

Chair
Ir. Dr. Khairul Salleh Basaruddin

Co-Chair
Dr. Mohd. Afendi Bin Rojan

Secretary
Dr. Mohd. Sani Mohamad Hashim
Dr.Ishak Ibrahim

Technical/ Publication
Dr. Nur Saifullah Kamarruddin
Dr. Tan Wei Hong
Dr. Shah Fenner Khan Mohamad Khan

Bursary
Dr. Liyana Tajul Lile

Arrangement.
Mohd Hafif Basha Mohamad Jamel Basha

Secretariat.
Dr. Mohd Ridzuan Mohd Jamir

Advisory Committees

Prof. Dr. Siva Prasad, India
Prof. Kenneth Dalgarno, U.K.
Prof. Dr. Mohd Nasir Tamin, UTM
Prof. Dr. Mohd Zulkifly Abdullah, USM
Prof. Dr. Mohd Hamdi Abd Shukor, UM
Assoc. Prof. Dr. Belal Yousif, Australia
Assoc. Prof. Dr. Ming Wang Fu, Hong Kong
Assoc. Prof. Dr. Nor Azwadi Che Sidik, UTM
Assoc. Prof. Dr. Mahadzir Ishak@Muhammad, UMP
Assoc. Prof. Dr. Mohd Fadzli Abdollah, UTeM
Assoc. Prof. Dr. Md Saidin Wahab, UTHM
Assoc. Prof. Dr. Ahmad Yasir Md Said, UniKL
Assoc. Prof. Dr. Mohamed Abd. Rahman, UIA
Assoc. Prof. Dr. Shaharin A. Sulaiman, UTP
Assoc. Prof. Ir. Dr. Mohamed Thariq Hameed Sultan, UPM
Assoc. Prof. Ir. Dr. Mohd Shukry Abdul Majid, UniMAP
Dr. Ir. Haftirman, Indonesia
Ir. Dr. Ruslizam Daud, UniMAP
Ir. Dr. Nasrul Amri Mohd Amin, UniMAP

IOP Conf. Series: Journal of Physics: Conf. Series **908** (2017) 011004 doi:10.1088/1742-6596/908/1/011004

No.	Session	Registration Name	Authors	Title	1st Day Attend	2nd Day Attend
1	A1	Dr Nik Mizamzul Mehat	Nik Mizamzul Mehat and Shahrul Kamaruddin	Evaluation on the feasibility of using bamboo fillers in plastic gear manufacturing via the Taguchi optimization method	/	
2	A1	Assoc. Prof Ir Dr Mohd Syukry	K.H. Lim, M.S. Abdul Majid, Ridzuan M.J.M., K.S. Basaruddin and M. Afendi	Effect of Nano-clay Fillers on Mechanical and Morphological Properties of Napier/Epoxy Composites	/	
3	A1	Mr Mohammad Asraf Alif b.Ahmad	M.A. A. Ahmad, M.S. Abdul Majid, Ridzuan M.J.M., Ahmad Firdaus A.Z. and N.A.M. Amin	Tensile Properties of Interwoven Hemp/PET (Polyethylene Terephthalate) Epoxy Hybrid Composites	/	
4	A1	Dr Azwan Iskandar Azmi	Nurdiyanah Nasuha Shah Fenner Khan, Azwan Iskandar Azmi and Chye Lih Tan	A Review on Mode-I Interlaminar Fracture Toughness of Fibre Reinforced Composites	/	
5	A1	Ir Dr Khairul Salleh bin Basaruddin	S.Z. Halim, K.S. Basaruddin, I. Ibrahim, M.S. Abdul Majid and M.J.M. Ridzuan	Determination of Effective Elastic Properties of Metal Matrix Composite with Damage Particulates using Homogenization Method	/	
6	A1	Dr Shamsul bin Kamaruddin	Shamsul Kamaruddin, Ai Bao Chai, Chuan Li Lim and Jee Hou Ho	Characterisation of Static and Dynamic Responses of Natural Rubber Bio-Composites	/	/
7	A1	Dr Shah Fenner Khan (Presenter: Rakhmad Arief Siregar)	Rakhmad Arief Siregar, Shah Fenner Khan and Khairul Umurani	Design and Development of Injection Moulding Machine for Manufacturing Laboratory	/	
8	A2	Assoc. Prof Dr Shariman Abu Bakar	Mohamad Syafiq Abdul Khadir, Shahriman Abu Bakar and Mohd Sani Mohamad Hashim	Analysis of Airbox Performance Improvement for Modenas Model by Geometrical Analysis	/	
9	A2	Dr Md Tasyrif bin Abdul Rahman	Aminurasyid Norizan, Md Tasyrif Abdul Rahman, Nasrul Amri Mohd Amin, Muhammad Haniff Naqiuddin Ismail, Azarul Fahmin Abdul Hamid, Mohd Hafif Basha Mohamad Jamel Basha and Ahmad Syayuthi Abdul Rahman	Study of Intake Manifold for Universiti Malaysia Perlis Automotive Racing Team Formula Student Race Car	/	
10	A2	Dr Narahari Marneni	Narahari Marneni	Buoyancy Induced Couette-Poiseuille Flow in a Vertical Microchannel	/	/
11	A2	Mrs Noor Saffreena binti Hamdan	Noor Saffreena	Dual direction blower system powered by solar energy to reduce car cabin temperature in open parking condition	/	/
12	A2	Mrs Nurul Izzatatie binti Ishak	Izzatie Ni, Mohd Hafif Basha, Yoshimitsu Uemura, Mohd Sani Mohamad Hashim, Nasrul Amri Mohd Amin and Mohd Fazli Hamid	Co-pyrolysis of rice straw and polyethylene terephthalate (PET) using fixed bed drop type pyrolyzer	/	
13	A2	Ms Aqilah binti Che Sulaiman	Aqilah Che Sulaiman, Nasrul Amri Mohd Amin, Mohd Saifizi Saidon and Mohd Shukry Abdul Majid	Preliminary study of TEC application in cooling system	/	
14	A2	Assoc. Prof Pushpa B V	Pushpa B V, Prasanna B M R, Younghae Do and Sankar M	Numerical study of double-diffusive convection in a vertical annular enclosure with a baffle	/	

No.	Session	Registration Name	Authors	Title	1st Day Attend	2nd Day Attend
15	A3	Mr Mohammed Mohammed	Ilyas A. S. Salem, Rozyanty Rahman, Bashir. O. Betar, Tijjani Adam and Mohammed Mohammed	Study of the effect of surface treatment of kenaf fiber on chemical structure and water absorption of kenaf filled unsaturated polyester composite	/	
16	A3	Mr Mohammed Mohammed	Ilyas A. S. Salem, Rozyanty Rahman, Bashir. O. Betar, Tijjani Adam and Mohammed Mohammed	Study of the effect of surface treatment of kenaf fiber on mechanical properties of kenaf filled unsaturated polyester composite	/	
17	A3	Mr Muhammad Zikri B.Zakariya	Zikri M.Z., Ridzuan M.J.M., M.S. Abdul Majid, M. Afendi and K.S. Basaruddin	Effect of glass fibre loading and moisture absorption on the tensile properties of hybrid Napier/glass reinforced epoxy composites	/	
18	A3	Mr Syafiq Farhan Shamsuri	Syafiq Farhan Shamsuri, Ruslizam Daud, Mohd Afendi, Khairul Salleh Basaruddin, Ahmad Burhanuddin Abdullah and Rozita Hassan	Energy release rate analysis on the interface cracks of enamel-cement-bracket fracture using virtual crack closure technique	/	
19	A3	Mr Wan Mohamad Adlan	W.M.A. W. Ramli, Ridzuan M.J.M., M.S. Abdul Majid, M.N. Rahman Y. and Azduwin K.	Effect of elevated temperatures on flexural strength of hybrid Napier/glass reinforced epoxy composites	/	
20	A3	Ms Nur Hafizah bt Zunaidi	Nur Hafizah Zunaidi, Wei Hong Tan, Muhd Shukry Abdul Majid and Eng Aik Lim	Effect of Physical Properties of Natural Fibre on the Acoustic Absorption Coefficient	/	
21	A3	Mr Gurubaran a/l Panerselvan	Gurubaran Panerselvan, M. Afendi, Nur Fareisha Mohd Azmi, Haftirman Idrus and M.S. Abdul Majid	Fatigue Life Investigation of UIC 54 Rail Profile for High Speed Rail	/	
22	A4	Dr William Pao King Soon	William Pao, Lian-Tat Hon, Ahmed Saieed and Sam Ban	Numerical Solution of Two-phase Slug Flow with Liquid Carryover in Different Diameter Ratio T-junction	/	/
23	A4	Mr Syed Samar Abbas	S S Abbas, M S Nasif, M A M Said and S K Kadhim	Numerical investigation on effect of aortic root geometry on flow induced structural stresses developed in a bileaflet mechanical heart valve	/	/
24	A4	Dr Nik Mizamzul Mehat(presenter: amirul aliff jamaluddin)	Amirul Alif Jamaludin, Nik Mizamzul Mehat and Shahrul Kamaruddin	Investigating the effects of PDC cutters geometry on ROP using the Taguchi technique	/	
25	A4	Assoc. Prof Dr Shariman Abu Bakar	Mohamad Syafiq Abdul Khadir, Shahriman Abu Bakar and Mohd Sani Mohamad Hashim	Optimization of Cam Performance Curve Using B-Spline Curve	/	

No.	Session	Registration Name	Authors	Title	1st Day Attend	2nd Day Attend
26	A4	Prof. Zhigang XU	Zhi-Gang Xu, Deyu Yang and Wei-Dong Shen	Functional design of heat exchanging for pneumatic vehicles	/	
27	A4	Prof. Zhigang XU	Zhi-Gang Xu, De-Yu Yang and Wei-Min Liu	Key issues in theoretical and functional pneumatic design	/	
28	A4	Prof. Zhigang XU	Zhi-Gang Xu, Wei-Min Liu and Tao-Tao Liu	Functional-to-form mapping for assembly design automation	/	
29	A4	Dr Azwan Iskandar Azmi	Azwan Azmi, Ahmad Zulfa Syahmi Ahmad Norllil, Mohd Naquib Nakiyudin, Tan Chye Lih, Ahmad Fairuz Mansor and Ahmad Nabil Mohd Khalil	Effects of machining conditions on the specific cutting energy of carbon fibre reinforced polymer composites	/	
30	A5	Mr Mohammed Mohammed	Mohammed Mohammed, Rozyanty Rahman, Bashir. O. Betar, Tijjani Adam, Azlin Fazlina Osman, Ilyas A. S. Salem, Omar Dahham and Marwa Al-Samarrai	The Weathering Effect in Natural Environment on Kenaf Bast and Unsaturated Polyester Composite		/
31	A5	Mr Mohammed Mohammed	Mohammed Mohammed, Rozyanty Rahman, Bashir. O. Betar, Tijjani Adam, Azlin Fazlina Osman, Ilyas A. S. Salem, Omar Dahham and Marwa Al-Samarrai	Effluence of Weathering on Thermal Properties of Kenaf Bast /Glass Fiber Polymer Composite		/
32	A5	Mr Mohammed Izzuddeen	Mohammed Izzuddeen Mohd Yazid, Supri A Ghani, Tan Soo Jin, Azlin Fazlina Osman and Siti Hajar Mohd Din	Utilization of Polyaniline (PAni) as Conductive Filler on Poly (Ethylene Oxide) / Poly (Vinyl Chloride) Films: Effects of Naphthalene as Surface Modifier on Electrical Conductivity	/	/
33	A5	Mrs Maslinda Abu Bakar@Yahaya	Maslinda A.B., M.S. Abdul Majid, Ridzuan M.J.M and Syayuthi A.R.A.	Water absorption behaviour of hybrid interwoven cellulosic fibre composites		/
34	A5	Ms Siti Hajar binti Mohd Din	Siti Hajar Mohd Din, Supri A. Ghani and Mohammed Izzuddeen Mohd Yazid	Effect of Graphite Loading on the Electrical and Mechanical Properties of Poly (Vinyl Chloride)Poly (Ethylene Oxide) Polymer Films	/	/
35	A5	Mr Faisal Abrar b Syamsul Bahri	Abrar F.S.B., M.S. Abdul Majid, Ridzuan M.J.M. and Syayuthi A. R.A.	Impact responses, compressive and burst tests of glass reinforced epoxy (GRE) composite pipes		/
36	A5	Ms Wanna a/p Soh Bua Chai	S B C Wanna, K S Basaruddin, M H Mat Som, M S Mohamad Hashim, R Daud and M S Abdul Majid	Prediction on fracture risk of femur with Osteogenesis Imperfecta using finite element models: Preliminary Study		/
37	A6	Mr Muhammed Abdullah Mutafi	M A Mutafi, M S Nasif, W Pao and F B Ismail	Numerical investigation on the effect of balcony open and solid upstand on smoke contamination in atrium balconies	/	/

No.	Session	Registration Name	Authors	Title	1st Day Attend	2nd Day Attend
38	A6	Ir. Ramesh Ramasamy	Ramesh Ramasamy, Zainah Ibrahim and Hwa Kian Chai	Numerical Investigations of Internal Stresses on Carbon Steel Based on Ultrasonic LCR Waves	/	/
39	A6	Dr Nik Mizamzul Mehat	Nik Mizamzul Mehat, Syahir Mohd Kassim and Shahrul Kamaruddin	Investigation on the effects of processing parameters on shrinkage behaviour and tensile properties of injection moulded plastic gear via the Taguchi method		/
40	A6	Ms Noor Shahirah binti Mohd Hussin	Noor Syahirah Mohd Hussin, Salem Awadh Nasser Gamil, Nasrul Amri Mohd Amin, Muhammad Juhairi Aziz Safar, Mohd Shukry Abdul Majid and Nashrul Fazli Mohd Nasir	Design and Analysis of Hydraulic Ram Water Pumping System		/
41	A6	Ir Dr Khairul Salleh bin Basaruddin	A.H. Abd Razak, K.S. Basaruddin, A.F. Salleh, W.M.R. Rusli, M.S. Mohamad Hashim and R. Daud	Finite Element Modelling of Plantar Fascia Response during Running on Different Surface Types		/
42	A6	Muhammad Faiz Hilmi bin Rani	Muhammad Faiz Hilmi Rani, Zuradzman Mohamad Razlan, Shahriman Abu Bakar, Chan Kah Yong, Azizi Harun, Mohd Sani Mohamad Hashim, Mohd Khairul Faizi Abd Rahman, Ishak Ibrahim, Nur Saifullah Kamarrudin, Mohd Asrul Md Saad, Zunaidi Ibrahim, Wan Khairunizam Wan Ahmad and Hazry Desa	Design of refrigeration system using refrigerant R134a for macro compartment		/
43	A6	Dr Shah Fenner Khan	Tai Yee Chen, Shah Fenner Khan and Wei Hong Tan	Influence of Processing Methods on Flexural Strength in Interwoven Hemp/ PET Reinforced POM Hybrid Composite.		/
44	A6	Mr Mohammad Ikram	Muhammad Ikram Md Fazli, Azli Yahya, Ardiyanshah Shahrom, Sophan Wahyudi Nawawi, Muhammad Rezza Zainudin and Muhammad Syazani Nazarudin	Finite Element Analysis of 2-Station Hip Simulator	/	/
45	A6	Dr Mohd Sani bin Mohamad Hashim	Mohammad Nur Aiman Zaidie, Mohd Sani Mohamad Hashim, Md Tasyrif Bin Abdul Rahman, Mohd Hafif Basha, Ishak Ibrahim, Nur Saifullah Kamarrudin and Shahriman Abu Bakar	Analysis of a Front Suspension System for UniART FSAE Car Using FEA		/

IOP Conf. Series: Journal of Physics: Conf. Series **908** (2017) 011004 doi:10.1088/1742-6596/908/1/011004

No.	Session	Registration Name	Authors	Title	1st Day Attend	2nd Day Attend
46	A7	Hassan Sheikh Hassan Ba Hamid(*on behalf Mr.Amares Singh)	Amares Singh, Endra Sujatmika, Tan Wei Hong and Rajkumar Durairaj	A Review: Characteristics of Noise Absorption Material	/	/
47	A7	Mr Mohammed Izzuddeen	Mohammed Izzuddeen Mohd Yazid, Supri A Ghani, Azlin Fazlina Osman, Siti Hajar Mohd Din and Tan Soo Jin	The Effect on Poly (Ethylene Oxide) / Poly (Vinyl Chloride) / Polyaniline (PAni) Films by Ethylene Dimethacrylate as Surface Modifier: Electrical Conductivity and Characterization	/	/
48	A7	Mr Ang Jia Yi	Ang Jia Yi, M.S. Abdul Majid, A. Mohd Nor, Ridzuan M.J.M. and S. Yaacob	Failure prediction of ±55° glass/epoxy composite pipes using system identification modelling		/
49	A7	Ms Fatinah Syafiqah Tajudin	Fatinah T.S., M.S. Abdul Majid, Ridzuan M.J.M., Ahmad Firdaus A.Z. and N.A.M. Amin	Tensile properties of compressed moulded Napier/glass fibre reinforced epoxy composites		/
50	A7	Ms Izzawati Basirom	Izzawati Basirom, Ruslizam Daud, Mohd Afendi, Mohd Shukry Abdul Majid, Mohd Yazid Bajuri and Noor Alia Mohd Zain	Stress Analysis of Implant-Bone Fixation at Different Fracture Angle		/
51	A7	Dr Tan Wei Hong	Wei Hong Tan, Jun Xiong Cheah, Eng Aik Lim, Hun Guan Chua and Chu Yee Khor	Vibration Analysis on Compact Car Shock Absorber		/
52	A7	Ms Revati radakisnin	Revati R., M.S. Abdul Majid, Mohd Ridzuan Mohd Jamir, Normahira M, Nashrul Fazli Mohd Nasir and Cheng Ee Meng	Biodegradation of PLA- Pennisetum purpureum based biocomposite scaffold		/
53	A8	Ms Nur Shafiqah binti Zhubir	Syafiqah Z., N.A.M. Amin, Irwan Y.M., M.S.A. Majid and N.A. Aziz	Simulation Study of Water Cooled Photovoltaic Panel Using ANSYS		/
54	A8	Ms Nur Syazwani Binti Abdul Aziz	Nursyazwani Abdul Aziz, Nasrul Amri Mohd Amin, Mohd Shukry Abdul Majid, Azman Hussin and Syafiqah Zhubir	Experimental investigation on AC unit integrated with sensible heat storage (SHS)		/
55	A8	Prof Dr B M R Prasanna	Sankar M, Jagadeesha R D, Prasanna B M R and Younghae Do	Natural convection in an inclined parallelogrammic porous enclosure under the effect of magnetic field		/
56	A8	Mr Ahmad Fairuz Bin Mansor	Ahmad Fairuz Mansor, Muhammad Syamil Zakaria, Dr Azwan Iskandar Azmi, Ahmad Nabil Khalil and Nur Atikah Musa	A Study of Energy Consumption in Turning Process using Lubrication of Nanoparticles Enhanced Coconut Oil (NECO)		/
57	A8	Dr Mohd Sani bin Mohamad Hashim	Yi Fei Ng, Mohd Sani Mohamad Hashim, Muhammad Amin Zulkifli Din, Nur Saifullah Kamaruddin, Mohd Hafif Basha, Nasrul Amri Mohd Amin and Md Tasyrif Abdul Rahman	Effect of Various Bending Angles on a Passive Light Pipe for Eco-Daylighting Systems		/

IOP Conf. Series: Journal of Physics: Conf. Series **908** (2017) 011004 doi:10.1088/1742-6596/908/1/011004

No.	Session	Registration Name	Authors	Title	1st Day Attend	2nd Day Attend
58	A8	Dr Md Tasyrif bin Abdul Rahman	Azarul Fahmin Bin Ab Hamid, Md Tasyrif Bin Abdul Rahman, Shah Fenner Khan, Muhajir Bin Ab Rahim, Norasmadi Bin Abdul Rahim, Mohd Hafif Basha Bin Mohamad Jamel Basha, Muhammad Hanif Naqiuddin Bin Ismail, Aminurasyid Bin Norizan, Mohd Sani Bin Mohamad Hashim and Ahmad Syayuthi Bin Abdul Rahman	Connected Car: Engines Diagnostic via Internet of Things (IOT)		/
59	A8	Assoc Prof Dr Shahriman bin Abu Bakar	Abdul Rahim Mahayadin, Shahriman Abu Bakar, Mohd Sani Mohamad Hashim, Zuradzman Mohamad Razlan, Mohd Khairul Faizi Abd Rahman, Azizi Harun, Nur Saifullah Kamarrudin, Ishak Ibrahim, Mohd Asrul Md Saad and Muhammad Faiz Hilmi Rani	Efficient methodology of route selection for driving cycle development		/
60	A9	Mr Mohd Hanif Mohd Pisal	Mohd Hanif Mohd Pisal, Supri A. Ghani, Rozyanty Rahman and Tan Soo Jin	Effect of Oxidation Agent on Wood Biomass in Ethylene Vinyl Acetate Conductive Polymer: Tensile Properties, Tensile Fracture Surface and Electrical Properties	/	/
61	A9	Ms Khalilah Khasnulhadi	Khalilah Khasnulhadi, Ruslizam Daud, Fauziah Mat, Khairul Salleh Basaruddin, Mohd Hafis Sulaiman and Siti Noor Fazliah Mohd Noor	Influence of ceramic dental crown coating substrate thickness ratio on strain energy release rate		
62	A9	Mr Ahmad Syayuthi	Syayuthi A.R.A., M.S. Abdul Majid, Mohd Ridzuan Mohd Jamir and K.S. Basaruddin	Effect of stress ratios on the fatigue behaviour of glass/epoxy composite		/
63	A9	Ir Dr Khairul Salleh bin Basaruddin	M F Hazman, K S Basaruddin, N S Kamarrudin, M S Abdul Majid and M Afendi	Effective Elastic Constants of Corrugated Core Sandwich Plate Microstructure Considering Imperfection in Adhesive Bonding		/
64	A9	Ir Dr Khairul Salleh bin Basaruddin	M Thomas, K S Basaruddin, M J A Safar, S F K Mohamad Khan and I Ibrahim	Homogenized Properties of Porous Microstructure: Effect of Void Shape and Arrangement		/
65	A9	Mr Mohd Khairul Faizi bin Abd Rahman	Faizi M.K., Mohd Shukry Abdul Majid, Shahriman A.B., Z.A. Ahmad, Shamsul B.M.T. and Y.G. Ng	The Effect of Alkaline Treatments Soaking Time on Oil Palm Empty Fruit Bunch (OPEFB) Fibre Structure	/	/
66	A9	Ir Dr Mohd Noor Arib Bin Md Rejab	Mohd Noor Arib Md Rejab, Shazmin Aniza Abdul Shukor, Mohd Sofian Mohammad Rosbi, Jawaid I. Inayat-Hussain, Nazirah Ahmad and Asyraf Ismail	Vibration Power Flow Method as an Additional Method to the Transmissibility Method in Quantifying the Effectiveness of Elastomeric Mount		/
67	A10	Mr Mohd Fadzil Faisae	Nur Hairunnisa Kamarudin and Mohd Fadzil Faisae Ab Rashid	Assembly Line Balancing with Resource Constraints using New Rank-Based Crossovers	/	/
68	A10	Mr Mohammad Rezza	Muhammad Rezza Zainudin, Azli Yahya, Muhammad Ikram Md Fazli, Ardiyansyah Syahrom, Fauzan Khairi Che Harun and Muhammad Syazani Nazarudin	Design of Biped Hip Simulator Using SolidWorks	/	/

IOP Conf. Series: Journal of Physics: Conf. Series **908** (2017) 011004 doi:10.1088/1742-6596/908/1/011004

No.	Session	Registration Name	Authors	Title	1st Day Attend	2nd Day Attend
69	A10	Mr Wan Mohd Syafiq bin Wan Sulong	W.M. Syafiq, M. Afendi, R. Daud, M.N. Mazlee, M.S. Abdul Majid and Y.S. Lee	Mechanical properties of friction stir welded butt joint of steel/aluminium alloys: effect of tool geometry		/
70	A10	Nurul Farhanah Azman	Nurul Farhanah Azman, Syahrullail Samion and Ahmad Ajruddin Mahzun	Evaluation of RBD Palm Stearin as Alternative Lubricant for Cold Forward Extrusion Process	/	/
71	A10	Dr. Shah Fenner	Muhammad Aizat Yazid and Shah Fenner Khan	Fabrication of Mandible Fracture Plate by Indirect Additive Manufacturing		/
72	A10	Mr Mohammed Mohammed	Tijjani Adam and Th. S. Dhahi	Fabrication and Electrical characterization of planner si-nanogap		/
73	A10	Jamali bin Md Sah	Jamali Md Sah, Khairul Azwan Ismail, Zahari Taha and Muhammad Aiman Mohd Fozi	Development of Matlab Simulink model for dynamics analysis of passive suspension system for lightweight vehicle		/

IOP Conf. Series: Journal of Physics: Conf. Series **908** (2017) 011005 doi:10.1088/1742-6596/908/1/011005

Peer review statement

All papers published in this volume of *Journal of Physics: Conference Series* have been peer reviewed through processes administered by the proceedings Editors. Reviews were conducted by expert referees to the professional and scientific standards expected of a proceedings journal published by IOP Publishing.

IOP Conf. Series: Journal of Physics: Conf. Series **908** (2017) 012001 doi:10.1088/1742-6596/908/1/012001

Study of the effect of surface treatment of kenaf fiber on chemical structure and water absorption of kenaf filled unsaturated polyester composite

I A S Salem[1], A R Rozyanty[1], B O Betar[2],T Adam[3,4], M Mohammed[1] and A M Mohammed[5]

[1]Center of Excellence Geopolymer and Green Technology, School of Materials Engineering, Universiti Malaysia Perlis (UniMAP), 01007, P.O Box 77, D/A PejabatPosBesar, Kangar, Perlis, Malaysia.
[2] Nanotechnology & Catalysis Research Centre (NANOCAT), University of Malaya, 50603 Kuala Lumpur, Malaysia.
[3] Faculty of Technology, Universiti Malaysia Perlis (UniMAP), KampusUnicitiAlam Sg. Chuchuh, 02100 Padang Besar (U), Perlis, Malaysia.
4 Institute of Nano Electronic Engineering (INEE), Universiti Malaysia Perlis (UniMAP), 01000 Kangar, Perlis, Malaysia.
[5]College of education for pure science Ibn-Alhaitham University of Baghdad,

rozyanty@unimap.edu.my

Abstract. In this research, unsaturated polyester/kenaf fiber (UP/KF) composites was prepared by using hand lay-up process. The effect of surface treatment of kenaf fiber on mechanical properties of kenaf filled unsaturated polyester composites were studied. Different concentrationsof stearic acid (SA) were applied, i.e. 0, 0.4, and 0.8 wt%. The Fourier transform infrared (FT-IR) spectra of kenaf fiber shows high intensity of the peak around 3300-3400 cm^{-1}, which is attributed to the hydrogen bonded O-H stretching. However, the treated kenaf fiber with stearic acid shows the elimination of O-H group and this peak is vanished. This is due to the reaction of (-COOH) group of stearic with (-OH) group of kenaf fiber. The results of water absorption study revealed that increasing the loading of KF in the composite will result is increasing the tendency to absorb water. However, the absorption was significantly decreased after treatment with stearic acid as well as the time to reach to the equilibrium state.

1.Introduction

Recently, there is a drastic growing attention in utilization of natural fibers materials as reinforcingcomponent in thermoplastics as well as thermosets.Natural fibers seem to be a superior material which has emerged as an abundant and sustainable replacement for the nonrenewable and expensive synthetic fibers. The advantages of naturalfibres overtraditional reinforcing materials such as lower density, less abrasiveacceptable specific strength and stiffness, greater deformability and less irritation to skin and respiratory [1].Several natural fibers have been used as a reinforcement in composites materials for advanced applications such as banana, sisal, oil palm, jute, kenaf and coir. Among different types of fibers, kenaf fiber has been widely explored over the past several years [2].

Kenaf (Hibiscus cannabinus, family *Malvaceae*L.) is a perennial plant herb. Kenaf is a plant heat-season annual lane and can be used for several industrial applications. Kenaf fiber has several The attractive features over other traditional reinforcement materials, as such it has gained research interest as a natural filler in polymer composites. The availability of kenaf fiber in Asia is more and it is abundantly available, environmental friendly, renewable, and cheap. Moreover, it has advantages the absence of silica content, which is criticalin the reduction of abrasiveness to processing equipment. The biodegradability of kenaf fiber has a great contribution to a healthy ecosystem while their low cost and high performance satisfies the economic concern of manufacturers [3,4].

The greatest challenge in working with natural fiber reinforced plastic composites is their large variation in properties and characteristics with respect to different condition [2]. Bio composite's properties are influenced by a number of variables, including the fiber type, environmental conditions (where the plant fibers are sourced), processing methods, and any modification of the fiber. Recently there has been a surge of interest in the industrial applications of composites containing biofibers reinforced with biopolymers [5]. Another problem is incompatibility to hydrophilic thermoplastic and thermosetmatrix weak interfacial strength between natural fiber surfaces and a polymer matrix.This is due to the chemical structure of natural fibers (such as kenaf) which is mainly has hydrophilic moiety (mainly OH), which will cause high polar characteristic. Therefore, it is important to enhance compatibility between the fiber and the polymer matrix in order to obtain a uniform composites with good properties. Several chemical treatments can be done to natural fibers in order to obtain better compatibility [6,7].

Therefore, a number of chemical and physicalsurface modification studies on a variety of natural fibers have been devoted to understand andimprove the interfacial strength between natural fibers and the polymer matrix. Various methods such as corona treatment [8], plasma treatment [9], mercerization [10], heat treatment [11], graft copolymerization [12,13] and silane treatment [10,14] have been reported to investigate the compatibility of natural fiber in polymeric composites, in most cases the results were positive [15-17]. However, these treatments have many disadvantages such as the use of expensive equipment or the use of expensive chemicals. A very attractive treatment is to modify the surface of natural fibers with fatty acids such as stearic acid[18-22]. The principle of this treatment is the carboxyl group of stearic acid reacts with the hydroxyl groups of the fiber through an esterification reaction and, thus, reduces the hydroxyl groups number available to boned with water molecules. One more advantage of treatment with stearic acid is that it is not sensitive to oxidation during the processing temperatures of natural fiber/polymer composites [22-26].

Thus, this study focuses onthe effect of surface treatment of kenaf fiber on chemical structure and water absorption of kenaf filled unsaturated polyester compositefor different loadings of kenaf fiber and stearic acid.

2.Experimental

2.1. Materials

Kenaf fiber was obtained from RahamatullahSdn. Bhd, Malaysia. The kenaf in this study were used in the form of mat. Unsaturated polyesters resin was supplied by Castmesch Technologies Sdn. Bhd. MEKP (solution in dimethyl phthalate) used was from KaumjungAkzoNobel peroxide Ltd by trade name Butanox M60. Stearic acid was purchased from Acidchem International Sdn. Bhd, Malaysia.

2.2. Treatment with stearic acid

In this study, treatment of kenaf fiber by using stearic acid was carried out. Prior to treatment, kenaf fiber mat was cut to 20 cm×20 cm dimensions. Different concentrations of stearic acid were applied, i.e. 0, 0.4, and 0.8 wt%. Stearic acid was added to water and warm up at 75 °C to form a suspension. Kenaf fiber was immersed in the suspensions containing stearic acid and stirred for 60 min at 70 °C. The treated kenaf fibers were then washed with distilled water until pH=7 by using pH meter. The treated kenaf fiber layers were placed in a conventional oven to dry at 80°C overnight.

IOP Conf. Series: Journal of Physics: Conf. Series **908** (2017) 012001 doi:10.1088/1742-6596/908/1/012001

2.3. Composite preparation

Initially kenaf mat was cut to 20 cm×20 cm dimensions and treated with different concentrations of stearic acid, this followed by compressing the untreated and treated kenaf fibers with stearic acid using compress machine at 90 °C.The composites formulation is presented in Table 1. Then prior putting to the mold and hand lay-up resin to bottom of the kenaf, then the kenaf was put inside the mold. Resin impregnated into untreated and treated kenaf via hand lay-up method and 2% of MEKP was added as a hardener for all the samples.This followed by separating it from mold using hack saw. The prepared composite was measured and cut to 2.5×20 cm about 13 samples were obtained.

Table 1. Composites formulation table.

No.	Kenaf mats %	SA %	UP %
UP	0	0	100
KF1	10	0	90
KF2	20	0	80
KF3	30	0	70
KF4	40	0	60
KF1-SA0.4	10	0.4	90
KF2-SA0.4	20	0.4	80
KF3-SA0.4	30	0.4	70
KF4-SA0.4	40	0.4	60
KF1-SA0.8	10	0.8	90
KF2-SA0.8	20	0.8	80
KF3-SA0.8	30	0.8	70
KF4-SA0.8	40	0.8	60

2.4. Fourier Transform Infrared Spectroscopy (FT-IR)

Infrared spectra were recorded using a Perkin- Elmer Spectrum 65 Infrared Spectrometer with 10 scans at 4 cm-1 resolutions. Kenaf fiber without treatment, with 0.4% treatment with stearic acid and 0.8% treatment with stearic acid were analyzed directly using an ATR technique within the wave number range of 4000-650 cm^{-1}.

2.5. Water absorption test

Water uptake was determined by soaking the sample in water and measuring the weight periodically. The samples were cut into 25 mm length, 12.7 mm width and 7 mm thickness in according to ASTM D5229. The samples were dried until consistent weight was obtained prior to water absorption test. The samples were then immersed in distilled water at room temperature and the weight change monitored as a function of time. All the experiments were done in triplicates The absorption behavior weighed were studied at every 24 hours and up to 648 hours (27 days) within 5 min of removing from distilled water. The formula used to calculate the water absorbency is shown in Equation 1.

$$\text{Percentage of water absorbency} = (W_1 - W_o)/W_o \times 100\ \% \quad (1)$$

Where W_0= Initial weight of the sample before immersion inside water and $W1$ = Weight of wet sample.

3. Results and discussion

FTIR analysis was carried out to study the characteristics of the kenaf fibers, before and after treatment with SA. FT-IR spectrum of untreated KF is shown in Figure 1. The broad peak at 3373 cm^{-1} is attributed to the hydrogen bonded O-H stretching, whereas the peak at 2923 cm^{-1} was attributed to the characteristic of C-H asymmetric and symmetric stretching. This is related to the aliphatic groups in cellulose and hemicellulose. Another peak at 1739 cm^{-1} corresponds to ester carbonyl vibrations from the acetyl, feruloyl and p-coumaryl groups in lignin. A stretching peak detected at 1633 cm^{-1} is attributed to the carbonyl group of the acetyl ester in hemicellulose and the carbonyl aldehyde in lignin. The broadpeak at 1252 cm^{-1} that corresponds to C–O vibration [16].

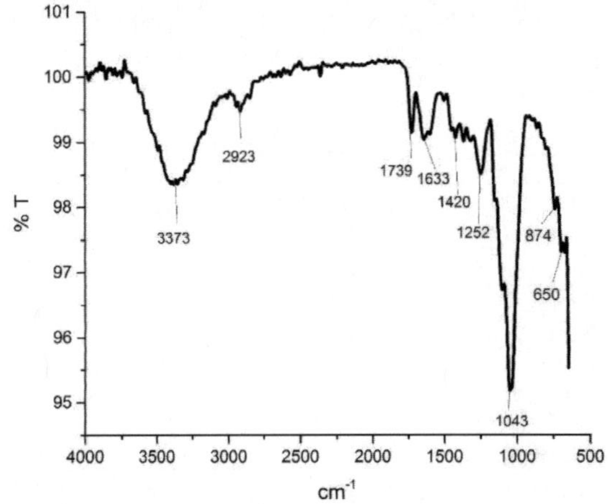

Figure 1.FT-IR spectrum of untreated KF.

The Fourier transform infrared (FT-IR) spectrums of treated KF with 0.4% and 0.8% of SA are shown in Figure 2 and 3, respectively. As can be seen, the intensity of the peak around 3300-3400 cm^{-1}, which is attributed to the hydrogen bonded O-H stretching, is vanished after treatment of KF with SA. The KF treated with SA have a hydrophobic chain of stearic acid molecules which imparts water resistance to the composite. Absorption peak appeared at 2925 and 2853 cm^{-1} are attributed to C-H asymmetric and symmetric of methylene group. The peak at 1715 cm^{-1} is the characteristic band for carbonyl (C=O) stretching. The peak at 1461 cm^{-1} is attributed to C-H bending frequency of hemicellulose. The peak at 1102 cm^{-1} is attributed to C-O-C stretching vibration of cellulose [17].

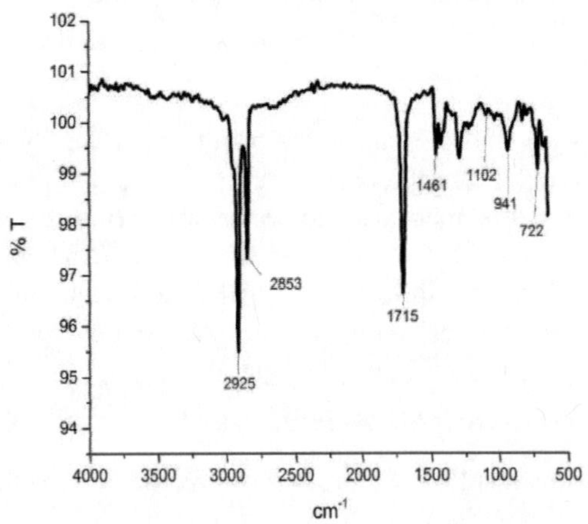

Figure 2.FT-IR spectrum of treated KF with 0.4% of SA.

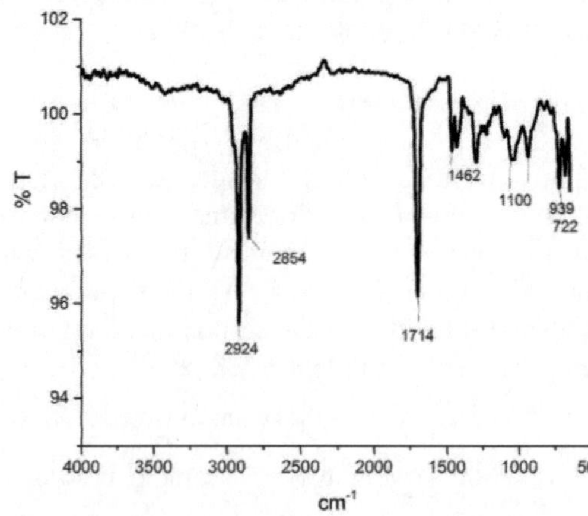

Figure 3.FT-IR spectrum of treated KF with 0.8% of SA.

The treatment of KF with SA undergoes the reaction which is presented in Figure 4. The principle of this treatment is that the carboxyl group (–COOH) in SA reacts with the hydroxyl groups of the fiber through an esterification reaction and, thus, this treatment reduces the number of hydroxyl groups available for bonding with water molecules. Furthermore, the long hydrocarbon chain of stearic acid (18 carbon atoms) provides an extra protection from water since it is itself quite hydrophobic.

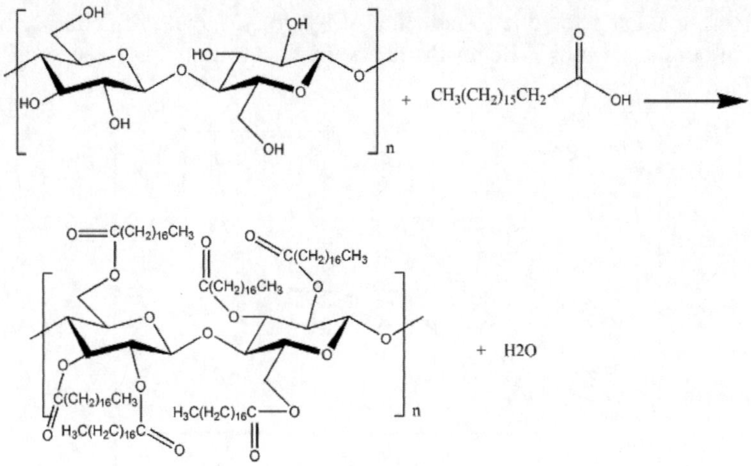

Figure 4. Schematic Illustration of the Reactions SA with KF during treatment.

When composites specimens were immersed into water, the water molecules penetrate and inducing the weight gain. The weight gain was recorded every one day (24 hours) in order to have a better understanding of water absorbency mechanism.

The water absorption of untreated UP/KF composites, UP/KF composites treated with 0.4 wt% of SA and UP/KF composites treated with 0.8 wt% of SA are depicted in Figures 5, 6 and 7, respectively.

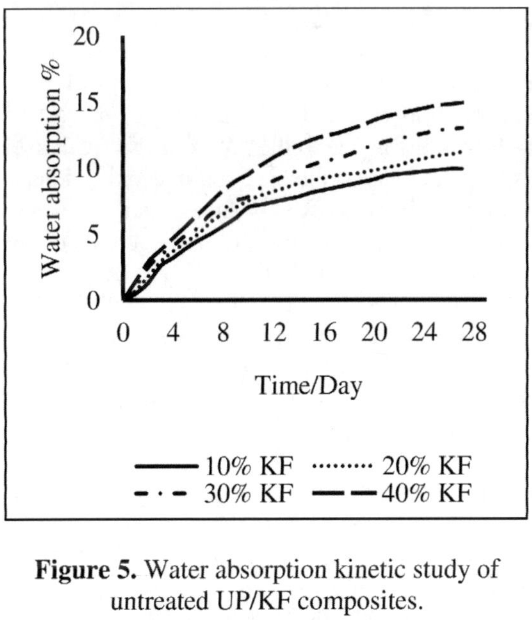

Figure 5. Water absorption kinetic study of untreated UP/KF composites.

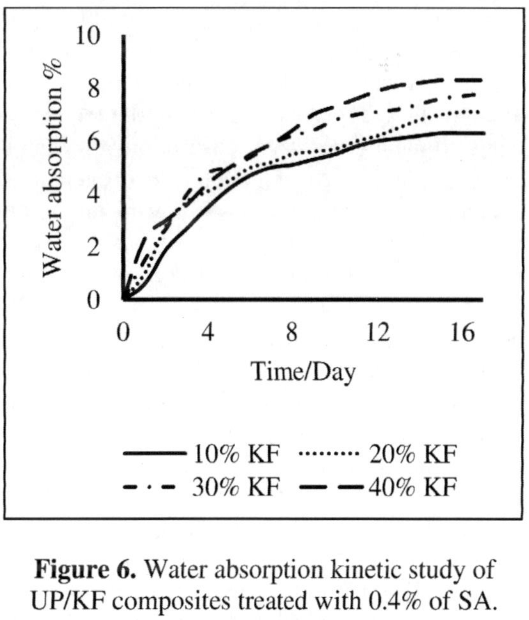

Figure 6. Water absorption kinetic study of UP/KF composites treated with 0.4% of SA.

It can be seen from the figures that all the composites show similar pattern of water absorption, which is at the beginning of the curve (within 10 days for untreated UP/KF composites, 5 days for UP/KF composites treated with 0.4 wt% of SA and 4 days for UP/KF composites treated with 0.8 wt% of SA) the weight increased sharply indicating the rapid water penetration into the composite materials. Then, the water abruption slows down to reach equilibrium stage, i.e. the composite was unable to absorb additional water. This can be due to the presence of KF, which is hydrophilic, which allow the composite to absorb water. However, the effect of SA treatment is clearly obvious from the figures. The reduction in water absorption after treatment with SA can be attributed to the improvement in interfacial adhesion due to reducing of hydroxyl moiety of cellulose in the surface of KF, which has been confirmed by FT-IR results. This would avoid an easy access of water molecule into the treated KF composites.

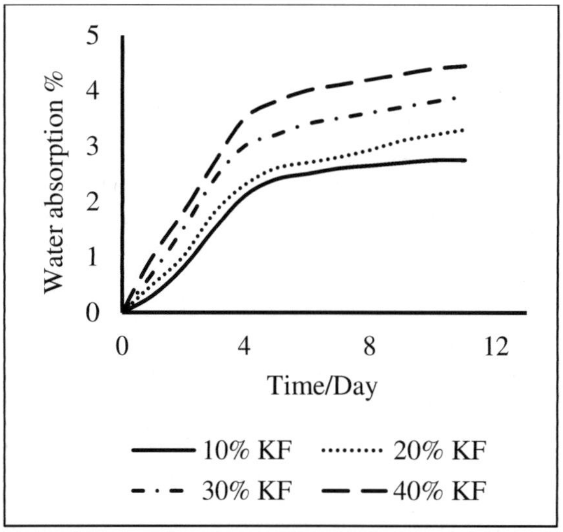

Figure 7. Water absorption kinetic study of UP/KF composites treated with 0.8% of SA.

The water absorbency equilibrium study of UP/KF composites are depicted in Figure 8. It was found that water absorption % of UP was only 1.5% after one day and it remains constant. This can be due to the fact that UP is hydrophobic in nature. The results reveal that increasing the loading of KF in the composite will result is increasing the tendency to absorb water, and 40 wt% loading of untreated KF showed the highest water absorption % which is 14.9 %. This is due to the nature of KF which is hydrophilic that lead to formation of hydrogen bonding between water molecule and hydroxyl moiety of cellulose, hemicellulose and lignin in the cell wall of KF. Thus, as KF loading increases in the composite, the number of hydrogen bonds between the organic molecules and water molecule increase and absorb higher amount of water. water absorption % increased with time and it reached to equilibrium state after 27, 17, and 11 days of conducting the experiment for untreated UP/KF composites, UP/KF composites treated with 0.4 wt% of SA and UP/KF composites treated with 0.8 wt% of SA, respectively. The difference in equilibrium time and water absorption % can be attributed to the effect of treatment of KF with SA as the tendency to absorb water is significantly lower in comparison to that of untreated fibers as the concentration of SA increased. This phenomenon is complementary match with FT-IR results.

It can be seen from the figure that increasing the loading of KF in the composite will result is increasing the tendency to absorb water, and the highest water absorption % is for 40 wt% of KF loading was 14.9%. When composites specimens were immersed inside water, the water molecules penetrate and inducing the weight gain. For all the composites, at the beginning of the curve, the water absorption % increased sharply indicating the rapid water penetration into the composite materials. However, after 10-12 days, the water absorption % was slow indicating that the tendency is less to absorb water as it was reaching the equilibrium state.

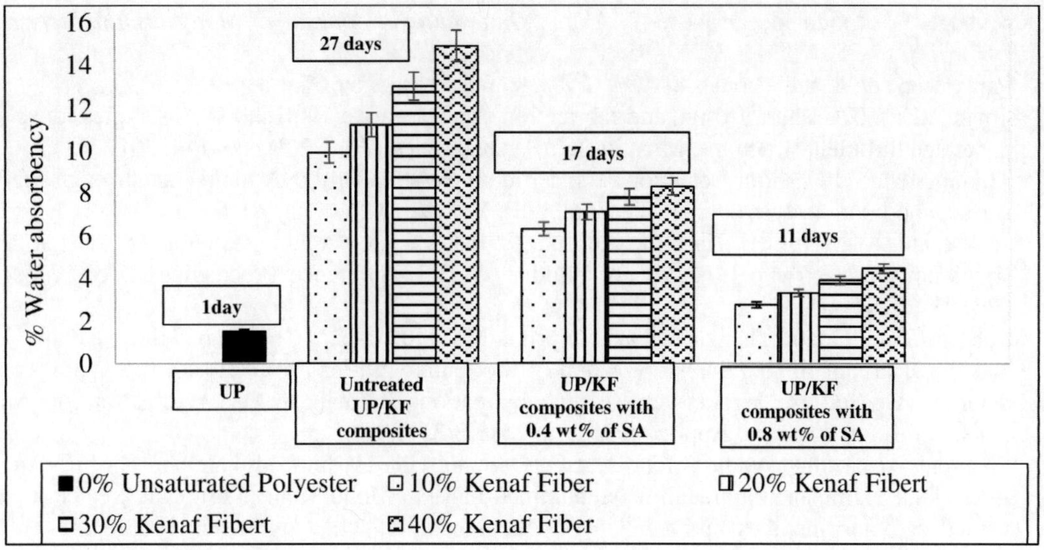

Figure 8. Water absorption equilibrium study of UP/KF composites.

4. Conclusion

The major challenge for natural fiber matrix its affinity to water, it has high absorption to moisture which have serious consequences to its properties.FT-IR analysis shows the elimination of OH group from kenaf fiber after treatment with stearic acid. The water absorption decreased with increasing of stearic acid concentration because stearic acid has modified kenaf fiber into hydrophobic material.

References

[1] Edeerozey A M M, HarizanM A, Azhar A B andAriffiM I Z 2007*Mater. Lett.*612023
[2] Akil H M, Omar M F, Mazuki A A M, Safiee S, Ishak Z A M and Abu Bakar A 2011 *Mater. Des.* 32 4107
[3] Facca AG, Kortschot M T and Yan N 2006 *Composites Part A*37 1660
[4] Raju G, Said MMand Ahmad M A 2008 *J. Rubb. Res.*11187
[5] Bismarck A, Askargorta IA, Springer J, Lampke T, Wielage B, Stamboulis A,ShenderovichI and Limbach HH2004 *Polym. Compos.*23 872
[6] Colom X, Carrasco F, Pages P andCanavate J 2003 *Comp. Sci. and Tech*63 161
[7] Lee SG, Choi SS, Park WH and Cho D 2003*Macromol. Syp.*19789
[8] Belgacem M N, BataillePandSapeiha S 1994 *J ApplPolymSci*53 379
[9] Felix J M, Carlsson C M G andGatenholm P 1994 *J AdhesSciTechnol*8 3043
[10] Bisanda E T N and Ansell M P 1991 Compos *SciTechnol*41 165–178.
[11] Sapieha S, Pupo J F and Schreiber H 1989 *J ApplPolymSci*37 233
[12] Felix J M and Gatenholm P 1991 *J ApplPolymSci*42 609
[13] Maldas D, Kokta B VandDaneault C 1989 *J ApplPolymSci*37751
[14] Mieck K P, Nechwatal A and Knobelsdorf C 1995 *Die AngewandteMakromolekulareChemi* 224 73
[15] GessJ M andRodriguez J M 2005*The sizing of Paper*(Atlanta: Tappi Press)

[16] MbadaN I., Aponbiede O, Ause T, Alabi A*International Journal of Materials Engineering*6 8
[17] Paul A, Joseph K and Thomas S 1997 *Compos. Sci. Technol.*51 67
[18] Tijjani Adam, U Hashim, Design and fabrication of micro-mixer with short turns angles for self-generated turbulent structures, Microsystem Technologies, Vol. 22(2) 433-440, 2016
[19] Tijjani Adam, U Hashim, Detection of specific single-stranded DNA molecules through SiNW surface modulation, Microsystem Technologies, Vol. 22(2) 269-273, 2016
[20] A Wesam Al-Muftil, U Hashim, Ausaila Fraihat, Tijjani Adam Studying on design and simulation of electronic IR guide for blindness, AIP Conference Proceedings, Vol 1775 (1) 030103, 2016
[21] Subash CB Gopinath, Thangavel Lakshmipriya, MK Md Arshad, CH Voon, Tijjani Adam, Uda Hashim, Harbant Singh, Suresh V Chinni, Shortening full-length aptamer by crawling base deletion–Assisted by Mfold web server application, Journal of the Association of Arab Universities for Basic and Applied Sciences, 2016/8/25
[22] Shahidah Arina Shamsuddin, Mohd Nazree Derman, Uda Hashim, Muhammad Kashif, Tijjani Adam, Nur Hamidah Abdul Halim, Muhammad Faheem Mohd Tahir, Nitric acid treated multi-walled carbon nanotubes optimized by Taguchi method, AIP Conference Proceedings, Vol, 1756 (1) 090002, 2016
[23] Tijjani Adam, U Hashim, Highly sensitive silicon nanowire biosensor with novel liquid gate control for detection of specific single-stranded DNA molecules, Biosensors and Bioelectronics. 67(2015) 656–661
[24] Tijjnai Adam U. Hashim and Th S. Dhahi, Silicon nanowire formed via shallow anisotropic etching, Si-ash-trimming for specific DNA and electrochemical detection, Chinese Physics B Vol. 24, No. 6 (2015) 06810
[25] Tijjani Adam, U Hashim, Design and fabrication of micro-mixer with short turns angles for self-generated turbulent structures, Microsystem Technologies, 1-8, 2015
[26] M Wesam Al-Mufti, U Hashim, M Rahman, T Adam, M Arshad, Studying Effect Dimensions of Design and Simulation Silicon Nanowire Filed Effect Biosensor , Applied Mechanics and Materials 754, 854-858, 2015

Study of the effect of surface treatment of kenaf fibre on mechanical properties of kenaf filled unsaturated polyester composite

I A S Salem[1], A R Rozyanty[1], B O Betar[2],T Adam[3,4], M Mohammed[1]and A M Mohammed[5]

[1]Center of Excellence Geopolymer and Green Technology, School of Materials Engineering, Universiti Malaysia Perlis (UniMAP), 01007, P.O Box 77, D/A PejabatPosBesar, Kangar, Perlis, Malaysia.
[2] Nanotechnology & Catalysis Research Centre (NANOCAT), University of Malaya, 50603 Kuala Lumpur, Malaysia.
[3] Faculty of Technology, Universiti Malaysia Perlis (UniMAP), KampusUniciti Alam Sg. Chuchuh, 02100 Padang Besar (U), Perlis, Malaysia.
4 Institute of Nano Electronic Engineering (INEE), Universiti Malaysia Perlis (UniMAP), 01000 Kangar, Perlis, Malaysia.
[5]College of education for pure science Ibn-Alhaitham University of Baghdad,

rozyanty@unimap.edu.my

Abstract. In this research, unsaturated polyester/kenaf fiber (UP/KF) composites was prepared by using hand lay-up process. The effect of surface treatment of kenaf fiber on mechanical properties of kenaf filled unsaturated polyester composites were studied. Different concentrationsof stearic acid (SA) were applied, i.e. 0, 0.4, and 0.8 wt%. Tensile strength of untreated UP/KF composites was found to be higher for 40 wt% loading of kenaf fiber. The highest tensile strength value was obtained after treatment with 0.4 wt% concentration of stearic acid at 56 MPa and tensile modulus was at 2409 MPa. From the flexural strength result obtained, it is clearly seen that 40 wt% loading of kenaf fiber and treatment with 0.4 wt% concentration of stearic acid give the highest value at 72 MPa and flexural modulus at 3929 MPa.

1.Introduction

The demand of high performance materials has led to extensive research and development of new and improved materials, such as composite materials. In recent years, natural fibers seem to be a superior material which has emerged as an abundant and sustainable replacement for the nonrenewable and expensive synthetic fibers. Several natural fibers have been used as a reinforcement in composites materials for advanced applications such as banana, sisal, oil palm, jute, kenaf and coir. Among different types of fibers, kenaf fiber has been widely explored over the past several years [1-3].

Kenaf (Hibiscus cannabinus, L. family Malvacea) has been discovered to be one of the main resource of natural fibers to be used for composites, and several other industrial applications. Kenaf fiber has several advantages over other traditional reinforcement materials, as such it has attracted the attention of researchers globally. The availability of kenaf fiber in Asia is more and it is abundantly available, environmental friendly, renewable, and cheap. Moreover, it has advantages over the synthetic fibers (glass, carbon and aramid fibers) such as light weight and full biodegradable. The biodegradability of kenaf fiber has a great contribution to a healthy ecosystem while their low cost and high performance satisfies the economic concern of manufacturers [4,5].

However, using kenaf also gives some disadvantages which occur during the processing stage of adding kenaf fiber into a polymer matrix. The lack of good interfacial adhesion between the fiber and polymer matrix, which results in poor mechanical properties in the final products.

Due to the chemical structure of natural fibers (such as kenaf) which is mainly has hydrophilic moiety (mainly OH), which will cause high polar characteristic. Therefore, it is important to enhance compatibility between the fiber and the polymer matrix in order to obtain a uniform composites with good properties. Several chemical treatments can be done to natural fibers in order to obtain better compatibility [6].

There were several attempts in the past years to modify the surface of natural fibers in order to enhance adhesion with the polymeric matrix. Various methods such as corona treatment [7], plasma treatment [8], mercerization [9], heat treatment [10], graft copolymerization [11,12] and silane treatment [9,13] have been reported to investigate the compatibility of natural fiber in polymeric composites, in most cases the results were positive[14-17]. However, these treatments have many disadvantages such as the use of expensive equipment or the use of expensive chemicals[18]. A very attractive treatment is to modify the surface of natural fibers with fatty acids such as stearic acid[19-24]. The principle of this treatment is the carboxyl group of stearic acid reacts with the hydroxyl groups of the fiber through an esterification reaction and, thus, reduces the hydroxyl groups number available to boned with water molecules[25-27]. One more advantage of treatment with stearic acid is that it is not sensitive to oxidation during the processing temperatures of natural fiber/polymer composites [28].

In this study, bio-composites comprising kenaf fiber and unsaturated polyester were prepared by using hand lay-up process. The effect of surface treatment of kenaf fiber on mechanical properties of kenaf filled unsaturated polyester composite were studied with different loadings of kenaf fiber and stearic acid.

2.Experimental

2.1. Materials

Kenaf fiber was obtained from RahamatullahSdn. Bhd, Malaysia. The kenaf in this study were used in the form of mat. Unsaturated polyesters resin was supplied by Castmesch Technologies Sdn. Bhd. MEKP (solution in dimethyl phthalate) used was from KaumjungAkzoNobel peroxide Ltd by trade name Butanox M60. Stearic acid was purchased from Acidchem International Sdn. Bhd, Malaysia.

2.2. Treatment with stearic acid

In this study, treatment of kenaf fiber by using stearic acid was carried out. Prior to treatment, kenaf fiber mat was cut to 20 cm×20 cm dimensions. Different concentrations of stearic acid were applied, i.e. 0, 0.4, and 0.8 wt%. Stearic acid was added to water and warm up at 75 °C to form a suspension. Kenaf fiber was then immersed in the suspensions containing stearic acid and stirred for 60 min at 70 °C. The treated kenaf fibers were then washed with distilled water until pH=7. The treated kenaf fiber layers were placed in a conventional oven to dry at 80°C overnight.

2.3.Composite preparation

Initially kenaf mat was cut to 20 cm×20 cm dimensions and treated with different concentrations of stearic acid, this followed by compressing the untreated and treated kenaf fibers with stearic acid using compress machine at 90 °C.The composites formulation is presented in Table 1. Then prior putting to the mold and hand lay-up resin to bottom of the kenaf, then the kenaf was put inside the mold. Resin impregnated into untreated and treated kenaf via hand lay-up method and 2% of MEKP was added as a hardener for all the samples.This followed by separating it from mold using hack saw. The prepared composite was measured and cut to 2.5×20 cm about 13 samples were obtained.

Table 1. Composites formulation table.

No.	Kenaf mats %	SA %	UP %
UP	0	0	100
KF1	10	0	90
KF2	20	0	80
KF3	30	0	70
KF4	40	0	60
KF1-SA0.4	10	0.4	90
KF2-SA0.4	20	0.4	80
KF3-SA0.4	30	0.4	70
KF4-SA0.4	40	0.4	60
KF1-SA0.8	10	0.8	90
KF2-SA0.8	20	0.8	80
KF3-SA0.8	30	0.8	70
KF4-SA0.8	40	0.8	60

2.4. Tensile test

Tensile test was carried out by using an Instron Universal Testing Machine (Model 5569), with crosshead speed of 2 mm/min and sample size of 140×23 mm. The tensile strength and modulus of the kenaf fiber composites before and after treatment of kenaf were determine according to ASTM D3039.

2.5. Flexural test

Flexural test was conducted using an Instron Universal Testing Machine (Model 5569) equipped with a 1 kN load cell. Flexural strength and modulus were obtained at constant crosshead speed of 1 mm/min and sample size of 140*23 mm. The flexural strength and modulus of the kenaf composites before and after treatment were determined using three-point bending test method following D7264 test method. A span of 140 mm maintaining a span to depth ratio of 23:1. The load was placed midway between the supports. Each sample was loaded until the core broke and their average is reported.

3. Results and discussion

Tensile strength of unsaturated polyester/ kenaf fiber (UP/KF) composites before and after treatment of kenaf with stearic acid (SA) are shown in Figure 1. The tensile strength results of UP by itself was very low, which is 26.734 MPa. KF/UP composites show a better tensile strength property. The tensile strength results of untreated KF/ UP composites show low tensile strength value at low loading of KF, i.e. 31.804 MPa of 10 wt% loading of KF. However, the tensile strength value was increased at higher loading of KF, i.e. 35.664 MPa for 40 wt% loading of KF. This attributed to KF which is in mat formas a reinforcement in UP composite although at higher loading.

The tensile strength property increased significantly after treatment of KF with SA. The treatment of KF with 0.4 wt% of SA showed higher tensile property as compared to that of without treatment with SA. However, the tensile property decreased thereafter treatment with 0.8% of SA, but it is still higher as compared to the composites prepared without treatment with SA. The highest tensile strength property is the composite with 40% loading of KF and 0.4 wt% treatment with SA which is 56.893 MPa.

The enhancement in properties is probably due to the higher polymer- fiber interactions. The increase in tensile strength for the UP /SA treated with stearic acid can be attributed to many factors;

such as formation of bonds between UP and KF fiber, better dispersion of filler, and by higher wettability of KF by the UP matrix, which enhances the orientation and dispersion of KF in the polymer (Colom et al., 2003). The treatment of KF would also increase the wettability and matrix interphase region, hence improve the tensile properties of the composite. The treatment helps in improving surface interaction between resin and fiber due to the increase of surface roughness resulting in superior mechanical properties. As SA concentration increased to 0.8 wt%, it can be seen that the tensile strength of UP/KF composites decreased. This can be attributed to a weakening of the cohesive strength of the fiber as a result of higher concentration of treatment with SA.

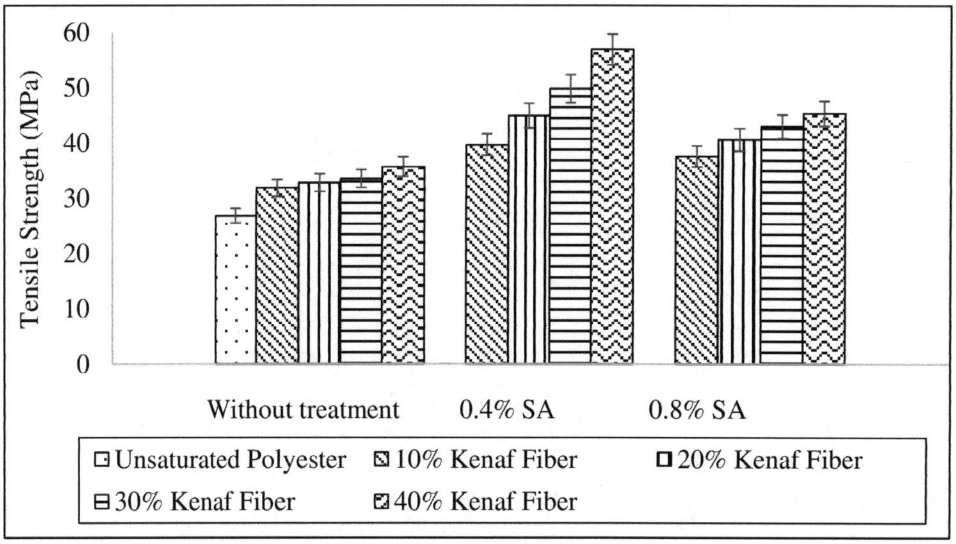

Figure 1. Tensile strength of UP/KF composites before and after treatment of KF with SA.

Figure 2 shows tensile modulus of UP/KF composites before and after treatment of KF with SA. The tensile modulus results of UP by itself was very low, which is 1175.99 MPa. UP/KF composites show a better tensile modulus property. This is due to the increase of stiffness that inherent from KF. The tensile modulus results of untreated KF/ UP composites show low tensile modulus value at low loading of KF, i.e. 1362.31 MPa of 10 wt% loading of KF. However, the tensile modulus value was increased at higher loading of KF, i.e. 1906.88 MPa for 40 wt% loading of KF. In other words, the tensile modulus of untreated UP/KF was increased by increasing the loading of KF up to 40 wt% loading of KF.

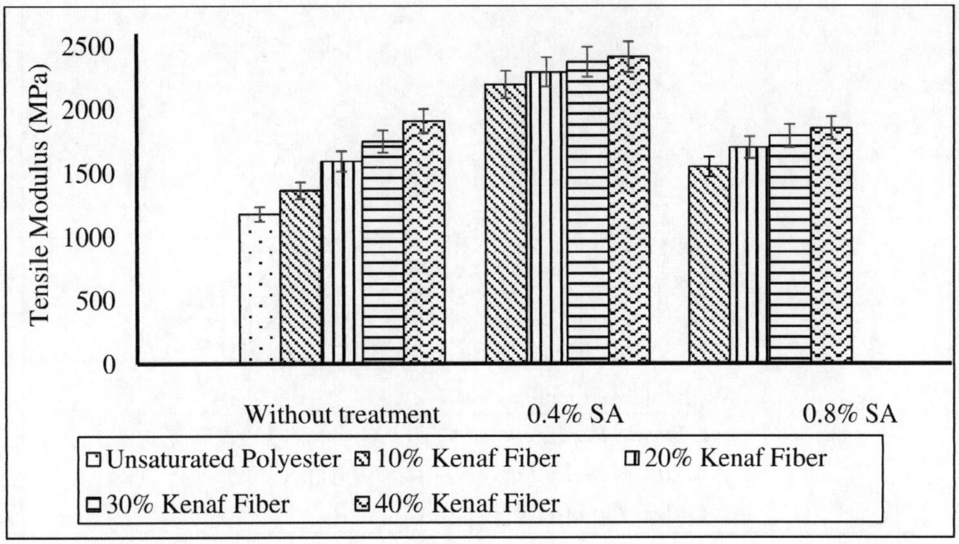

Figure 2. Tensile modulus of UP/KF composites before and after treatment of KF with SA.

The tensile modulus property increased significantly after treatment of KF with SA. The treatment of KF with 0.4 wt% of SA shows higher tensile modulus as compared to that of without treatment with SA. However, the tensile property decreased thereafter with treatment with 0.8 wt% SA. The highest tensile modulus value is the composite with 40 wt% loading of KF and 0.4 wt% treatment with SA which is 2409.84 MPa. This is attributed to the better stress transfer from matrix to the treated KF due to the SA treatment that improved the stress transfer and led to uniform stress distribution in the composite. Tensile modulus also showed the same trend as tensile strength. The tensile modulus of UP/KF treated with SA decreased due to degradation of KF. However, the tensile modulus of 0.8 wt% SA treated composites were higher than untreated UP/KF composites.

Figure 3 shows the elongation at break of UP/KF composites before and after treatment of KF with SA. The elongation at break results of UP by itself was high, which is 2.3%. It can be seen clearly that the addition of KF showed significant decrease of the elongation at break of the composites. The reason of the decrease in elongation at break of the composites was due to the lower elongation at break of KF compared to UP matrix. According to Bismark et al., the elongation at break of KF is 1.6 % while and UP is around 2 %. Generally, increasing the fiber content in UP led to a lower elongation at break of its composite. In addition, as the fiber content increased to 40 wt%, the UP was insufficient to wet the fiber entirely and led to poor interfacial bonding between the fiber and the matrix. When forced is applied, the composite had tendency to fail rather to elongate. This result indicates that the polyester effectively transferred the tensile force to the kenaf fiber. The reinforcement of kenaf fiber eventually absorbed the tensile force.

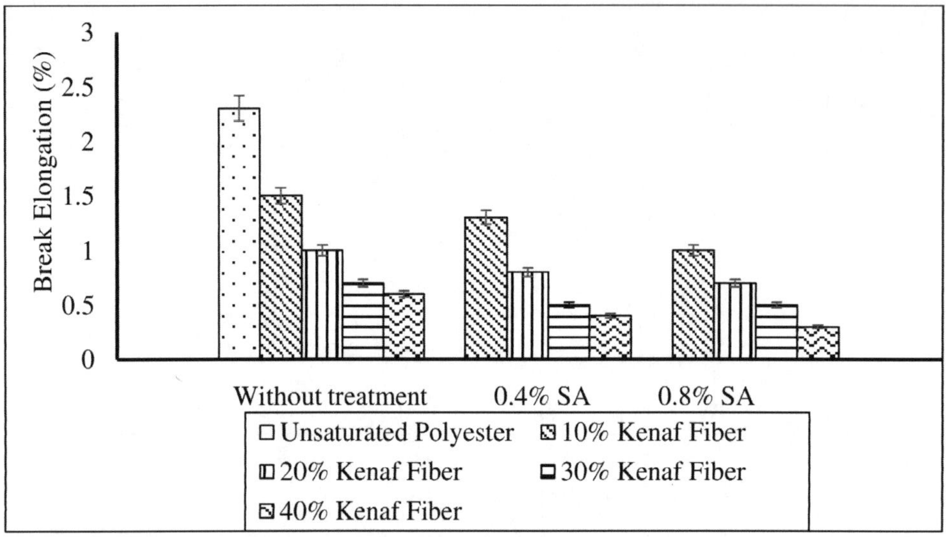

Figure 3. The elongation at break of UP/KF composites before and after treatment of KF with SA.

The elongation at break property slightly decreased after treatment of KF with SA. The treatment of KF with 0.4 wt% of SA shows lower elongation at break as compared to that of without treatment with SA. However, the elongation at break decreased thereafter with treatment with 0.8 wt% SA. The elongation at break value of 40 wt% loading of KF and 0.4% treatment with SA which is 0.4% and it is 0.3% for the same content of fiber for 0.8 wt% treatment with SA. This decrement is attributed to the creation ofa certain interface and the rigidity of the fibers are responsible of this decrease.

Flexural tests were also carried out to have a better understanding of the mechanical properties of the composites before and after treatment of KF with SA. Figure 4 shows the flexural strength of UP/KF composites before and after treatment of KF with SA. UP shows a flexural strength value of 40.48 MPa, however this value was increased after incorporation KF with UP. The results show an increment in the flexural strength value by increasing the loading of KF up to 40 wt%.

An improvement of the flexural strength was observed after treatment of KF with 0.4% of SA and when KF loading increased up to 40 wt%. This reveal a clear evidence that treatment of KF with 0.4 wt% of SA gives better compatibility between KF and its matrix. However, the flexural strength value decreased after treatment of KF with 0.8 wt% of SA. This can be due as we increase the SA concentration, KF became degraded.

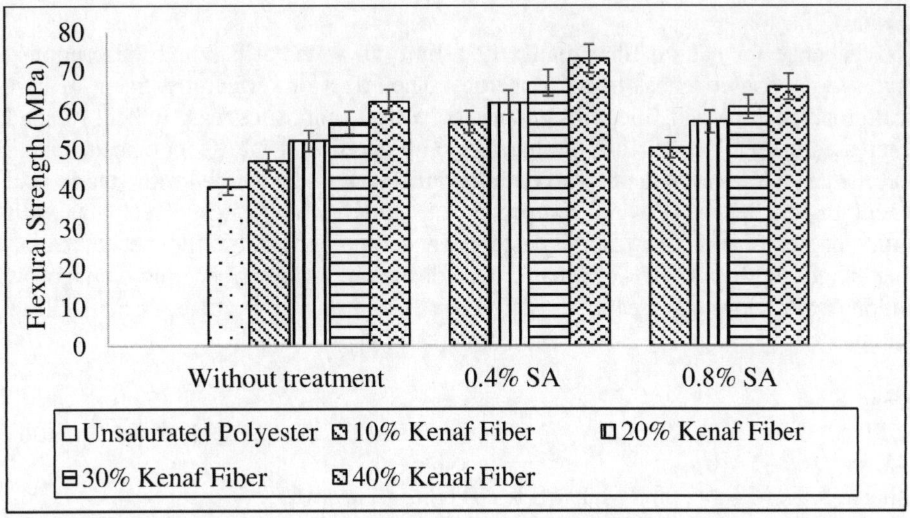

Figure 4. Flexural strength of UP/KF composites before and after treatment of KF with SA.

Figure 5 shows the flexural modulus of UP/KF composites before and after treatment of KF with SA. The flexural modulus showed a similar trend as tensile modulus. The flexural modulus increased by increasing the loading of KF up to 40 wt% and it exhibits 3781.62 MPa with 40 wt% loading of KF. After treatment with 0.4 wt% and 0.8 wt% of SA, the flexural modulus was higher for treatment with 0.4 wt% of SA as compared to that of 0.8% of SA. The reason can be due to achievement of higher compatibility and interfacial adhesion between KF and its matrix by treatment with 0.4 wt% of SA. This finding is in agreement with a study by Huda et al., 2007, which also found the improvement in flexural modulus of the PLA/kenaf fiber composites when the fiber was treated with chemical treatment. They found that the increment is influenced by the good compatibility between fiber and matrix.

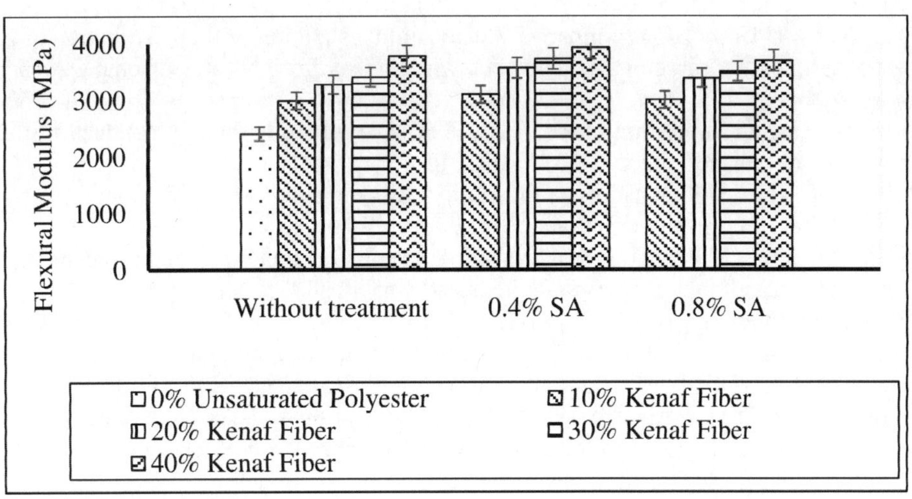

Figure 5. Flexural modulus of UP/KF composites before and after treatment of KF with SA.

4. Conclusion

The major challenge for natural fiber matrix its affinity to water, it has high absorption to moisture which have serious consequences to its properties. The effect of surface treatment of kenaf fiber on mechanical properties of kenaf filled unsaturated polyester composites was studied. As the kenaf fiber loading increased from 10 to 40 wt%, the mechanical properties of UP/KF composites increased. This shows that the kenaf fiber acts as reinforced material. Treated kenaf fiber with stearic acid shows an improvement in tensile and flexural properties of UP/KF composites. Treatment with 0.4 wt% concentration of stearic acid gives the highest mechanical properties, but the mechanical properties of kenaf fiber treated with 0.8 wt% of stearic acid decreased as the fiber was damaged with higher concentration of SA. However, treatment with 0.8 wt% concentration of stearic acid still gives higher mechanical properties comparing to that of untreated UP/KF.

References

[1] Akil H M, Omar M F, Mazuki A A M, Safiee S, Ishak Z A M and Abu Bakar A 2011 *Mater. Des.* **32** 4107
[2] Acharya S K., Mishra P and Mehar S K 2011*Bioresources***6**3155
[3] RajuG, SaidMMandAhmadM A 2008 *J. Rubb. Res.***11**187
[4] Lei Y, Wu Q, Yao F and Xu Y 2007*Composites Part A***38** 1664
[5] AbdullahH H, Asa'ariA Z M, ZawawiN I M, AbdullahL Cand ZakariaS 2013 *BioResources***8** 4088
[6] Colom X, Carrasco F, Pages P andCanavate J 2003 *Comp. Sci. and Tech***63** 161
[7] Belgacem M N, BataillePandSapeiha S 1994 *J ApplPolymSci***53** 379
[8] Felix J M, Carlsson C M G andGatenholm P 1994 *J AdhesSciTechnol***8** 3043
[9] Bisanda E T N and Ansell M P 1991 Compos *SciTechnol***41** 165–178.
[10] Sapieha S, Pupo J F and Schreiber H 1989 *J ApplPolymSci***37** 233
[11] Felix J M and Gatenholm P 1991 *J ApplPolymSci***42** 609
[12] Maldas D, Kokta B VandDaneault C 1989 *J ApplPolymSci***37**751
[13] Mieck K P, Nechwatal A and Knobelsdorf C 1995 *Die AngewandteMakromolekulareChemi* **224** 73
[14] GessJ M andRodriguezJ M 2005*The sizing of Paper*(Atlanta: Tappi Press)
[15] AW Al-Mufti, U Hashim, M Rahman, T Adam An Investigation of the Distribution of Electric Potential and Space Charge in a Silicon Nanowire, Zeitschrift für Naturforschung A 69 (10-11), 597-605, 2014
[16] T Adam, U Hashim Novel Iron-oxide Catalyzed CNT Formation on Semiconductor Silicon Nanowire Current nanoscience 10 (5), 695, 2014
[17] T Adam, U Hashim Silicon nanowire fabrication: Silicon trimming via shallow anisotropic etching, Microelectronics International 31 (2), 78-85, 2014
[18] ME Ali, U Hashim, S Mustafa, YB Che Man, T Adam, Q Humayun Nanobiosensor for the detection and quantification of pork adulteration in meatball formulation, Journal of Experimental Nanoscience 9 (2), 152-160, 5, 2014
[19] M Wesam Al-Mufti, U Hashim, T Adam The State of the Arts: Simulation of Nanostructures Using COMSOL Multiphysics, Advanced Materials Research 832, 206-211, 2014
[20] V Perumal, U Hashim, T Adam Mask design and fabrication of Micro/Nanowire biochip for reliable and repeatability pattern transfer, Advanced Materials Research 832, 79-83, 2014

[21] Tijjani Adam, U Hashim, Design and fabrication of micro-mixer with short turns angles for self-generated turbulent structures, Microsystem Technologies, Vol. 22(2) 433-440, 2016

[22] Tijjani Adam, U Hashim, Detection of specific single-stranded DNA molecules through SiNW surface modulation, Microsystem Technologies, Vol. 22(2) 269-273, 2016

[23] A Wesam Al-Muftil, U Hashim, Ausaila Fraihat, Tijjani Adam Studying on design and simulation of electronic IR guide for blindness, AIP Conference Proceedings, Vol 1775 (1) 030103, 2016

[24] Subash CB Gopinath, Thangavel Lakshmipriya, MK Md Arshad, CH Voon, Tijjani Adam, Uda Hashim, Harbant Singh, Suresh V Chinni, Shortening full-length aptamer by crawling base deletion–Assisted by Mfold web server application, Journal of the Association of Arab Universities for Basic and Applied Sciences, 2016/8/25

[24] Shahidah Arina Shamsuddin, Mohd Nazree Derman, Uda Hashim, Muhammad Kashif, Tijjani Adam, Nur Hamidah Abdul Halim, Muhammad Faheem Mohd Tahir, Nitric acid treated multi-walled 'carbon nanotubes optimized by Taguchi method, AIP Conference Proceedings, Vol, 1756 (1) 090002, 2016

[26] Tijjani Adam, U Hashim, Highly sensitive silicon nanowire biosensor with novel liquid gate control for detection of specific single-stranded DNA molecules, Biosensors and Bioelectronics. 67(2015) 656–661

[27] Tijjnai Adam U. Hashim and Th S. Dhahi, Silicon nanowire formed via shallow anisotropic etching, Si-ash-trimming for specific DNA and electrochemical detection, Chinese Physics B Vol. 24, No. 6 (2015) 06810

[28] Tijjani Adam, U Hashim, Design and fabrication of micro-mixer with short turns angles for self-generated turbulent structures, Microsystem Technologies, 1-8, 2015

[29] M Wesam Al-Mufti, U Hashim, M Rahman, T Adam, M Arshad, Studying Effect Dimensions of Design and Simulation Silicon Nanowire Filed Effect Biosensor , Applied Mechanics and Materials 754, 854-858, 2015

The weathering effect in natural environment on kenaf blast and unsaturated polyester composite

M Mohammed[1] , A R Rozyanty[1], B O Beta[4], T Adam[2,3], A F Osman[1], I A S Salem [1] ,O S Dahham[2], M N Al-Samarrai[2] and A M Mohammed [5]

[1]Center of Excellence Geopolymer and Green Technology, School of Materials Engineering, Universiti Malaysia Perlis (UniMAP), 01007, P.O Box 77, D/A PejabatPosBesar, Kangar, Perlis, Malaysia.

[2]Faculty of Technology, Universiti Malaysia Perlis (UniMAP), KampusUnicitiAlamSg. Chuchuh, 02100 Padang Besar (U), Perlis, Malaysia

[3]Institute of Nano Electronic Engineering (INEE), Universiti Malaysia Perlis (UniMAP), 01000 Kangar, Perlis, Malaysia

[4]Research Center (NANOCAT),University of Malaya, Kuala Lumpur 50603 Malaysia

[5]College of education for pure science Ibn-Alhaitham University of Baghdad,

rozyanty@unimap.edu. my

Abstract. A Kenaf composite was prepared by using hand lay-up process. The effect of weather on mechanical, morphology and thermal properties of kenaf composite were studied. Tensile strength of kenaf compositewas found to be 60MPa. Unfortunately, tensile strength of thecomposite starts to decrease after the first weathering month through to the weathering periods with constant reduction of tensile, at the end of the wreathing period.,almost 85% the composite mechanical behaviour is lost. From the mechanical properties result obtained it clearly seen that natural fiber and their composites are not able to stand environmental condition because they have poor wettability, incompatibility with some polymeric matrices and high moisture absorption. Due to the high moisture absorption properties, there are formations of void in interfacial adhesion between fiber and matrix which can reduce the mechanical properties of composite such as flexural strength or flexural modulus this clearly supported by SEM results. Fortunately some modification can do towards improving the mechanical properties and it is good enough to achieve the high performance of the composite with proper system formulation during the modification processand the result of reinforced kenaffiber will be presented in our next publication.

1. Introduction

Considerable interest has been shown research community to work with kenaf and Kenaffiber is unique and potentially reliable natural material that occupied the mind scientist engineered and the development of high-performance engineering products made from natural resources is increasing worldwide, due to renewable and environmental issues[1]. Among the many different types of natural resources, kenaf plants have been extensively exploited over the past few years[2]. Kenaffiber is extracted from bastfiber of kenaf plants and properties of kenaffiber composite are comparable to conventional fiber composites, moreover, kenaffiber composite can be produced using conventional fiber composite manufacturing and kenaffiber composite have a bright future due to its renewability and eco-friendly. Hence this work is presenting reinforcing of glass fiber with kenaffiber [3]. The quest to develop high-performance materials made from natural obtained from resources is increasing worldwide, especially the material with integrity to be used in much application[1,4].

The greatest challenge in working with natural fiber reinforced plastic composites is their large variation in properties and characteristics with respect to different condition [5[. Bio composite's properties are influenced by a number of variables, including the fiber type, environmental conditions (where the plant fibers are sourced), processing methods, and any modification of the fiber. Recently there has been a surge of interest in the industrial applications of composites containing biofibers reinforced with biopolymers [6]. In the other hand, there is a growing trend to use bio fibers as fillers and/or reinforces in plastics composites[7]. This is due their flexibility during processing, highly specific stiffness, and low cost (on a volumetric basis) makes them attractive to manufacturers[8]. This century has witnessed ever-increasing demands for the utilization of plastics as important raw materials, more than 80% of which are thermoplastics[9-15]. However, using kenaf constitute some disadvantage which occur during the addition of natural fibers, including kenaffiber into a polymer matrix, is the lack of good interfacial adhesion between the two components, which results in poor properties in the final product[3].

Another problem is affected by humidity aging is widely recognized as one of the main causes of long-term failure of an organic matrix exposed to the atmosphere; or in contact with an aqueous media[15,16]. There are several recognized modes of humidity aging, namely plasticization of the matrix; differential swelling related to concentration gradients, linked the degradation of the macromolecular [17-25].Kenaffibers, as reinforcement, have recently attracted the attention of researchers because of their advantages over other established materials[26].

Moreover, They are environmentally friendly, fully biodegradable, abundantly available, renewable and cheap and have low density[27-33]. Plant fibers are light weight compared to glass, carbon and aramid fibers. The biodegradability of plant fibers can contribute to a healthy ecosystem while their low cost and high performance fulfills the economic interest of industry[35,36]. When natural fiber-reinforced plastics are subjected, at the end of their life cycle, to combustion process or landfill, the released amount of CO_2 of the fibers is neutral with respect to the assimilated amount during their growth[37]. The abrasive nature of fiber is much lower which leads to advantages in regard to technical process and recycling process of the composite materials in general [37-40]. Natural fiber-reinforced plastics, by using biodegradable polymers as matrices, are the most environmental friendly materials, which can be composed at the end of their life cycle [41].

Natural fiber composites are used in place of glass mostly in non-structural applications [42]. A number of automotive components previously made with glass fiber composites are now being manufactured using environmentally friendly composites [43].Glass fibers (GF) are the most common reinforcement for polymeric matrix composites [44]. Their principal advantages are the relationship between their high tensile strength, high chemical resistance, and insulating properties [45]. The disadvantages are low tensile modulus, relatively high specific gravity, sensitivity to abrasion during handling, low fatigue resistance, and high hardness and the compositional difference and low manufacturing cost make it more attractive materials[46].The best way to increase the material efficiency without sacrificing environmental safety is to employ natured fiber reinforced composite materials[47-49]. Producing the material with the focus on improvement aspects is very important in various applications [50]. The goals are to increase the performance of the materials and also to find the solution to reduce the cost hence able to reduce the production cost[50-54]. The latest incorporation natural material develops plastic based material has resulted in higher material properties and more possibilities use in many different applications [55-60]. However the use glass fiber becomes at higher cost than long natural fiber reinforced polymer, additional costs incurred by using glass include the added compounding and using specialized equipment to process [50, 61].

2. Methods

Kenaffiber was obtained from RahamatullahSdn. Bhd, Malaysia. All the kenaf mats in this study were used without any surface treatment. A standard unsaturated polyesters resin was supplied by CastmeschTechnilogiesSdn. Bhd.MEKP(solution in dimethyl phthalate) used was fromKaumjungAkzoNobelperoxide Ltd by trade name Butanox M60. Initially kenaf mat was cut to 20×20 cm dimensions (four layers), this followed by compressing them using compress machine at 90^0C for ten minute. This process was done to remove air between the kenaffibers and to uniform thickness. The first layer of kenaffiber was put in the mold and hand lay up the resin (unsaturated polyester +hardener (2 %)) and with ratio of 30:70 wt% (fiber to resin) to it. Resin impregnated into kenaf fiber mat via hand lay-up method. Where in impregnated into 3 layers to make purekenaf is the topcomposite after that compressed the composites for 10min without heating at 5MPa and allowed it to cure at room temperature for 24hrs. This followed by separating it from mold using hack saw. The prepared composite was measured and cut to 2.5 × 20 cm about 8 samples were obtained. The was followed by preparing the sample for test and the first test was tensile and in this case, the sample need to measure with the gauge length of 140mm to be ready for tensile test (the speed 2 mm/min). To get the properties such as ultimate tensile, strength, yield strength, break elongation and modulus of elasticity and also do scanning electron microscopic test to observed degradation process that will happen in natural fiber and matrix . We compared the composite in room temperature and composite after exposure to weathering by two weeks for three monthsIf a figure has parts these should be labelled as (a), (b), (c) etc on the actual figure. Parts should not have separate captions.

3. Results and Discussion

The preparation of kenaf fiber was completed successfully and kenaf fiber composite showed very tensile properties. This is due to the improvement in adhesion between fibers to polyester matrix.The composites with the kenaf exhibited higher tensile and module and strength compared to other material. The strength is decrease when they are exposed to weather. As expected, the results of the tensile tests, in terms of tensile strength, tensile modulus breaking elongation show similar trend. Fives samples (pure kenaf composite fiber hybrid composite with different exposure were tested (before exposure, 1 month of exposure, 2 months of exposure, 3 month of exposure in an interval of 2 weeks) were tested.

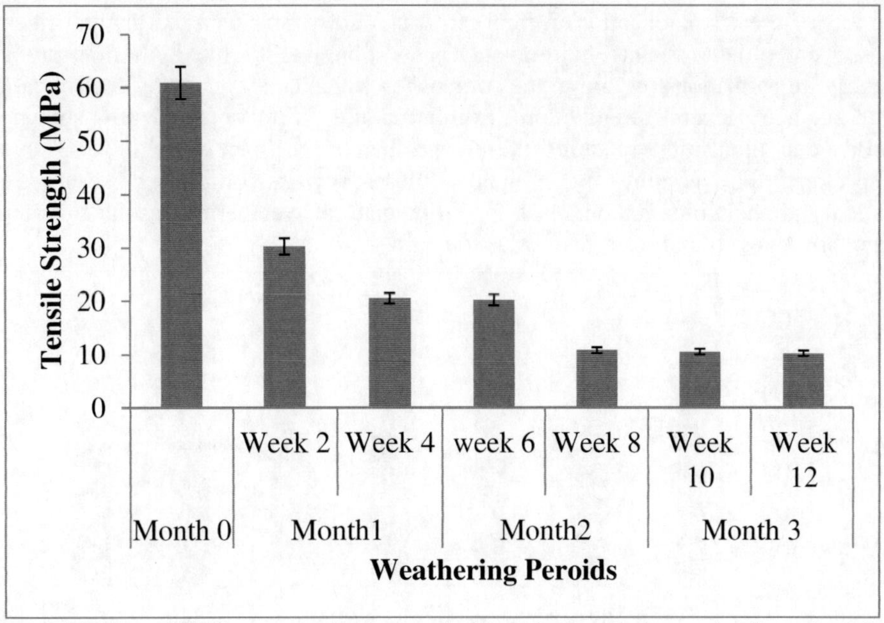

Figure1. Effect of different weathering periods on the tensile strength of purekenaf composite.

The degradation process due to the weathering is usually affecting the mechanical properties of the composite. Figure 1, show the tensile strength of the pure kenaf composites.Unexposed pure kenaf composite show approximately 60.8 MPastrength and is very important to note that, decrement trend of tensile strength of kenaf composite. From the results, it can be seen that the kenaf composite has significant decrement for before exposure compared to the after exposure,due to natural plant fibers are susceptible to biodegradation thus their composites face high risk of degradation when subjected to outdoor application When natural fiber composite exposed to moisture, water penetrates and attached ontohydrophilic groups of fiber establishing intermolecular hydrogen bonding with fibersand reduced interfacial adhesion of fiber to matrix this lead to poor transfer stress from matrix to fiber will result to lower tensile strength of pure kenaf composite .

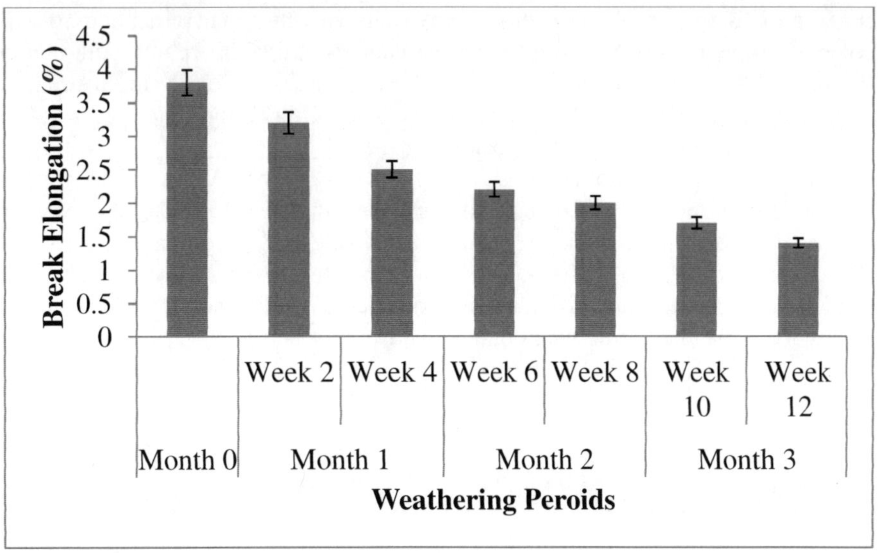

Figure 2. Effect of different weathering period on the elongation at break of pure kenaf composite.

From the results of break at elongation in figure 2, it can clearly be seen that natural fiber and their composites are not able to stand environmental condition due to the high moisture absorption properties, there are formations of void in the composites which can reduce the mechanical properties of composite such as flexural strength or flexural modulus. Similar trend also shown by many researchers.Break at elongation modulus of the weathered specimen were lower than those not weathered specimen. This is common for lignocellulose polymer composites because exposure to photo-oxidation and thermo-oxidation reaction during natural weathering results in a more brittle composite and brittleness of polymer matrix due to chain scission.

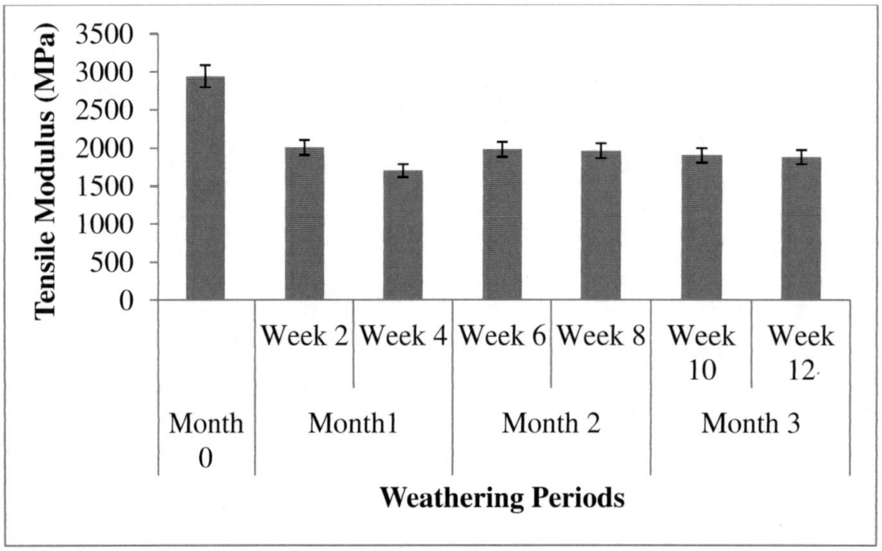

Figure 3. tensile modulus of the composite with weathering periods.

From the figure 3 the tensile modulus shows the same trend with tensile strength.
Incorporation of fiberbefore weathering has better the tensile modulus of kenaffibercomposite to 3000 MPa compared to after weathering kenaf composite 2000 MPa. When expose to environment, it can be seen that the tensile modulus decrease for kenaf composite. Kenaffiber provide stiffness to the composite but when expose for a longer time to environment, fiber-matrix debonding will occur in pure kenaf composite due to the weathering effect. Besides moisture absorption, both polymer matrix and kenaffiber in pure kenaf composite will experience solar degradation during natural weathering. Photo degradation upon ultraviolet (UV) exposure is expected to be the major cause of changes to surface appearance and chemistry of the composite.
Morphological study of the pure kenaf composite were performed by SEM for the determination of the mode or cause of failure and to fully reveal the mechanical behavior that was seen in the above section. The effect of weather on properties of the composite and pure kenaf can be evaluated by comparing the morphologies of the pure kenaf composite with different weathering periods. All the possible effects on composite properties must be considered which indicates such behaviour. The micrographs with 500X magnification were obtained and shown in Figure 4.

IOP Conf. Series: Journal of Physics: Conf. Series **908** (2017) 012003 doi:10.1088/1742-6596/908/1/012003

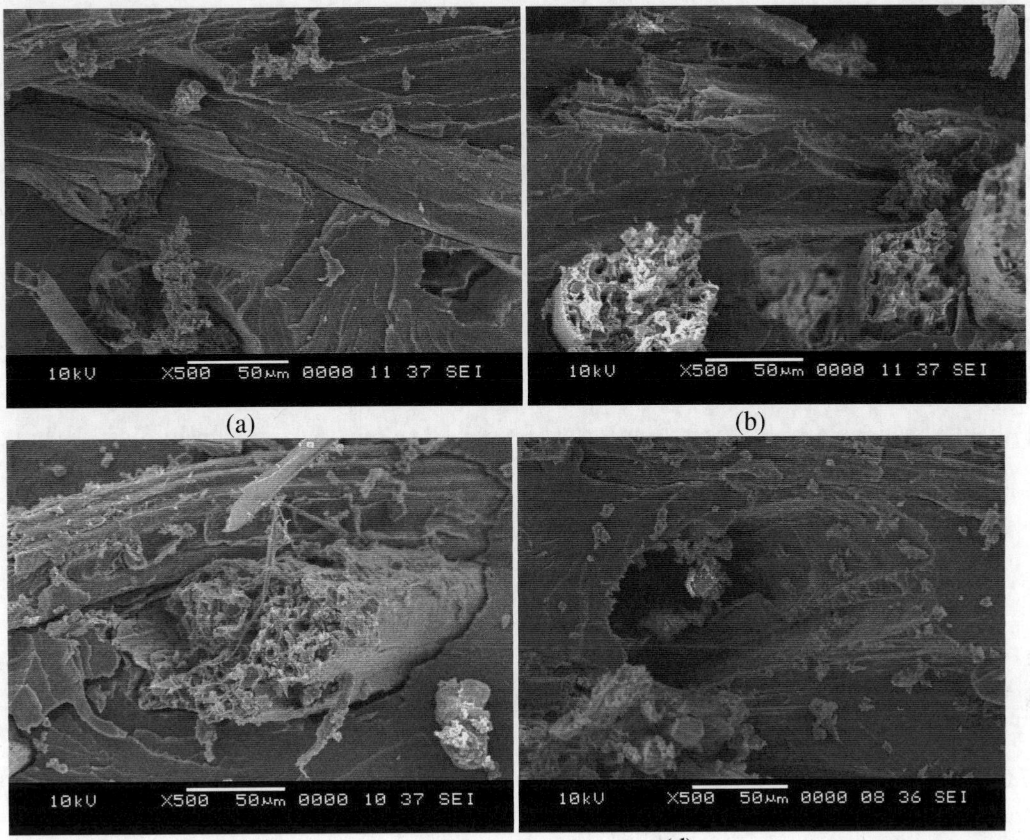

(c) (d)

Figure 4. Micrograph of thekenaf composite by SEM at 500X magnification (a) kenaf composite, (b) kenaf composite after one month of exposure (c) kenaf composite after two months exposure (d) kenaf composite after three month of exposure.

As shown in Figure 4, the kenaf composite show very good and smooth morpholgy compared to other three samples, as the weathering periods increased the void in the kenaf increased compared to pure kenaf before exposure. Based on the micrographs, weathering cause the kenaf degrade due to obsorption of the mosture, this equally revealed in the tensile behavior discussed in this above section. As the weathering periods increase the kenaf dislocation increase. Natural materials tend to show very low mechanical propeties because they do not plastically deformed and in agreement with tensile, flexural results. decrease of the mechanical propeties of samples is due to the formation of void in within fiber molecular bond. Micrographs figure 4(c) and figure 4(d) show obvious picture where there are some voids in the fiber molecular chain. It can be seen from the micrographs that, the size of voids were increased with the increase in weahtering exposure. Consequently, it has caused the reduction in mechanical properties as discussed before. Micrograph shows some obvious cracked . This indicates weak interfacial interaction between natural fibre because there were micro spaces existing at the interfacial area between fibres within the same system due moisture.

4. Conclusion

The major challenge for natural fiber matrix its affinity to water, it has high absorption to moisture which have serious consequences to it properties. Al though, incorporating reinforcement in kenaf play an important role in improving the mechanical and thermal properties this is could be due to the maintaining of structural integrity and bonding of the fibre from the matrix, resulting from the prevention of shear stress at the interface. However, this gradually affected due to absorbed moisture with weathering. In in conclusion there is tensile and break elongation reduction due to long term exposure to weather and the effect of weather on tensile properties of kenaf compositesthe modulus is 10MPa compared with tensile strength 60MPa before weathering. As mentioned mechanical behaviour of materials starts to decrease after the first weathering month through to the weathering periods with constant reduction tensile strength.

Acknowledgment

The author thanks to LembagaKenafdanTembakau Negara (LKTN) and UniversitiMalaysia Perlis (UniMAP) for facilities and support.

References

[1.] Abdul Malek Ya'acob1, Azhar Abu Bakar ,Hanafi Ismail and KhairulZamanDahlan Hybrid Composites: Study on Un-Treated Kenaf/Glass Fibre Properties *Pertanika J. Sci. & Technol.19* (2): (2011) 373– 381

[2.] Acharya S.K., Mishra P. and Mehar S.K. Effect of surface treatment on the mechanical properties of bagasse fiber reinforced polymer composite, *Bioresources.,6*, 3(2011) 155-3165

[3.] Aji, I.S. Sapuan,.S.M. Zainudin E.S. and Abdan K. (2009)Kenaffibres as reinforcement for polymeric composites: a review, *International Journal of Mechanical and Materials Engineering (IJMME), Vol. 4 No. 3, 239-248*

[4.] Aji. I. S., S.M. Sapuan, E.S. Zainuddin, K. Abdan (2009) Kenaffibres as reinforcement for polymeric composites: a review. . *International Journal of Mechanical and Materials Engineering (IJMME)*, Vol.4, No.3, 239-248

[5.] Akil H.M., I.M. De Rosa, C. Santulli, F. Sarasini. (2010) Flexural behaviour of pultruded jute/glass and kenaf/glass hybrid composites monitored using acoustic emission. *Materials Science and Engineering A,* 527, 2942-2950,

[6.] Alexander Bismarck, IbonAranberri-Askargorta, Jürgen Springer, T. L., Bernhard Wielage, A. S., IljaShenderovich&Limbach, H.-H.(2004) Surface characterization of flax, hemp and cellulose fibers; Surface properties and the water uptake behavior *Polymer Composites*, *23*, 872-894.

[7.] AmalNassar, EmanNassar (2013) Study on Mechanical Properties of Epoxy Polymer Reinforced with NanoSiC particles, *Nanoscience and Nanoengineering1*(2): 89-93, 2013

[8.] Anuar, H. and A. Zuraida, (2011). Improvement in Mechanical Properties of Reinforced Thermoplastic Elastomer Composite with KenafBastFibre, *Composites Part B: Engineering* 42, p. 462.

[9.] Arun K. V., S. Basavarajappa, B.S. Sherigara. (2010) Damage characterization of glass textile fabric polymer hybrid composites in sea water environment.*Materials and Design31*, pp 930 – 939.

[10.] Ayensu A. Interfacial debonding of natural fibre reinforced composites (2000).*QuartSci Vision*;6:25–34.

[11.] Aziz, S. H. & Ansell, M. P. (2004) The effect of alkalization and fibre alignment on the mechanical and thermal properties of kenaf and hemp bastfibre composites: part 2 - cashew nut shell liquid matrix. *Composites Science and Technology, 64*, 1231-1238.

[12.] Aziz, S. H., Ansell, M. P., Clarke, S. J. &Panteny, Y, S. R. (2005) Modified polyester resins for natural fibre composites. *Composites Science and Technology, 65*, 525-535.

[13.] Bernard, M., A. Khalina, A. Ali, R. Janius, M. Faizal, K.S. Hasnah, and A.B. Sanuddin, (2011). The Effect of Processing Parameters on The Mechanical Properties of KenafFibre Plastic Composite ,*Materials& amp; Design 32*, p. 1039.

[14.] Bikiaris, D. &Panayiotou, C. (1998). LDPE/starch blends compatibilized with PE-*g*-MA copolymers. *J. Appl. Polym. Sci.*, *70*, 1503–1521.

[15.] Bledzki A.K., Mamuna A.A. and Volk J., Physical (2010) chemical and surface properties of wheat husk, rye husk and soft wood and their polypropylene composites *Composite. Part A*, 41, 480-488

[16.] Bras J, Hassan ML, Bruzesse C, Hassan EA, El-Wakil NA, DufresneA.(2010) Mechanical, barrier, and biodegradability properties of bagassecellulose whiskers reinforced natural rubber nanocomposites.*Industrial Crops and Products32*, 627–33.

[17.] Chollakup R., Tantatherdtam R., Ujjin S. and Sriroth K. (2011) Pineapple leaf fiber reinforced thermoplastic composites, *J. Appl. Polym. Sci., 119*, 1952-1960

[18.] Coutinho, F.M.B., Costa, T.H.S. &Carvalho, D.L. (1997) Polypropylene-wood fiber composites: Effect of treatment and mixing conditions on mechanical properties. *J .Appl. Polym.Sci, 65*, 1227-1235.

[19.] Dhakal, H. N., Zhang, Z. Y. & Richardson, M. O. W. (2007) Effect of water absorption on the mechanical properties of hemp fibre reinforced unsaturated polyester composites. *Composites Science and Technology, 67*, 1674-1683.

[20.] El-Sayed, A. A., El-Sherbiny, M. G., Abo-El-Ezz, A. S. &Aggag, G. A. (1995) Friction and wear properties of polymeric composite materials for bearing applications. *Wear, 184*, 45-53.

[21.] Facca A.G., Kortschot MT & Yan N. (2006) Predicting the elastic modulus of natural fibre reinforced thermoplastics. *Comp Part A-Ap. Sci. and Manufacturing, 37*, 1660-1671.

[22.] Fiorea, V., Bellab, G. D.,&Valenzaa, A. (2015). The effect of alkaline treatment on mechanical properties of kenaf fibers and their epoxy composites.*Composites Part B: Engineering.68.*14–21

[23.] Gassan, J. &Cutowski V. S. (2000).Effect of corona discharge and UV treatment on the properties of jute-fibre epoxy composite.*ComposSciTechnol, 60*, 2857-2863.

[24.] Gomes, A., Goda, K. &Ohgi, J. (2004).Effects of Alkali Treatment to Reinforcement on Tensile Properties of Curaua Fiber Green Composites.*JSME International Journal Series A Solid Mechanics and Material Engineering, 47*, 541-546.

[25.] Hill, C. A. S., Abdul, H. P. S. & Khalil. (2000). The effect of environmental exposure upon the mechanical properties of coir or oil palm fiber reinforced composites. *Journal of Applied Polymer Science, 77*, 1322-1330.

[26.] Huda, M. S., Drzal, L. T., Mohanty, A. K. and Misra, M. (2006). Chopped Glass and Recycled Newspaper as Reinforcement in Injection Molded Poly(lactic acid) (PLA) Composites: A Comparative Study. *Composites Science and Technology, 66* 1813-1824.

[27.] Iannace, S., Nocilla, G. &Nicolais, L. (1999).Biocomposites based on sea algae fibers and biodegradable thermoplastic matrices. *Journal of Applied Polymer Science, 73*, 583-592.

[28.] Irene S. Fahim, Salah M. Elhaggar, HatemElayat.(2012). Experimental Investigation of Natural Fiber Reinforced Polymers, *Materials Sciences and Applications*, 2012, 3, 59-66

[29.] Jochen, G., Andrzej, K.,&Bledzki. (1999). Possibilities for improving the mechanical properties of jute/epoxy composites by alkali treatment of fibers.*Composites Science and Technology. 59,* 1303–1309.

[30.] Joseph PV, Rabello MS, Maattoso LHC, Joseph K, Thomas S. (2002). Environmental effects on the degradation behavior of sisal fibre reinforced polypropylene composites. *Compos SciTechnol*;*62*:1357–72.

[31.] Joshi, S.V. DrzalL.T. ,Mohanty A.K., Arora S. (2004) Are natural fiber composites environmentally superior to glass fiber reinforced composites, *Composites: Part A 35* 371–376

[32.] Karmakar AC, Hoffmann A, Hinrichsen G. (1994) Influence of water uptake on the mechanical properties of jute fibre reinforced polypropylene. *J ApplPolymSci*;54:1803–7.

[33.] Karmakar, A. C., Hoffmann, A. G. H. (1994.) Influence of Water Uptake on the Mechanical Properties of Jute Fibre Reinforced Polypropylene. *Journal of Applied Polymer Science, 54*, 1803-1807

[34.] Mansur, M. A. and M. A. Aziz. (1983). Study of Bamboo-Mesh Reinforced Cement Composites. *Int. Cement Composites and Lightweight Concrete,, 5*, 165–171.

[35.] Maria, L. S., Corneliu, C., Teodorescu-D.H.,Janos, T. (2012). Properties of Advanced New Hemp Fiber Materials Used inAutomotiveEngineering.*Advances in Production, Automation and Transportation Systems. 978-1-61804.*193

[36.] Mariatti, W.L. Lai, M. and M. Jani. (2008) The properties of woven kenaf and betal palm (Areca catechu) reinforced unsaturated polyester composites. *Polymer Plastic technology and Engineering*, *47*, 1193-1199.

[37.] MohdSuhairilMeon,,MuhamadFauzi Othman, Hazran Husain, Muhammad FairuzRemeli, MohdSyaharMohdSyawal (2012). Improving tensile properties of kenaf fibers treated with sodium hydroxide, *Procedia Engineering 41*(2012) 1587 – 1592

[38.] MohdYuhazri, Y., Phongsakorn, P.T., HaeryipSihombing, Jeefferie A.R., PuvanasvaranPerumal, Kamarul, A.M., and KannanRassiah.(2012). Mechanical Poperties of Kenaf/polyester composites. *International Journal of Engineering & Technology IJET-IJENS*, 11(1), 127-131

[39.] Omar F., Andrzej K. Bledzki Hans-Peter F., Mohini S. (2012). Biocomposites reinforced with natural fibers: 2000–2010, *Progress in Polymer Science 37*, 1552– 1596

[40.] Poostforush M., Al-Mamun M. and Fasihi M. (2013). Investigation of Physical and Mechanical Properties of High Density Polyethylene/Wood Flour Composite Foams,*Res. J. Engineering Sci*., 2(1), 15-20

[41.] Prachayawarakorn, J. et al. (2010). Property improvement of thermoplastic mung bean starch using cotton fiber and low-density polyethylene.*Starch – Stärke*, 62(8), 435–443.

[42.] Prachayawarakorn, J., Sangnitidej, P. &Boonpasith, P. (2010). Properties of thermoplastic rice starch composites reinforced by cotton fiber or low-density polyethylene. *Carbohydr.Polym.*, 81(2), 425–433.

[43.] Rashdi, A. A. A., S. M. Sapuan, M. M. H. M. Ahmad &Khalina, A. (2 0 0 9) Water absorption and tensile properties of soil buried kenaffibre reinforced unsaturated polyester composites (KFRUPC). *Journal of Food, Agriculture & Environment, 7*, 9 0 8 - 9 1 1.

[44.] Sharma, N. et al. (2001). A study on the effect of pro-oxidant on the thermo-oxidative degradation behaviour of sago starch filled polyethylene. *Polym.Degrad. Stab.*, 71, 381–393.

[45.] Stamboulis A, Baillie CA, Peijs T. (2001) Effects of environmental conditions on mechanical and physical properties of flax fibers.*Compos Part A ApplSciManuf* 2001;32:1105–15

[46.] SuhailahMohdSukri, NorLiyanaSuradi, AgusArsad, Abdul RazakRahmat and Azman Hassan. (2012). Green composites based on recycled polyamide-6/recycled polypropylene kenaf composites: mechanical, thermal and morphological properties. *Journal of Polymer Engineering*. 32, 291-299

[47.] SupraneeSangthong, ThirawudhPongprayoon, NantayaYanumet. (2009) Mechanical property improvement of unsaturated 5.polyester composite reinforced with admicellar-treated sisal fibers. *Composites: Part A* 40, pp 687–694.

[48.] Tan, C., Ahmad, I., Heng, M. (2011). Characterization of polyester composites from recycled polyethylene terephthalate reinforced with empty fruit bunch fibers, *Materials and Design*.32 (8-9): 4493-4501

[49.] Threepopnatkul, P N. Kaerkitcha, N. Athipongarporn. (2009) Effect of surface treatment on performance of 7.pineapple leaf fiber–polycarbonate composites. *Composites: Part B 40*, pp 628–632.

[50.] Zampalonib, M.,Pourboghrata, F., Yankovicha, S.A., Rodgersa, B.N., Moorea, J., Drzalb,L.T and Mohantyc, A.K. (2007). Kenaf natural fiber reinforced polypropylene composites: A discussion on manufacturing problems and solutions. *Composites Part A: Applied Science and Manufacturing*. *38*. 1569–1580.

[51.] Zou Yi, Shah Huda, Yiqi Yang. (2010). Lightweight composites from long wheat straw and polypropylene web. 3. *Bioresource Technology 101*, pp 2026–2033.

[52.] T Adam, U Hashim, TS Dhahi, KN Khor, PS Chee, PL Leow, ELECTROCHEMICAL ETCHING: An Ultrasonic Enhance Method of Silicon Nano Porous Fabrication, Wulfenia Journal 20 (1), 45-55, 2013

[53.] Tijjani Adam, H Uda, M Eaqub, PL Leow, The electroosmosis mechanism for fluid delivery in PDMS multi-layer microchannels, American Scientific Publishers All rights reserved.2013
[54.] U Hashim, T Adam, J Lung, P Ling, Fabrication of Microchannel and Micro Chamber for Microfluidic Lab-on-Chip.Australian Journal of Basic & Applied Sciences 7 (1)2013
[55.] T Adam and U. Hashim Light Observation in Polymer: A Study of Silicon-Based Organic Polymer Using Visible Spectroscopy, Australian Journal of Basic and Applied Sciences 7 (1), 76-80, 2013.
[56.] U Hashim, MW Al-Mufti, T Adam, Silicon Nanowire Geometry: Investigation of Interaction Site Potential in Semiconductor-DNA Interaction, Australian Journal of Basic and Applied Sciences 7 (5), 242-245, 2013
[57.] MW Al-Mufti, U Hashim, T Adam, Simulation of Nano lab on chip devices by using COMSOL MultiphysicsJournal of Applied Sciences Research 9 (2), 1056-1061, 2013
[58.] T Adam, U Hashim, KL Foo, TS Dhahi, T Nazwa, Technology development for nano structure formation: Fabrication and characterization, Advanced Science Letters 19 (1), 132-137, 2013
[59.] T Adam, U Hashim, PL Leow, KL Foo, PS Chee, Selection of optimal parameters in fabrication of poly (dimethylsiloxane) microfluidics using taguchi method, Advanced Science Letters 19 (1), 32-36,2013
[60.] T Adam, U Hashim, KL Foo, Microfluidics design and fabrication for life sciences application, Advanced Science Letters 19 (1), 48-53, 2013
[61.] T Adam, U Hashim, ME Ali, PL Leow, The electroosmosis mechanism for fluid delivery in PDMS multi-layer microchannel, Advanced Science Letters 19 (1), 12-15, 2013

IOP Conf. Series: Journal of Physics: Conf. Series **908** (2017) 012004 doi:10.1088/1742-6596/908/1/012004

Influence of weathering effect in natural environment on thermal properties hybrid kenaf blast/glass fibre and unsaturated polyester composite

M Mohammed[1], A R Rozyanty[1], B O Beta[4], T Adam[2,3], A F Osman[1] , I A S Salem [1] ,O S Dahham[2] , M N Al-Samarrai[2] and A M Mohammed [5]
[1]Center of Excellence Geopolymer and Green Technology, School of Materials Engineering, Universiti Malaysia Perlis (UniMAP), 01007, P.O Box 77, D/A PejabatPosBesar, Kangar, Perlis, Malaysia.
[2]Faculty of Technology, Universiti Malaysia Perlis (UniMAP), KampusUniciti AlamSg. Chuchuh, 02100 Padang Besar (U), Perlis, Malaysia
[3]Institute of Nano Electronic Engineering (INEE), Universiti Malaysia Perlis (UniMAP), 01000 Kangar, Perlis, Malaysia
[4]Research Center (NANOCAT),University of Malaya, Kuala Lumpur 50603 Malaysia
[5]College of education for pure science Ibn-Alhaitham University of Baghdad,

rozyanty@unimap.edu. my

Abstract. Unprecedented growing on environmental concern has put research on completive driven effort to quest for new material in various application, the effort toward producing thermally stable polymer is ever gaining considerable interest. Thus, this study proposed the integration of glass fiber with kenaf based polymer to improve thermal properties. Based on the TGA and DSC results, the composites show slow and steady initial weight loss until major weight loss at 360°C.Thus, with incorporation of glass fiber extend region of degradation until 260-360 °Cshow no exothermic or endothermic changes, this reflected that the composites thermally stability have been improved.

1. Introduction

The researchers across the globe have begun to focus attention on increasing demand for environmentally friendly materials [1-7]. Besides ecological considerations natural fibers exhibit many advantageous properties which promote the replacement of synthetic fibers in polymer composites, offer high potential for an outstanding reinforcement in lightweight structures [8-16]. Natural fibers are derived from a renewable resource and do not require a large energy requirement to process and biodegradable[17-25]. These fibers also offer significant cost advantages and therefore the utilization of lightweight, lower cost natural fibers such as jute, flax, hemp, sisal, abaca, coir offer the potential to replace a large segment of the synthetic fibers in numerous applications [25-35]. In recent years this natural fiber composites have been used in many engineering applications due to natural properties that they offer such a variety of lightweight, reform, cheap and environmentally friendly[36-40]. Natural fiber composites are used in various industries such as automotive, sporting goods, marine, electrical, construction and household appliance[41-51].Kenaf, sisal, coir, banana, hemp, pulp, wood flour, palm oil, pineapple leaf and coir is the main natural fiber used as reinforcement [50-57] .Kenaf fiber provides the hardness and strength values are high [48, 58]. They also have a higher aspect ratio which makes them suitable for use as reinforcement in polymer composites is a perennial plant herb [58-61]. Kenaf is a plant heat-season annual lane. The attractive feature of kenaf fiber is inexpensive, lightweight, Reform, biodegradability and certain properties of high mechanical. Kenafcontainsbast fiber containing 75% cellulose and 15% lignin and offers the advantage of being environmentally friendly and safe environment (Mansur & Aziz, 1983). Kenaf fibers have superior flexural strength and excellent tensile strength that makes the kenaf good candidate for many applications [48]. However, it has several drawbacks, including high moisture absorption, which have a negative impact on the mechanical properties [36-40]

2. Methods

Initially kenaf mat was cut to 20×20 cm dimensions (three layers), this followed by compressing them using compress machine at 90 0C. Two layers of fiber glass 20×20cm were also prepared. The first layer of fiber glass was put in the mold and hand lay up the resin (unsaturated polyester +hardener (2 %)) and with ratio of 30 wt %:70 wt % (fiber to resin) to it. Then prior putting to the mold and hand lay-up resin to bottom of the kenaf, then the kenaf was put inside the mold. Resin impregnated into kenaf/glass fiber mat via hand lay-up method. This processes is divided into two stages (1) for fabrication of layer of kenaf whereas in impregnated into 4 layer to make kenaf (UP) composite (2) fabrication of hybrid kenaf/glass (UP) composite impregnated resin into 3 layers of kenaf and 2 layers of glass fiber after that compressed for 10 min without heating at 5 MPa and allowed it to cure at room temperature for 24 hrs. This followed by separating it from mold using hack saw. The prepared composite was measured and cut to 2.5 × 20 cm about 8 samples were obtained. The was followed by preparing the sample for test. We compared the composite in room temperature and composite after exposure to weathering by two weeks for three months

3. Results and Discussion

Thermogravimetric analysis was carried out to observe distinct process of weight loss occurring on natural fibers and their composites at different temperature. Figures 1 &2 show the thermogravimetric analysis curves of pure kenaf composited and hybrid kenaf/glass fiber composites, respectively, before and after exposed to weathering.From figure 1, pure kenaf composites exposed and unexposed to weathering displayed three stages of weight loss. The first stage occurs at temperature range from 50-110 ^{o}C due to release of moisture which was absorbed by natural fiber. At the second stage, the major degradation occurred with temperature range at about 200-300 ^{o}C, which was related to degradation of hemi cellulose and cellulose of natural fiber. The third stage of weight loss occurred at temperature range at about 460-600 ^{o}C, which was related to degradation of lignin (Campos A, et al,. 2011). All this values mentioned prior lines will increased in hubris kenaf/glass fiber composites due to high thermal stability of glass fiber existed in the composites, which will act as a shield in the composite for burning process (Yao et al., 2008).

IOP Conf. Series: Journal of Physics: Conf. Series **908** (2017) 012004 doi:10.1088/1742-6596/908/1/012004

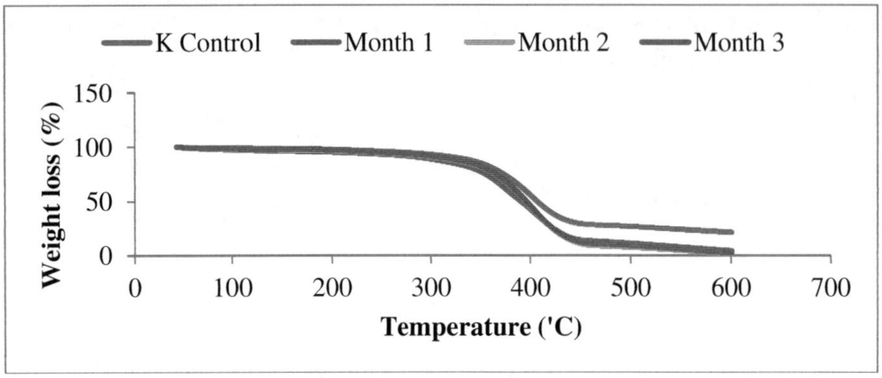

Figure 1. TGA curve Thermogravimetric curves of pure kenaf composite before and after exposed to weathering.

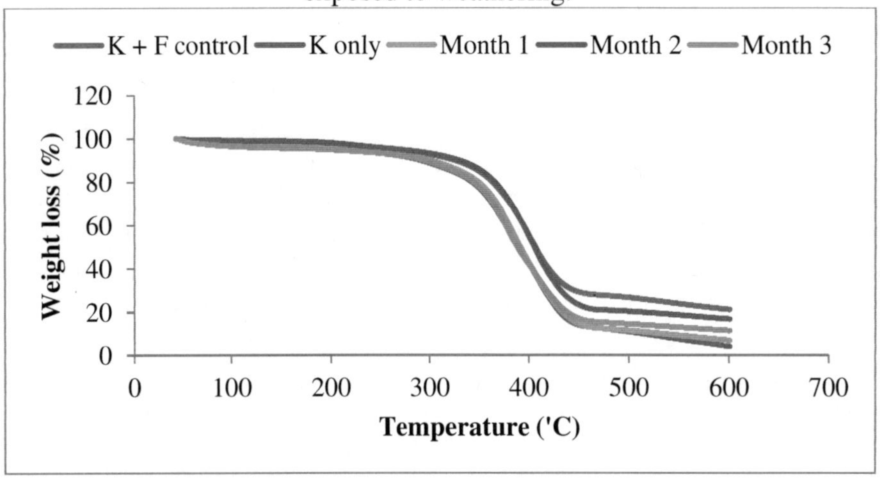

Figure 2. TGA curve Thermogravimetric curves of kenaf /glass fiber hybrid composite before and after exposed to weathering.

By increasing the weathering period, TGA curves are shifted to lower temperatures, showing a decrease in the thermal stability for both unsaturated polyester and its composites. The decrease in thermal stability in polymer matrix could be due the reduction of molecular weight of unsaturated polyester. For the composites, TGA transition were shifted to slightly lower temperatures after increasing the exposure period. This could be the result of unsaturated polyester chain scission along with degradation of both fiber and fiber matrix interfacial bonding (M. D. H. Beg and K. L. Pickering, 2008).

It can be noticed in figure 4.9 that the slop of TGA curves for all the composites after exposure to weathering is almost similar. This can be due to degradation process and moisture absorption have reached to the saturation stage after the first month of exposure to weathering.Derivative thermogravimetric is used to determine the weight loss and identify the decomposition of the composites at different temperature.Figures 3 and 4 show the derivative thermogravimetric curves of pure kenaf composites and hybrid kenaf/glass fiber composites, respectively, before and after to weathering.

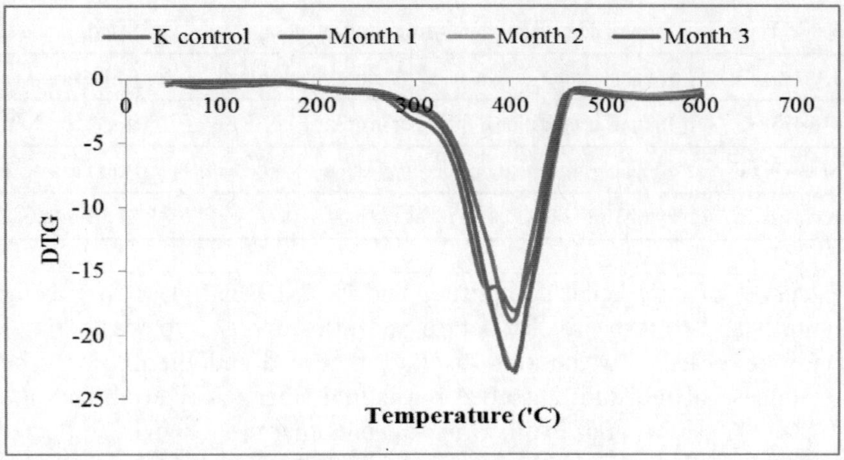

Figure 3. DTG curve Thermogravimetric curves of pure kenaf composite before and after exposed to weathering.

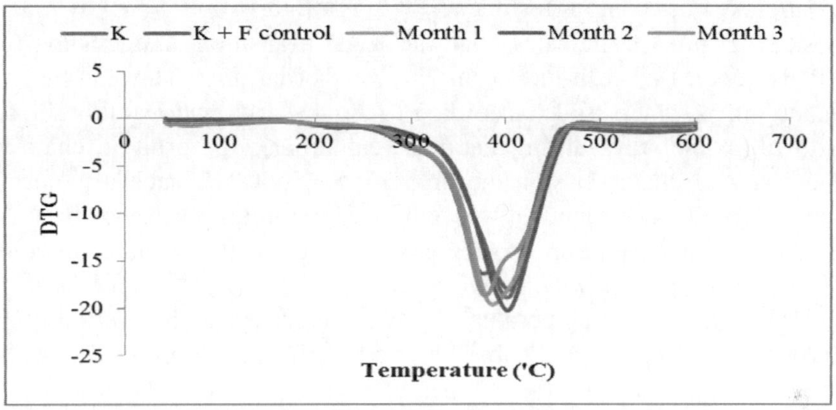

Figure 4. DTG curve Thermogravimetric curves of kenaf/glass fiber hybrid composite before and after exposed to weathering.

Table 4 shows the three stages of weight loss during the weight loss process with prolonged temperatures.All the values presented in the table increase in the hybrid kenaf/glass fiber composites because of existing of glass fiber in the composites which has high thermal stability. This glass fiber will act like a protector to the composites against burning process (Yao et al., 2008)

Table 1. Weight loss stages of the composites.

Stage 1	50- 110 oC	Evaporation of moisture in the fiber	(P.Methacanon et al., 2010)
	400 oC	Thermal decomposition of hemicellulose and glycosidic links of cellulose	(Araújo et al., 2008)
	350-400 oC	Characteristic of low molecular weight components such as hemicelluloses	(Lee et al ,. 2006)
	320-400 oC	Pyrolysis of hemicelluloses	(Suardana et al., 2011)
Stage 2	300-400 oC	Thermal decomposition of hemicellulose	(P.Methacanon et al., 2010)
	460 oC	Thermal decomposition of α-cellulose	(Manfredi et al., 2006)

	400-460 oC	Thermal degradation and pyrolysis of cellulose	(Suardana et al., 2011)
Stage 3	450-480 oC	Weight loss due to lignin and cellulose degradation	(P.Methacanon et al., 2010)
	440-475 oC	Thermal degradation of polymer matrix	(Lee et al ,. 2006)
	Near 450 oC	Weight loss due to lignin decomposition	(Lee et al ,. 2006)
	160-600 oC	Pyrolysis of lignin	(Suardana et al., 2011)

DSC curves of pure kenaf composites and hybrid kenaf/glass fiber composites are shown in Figure 5 and 6, respectively, before and after exposed to weathering.Pure kenaf composites and unexposed to weathering display one broad endothermic peak between 50-100 ^{o}C due to release of moisture absorbed by natural fiber (Ray et al,.2002). The region between 100-200 ^{o}C shows no exothermic or endothermic changes reflected that the composites are thermally stable. It was reported that natural fibers lignin degrade at temperature around 200 ^{o}C while the other constituents such as hemicellulose and cellulose degrade at higher temperature (Aziz, S.H. & Ansell, M.P. 2004). All these values increased in the hybrid kenaf/glass fiber composites due to existing of glass fiber which will act as a mask to its composites. Figure 3 indicates that the glass transition temperature (Tg) of the weathered UP also decreased with increasing the weathering time. This can be probably due to photo-oxidation of polymer itself. According to Abu-Sharkh and Hamid (2004), exposure of polymers to ultraviolet rays during natural weathering will result in chain scission of polymer. As a result of chain scission the properties of polymer such a molecular weight, chain symmetry, flexibility, mechanical strength and intermolecular forces have deteriorated (Zhu et al., 2006). Thus resulting in decrease of Tg of the weathered composites in comparison with the composites which have been reinforced by glass fiber.As seen in Figure 4 which is the glass transition temperature, Tg was increased with addition of glass fiber. Wang et al. (2005) reported that glass fiber plays a significant role in the thermal stability of natural fiber composites wherein the temperature causes a direct thermal expansion or contraction. Glass fiber has high content of inorganic silica. During pyrolysis process organic components such as cellulose and hemicelluloses decomposed and gasified leaving silica which can play a role as a heat shield (Yao, et al,.2008).

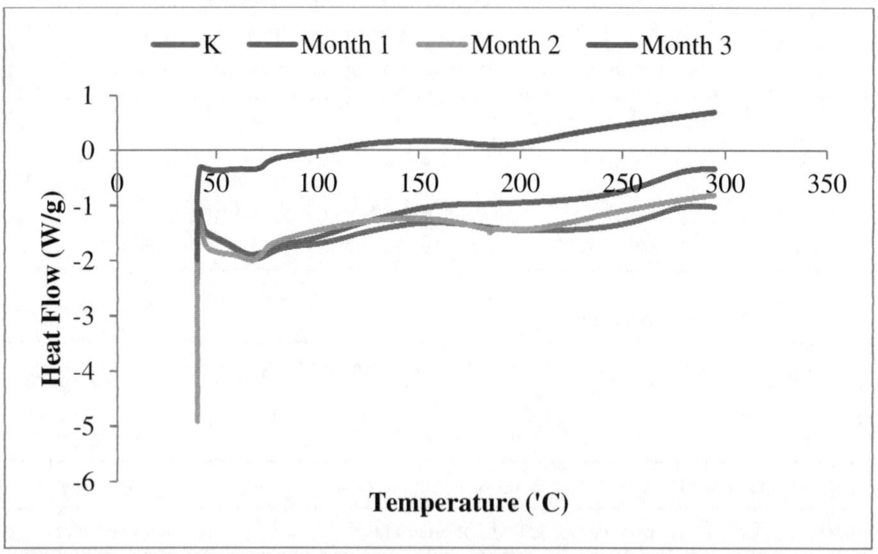

Figure 5. DSC curve of pure kenaf composite before and after exposed to weathering.

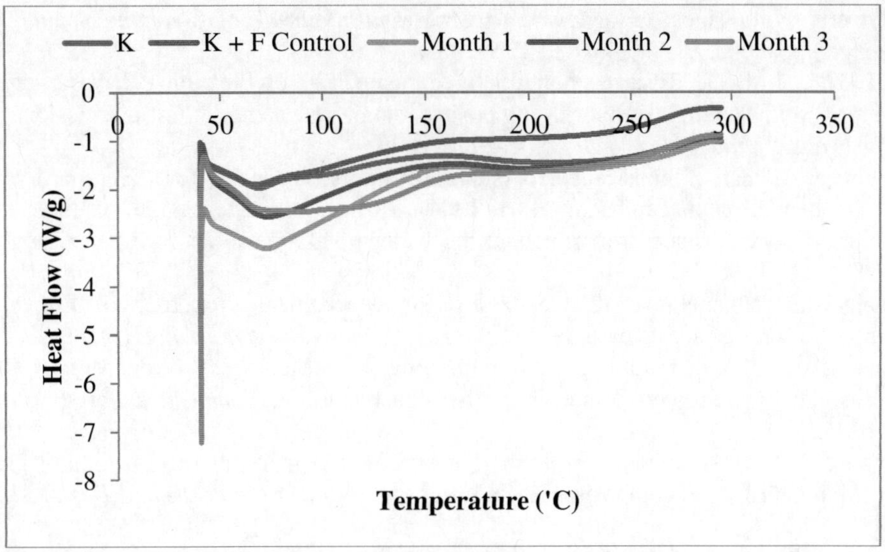

Figure 6. DSC curves of kenaf/glass fiber hybrid composite before and after exposed to weathering.

4. Conclusion

The study demonstrate the study on the effect of weathering on thermal properties of kenaf hybrid we compared both pure kenaf and hybrid kenaf-glass fiber before and after exposure, the composite shows slow and steady initial weight loss until major weight Loss is detected at approximately temperature 390^{0}C. While as the weathering increase, the weight Loss started as low as 290^{0}C. it was equally found that, the weakness in the molecular interfacial bond cause by moisture absorption. The glass transition temperature also is affected by weather for both composites, pure kenaf composites shows lower value of glass transition temperature compared with kenaf /glass fiber hybrid composites . This is due to photo-oxidation of weathered UP during natural weathering will results in chain scission of polymer . As a consequence the properties of polymer including chain symmetry ,flexibility and intermolecular forces have deteriorated. Thus study has revealed the effect of the weather to pure kenaf thermal properties and improve could be achieved with the integration of GF.

5. Acknowledgment

The author thanks to LembagaKenafdanTembakau Negara (LKTN) and UniversitiMalaysia Perlis (UniMAP) for facilities and support.

References

[1.] Abdul Malek Ya'acob1, Azhar Abu Bakar ,Hanafi Ismail and KhairulZamanDahlan Hybrid Composites: Study on Un-Treated Kenaf/Glass Fibre Properties *Pertanika J. Sci. & Technol.19* (2): (2011) 373– 381

[2.] Acharya S.K., Mishra P. and Mehar S.K. Effect of surface treatment on the mechanical properties of bagasse fiber reinforced polymer composite, *Bioresources.,6*, 3(2011) 155-3165

[3.] Aji, I.S. Sapuan,.S.M. Zainudin E.S. and Abdan K. (2009)Kenaffibres as reinforcement for polymeric composites: a review, *International Journal of Mechanical and Materials Engineering (IJMME), Vol. 4 No. 3, 239-248*

[4.] Aji. I. S., S.M. Sapuan, E.S. Zainuddin, K. Abdan (2009) Kenaffibres as reinforcement for polymeric composites: a review. . *International Journal of Mechanical and Materials Engineering (IJMME)*, Vol.4, No.3, 239-248

[5.] Akil H.M., I.M. De Rosa, C. Santulli, F. Sarasini. (2010) Flexural behaviour of pultruded jute/glass and kenaf/glass hybrid composites monitored using acoustic emission. *Materials Science and Engineering A,* 527, 2942-2950,

[6.] Alexander Bismarck, IbonAranberri-Askargorta, Jürgen Springer, T. L., Bernhard Wielage, A. S., IljaShenderovich&Limbach, H.-H.(2004) Surface characterization of flax, hemp and cellulose fibers; Surface properties and the water uptake behavior *Polymer Composites*, *23*, 872-894.

[7.] AmalNassar, EmanNassar (2013) Study on Mechanical Properties of Epoxy Polymer Reinforced with NanoSiC particles, *Nanoscience and Nanoengineering1*(2): 89-93, 2013

[8.] Anuar, H. and A. Zuraida, (2011). Improvement in Mechanical Properties of Reinforced Thermoplastic Elastomer Composite with KenafBastFibre, *Composites Part B: Engineering* 42, p. 462.

[9.] Arun K. V., S. Basavarajappa, B.S. Sherigara. (2010) Damage characterization of glass textile fabric polymer hybrid composites in sea water environment.*Materials and Design31*, pp 930 – 939.

[10.] Ayensu A. Interfacial debonding of natural fibre reinforced composites (2000).*QuartSci Vision*;6:25–34.

[11.] Aziz, S. H. & Ansell, M. P. (2004) The effect of alkalization and fibre alignment on the mechanical and thermal properties of kenaf and hemp bastfibre composites: part 2 - cashew nut shell liquid matrix. *Composites Science and Technology, 64*, 1231-1238.

[12.] Aziz, S. H., Ansell, M. P., Clarke, S. J. &Panteny, Y, S. R. (2005) Modified polyester resins for natural fibre composites. *Composites Science and Technology, 65*, 525-535.

[13.] Bernard, M., A. Khalina, A. Ali, R. Janius, M. Faizal, K.S. Hasnah, and A.B. Sanuddin, (2011). The Effect of Processing Parameters on The Mechanical Properties of KenafFibre Plastic Composite ,*Materials& amp; Design 32*, p. 1039.

[14.] Bikiaris, D. &Panayiotou, C. (1998). LDPE/starch blends compatibilized with PE-*g*-MA copolymers. *J. Appl. Polym. Sci.*, *70*, 1503–1521.

[15.] Bledzki A.K., Mamuna A.A. and Volk J., Physical (2010) chemical and surface properties of wheat husk, rye husk and soft wood and their polypropylene composites *Composite. Part A*, 41, 480-488

[16.] Bras J, Hassan ML, Bruzesse C, Hassan EA, El-Wakil NA, DufresneA.(2010) Mechanical, barrier, and biodegradability properties of bagassecellulose whiskers reinforced natural rubber nanocomposites.*Industrial Crops and Products32*, 627–33.

[17.] Chollakup R., Tantatherdtam R., Ujjin S. and Sriroth K. (2011) Pineapple leaf fiber reinforced thermoplastic composites, *J. Appl. Polym. Sci., 119*, 1952-1960

[18.] Coutinho, F.M.B., Costa, T.H.S. &Carvalho, D.L. (1997) Polypropylene-wood fiber composites: Effect of treatment and mixing conditions on mechanical properties. *J .Appl. Polym.Sci, 65*, 1227-1235.

[19.] Dhakal, H. N., Zhang, Z. Y. & Richardson, M. O. W. (2007) Effect of water absorption on the mechanical properties of hemp fibre reinforced unsaturated polyester composites. *Composites Science and Technology, 67*, 1674-1683.

[20.] El-Sayed, A. A., El-Sherbiny, M. G., Abo-El-Ezz, A. S. &Aggag, G. A. (1995) Friction and wear properties of polymeric composite materials for bearing applications. *Wear, 184*, 45-53.

[21.] Facca A.G., Kortschot MT & Yan N. (2006) Predicting the elastic modulus of natural fibre reinforced thermoplastics. *Comp Part A-Ap. Sci. and Manufacturing, 37*, 1660-1671.

[22.] Fiorea, V., Bellab, G. D.,&Valenzaa, A. (2015). The effect of alkaline treatment on mechanical properties of kenaf fibers and their epoxy composites.*Composites Part B: Engineering.68.*14–21

[23.] Gassan, J. &Cutowski V. S. (2000).Effect of corona discharge and UV treatment on the properties of jute-fibre epoxy composite.*ComposSciTechnol, 60*, 2857-2863.

[24.] Gomes, A., Goda, K. &Ohgi, J. (2004).Effects of Alkali Treatment to Reinforcement on Tensile Properties of Curaua Fiber Green Composites.*JSME International Journal Series A Solid Mechanics and Material Engineering, 47*, 541-546.

[25.] Hill, C. A. S., Abdul, H. P. S. & Khalil. (2000). The effect of environmental exposure upon the mechanical properties of coir or oil palm fiber reinforced composites. *Journal of Applied Polymer Science, 77*, 1322-1330.

[26.] Huda, M. S., Drzal, L. T., Mohanty, A. K. and Misra, M. (2006). Chopped Glass and Recycled Newspaper as Reinforcement in Injection Molded Poly(lactic acid) (PLA) Composites: A Comparative Study. *Composites Science and Technology, 66* 1813-1824.

[27.] Iannace, S., Nocilla, G. &Nicolais, L. (1999).Biocomposites based on sea algae fibers and biodegradable thermoplastic matrices. *Journal of Applied Polymer Science, 73*, 583-592.

[28.] Irene S. Fahim, Salah M. Elhaggar, HatemElayat.(2012). Experimental Investigation of Natural Fiber Reinforced Polymers, *Materials Sciences and Applications*, 2012, 3, 59-66

[29.] Jochen, G., Andrzej, K.,&Bledzki. (1999). Possibilities for improving the mechanical properties of jute/epoxy composites by alkali treatment of fibers.*Composites Science and Technology. 59,* 1303–1309.

[30.] Joseph PV, Rabello MS, Maattoso LHC, Joseph K, Thomas S. (2002). Environmental effects on the degradation behavior of sisal fibre reinforced polypropylene composites. *Compos SciTechnol*;*62*:1357–72.

[31.] Joshi, S.V. DrzalL.T. ,Mohanty A.K., Arora S. (2004) Are natural fiber composites environmentally superior to glass fiber reinforced composites, *Composites: Part A 35* 371–376

[32.] Karmakar AC, Hoffmann A, Hinrichsen G. (1994) Influence of water uptake on the mechanical properties of jute fibre reinforced polypropylene. *J ApplPolymSci*;54:1803–7.

[33.] Karmakar, A. C., Hoffmann, A. G. H. (1994.) Influence of Water Uptake on the Mechanical Properties of Jute Fibre Reinforced Polypropylene. *Journal of Applied Polymer Science, 54*, 1803-1807

[34.] Mansur, M. A. and M. A. Aziz. (1983). Study of Bamboo-Mesh Reinforced Cement Composites. *Int. Cement Composites and Lightweight Concrete,, 5*, 165–171.

[35.] Maria, L. S., Corneliu, C., Teodorescu-D.H.,Janos, T. (2012). Properties of Advanced New Hemp Fiber Materials Used inAutomotiveEngineering.*Advances in Production, Automation and Transportation Systems. 978-1-61804.*193

[36.] Mariatti, W.L. Lai, M. and M. Jani. (2008) The properties of woven kenaf and betal palm (Areca catechu) reinforced unsaturated polyester composites. *Polymer Plastic technology and Engineering*, *47*, 1193-1199.

[37.] MohdSuhairilMeon,,MuhamadFauzi Othman, Hazran Husain, Muhammad FairuzRemeli, MohdSyaharMohdSyawal (2012). Improving tensile properties of kenaf fibers treated with sodium hydroxide, *Procedia Engineering 41*(2012) 1587 – 1592

[38.] MohdYuhazri, Y., Phongsakorn, P.T., HaeryipSihombing, Jeefferie A.R., PuvanasvaranPerumal, Kamarul, A.M., and KannanRassiah.(2012). Mechanical Poperties of Kenaf/polyester composites. *International Journal of Engineering & Technology IJET-IJENS*, 11(1), 127-131

[39.] Omar F., Andrzej K. Bledzki Hans-Peter F., Mohini S. (2012). Biocomposites reinforced with natural fibers: 2000–2010, *Progress in Polymer Science 37*, 1552– 1596

[40.] Poostforush M., Al-Mamun M. and Fasihi M. (2013). Investigation of Physical and Mechanical Properties of High Density Polyethylene/Wood Flour Composite Foams,*Res. J. Engineering Sci*., 2(1), 15-20

[41.] Prachayawarakorn, J. et al. (2010). Property improvement of thermoplastic mung bean starch using cotton fiber and low-density polyethylene.*Starch – Stärke*, 62(8), 435–443.

[42.] Prachayawarakorn, J., Sangnitidej, P. &Boonpasith, P. (2010). Properties of thermoplastic rice starch composites reinforced by cotton fiber or low-density polyethylene. *Carbohydr.Polym.*, 81(2), 425–433.

[43.] Rashdi, A. A. A., S. M. Sapuan, M. M. H. M. Ahmad &Khalina, A. (2 0 0 9) Water absorption and tensile properties of soil buried kenaffibre reinforced unsaturated polyester composites (KFRUPC). *Journal of Food, Agriculture & Environment, 7*, 9 0 8 - 9 1 1.
[44.] Sharma, N. et al. (2001). A study on the effect of pro-oxidant on the thermo-oxidative degradation behaviour of sago starch filled polyethylene. *Polym.Degrad. Stab.*, 71, 381–393.
[45.] Stamboulis A, Baillie CA, Peijs T. (2001) Effects of environmental conditions on mechanical and physical properties of flax fibers.*Compos Part A ApplSciManuf* 2001;32:1105–15
[46.] SuhailahMohdSukri, NorLiyanaSuradi, AgusArsad, Abdul RazakRahmat and Azman Hassan. (2012). Green composites based on recycled polyamide-6/recycled polypropylene kenaf composites: mechanical, thermal and morphological properties. *Journal of Polymer Engineering*. 32, 291-299
[47.] SupraneeSangthong, ThirawudhPongprayoon, NantayaYanumet. (2009) Mechanical property improvement of unsaturated 5.polyester composite reinforced with admicellar-treated sisal fibers. *Composites: Part A* 40, pp 687–694.
[48.] Tan, C., Ahmad, I., Heng, M. (2011). Characterization of polyester composites from recycled polyethylene terephthalate reinforced with empty fruit bunch fibers, *Materials and Design*.32 (8-9): 4493-4501
[49.] Threepopnatkul, P N. Kaerkitcha, N. Athipongarporn. (2009) Effect of surface treatment on performance of 7.pineapple leaf fiber–polycarbonate composites. *Composites: Part B 40*, pp 628–632.
[50.] Zampalonib, M.,Pourboghrata, F., Yankovicha, S.A., Rodgersa, B.N., Moorea, J., Drzalb,L.T and Mohantyc, A.K. (2007). Kenaf natural fiber reinforced polypropylene composites: A discussion on manufacturing problems and solutions. *Composites Part A: Applied Science and Manufacturing*. *38*. 1569–1580.
[51.] Zou Yi, Shah Huda, Yiqi Yang. (2010). Lightweight composites from long wheat straw and polypropylene web. 3. *Bioresource Technology 101*, pp 2026–2033.
[52.] T Adam, U Hashim, TS Dhahi, KN Khor, PS Chee, PL Leow, ELECTROCHEMICAL ETCHING: An Ultrasonic Enhance Method of Silicon Nano Porous Fabrication, Wulfenia Journal 20 (1), 45-55, 2013
[53.] Tijjani Adam, H Uda, M Eaqub, PL Leow, The electroosmosis mechanism for fluid delivery in PDMS multi-layer microchannels, American Scientific Publishers All rights reserved.2013
[54.] U Hashim, T Adam, J Lung, P Ling, Fabrication of Microchannel and Micro Chamber for Microfluidic Lab-on-Chip.Australian Journal of Basic & Applied Sciences 7 (1)2013
[55.] T Adam and U. Hashim Light Observation in Polymer: A Study of Silicon-Based Organic Polymer Using Visible Spectroscopy, Australian Journal of Basic and Applied Sciences 7 (1), 76-80, 2013.
[56.] U Hashim, MW Al-Mufti, T Adam, Silicon Nanowire Geometry: Investigation of Interaction Site Potential in Semiconductor-DNA Interaction, Australian Journal of Basic and Applied Sciences 7 (5), 242-245, 2013
[57.] MW Al-Mufti, U Hashim, T Adam, Simulation of Nano lab on chip devices by using COMSOL MultiphysicsJournal of Applied Sciences Research 9 (2), 1056-1061, 2013
[58.] T Adam, U Hashim, KL Foo, TS Dhahi, T Nazwa, Technology development for nano structure formation: Fabrication and characterization, Advanced Science Letters 19 (1), 132-137, 2013
[59.] T Adam, U Hashim, PL Leow, KL Foo, PS Chee, Selection of optimal parameters in fabrication of poly (dimethylsiloxane) microfluidics using taguchi method, Advanced Science Letters 19 (1), 32-36,2013
[60.] T Adam, U Hashim, KL Foo, Microfluidics design and fabrication for life sciences application, Advanced Science Letters 19 (1), 48-53, 2013
[61.] T Adam, U Hashim, ME Ali, PL Leow, The electroosmosis mechanism for fluid delivery in PDMS multi-layer microchannel, Advanced Science Letters 19 (1), 12-15, 2013

IOP Conf. Series: Journal of Physics: Conf. Series **908** (2017) 012005 doi:10.1088/1742-6596/908/1/012005

A Review: Characteristics of Noise Absorption Material

S Amares[1, 4], E Sujatmika[1,5], T W Hong[2] , R Durairaj[3] and H S H B Hamid[1]

[1] Faculty of Engineering and Built Environment, SEGi University, No. 9, Jalan Teknologi, Taman Sains Selangor, Kota Damansara PJU 5, 47810 Petaling Jaya, Selangor, Malaysia.
[2] School of Mechatronic Engineering, Universiti Malaysia Perlis (UniMAP), Pauh Putra Campus, 02600 Arau, Perlis, Malaysia.
[3] Lee Kong Chian Faculty of Engineering and Science, Universiti Tunku Abdul Rahman, Jalan Sungai Long, Bandar Sungai Long, 43000 Kajang, Selangor, Malaysia.

amaressinghgill@segi.edu.my, endra@segi.edu.my

Abstract. Noise is always treated as a nuisance to human and even noise pollution appears in the environmental causing discomfort. This also concerns the engineering design that tends to cultivate this noise propagation. Solution such as using material to absorb the sound have been widely used. The fundamental of the sound absorbing propagation, sound absorbing characteristics and its factors are minimally debated. Furthermore, the method in order to pertain sound absorbing related to the sound absorption coefficient is also limited, as many studies only contributes in result basis and very little in literature aspect. This paper revolves in providing better insight on the importance of sound absorption and the materials factors in obtaining the sound absorption coefficient.

1. Introduction

Noise is obviously related as nuisance endured by humans from the machineries and appliances of daily activities [1]. Sound is a propagation of disturbance in fluid or in solid, principally; a wave motion is set off when an element sets the nearest particle of air into motion, which then creates difference in pressure in the medium where the wave travels [2]. Noise and sound carries the same definition, however denoted with different perception and definition. Depending on the medium the sound wave travels, sound propagates at different speeds, related to the pressure difference [4]. Normally, the static air pressure is about 105Pa in which the audible sound pressure variations are minor ranging from 20μPa to 100Pa [5]. Such in that, the sound pressure of 20μPa corresponds to the average human's threshold of hearing, making it the par. As for principle of the sound detection, theory notes the vibrations caused by the sound pressure hits a surface and travels to been detected by the human ear [6] as in figure. 1. The normal static pressure appears hits the object causing pressure change in the air thus propagating a sound wave through the vibration back and forward by the object [7]. The sound being emitted in terms of sound waves in a longitudinal manner, which eventually calculated using frequency, *f* measured in Hertz, Hz. As researched by [3], frequency as the physic quantity can be used as the sound wave and enables calculation of the sound pressure level and also

Figure 1. Schematic diagram of reflected sound wave.

sound absorption coefficient. Theoretically, the frequency is produced by the vibration of the particles upon the travelling through the pressure wave or the sound wave [8]. Subsequently the particles moves in parallel direction to the wave movement, the reason of the sound wave being a longitudinal wave. These longitudinal waves occurs because of the creation of compressions and rarefactions within the air. As the sound wave travels through the air, two regions are formed, one of it at where the air molecules are crowded together in the high pressure regions making compression between the molecules in the air. Such action makes the molecules to push each other and enter another region of low pressure, where the rarefactions occur [9]. During the compression, individual molecule moves side to side, making the speed at which a compression propagates through the medium to be known, which is actually the wave/sound speed [10]. Closely related to the speed of sound is the wavelength denoted as λ, used to calculate the frequency. The air experiences in norm its own normal pressure, termed as P_0, and then as the sound waves travels through the air, the differences in the pressure occurs aforementioned, denoted as ΔP. Changes in the pressure can be determine by taking the reference pressure of P_0 which appears at the initial point, which can also be related to the pressures at other points. These changes in pressure can be at maximum P_{max} or at minimum P_{min} and these are the pressures that will be detected by human [11], where P_{max} or P_{min} are primarily the amplitudes. Apparently, as the wave travels, a certain cycle will be occurred due to displacement. The instantaneous displacement waveform is the cycle taken, and each cycle can be represented by period, which the time needed to complete one cycle [12] and knowing the time taken for the one cycle, t, the frequency, f can be determined [13].

2. Sound Absorption Coefficient (SAC)

Sound absorption coefficient is defined as ratio of absorbed energy to incident energy noting the amount of sound being absorbed by a material [14]. The higher the percentages, the better the absorption is, indicating most of the sound being absorbed and less is being reflected back and is portrayed in the work of [15]. As presented in figure 2, it is observed that there are three types of waves encountered by a certain material which are the incident, reflected and transmitted sound waves. Incident waves are those waves that are being projected to a certain material which then can either be reflected or transmitted [16]. This principally contributes in finding a material's capability of absorbing sound wave. With aid of equations 1, 2, 3, and 4 the sound absorption coefficient, α can be calculated [4].

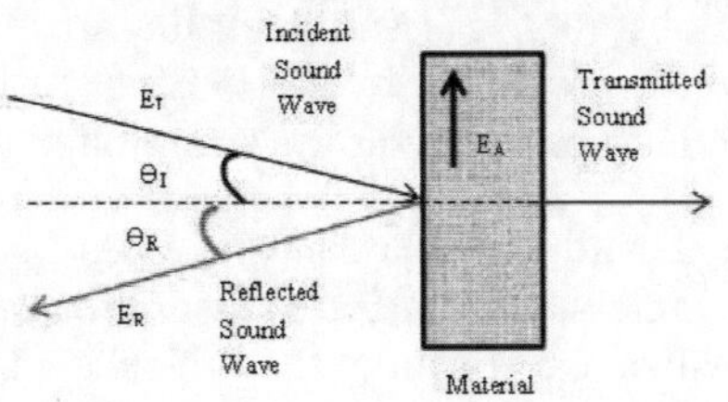

Figure 2. Schematic diagram of a material's absorption.

$$\alpha = E_A/E_I; \quad (1)$$

$$\alpha = 1 - E_R/E_I. \quad (2)$$

$$\alpha = 1 - |\, r \,|^2. \quad (3)$$

$$\alpha = I_{Abs}/I_I. \quad (4)$$

α = sound absorption coefficient E_I = incident energy, E_R = reflection energy, E_A = absorbed energy, r= incident reflection factor I_{Abs} = Sound intensity absorbed, I_I = Incident sound intensity

3. Sound Absorbing Material and Influencing Factors

Sound absorbing materials are classified on the ability of the material to absorb as much as sound wave and reflect as minimal as it could and at the same time transmit more of the waves. A material that could absorb and transmit more sound waves than it reflects, is considered a good sound absorbing material [17]. Factors such as the thickness, density and porosity influences a material's capability to absorb sound and will be discussed in this paper to provide better insight [18].

3.1. Material's Thickness

One of the factors that influence a material's sound absorption is the thickness. Though, the thickness of the materials is only relevant or has direct relationship at low frequency range (100-2000Hz) and is insignificant for high frequency (>2000Hz) [19]. Increase in the thickness provides better absorption of the wave and reflect less energy [20] Study by [21] shows that as the thickness of the samples increases, the sound absorption coefficient increases too as in table 1. The reason is at low frequency, waves have higher wavelength, which means the thicker material contributed in better absorption [22].The result shows that the thickness influences better sound absorption and is the reason of the high wavelength mentioned [23]. In that study, biopolymer foams were coated with three different materials comprising cotton fabric, single knitted jersey fabric and polyester fabric that been layered with four panels. It was found that taking the result at 2000Hz, the sound absorption coefficient increases by an average 58% comparing the first and fourth panel. This indicates that as the thickness increases, the absorption increases [24]. Findings by [25] also proved that thicker elastomeric foam based on synthetic rubber produces better absorption. Obvious trend of increase was noted upon increase in the thickness of the material. Comparing different thicknesses, the thicker elastomeric foam produced better sound absorption from the frequency to higher frequency. Although thicker material does not prevail at higher frequency due to the coincidence dip phenomenon, it was still noticeable that at 2000Hz, the 50mm material had almost 20% better absorption than the 10mm material. Higher wavelength by a low frequency needs a thicker material to ensure contact and making the absorption better [26]. Crucial point to be noted is the materials thickness should be one tenth of the wavelength in order to act as an effective sound absorber [27] and is on the recommended condition for the lower frequency (<2000Hz). Further study providing by [28] using polyurethane resin added with fire retardant was used with four different thicknesses, as the data plotted in figure. 3.

Thicker polyurethane resin added with fire retardant produces better sound absorption coefficient similarly reported by [21], nevertheless, shift in better absorption occurs towards thinner material as frequency increases. This provides clarification that thicker material contributes better sound absorption for low or medium range frequency (<2000Hz), supported [29]. Reason for this observation is because standards clarified that the thickness of the materials should be at least one tenth of the wavelength for that particular frequency in order to obtain better absorption [30]. Thus, if the frequency of a material is at 1000Hz, the wavelength will be 0.34m, which then the thickness shall be about one tenth of the wavelength and in this study, it is 0.03m. Such that, the thickness must be about 0.03m and if higher it would have no profound effect to absorption.

Table 1. Sound absorption coefficient influenced by thickness of plywood panel and fibreglass board.

Material	Frequency (Hz)					
	125	250	500	1000	2000	4000
Wood: Plywood panel(10 mm thick)	0.28	0.22	0.17	0.09	0.1	0.11
Fibreglass board(25 mm thick)	0.06	0.20	0.65	0.9	0.95	0.98

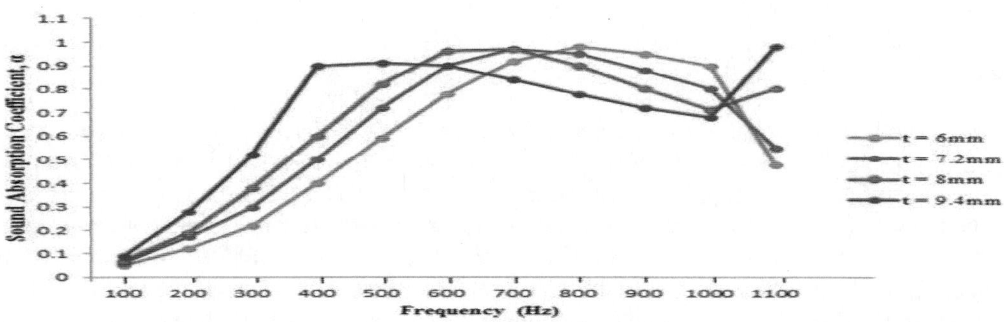

Figure 3. Sound absorption of polyurethane resin added with fire retardant with various thicknesses.

3.2. Material's Density

The density of the material is another factor and is defined as mass per unit volume. Density influences the acoustic impedance as the impedance determines the reflection of materials, equation 5. It was stated by [31] that a high density material due of mass increase increases the sound absorption and provided by [32] in table 2, the reason is justified.

$$Z = \rho cs \frac{A+B}{A-B} \quad (5)$$

ρ = density of material, c = speed of sound, s = cross sectional area of material, A + B = maximum of wave pattern, A - B = minimum of wave pattern.

Research by [33] compared the 25% percentages rice husk mass added to polyurethane with the bare polyurethane and concluded the prior had better absorption of sound with a frequency ranging 200-1000Hz that later decreases at 2000Hz as in figure 4. Contribution of this 25% rice husk in of increase in mass subsequently increasing the density. However, as for the decrease in the absorption at a higher frequency is related to the thickness of material that will be the dominant factor at higher frequencies. Demonstrated also by [34], this occurrence of decrease in the sound absorption at higher frequency is known as coincidence dip phenomenon. Critical frequency (1 kHz-4 kHz) strictly limits the sound absorption ability of a material and when the incident sound wave is in phase with the reflected wave, the coincidence dip phenomenon occurs, similarly reported by [35]. A material with high density has

Table 2. Sound absorption coefficient on different densities (RW = Rockwool).

Material	Densities (kg/m^3)	Frequency (Hz)					
		200	250	500	1K	2K	4K
RWA45	45	0.20	0.50	0.85	1.00	1.00	1.00
RW3	60	0.11	0.60	0.96	0.94	0.92	0.82
RW5	100	0.10	0.40	0.80	0.90	0.90	0.90

a larger surface are per unit volume and is effective in sound absorbing associating with fibre containing porous materials [37]. The increase in fibres surface area prompts more energy loss due to high friction loss, making more incident waves being converted to heat energy [19]. Research by [37] clarified that as the fibre size decreases, it is easier for the fibres to move among them making the airflow resistance higher contributing in increase of friction [38]. Few recommendations in the size by stating that 1.5 to 6 denier per filament (dpf) produces better sound absorption than coarser fibres. Similarly, 1.25dpf nonwoven fibres produces better sound absorption coefficient than 7dpf.

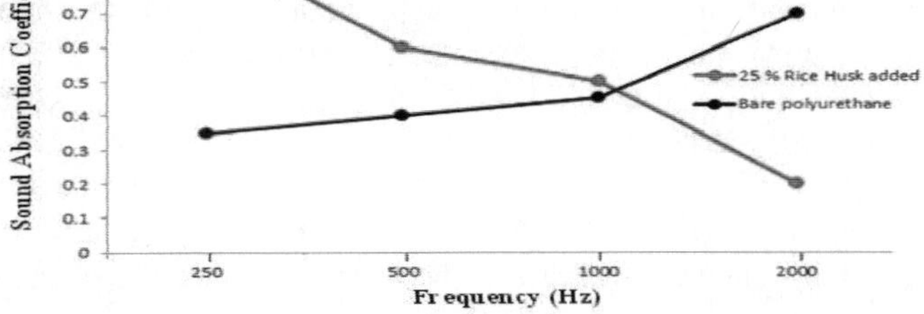

Figure 4. Sound absorption of 25% husk rice added with polyurethane and pure polyurethane [33].

Figure. 5 shows the result based on the nonwoven fabric's density and its influence in sound absorption, and it was evident that increase in density increases the absorption coefficient. Primarily, the increase in the density increases the thickness of the fibre. Henceforth, this contributes in more contact of the sound wave and the fibre, increasing the friction of airflow resistance and released as heat energy [34].

3.3. Material's Porosity

This factor is closely related with material's configuration involving the volume of the voids/holes to the total volume [39] shown in equation 6. The presence of pores or voids plays a crucial part as they act as the medium of sound wave dissipation. Figure 6 shows a schematic diagram of types of voids present in the material as porous material's configuration are such containing pores, granular, fibrous or cellular types. However, the principle of sound wave reaching the pores is similar for each type and this influenced by the existing voids. Upon contact, the channels or pores contains air molecules will vibrate and losses energy due to energy of the air molecules by the sound wave being converted to heat due to thermal and viscous losses at the walls of the interior pores and channels. The isothermal changes are noted as domain at low frequencies while at high frequencies, adiabatic acts as domain factor [39].

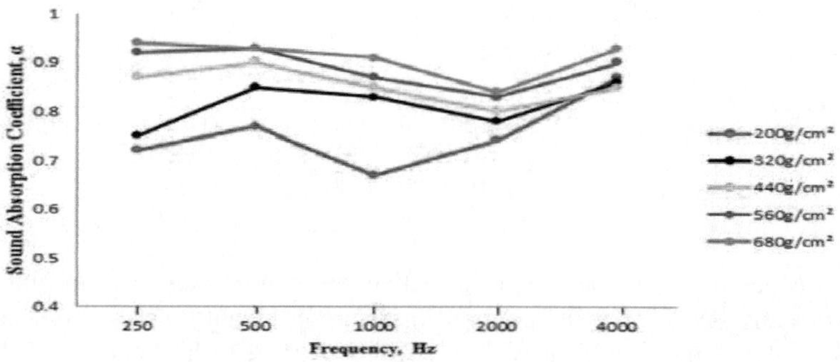

Figure 5. Sound absorption of nonwoven fabric.

Ultimately, there must be a sufficient transfer of energy with an acceptable range of porosity, meaning an open pores with continuous channels prevails better sound absorbing, because of the multiple reactions between the sound wave and the walls of the pores. Eventually more energy will be transformed in to heat energy [31]. As found out by [8], micro-perforated panel (MPP) were used with different cavity/pores diameters and upon two different diameters of 30 and 60mm, the later the as better sound absorption coefficient. Different air gaps diameter behind the sample of polyurethane foam (fixed diameter = 25mm) was tested as air gaps behind the sample acts as cavities that allows sound that has been transmitted to flow and hit the rigid back plate and encounter the sample after reflection[40]. In other word, the cavity acts as open pores. The diameter of 25mm gap produces better sound absorption than the smaller gaps almost 25% better comparing to the 5mm diameter. This is only applicable up to 1250Hz of frequency which later had reduction in absorption for the 25mm diameter gaps as the frequency increases. The reason is at high frequency, the wavelength is short and the diameter of the gaps has no profound effect. This proves that as the increase in frequency occurs, the wavelength becomes shorter needing thicker of materials to absorb more incident energy [17]. Contradicting to that, studied that the perforation panels should be having smaller diameters to allow better absorption [41]. However, the perforations are usually panels that act as backing for absorbent material and is applicable for a low frequency range only.

$$Porosity\ (H){:}\frac{Va}{Vm} \tag{6}$$

Va = Volume of air in voids, Vm = Total volume of material

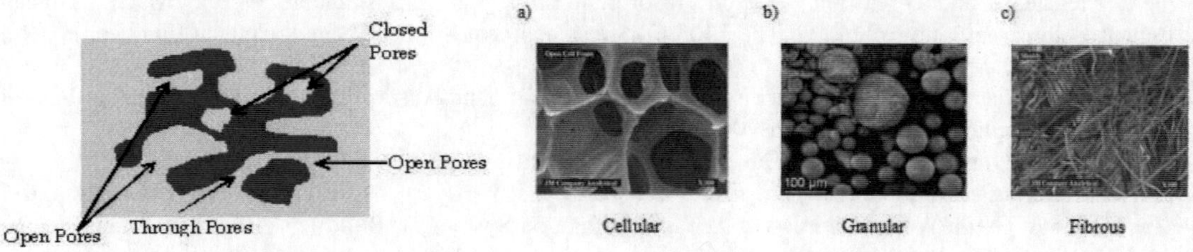

Figure 6. Schematic diagram of porous materials with pores/voids [39].

4. Conclusion

This paper aims to provide information as a literature study as the concept of sound and noise were deliberated. Adding to that, the material's characteristics were further explained which then linked to the test method to figure or obtain the sound absorption coefficient. Obtaining such information will provide better understanding of the parameter needed to select a material based on the sound absorption criteria. Conventionally, the major confusion arises in this area of study is to link each parameter and combine them to produce or even study the acoustical aspects. Nevertheless, this paper solves such problem and provides simpler yet sufficient linkage of information in this field of study with the method of solving the sound absorption coefficient.

References

[1] N. Indrianti, N. B. Biru and T. Wibawa, "The development of compressor noise barrier in the assembly area: (Case study of PT Jawa Furni Lestari)," Procedia CIRP, pp. 705-710, 2016.
[2] C. H. Hansen, "Fundamental of Acoustics," University of Adelaide, Adelaide, 2001.
[3] F. Jacobsen, T. Poulsen, J. H. Rindel, A. C. Gade and M. Ohlrich, "Fundamental of Acoustic and Noise Control," Technical University of Denmark, 2011.
[4] P. N. Breysse and P. S. Lees, "Noise," The Johns Hopkins University, 2006.
[5] J. A. Scales and R. Snieder, "What is noise?," Geophysics, vol. 63, no. 4, p. 1122–1124, 1998.
[6] A. R. M. and . S. F. , "An overview of automobile noise and vibration control," Noise Notes, vol. 13, no. 1, pp. 10-19, 2014.
[7] D. Zhou, Z. Luo, M. Fang, J. Jiang, H. Chen, D. Sha and M. Lu, "Numerical study of the movement of fine particle in sound wave field," Energy Procedia, pp. 2415-2420, 2015.
[8] M. Crocker and J. Arenas, Handbook of Noise and Vibration Control., M.J. Crocker (Ed.) ed., New York: John Wiley and Sons, 2007.
[9] S. R. and A. H. , An Introduction to Acoustics, Eindhoven: Eindhoven University of Technology, 2016.
[10] M. Bruneau, "Fundamentals of Acoustics," ISTE Ltd , Great Britain, 2006.
[11] S. Errede, "Acoustical Physics of Music," UIUC, Illinois, 2012.
[12] ASHRAE, Fundamentals, ASHRAE, 2013.
[13] R. T. Muehleisen, "Measurement of the acoustic properties of acoustic absorbers.," Ilinois Institute of Technology, Illinois, 2007.

[14] M. R. Bai, Y.-Y. Lo and Y. S. Chen, "Impedance measurement techniques for one-port and two-port networks.," The Journal of Acoustical Society of America, p. 2279, 2015.
[15] M. A. Kuczmarski and J. C. Johnston, "Acoustic Absorption in Porous Materials," NASA, Ohio, 2011.
[16] K. Ghorbani, H. Hasani, M. Zarrebini and R. Saghaf, "An investigation into sound transmission loss by polypropylene needle-punched nonwovens," Alexandria Engineering Journal, pp. 907-914, 2016.
[17] O. Doutres and N. Atalla, "Sound Absorption Properties of Functionally Graded Polyurathane Foams," New York, 2012.
[18] F. D'Alessandro and G. Pispola, "Sound Absorption Properties of Sustainable Fibrous Materials in an Enhanced Reverberation Room," in The 2005 Congress and Exposition of Sound Control Engineering, Rio de Janeiro, 2005.
[19] H. S. Seddeq, "Factors Influencing Acoustic Performance of Sound Absorptive Material," Australian Journal of Basic and Applied Sciences, pp. 4610-4617, 2009.
[20] K. Limited, "Acoustic Performance Guide," Kingspan Limited, 2005.
[21] Z. Azkorra, G. Perez, J. Coma, L.F. Cabeza, S. Bures, J.E. Alvaro, A. Erkoreka and M. Urrestarazu, "Evaluation of Green Walls as a Passive Acoustic Insulation System for Buildings.," Applied Acoustics, pp. 46-56, 2015.
[22] N. Q. A. Adnan and A. Z. M. Rus, "Sound Absorption of Laminated Biopolymer Foam and Epoxy Foam," Key Engineering Materials, pp. 291-295, 2014.
[23] N. N. Mat Hassan and A. Z. M. Rus, "Influences of Thickness and Fabric for Sound Absorption of Biopolymer Composite," Applied Mechanics and Materials, pp. 102-107, 2013.
[24] E. Julia, J. Segura, A. Nadal, J. M.Gadea and J. E. Crespo, "Study of Sound Absorption of Multilayer Panels Made from Ground Tyre Rubbers," Annals of the Oradea University, pp. 147-150, 2013.
[25] A. "High Performance Acoustic Insulation Material for a Quieter Environment," Armacell Advanced Insulation, 2011.
[26] M. M. Al Bakri Abdullah, L. Jamaludin, A. Abdullah, R. Abd Razak and K. Hussin, "Influence of Multilayer Textile Biopolymer Foam Doped with Titanium Dioxide for Sound Absorption Materials," Key Engineering Materials, Vols. 594-595, pp. 750-754, 2013.
[27] M. Coates and M. Kierzkowski, "Lighter, Thinner and More Absorbent," Technical Textile International, 2002.
[28] F. A. F. D. A. and S. Schiavoni, "Sound Absorbing Properties of Materials Made of Rubber Scrumbs," Paris, 2008.
[29] C. T. Corporation, "Noise Control for Buildings," CertainTeed Corporation, Pennsylvania, 2011.
[30] I. Standard (ISO), "Acoustics - Determination of Sound Absorption Coefficient and Impedance in Impedance Tubes". United States of America Patent ISO 10534-2, 1998.
[31] S. J. Wertel , "Experimental Analysis of Noise Reduction Properties of Sound Absorbing Foam," University of Wisconsin-Stout , 2000.
[32] R. Limited, "RW Semi Rigid & Rigid Slabs," Rockwool Limited, New York, 2015.
[33] S. M. A. A. Zaidi, M. . M. Y. and M. I. , "Investigation on Sound Absorption of Rice-Husk Reinforced Composite," Kuantan, 2009.
[34] F. Shahani, P. Soltani and M. Zarrebini, "The Analysis of Acoustic Characteristics and Sound Absorption Coefficient of Needle Punched Nonwoven Fabrics," Journal of Engineered Fibers and Fabrics, vol. 9, no. 2, pp. 84-92, 2014.
[35] M. I. A. Abdalla , "Treatment the Effects of Studio Wall Resonance and Coincidence Phenomena for Recording Noisy Speech Via FPGA Digital Filter," Journal of Telecommunication, vol. 2, no. 2, pp. 42-48, 2010.
[36] W. D. Callister Jr, Materials Science and Engineering, John Wiley & Sons Inc., 2006.
[37] Y. E. Lee and C. W. Joo, "Sound Absorption Properties of Thermally Bonded Nonwovens based on Composing Fibers and Production Parameters," Journal of Applied Polymer Science, p. 2295–2302, 2004.
[38] W. Lauriks, D. Dan, A. Dijckmans and G. Vermier, "Characterisation of Sound Absorbing Material," in 10th Acoustic Congres of France, Lyon, 2010.
[39] P. Soltani and M. Zerrebini, "The Analysis of Acoustical Characteristics and Sound Absorption Coefficient of Woven Fabrics.," Textile Research Journal, vol. 82, pp. 875-882, 2012.

[40] M. Bilova and E. Lumitzer, "Acoustical Parameter of Porous Material and their Measurement," Acta Technica Corviniensis, pp. 39-42, 2011.
[41] J. Sikora and J. Turkiewicz, "Sound Absorption Coefficient of Granular Material," Journal of Mechanics and Control, pp. 149-157, 2010.

Utilization of Polyaniline (PAni) as Conductive Filler on Poly (Ethylene Oxide) / Poly (Vinyl Chloride) Films: Effects of Naphthalene as Surface Modifier on Electrical Conductivity

Mohammed Izzuddeen Bin Mohd Yazid[1], Supri A Ghani[2], Tan Soo Jin[2], Azlin Fazlina Osman[1] and Siti Hajar Mohd Din[1]

[1] School of Materials Engineering, Universiti Malaysia Perlis, Kompleks Pusat Pengajian Jejawi 2, 02600 Arau, Perlis, Malaysia.
[2] Faculty of Engineering Technology, Universiti Malaysia Perlis (UniMAP), Sungai Chuchuh, Padang Besar, Perlis, Malaysia.

moizzuddeen@gmail.com

Keywords: poly (vinyl chloride), poly (ethylene oxide), polyaniline, naphthalene, electrical conductivity

Abstract: The aim of the study was to utilize polyaniline (PAni) as conductive filler in poly (ethylene oxide) / poly (vinyl chloride) (PEO/PVC) films. Naphthalene was used as surface modifier to increase the properties of PEO/PVC/PAni films. The electrical conductivity of PEO/PVC/PAni film improved with higher loading of PAni until it reached the percolation threshold at 10 wt%. Moreover, the conductivity of the films also improved with the addition of naphthalene compared to the films without naphthalene. These enhancements were due to the improved interfacial bonding between PEO/PVC blend and PAni, which were supported by scanning electron microscopy (SEM) analysis. In addition, no new chemical bonding was created with the incorporation of naphthalene as proved by the FTIR analysis.

1. Introduction

Polymers typically do not conduct electricity and are used universally as an insulator. However, since the discovery of polyacetylene by Shirakawa and his fellow researchers, intrinsically conducting polymers (ICPs) such as polyaniline, polypyrrole and polythiophene have change the conventional perspective on polymers. Among numerous intrinsically conducting polymers available, polyaniline (PAni) has gain a great reputation as one of the most promising intrinsically conducting polymers (ICPs) due to its manageable electrical conductivity through protonation/deprotonation method, easiness to synthesis, good environmental stability and relatively inexpensive reagents [1].

However, despites all that, PAni still possess several issues that need to be solves to fulfill its high potential. PAni blending with commercially available polymers are one of the frequently practiced methods to overcome the drawbacks of PAni such as the processing difficulties and relatively poor mechanical properties [2,3]. This methods is advantageous when it comes to designing polymer products with wide variety of characteristics. Research on the blending of PAni with conventional polymers have been reported and have been proved to balance the good mechanical properties of polymers with good electrical conductivity of PAni [4,5]. In this study, poly (ethylene oxide) (PEO) and poly (vinyl chloride) (PVC) blend were chosen as the matrix to compliment PAni as the filler. The PEO/PVC/PAni films were fabricated via solution casting as this method offers cheap, easy and efficient processing.

PVC is one of the most utilized thermoplastic thanks to its application versatility, biocompatibility and chemical stability. Interestingly, for several decades, researchers have been developing and studying the ways to improve electrical conductivity of PVC. Such reports suggest several methods including blending with intrinsically conducting polymers (ICPs), coating of conductive layers onto PVC surfaces, and also incorporating liquid anti-static agents or conductive inorganic filler into PVC [6-9]. Otherwise, PEO which is also used as a matrix in this paper, have been repeatedly reported as a promising material in this field. PEO is well known for its ability to dissolve large amount of inorganic salts homogeneously [10] and the ether oxygen of PEO also made it useful to hold large dipole moment. These features made PEO stand out from the rest as it facilitated the transfer of conductivity from intrinsically conducting polymers (ICPs) to PEO with ease. One significant criteria to consider when designing a new polymer is miscibility. Researchers had analyzed the miscibility of PEO and PVC and they all came out with the same conclusion which confirmed that PEO and PVC formed a miscible blend [11,12].

In contrast, filler incorporation into matrix usually disrupted the interfacial adhesion between matrix and filler. In order to overcome this drawback, several techniques such as chemical treatment or surface modifier were applied. These practices altered the surface of the filler or establishing new bonds respectively which in turn improved the electrical conductivity of the composites [13]. Naphthalene were used as a surface modifier on the properties of PEO/PVC/PAni films. In this paper, the effect of PAni loading and naphthalene as the surface modifier were discussed according to the electrical conductivity properties, morphology and Fourier Transform Infrared (FTIR) study.

2. Materials

PAni with the particle size of 21 μm was obtained in the form of emeraldine salt with 20 wt% of carbon black from HmBG Co. Inc. PEO (M_w of 100,000) and PVC (M_w of 220,000) used in this study were acquired from AR Alatan Sdn. Bhd., Kedah, Malaysia. The analytical grade of tetrahydrofuran (M_w of 72.11) was available through Sigma Aldrich. Dioctyl terephthalate (M_w of 390) used in this study as a plasticizer and were obtained via AR Alatan Sdn. Bhd. Naphthalene was acquired from a local shop in Kangar, Perlis, Malaysia.

3. Methods

3.1 Samples Preparation

To prepare the PEO/PVC/PAni films, required amount of PEO and PVC were dissolved individually in THF. The solution were stirred at ambient temperature until it became homogeneous. Later, DOTP and PAni and were integrated into the solution and were stirred with magnetic stirrer at 400 rpm for 4 hours. Noted for PEO/PVC/PAni/Naphthalene films, naphthalene were added together with DOTP and PAni at this stage. Lastly, the solution were left to evaporate at room temperature in fume cupboard for the film to form. The same procedures was repeated to prepare other films with different filler loading as displayed in Table 1.

Table 1: Formulations of the PEO/PVC/PAni films and PEO/PVC/PAni/Naphthtalene films with different PAni loading.

Conductive films code	PEO/PVC (40:60) (wt%)	DOTP (wt%)	PAni (wt%)	Naphthalene (wt%)
PEO/PVO	85	15	-	-
PEO/PVC/PAni -2.5	82.5	15	2.5	-
PEO/PVC/PAni -5	80	15	5	-
PEO/PVC/PAni -7.5	77.5	15	7.5	-
PEO/PVC/PAni -10	75	15	10	-
PEO/PVC/PAni/Napthalene-2.5	76.5	15	2.5	6
PEO/PVC/PAni/Naphthalene-5	74	15	5	6
PEO/PVC/PAni/Naphthalene-7.5	71.5	15	7.5	6
PEO/PVC/PAni/Naphthalene-10	69	15	10	6

3.2 Electrical Conductivity Test

For the electrical conductivity measurement, 4 point probe technique was executed via Keithley Model 4200 semiconductor characterization system. The value were determined via the relationship with resistivity as displayed in Eq. (1) and Eq. (2):

$$\rho = R\left(\frac{w \times t}{l}\right) \qquad (1)$$

where R = resistance of the films, w = width, t = thickness, l = length between the metal probe contact, and the conductivity, σ, was calculated using the Eq. (2):

$$\sigma = \frac{1}{\rho} \qquad (2)$$

3.3 Scanning Electron Microscopy (SEM)
The morphology of the conductive films were analyzed via JEOL JSM-6460LA. The films were cut into square shaped sample to fit the dimension of the SEM aluminium stubs to determine the samples morphology. The samples must be coated first before the examinations can be carried out. The samples were coated with thin palladium layer to prevent electrostatic charges during the analysis. The morphology were examined at 400x magnification and 10 kV.

3.4 Fourier Transform Infrared (FTIR) Spectroscopy
FTIR spectra of the films were examined using attenuated total reflectance (ATR) technique via Perkin-Elmer Spectrum 400 Series equipment. The testing were operated from 650 cm^{-1} to 4000 cm^{-1} at room temperature.

4. Results and Discussion

4.1 Electrical Conductivity
Fig. 1 showed the electrical conductivity versus PAni loading of PEO/PVC/PAni films without and with naphthalene. The electrical conductivity was found to increase with the addition of PAni from 2.5 wt% to 10 wt%. The increment of PAni loading were related to the conducting path that occurred as PAni particles attracted to each other. This phenomenon resulted in improved efficiency of the electrons to flow in the films thus increased the conductivity. Merlini et al. also explained that PAni introduction in PAni-coated coconut fibers lead to increased electrical conductivity due to higher conducting pathway formed in the composites [14].

The incorporation of naphthalene, on the other hand, enhanced the electrical conductivity of PEO/PVC/PAni/Naphthalene films. The observations suggest that naphthalene improved the dispersion of PAni which subsequently enhanced the formation of conductive pathway in the films. Report by Kim and Jo, [15] analyzed the effect of naphthalene on electrical conductivity and their study concluded that a compatibilizer consisting naphthalene enhanced the dispersion of filler in the matrix. Moreover, the benzene ring in naphthalene structure also aided the interaction between PEO/PVC phase and and PAni which also helped to increase the electrical properties of the films.

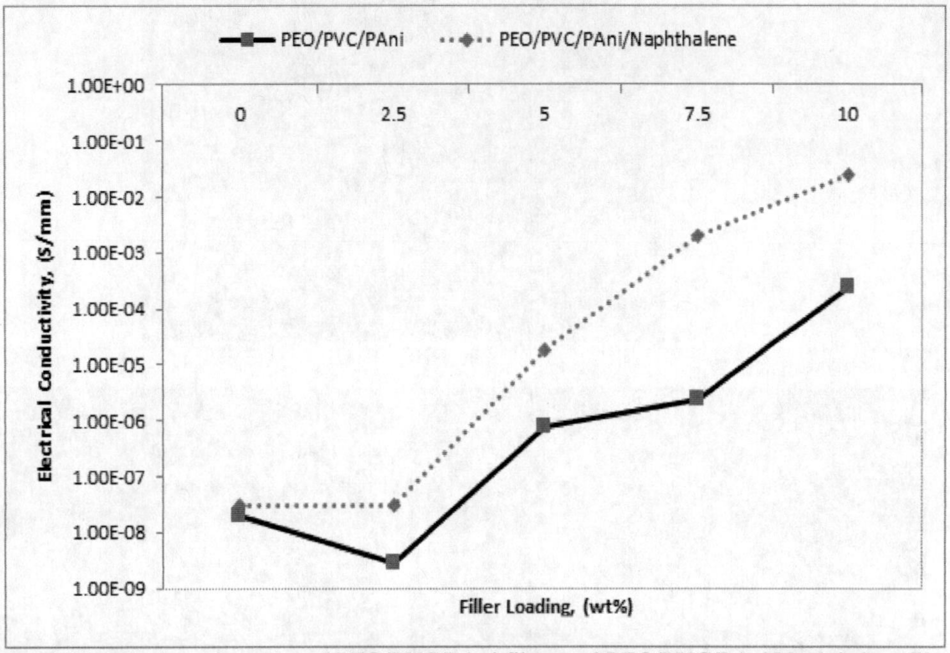

Figure 1. Electrical conductivity of PEO/PVC/PAni films and PEO/PVC/PAni/Naphthalene films at different filler loadings.

4.2 Morphology analysis

The morphologies of the PEO/PVC/PAni films without and with naphthalene were recorded in SEM micrographs displayed in Fig. 2a-e. First, the morphology of PEO/PVC blend presented a smooth surface which pointed out that PEO and PVC formed a miscible blends. Moreover, from this micrograph, it can be understood that at 5 wt% of PAni loading, PAni were properly dispersed (Fig. 2b and d) while at higher filler loading (10 wt%), the SEM micrographs showed large agglomerates of PAni in the PEO/PVC blend (Fig. 2c and e). In short, at higher filler loading, PAni have a habit to agglomerate as it became harder to disperse in congested spaces.

On the other hand, the SEM micrographs of PEO/PVC/PAni/Naphthalene films (Fig. 2d and e) revealed better surface morphology than PEO/PVC/PAni films (Fig. 2b and c). The SEM micrographs presented good adhesion and fine dispersion between PEO/PVC blend and PAni filler which were verified by rough surface and less occupancy of the filler on the matrix surfaces. Furthermore, naphthalene developed strong chemical linkage between PEO/PVC phases and PAni to boost the interaction of matrix and filler which in turned formed good interfacial adhesion.

IOP Conf. Series: Journal of Physics: Conf. Series **908** (2017) 012006 doi:10.1088/1742-6596/908/1/012006

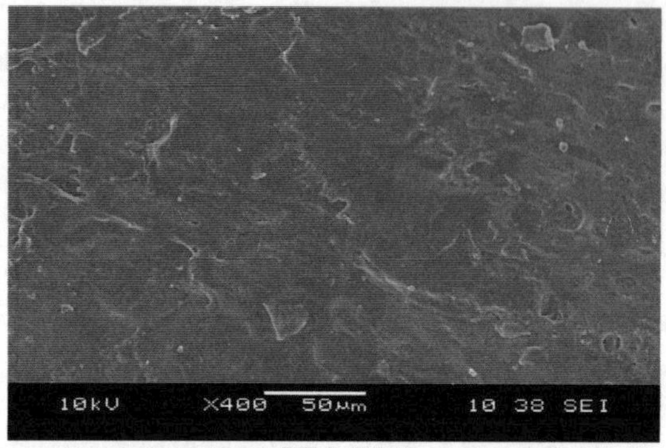

a) PEO/PVC

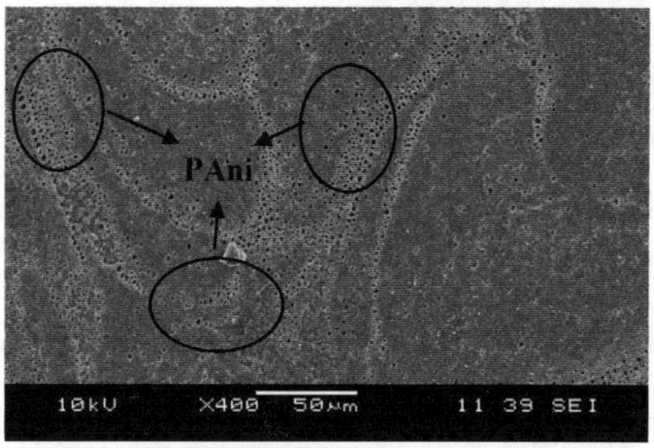

b) PEO/PVC/PAni – 5

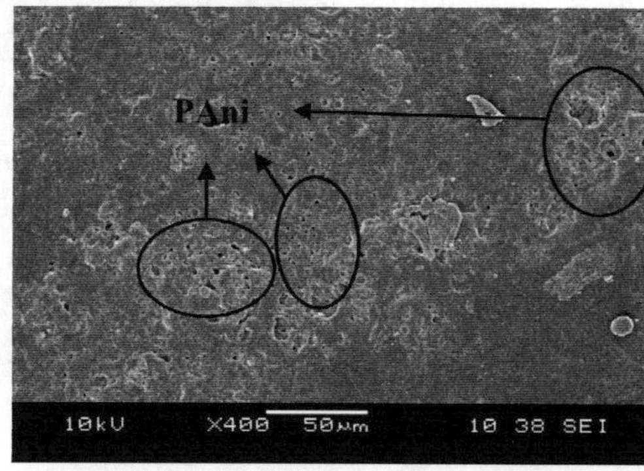

d) PEO/PVC/PAni-5/Naphthalene

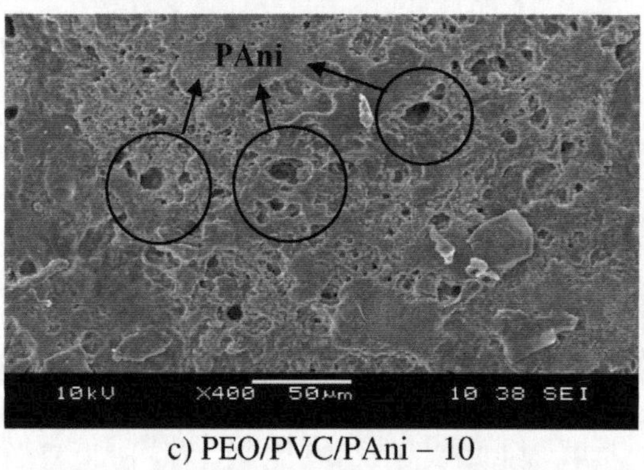

c) PEO/PVC/PAni – 10

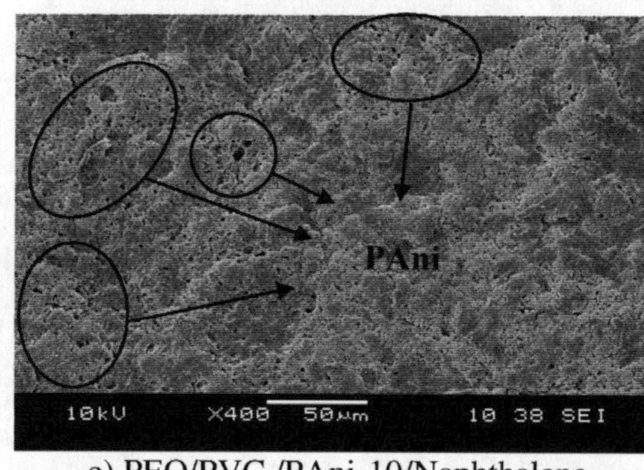

e) PEO/PVC /PAni-10/Naphthalene

Figure 2. (a-e) SEM morphology of PEO/PVC/PAni films and PEO/PVC/PAni/Naphthalene films at different filler loadings.

4.3 Spectroscopy Infrared Analysis

Fig. 3 represented the FTIR spectra of PEO/PVC/PAni films without and with naphthalene. The peaks at 2885.91 cm^{-1} and 2886.33 cm^{-1} were assigned to CH_2 asymmetric stretching which derived from PEO and PVC matrix. The band at 1715.79 cm^{-1} and 1771.13 cm^{-1} corresponded to the C=O stretching of PEO. The peaks at 1466.98 cm^{-1} and 1465.95 cm^{-1} were indicated to CH_2 deformation while the absorption peak at 1342.28 cm^{-1} were attributed to CH_2 vibration. Strong vibration of C-O structure was observed at 1278.77 cm^{-1} and 1274.90 cm^{-1}. In addition to the above peaks, the spectrum recorded the C-O-C stretch at 1147.90 cm^{-1}, 1103.54 cm^{-1}, 1099.61 cm^{-1} and 1061.38 cm^{-1} and also the characteristics peaks of gauche conformation of [-CH_2-CH_2-] group from PEO at 962.69 cm^{-1}, 960.96 cm^{-1} and 842.82 cm^{-1} and 842.82 cm^{-1}. The peak at 732.59 cm^{-1} was attributed to the C-Cl stretching of PVC. In short, the FTIR spectra validated that the addition of naphthalene in

PEO/PVC/PAni/Naphthalene did not establish new functional group as naphthalene only created physical interaction between PEO/PVC blend with PAni. Fig. 4 displayed the proposed mechanism between matrix and filler with the presence of naphthalene.

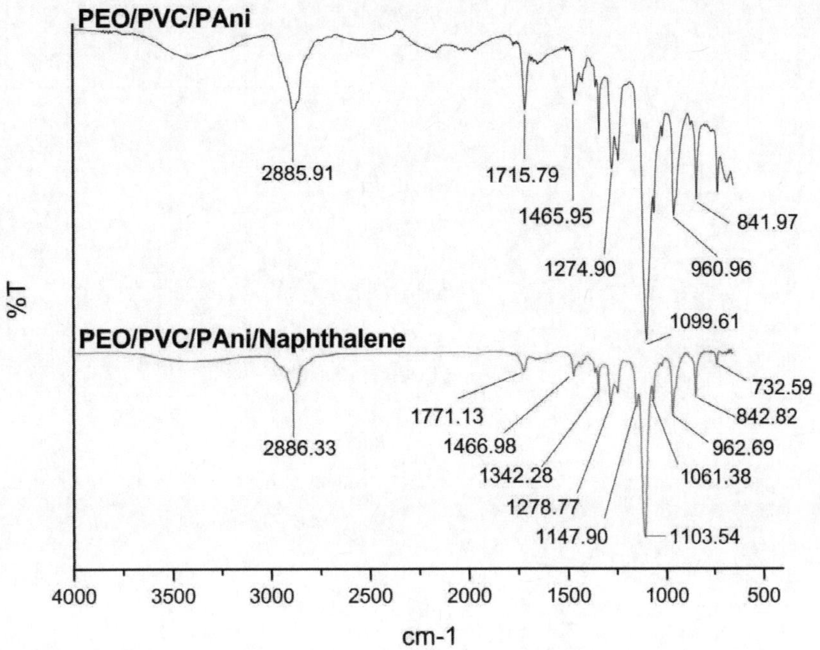

Figure 3.: FTIR spectra of PEO/PVC/PAni films and PEO/PVC/PAni/Naphthalene films.

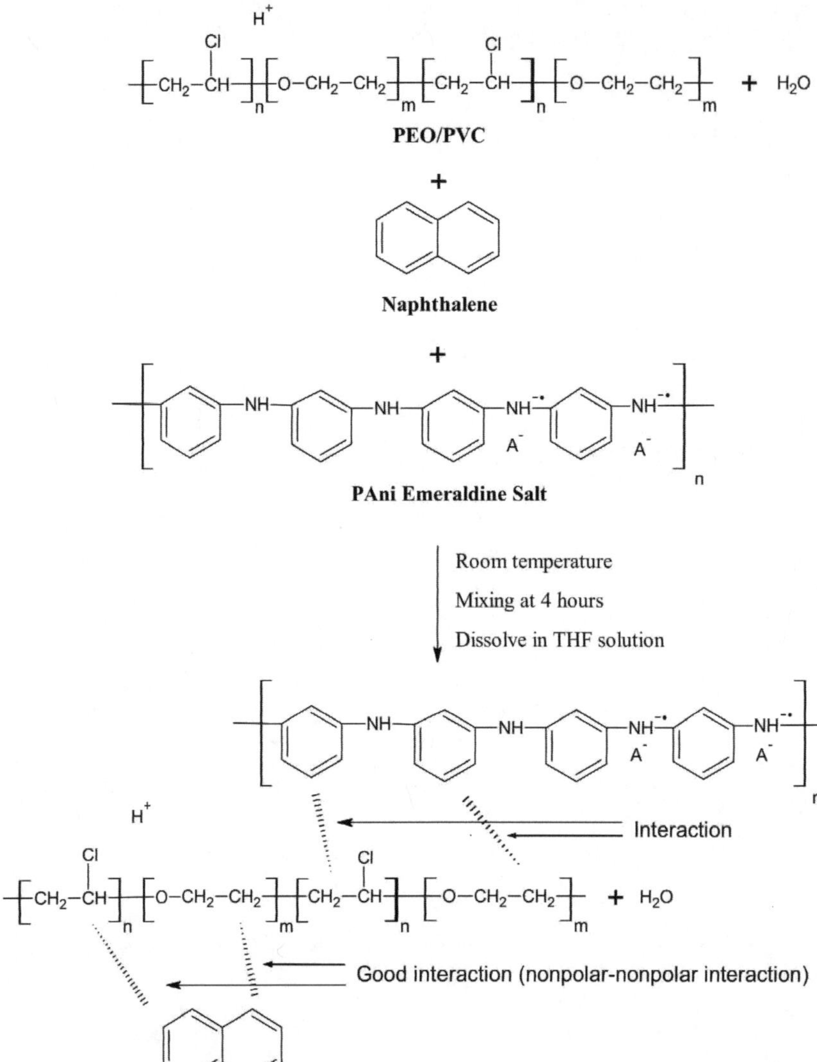

Figure 4.: Proposed interaction of naphthalene with PAni and PEO/PVC matrix.

5. Conclusion

We executed the current research to relate the addition of PAni loading and naphthalene into the PEO/PVC blend by using electrical conductivity, SEM morphology and FTIR spectra. Electrical conductivity test revealed that the incorporation and PAni and naphthalene into the films increased the electrical conductivity values. PEO/PVC/PAni/EDMA films showed superior dispersion of filler in matrix compared to PEO/PVC/PAni films, as demonstrated by the SEM micrographs. The interaction of naphthalene between the PEO/PVC matrix and PAni filler did not formed any new functional group as sanctioned by the FTIR spectra.

6. Acknowledgements

Author would like to record his appreciation to the Ministry of Higher Education Malaysia for the funds given which facilitate in accomplishing the study.

References

[1] MacDiarmid AG 2001 Angew. Chem. Int. Ed. **40** 2581-2590
[2] Varesano A, Dall'Acqua L and Tonin C 2005 Polym. Degrad. Stab. **89** 125-132
[3] Souza FG, Pinto JC, de Oliveira GE and Soares BG 2007 Polym. Test. **26** 720-728
[4] Supri AG, Tan SJ, Ismail H and Teh PL 2014 J. Vinyl. Addit. Techn. **20** 122-130
[5] Chen X, Zhou Z, Lv W, Huang T and Hu S 2009 Mater. Chem. Phys. **115** 258-262
[6] Saad ALG, Aziz HA and Dimitry OIH 2004 J. Appl. Polym. Sci. **91** 1590-1598
[7] Vadukumpully S, Paul J, Mahanta N and Valiyaveettil S 2011 *Carbon* **49** 198-205
[8] Ouyang M and Chance CM 1998 *Polym.* **39** 1857-1862
[9] Saad ALG, Hussien LI, Ahmed MGM and Hassan AM 1998 J. Appl. Polym. Sci. **69** 685-693
[10] Ito K, Tominaga Y and Ohno H 1997 Electrochim. Acta **42** 1561-1570
[11] da Silva Neiro SM, Dragunski DC, Rubira AF and Muniz EC 2000 Eur. Polym. J. **36** 583-589
[12] Ramesh S and Arof AK 2009 Int. Polym. Proc. **24** 354-358
[13] Supri AG, Hajar MS, Hanif MPM 2016 Polym. Bull. **73** 2831-2841
[14] Merlini C, Barra GM, Schmitz DP, Ramôa SD, Silveira A, Araujo TM and Pegoretti A 2014 Polym. Test. **38** 18-25
[15] Kim KT and Jo WH J Polym. Sci. A Polym. Chem. **48** 4184-4191

IOP Conf. Series: Journal of Physics: Conf. Series **908** (2017) 012007 doi:10.1088/1742-6596/908/1/012007

The effect on poly (ethylene oxide) / poly (vinyl chloride) / polyaniline (PAni) films by ethylene dimethacrylate as surface modifier: electrical conductivity and characterization

M I M Yazid[1], S A Ghani[2], A F Osman[1], S H M Din[1] and T S Jin[2]

[1] School of Materials Engineering, Universiti Malaysia Perlis, Kompleks Pusat Pengajian Jejawi 2, 02600 Arau, Perlis, Malaysia.
[2] Faculty of Engineering Technology, Universiti Malaysia Perlis (UniMAP), Sungai Chuchuh, Padang Besar, Perlis, Malaysia.

moizzuddeen@gmail.com

Abstract: In this paper, the electrical conductivity and the characterizations of poly (ethylene oxide) / poly (vinyl chloride) / polyaniline (PEO/PVC/PAni) films with and without the presence of Ethylene Dimethacrylate (EDMA) were investigated. The films were prepared using solution casting method using tetrahydrofuran as the solvent. The results indicated that the electrical conductivity of PEO/PVC/PAni films increased as the PAni loadings increased while PEO/PVC/PAni films with EDMA showed higher electrical conductivity than PEO/PVC/PAni films. PEO/PVC/PAni films with EDMA showed a value as high as 0.68 x 10^{-4} S/mm as it enhanced the interfacial adhesion between the matrix and the filler as proved by the SEM morphology. Meanwhile, the FTIR analysis confirmed there are no new chemical bonding occurred with the introduction of EDMA into the films.

1. Introduction

Intrinsically conducting polymers (ICPs) have caught many recognition due to their exclusive physicochemical properties and huge potential in applications [1-6]. Among numerous ICPs, polyaniline (PAni), polypyrrole (PPy) and polythiopene (PTh) are favorable to be applied in practical appliance as they acquire a combination of unique chemical and physical processes besides enough stability at normal processing conditions [1,2,7]. PAni itself has been extensively studied as it holds many advantages compared to many other ICPs such as ease of synthesis, high electrical conductivity, decent environmental stability, fast oxidation-reduction reactions and relatively cheap [8].

However, PAni still has some deficiencies that need to overcome which hindered their commercial potential. Poor mechanical properties and difficulties in processing are some of the drawbacks that require attention [9]. Various approaches have been reported to solve these shortcomings in the last three decades. One of the favorable way to counter these issues is by polymer blending [10]. Synthesis of PAni with poly (vinyl chloride) (PVC) for example, can be reflected as versatile and efficient method for exploiting PAni and reducing its shortcoming [11, 12]. Likewise, the potential of poly (vinyl chloride) (PVC) and poly (ethylene oxide) (PEO) as matrix will be analyzed in this study with PAni as a function of conductive filler. In addition, solution casting which offers simplicity in processing and also cost effective is chosen as the preferred fabrication method.

Studies on PEO and PVC blend have been reported as the blend offers one important criteria in polymer blend, which is miscibility [13-16]. PVC is one of the major commercial polymer which provides low cost, chemical stability, application versatility and also biocompatible [17]. Furthermore, PVC has competent compatibility with many types of plasticizers besides easiness in processability. On the other hand, PEO is one of the most frequently studied polymer in this field as PEO can dissolve huge capacity of inorganic salts homogeneously [18] and ether oxgens in PEO makes it able to hold large dipole moment. These unique properties of PEO is very useful in helping PAni to transfer its electrical conductivity.

On the other hand, the integration of filler into matrix will typically resulting in deficiencies of matrix/filler interfacial adhesion. This phenomenon affected the properties of composites and common method to enhance the interfacial adhesion is by utilizing surface modifier or chemical treatment. These methods helps to modify the composites surface and consequently improve the transfer of electrical conductivity in the composites [19]. Ethylene dimethacrylate (EDMA) were used in this paper to represent the effect of surface modifier on the properties of PEO/PVC/PAni films. In short, the effect of PAni and EDMA were analyzed on PEO/PVC/PAni films according to the electrical conductivity properties, morphology and Fourier Transform Infrared (FTIR) study.

2. Materials

PAni in emeraldine salt form with 20 wt% of carbon black has the particle size of 21 μm was brought from HmBG Co. Inc. PEO with 100,000 g/mol molecular weight and PVC powder with molecular weight of 220,000 g/mol were applied in this study as the matrix were obtained from AR Alatan Sdn. Bhd., Kedah, Malaysia. The solvent used, tetrahydrofuran, with molecular weight of 72.11 g/mol was analytical grade from Sigma Aldrich. Dioctyl terephthalate with molecular weight of 390 g/mol was used as a plasticizer was provided by AR Alatan Sdn. Bhd., Kedah, Malaysia. Ethylene dimethacrylate with the density of 1.05 g/mL and molecular weight of 198.22 g/mol was obtained from Fisher science.

3. Methods

3.1 Samples Preparation

Two types of films were prepared, PEO/PVC/PAni films and PEO/PVC/PAni films with EDMA were fabricated using solution casting technique. The ratio of PEO/PVC was fixed at 40/60, and these materials were incorporated with PAni loading from 2.5 wt% to 10 wt%. First, PEO and PVC were dissolved in THF in a separate beaker. The fully dissolved PEO and PVC were then mixed and stirred at 400 rpm. Afterward, DOTP, EDMA and 2. 5 wt% of PAni were poured into the PEO/PVC solvent and were continuously stirred for another 4 hours to reach homogenous solution. Lastly, the solvent were poured onto a mold and left to dry under controlled conditions. Similar procedures were used to fabricate another films with PAni loadings of 5 wt%, 7.5 wt% and 10 wt%. The formulations for PEO/PVC/PAni films and PEO/PVC/PAni/EDMA films are shown in Table 1.

Table 1. Formulations of the PEO/PVC/PAni and PEO/PVC/PAni/EDMA films with different PAni. loading.

Conductive films code	PEO/PVC (40:60) (wt%)	DOTP (wt%)	PAni (wt%)	EDMA (wt%)
PEO/PEO	85	15	-	-
PEO/PVC/PAni -2.5	82.5	15	2.5	-
PEO/PVC/PAni -5	80	15	5	-
PEO/PVC/PAni -7.5	77.5	15	7.5	-
PEO/PVC/PAni -10	75	15	10	-
PEO/PVC/PAni /EDMA-2.5	76.5	15	2.5	6
PEO/PVC/PAni /EDMA-5	74	15	5	6
PEO/PVC/PAni /EDMA-7.5	71.5	15	7.5	6
PEO/PVC/PAni/EDMA-10	69	15	10	6

3.2 Electrical Conductivity Test

4 point probe technique was performed via Keithley Model 4200 semiconductor characterization system to determine the electrical conductivity value. The value of the electrical conductivity were figured out via the relationship with resistivity such illustrated using Eq. (1) and Eq. (2):

$$\rho = R\left(\frac{w \times t}{l}\right) \quad (1)$$

where R = resistance of the films, w = width, t = thickness, l = length between the metal probe contact, and the conductivity, σ, was calculated using the Eq. (2):

$$\sigma = \frac{1}{\rho} \quad (2)$$

3.3 Scanning Electron Microscopy (SEM)

The morphology of the PEO/PVC/PAni films and PEO/PVC/PAni/EDMA films were carried out by using model JOEL JSM-6460LA. The samples undergo a thin palladium layer of 20 nm sputtering coating. The surface of the films were coated to inhibit electrostatic charges during the analysis. The morphology were examined at 400x magnification and 10 kV.

3.4 Fourier Transform Infrared (FTIR) Spectroscopy

FTIR spectra of the films were established with Perkin-Elmer Spectrum 400 Series equipment using attenuated total reflectance (ATR) technique. The spectra of % transmittance versus wavelength (cm^{-1}) were analyzed at room temperature with scanning range of 650 cm^{-1} to 4000 cm^{-1}.

4. Results and Discussion

4.1 Electrical Conductivity

Figure 1 displays the effect of PAni loadings on the electrical conductivity of PEO/PVC/PAni films and PEO/PVC/PAni/EDMA films. Increasing of PAni loading on the PEO/PVC matrix improved the films conductivity and lead to percolation threshold at 10 wt% of PAni loading. This could be explained by the ability of intrinsically conducting polymer, PAni, to transmit the electrical conductivity into the PEO/PVC matrix. Moreover, as the filler loading increase, the gap between PAni particles will become narrower which will subsequently create conductive paths. As a results, electrons will effortlessly transmitted in films and enhanced the electrical conductivity of PEO/PVC/PAni films and PEO/PVC/PAni/EDMA films. In a study reported by Merlini et al., the presence of PAni in PAni-coated coconut fibers exhibited significant rise in electrical conductivity as PAni provides the conductivity through its conjugated backbone [20].

At a similar composition, PEO/PVC/PAni/EDMA films exhibited higher electrical conductivity than PEO/PVC/PAni films. The improved conductivity of the films is credited to the enhanced distribution of PAni in the films as a results of EDMA incorporation. The results is in line with Castillo- Castro et al. as they reported that the addition of 5 wt % polyethylene-graft-maleic anhydride (PEgMA) as a coupling agent improve the conducting paths organization of PAni in low density polyethylene / n-dodecylbenzene sulfonate doped polyaniline films [21].

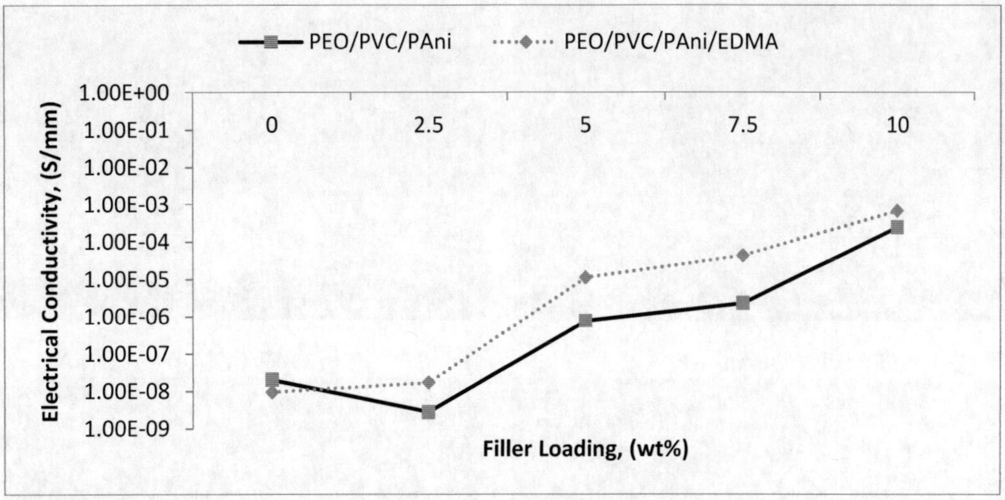

Figure 1. Electrical conductivity of PEO/PVC/PAni films and PEO/PVC/PAni/EDMA films at different filler loadings.

4.2 Morphology analysis

Figure 2 (a-e) explains the morphologies of PEO/PVC/PAni films with the effect of different loading of PAni and the effect of EDMA presence to the films. Figure 2a proves that the blend of PEO and PVC is miscible as the figure exhibits a smooth surface as expected. The founding is in line with other researchers such as Ramesh et al. that claimed PEO and PVC produced a miscible blend. Figure 2b and 2c present the effect of filler loading on the films with 5 wt% and 10 wt% PAni loading respectively [15]. PEO/PVC/PAni-5 wt% (Figure 2b) displays a better distribution of PAni and less visible PAni agglomeration compared to PEO/PVC/PAni-10 wt% (Figure 2c). PAni agglomeration at higher loadings is understandable as fillers have a tendency for filler-filler interaction rather than matrix-filler interaction [22]. Meanwhile, the addition of EDMA into the films (Figure 2d and Figure 2e) demonstrates lower filler agglomerations and better filler distribution at high PAni loading compared to the films without EDMA addition (Figure 2b and Figure 2c). EDMA improves the interfacial adhesion between the matrix and the filler and rearrange the conductive path in the films which explains the improvement on electrical conductivity of PEO/PVC/PAni/EDMA films.

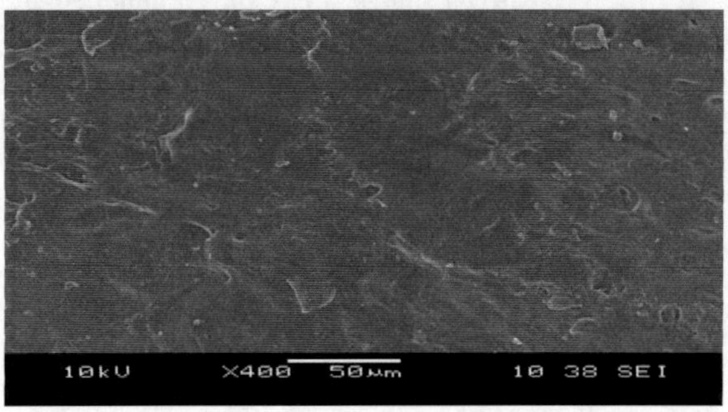

a) PEO/PVC

IOP Conf. Series: Journal of Physics: Conf. Series **908** (2017) 012007 doi:10.1088/1742-6596/908/1/012007

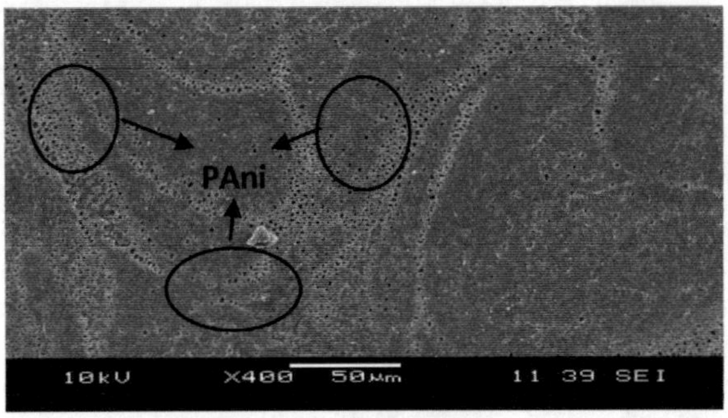

b) PEO/PVC/PAni – 5

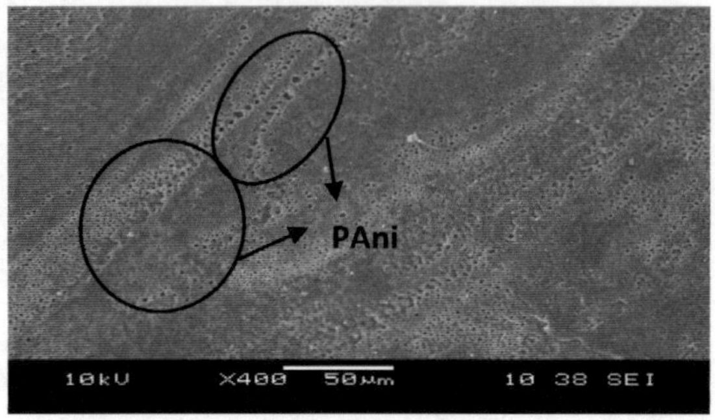

d) PEO/PVC/PAni–5/EDMA

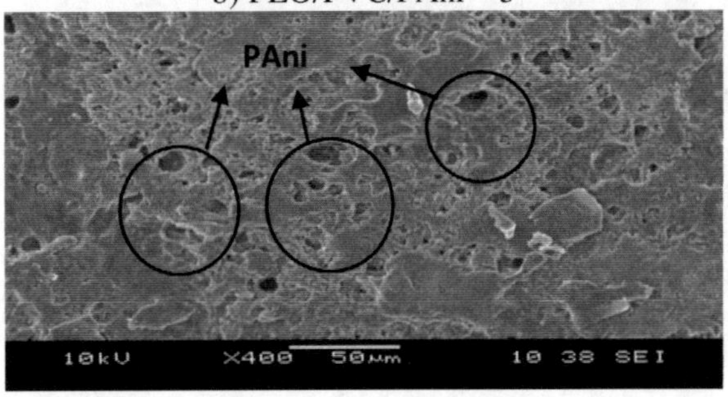

c) PEO/PVC/PAni – 10

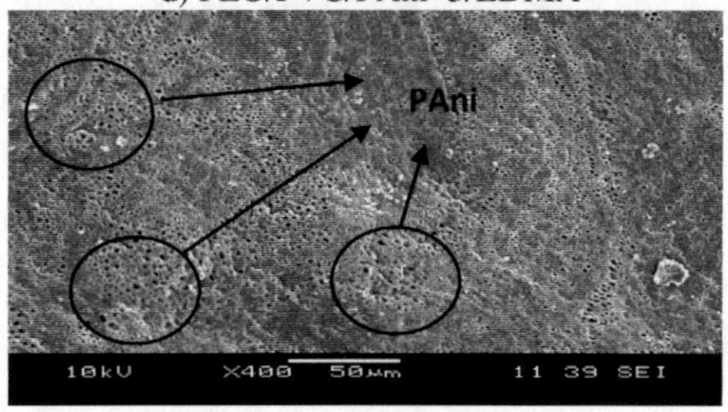

e) PEO/PVC/PAni–10/EDMA

Figure 2 (a-e). SEM morphology of PEO/PVC/PAni films and PEO/PVC/PAni/EDMA films at different filler loadings.

4.3 Spectroscopy Infrared Analysis

The FTIR spectra of PEO/PVC/PAni films and PEO/PVC/PAni/EDMA films are presented in Figure 3. The spectra shows no striking peak from 1750 to 2800 cm^{-1} but C-N stretching of secondary aromatic amine is detected at the medium intensity band of 1465 cm^{-1}. On the other hand, a peak at 1278 cm^{-1} is spotted which signals to the PAni polaronic structure. The bands corresponding to the in-plane and out-of-plane C-H stretching appears at 962 cm^{-1} and 842 cm^{-1} while the band at 731 cm^{-1} are related to The C-Cl stretching. The FTIR spectra indicates that the presence of EDMA in films did not generate any new functional group. It shows that EDMA only enhanced the matrix-filler interfacial adhesion physically without creating any chemical interactions.

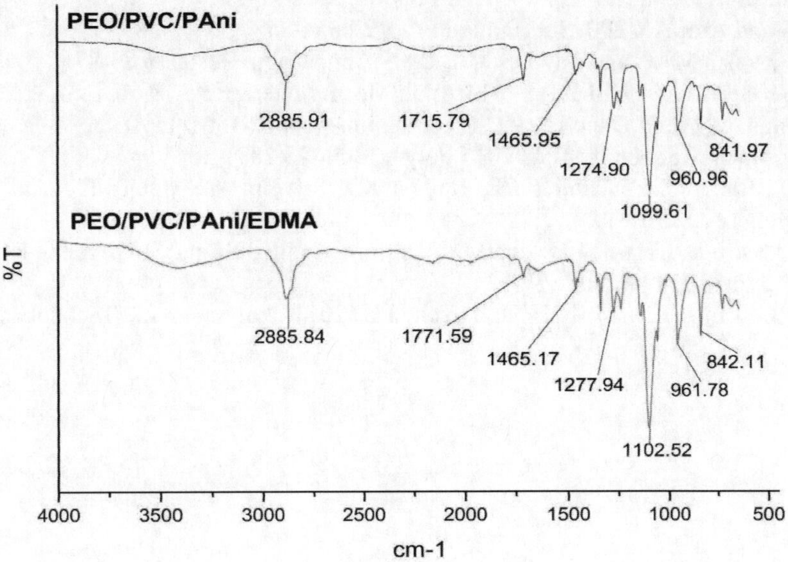

Figure 3. FTIR spectra of PEO/PVC/PAni films and PEO/PVC/PAni/EDMA films.

5. Conclusion

The addition of PAni and EDMA in PEO/PVC/PAni films significantly increased the electrical conductivity of the films. PEO/PVC/PAni films with EDMA exhibit better dispersion of PAni in the PEO/PVC matrixes compared to the films without EDMA, as evidenced by the morphological study using SEM. The FTIR spectra also confirms that there is no new functional group created with the addition of EDMA in the films.

6. Acknowledgements

Author would like to record his appreciation to the Ministry of Higher Education Malaysia for the funds given which facilitate in accomplishing the study.

7. References

[1] Jaymand M 2013 Prog. Polym. Sci. **38** 1287-1306
[2] Jaymand M, Hatamzadeh M and Omidi Y 2015 Prog. Polym. Sci. **47** 26-69
[3] James R, Nagarale RK, Sachan VK, Badalucco C, Bhattacharya PK and Kumbar SG 2014 Polym. Adv. Technol. **25** 1439-1445
[4] Vonlanthen D, Lazarev P, See KA, Wudl F and Heeger AJ 2014 Adv. Mater. **26** 5095-5100
[5] Lee K, Cho S, Park, SH, Heeger AJ, Lee CW and Lee SH 2006 Nature **441** 65-68
[6] Massoumi B, Davtalab S, Jaymand M and Entezami AA 2015 RSC Adv. **5** 36715-36726
[7] Hatamzadeh M, Mohammad-Rezaei R and Jaymand M 2015 Mater. Sci. Semiconduct. Proces. **31** 463-470
[8] Massoumi B, Mohammad- Rezaei R, and Jaymand M 2016 2014 Polym. Adv. Technol. **27** 1056-1063
[9] Varesano A, Dall'Acqua L and Tonin C 2005 Polym. Degrad. Stab. **89** 125-132
[10] Souza FG, Pinto JC, de Oliveira GE and Soares BG 2007 Polym. Test. **26** 720-728
[11] Heydari M, Najafi Moghadam P, Fareghi AR, Bahram M and Movagharnezhad N 2015 Polym. Adv. Technol. **26** 250-254
[12] Nazari M., Agbolaghi S, Abbaspoor S, Gheybi H and Abbasi F 2015 Macromolecules **48** 8947-8957
[13] da Silva Neiro SM, Dragunski DC, Rubira AF and Muniz EC 2000 Eur. Polym. J. **36** 583-589

[14] MacCallum JR and Smith JSG 2000 Eur. Polym. J. **36** 491-494
[15] Ramesh S and Arof AK 2009 Int. Polym. Proc. **24** 354-358
[16] Silva EP, Kaplan MAC and Tavares MIB 2007 Ann. Magn. Reson. **6** 83-87
[17] Hajar MS, Supri A and Jalilah AJ 2014 Appl. Mech. Mater. **695** 348
[18] Ito K, Tominaga Y and Ohno H 1997 Electrochim. Acta **42** 1561-1570
[19] Supri AG, Hajar MS, Hanif MPM 2016 Polym. Bull. **73** 2831-2841
[20] Merlini C, Barra GM, Schmitz DP, Ramôa SD, Silveira A, Araujo TM and Pegoretti A 2014 Polym. Test. **38** 18-25
[21] Castillo- Castro D, Castillo- Ortega MM, Herrera- Franco PJ and Rodríguez- Félix DE 2011 J. Appl. Polym. Sci. **119** 2895-2901
[22] Supri AG, Tan SJ, Ismail H and Teh PL 2011 Polym-Plast. Technol. **50** 898-906

IOP Conf. Series: Journal of Physics: Conf. Series **908** (2017) 012008 doi:10.1088/1742-6596/908/1/012008

Effect of oxidation agent on wood biomass in ethylene vinyl acetate conductive polymer: tensile properties, tensile fracture surface and electrical properties

M P M Hanif[1], A G Supri[2], A R Rozyanty[1,3], S J Tan[2,3],

[1]School of Materials Engineering, Universiti Malaysia Perlis (UniMAP), Kompleks Taman Muhibbah, Jejawi 2, 02600 Arau, Perlis, Malaysia.
[2]Faculty of Engineering Technology, Universiti Malaysia Perlis (UniMAP), P.O Box 77, D/A Pejabat Pos Besar, 01000 Kangar, Perlis, Malaysia.
[3]Center of Excellence Geopolymer and Green Technology, School of Materials Engineering, Universiti Malaysia Perlis (UniMAP), Kompleks Taman Muhibah, Jejawi 2, 02600, Perlis, Malaysia.

anepisal@gmail.com

Abstract: The wood fiber (WF) type of Pulverised Wood Filler obtained by combustion process at temperature under 700 °C for 3 hours was characterized and coated with ferric chloride ($FeCl_3$) by ethanol solution. Both carbonized wood fiber (CWF) and carbonized wood fiber-ferric chloride (CWF-$FeCl_3$) were used as filler in ethylene vinyl acetate (EVA) conductive polymer. The filler was coated with $FeCl_3$ to enhance the properties of the CWF to achieve progressive mechanical and electrical properties. The CWF and CWF-$FeCl_3$ loading were varied from 2.5 to 10.0 wt%. EVA/CWF and EVA/CWF-$FeCl_3$ conductive polymer were processed by using Brabender Plasticoder at 160 °C with 50 rpm rotor speed for 10 min. The mechanical properties were investigated by tensile testing and the tensile fractured surface of conductive polymers was analyzed by scanning electron microscopy (SEM) analysis. Then, the electrical conductivity of conductive polymer was determined by four-point probe I-V measurement system. The EVA/CWF-$FeCl_3$ conductive polymer showed greater electrical conductivity and tensile strength but lower elongation at break than EVA/CWF conductive polymer. SEM morphology displayed rougher surface between CWF-$FeCl_3$ and EVA phases compared to EVA/CWF conductive polymer.

1. Introduction

Lately, conductive polymer composite acquired from the addition of various carbon blacks to polymer matrices was reported[1-3]. Polymer electrolytes have drawn attention due to their achievable use as electrolytes in solid-state electrochemical devices, for example like super capacitors, batteries, fuel cells, electrochemical windows, sensors, etc.[4]. The possibility of applying carbon black filled composites in various electronics application as an electrode for electric double layer capacitors also had been previously reported especially particulate fillers of a carbon which are broadly used as in polymer industries[5]. This carbon black are obtained from crude oil hence, is relatively expensive. Therefore, it is important to generate other possible substitute fillers from renewable resources like agricultural wastes.

Rice husks, wood remains, nut shells, manure, and crop residues are regarded as agricultural waste, but recently such solid waste have been transformed into biochar for the purpose of carbon sequestration. In addition, transportation and handling of lignocellulosic wood biomass is often challenging as they have low bulk density, in the range of 200 to 600 kg/m^3[6]. Biochar is commonly defined as charred organic matter, produced with the intent of being deliberately added to soil to improve its agronomic properties. On average, one ton of dry biomass can create 400 kg of biochar containing 80 to 90 % pure carbon, reported by Lehmann et. al., at temperature 300 to 700 °C, under low oxygen concentration[7]. The study conducted by Milla et. al. found that SEM-EDX analysis for wood biochar demonstrates high carbon content and a large degree of macro-porosity in the 1 to 10 micron scale, with contents of potassium (K), carbon (C), chlorine (Cl) and calcium (Ca)[8].

The thermoplastic polymer materials currently have been selected as matrices for bio-fibers. The most frequently used thermoplastics polymer in this area are polyethylene (PE), polypropylene (PP), polystyrene, etc.[9]. However, thermoplastic copolymer such as acrylonitrile butadiene styrene (ABS), poly vinyl chloride (PVC), ethylene vinyl acetate (EVA), etc. are used for advanced application because of customizable chemical structure. EVA copolymer is a polymer that has low temperature flexibility, toughness, good adhesion to a wide variety of materials and recognized as 'rubber-like' in softness. EVA copolymers are naturally polar due to the presence of vinyl acetate (VA) group which causing the polarity of EVA increases with higher VA content[10]. Moreover, an increase in VA content provides better flexibility and adhesion and low temperature performance, but increase of ethylene content promotes mechanical strength and block resistance[11].

The CWF acted as conductive fillers that provide dramatically increment in electrical conductivity at the very low filler content. Meanwhile, salt is added into solvent such as water to form an electrolyte. Individual atomic components are split by the force applied on the solute to hold ion apart in a chemical dissociation process in which the solution applies force[12]. Ferric chloride ($FeCl_3$) is a chemical compound at industrial scale with iron in the oxidation state of +3. The advantages of metal salt catalyst are ease of volatilization, availability, non-toxicity, low price, and easy in handling, while hold in combination with the toluene precursor, which is less toxic than benzene[13]. In this research, the compatibility of EVA filled carbonized wood fiber using addition of oxidation agent was studied.

2. Methods

2.1 Materials

EVA grade 2518 CO contains 18.1% of vinyl acetate (VA) pellets with density of 0.94 g/cm^3, processing temperature of 160 °C, and melt flow rate of 2.5 g/10 min was supplied by The Polyolefin Company (Singapore) Pte Ltd. The wood fiber with size of 199 μm (average diameter of the filler) and the types of Pulverised Wood Filler with mixed wood grade was obtained from PolyComposite Sdn. Bhd., Selangor. The metallic salt, ferric chloride ($FeCl_3$ with molar mass of 162.2 g/mol and density of 2.9 g/cm^3) and chemical, ethanol (C_2H_5OH with molar mass of 46.07 g/mol and density of 0.79 g/cm^3) was brought from AR Alatan Sdn. Bhd., Kedah.

2.2 Preparation of fillers CWF and CWF-$FeCl_3$

The wood fiber (WF) was ground and sieved using 150 μm and then was carbonizing using a small-scale reactor at 700 °C (3 hours) in a muffle furnace. After that, CWF with diameter size in the range of 41 to 110 μm (average is around 85 μm) were used as conductive filler in compounding A. For compounding B, CWF was coated by using 6% of ferric chloride ($FeCl_3$) under room temperature. Firstly, the 11 g of $FeCl_3$ was transferred into 500 ml of ethanol in a beaker. After the mixture was homogenous, the CWF were mixed together by using mechanical stirrer model 5430 with rotor speed of 500 rpm (24 hours). Then, filler was filtered and left to dry in the oven at 50°C (24 hours) to eliminate ethanol residue. After that, the modified CWF- $FeCl_3$ with diameter size in the range of 45 to 100 μm (average is around 75 μm) were used as filler in conductive polymer. The percentages of atomic mass of both filler are presented in Table 1.

Table 1. Atomic mass (%) from SEM-EDX spectra for CWF and CWF-$FeCl_3$.

Element	CWF	CWF-$FeCl_3$
Carbon (C)	74.20	79.70
Oxygen (O)	23.32	18.88
Magnesium (Mg)	0.55	0.26
Silica (S)	0.13	0.07
Potassium (K)	0.60	0.01
Platinum (Pt)	1.20	0.39
Chloride (Cl)	-	0.52
Ferum (Fe)	-	0.18

2.3 Conductive polymer composites preparation

The EVA/CWF and EVA/CWF-$FeCl_3$ conductive polymer with different ratio of filler loadings were carried out by using Brabender Plasticorder and was setup at temperature of 160 °C (rotor speed of 50 rpm). Firstly, the EVA pellet was loaded into the mixing chamber and preheated for 2 minutes. After that, CWF or CWF-$FeCl_3$ was added into the chamber and mixed for another 8 minutes. The molten conductive polymer was removed from the chamber. Then, to produce the conductive polymer in plate form, the hydraulic hot press machine (GT 7014A) was used. The hot and cool press was set at the temperature of 160 °C and 20°C. 1 mm empty mold was heated for 5 min before used. Then, the conductive polymer were put into the mold, preheated and compressed partially for 10 minutes. Once the conductive polymers started soften, they were fully compressed for 4 min. Finally, the composites were compress for 3 minutes for cooling process. The formulations of EVA/CWF and EVA/CWF-$FeCl_3$ conductive polymer are shown in Table 2.

Table 2. Formulations of EVA/CWF and EVA/CWF-$FeCl_3$ conductive polymer composites at different fiber loading.

Materials	EVA/CWF (wt%)	EVA/CWF-$FeCl_3$ (wt%)
EVA	100	100
CWF	2.5, 5.0, 7.5, 10	-
CWF-$FeCl_3$	-	2.5, 5.0, 7.5, 10

2.4 Characterization and measurements

Tensile properties in which tensile strength and percentage elongation at break were conducted based on ASTM D638 using Universal Testing Machine Instron 5569 at a capacity of the machine was 5kN, crosshead speed 50 mm/min and gauge length 50 mm. An average of five dumbbell shaped type four were used and the shaped specimens were conditioned at room temperature before the testing is conducted. Then, tensile fracture surface of conductive polymer were carried out using a Scanning Electron Microscopy (SEM) model JOEL JSM-6460LA, where, surface of the samples were coated with a thin platinum layer of about 20 μm in thickness using Auto Fine Coater to avoid electrostatic charges during examination. The elemental composition of CWF and CWF^{FeCl3} was determined using an Energy Dispersive X-Ray Spectroscopy (EDX). Similar specimen for microstructural analysis was used for elemental composition analysis. Finally, the electrical properties of conductive polymer were measured using four-probe I-V measurement system model Keithley Model 4200 Semiconductor Characterization System with voltage varied from 0 until 10 V. The resistivity (ρ) and electrical conductivity (σ) can be calculated using the equation 1 and 2 below:

$$\rho = \left(\frac{\pi}{ln2}\right) t \left(\frac{V}{I}\right) \quad (1)$$

$$\sigma = \frac{1}{\rho} \quad (2)$$

Where; t is sample thickness, V is voltage difference in charge between two points and I is the rate at which charge is flowing.

3. Results and Discussion

3.1 Tensile properties

Fig 1 presents the effect of CWF loading on the tensile strength of the EVA/CWF and EVA/CWF-$FeCl_3$ conductive polymer at different fiber loading. The results showed that the tensile strength of the EVA/CWF conductive polymer decreased slightly as the fiber loading increased due to the incompatibility of the CWF in the EVA phases. This incompatibility reduced the tensile strength because fractures were initiated at the weak interface between the composites. It can be seen from Fig 1 that under a similar fiber loading, the EVA/CWF-$FeCl_3$ conductive polymer had a greater tensile strength than the EVA/CWF conductive polymer. This can be explained that incorporating of $FeCl_3$ into EVA/CWF conductive polymer had improved the interfacial interaction or van der waal's forces between CWF and EVA phases. This was due to CWF-$FeCl_3$ induced carbonyl group of vinyl acetate in EVA to form temporary induced dipole molecules which provide good interfacial interaction between CWF-$FeCl_3$ and EVA matrix. Hence, good interfacial interaction allows applied load to be transfer from EVA matrix to CWF-$FeCl_3$ and dissipated throughout the matrix. The morphology of tensile fracture surface by using SEM proved the enhancement in mechanical properties was due to the improvement in the interfacial bonding between CWF and EVA matrix as presented in Figure 3.

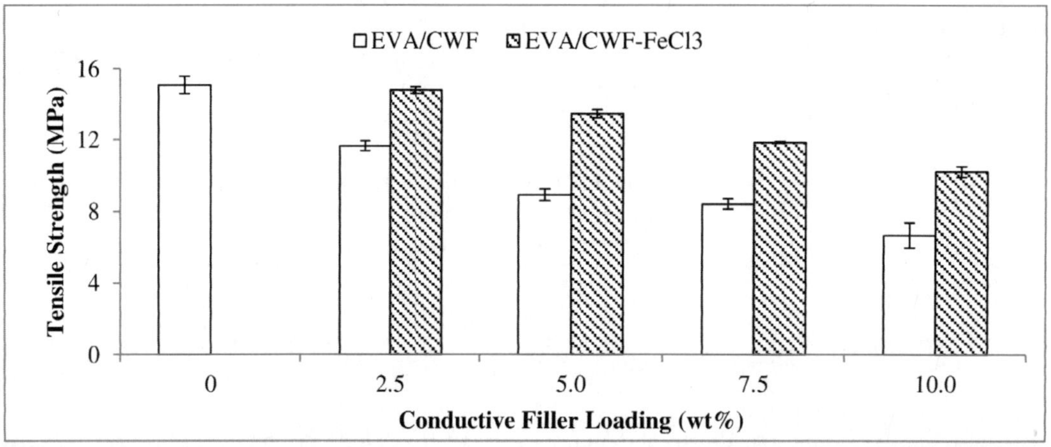

Figure 1. Tensile strength vs. conductive filler loading of EVA/CWF-$FeCl_3$ conductive polymer.

Figure 2 illustrates the effect of CWF loading on elongation at break of EVA/CWF and EVA/CWF-$FeCl_3$ conductive polymer. Both of the elongation at break of EVA/CWF and EVA/CWF-$FeCl_3$ conductive polymer decreased with increasing of filler loading. The addition of greater CWF loading into the conductive polymer tended to cause failure to occur earlier than lower CWF loading. This was due to the plasticity of the polymer matrix, which decreased the chain mobility as the conductive polymer become stiffer. Besides that, weaker interfacial region between the EVA copolymer and CWF were formed when CWF loading increased. Thus, cracks could travel more easily through these weak interfacial regions and caused the elongation at break of EVA/CWF conductive polymer to reduce. At the similar CWF loading, the EVA/CWF-$FeCl_3$ conductive polymer exhibited lower elongation at break compared to EVA/CWF conductive polymer. This was due to good interfacial interaction between the filler and matrix had increased the stiffness of the conductive polymer and thus reduced the elongation at break of the conductive polymer[14].

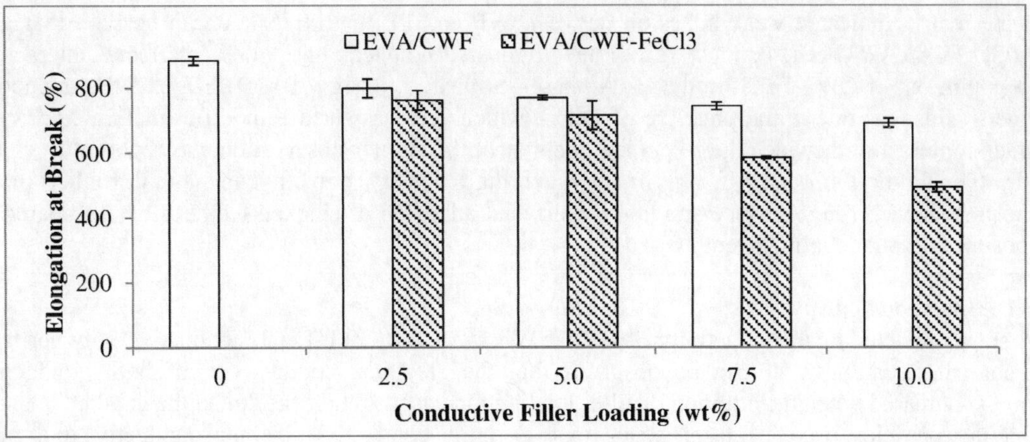

Figure 2. Elongation at break vs. conductive filler loading of EVA/CWF-$FeCl_3$ conductive polymer.

3.2 Tensile fracture surface

The influence of the CWF loading of EVA/CWF and EVA/CWF-$FeCl_3$ conductive polymer was observed in Fig 3(A)-(G) from the tensile fracture surface of conductive polymer. The micrograph of the tensile facture surface of EVA is presented in Fig 3(A).

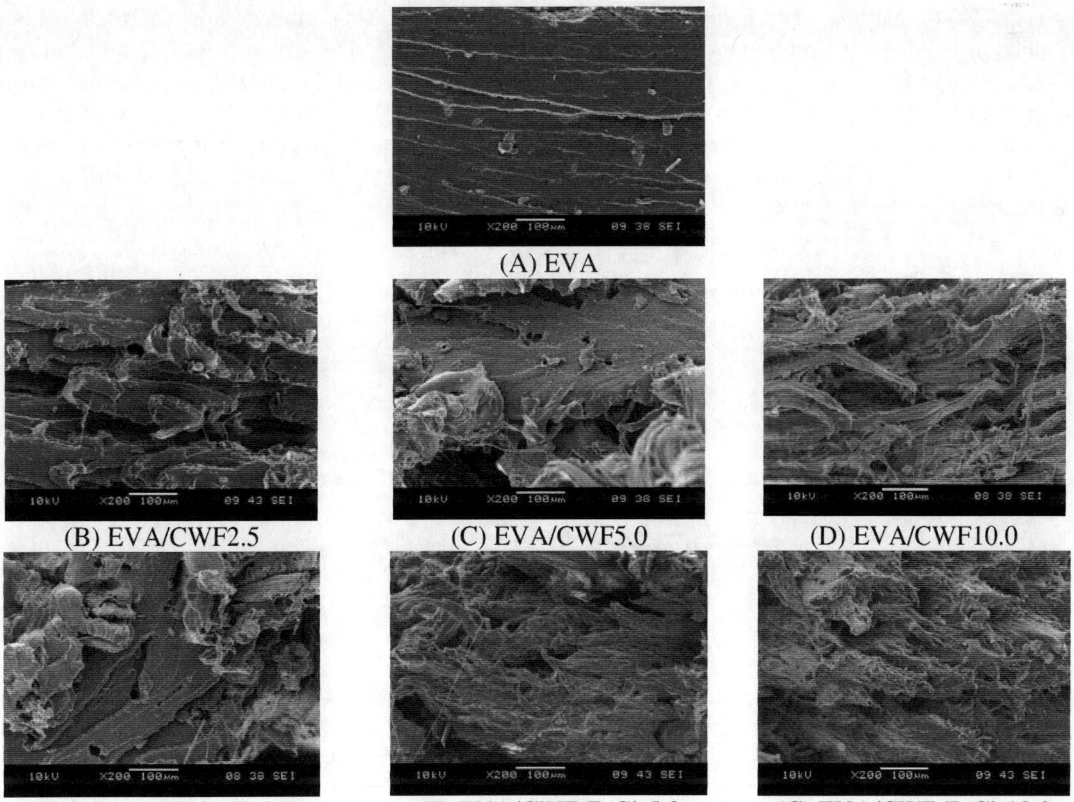

(A) EVA

(B) EVA/CWF2.5 (C) EVA/CWF5.0 (D) EVA/CWF10.0

(E) EVA/CWF-$FeCl_3$2.5 (F) EVA/CWF-$FeCl_3$5.0 (G) EVA/CWF-$FeCl_3$10.0

Figure 3. The SEM micrographs of tensile facture surface of EVA/CWF and EVA/CWF-$FeCl_3$ conductive polymer at different fiber loading.

It can be seen that the micrograph exhibited a smooth surface. As the filler loading increases, the micrographs show weak interfacial adhesion by the presence of voids from the matrix surface. According to Danyadi et. al., the small numbers of holes which related to debonding or fiber pull out from the matrix indicated weak adhesion between WF and PP matrix[15]. It can be seen that, Fig 3(E)-(G) EVA/CWF-$FeCl_3$ conductive polymer indicated that tear line, good interfacial interaction and less pull out of CWF-$FeCl_3$ in the EVA phases compared to Fig 3(B)-(D) EVA/CWF conductive polymer. This was due to the presence of $FeCl_3$, which acted as surface modifier that enhanced the interfacial interaction between the fillers and the matrix. A similar observation was reported by Supri et. al., the addition of benzyl urea in recycled high density polyethylene/natural rubber blend composites showed rough surface and good interfacial adhesion of chicken feather fiber compared to composites without coupling agent[16].

3.3 Electrical conductivity

Fig 4 shows the electrical conductivity of EVA/CWF and EVA/CWF-$FeCl_3$ conductive polymer with different filler loadings. It can be observed that the electrical conductivity of both conductive polymers increased when the addition of filler loading increased. This was due to the gap between the conducting particles become closer hence increased the conductivity of the conductive polymer. When filler loading increased, the electrons are able to flow throughout filler and matrix because of CWF which having role as networks for the electron to flow. Das et al., reported that the addition of more polyaniline/dodecylbenzene sulfonic acid (PAni/DBSA) in EVA matrix contributed to the increasing of conductivity[17]. However, at lower content of conductive fillers, the distance of conductive fillers particles is very broad, and the conducting pathways are discontinuous within EVA matrix which may bring to low conductivity. Low amount of conductive filler dispersed to homogeneous matrix would not enhance the conductivity until the conductive films was highly loaded with CWF so that the aggregates would be interacted with each other. For EVA/CWF-$FeCl_3$ conductive polymer, the principle of electric conduction occurred when the incorporation of $FeCl_3$ as oxidation agent filler that will be coated onto CWF. This was due to the metal iron particles allowed the ease of electron transfer via conduction. The $FeCl_3$ acts as a medium to transfer the electron in conductive polymer composite due to their formation. Supported research by Suherman et al. stated that the effect of graphite loading on the conductivity of graphite/epoxy nanocomposites will increase the electrical conductivity[18].

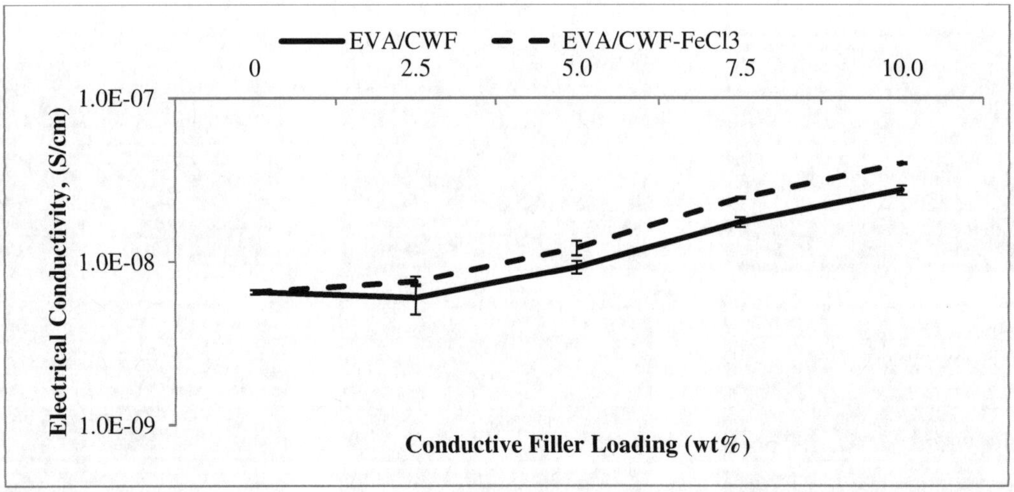

Figure 4. Electrical conductivity vs. conductive filler loading of EVA/CWF-$FeCl_3$ conductive polymer.

4. Conclusion

The results show an increment in tensile strength but decrement in ductility of EVA/CWF-$FeCl_3$ conductive polymer in comparison with EVA/CWF conductive polymer. The tensile fracture studies proved better filler distribution on the surface of EVA/CWF-$FeCl_3$ conductive polymer when compared to EVA/CWF conductive polymer. However, the electrical conductivity increased progressively with the addition of $FeCl_3$ in EVA/CWF-$FeCl_3$ conductive polymer than the electrical conductivity in EVA/CWF conductive polymer.

5. Acknowledgements

The authors gratefully acknowledge the financial support for this research from Skim Latihan Tenaga Pengajar Akademik (SLTPA) under Universiti Malaysia Perlis (UniMAP) and the Ministry of Education Malaysia.

6. References

[1] M. Krzesinska, J. Zachariasz, and A. I. Lachowski, "Eco-composite developed using biomorphous stiff skeleton of carbonised yucca and furfuryl alcohol as a filler," vol. 43, ed: Springer, 2008, pp. 5763-5771.

[2] M. S. Dandekar, G. Arabale, and K. Vijayamohanan, "Preparation and characterization of composite electrodes of coconut-shell-based activated carbon and hydrous ruthenium oxide for supercapacitors," vol. 141, ed: Elsevier, 2005, pp. 198-203.

[3] Y.-J. Kim, B.-J. Lee, H. Suezaki, T. Chino, Y. Abe, T. Yanagiura, *et al.*, "Preparation and characterization of bamboo-based activated carbons as electrode materials for electric double layer capacitors," vol. 44, ed: Pergamon, 2006, pp. 1592-1595.

[4] F. M. Gray, *Solid polymer electrolytes*: VCH New Tork etc., 1991.

[5] W. Zhang, X. Zhang, M. Liang, and C. Lu, "Mechanochemical preparation of surface-acetylated cellulose powder to enhance mechanical properties of cellulose-filler-reinforced NR vulcanizates," vol. 68, ed: Elsevier, 2008, pp. 2479-2484.

[6] P. Mitchell, J. Kiel, B. Livingston, and G. Dupont-Roc, "Torrefied Biomass-A Foresighting Study into the Business Case for Pellets from Torrefied Biomass as a New Solid Fuel," *All Energy,* vol. 24, p. 2007, 2007.

[7] J. Lehmann, C. Czimczik, D. Laird, and S. Sohi, "Stability of biochar in soil," *Biochar for environmental management: science and technology,* pp. 183-206, 2009.

[8] O. Varela Milla, E. B. Rivera, W. J. Huang, C. Chien, and Y. M. Wang, "Agronomic properties and characterization of rice husk and wood biochars and their effect on the growth of water spinach in a field test," vol. 13, ed: SciELO Chile, 2013, pp. 251-266.

[9] R. Malkapuram, V. Kumar, and Y. S. Negi, "Recent development in natural fiber reinforced polypropylene composites," ed: SAGE Publications, 2008.

[10] I. Ali, R. Elleithy, S. M. Al-Zahrani, and M. E. A. Mohsin, "Viscoelastic, thermal, and morphological analysis of HDPE/EVA/CaCO3 ternary blends," vol. 67, ed: Springer, 2011, pp. 1961-1978.

[11] R. Tlili, V. Cecen, I. Krupa, A. Boudenne, L. Ibos, Y. Candau, *et al.*, "Mechanical and thermophysical properties of EVA copolymer filled with nickel particles," vol. 32, ed: Wiley Online Library, 2011, pp. 727-736.

[12] S. R. Majid and A. K. Arof, "Proton-conducting polymer electrolyte films based on chitosan acetate complexed with NH4NO3 salt," vol. 355, ed: Elsevier, 2005, pp. 78-82.

[13] H. Hou, A. K. Schaper, Z. Jun, F. Weller, and A. Greiner, "Large-scale synthesis of aligned carbon nanotubes using FeCl3 as floating catalyst precursor," vol. 15, ed: ACS Publications, 2003, pp. 580-585.

[14] K. Ahmed, N. Z. Raza, F. Habib, M. Aijaz, and M. H. Afridi, "An investigation on the influence of filler loading and compatibilizer on the properties of polypropylene/marble sludge composites," vol. 19, ed: Elsevier, 2013, pp. 1805-1810.

[15] L. Danyadi, J. Moczo, and B. Pukanszky, "Effect of various surface modifications of wood flour on the properties of PP/wood composites," vol. 41, ed: Elsevier, 2010, pp. 199-206.

[16] A. G. Supri and H. Ismail, "Recycled high density polyethylene/natural rubber/chicken feather fibers (RHDPE/NR/CFF) composites: The effect of fiber loading and benzyl urea," vol. 52, ed: Taylor & Francis, 2013, pp. 1316-1322.
[17] N. C. Das, S. Yamazaki, M. Hikosaka, T. K. Chaki, D. Khastgir, and A. Chakraborty, "Electrical conductivity and electromagnetic interference shielding effectiveness of polyaniline-ethylene vinyl acetate composites," vol. 54, ed: Wiley Online Library, 2005, pp. 256-259.
[18] H. Suherman, A. B. Sulong, and J. Sahari, "Effect of filler loading concentration, curing temperature and molding pressure on the electrical conductivity of CNTS/graphite/epoxy nanocomposites at high loading of conductive fillers," vol. 5, ed: University of Malaya, 2010, pp. 74-79.

Evaluation on the feasibility of using bamboo fillers in plastic gear manufacturing via the Taguchi optimization method

N M Mehat[1] and S Kamaruddin[2]

[1]Department of Mechanical Engineering Technology, Faculty of Engineering Technology, Universiti Malaysia Perlis (UniMAP), Uniciti Alam Campus, Sungai Chuchuh, Padang Besar, Perlis, Malaysia.
[2]Mechanical Engineering Department, Universiti Teknologi PETRONAS, 32610 Seri Iskandar, Perak Darul Ridzuan, Malaysia

nikmizamzul@unimap.edu.my

Abstract. An increase in demand for industrial gears has instigated the escalating uses of plastic matrix composites, particularly carbon or glass fibre reinforced plastics as gear material to enhance the properties and limitation in plastic gears. However, the production of large quantity of these synthetic fibres reinforced composites has posed serious threat to ecosystem. Therefore, this work is conducted to study the applicability and practical ability of using bamboo fillers particularly in plastic gear manufacturing as opposed to synthetic fibres via the Taguchi optimization method. The results showed that no failure mechanism such as gear tooth root cracking and severe tooth wear were observed in gear tested made of 5-30 wt% of bamboo fillers in comparing with the unfilled PP gear. These results indicated that bamboo can be practically and economically used as an alternative filler in plastic material reinforcement as well as in minimizing the cost of raw material in general.

1. Introduction

Gears have been in used for more than three thousand years and are commonly utilized in power and motion transmission under different loads and speeds. A variety of cast irons, powder metallurgy materials, and nonferrous alloys are used in gears. To date, with advance in the development of plastic materials, and with a wide spectrum of fiscal and practical properties available, plastics have become one of the most sought-after materials in the world today. Plastics have quickly become common throughout the modern world due to their intrinsic properties such as lightweight, versatility, and durability [1-2]. By possessing these tremendous range properties, plastics have become a good candidate for replacing the primary materials used for gear manufacturing and the demand for plastic gears has significantly increased and indubitably continues in the future [3].

An increase in demand for industrial plastic gears has resulted in increased use of plastic-matrix composites, particularly carbon or glass fibre-reinforced plastics as gear materials, to enhance the mechanical strength and the thermal resistance of plastic gears. Nevertheless, the production of large quantities of these synthetics fibre-reinforced composites poses a serious threat to the ecosystem. These fibres or fillers, especially glass, have become a waste management issue due to their non-biodegrability. Glass fibre filled plastic waste can take decades, and probably centuries, to decompose and constitute an environmental hazard. The disposition or the recycling of glass fibre reinforced composites is not easy and safe for environment, since even after recycling these thermoplastic composites, very small amounts of them can be incinerated.

As a consequence, researcher have been looking into the possibility of using natural fillers as reinforcement in plastic gear manufacturing. The use of natural fibres, as opposed to synthetic fibres such as carbon and glass fibres, has been widely recognized due to the positive effect on the environment with respect to ultimate disposability, not to mention the rising prices of raw materials used for manufacturing standard plastics. In addition, natural fibres have properties including low cost, light weight, easy avaibility, high strength-to-weight-ratio and also, less wear and tear in processing machinery [4-6].

Therefore, in this study, an attempt is given on the possibility of using natural filler especially the bamboo filler of the Bambusa Bluemeana species, which grows abundantly in Malaysia for reinforcement in plastic gear manufacturing. The Taguchi method is adopted in designing the experiments for gear performance testing. The wear behaviour of the gear produced by various loading rate of bamboo fillers are investigated in this study.

2. Gear performance testing procedures

The flowchart of the gear performance testing procedures is illustrated in Figure 1 and each stage of the experimental procedures is elaborated in the following section.

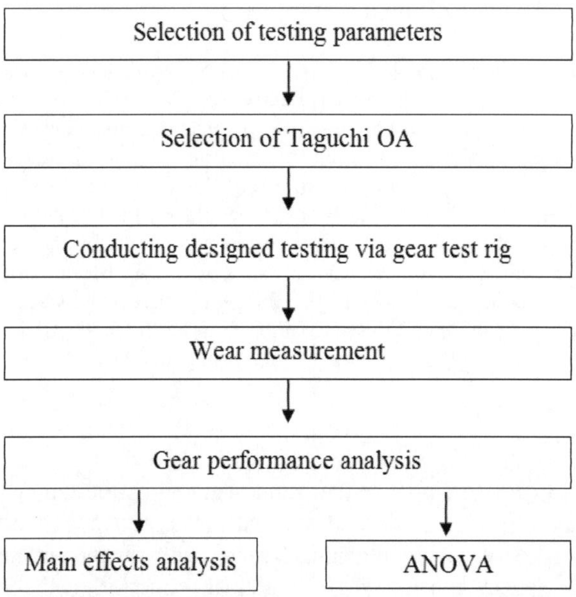

Figure 1. Gear performance testing procedures.

2.1. Selection of testing parameters

Five parameters control including percentage of bamboo loading, gear meshing gap, speed, load and running time were selected for conducting the gear performance test. Each parameter of gear meshing gap, speed, load and running time was set at four levels. As in this research, the percentage of bamboo filler incorporated to the PP gears varied from 0 wt%, 5 wt%, 10 wt%, 15 wt%, 20 wt%, 25 wt%, 30 wt% (refer Table 3.3), hence seven levels were set for bamboo loading (parameter A). Table 1 shows the parameters control and their levels.

Table 1. Parameters control and their levels in the gear performance test.

Column	Factors	Level						
		1	2	3	4	5	6	7
A	Bamboo loading (%)	0	5	10	15	20	25	30
B	Gear meshing gap (mm)	3.0	3.1	3.2	3.3			
C	Speed (Hz)	12	13	14	15			
D	Load (%)	0	5	10	15			
E	Running time (min)	30	40	50	60			

2.2. Selection of Taguchi orthogonal arrays

After determining the number of control parameters and their levels, an appropriate OA was established for laying out the design of experiment in testing the performances of unfilled PP gears and bamboo-based PP composite gears at various bamboos loading percentage related to wear failure. The selection of an appropriate OA depends on the total DOF of the control parameters. In this study, there were four control parameters (meshing gap, speed, load and running time) with four levels of each and one control parameter (bamboo loading) with seven levels. As DOF was calculated by subtracting the number of levels with one (DOF = number of levels-1), therefore for four control parameters with four levels each, the total DOF is equal to twelve and for control parameters with seven levels, the DOF is six. Hence, the total DOF required for all control parameters is 18 DOF. According to Roy (1990) [7], the total DOF of selected OA must be greater than or equal to the total DOF required for the experiment. There are some standard OAs which treats four levels factors such as L16, L32 and L64. However, in selecting an OA for mixed levels, an L32 OA was chosen to lay out the design of experiment in conducting the gear performance test in this study. Referring to Table 2, percentage of bamboo loading (A) was assigned to column 2 varied from 0 to 30 wt %, and all the four testing parameters with four levels were assigned to columns 3, 4, 5, and 6. The remaining five columns (Column 1, 7, 8, 9, and 10) were not used when running the experiments. The details of assigning the columns for the parameters studied can be found in Roy (1990) [7] and Roy (2001) [8]. There were total 32 trials need to be conducted under different combination of test parameters as shown in Table 3.9, however in this study, gear tests were conducted only up to 28 trials to be adjusted with percentage of bamboo loading varied from 0 to 30 wt %. All the gear tests were conducted under unlubricated dry conditions via gear test rig.

2.3. Test rig details

Figure 2 shows the gear test rig used for conducting gear performance test. In this test rig, the test gear was driven using a DC motor and was run at selected speed up to 1500 rpm. Test gear mates with an identical gear, was connected to the DC generator. The required test torque was introduced by loading the rheostat connected to the generator. Speed and digital counter were suitably placed to monitor the speed and number of cycles run, respectively. A pair of similar gears was allocated and continuously run following the designed L32 OA (refer to Table 3.9) for each trial. All the gear tests were conducted at the laboratory conditions (room temperature: 25 °C and humidity: 64 %). The wear performance of gear made from unfilled PP and bamboo-based PP composite gears were then systematically studied.

Table 2. The L_{32} OA used in the gear performance test.

Trial No.	1	2 A	3 B	4 C	5 D	6 E	7	8	9	10
1	1	1	1	1	1	1	1	1	1	1
2	1	1	2	2	2	2	2	2	2	2
3	1	1	3	3	3	3	3	3	3	3
4	1	1	4	4	4	4	4	4	4	4
5	1	2	1	1	2	2	3	3	4	4
6	1	2	2	2	1	1	4	4	3	3
7	1	2	3	3	4	4	1	1	2	2
8	1	2	4	4	3	3	2	2	1	1
9	1	3	1	2	3	4	1	2	3	4
10	1	3	2	1	4	3	2	1	4	3
11	1	3	3	4	1	2	3	4	1	2
12	1	3	4	3	2	1	4	3	2	1
13	1	4	1	2	4	3	3	4	2	1
14	1	4	2	1	3	4	4	3	1	2
15	1	4	3	4	2	1	1	2	4	3
16	1	4	4	3	1	2	2	1	3	4
17	2	5	1	4	1	4	2	3	2	3
18	2	5	2	3	2	3	1	4	1	4
19	2	5	3	2	3	2	4	1	4	1
20	2	5	4	1	4	1	3	2	3	2
21	2	6	1	4	2	3	4	1	3	2
22	2	6	2	3	1	4	3	2	4	1
23	2	6	3	2	4	1	2	3	1	4
24	2	6	4	1	3	2	1	4	2	3
25	2	7	1	3	3	1	2	4	3	2
26	2	7	2	4	4	2	1	3	4	1
27	2	7	3	1	1	3	4	2	2	4
28	2	7	4	2	2	4	3	1	1	3
29	2	8	1	3	4	2	4	2	1	3
30	2	8	2	4	3	1	3	1	2	4
31	2	8	3	1	2	4	2	4	3	1
32	2	8	4	2	1	3	1	3	4	2

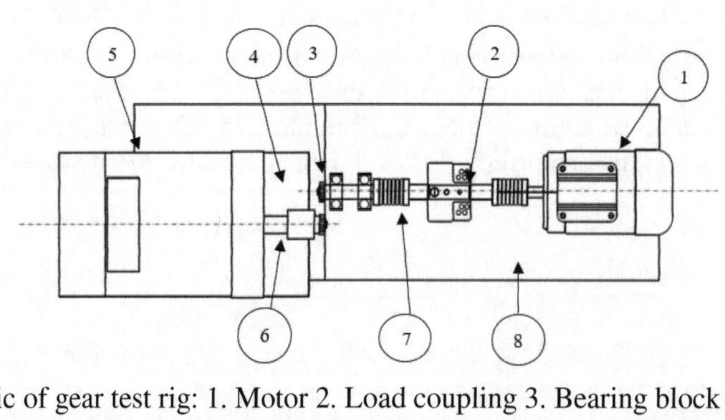

Figure 2. Schematic of gear test rig: 1. Motor 2. Load coupling 3. Bearing block 4. Test spur gear 5. Generator 6. Standard spur gear 7. Torque sensor 8. Base plate

2.4. Wear measurement

Severe wear of the gear tooth is one of the gear failure modes, which leads to a gradual increase in sound and vibration of the assemble unit. For extensive investigation of wear behaviour in plastic gear, a wear measurement was carried out to determine the amount of materials removed or worn away after the gears have been tested for a period of running time as specified in Table 3.8. In this research, the material worn away is expressed as weight (mass) loss. Weight measurement was taken from both driving and driven gear by using Shimadzu precision balance (model ATY224) just before and after each test to evaluate the weight loss. The difference in weight before and after gear performance test represents the weight loss in gram (g) caused by wear. For further understanding the wear failure mechanism, the optical photograph of the tested gears was also observed.

2.5. Gear performance analysis

All the experimental results obtained from the wear measurement of each gear made from unfilled PP and bamboo-based PP composites were then systematically and statistically analyzed by main effects analysis and ANOVA.

3. Results and discussion

3.1 Main effects analysis

By using the results in Table 4, the main effects analysis of testing parameters including the bamboo loading percentage (A), the gear meshing gap (B), the speed (C), the load (D), as well as the running time (E), are presented in the form of a graphical display to demonstrate the impact of variation in testing parameters on wear behaviour of the gear produced (see Figure 3).

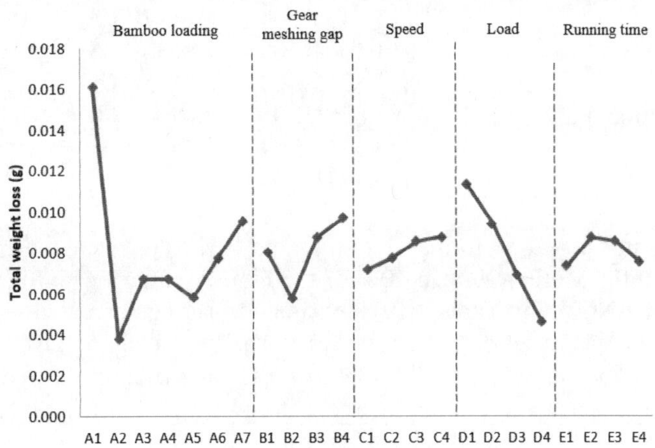

Figure 3. Main effects plot.

Figure 3 clearly shows that the total weight loss of the meshing gear studied is significantly affected by the adjustments of the testing parameters. In this case, the variations in the gear meshing gap, the speed, the load, and the running time, as well as the percentage of bamboo filler loading in the gears, have different effects on the wear behaviour of the meshing gears. Any changes in parameters can led to an improvement or a deterioration in gear quality. Plastic gears are known to fail in a number of ways, including wear, which needs to be minimized during the gear running. It is important to understand the wear characteristics, particularly the way in which such gear wear is related to the contact conditions during gear running [9-10]. From Figure 4.22, it can be concluded that the percentage of bamboo loading in the gear is significant parameter, because the slope gradient is very big. Compared to that in an unfilled PP gear, the addition of 5 wt% of bamboo filler has a great impact in minimizing the wear during the gear running.

IOP Conf. Series: Journal of Physics: Conf. Series **908** (2017) 012009 doi:10.1088/1742-6596/908/1/012009

Figure 3 also shows that lower speeds minimize the wear behaviour in the gear meshing. With loading, the opposite trend occurs. The wear of the meshing gear decreases with higher loads. For the meshing gap and the running time, the total weight loss for both running gears at 3.1 mm and 30 minutes of running time, respectively, showed the least amount of wear. Repeated gear tooth loading contributes to subsequent deformation of gear teeth, due to hysteresis heating during the gear running. Friction between the mating gear teeth resulting from the gear meshing gap selected during the gear running generates the heat and propagates hysteresis effects. This heat, generated due to friction as well as hysteresis loss, also increases with increasing gear rotational speed, which then results in wear failure gear teeth and on the surface.

3.2 Analysis of variance (ANOVA)

For further investigation into the effect of different testing parameters on the wear behaviour of the optimized unfilled PP and bamboo-based PP gears produced, the ANOVA is also performed. The relative contributions of the gear meshing gap, the speed, the load, the running time, as well as the bamboo loading percentage, to the wear behaviour of the gears, are measured. By adopting similar procedures to those followed in the previous experiments, the quantities such as DOF, sums of squares, variances, F-Ratios, and percentage contributions, are computed and presented in Table 3.

Table 3. ANOVA table for the grey relational grade of nine comparability sequences.

Column	Parameters	DOF	Sum of squares	Variances	F-Ratio	Percent
A	Bamboo loading (%)	6	3.7650E-04	6.2750E-05	1.5800	37.85
B	Gear meshing gap (mm)	3	5.8430E-05	1.9477E-05	0.4904	5.87
C	Speed (Hz)	3	1.1604E-05	3.8680E-06	0.0974	1.17
D	Load (%)	3	1.8046E-04	6.0155E-05	1.5146	18.14
E	Running time (min)	3	1.0261E-05	3.4204E-06	0.0861	1.03
All others/error		9	3.5745E-04	3.9717E-05		35.94
Total		27	9.9471E-04			100.00

In analysing the results in the gear performance testing, the 10% rule, where a parameter is considered insignificant when its influence is less than 10% of the highest parameter influence, is implemented [8]. From the results of ANOVA in Table 3.19, the bamboo loading percentage appears to be the key parameter affecting the wear in an engaged gear pair. The percentage contribution of the bamboo loading percentage is 37.85%, the highest of all the process parameters, followed by the load and the gear meshing gap. While, speed and running time are insignificant, as their percentages are less than 10% of the highest parameter influence, which is 3.785%.

Table 4. Wear measurement results.

Test	Testing parameters					Weight loss (g)		
	Bamboo loading (%)	Gear meshing gap (mm)	Speed (Hz)	Load (%)	Running time (min)	Driver Gear	Driven Gear	Total
1	0	3.0	12	0	30	0.0116	0.0096	0.0212
2	0	3.1	13	5	40	0.0115	0.0018	0.0133
3	0	3.2	14	10	50	0.0159	0.0067	0.0226
4	0	3.3	15	15	60	0.0022	0.0051	0.0073
5	5	3.0	12	5	40	0.0014	0.0006	0.0020
6	5	3.1	13	0	30	0.0033	0.0018	0.0051
7	5	3.2	14	15	60	0.0017	0.0010	0.0027
8	5	3.3	15	10	50	0.0027	0.0025	0.0052
9	10	3.0	13	10	60	0.0013	0.0005	0.0018
10	10	3.1	12	15	50	0.0019	0.0016	0.0035
11	10	3.2	15	0	40	0.0081	0.0065	0.0146
12	10	3.3	14	5	30	0.0035	0.0035	0.0070
13	15	3.0	13	15	50	0.0021	0.0015	0.0036
14	15	3.1	12	10	60	0.0021	0.0016	0.0037
15	15	3.2	15	5	30	0.0025	0.0024	0.0049
16	15	3.3	14	0	40	0.0059	0.0087	0.0146
17	20	3.0	15	0	60	0.0052	0.0051	0.0103
18	20	3.1	14	5	50	0.0022	0.0014	0.0036
19	20	3.2	13	10	40	0.0022	0.002	0.0042
20	20	3.3	12	15	30	0.0025	0.0026	0.0051
21	25	3.0	15	5	50	0.0071	0.0066	0.0137
22	25	3.1	14	0	60	0.0033	0.0024	0.0057
23	25	3.2	13	15	30	0.0026	0.0019	0.0045
24	25	3.3	12	10	40	0.0042	0.0029	0.0071
25	30	3.0	14	10	30	0.0024	0.0012	0.0036
26	30	3.1	15	15	40	0.0027	0.0027	0.0054
27	30	3.2	12	0	50	0.0041	0.0035	0.0076
28	30	3.3	13	5	60	0.0109	0.0105	0.0214

4. Conclusion

The Taguchi technique's OA and main effects analysis was recommended as a way of efficiently optimizing the cutters geometry in studying the feasibility of using bamboo fillers in plastic gear manufacturing related to wear behaviour. The main effects plot showed that A^1, B^4, C^4, D^1 and E^2 was identified as the optimal parameters setting that statistically results in minimum wear behaviour in plastic gear testing performance. The bamboo loading percentage appears to be the key parameter affecting the wear in an engaged gear pair with the percentage contribution of 37.85%. The results indicated that bamboo can be potentially used as a practical and economical filler in plastic gear material reinforcement as well as in minimizing the cost of raw material in general.

5. References

[1] Fortelny I, Michalkova D, and Krulis Z 2004 *Polymer Degradation and Stability* **85(3)** 975.
[2] Gu F, Hall P and Miles N J 2014 *Material & Design* **62** 189.
[3] Hasl C, Liu H, Oster P, Tobie T, Stahl K and Forschungsstelle fuer Zahnraeder und Getriebebau (Gear Research Centre) 2017 *Mechanism and Machine Theory* **111** 152.
[4] Ren D, Yu Z, Li W, Wang H and Yu Y 2014 *Industrial Crops and Products* **62** 94.
[5] Tausif M, Ahmad F, Hussain U, BAsit A and Hussain T 2015 J. of *Cleaner Production* **89** 110.
[6] Nirmal U, Hashim J and Megat Ahmad M H H 2015 *Tribology International* **83** 77.
[7] Roy R K 1990 *A Primer on The Taguchi Method* (New York: Hanser)
[8] Roy R K 2001 *Design of Experiment Using the Taguchi Approach: 16 Steps to Product and Process Improvement* (New York: John Wiley and Sons).
[9] Mao K, Li W, Hooke C J and Walton D (2010) *Tribology International* **43** 433.
[10] Liu X, Yang Y and Zhang J 2016 *Tribology International* **101** 383.

Effect of nano-clay fillers on mechanical and morphological properties of Napier/epoxy Composites

K H Lim, M S A Majid, M J M Ridzuan, K S Basaruddin, M Afendi

School of Mechatronic Engineering, Universiti Malaysia Perlis, Pauh Putra Campus, 02600 Arau, Perlis, Malaysia

shukry@unimap.edu.my

Abstract. The effect of nano clay filler on the mechanical and morphological properties of Napier/epoxy composites was investigated. Neat, 2 wt%, 3 wt%, 4 wt% and 5 wt% of Montmorillonite (MMT) nano clay filled Napier/epoxy composites were fabricated by vacuum infusion technique. These specimens were tested in the three points bending according to the ASTM D790. The flexural stress-strain curve, flexural strength, flexural modulus and strain to failure were then discovered based on the flexural test results. The results revealed that flexural strength and flexural modulus increased when a particular amount of nano clay was added to the epoxy matrix. 3 wt% of nano clay filler yielded the highest flexural strength with an improvement of 163% when compared to the neat Napier/epoxy composites. Moreover, a maximum of 180% increases in flexural modulus was registered at 5 wt% of nano clay filler. The enhanced properties of nano clay filled composites were highly achieved due to better dispersion and distribution of nano clay in the epoxy resin as well as an increase on the interfacial bonding. Using Scanning Electron Microscopy (SEM), morphological analysis was conducted to observe the fracture surfaces of the specimens after the flexural test. Overall, the presence of nano clay filler loading with a range of 3 wt% to 5 wt% in the Napier/epoxy composites shows the significant improvement in mechanical and morphological properties.

1. Introduction

Over the past decade, the adoption of natural fibres is gradually increasing as reinforcing resources in particular applications for composites structural industries all over the world. A lot of industrial and structural applications such as aerospace, automotive, marine, construction and packaging are raising utilisation of natural fibre reinforced composites due to their cost-effective, particularly high specific strength, toxic free, renewability, low density and biodegradable behaviour [1]. Currently, some research study on extracting fibres from Napier grass which also known as 'Pennisetum purpureum' or elephant grass. It is looking like bamboo, yellowish colour stem and highly sustainable in Malaysia. Napier grass fibres have been successfully proved that it has potential to use as alternative resources in reinforcement composite materials [2-6]. Epoxy resins are commonly selected in fabrication natural fibres reinforced composites as matrix materials. They are one of the outstanding resins amongst the class of thermosetting resins. Their primary function is to enhance the structural adhesion when reinforcing together in natural fibres and composite materials [7]. In recent year, a new interest is generated to improve the performance of the composite by adding nano clay filler into the matrix system. Montmorillonite (MMT) clay contains 25-30 wt. % trimethyl stearyl ammonium that able to improve flexural strength and modulus, reduce crack propagation and achieve better surface modification [8-10]. Consequently, this study aims to determine the effect of nano clay filler on

IOP Conf. Series: Journal of Physics: Conf. Series **908** (2017) 012010 doi:10.1088/1742-6596/908/1/012010

mechanical and morphological properties of Napier fibres reinforced epoxy composites. The various amounts of nano clay filler are added to an epoxy resin to improve the mechanical behaviour and morphological performance of the Napier epoxy composites. Moreover, Napier epoxy composites are going to fabricate using vacuum infusion technique to generate superior performance of composites. Overall, the flexural test is going to carry out in this paper to investigate the effect of nano clay filler on mechanical performance of Napier reinforced epoxy composites. Also, the fracture behaviour obtained on the surface of specimens is achieved through FESEM observations.

2. Materials and Methods

2.1 Extraction of fibres

The Napier grass stems were obtained from Kodiang, Perlis, in northern peninsular Malaysia. The Napier grass fibres were being extracted by using a simple manual after few weeks of water retting process. In the beginning, the Napier stems were cleaned to remove the impurities on its surface and then the leaves, thorn node and root portion were cut off since only the stem part was required. Next, the stems were crushed with a hammer between the stem internodes to crack the thick and harder stem wall and enable water to diffuse into the inner stem. The stems then were chopped into the small portions with 2 or 3 segments between stems. The cracked-stems were submerged in water tanks filled with water for four weeks and monitored daily. After that, the soaked Napier fibres were extracted manually and made sure all the impurities such as hemicelluloses, lignin or waxy component removing from the surface of the fibres. Lastly, the extracted raw fibres were going to put under the sun for drying process for few days to eliminate the excess moisture inside.

2.2 Preparation of nano clay-epoxy mixture

The various amounts of nano clay filler such as 2 wt%, 3 wt%, 4 wt% and 5 wt% were prepared for fabricating Napier/epoxy composites. Firstly, the nano-clay filler was directly added into epoxy resin. Manual hand mixing method was used for half an hour to reduce the nano-clay filler agglomeration and improvise their dispersion. After that, the epoxy hardener at a ratio of 3:1 by weight (epoxy: hardener) was then slowly added to the nano-clay filled epoxy resin and stirred until it becomes a bit gelling.

2.3 Composite fabrication

Neat, 2 wt%, 3 wt%, 4 wt% and 5 wt% nano clay filled Napier/epoxy composites were fabricated through vacuum infusion technique. First and foremost, the extracted raw Napier fibres were oven dried at 55 °c for 30 min to remove excess moisture. The composite samples were fabricated with 75% of nano clay-epoxy resin and 25% of Napier fibre weight content (with ratio 3: l). Next, the glass surface was being polished by wax. Few-layer of Napier fibres were prepared and then positioned on the grass surface. The black enka channel was then put and taped at the two end sides of the Napier fibres. The Napier fibres were then surrounded by adhesive clay as the area for infusion purpose. After that, the resin feed spiral and vacuum hose were sealed and positioned at the two centre sides by using a sealant tape. The transparent Perspex was covered and taped on the Napier fibres which acted as a mould for fabricating composite. The vacuum bag was being taped and sealed on the area surrounding by a sealant tape. The vacuum pump was then connected and switched on to withdraw the air inside for around half an hour. Then the epoxy resin and hardener was mixed and prepared. Furthermore, the prepared resin was then infused and flow slowly onto the surface of Napier fibres. While the resin was distributed until the end of the edge, the excess resin flowed into an installed resin trap. The process was continued run at room temperature for around 6 hours to ensure the epoxy resin with nano clay filler full covered onto the Napier fibres. Lastly, the Napier fibre epoxy composite was done fabricating and stored for fully dried and hardened.

2.4 Flexural testing

Three-point bending test was carried out by using INSTRON micro-tester according to ASTM D-790 standards at a constant crosshead speed of 1 mm/min. The specimens were cut into dimensions of rectangular shape (12.7cm x 1.27cm x 0.3cm) by using hand Dremel. Eight specimens were tested for each type of composites of different filler contents. The flexural stress-strain curve, flexural strength and modulus, as well as the strain to failure were determined. In a three-point bending test, the flexural stress was given by $\sigma_f = (3PL)/(2bd^2)$ where P is maximum load (N), L is a support span length (mm), b is the width of the specimen (mm), and d is the thickness of specimen (mm), respectively. While flexural strain was calculated by $\varepsilon_f = (6Dd)/(L^2)$ where D is a maximum deflection of the centre of the beam (mm). Besides, the modulus was given by $E = (mL^3)/(4bd^3)$ where m is the initial slope of the load-deflection (N/mm) and could also be determined using the stress-strain curve.

2.5 Morphological testing

The morphology properties of the Napier fibre epoxy composited is observed by using scanning electron microscope (SEM). It was used to investigate the fracture surface of the specimens, the distribution of nano clay filler within the epoxy matrix, agglomeration of nano clay filler and fibre pull-out failure. The microscope was acquired with 5-15 kV accelerating voltages, and the magnification was focused in a range of 50 to 200 times. Thus, the specimens must cut into a small size with diameter 10mm x 10mm. Before scanning, the specimens have to be consistently coated with thin layer of platinum to prevent charging and caused a white spot on the scanned pictures.

3. Results and Discussion

3.1 Flexural Test properties

Table 3.1 shows the mean value of the *ultimate* flexural load and maximum extension of different loading of 2%, 3%, 4% and 5% of nano clay filled Napier/epoxy composites. Throughout, the results for neat Napier/epoxy composite are included for comparison. It can be seen that 3% of nano clay-filled Napier/epoxy composite showed most improved ultimate load capacity than other nano-clay filled composite. There is an increase of over 125.33%. The 2 wt%, 4 wt% and 5 wt% of nano clay filler displayed the ultimate load of 46.30N, 66.34N and 78.83N respectively. This indicates that nano clay filler further acts as a load bearing reinforcement in the Napier/ epoxy composites. Besides, 5% of nano clay-filled Napier/epoxy composite has the smallest deflection with the value of 3.78mm only which is a reduction of 64% compared to pure Napier/epoxy composite. However, 2 wt%, 3 wt% and 4 wt% of nano clay filler exhibited maximum extension of 8.85mm, 6.07 and 4.43mm respectively. This suggested that the addition of nano clay filler might reduce the elasticity and flexibility of the composites due to the agglomeration of nano clay and thus attributed to the weak intercalated structure in the Napier/epoxy composites.

Table 3.1. Ultimate Load and Maximum Extension of Composites.

wt% of Nanoclay in Epoxy	Ultimate Load (N)	Max. Extension(mm)
Neat	38.13	10.56
2%	46.30	8.85
3%	85.92	6.07
4%	66.34	4.43
5%	78.83	3.78

The flexural stress-strain trend of the neat, 2 wt%, 3wt%, 4 wt% and 5 wt% Nanoclay filled Napier/epoxy composite is illustrated in Figure 3.1(a). It clearly showed that the composites undergo elastic region and plastic deformation before fracture. After the yield point, the composite did experience slight plastic strain, before it decreased sharply and fractures. Neat Napier/epoxy composites, shows distinct yielding before fractures at 3.5% strain while the 2 wt%, 3wt%, 4 wt% and 5 wt% nano clay filled Napier/epoxy composite indicate lower strain to failure at ranges 2% to 3%.

The trend of this stress-strain curve shows similar characteristics to those of brittle materials. Therefore, the results suggest that the composites become stiffer and more brittleness like as the nano-clay filler contents increases.

The flexural strength of nano clay filled Napier/epoxy composites with 2 wt% to 5 wt% filler contents exhibited significant improvement than that of neat Napier/epoxy composites as shown in Figure 3.1(b). It is suggested that the presence of nano-clay filler is creating strong bonding between Napier fibres and epoxy resin. Thus, it resulted in increases the capability of the epoxy composite to sustain the stress during bending test. The flexural strength of filled epoxy composite increases with a rise in nano clay contents from neat (0 wt%) to 3 wt% before went slightly decreases with further increase in filler loading. The addition of 3 wt% of nano clay filler exhibits the highest flexural strength with the value 57.72 MPa and with an improvement of approximately 163% comparing to the neat Napier/epoxy composites, followed by 5 wt%, 4 wt%, and 2 wt% and without nano clay filler respectively. The improvement was achieved due to the interface interaction and created exfoliated structure in the composite with the addition of nano clay filler. Also, with a further increase in the filler loading of nano clay to 4 wt% and 5 wt%, the flexural strength Napier/epoxy composite dropped to 43.77 MPa and 48.599 MPa respectively. In Solhi et al.'s [11] and Zukas et al.' [12] studies, results showed that by increasing the number of nanoparticles to more than a particular point, the flexural strength faced a significant reduction. Nevertheless, the presence of higher amounts of nano clay up to 5 wt% gradually decreased the flexural strength due to the agglomeration of the nano clay fillers at higher ratios and consequently embrittlement of the matrix [13].

Figure 3.1(c) showed the flexural modulus for various nano clay-filled loading Napier/Epoxy composite. There was an increment in the flexural modulus as the nano-clay filler loading increases. The flexural modulus of 5 wt% nano clay filled Napier/epoxy composites ranked the highest flexural modulus of approximately mean value of 3.2GPa. There was over 180% of improvement compared to neat Napier/epoxy composites with the value of 1.12GPa. It ranked as the highest flexural modulus due to the formation of exfoliated structure where strong interfacial interaction between the epoxy matrix and nano clay filler [14]. Besides, the flexural modulus for 2 wt%, 3 wt% and 4 wt% of nano clay filler also displayed significant enhancement with 42%, 132% and 130% respectively. The presence of nano clay filler in the epoxy matrix contributed in stronger the bonding via rearranging the chain structure of the epoxy, and it enables to tighten the chain for preventing the chain free to move. Thus, it significantly improves the flexural modulus as consequence of the addition of nano clay filler in the epoxy metric. Based on Chisholm et al.'s study, a significant enhancement was investigated in flexural modulus of epoxy metric by adding nanoparticles due to it able provided higher surface energy [15]. Besides, the highly dispersed of nano clay filler could provide well bonding with epoxy matrix and thus the efficiency of stress transferring boosted up subsequently, leading to ameliorate the flexural modulus.

Figure 3.1(d) presented the strain to failure with the different nano clay filler content for Napier/epoxy composites. The figure showed that the strain to failure values decreased with increases the amount of nano clay filler content of the composites. This suggested that addition of nano clay filler seem to change the epoxy matrix to be more brittle behaviour, which was reflected by minimised strain to failure values of the composites. The results show that the maximum strain to failure values occurred at neat Napier/epoxy composite at 3.08% while the minimum strain to failure values happened to 5 wt% of nano clay filled composite at 2.21% strain. Besides, the strain-to-failure for 2 wt%, 3 wt% and 4 wt% of nano clay filler contents yield reduced strain at 3.0%, 2.98% and 2.73% respectively. The drop in strain to failure values of the nano clay filled Napier/epoxy composite could be described as the presence of the nano clay filler in the epoxy matrix would change their properties from elastic behaviour to brittle behaviour.

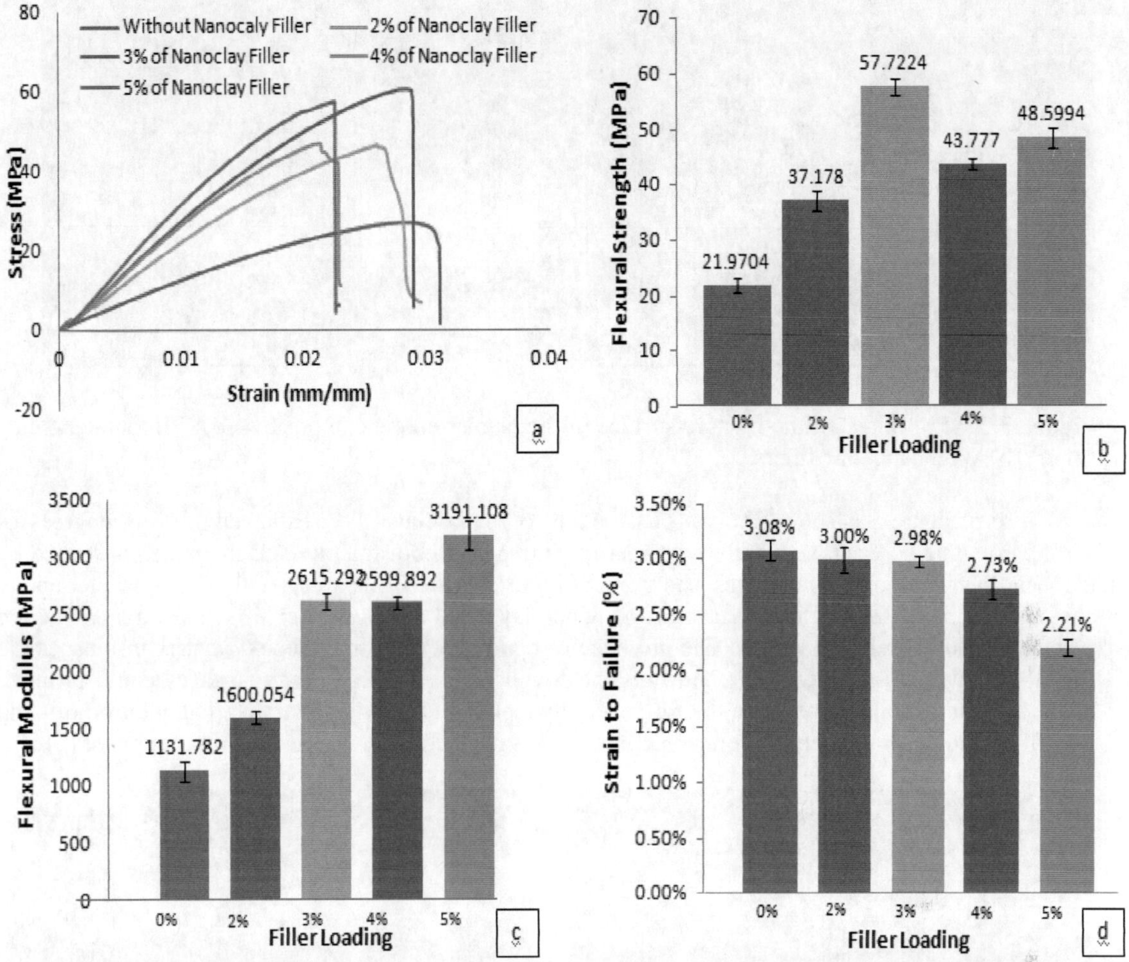

Figure 3.1. (a) Graph of Flexural Stress-Strain, (b) Histogram of Flexural Strength, (c) Histogram of Flexural Modulus, (d) Histogram of Strain to Failure for Various Nanoclay-filled Loading Napier/Epoxy Composite.

3.2 Morphological Properties on Facture Surface

The fracture surface observation of Napier/epoxy composites with and without nano clay filler through SEM are shown in Figure 3.2. It revealed that the fracture surface of composites without nano clay filler is clear and with a smooth surface. It can be observed that there are a large number of tiny white particles inspected on the fracture surface of the composites. The increased nano clay filler contents in the reinforced Napier/epoxy composites can be evident from the increased of the tiny white spot on within the epoxy matrix. In order to achieve the better mechanical performance of the composites, the nano-clay filler should be well distributed in the epoxy matrix. Moreover, for the 3 wt% and 5 wt% of nano clay loading, the SEM image displayed that the nano-clay filler is quite well distributed throughout the Napier fibres and epoxy matrix and also distributed equally without any agglomeration observed. Thus, it brings significant advantages to the flexural strength and modulus properties.

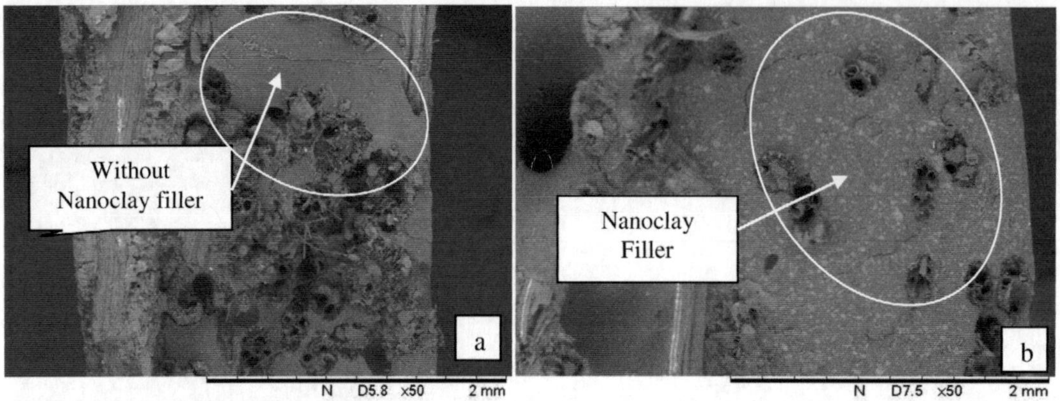

Figure 3.2. SEM image of the fracture surface of the Napier/epoxy composites, (a) Without nano clay fillers, (b) With nano clay fillers.

From Figure 3.3 shows the various loading of nano clay filler displayed varying degrees of surface roughness. It can be clearly seen that a relatively smooth fracture surface is demonstrated by the neat Napier/epoxy composites shown in Figure 3.3(a). Compared to the neat Napier/epoxy composite sample, the fracture surface of the nano clay filled composites exhibited a rougher fracture surface as shown in Figure 3.3(b). The presence of nano clay filler in the epoxy matrix influences the fracture path to be more complex and thus the rough surface. These rough surfaces could promote better interfacial bonding between the fibres and the epoxy resin. It is illustrated that addition of nano clay filler into the Napier/epoxy composite interphases resulted in an improvement in toughness [16].

Figure 3.3. SEM image of the roughness surface of the Napier/epoxy composites, (a) Without nano clay fillers, (b) With nano clay filler.

Besides, Figure 3.4 exhibited that a huge number of agglomerates which was observed in SEM micrograph with 200 x magnifications. This kind of observation can justify the reduction of mechanical properties at nano clay loading of beyond 3 wt%. However, the accumulation of the nano clay filler in the composites causes the stress concentration and consequently crack propagation occurred. Therefore, higher agglomeration clay can be the critical reason for crack propagation, and it affects to the reduction in flexural strength as well as matrix fracture. [17-18].

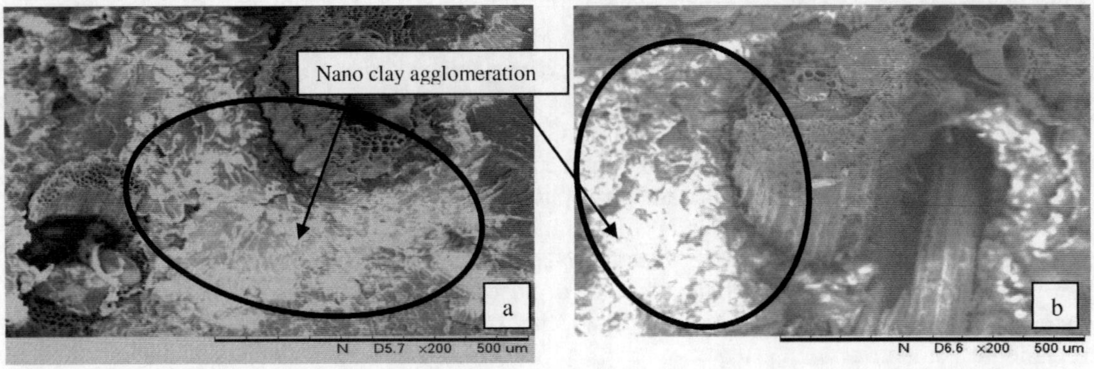

Figure 3.4. SEM image of the agglomeration of nano clay on the fracture surface of the Napier/epoxy composites at a various wt% of nano clay contents.

As shown in the SEM image of Figure 3.5(a), the neat Napier/epoxy composite demonstrates the fracture took place along the bonding interface between the epoxy resin and Napier fibres. The Napier fibres are completely separate from the fracture surface that proves the poor bonding between the Napier fibre and epoxy matrix and consequently, more Napier fibres pull out. In contrast, with the presence of nano clay filler reinforced composites, the Napier fibres pull-out the phenomenon has been significantly reduced, and thus improve the consolidation between the epoxy matrix and the Napier fibres as shown in figure 3.5(b). The 5 wt% nano clay filled composites were observed with the least fibres pull-out under 100 x magnifications. As a result, it suggests an enhancement of adhesion capability between the Napier fibres and epoxy matrix with the presence of the nano clay filler. Some research mentioned that nano clay filler could improvise the cohesion properties between the adjacent and the interlayer of composite layers [13].

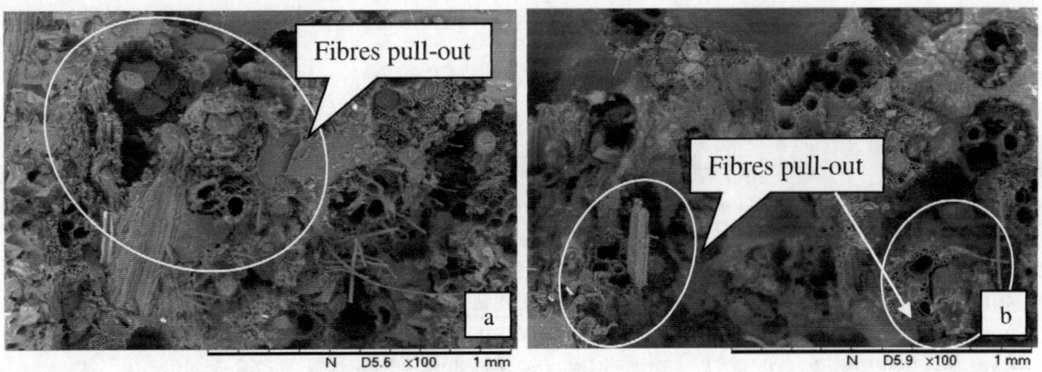

Figure 3.5. SEM image of the fibre pull-out on the fracture surface of the Napier/epoxy composites, (a) Without nano clay fillers, (b) With nano clay fillers.

4. Conclusion

The effect of nano clay filler loading on mechanical and morphological properties of Napier/epoxy composites was investigated. Based on the experimental results, it can be figured out that the reinforcement of nano clay filler in Napier/epoxy composites has a significant impact on the flexural properties compared to neat Napier/epoxy composite. However, for the fracture mode, it is no longer ductile with plastic deformation, but showing becoming brittleness like behaviour. The highest tensile strength was ranked by 3 wt% of nano clay filler with the value 57.72 GPa. However, there is a reduction in flexural strength for further increase of nano clay filler loading. It was attributed to the intercalated structure in the Napier/epoxy composite with above 3 wt% of nano clay filler was loaded. This study demonstrates that the flexural modulus is improved with increasing of the nano clay filler content up to 5 wt%. This is due to the formation of exfoliated structure where strong interfacial interaction between the epoxy matrix and nano clay filler. From the observation under SEM, the better intercalation of nano clay in the epoxy composites was examined in the specimen with 3 wt% and 5 wt% of nano clay filler loadings. By comparison, the fracture surface of 5 wt% of nano clay filler composite is rougher, and it ranked as the highest in flexural modulus. Also, fibre pull-out failure was observed on the flexural fracture surfaces of the neat Napier/epoxy composite. With the addition of nano clay filler, it could help to reduce the fibre pull-out failure problem due to the enhancement of adhesion capability between the Napier fibres and epoxy matrix. Overall, the SEM shows an improvement in interfacial bonding between the Napier fibres and epoxy matrix upon the nano-clay filler was added [19]. As concluded, the presence of nano clay filler loading with a range of 3 wt% to 5 wt% in the Napier/epoxy composites shows the significant improvement in mechanical and morphological properties.

5. Acknowledgements
The authors would like to express my gratitude and special thanks to University Malaysia Perlis (UniMAP) in providing researching platform. Besides, the authors are also thankful to all fellows for providing their valuable suggestions and guidance during this work.

6. References
[1] D. H. H. Cheung, M. P. Ho, K. T. Lau and F. Cardona: 'Natural fibre-reinforced composites for bioengineering and environmental engineering applications', Compos. Part B Eng., vol. 40(7), pp. 655–663, 2009.
[2] Kovier K, "Fiber reinforced concrete". The Cement and Concrete institute, Midrand, 2001.
[3] M.J.M. Ridzuan, M.S. Abdul Majid*, M. Afendi, S.N. Aqmariah Kanafiah, J.M. Zahri, and A.G. Gibson, "Characterisation of natural cellulosic fibre from Pennisetum purpureum stem as potential reinforcement of polymer composites," Mater. Des., vol. 89, pp. 839–847, Jan. 2016.
[4] M. Haameem J.A., M.S. Abdul Majid, M. Afendi, H.F.A. Marzuki, I. Fahmi, and A.G. Gibson, "Mechanical properties of Napier grass fibre/polyester composites," Compos. Struct., vol. 136, pp. 1–10, Feb. 2016.
[5] M. Haameem J.A., M.S. Abdul Majid*, M. Afendi, H.F.A. Marzuki, E.A. Hilmi, I. Fahmi, and A.G. Gibson, "Effects of water absorption on Napier grass fibre/polyester composites," Compos. Struct., vol. 144, pp. 138–146, Jun. 2016
[6] M.J.M. Ridzuan, M.S. Abdul Majid*, M. Afendi, K. Azduwin, N.A.M. Amin, J.M. Zahri, and A.G. Gibson, "Moisture absorption and mechanical degradation of hybrid Pennisetum purpureum/glass–epoxy composites," Compos. Struct., vol. 141, pp. 110–116, May 2016.
[7] Amir, W.W., Jumahat, A. and Kalam, A., Characterization of Nanoclay-Modified Epoxy Polymers Using X-Ray Diffraction Analysis. Applied Mechanics and Materials, 2015, pp. 175-180.
[8] Mousavinasab SM, Atai M, Alavi B., "To Compare the Microleakage Among Experimental Adhesives Containing Nanoclay Fillers after the Storages of 24 Hours and 6 Month," Open Dent J, vol. 5, pp. 52-7, 2011.
[9] Paul DR, Robeson LM. Polymer nanotechnology: Nanocomposites. Polymer (Guildf), 2008, pp. 3187-3391.
[10] M.S. Fartini, M.S. Abdul Majid*, M.J.M. Ridzuan · N.A.M. Amin, Geoff Gibson, Compressive properties of Napier (Pennisetum Purpureum) filled polyester composites. Plastics, Rubber and Composites: Macromolecular Engineering, Plast. Rubber Compos., vol. 45, no. 3, pp. 136-146, April. 2016.
[11] Solhi L, Atai M, Nodehi A, Imani M, "A novel dentin bonding system containing poly (methacrylic acid) grafted nanoclay: synthesis, characterization and properties," Dent Mater, vol. 28, pp. 1041-1050, 2012.
[12] Zukas T, Jankauskaite V, Žukienė K, "The Influence of Nanofillers on the Mechanical Properties of Carbon Fibre Reinforced Methyl Methacrylate Composite," Mater Scienc, vol. 18, pp. 250-255, 2012.

[13] K. Krushnamurty, I. Srikanth, B. Rangababu, S. K. Majee, R. Bauri, Ch. Subrahmanyam, "Effect of nanoclay on the toughness of epoxy and mechanical, impact properties of E-glass-epoxy composites," Advanced Materials, vol. 6 , pp. 684-689, March 2015.
[14] 6. S. S. Ray and M. Okamoto, "Polymer/layered silicate nanocomposites: a review from preparation to processing," Progress in Polymer Science, vol. 28, no. 11, pp. 1539–1641, 2003.
[15] Chisholm N, Mahfuz H, Rangari VK, Ashfaq A, Jeelani S, "Fabrication and mechanical characterization of carbon/ SiC-epoxy nanocomposites," Composite Structural, vol. 67, pp. 115-139, 2005.
[16] Kusmono and Zainal ArifinMohd Ishak, "Effect of Clay Addition on Mechanical Properties of Unsaturated Polyester/Glass Fiber Composites," International Journal of Polymer Science, Oct 2013.
[17] Ikejima I, Nomoto R, McCabe JF, "Shear punch strength and flexural strength of model composites with varying filler volume fraction, particle size and silanation," Dent Mater, vol. 19, pp. 206-211, 2003.
[18] Nunes MF, Swift EJ, Perdigão J, "Effects of adhesive composition on microtensile bond strength to human dentin," Am J Dent, vol. 14, pp. 340-343, 2001.
[19] Mohammad Hossein Pol, G.H. Liaghat, "Studies on the Mechanical Properties of Composites Reinforced with Nanoparticles," Polymer Composites, vol.10, pp. 206-211, 2017.

Tensile properties of interwoven hemp/PET (Polyethylene Terephthalate) epoxy hybrid composites

M A A Ahmad, M S A Majid, M J M Ridzuan, A Z A Firdaus and N A M Amin

School of Mechatronic Engineering, Universiti Malaysia Perlis, Pauh Putra Campus, 02600, Arau, Perlis, Malaysia.

shukry@unimap.edu.my

Abstract. This paper describes the experimental investigation of the tensile properties of interwoven Hemp/PET hybrid composites. The effect of hybridization of hemp (warp) with PET fibres (weft) on tensile properties was of interest. Hemp and PET fibres were selected as the reinforcing material while epoxy resin was chosen as the matrix. The interwoven Hemp/PET fabric was used to produce hybrid composite using a vacuum infusion process. The tensile test was conducted using Universal Testing Machine in accordance to the ASTM D638. The tensile properties of the interwoven Hemp/PET hybrid composite were then compared with the neat woven hemp/epoxy composite. The results show that the strength of hemp/PET with the warp direction was increased by 8% compared to the neat woven hemp composite. This enhancement of tensile strength was due to the improved interlocking structure of interwoven Hemp/PET hybrid fabric.

1. Introduction

The increasing environmental issues due to the used of petroleum based product and fuels lead to the development of natural fibre composite. Fibre reinforced composite are low cost, easier to handle and have good mechanical properties. Fibre can be easily obtained from the natural environment or readily available such as hemp, wood, carbon fibre, and fibre glass[1]. Iqbal et. al provides an explanation on how to utilise textile materials and fabrics manufacturing methods such as weaving, knitting and braiding[2]. The mechanical properties of the fibre reinforced composite depend on the fibre geometry, hybridization, and weaving architecture. These parameters will increase their performance of high strength application. Besides, a knitted or woven fibre provides equally distributed strength through the transverse direction and longitudinal direction [3].

Woven are made by interlacing fibre bundles (yarn) to form a fabric layer [4]. It gives high packing density and excellent dimensional stability. Further studied have been done to improve more on the strength of composite by producing woven and hybrid composite. Hybrid composite refers to more than one reinforcing material used in a polymer matrix, develop to achieve properties that cannot be made by a single fibre [3]. The hybrid will make each fibre improve the weakness of their properties, thus strengthen the structure of composite. Review by Swollfs shows that hybridization is a method to increase the toughness of fibre reinforced composite and three leading hybrid configurations to combine the fibres are interlayer or layer-by-layer configuration, intralayer or yarn-by-yarn and intra-yarn or fibre-by-fibre [5].

An experimental study involving interwoven of kenaf/PET reinforced POM by Dan-Mallam showed that the high resistance to water uptake due to the interlacing of the hydrophobic PET fibre in the interwoven increased the composite strength [3]. Venkateshwaran et. al [6] indicates that the addition of sisal fibre to banana/epoxy composite will bring the hybrid effect that eventually improves the mechanical properties of the composite. Other than that, the similar result is shown by Yahaya et. al [7] in his investigation of tensile, flexural, and impact properties of woven kenaf/Kevlar reinforced epoxy composite. The primary objective of this paper is to investigate the effect of hybridization of interwoven Hemp/PET hybrid composite on their tensile properties.

2. Experimental Methods

2.1 Woven Fabric
Hemp and PET yarn were used as reinforcement material that supplied by Innovative Pultrusion Sdn. Bhd. The fabric was weaved using a locally constructed wooden frame with a size of 400×400 mm. It consists of sets of warp yarn guider at the both sides of the wood. The weaving process starts with the arranging warp yarn along the nail. Then the weft yarn was arranged passing over and underneath the warp yarn until the fabric was produced [8]. Figure 1 shows the weaving process and fabric.

(a)

(b)

Figure 1. (a) weaving process, (b) fabric.

2.2 Composite Fabrication
The composite was produced by the vacuum infusion process as shown in Figure 2. The vacuum infusion machine consists of the vacuum pump, pressure gauge, suction hose, and container for residual resin collection. Firstly, the fabric was placed on top of the transparent glass table. Peel ply was used to cover on top of it for the removal after curing. Netting was used to disperse the resin thoroughly on the fabrics during the suction process. Then vacuum bagging was used on the top of the sample. The resin flows through the hose by evacuating the mould of fabric into the residual resin container[9].

EpoxeAmite was mixed with a slow hardener in a weight ratio of 100g epoxy: 28.4g hardener according to the manufacturer suggestion. The infused composite was removed from the mould and cured for a complete 12 hour at room temperature (23°C). The vacuum pump was running about 6 hours before switch off. This is to ensure that the resin will disperse and filled thoroughly on the fabric during the curing process. H/H composite and H/P hybrid composite were prepared in this experiment. Abbreviations representing the composites types are listed in Table 1.

Table 1. Symbol used to represent different types of composite prepared.

Type of composite	Abbreviation
Woven hemp	H/H
Interwoven Hemp/PET along hemp direction	P/H
Interwoven Hemp/PET along PET direction	H/P

Figure 2. Vacuum infusion process.

2.3 Tensile Testing

The tensile test samples were cut with a Dremel 4000 tool according to the ASTMD638 [10]. Fifteen samples were used in this experiment which are five samples for H/H composites and ten samples for the H/P composite with two different directions of fibre. Universal Testing Machine with a crosshead speed 1mm/min and 100kN strength was used for the tensile test. The extensometer was used to measure the strain as shown in Figure 3.

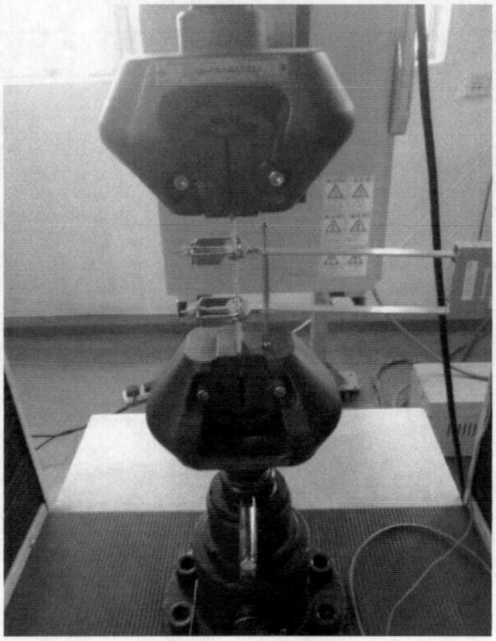

Figure 3. Tensile test.

3. Results and Discussion

The results show that the maximum tensile strength and modulus elasticity of P/H hybrid, H/H, and H/P composites as shown in Figure 4 and 5. The tensile strength of P/H hybrid composite increased by 4% compared to the H/H composite which was recorded about 47 MPa and 51MPa respectively. However, the strength showed by H/P hybrid composite is the lowest strength about 35 MPa. This leads to the significant of the hybridization to increase the strength of fibre in the composite.

The modulus of elasticity of the composite shows similar responses to their tensile strength. The P/H hybrid composite give the highest modulus of elasticity recorded about 2.3 GPa compared to the H/H and H/P composite with 2GPa and 1.8 GPa respectively. The modulus is said to link with the stretching behaviour between the fibre and matrix in the composite. This is because the break and stretch of PET fibre when it approaches the breaking stress. Eventually, the sound characteristic PET fibre will help to enhance the strength of fibre and composite[8].

In addition, a better interfacial adhesion during the vacuum infusion process was lead to improving the strength of the composite. The P/H hybrid composite exhibit higher strength and elasticity modulus because of the fibre arrangement. Other than that, the stable network structure of interwoven hemp and PET fibres will improve the interlocking between fibre and matrix of the composite. Thus, it will prevent for fibre pull out while conducting the test.

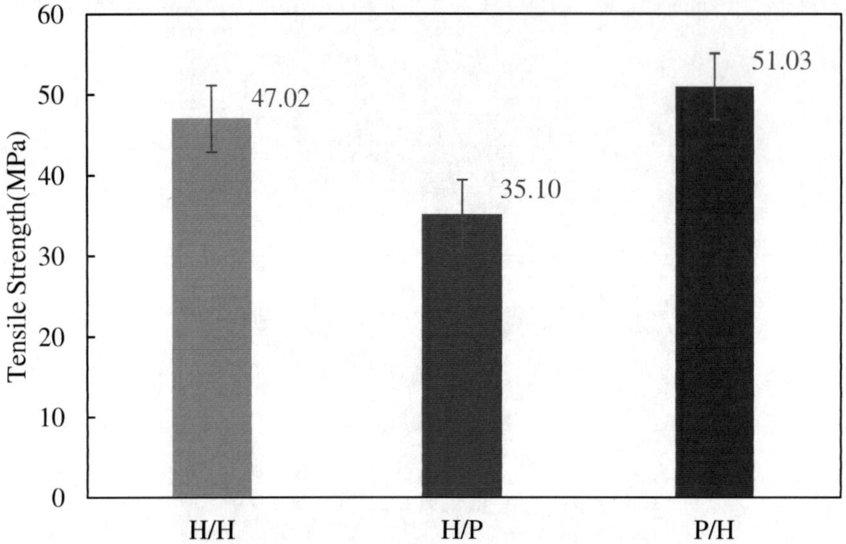

Figure 4. Comparison of the tensile strength of woven and interwoven composite.

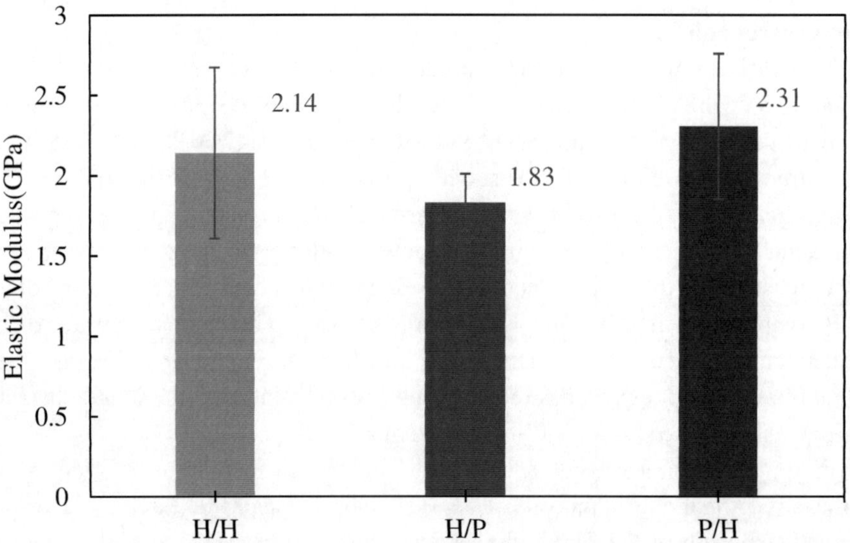

Figure 5: Comparison of the modulus elastic of woven and interwoven composite.

4. Conclusion

The mechanical properties of H/H, H/P and P/H hybrid composites were investigated through the tensile test. The results showed the tensile strength and elastic modulus of P/H hybrid composite were higher than H/H composite. The interfacial adhesion was improved with a combination of hemp and PET fibres, thus prevent for the fibre pull out during the test. Thus, the higher load is needed to break the composites structure.

5. Acknowledgement

The authors would like to express gratitude and special thanks to University Malaysia Perlis (UniMAP) in providing researching platform well as providing the equipment and financial assistance for this project. Besides, the authors are also thankful to all fellows for providing their valuable suggestions and guidance during this investigation.

6. References

[1] E. Charlton, "Basics-Of-Composite-Materials," *about money.com*, 2016. [Online]. Available: http://composite.about.com/od/aboutcompositesplastics/u/Basics-Of-Composite-Materials.htm.

[2] M. Iqbal, M. Islam, J. Ananda, and K. Lau, "Potentiality of utilising natural textile materials for engineering composites applications," *J. Mater.*, vol. 59, pp. 359–368, 2014.

[3] Yakubu Dan-mallam, TanWei Hong, and Mohd Shukry AbdulMajid, "Mechanical Characterization and Water Absorption Behaviour of Interwoven Kenaf / PET Fibre Reinforced Epoxy Hybrid Composite," *Int. J. Polym. Sci. vol. 2015* (2015), Article ID 371958.

[4] F. C. Suhad D. Salman, Mohaimin J. Sharba, Z. Leman, M.T.H. Sultan, M.R. Ishak, "Physical , Mechanical , and Morphological Properties of Woven Kenaf / Polymer Composites Produced Using a Vacuum Infusion," *Int. J. Polym. Sci., vol. 2015,* (2015), Article ID 894565.

[5] Y. Swolfs, L. Gorbatikh, and I. Verpoest, "Composites : Part A Fibre hybridisation in polymer composites : A review," *Compos. PART A*, vol. 67, pp. 181–200, 2014.

[6] N. Venkateshwaran, a. ElayaPerumal, a. Alavudeen, and M. Thiruchitrambalam, "Mechanical and water absorption behaviour of banana/sisal reinforced hybrid composites," *Mater. Des.*, vol. 32, no. 7, pp. 4017–4021, 2011.

[7] R. Yahaya, S. Sapuan, M. Jawaid, Z. Leman, and E. Zainudin, "Mechanical performance of woven kenaf-Kevlar hybrid composites," *J. Reinf. Plast. Compos.*, vol. 33, no. 24, pp. 2242–2254, 2014.

[8] Y. Dan-Mallam, M. Z. Abdullah, and P. S. M. M. Yusoff, "The effect of hybridization on mechanical properties of woven kenaf fiber reinforced polyoxymethylene composite," *Polym. Compos.*, vol. 35, no. 10, pp. 1900–1910, Oct. 2014.

[9] A. B. Maslinda, M. S. A. Majid, M. J. M. Ridzuan, M. Afendi, and A. G. Gibson, "Effect of water absorption on the mechanical properties of hybrid interwoven cellulosic-cellulosic fibre reinforced epoxy composites," *Compos. Struct.*, vol. 167, pp. 227–237, 2017.

[10] A. D638-10, "Standard Test Method for Tensile Properties of Plastics 1," no. ASTM International, pp. 1–16, 2012.

Failure prediction of ±55° glass/epoxy composite pipes using system identification modelling

A J Yi[1], M S A Majid[1], A M Nor[1], M J M Ridzuan[1] and S Yaacob[2]

[1]School of Mechatronic, Pauh Putra Campus, Universiti Malaysia Perlis, Arau,02600,Perlis
[2]Universiti Kuala Lumpur, Malaysia Spanish Institute, Kulim Hi-Tech Park 09000, Kulim,Kedah

jiayi89@msn.com

Abstract: Black-box modelling using system identification method to predict the performance of glass fibre reinforced epoxy (GRE) composite pipe under multiaxial loading stress ratio is presented. In this study, both linear and nonlinear models were derived namely; linear time-invariant parametric model and artificial neural network model. The models derived are to approximate the pure hydrostatic loading performance using GRE pipes with winding angles of ±55°. Three different linear model structures were derived, and the best fit model achieved at 96.64% of best fit. On the other hand, the Artificial Neural Network (ANN) modelling showed better accuracy with the best fit of 99.82%. Finally, the point of failure at which first damage takes place predicted by the models derived was validated using experimental data.

1. Introduction

Glass fibre reinforced epoxy (GRE) composite pipes possess unique characteristic of high strength, high stiffness, light weight and corrosion resistance. It has been widely used in oil and gas industries as an alternative to conventional steel pipes [1]. The qualification of GRE composites pipe is conducted through international standard ASTM D2992 [2] in order to ensure its quality and safety. Based on the standard, GRE composite pipe has to withstand a minimum life span service of 20-50 years. This requirement has become a primary design requirement for quality control. To meet these design requirement, a full-scale test of GRE pipes under long-term hydrostatic pressure loading test up to 10,000 hours was established [3]. This quality testing method required a duration of the testing up to one year or more, which will be an increased cost and time for industries. Therefore, to ensure smooth qualifying procedure of the GFE composite pipes without wasting unnecessary cost and time, pre-qualifying modelling is essential. Efficient modelling can significantly provide a faster and cheaper solution to predict the performance of the composite pipe before the actual qualifying procedure [6].

In recent research studies, several researchers have been working on a model to provide realistic predictions using the finite element modelling method. R.M. Guedes has studies on a stress-strain analysis of a cylindrical pipe subjected to a transverse load and large deflections [4]. The result discusses the deformation theory on a high strength compared with the stiffness of the new material. The strain experimental measured and the final element analysis was performed to verify the phenomenon. A simple approach based on finite deformation theory was proposed and assessed.

Of late, Roham Fafiee et al. has also been looking into the theoretical modelling of fatigue phenomenon in composite pipes [5]. The modelling procedure has been divided into few phases to predicting the fatigue lifetime of the composite pipe. Their results suggested a good agreement between theoretically predicted fatigue lives of an industrial pipe with it long-term behaviour operation estimated test.

Besides that, Roham Fafiee and Ali Amini investigated on the modelling, and experimental evaluation of functional failure pressures in glass fibre reinforced polyester pipes [7]. A progressive modelling has been developed for the internal pressure weepage prediction which consists of failure evaluation, material degradation and stress analysis. The result shows that proper selection of the winding angle can overcome the adverse effect of fibre volume fraction.

Recently, most of the modelling methods are based on the theoretical analysis of the composite pipe. Hence, derivation of a mathematical model using identification method can be relativity new to the composite modelling. Based on the Li Fu research on the research survey of system identification method [8]. The study has mentioned that various types of identification technique which include system identification, neural network or fuzzy logic can be used to analyses complex model.

A.G Gibson and his team have conducted research on the qualification and lifetime modelling of fibreglass pipe [9]. They have been using the UEWS test on the cyclic fatigue test on fibreglass to model out the relationship between cyclic and static fatigue behaviour. By using the Miner's law, it has found that it is effective in modelling the damage that can be related to the matrix cracking.

Abdul Majid and co-worker have modelled the strain responses of glass/epoxy pipes under various stress ratios. They managed to produce a model to predict the non-linear stress-strain response caused by the fatigue static and cyclic loading stress test. The prediction from the model is found to be in good agreement with the data from the multiaxial UEWS tests of ±55° filament wound GRE pipes [10]. They also have been working on the modelling of the general lifetime damage model for the GRE pipes which the model produced in a good agreement between the experimental data and the modelled developed [11]. Farhana et al. performed research on a new vibration based non-destructive testing for prediction glass fibre/matrix volume fraction in composites using a neural network model [12]. The neural network is used to determine the percentage of the fibre volume fraction. The results showed a good agreement with the standard destructive test for volume fraction determination which between the range of 90-98%.

The present research is intended to develop and validate mathematical models to predict the failure of the composite pipe. The model development will be based on system identification methods. The model will able to help to predict the behaviour of the composite pipe which will reduce the time and cost when developing new composite pipes.

2. Methodology

2.1. Glass Fibre Reinforced Epoxy Pipes (GRE)

The composite pipes with ±55° winding angle were fabricated through filament winding process at Industrial Research Institute of Malaysia (SIRIM), Kulim, Kedah. Standards E-glass fibres with a linear density of 1200 tex and DER-331 epoxy were used to create the fibre-matrix interface. The glass fibres were passing through a bath of epoxy resin maxis with hardener, and the wet fibre was wounded onto a mandrel of 100 mm in diameter. The pipe then cured around 2 hours in an oven. The specification of the GRE pipe is given in Table 1.

Table 1. Composite pipes specifications.

Winding Angle	**±55°**
Average Length(mm)	1050
Average Diameter(mm)	105
Density(kg/m^2)	1150
Average wall thickness(mm)	4.34

2.2 Experimental Procedure

A test rig was developed to undergo the cyclic loading test under multiple stress ratio loading tests. In this research, the focus of the experimental will only be on the pure hydrostatic (2H: 1A) loading test. The full schematic diagram of the automated pressure test rig for the pure hydrostatic setup is shown in Figure 1.

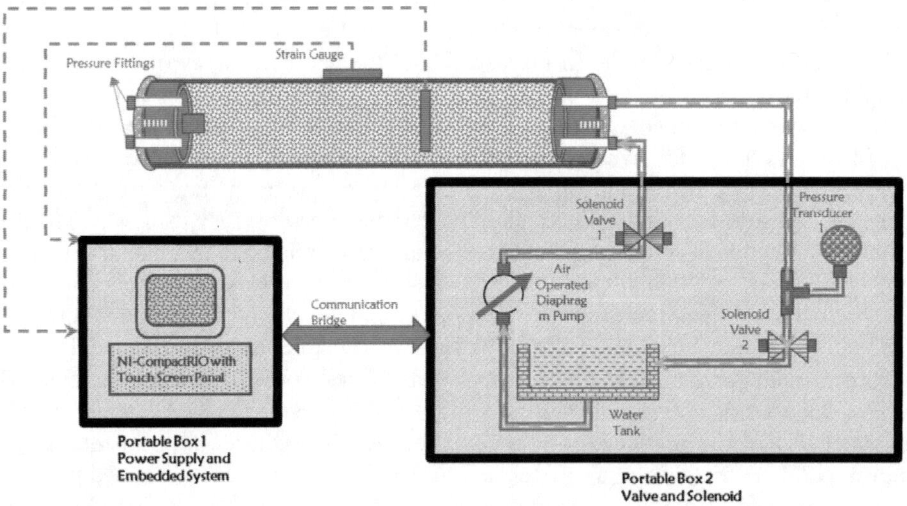

Figure 1. Schematic Diagram of the Automated Pressure Test Rig.

First, the pipe was installed with two custom design end-fittings with water filled inside it. The end fitting is sealed with O-rings to ensure that the leaks will not happen at the end fitting. To ensure no outward movement of the end fitting, serrated wedges placed in between the pipe wall to lock the movement. Before proceeding with the pressure testing, the composite pipe must be inspected to ensure that bubble trap inside the pipe is fully removed. Next, the pipe was placed in free-free conditions by hanging onto the test rig using lubricated rubber ring to reduce external disturbance. The strain gauge will be mounted onto the composite pipe using some special epoxy. The strain gauge was used to monitor the deformation takes place when the test is being conducted. The pipe fitting will be installed at the final stage depends on the test ratio needed.

The input signal is designed following the cyclic pressure loading which is controlled automatically using high-pressure solenoid valves. The input cyclic pressure was determined to be started with the pressure increment by 10% of the total expected failure pressure. The cycle of each pressure will last for ten cycles which consist of 60 seconds of the pressurised and non-pressurized condition. The test will be conducted until the leakage or weepage is observed from the outer surface of the pipe wall. The whole process was carried out automatically using the Ni-Compact RIO devices which including the increment of the input signal till the data collection. Visual inspection of the entire test will be done manually to determine the stopping time. All the data recorded from the experiment will be used to produce the model using identification method. A flow chart illustrated in Figure 2 presents the whole process of this research.

IOP Conf. Series: Journal of Physics: Conf. Series **908** (2017) 012012 doi:10.1088/1742-6596/908/1/012012

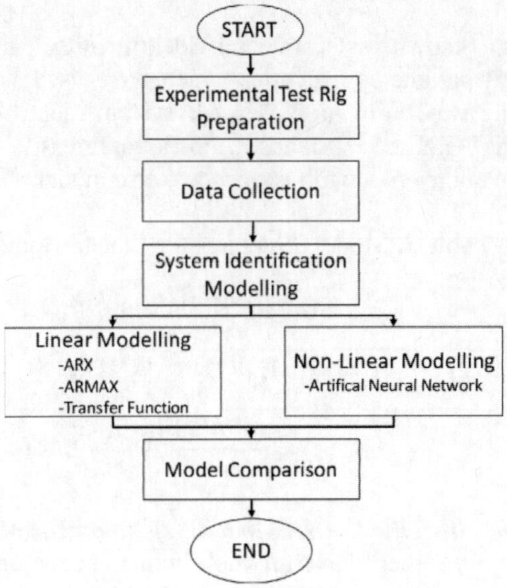

Figure 2. Flow Chart for Identification Modelling.

2.3. Data Preparation

The modelling need the input and output data for construction, preparation of the data become an important role in determining the accuracy of the model. The data set from the experimental study is given in Table 2, which consist of the input pressure and the corresponding strain reading. A total number of 40562 data were prepared in each experiment. Before the modelling, a smoothing process will be conducted towards the data to remove unnecessary noise.

Table 2. Sample Data Collected.

Pressure 1 (bar)	Pressure 2 (bar)	Hoop Strain	Axial Strain
0.837809	0.614225	0.000029	0.000012
0.886995	0.659944	0.000028	0.000015
0.941061	0.715936	0.000033	0.000013
0.999558	0.770901	0.000042	0.000019
1.059467	0.830297	0.000042	0.000007
1.115973	0.893481	0.000039	0.000019

2.4. Identification Method

Identification modelling is a method which uses the input and the output of the system to construct a mathematical modelling. This approach has been developed and used by L.A Zadeh since 1962 [13]. There are several methods which can be employed in the identification modelling. For example, least square method, gradient correction process is among the few that are commonly used. Some advanced method such as neural network and fuzzy logic has slowly become a trend in the modelling world [14]. However, each different method consists of pros and cons of its usage. In this study, both linear and nonlinear model consist of system identification based method and artificial neuro network modelling will be used.

IOP Conf. Series: Journal of Physics: Conf. Series **908** (2017) 012012 doi:10.1088/1742-6596/908/1/012012

2.4.1. System identification

System identification is also known as parameter identification. In constructing the system identification, there are several parameter structures which were used which given in Table 3. The system identification modelling was built via MATLAB system identification toolbox (SIT) which consists of those structures. The ARX model stands for the autoregressive moving average model, and ARMAX is autoregressive moving average with exogenous term model.

Table 3. Model Structures for Linear Model.

ARX MODEL	A(z)y(t)=B(z)u(t−nk)+e(t)	(1)
ARMAX MODEL	A(z)y(t)=B(z)u(t−nk)+C(z)e(t)	(2)
TRANSFER FUNCTION	$Y(s) = \frac{num(s)}{den(s)} u(s) + E(s)$	(3)

2.4.2. Artificial Neuron Network Modelling (ANN Modelling)(Nonlinear Model)

Artificial Neuron Network is the other powerful tools which have the ability to construct some complex model [19]. By using the mathematical structure which shown in Equation **Error! Reference source not found.**, the construction of the AI model can be processed by using the input-output of the data.

$$y_{output} = (\textstyle\sum_{i=0}^{k} W_i * I_i) \quad (4)$$

where y_{output}=value propagates to next layer
W = weight of the neuron
I = neuron value gain

3. Results and Discussion

3.1. Comparison of the Model Structure

For comparison purposes, all three linear model structures were analysed, and the 1st order mathematical models are tabulated in Table 4. Three models where produce and their accuracy according to the best fit are compared. According to the result, the ARX model achieved the best fit of 86.39% followed by the ARMAX with the best fit of 87.13%. Overall, the highest best fit produce by the system identification modelling is the transfer function structure model at 96.64%. The results show that ARMAX model has the better fit than ARX. This is due to the ARMAX model structure has included the properties of the error term which is lag inside the ARX model. For the ARX structure model, the noise entered directly into the structure parameter which introduced errors within the design model. Inside the ARMAX model, the noise is separated and resulted in the better model closer representation of the physical design.

Transfer function structure model are the model which using the Laplacian transform pole and zero. This method will produce a continues equation based on the S-plane model. On the other hand, the ARX and ARMAX model are using the linear time invariant model which produce model base on the Z-transform. Therefore, the model produces on the continues plane (S-plane) will obtain a better best fit result compare with a discrete model (Z-Transform). This is due to the discrete model normally will lose part of it data when the model is being digitalize.

Table 4. 1st Order Models of GRE Composite Pipes.

Pipe Winding Angle and Estimator	Model Obtain for Hoop Strain	Best Fit (%)	Model Obtain for Axial Strain	Best Fit (%)
55°				
ARX	$\frac{2.676\times10^{-5}}{1-0.6926Z^{-1}}u(t)$	86.39	$\frac{2.277\times10^{-6}}{1-0.8742Z^{-1}}u(t)$	66.07
ARMAX	$\frac{2.902\times10^{-5}}{1-0.67Z^{-1}}u(t)+\frac{1+0.9434Z^{-1}}{1-0.67Z^{-1}}e(t)$	87.13	$\frac{2.398\times10^{-6}}{1-0.8675Z^{-1}}u(t)+\frac{1+0.9405Z^{-1}}{1-0.8675Z^{-1}}e(t)$	67.12
Transfer Function	$\frac{8.317\times10^{-5}S+9.242\times10^{-9}}{S+6.982\times10^{-5}}$	96.64	$\frac{1.981\times10^{-5}S+2.653\times10^{-8}}{S+0.001653}$	86.1

The other factor which will impact on the accuracy of the model will be the decision of the number of order. The number of the order will be best to describe as the power factor is a mathematical equation. Table 5 has shown the best fit that each model until 3rd order model. This is because most of the model can be achieved in this region. Moreover, the model will be too complex to handle if the number of order is too high. From Table 5, The 1st order model produced by transfer function will be chosen to represent the model. Although it has shown that the best fit is the 3rd order model produced by the transfer function structure, due to the small increment the 1st order structure will be used.

Table 5. Accuracy of the Model for different Pole and Zero.

Pipe Winding Angle and Estimator	Best Fit (%) Hoop Strain						Best Fit (%) Axial Strain					
Number Of Order(P/Z)	1-1	2-1	2-2	3-1	3-2	3-3	1-1	2-1	2-2	3-1	3-2	3-3
55°												
ARX	86.39	73.45	93.12	77.15	93.21	92.61	66.07	70.42	84.88	64.77	84.88	85.3
ARMAX	87.13	80.35	92.92	81.28	92.88	92.89	67.12	72.98	84.68	70.38	85.55	86.03
Transfer Function	96.64	89.4	97.22	85.94	89.31	97.5	86.1	85.01	81.73	82.65	85.89	86.15

Figure 3 illustrates the nonlinear model structure which was generated to represent the system. From the figure, the ANN model structure consists of 1 layer with three neurones. The structure value for all weight and bias which represent the system model is shown in Figure 3. From the calculation using the system, the ANN Model has achieved the best fit of 99.82%.

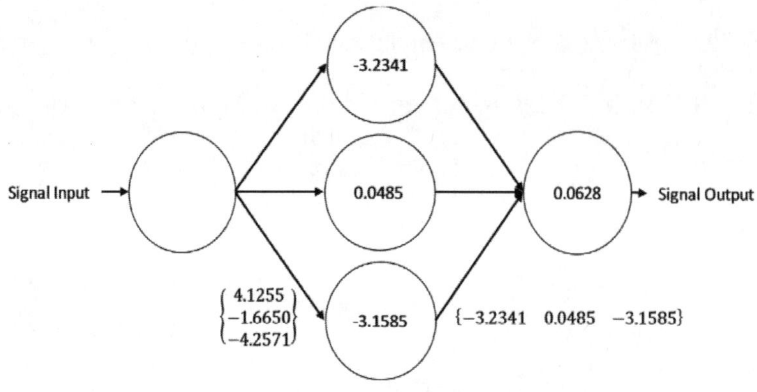

Figure 3. Model Structure of ANN Model.

3.2. Model Comparison

The developed mathematical model is expected to determine the onset of ply failure for the composite pipes. Figure 4 below shows the hoop strain versus pressure results for both; identification model and the experimental result. There were three plots presented; two were predictions from the system identification model and artificial neural network models and another is the experimental findings as reported by M.S. Abdul-Majid et al. [15]. The prediction from system identification model is computed using equation (5) while artificial neural network model was developed from the output obtained from the model structure shown in Figure 3

$$F(x) = \frac{8.317 \times 10^{-5}S + 9.242 \times 10^{-9}}{S + 6.982 \times 10^{-5}} \quad (5)$$

From Figure 4, the graph shows that the linear model and the non-linear model are able to predict the behavior of the system. But the transfer function model which is the linear model were only able to predict the behavior of the system up to the point where the crack happens. It has shown that the models are only good in predicting the linear region. Since the system which under testing will change its behavior to shows exponential like behavior, the ANN model has shown a much better performance in predicting the behavior of the system. Thus, the ANN model is better in modelling since it represents a nonlinear model structure.

By comparing both the linear and on-linear model, the onset of first ply failure is suspected of taking place when the stress-strain curve deviates from the linear responses and exceeds a defined threshold level as to suggest permanent deformation occurred within the ply structure. From Figure 4, it could be suggested that the model did show some agreement to a degree to the experimental data. The prediction from both models shows good prediction, particularly in the linear region. The predictions from system identification model start to show deviations from the straight linear relationship at roughly about 40 bars.

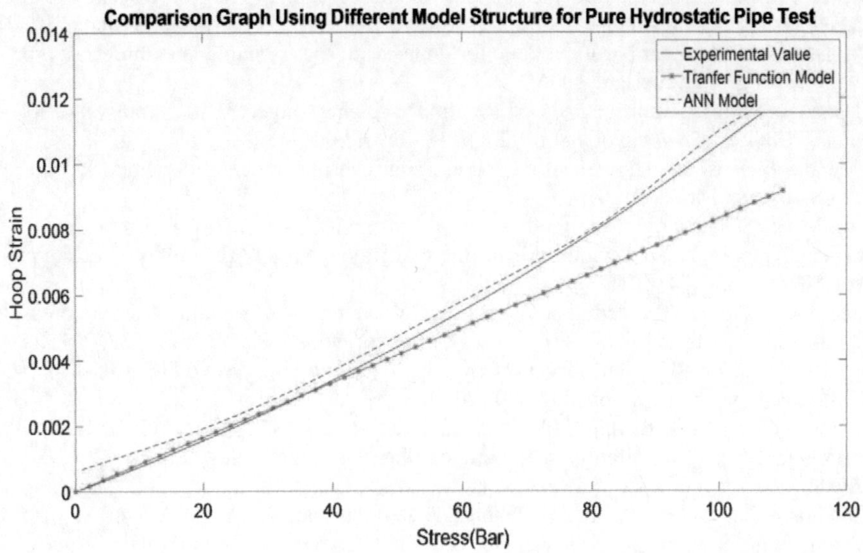

Figure 4. Comparison between linear and non-linear model for stress-hoop strain response and the experimental data.

The results imply that the models provide a reasonable prediction of the onset first ply failure before functional failure such as weepage takes place albeit the variance in accuracy. The result seems to suggest that the system identification model has detected the damage by comparing with the ANN Model structure. This is because the system identification is a linear model compared to the ANN Model which is more robust. The ANN Model produced inclines to adapt the characteristic of the composite pipes thus indicates the non-linearity. Nonetheless, both models showed acceptable predictions of the onset of the damage for the GRE pipes when comparing together.

5. Conclusion

In conclusion, two different model types were developed to predict the onset of damage in the ply structure of GRE composite pipes, namely; linear time-invariant and ANN nonlinear models. The system identification based model was performed by using three different model structures, and the best fit the model achieved at 96.64% of best fit. On the other hand, the ANN modelling method produced the best fit of 99.82%. The models generated are then compared to the experimental data and presented in the hoop strain-stress (bar) plot. Both prediction models show reasonable agreement with experimental findings though with variance in accuracy levels. The results suggest that the models can be used as a preliminary/early procedure to predict the onset of damage in the GRE pipes before experimental means is applied.

6. Acknowledgements

The authors would like to express the gratitude for those involved in this research. Also, the authors are grateful to Universiti Malaysia Perlis for providing all the access to the facilities and financial support for this investigation.

7. References

[1] S. N. Fitriah, M. S. Abdul Majid, M. J. M. Ridzuan, R. Daud, A. G. Gibson, and T. A. Assaleh, "Influence of hydrothermal ageing on the compressive behaviour of glass fibre/epoxy composite pipes," *Compos. Struct.*, vol. 159, pp. 350–360, 2017.

[2] "ASTM D2992-06. Standard test method for obtaining hydrostatic or pressure design basis for reinforced thermosetting resin pipe and fitting. American Society for Testing and Materials; 2006."

[3] "ASTM D5365-06. Standard test method for long-term ring bending of fibreglass pipe. American Society for Testing and Materials; 2006." .

[4] R. M. Guedes, "Stress-strain analysis of a cylindrical pipe subjected to a transverse load and large deflections," *Compos. Struct.*, vol. 88, no. 2, pp. 188–194, 2009.

[5] R. Rafiee and F. Elasmi, "Theoretical modeling of fatigue phenomenon in composite pipes," *Compos. Struct.*, vol. 161, pp. 256–263, 2016.

[6] A. Hawa, M. S. Abdul Majid, M. Afendi, H. F. A. Marzuki, N. A. M. Amin, F. Mat, and A. G. Gibson, "Burst strength and impact behaviour of hydrothermally aged glass fibre/epoxy composite pipes," *Mater. Des.*, vol. 89, pp. 455–464, 2016.

[7] R. Rafiee and A. Amini, "Modeling and experimental evaluation of functional failure pressures in glass fiber reinforced polyester pipes," *Comput. Mater. Sci.*, vol. 96, no. PB, pp. 579–588, 2015.

[8] L. Fu and P. Li, "The Research Survey of System Identification Method," *2013 5th Int. Conf. Intell. Human-Machine Syst. Cybern.*, pp. 397–401, 2013.

[9] A. G. Gibson, M. S. Abdul Majid, T. A. Assaleh, J. M. Hale, A. Fahrer, C. A. P. Rookus, and M. Hekman, "Qualification and lifetime modelling of fibreglass pipe," *Plast. Rubber Compos.*, vol. 40, no. 2, pp. 80–85, 2011.

[10] M. S. Abdul Majid, A. G. Gibson, M. Hekman, M. Afendi, and N. A. M. Amin, "Strain response and damage modelling of glass/epoxy pipes under various stress ratios," *Plast. Rubber Compos.*, vol. 43, no. January 2016, pp. 290–299, 2014.

[11] M. S. A. Majid, M. Afendi, R. Daud, N. a. M. Amin, A. Mohamad, E. M. Cheng, a. G. Gibson, and M. Hekman, "General Lifetime Damage Model for Glass Fibre Reinforced Epoxy (GRE) Composite Pipes under Multiaxial Loading," *Key Eng. Mater.*, vol. 594–595, no. January 2014, pp. 624–628, 2013.

[12] N. I. E. Farhana, M. S. Abdul Majid, M. P. Paulraj, E. Ahmadhilmi, M. N. Fakhzan, and A. G. Gibson, "A novel vibration based non-destructive testing for predicting glass fibre/matrix volume fraction in composites using a neural network model," *Compos. Struct.*, vol. 144, pp. 96–107, 2016.

[13] L. A. Zadeh, "From Circuit Theory to System Theory," *Proc. IRE*, vol. 50, no. 5, pp. 856–865, 1962.

[14] A. J. Yi, M. S. A. Majid, A. M. N, and S. Yaacob, "Microcontroller-Based For System Identification Tools Using Least Square Method For Rc Circuits," *J. Teknol.*, 2015.

[15] M.S. Abdul Majid, T.A. Assaleh, A.G. Gibson, J.M. Hale, A. Fahrer, C. A. P. Rookus, and M. Hekman, "Ultimate elastic wall stress (UEWS) test of glass fibre reinforced epoxy (GRE) pipe," Compos. Part A Appl. Sci. Manuf., vol. 42, no. 10, pp. 1500–1508, Oct. 2011.

Tensile properties of compressed moulded Napier/glass fibre reinforced epoxy composites

T S Fatinah, M S Abdul Majid[1], M J M Ridzuan, T W Hong, N A M Amin and M Afendi

[1]School of Mechatronic Engineering, Universiti Malaysia Perlis, Pauh Putra Campus, 02600, Arau, Perlis, Malaysia.

shukry@unimap.edu.my

Abstract. This paper describes the experimental investigation of the tensile properties of compressed moulded Napier grass fibres reinforced epoxy composites. The effect of treatment 5% sodium hydroxide (NaOH) concentrated solution and hybridization of Napier with CSM E-glass fibres on tensile properties was also studied. The untreated and treated Napier fibres with 25% fibre loading were fabricated with epoxy resin by a cold press process. 7% fibre loading of CSM glass fibre was hybrid as the skin layer for 18% fibre loading of untreated Napier grass fibre. The tensile tests were conducted using Universal Testing Machine in accordance with ASTM D638. The tensile properties of the untreated Napier/epoxy composites were compared with treated Napier/epoxy and untreated Napier/CSM/epoxy composites. The results demonstrated that the tensile performance of untreated Napier fibre composites was significantly improved by both of the modification; alkali treatment and glass fibre hybridization. Napier grass fibres showed promising potentials to be used as reinforcement in the polymer based composites.

Keywords: Natural fibre composites; hybrid composites; tensile properties; compressed moulding.

1. Introduction

The increase in environmental awareness in the general public and community interest, the new environmental regulations and unsustainable consumption of non-renewable petroleum-based materials, have resulted in the promotion of environmentally cautious materials. Natural fibre is regarded as green materials which possess relatively excellent mechanical properties compared to synthetic fibre [1]. Natural fibres in the simple definition are fibres that are not artificial or human-made. They can be harvested from plants or animals [2]. Amongst other attractive properties of natural fibres which make them an attractive alternative to the traditional material are low cost, low density, excellent relative specific tensile strength, non-abrasive, non-irritation, reduced energy consumption thus lower the carbon footprint, less risk to health, renewability, recyclability and biodegradability [3]. Nowadays, natural fibres are used in various applications such as structural materials, particle boards, insulating materials, biomedical and for other biopolymers applications. Besides that, natural fibre composites also have been used in automotive industry for interior components such as door panels, instrument panels, armrests, headrests and seat shells [4]. The natural fibres extracted from banana, bamboo, coir, sisal, ramie, flax, jute, date palm and many more were at present used as reinforcements in polymer matrix composites. The effectiveness of the fibre reinforced composites is highly depended on fibre–matrix interface and its capacity to transfer the stress from the matrix to the reinforcing fibre. To obtain the optimal mechanical properties, several important things must be identified, such as the fraction of the fibres, fibre orientation, fibre aspect ratio, fibre length, fibre-matrix adhesion, and stress transfer at the interface [5]. Many research works were published to represent the importance of newly identified natural fibres with the optimised fibre length and weight or volume fractions. Usually, high fibre loading is needed to attain excellent properties of natural fibre polymer composites [6]. Moreover, Shinoj et al. [7] noticed that the rise in fibre content causes improvement in the tensile properties of the composites. Furthermore, Facca et al. [8] reported that the tensile strengths of 40-mesh hardwood fibres reinforced HDPE composites increased gradually, and up to a maximum of 25% of the fibre loading by volume before dropped back as the fibre loading further increased above 25%.

However, the major shortcoming of natural fibres as reinforcing materials for polymer based composites is their weak interfacial bonding due to the poor compatibility of the hydrophilic nature of the fibres with hydrophobic polyolefin polymers [9]. Thus it is essential that the fibre–matrix adhesion is to be improved and at the same time to reduce the moisture content of the fibre. This can be achieved by fibre surface modification via physical or chemical treatments. The use of different physical treatment methods (corona discharge, cold plasma, x-ray and UV bombardment) and chemical treatment methods (mercerization (alkali), grafting, acrylation, permanganate, acetylation, silane and peroxide) causes a reduction in moisture gain as well as changes in the fibre surface [10–14]. Alawar et al. [12] have studied the effect of different chemical treatment processes on date palm fibre surrounding the stems of date palm tree. From their work, it has been concluded that 5% NaOH treated date palm fibre shows optimum mechanical properties. Meanwhile, Prasad et al. [15] found that the alkali and acrylic acid treatment and the use of (maleic anhydride grafted LDPE) MA-g-LDPE as a compatibilizer improved the mechanical properties of the LDPE composites. Maheswari et al. [16] found that the mechanical (tensile and flexural) properties improved significantly with the addition of alkali combined with silane-treated fibre composites. Besides that, Maheswari et al. [17] reported that the tensile and thermal properties for the alkali and silane treated fibres were found to be improved by polymer coating

Meanwhile, Pothan et al. [18] suggested one of the modified solutions in most of the limitations is the hybridization of natural fibres with synthetic fibres. Natural and synthetic fibres can be used in the same matrix to produce hybrid composites that take full advantage of the best properties of the constituents, and thereby an optimal, superior but economical composite can be attained. In general, hybridization of synthetic–natural fibres hybrid composites is performed with the aims to reduce the use of synthetic fibres [19]. Mishra et al. [20] proved that addition of glass fibres in small amounts to the sisal and pineapple fibre–polyester composites increased the mechanical properties of the hybrid composites also the use of alkali treated fibres with glass fibres further enhanced the properties with reduced moisture absorption of the composites. Not only that, Noorunnisa et al. [21] found that the mechanical properties of sisal fibre reinforced polyester composites were improved by adding the carbon glass fibres in the sisal composites. Moreover, Idicula et al. [22] indicated that the natural fibre with glass allows a significantly better heat transportability for the composites. Carvalho [23] reported that the addition of 2% and 3% of short jute fibre to the brittle cement matrix changed its stress-strain behaviour significantly under direct tension as well as its load-deflection response under bending load. The energy capacity of the composites was increased expressively under both bending and direct tension load.

The main objective of this paper is to explore the potential of Napier grass fibre as reinforcing materials in polymer composites. Some work had been conducted toward the Napier grass fibre; Fartini et al. [24] studied compressive properties of Napier (*Pennisetum Purpureum*) filled polyester composites. Reddy et al. [25] analysed chemical composition, and structural characterization of Napier grass fibres while Phitsuwan et al. [26] evaluated structural features and enzymatic digestibility of Napier grass fibre treated with aqueous ammonia. Reddy et al. [27] also investigated the properties of Napier grass pulp as handmade paper sheets. This study involving the treatment of 5% NaOH solution to the Napier grass fibres. This article also involving the fabrication of 25% of fibre-loading of untreated and treated Napier grass fibre reinforced epoxy resin composites and the CSM glass fibre will be utilised as the skin layer of the untreated hybrid Napier fibre composites. The tensile test was then performed to study the tensile performance for the untreated Napier/epoxy composites, 5% NaOH treated Napier/epoxy composites and untreated Napier/CSM/epoxy composites.

2. Experimental Methods

2.1 Fibre Extraction

Napier grass stems were supplied from a local farm, located at Beseri, Perlis, Malaysia. The water retting process was utilised to extricate the fibre strand from the Napier grass stem. The stems were crushed by a mallet to allow the water to seep into the stems and accelerates the water absorption and thus decomposition. Figure 1 (a) shows the stems were submerged in a tank filled with running tap water for three to four week as by the time the Napier grass fibre strands become easy to separate from the grass internodes. The fibre was extricated physically and dried under the sun for few days to ensure for removal of moisture as shown in Figure 1 (b). Then the Napier grass fibre strands were dried in an air-circulating oven at 60ºC for 2 hours to ensure maximum moisture removal.

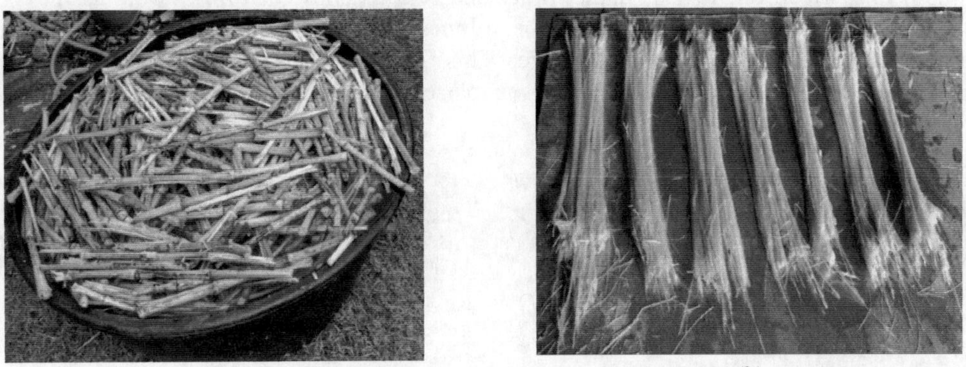

(a) (b)

Figure 1. (a) Water retting process, (b) Napier grass fibres

2.2 Alkaline treatment

The Napier grass fibres were treated with 5% of aqueous sodium hydroxide (NaOH) solutions at room temperature for 6 hours as shown in Figure 2 (a), maintaining a liquor ratio of 40:1, which is 40gram of solution for 1gram of fibre to remove the hemicellulose and other greasy materials. After that, the fibres were washed and rinsed using distilled water to remove the remaining traces of NaOH solutions. It also helps to reduce the stress on the fibres during treatment. Next, the fibres were dried under the sun for few days as shown in Figure 2 (b). Moreover, the fibres were oven dried at 60ºC for 1 hour.

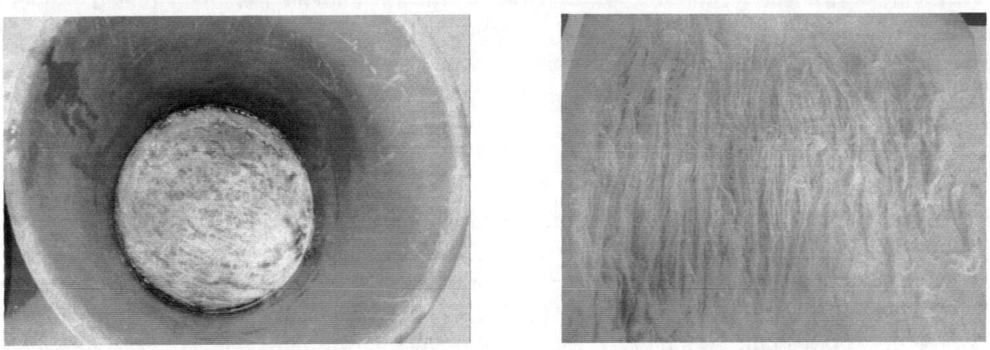

(a) (b)

Figure 2. (a) Soaking process, (b) treated Napier grass fibre

2.3 Composite Fabrication
The composite was produced by the cold press process as shown in Figure 3 (a). Firstly, the male and female section of the mould was cleaned and coated with the releasing agent for easy removal of the specimens. EpoxeAmite was mixed with a slow hardener in a weight ratio of 100g epoxy: 28.4g hardener and they were gently stirred, avoiding the formation of air bubbles, until the mix was entirely homogeneous. After that, some of the resin was poured first into the mould base, followed by placing the untreated Napier fibres randomly oriented into the mould and the remaining resin was poured slowly over the fibres. Initially, the fibres were oven dried for 30 minutes at 65°C. It is often necessary to preheat the natural fibres to reduce the moisture before processing the composites. A roller was used to expel entrapped air and to spread the resin all over the fibres uniformly. Then a load (5-8 kg) was placed on the top of the mould and compressed (0.6-2 MPa) to remove all excess resin and attain designated dimensions. The composite plates (300 x 300 x 3.5 mm) as shown in Figure 3 (b) were then left to cure at room temperature and removed from the mould after 24 h. Eventually, 240 g of epoxy resin and 26 g of Napier fibres were used to produce both untreated and treated composites plates of 25% fibre loading. However, 19 g of Napier fibres and 50 g of CSM glass fibres were used to produce the hybrid composites of 25% fibre loading. The volume fraction fibre for hybrid composite becomes 18% of Napier fibre and 7% of CSM glass fibre. This is because the CSM glass fibre was applied to the skin layer for untreated Napier fibres and was placed at the top and bottom of the Napier fibres during their hybrid composite plates fabrication.

(a)

(b)

Figure 3. (a) Cold press process, (b) Napier grass fibre reinforced epoxy composite plates

2.4 Tensile Testing
The composite plates were cut with a Dremel 4000 tool according to the ASTMD638 requirement as shown in Figure 4 (a). The tensile test was performed toward untreated Napier, treated Napier and hybrid untreated Napier/CSM composites with a crosshead speed 1mm/min using 100kN Universal Testing Machine. The ultimate strength, the elongation and the modulus of elasticity of the composites were calculated from the average of 5 samples for each composite. The extensometer was used to measure the strain as shown in Figure 4 (b).

3. Results and Discussion
The results of the tensile test conducted are presented in Figure 5 and 6 respectively. All the stress-strain curves for the three types of the Napier composites is similar to the brittle materials [28,29] The ultimate strength for the 25% of fibre loading of untreated Napier/epoxy composites was recorded about 22.78 MPa and modulus elasticity at 3.42 GPa. In contrast, Hameem et al. [30] reported the ultimate strength and modulus elasticity for 25% of fibre loading for untreated Napier fibre reinforced polyester composites are at 18.17 MPa and 2.00 GPa respectively. This indicates that the strength of the untreated Napier/epoxy is 25% stronger than the untreated Napier/polyester composites. The modulus elasticity of Napier/epoxy composites is almost 71% higher than Napier/polyester composites. This is due to the mechanical properties strength of epoxy resin is noticeably stronger

compared to polyester resin. Gopinath et al. [31] reported that the jute reinforced epoxy composite exhibited better mechanical properties than jute-polyester composites. Also, Satishkumar et al. [32] also found the maximum tensile strength and modulus of randomly oriented chopped snake grass fibre reinforced isophthalic polyester composites at 25% volume fraction fibre loading and reported that the volume fraction significantly influenced the mechanical properties of composites. Nirmal and Premkumar [33] concluded that the tensile strength of the composite increased up to 25% of the volume fraction of the fibres and further for the increase in the volume fraction of fibre the mechanical properties were decreased.

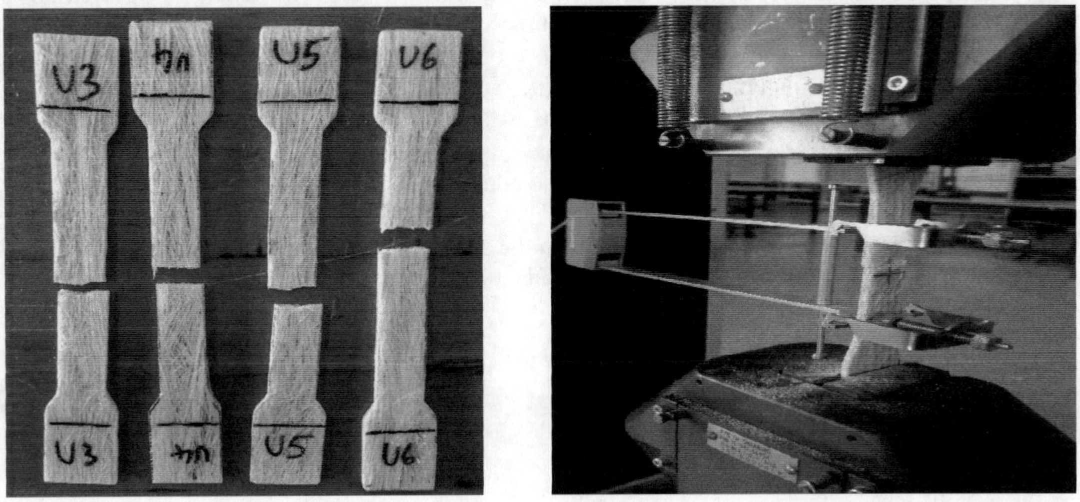

(a) (b)

Figure 4. (a) Tensile dog bone specimens, (b) tensile test

Meanwhile, the maximum ultimate strength and modulus elasticity for the treated Napier reinforced epoxy composites were computed at 40.06 MPa and 3.86 GPa respectively. The increase of strength and modulus between 5% NaOH Napier grass fibre and untreated Napier grass fibre composites were about 75 % and 12% respectively. Besides that, the elongation for treated Napier epoxy composites which at 0.0217 mm was two times higher than the elongation for untreated Napier epoxy composites which at 0.00995 mm. The effect of alkali treatment toward Napier fibre was very significant in term of strength and elongation compared to modulus elasticity. This is because the alkali treatment helps to extract the non-cellulosic components (hemicellulose, lignin, pectin, waxes and impurities) and some part of amorphous cellulose which is responsible for poor surface wetting and inefficient fibre-matrix interaction [34].Removal of these components increases the surface roughness and reduces its hydrophilic nature, thus generates a larger contact surface area and leads to better mechanical interlocking between Napier fibres and epoxy matrix which provides better mechanical properties. Amroune et al. [35]reported that the tensile tests show that use of post-processing NaOH-based chemical treatments towards date palm fruit branches (Phoenix dactylifera L.) allow a significant increase in stress at failure and Young's modulus with a profound influence on strain at failure. Also, Hossain et al. [36] concluded that fibres treated with 5% alkali solution yielded the best mechanical properties.

Not only that, the ultimate strength and modulus elasticity for untreated Napier fibres hybrid CSM glass fibres composites showed the highest result among these composites which at 44.71 MPa and 4.09 GPa respectively. The strength of hybrid Napier/CSM glass composite was 96% stronger than untreated Napier composites and only 11% stronger than 5% NaOH treated Napier fibre composites. While for the modulus elasticity of untreated Napier /CSM glass fibre composites showed about 19% higher than untreated Napier composites and only 6% higher than treated Napier composites. However, the elongation of untreated Napier/CSM glass fibre composites was 0.0311 mm which three times higher than untreated Napier composites and 43% higher than treated Napier composites. The elongation of Napier hybrid fibre showed biggest effect compare than the strength and modulus elasticity toward the untreated and treated Napier/epoxy composites probably due to a little ductility behaviour of the CSM glass fibres. This indicates that Napier grass fibre hybrid with CSM glass fibres composites gives significant effect toward the untreated Napier composites, but small impact toward the 5% NaOH treated Napier composites. Heckadka et al. [37] concluded that combination of three plain weave mats demonstrated better tensile, flexural, and interlaminar shear properties. However, the arrangement of three chopped strand mats resulted in inferior properties except for impact strength. However, Tshai et al. [38] reported that the inclusion of chopped strand E-glass fibres in Hybrid polylactide acid (PLA) composites reinforced with palm empty fruit bunch (EFB) improved the tensile and flexural performance of the hybrid composites. Besides that, hybrid composite with Kevlar as outer layers display better mechanical properties as compared to hybrid composites with kenaf as the outer layer was demonstrated by Yahaya et al. [39]. Vijaya et al. [40] also agreed that glass fibre is used to laminate the composite on the top and bottom layer, which improves the surface finish and adds strength to the composites. They also suggested that, generally, the % elongation of single fibre composite is less than that of the hybrid composite which indicates that the hybrid composite tolerates more strain than the single fibre composites before failure. Thus it can be concluded that the hybrid composite is more ductile than the neat composite.

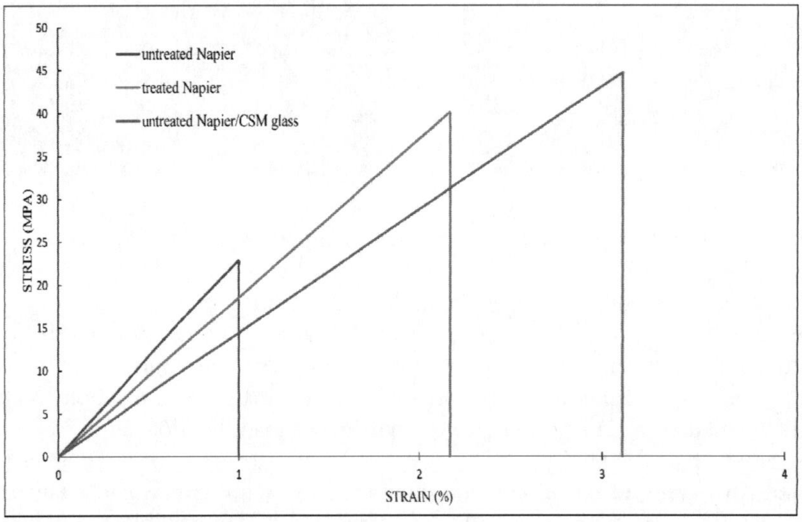

Figure 5. Comparison of the tensile strength of untreated Napier, treated Napier and hybrid untreated Napier/CSM glass fibre composites.

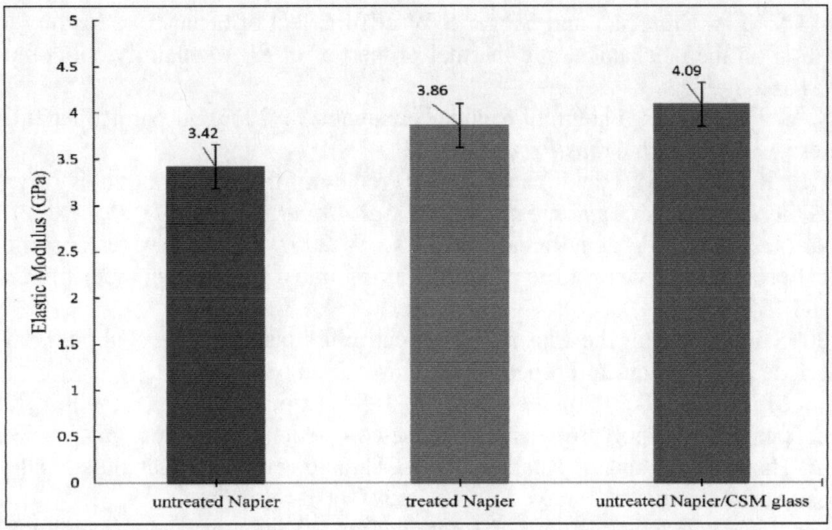

Figure 6: Comparison of the modulus elastic of untreated Napier, treated Napier and untreated Napier/CSM glass fibre composites

4. Conclusion

The mechanical properties of untreated Napier fibre, 5% NaOH treated Napier fibre and untreated Napier hybrid with CSM glass fibre reinforced epoxy composites were investigated through the tensile test. The result showed that Napier fibre with 25% of fibre loading exhibit efficient load transfer between the fibre and the matrix. The modified Napier grass fibres reinforced epoxy composites by both alkali treatment, and E-glass fibre hybridization gives a significant increase in ultimate strength, elastic modulus and fibre elongation behaviour of Napier fibre reinforced epoxy composites. The results of tensile strength and the modulus elasticity between Napier hybrid CSM glass fibres and 5% NaOH treated Napier fibre composites did not exhibit much differences gap. However, the modification of Napier grass fibres between hybridization with CSM glass fibres as the skin layer for Napier fibre and treatment of 5% NaOH concentration demonstrated a significant distinction regarding fibre elongation. Napier grass fibres appear to have bright future as an alternative reinforcement for fibre reinforced polymer composites.

5. Acknowledgement

The authors sincerely appreciate the financial support from the Ministry of Education, Malaysia through the Research Acculturation Grant Scheme (RAGS) 2014 under the project number 9018-00072. Also thanks to School of Mechatronic Engineering, Universiti Malaysia Perlis for the use of its facilities.

6. References

[1] May-Pat A, Valadez-González A and Herrera-Franco P J 2013 Effect of fiber surface treatments on the essential work of fracture of HDPE-continuous henequen fiber-reinforced composites *Polym. Test.* **32** 1114–22

[2] Ticoalu a, Aravinthan T and Cardona F 2010 A review of current development in natural fiber composites for structural and infrastructure applications *South. Reg. Eng. Conf.* 1–5

[3] Malkapuram R, Kumar V and Yuvraj Singh Negi 2009 Recent Development in Natural Fiber Reinforced Polypropylene Composites *J. Reinf. Plast. Compos.* **28** 1169–89

[4] Satyanarayana K G, Arizaga G G C and Wypych F 2009 Biodegradable composites based on lignocellulosic fibers-An overview *Prog. Polym. Sci.* **34** 982–1021

[5] Hepworth D G, Bruce D M, Vincent J F V and Jeronimidis G 2000 Manufacture and mechanical testing of thermosetting natural fibre composites *J. Mater. Sci.* **35** 293–8

[6] Tawakkal I S M A, Cran M J and Bigger S W 2014 Effect of kenaf fibre loading and thymol concentration on the mechanical and thermal properties of PLA/kenaf/thymol composites *Ind. Crops Prod.* **61** 74–83
[7] Shinoj S, Visvanathan R, Panigrahi S and Kochubabu M 2011 Oil palm fiber (OPF) and its composites: A review *Ind. Crops Prod.* **33** 7–22
[8] Facca A G, Kortschot M T and Yan N 2006 Predicting the elastic modulus of natural fibre reinforced thermoplastics *Compos. Part A Appl. Sci. Manuf.* **37** 1660–71
[9] Dhakal H N, Zhang Z Y and Richardson M O W 2007 Effect of water absorption on the mechanical properties of hemp fibre reinforced unsaturated polyester composites *Compos. Sci. Technol.* **67** 1674–83
[10] Sun D 2005 Investigating the Plasma Modification of Natural Fiber Fabrics-The Effect on Fabric Surface and Mechanical Properties *Text. Res. J.* **75** 639–44
[11] Belgacem M N, Bataille P and Sapieha S 1994 Effect of corona modification on the mechanical properties of polypropylene/cellulose composites *J. Appl. Polym. Sci.* **53** 379–85
[12] Alawar A, Hamed A M and Al-Kaabi K 2009 Characterization of treated date palm tree fiber as composite reinforcement *Compos. Part B Eng.* **40** 601–6
[13] Mwaikambo L Y and Ansell M P 2002 Chemical modification of hemp, sisal, jute, and kapok fibers by alkalization *J. Appl. Polym. Sci.* **84** 2222–34
[14] Xie Y, Hill C A S, Xiao Z, Militz H and Mai C 2010 Silane coupling agents used for natural fiber/polymer composites: A review *Compos. Part A Appl. Sci. Manuf.* **41** 806–19
[15] Prasad N, Agarwal V K and Sinha S 2016 Banana fiber reinforced low-density polyethylene composites: effect of chemical treatment and compatibilizer addition *Iran. Polym. J.* **25** 229–41
[16] Maheswari C U, Reddy K O, Muzenda E, Shukla M and Rajulu A V 2013 Mechanical Properties and Chemical Resistance of Short Tamarind Fiber/Unsaturated Polyester Composites: Influence of Fiber Modification and Fiber Content *Int. J. Polym. Anal. Charact.* **18** 520–33
[17] Maheswari C U, Reddy K O, Muzenda E and Rajulu A V 2012 Tensile and Thermal Properties of Polycarbonate-Coated Tamarind Fruit Fibers *Int. J. Polym. Anal. Charact.* **17** 578–89
[18] Pothan L A, Cherian B M, Anandakutty B and Thomas S 2007 Effect of Layering Pattern on the Water Absorption Behavior of Banana Glass Hybrid Composites *J. Appl. Polym. Sci.* **105** 2540–8
[19] Joshi S V., Drzal L T, Mohanty A K and Arora S 2004 Are natural fiber composites environmentally superior to glass fiber reinforced composites? *Compos. Part A Appl. Sci. Manuf.* **35** 371–6
[20] Mishra S, Mohanty A K, Drzal L T, Misra M, Parija S, Nayak S K and Tripathy S S 2003 Studies on mechanical performance of biofibre/glass reinforced polyester hybrid composites *Compos. Sci. Technol.* **63** 1377–85
[21] Khanam P N, Khalil H P S A, Jawaid M, Reddy G R, Narayana C S and Naidu S V 2010 Sisal/Carbon Fibre Reinforced Hybrid Composites: Tensile, Flexural and Chemical Resistance Properties *J. Polym. Environ.* **18** 727–33
[22] Idicula M, Boudenne A, Umadevi L, Ibos L, Candau Y and Thomas S 2006 Thermophysical properties of natural fibre reinforced polyester composites *Compos. Sci. Technol.* **66** 2719–25
[23] Carvalho L V, Evangelista A C J and Toledo Filho R D 2012 Mechanical Behavior of Jute Fiber-Cement Based Composites *Key Eng. Mater.* **517** 469–76
[24] Fartini M S, Majid M S A, Ridzuan M J M, Amin N A M, Gibson A G, Compressive properties of Napier (*Pennisetum Purpureum*) filled polyester composites *Plast. Rubber Compos. Macromol. Eng.* **45** 136–46
[25] Reddy K O, Maheswari C U, Shukla M and Rajulu A V 2012 Chemical composition and structural characterization of Napier grass fi bers *Mater. Lett.* **67** 35–8
[26] Phitsuwan P, Charupongrat S, Klednark R and Ratanakhanokchai K 2015 Structural features and enzymatic digestibility of Napier grass fibre treated with aqueous ammonia *J. Ind. Eng.*

IOP Conf. Series: Journal of Physics: Conf. Series **908** (2017) 012013 doi:10.1088/1742-6596/908/1/012013

Chem. **32** 360–4

[27] Reddy K O, Maheswari C U, Shukla M and Muzenda E 2014 Preparation , Chemical Composition , Characterization , and Properties of Napier Grass Paper Sheets Properties of Napier Grass Paper Sheets *Sep. Sci. Technol.* **49** 1527–34

[28] Satyanarayana K G, Sukumaran K, Mukherjee P S, Pavithran C and Pillai S G K 1990 Natural fibre-polymer composites *Cem. Concr. Compos.* **12** 117–36

[29] Sreenivasan V S, Ravindran D, Manikandan V and Narayanasamy R 2011 Mechanical properties of randomly oriented short Sansevieria cylindrica fibre/polyester composites *Mater. Des.* **32** 2444–55

[30] Haameem M J A, Majid M S A, Afendi M, Marzuki H F A, Fahmi I and Gibson A G 2016 Mechanical properties of Napier grass fibre / polyester composites *Compos. Struct.* **136** 1–10

[31] Gopinath A, M S K and Elayaperumal A 2014 Experimental Investigations on Mechanical Properties Of Jute Fiber Reinforced Composites with Polyester and Epoxy Resin Matrices *Procedia Eng.* **97** 2052–63

[32] Sathishkumar T P, Navaneethakrishnan P and Shankar S 2012 Tensile and flexural properties of snake grass natural fiber reinforced isophthallic polyester composites *Compos. Sci. Technol.* **72** 1183–90

[33] Nirmal K J and Premkumar D 2015 Assessing of Mechanical Properties of Natural Fiber Reinforced Polymer Matrix Hybrid Composites *Appl. Mech. Mater.* **766–767** 199–204

[34] Venkateshwaran N, Elaya Perumal A and Arunsundaranayagam D 2013 Fiber surface treatment and its effect on mechanical and visco-elastic behaviour of banana/epoxy composite *Mater. Des.* **47** 151–9

[35] Amroune S, Bezazi A, Belaadi A, Zhu C, Scarpa F, Rahatekar S and Imad A 2015 Tensile mechanical properties and surface chemical sensitivity of technical fibres from date palm fruit branches (Phoenix dactylifera L.) *Compos. Part A Appl. Sci. Manuf.* **71** 98–106

[36] Hossain M K, Karim M R, Chowdhury M R, Imam M A, Hosur M, Jeelani S and Farag R 2014 Comparative mechanical and thermal study of chemically treated and untreated single sugarcane fiber bundle *Ind. Crops Prod.* **58** 78–90

[37] Heckadka S S, Nayak S Y, Narang K and Pant K V 2015 Chopped Strand / Plain Weave E-Glass as Reinforcement in Vacuum Bagged Epoxy Composites *J. Mater.* **2015**

[38] Tshai K Y, Chai A B, Kong I, Hoque M E and Tshai K H 2014 Hybrid Fibre Polylactide Acid Composite with Empty Fruit Bunch: Chopped Glass Strands *J. Compos.* **2014**

[39] Yahaya R, Sapuan S M, Jawaid M, Leman Z and Zainudin E S 2015 Effect of layering sequence and chemical treatment on the mechanical properties of woven kenaf-aramid hybrid laminated composites *Mater. Des.* **67** 173–9

[40] Vijaya Ramnath B, Junaid Kokan S, Niranjan Raja R, Sathyanarayanan R, Elanchezhian C, Rajendra Prasad A and Manickavasagam V M 2013 Evaluation of mechanical properties of abaca–jute–glass fibre reinforced epoxy composite *Mater. Des.* **51** 357–66

Effects of fibre loading and moisture absorption on the tensile properties of hybrid Napier/glass/epoxy composites

M Z Zikri, M J M Ridzuan, M S Abdul Majid, M Afendi and K S Basaruddin

Mechanical Engineering Programme, School of Mechatronic Engineering, Universiti Malaysia Perlis, Pauh Putra Campus, 02600 Arau, Perlis, Malaysia

ridzuanjamir@unimap.edu.my

Abstract. The primary objective of this study was to investigate the effect of moisture absorption on the mechanical degradation of hybrid Napier/glass-epoxy composites. The hybrid Napier/glass-epoxy composites plates were produced by the vacuum infusion method using epoxy resin as a matrix. The hybrid composite specimens were tested after following 50 h of water immersion. The moisture content decreased as the glass fibre content increased. The wet and dry hybrid composite samples were subjected to tensile tests. The incorporation of the glass fibre into the Napier grass fibre-epoxy composites enhanced their tensile strength and tensile modulus. The tensile strength and tensile modulus of the hybrid Napier/glass-epoxy composites (24/6–70 vol%) were 43 MPa and 3.2 GPa, respectively. However, the tensile strength and tensile modulus properties highly degraded under wet conditions.

Keywords: Hybrid composites; natural fibres; tensile properties; moisture absorption; fibre loading.

1. Introduction

Greater than before pressure from ecological campaigners, conservation of natural resources, and emergence of stringency of laws passed have results in promotion of natural materials with an attention on renewable materials. Composite manufacturers have been searching for plant based fibre such as flax, hemp, jute, sisal, kenaf, banana as alternative substitute for reinforcing materials. Natural fibre-reinforced polymer composites are preferred today as they are harvested from renewable resources, environmentally friendly, biodegradable, low density, lightweight, and recyclable compared to conventional synthetic fibres [1–3]. However, the nature of natural fibre itself which is hydrophilic, permits the composite to absorb a large quantity of moisture from the environment than that of the synthetic composites [4]. The high moisture in composites is known to cause dimensional variation and greatly affects the mechanical properties of the composites. Moisture allowed the formation of voids and inversely affected the fibre–matrix bonding which eventually result in a reduction of their mechanical properties.

Hybrid composites are fabricated by the joining of at least two sorts of fibre into a single matrix. The performance of these hybrid composites highly influenced by types the matrix used, length and shape of individual fibres, fibre–matrix interface bonding, and volume fraction of the natural/synthetic fibres[5]. There were few investigations on the previous studies on the impact of changing the quantity of fibre loading on the mechanical properties of hybrid composites comprising of natural fibres and glass fibre reported which amongst them are sisal/glass [6], kenaf/glass [7], okra/glass[8] and jute/glass [9] hybrid composites. Mishra et al. investigated the water take-up of hybrid composites is low compared to un-hybridized composites [6].

Mechanical degradation of the polymeric composites due to moisture absorption have been studied by several researchers [6,8,9]. The length of the fibres and the hybridization ratio utilized for the reinforcement are highly significant towards the mechanical properties and moisture absorption of and polymeric composites. Phan Braga and Magalhaes have studied that jute/glass composites, which consisted of a larger quantity of jute fibre have high moisture content than those that have a larger quantity of glass [9]. Recently, research conducted on Napier grass fibres suggested possible alternatives as reinforcement material in polymer composites [1,2,10].

The current work discusses the experimental investigation conducted to evaluate the mechanical degradation of hybrid Napier/glass-epoxy composites as results of exposure to moisture.

2. Materials and experimental procedure

2.1 Materials

The Napier grass fibre was acquired from a local plantation near Bukit Kayu Hitam, Kedah, Malaysia. In producing, hybrid composite, woven E-glass fibres were provided by a local provider, with the dimensions of 300 × 300 × 15 mm for each layer. EpoxAmite 100 series resin mixed with a hardener to a proportion of 3:1 is used for the matrix material.

2.2 Composite fabrication

The hybrid Napier/glass-epoxy composites were fabricated through in house-built vacuum infusion machine. The reinforcement plies were laminated over a glass mould, having dimensions of 300 × 300 × (3.2 ± 0.4) mm^3. The resin was infused into the lamination plies through near vacuum in-mould pressure (~10 mbar) during infusion process. Afterwards, the specimens were left for 12h at room temperature (25 °C) for curing process. Next, post-curing was proceeded in an oven for 2h at temperature of 80 °C, under air circulation [11]. Five different types of composite plates were produced at different volume fraction of the Napier and glass fibres. Their volume fraction contents are showed in Table 1. Specimens were then prepared to specific size and shape according to the standard for testing using a Dremel 4000 tool.

Table 1: Percentage (by volume) of the constituents of the hybrid composite samples

Composite sample	Epoxy resin (% volume)	Napier grass fibre content (% volume)	Glass fibre content (% volume)
S1	70	30	0
S2	70	24	6
S3	70	18	12
S4	70	12	18
S5	70	6	24

IOP Conf. Series: Journal of Physics: Conf. Series **908** (2017) 012014 doi:10.1088/1742-6596/908/1/012014

2.3 Moisture absorption test
Specimens of 76.2 × 25.4 × (3.2 ± 0.4) mm dimension were immersed in distilled water at room temperature, as per the standards provided by ASTM D570 [12]. The change of mass of the samples was recorded using an analytical balance at intervals of 10 h. The water take-up by the samples was measured in period, up to 50h immersion period. The moisture uptake, expressed as the percentage of weight gain, $\Delta\boldsymbol{M(t)}$, is given as:

$$\Delta\boldsymbol{M(t)} = \frac{\boldsymbol{M_t - M_0}}{\boldsymbol{M_0}} \times \mathbf{100} \quad \mathbf{(1)}$$

Where $\mathbf{M_0}$ and $\mathbf{M_t}$ represent the mass of the dry sample and the mass after time (wet), t, respectively. The relation between the moisture uptakes by the specimen as a function of immersion period then established. Next, the procedure was repeated for the dog-bone shaped specimens before proceeding with tensile testing. Afterwards, tensile test was conducted in order to determine the residual strength and elastic modulus following moisture exposure. The tests were conducted for five replicate for each type of the composite. The average results were recorded and reported.

2.4 Tensile testing
Dog-bone composite samples with dimensions of 165 × 19 × (3.2 ± 0.4) mm^3 prepared from the composite plates were tested in accordance with ASTM D638-10 [13]. The specimens were tested to failure under tension using a universal testing machine (Instron) with a load cell of 100 kN at a cross-head speed of 1 mm/min. An extensometer was attached to a 50 mm gauge length for strain measurement. The unidirectional *Napier grass* fibres of the samples were parallel to the loading direction throughout testing period. The tests were conducted repeated for five replicates for each type of composites and the average result was recorded. The tensile strength and tensile modulus of the specimens were then computed.

3. Results and discussion

3.1 Moisture absorption
Figure 1 shows the % of moisture absorption of five composite samples as a function of the immersion time. The % of moisture absorption of the composites was determined by computing the differences between the weight of the wet sample following immersion in water and that of the dry sample, using Eq. 2. Based on Figure 1, it can be observed that the moisture absorption increases along with the extended immersion time. Previous reports stated the similar results with different types of natural composites [14,15].Besides that, it can be clearly seen moisture absorption rate reduced as the glass fibre volume content increased. The results show that sample S1 absorbs the greatest moisture content of 7% after being soaked in water for a 50 h period. Increasing the glass fibre content in composite sample S2 showed that the moisture content of the sample S2 was half that of the Napier-epoxy composite sample(S1) when pursuing 50 h immersion period.

The moisture content obviously decreased as the glass fibre content of the samples increased: the S3 and S4 samples show the moisture contents of 3.78 and 3.06%, individually. Finally, the moisture content of samples S5 was 1.51% indicates a reduction of over 78% after following a 50h period of immersion time. The results also were proved from the similar observation reported by Salleh [15]. This is because the natures of the Napier grass itself which is hydrophilic, enabling the formation of microvoids within the resin matrix [16]. The observation differed from the hybrid composite which is the fibres are arranged in a closed packed manner, where the water-impermeable glass fibres act as barriers to avert contact between the water and Napier fibre [15]. Therefore, Napier-epoxy composite allowed high moisture absorption than of Napier/glass-epoxy composites. Other than that, a fibre/matrix interface of a composite that contains only natural fibres would be fully damaged because of moisture absorption effects [17]. This differs from the natural/synthetic hybrid composites, which are only less adversely affected due to their resistance towards water uptake [18]. In addition, the quantity of moisture absorbed by the epoxy is practically negligible because of its hydrophobic nature [16].

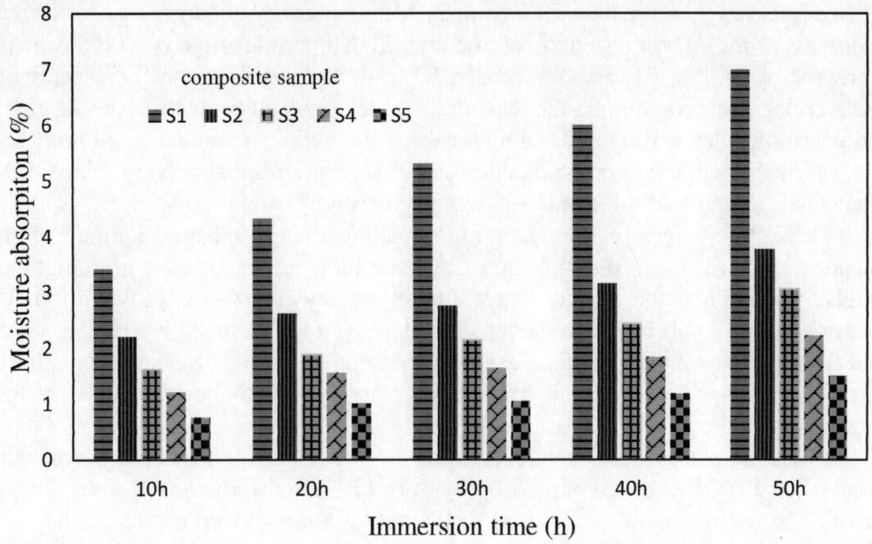

Figure 1. Moisture absorption responses of hybrid Napier/glass fibre-reinforced composites

3.2 Effect of fibre loading and moisture absorption on the tensile behaviour of hybrid Napier/glass-epoxy composites

Figure 2 illustrates the tensile strength performance of the dry and wet (immersed in water) hybrid Napier/glass-epoxy composites. From the figure, it can be clearly observed that the tensile strength of the Napier fibre/epoxy composite is greatly enhanced with the higher presence of the glass fibre. The tensile strengths of composite samples S1, S2, S3, S4 and S5 were 21, 43, 73, 109 and 131 MPa, respectively. The results indicates that 6 vol % content of glass fibre hybridization has improved the tensile strength of the composites by 105%.

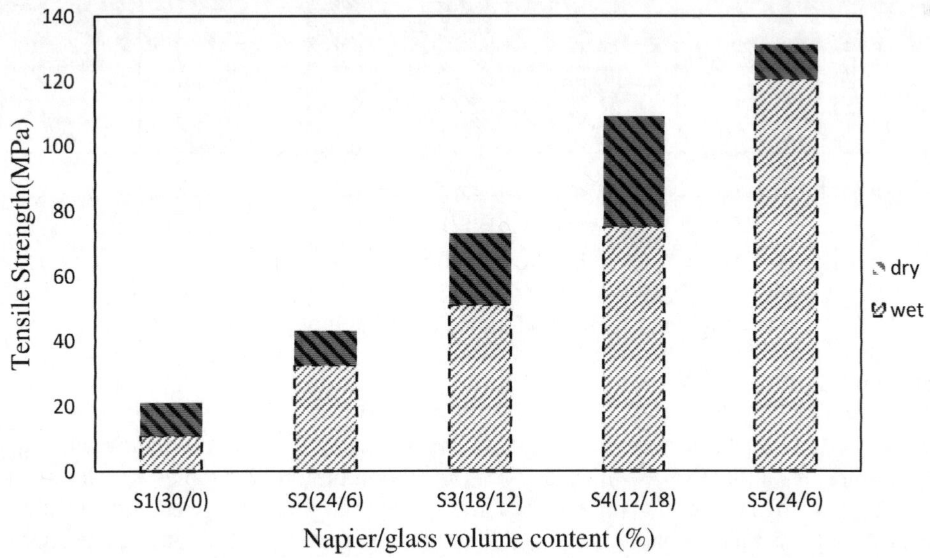

Figure 2. Tensile strength of dry and wet hybrid Napier/glass fibre-reinforced composites

Following 50 h of immersion period, the extensive matrix/interface cracking correlated with the moisture absorption of the composites has clearly influenced the tensile strength of the hybrid Napier/glass epoxy composites under axial loading. The tensile strength of the Napier-epoxy composites (S1) decreased by more than 50% from 21 MPa under dry conditions to 11 MPa under wet conditions. Moreover, the tensile strength of the hybrid Napier/glass-epoxy composite under wet condition decreased by 24, 30, 31, and 8% for the S2, S3, S4, and S5 samples, respectively. It was suspected that, under wet conditions the absorbed water molecules within the composites had weakened the intermolecular hydrogen bonding between the cellulose molecules in the fibre and the water molecules [16]. Hence, the interfacial adhesion between the fibre and the matrix was weakened. As a result, the tensile strength of the composite was decreased.

Figure 3 shows the effect of the glass fibre loading on the tensile modulus of the hybrid Napier/glass-epoxy composites. From the data obtained indicate that the reinforcing glass fibres resulted in higher stiffness of the hybrid composites either under dry or wet condition. Under dry conditions, it can be clearly observed that the tensile modulus of the *Napier fibre*-reinforced composite (S1) increased from 1.7 to 3.2 GPa with a glass fibre content of 6 vol% (S2) indicates an increase of 88%. Besides, the tensile modulus of the hybrid Napier/glass-epoxy composite samples of S3, S4, and S5 also increased by 5.3, 7.6, and 9.1 GPa, respectively.

Under the wet condition, the tensile modulus of the Napier fibre-epoxy composites (S1) decreased from 1.7 to 1.0 GPa, representing drop by over 45% after immersed in water for 50 h of the period. Moreover, the tensile modulus of the hybrid Napier/glass-epoxy composites sample of S2, S3, S4, and S5 decreased by 19, 23, 8, and 8 %, respectively under wet condition. Thus, the increases in the synthetic fibre content have found its capability to reduce the degradation due to moisture absorption. This shows that hybridization acts a significance impact in reducing the mechanical degradation of the hybrid fibre composites.

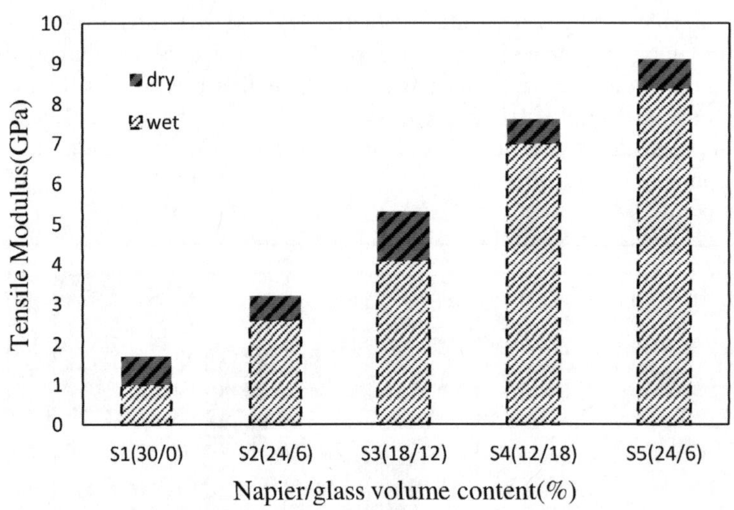

Figure 3. Tensile Modulus of dry and wet samples

4. Conclusions

The effects of fibre loading and moisture absorption on the mechanical degradation of hybrid Napier/glass/epoxy composites were investigated. The moisture absorption of the hybrid Napier/glass-epoxy composites found increased with the water-immersion period of samples and diminished as the glass fibre content in the samples increased. The tensile strength of the wet samples was recorded lower than the dry sample. Finally, the tensile strength increased as the glass fibre content of the composite increased and has a similar response towards their tensile modulus.

5. Acknowledgement
The authors are thanking Universiti Malaysia Perlis and the Ministry of Education, Malaysia, for providing financial support (FRGS No: 9003-00515) and the access for using the facilities during the course of the research conducted.

References

[1] Ridzuan M J M, Abdul Majid M S, Afendi M, Kanafiah S N A, Zahri J M and Gibson A G 2016 Characterisation of natural cellulosic fibre from Pennisetum purpureum stem as potential reinforcement of polymer composites *Mater. Des.* **89** 839–47

[2] Ridzuan M J M, Abdul Majid M S, Afendi M, Azduwin K, Kanafiah S N A and Dan-mallam Y 2015 The Effects of the Alkaline Treatment's Soaking Exposure on the Tensile Strength of Napier Fibre *Procedia Manuf.* **2** 353–8

[3] Ridzuan M J M, Abdul Majid M S, Afendi M, Kanafiah S N A and Nuriman M B M 2015 Effects of alkaline concentrations on the tensile properties of Napier grass fibre *Appl. Mech. Mater.* **786** 23–7

[4] Ridzuan, M.J.M.;Abdul Majid, M.S.: Afendi, M.; Rahman, Y.M.N. ; Azduwin K 2016 Thermal, composition and morphological properties of untreated and alkali-treated Napier fibres *J. Eng. Appl. Sci.* **11** 2485–9

[5] Ridzuan M J M, Majid M S A, Afendi M, Mazlee M N and Gibson A G 2016 Thermal behaviour and dynamic mechanical analysis of Pennisetum purpureum/glass-reinforced epoxy hybrid composites *Compos. Struct.* **152** 850–9

[6] Mishra S, Mohanty A K, Drzal L T, Misra M, Parija S, Nayak S K and Tripathy S S 2003 Studies on mechanical performance of biofibre/glass reinforced polyester hybrid composites *Compos. Sci. Technol.* **63** 1377–85

[7] Atiqah A, Maleque M A, Jawaid M and Iqbal M 2014 Development of kenaf-glass reinforced unsaturated polyester hybrid composite for structural applications *Compos. Part B Eng.* **56** 68–73

[8] Sule U 2014 Studies on the Properties of Short Okra/Glass Fibers Reinforcedepoxy Hybrid Composites *Int. J. Sci. Technoledge* **2** 260–5

[9] Braga R A and Magalhaes Jr. P A A 2015 Analysis of the mechanical and thermal properties of jute and glass fiber as reinforcement epoxy hybrid composites *Mater. Sci. Eng. C* **56** 269–73

[10] Obi Reddy K, Uma Maheswari C, Shukla M, Song J I and Varada Rajulu A 2013 Tensile and structural characterization of alkali treated Borassus fruit fine fibers *Compos. Part B Eng.* **44** 433–8

[11] Júnior J H S A, Júnior H L O, Amico S C and Amado F D R 2012 Study of hybrid intralaminate curaua/glass composites *Mater. Des.* **42** 111–7

[12] ASTM D570 1998 Standard Test Method for Water Absorption of Plastics *ASTM Int.* 25–8

[13] ASTM D638 2010 Standard test method for tensile properties of plastics *ASTM Int.* 1–16

[14] Chow C P L, Xing X S and Li R K Y 2007 Moisture absorption studies of sisal fibre reinforced polypropylene composites *Compos. Sci. Technol.* **67** 306–13

[15] Doan T T L, Brodowsky H and Mäder E 2007 Jute fibre/polypropylene composites II. Thermal, hydrothermal and dynamic mechanical behaviour *Compos. Sci. Technol.* **67** 2707–14

[16] Girisha C and Srinivas G R 2012 Sisal/Coconut Coir Natural Fibers – Epoxy Composites : Water Absorption and Mechanical Properties *Int. J. Eng. Innov. Technol.* **2** 166–70

Water absorption behaviour of hybrid interwoven cellulosic fibre composites

A B Maslinda[1,2], M S Abdul Majid[1], M J M Ridzuan[1] and AR.A. Syayuthi[1]

[1] School of Mechatronic Engineering, Universiti Malaysia Perlis, Pauh Putra Campus, 02600, Arau, Perlis, Malaysia
[2] Faculty of Engineering Technology, Universiti Malaysia Perlis, Uniciti Alam Campus, 02100, Padang Besar, Perlis, Malaysia

shukry@unimap.edu.my

Abstract. The present paper investigated the water absorption behaviour of hybrid interwoven cellulosic fibre composites. Hybrid composites consisting of interwoven kenaf/jute and kenaf/hemp yarns were prepared by an infusion manufacturing technique that used epoxy as the polymer matrix. Water absorption test was conducted as elucidated in ASTM D570 standard by immersing the composite samples in tap water at room temperature until reaching their water content saturation point. For each composite type, average from five samples was recorded and the percentage of water uptake against the square root of time was plotted. As the effect of hybridization, the water uptake, diffusion and permeability coefficient of the hybrid composites were lesser than the individual woven composites.

Keywords: Hybrid composites; natural fibres; water absorption; interwoven; cellulosic.

1. Introduction

Exploration on the natural resources fibres as a substitution to the petroleum-based fibres was rapid growth in the last few decades and nowadays, natural fibre reinforced polymer composites are widely used in the automotive industry, sports equipment and marine structure [1]. In comparison with the synthetic fibre, natural fibre possessed high specific strength, readily available at low cost, renewable and less abrasive to processing equipment and these features correspond well with the consumer demands and environmental sustainability. Cellulose based natural fibres that are widely studied and applied in composites engineering applications are flax, hemp, jute, ramie, kenaf and sisal [2].

The hybrid composite consists of at least two types of fibre in a single matrix system. Interest in hybridization started when there was increasing demand for a lighter material with improved toughness, especially for aircraft application. The early focus of hybridization is reducing the cost by combining carbon fibre with cheaper glass fibre. Through hybridization, the mechanical performance of the composites was improved which attributed by the changes in applied load distribution and failure development. Now, hybridization has been widely explored such as pseudo-ductility, ductile fibre and natural fibre hybrids composites [3].

Common types of reinforcement form in fibre reinforced composites are fibre, yarn and fabric. The arrangement of the reinforcement or layered affects the final performance of the composites and fabric structure, governs the properties rather than the others. Fabric can be classified as woven, knitted or non-woven. Woven fabric is interlacing between two sets of yarns along the longitudinal (warp) and transverse (weft) direction. Interloping between yarns at weft or warp direction defined the knitted fabric, and for a non-woven fabric, it is termed as the bonding of raw material (fibre) using mechanical, chemical or solvent [4].

In particular, all polymer composites absorb moisture, and the water molecules can act as a plasticiser by influencing the fibres, the matrix and the interface simultaneously, thus creating regions of poor transfer efficiency, which results in a reduction of mechanical properties. It is now well established that composite materials are sensitive to humidity through absorption of water leading to differential swelling between the fibres and the matrix. This situation is even more complicated in the case of natural fibres, which are known to exhibit a poor resistance to water absorption, due to their hydrophilic nature that is related [5]. As an extension to the research involving hybrid and woven structure, this research's primary focus is investigating the water absorption behaviour of hybrid composite made from interwoven kenaf/jute and kenaf/hemp yarns to observe their responses when subjected to moisture and at the same time predict their capability for outdoor applications.

2. Experimental procedure

2.1 Materials

As shown in figure 1, kenaf, jute and hemp fibre in a yarn form were used for the preparation of the woven fabric. All the fibre yarns were supplied by Innovative Pultrusion Sdn. Bhd. The EpoxAmite 100 series resin, provided by Castmech Technologies Sdn. Bhd was selected as the polymer matrix, mixed with a hardener in a ratio of 3:1 to form the binder for the composites preparation.

Figure 1. Fibre yarn; (a) jute, (b) kenaf and (c) hemp.

2.2 Woven fabric production

The woven fabric was fabricated using the manually constructed wooden frame. The wooden frame consists of nails at both of its side that served as the warp yarn guider. Following the plain weave pattern, the weft yarns were passed over and underneath the warp yarns that have been arranged earlier on the wooden frame to complete the woven fabric. The interlacing of kenaf yarns as the warp direction fibre with jute and hemp yarns as the weft direction fibre in the construction of interwoven kenaf/jute and kenaf/hemp hybrid composites is illustrated in figure 2.

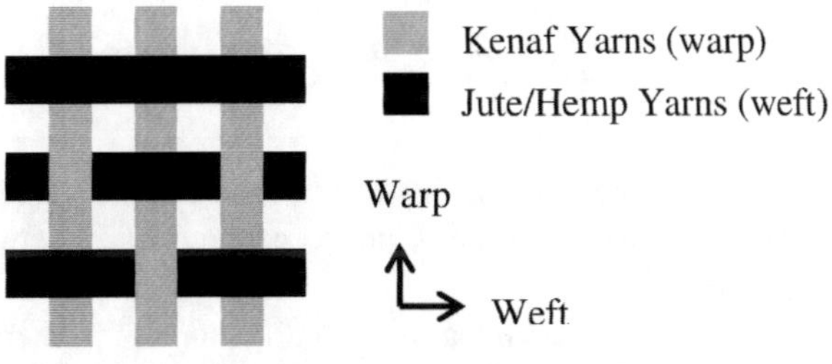

Figure 2. Warp and weft direction fibre of interwoven hybrid composites.

2.3 Composite fabrication
For comparison, five types of composites were prepared, and a symbol representing the composite types is as listed in Table 1. All the composites, with fibre weight content of 30% were prepared via vacuum infusion process using epoxy as the polymer matrix. Figure 3 shows the infused composites after moulding and curing for 24 hours at room temperature.

Table 1. Symbols used for representing the different types of composite prepared.

Type of Composite	Symbols
Woven Kenaf	KK
Woven Jute	JJ
Woven Hemp	HH
Interwoven Kenaf/Jute	KJ
Interwoven Kenaf/Hemp	KH

Figure 3. Infused woven composite.

2.4 Water absorption test
Water absorption test was carried as per instructed in the ASTM D570 standard. Mass of the dried samples, *Mo* were taken after oven dried at 100 °C for 10 minutes and left to cool at room temperature for 24 hours. Later, specimens were submerged in a container filled with tap water before being placed in the humidity chamber at 25 °C. To monitor the weight gain due to the water absorption, the specimens were withdrawn from the water, wiped dry to remove the surface moisture, and then weighted using a high accuracy 4-digit analytical balance. Samples are considered reached their equilibrium when the daily weight gain of the samples was less than 0.01%. For each type of the composites, five specimens were used, and the average result was recorded. Moisture content percentage, $\Delta M(t)$ was calculated using equation (1):

$$\Delta M(t) = \frac{M_t - M_0}{M_0} \times 100 \quad (1)$$

where M_o is the weight of the specimen before immersion and M_t representing the weight of the specimen after immersion at a specific time. The percentage of moisture absorption was plotted against the square root of time ($\sqrt{t}$, hours).

Assuming the absorption process is linear at the early stage of immersion; times are taken at the beginning of absorption process so that the weight change is expected to vary linearly with the square root of time. The kinetic parameter, diffusion coefficient, D (m^2/s) is computed from the slope of moisture content versus the square root of time by equation (2):

$$D = \pi\left(\frac{kh}{4m_\infty}\right)^2 \quad (2)$$

Where h is the thickness of the specimen, m_∞ is the maximum water uptake, and k is the initial slope of the linear portion of a plot of $M(t)$ versus $t^{1/2}$ curve.

The permeability of water molecules through the composite sample depends on the sorption of water by the fibre. Therefore, the sorption coefficients that are related to the equilibrium sorption of the principal penetrated is calculated by the equation (3):

$$S = M_\infty / M_t \quad (3)$$

where M_∞ and M_t are molar percentages of water uptake at infinite time and at time t. The permeability coefficient, P (m^2/s), which implies the net effect of sorption and diffusion, is given by equation (4):

$$P = D\,x\,S \quad (4)$$

where D is the diffusion coefficient (m^2/s), and S is the sorption coefficient.

When materials are known or suspected to contain any appreciable amount of water-soluble ingredients, the specimens, after immersion, shall be weighed, and then reconditioned for the same time and temperature as used in the original drying period. If the reconditioned weight is lower than the conditioned weight, the difference shall be considered as water-soluble matter lost during the immersion test. Percentage of soluble matter lost during immersion, calculated to the nearest 0.01% as equation (5):

$$\text{Soluble matter lost, \%} = \frac{conditioned\ weight - reconditioned\ weight}{conditioned\ weight} \times 100 \quad (5)$$

3. Result and discussion

Figure 4 shows the moisture absorption percentage as a function of the square root of time (hours) for various woven composites, which had been immersed in tap water at room temperature. From the water absorption curves, the absorbed water content increased with increasing immersion time. A similar finding was reported in a previous study where the water absorption effects of natural fibre reinforced polymer composites were investigated [6]. The water uptake process of the woven composites was linear at the beginning, especially for KK, JJ and HH composites, demonstrating the rapid water penetration into the composite materials. The water uptake then slowed down and approached saturation after a prolonged period.

From the graph, KK composites absorbed a higher amount of water, compared with JJ and HH composites. This was expected due to the higher cellulose and hemicellulose content in the kenaf fibre [2]. Cellulose and hemicellulose are the principal contributors to the moisture absorption, due to its open structure containing hydroxyl (OH) and acetyl (C_2H_3O) groups. When compared to the individually woven composites, a reduction of the water uptake was observed for KJ and KH hybrid composite. The KJ hybrid composite absorbed 46% less water than the KK and JJ composites. Meanwhile, the amount of water absorption of the KH hybrid composite was reduced by 64% and 58%, in comparison with the KK and HH composites, respectively. This reduction shows that the hybridization enhanced the water resistance properties of the kenaf, jute, and hemp fibres. A decrease in the water absorption properties of composites was also found when sisal fibre was added to a banana/epoxy composite [7]. Lignin protects the fibres from hydrothermal degradation due to its hydrophobic features, and theoretically, composites with fibre containing higher lignin content as a filler should present lower values of water uptake [7,8].

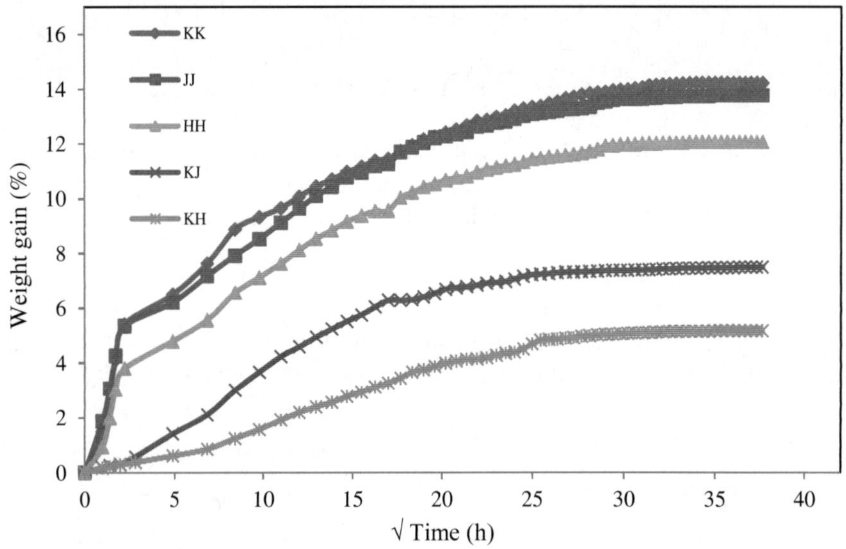

Figure 4. Water absorption curve for various woven composites.

It is essential to measure their diffusion, D (m^2/s) and solubility coefficient, S, when discussing the absorption behaviour of fibre or polymer as these two parameters determine the permeability coefficient of the polymer, P (m^2/s); which implies the net effect of the sorption and diffusion [7]. The diffusion coefficient, D is the velocity of water molecules transported into the composite structure, and also represent the ability of the water molecules to move among the polymer segments. The amount of water dissolved in a polymer is described by the sorption behaviour, S. Table 2 shows the diffusion and permeability study of the woven composite. The amounts of water uptake are corresponding well with the value of diffusion and permeability coefficient, where the individual woven composite have higher diffusion and permeability coefficient; compared with the interwoven hybrid composites.

Table 2. Summary of water absorption parameters for various woven composites.

Specimens	Percentage of water uptake at infinite time M_∞ (%)	Diffusion coefficients, D x 10^{-8} (m^2/s)	Sorption coefficient, S	Permeability coefficient, P x 10^{-7} (m^2/s)	k	n
KK	14	6.32	3.7	4.0	1.311	0.4823
JJ	13.8	6.29	4.9	3.1	1.5086	0.5927
HH	12	4.36	5	2.2	1.6957	0.6529
KJ	7.5	0.36	29	1.0	1.7819	0.7437
KH	5	0.15	5	0.075	1.7193	0.6039

Table 2 also presented the k and n, which are constants. k is a constant which indicates the interaction between the sample and water, while n indicates the mode of diffusion. Both constants were determined from the slope (n) and intercept (k) of $Mt/M\infty$ versus t in a log plot; see figure 5, which can be drawn from the experimental data according to equation (6):

$$\log\left(\frac{M_t}{M_\infty}\right) = log(k) + n\,log(t) \tag{6}$$

Where M_t is moisture content at time *t,* and M_∞ is moisture content at saturation. In particular, the value of *n* shows different behaviour among the three cases; for Fickian diffusion (Case I) $n = 0.5$, while for Case II $n = 1$ (and for super Case II, $n > 1$) and Case III (non-Fickian or anomalous diffusion) $0.5 < n < 1$. Usually, the water absorption pattern of natural fibre composites at room temperature is found to follow Fickian behaviour, whereas at the elevated temperature the absorption behaviour can be non-Fickian [9].

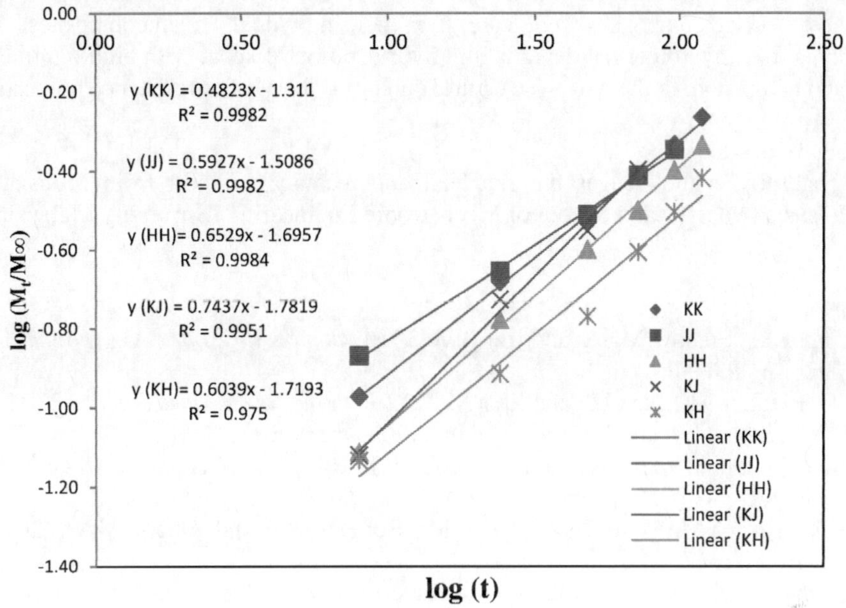

Figure 5. Log (M_t/M_∞) versus log (t) plot.

In contrast with the literature, the *n* values of the individual and interwoven hybrid composites, as presented in Table 2; are higher than 0.5, which indicates non-Fickian diffusion. The water uptake seems more sigmoidal in nature, could be due to the content of fibre, type of matrix, fibre/matrix bonding, and leaching out of resin particles [10, 11]. Calculated using equation (5), the non-Fickian diffusion mode of the woven composites also supported by the weight loss percentage of the water immersed samples, as presented in Table 3.

Table 3. Weight loss percentage of the water immersed samples after reconditioned.

Composite	Conditioned weight (*g*)	Reconditioned weight (*g*)	Weight loss (%)
KK	10.453	10.0966	3.4
JJ	11.1028	11.0503	0.5
HH	11.9097	11.3211	5
KJ	11.0924	11.0732	0.2
KH	11.1812	10.9275	2.3

4. Conclusion

As a conclusion, longer exposure to the aqueous environment increased the water uptake of the composites, and the water content reaches saturation after 1400 h of exposure. Loss of resin particles was observed for the water-immersed samples, which confirmed that the absorption pattern of the composites followed the non-Fickian behaviour. Woven kenaf composite (KK) had the highest water uptake, diffusion and permeability coefficient, attributed by the higher cellulose and hemicellulose content in kenaf fibre compared with jute and hemp fibre. The diffusion coefficient and water uptake of the interwoven hybrid composites were lesser than their individual woven composites. As the lignin protect the fibres from hydrothermal degradation, hybridization of kenaf with high lignin content fibre such as jute and hemp fibre in the woven structure; enhanced their water resistance properties.

5. Acknowledgments

The author gratefully acknowledges the financial and technical support from Ministry of Higher Education Malaysia (MOHE) and School of Mechatronic Engineering, University Malaysia Perlis.

6. References

[1] Pickering KL, Efendy MGA and Le TM 2016 *Composites Part A: Applied Science and Manufacturing* **83** 98
[2] Faruk O, Bledzki AK, Fink HP and Sain M 2012 *Progress in Polymer Science* **37** 1552
[3] Swolfs Y, Gorbatikh L and Verpoest I 2014 *Composites Part A* **67** 181
[4] Misnon MI, Islam MM, Epaarachchi JA and Lau K 2014 *Materials and Design* **59** 359
[5] Sethi S and Ray BC 2015 *Advances in Colloid and Interface Science* **217** 43
[6] Salleh Z, Taib YM, Hyie KM, Mihat M, Berhan MN and Ghani MAA 2012 *Procedia Engineering* **41** 1667
[7] Venkateshwaran N, Elayaperumal A, Alavudeen A and Thiruchitrambalam M 2011 *Materials and Design* **32** 4017
[8] Akil HM, Omar MF, Mazuki AAM, Safiee S, Ishak ZAM and Bakar AA 2011 *Material and Design* **32** 4107
[9] Chen H, Miao M and Ding X 2009 *Composite Part A* **40** 2013
[10] Espert A, Vilaplana F and Karlsson S 2004 *Composite Part A* **35** 1267
[11] Akil HM, Santulli C, Sarasini F, Tirillo J and Valente T 2014 *Composites Science and Technology* **94** 62

IOP Conf. Series: Journal of Physics: Conf. Series **908** (2017) 012016 doi:10.1088/1742-6596/908/1/012016

Energy release rate analysis on the interface cracks of enamel-cement-bracket fracture using virtual crack closure technique

S F Samshuri[1a], R Daud[1b], M A Rojan[1], F Mat[1], K S Basaruddin[1] and R Hassan[2]

[1] Fracture and Damage Mechanics SIG, School of Mechatronic Engineering, Universiti Malaysia Perlis 02600 Arau, Perlis, Malaysia.
[2]School of Dental Sciences, Health Campus,Universiti Sains Malaysia, 16150 Kubang Kerian, Kelantan.

[a] syafiqfarhansamshuri@gmail.com, [b] ruslizam@unimap.edu.my

Abstract. This paper presents the energy method to evaluate fracture behavior of enamel-cement-bracket system based on cement thickness. Finite element (FE) model of enamel-cement-bracket was constructed by using ANSYS Parametric Design Language (APDL). Three different thickness were used in this study, 0.05, 0.2, and 0.271 mm which assigned as thin, medium and thick for both enamel-cement and cement bracket interface cracks. Virtual crack closure technique (VCCT) was implemented as a simulation method to calculated energy release rate (ERR). Simulation results were obtained for each thickness are discussed by using Griffith's energy balance approach. ERR for thin thickness are found to be the lowest compared to medium and thick. Peak value of ERR also showed a significant different between medium and thick thickness. Therefore, weakest bonding occurred at low cement thickness because less load required to produce enough energy to detach the bracket. For medium and thick thickness, both increased rapidly in energy value at about the mid-point of the enamel-cement interface. This behavior occurred because of the increasing in mechanical and surface energy when the cracks are increasing. However, result for thick thickness are higher at mid-point compared to thin thickness. In conclusion, fracture behavior of enamel cracking process for medium most likely the safest to avoid enamel fracture and withstand bracket debonding.

1. Introduction

Bracket debonding occurred during tooth misalignment treatment due to a lot of factors such as incorrect bonding procedures, excessive load exerted from chewing food and unsuitable bracket material used [1]. Therefore, adhesive bonding between bracket and enamel surface should be at optimum state to avoid bracket bonding failure [2]. Two most common debonding type occurred during this failure were at enamel-cement interface and cement-bracket interface [3]. Studying fracture behavior of this two failure condition by using cement thickness as the parameter will lead to a beneficial outcome to improve orthodontist bracket treatment procedure. According to Hajizadeh et al. [4], various factor contributing to bracket debonding based on engineering finding such as tooth surface geometry, bracket base design, location and direction of loading exerted, and adhesive material and geometry. Stress analysis using finite element method (FEM) have been widely used to study bracket debonding. The main purpose of the stress analysis are to recognize the maximum stress region [3]–[5]. However, no fracture mechanics approach to study enamel-cement-bracket system in current research. Using fracture mechanics to study adhesive bonding and debonding was a challenge because of the stress condition along adhesive joints are quite complex [7]. The complexity of adhesive joints are related to different mechanical properties involve such as young modulus, shear modulus and poison ratio. More challenging factor to study adhesive joint was because of the oscillatory singularity located in the crack tip of the biomaterial interface [8]. Therefore, as suggested by Soderholm. [7], energy method was a better approach to eliminate the oscillatory stress factor to study the interfacial cracks because the result obtained are converged at a better range. Two method widely used to calculate the energy release rate (ERR) which are virtual crack closure technique (VCCT) and J – integral. Both method are comparable and validated with each other as stated in Valvo. [8]. VCCT required lower meshing element to compute 2D and 3D calculation compared to J – integral. This study aims to investigate the fracture behavior of interface cracks to determine the safest thickness range for cement layer of enamel-cement bracket system.

2. Finite element Modelling

Two-dimensional (FE) enamel-cement-bracket model was constructed by using ANSYS Parametric Design Language (APDL). Model geometry are configured according to Elaska et al. [6]. Material properties was set according to Table 1. Shear loading applied on enamel-cement-bracket model to simulate the masticatory process involved during bracket debonding. For this model, soft uniform loading of 10 N load applied on bracket base at bracket top side as shown in Figure 1.

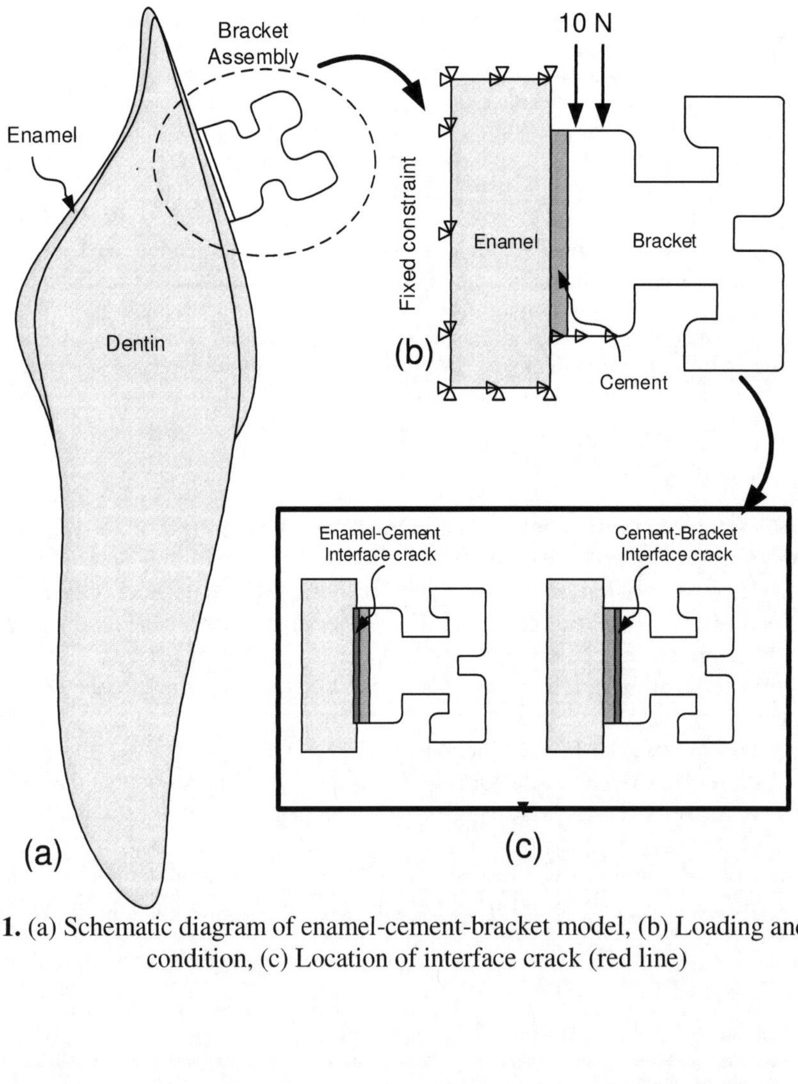

Figure 1. (a) Schematic diagram of enamel-cement-bracket model, (b) Loading and boundary condition, (c) Location of interface crack (red line)

In FE modeling, the nodes along enamel outer layer were fix at all direction to make an assumption of enamel as the fix layer. Then, lower area cement, and bracket base layer, fixed at y – direction to avoid cement layer from entering enamel layer. Meshing size 0.1 was used for enamel region near cement layer, bracket base and cement layer. Small meshing size used at the area close to cement layer to avoid meshing error because high stress at that region. The 8 nodes quadrilateral meshing element used to construct the meshing area with meshing size 0.15 is used for the area away from cement layer. For validation, meshing convergence was conducted. This study involved the assumption of homogenous, isotropic, and linear elastic.

Table 1. Material properties [3]

Material	Manufacturer	Elastic Modulus	Poisson's Ratio
Ceramic	Clarity, 3M Unitek, Monrovia, CA, USA	380000	0.29
Transbond XT	3M Unitek, Monrovia, CA, USA	5000	0.30
Enamel	-	84000	0.30

Interface fracture concerned in this study are enamel-cement interface and cement-bracket interface. Virtual crack closure technique (VCCT) was used to calculated energy release rate by implementing delamination concept on enamel-cement-bracket model. For this study, a simplified flat cement and bracket base geometry according literature was implemented [6]. The evaluation of ERR by using VCCT according to Valvo. [8] can be expressed as

$$G_{ERR} = G_I + G_{II} + G_{III} \tag{1}$$

In this FE formulation, the ERR of interface fracture is evaluated based on

$$G_{IF} = \lim_{x\to 0} \frac{1}{2S} \int_0^{\Delta a} F_{ix}\Delta u_{ix}(a)da + \int_0^{\Delta a} F_{iy}\Delta u_{iy}(a)da + \int_0^{\Delta a} F_{iz}\Delta u_{iz}(a)da \tag{2}$$

The total ERR, G_{IF} involves the opening of initial combined nodes at crack tip to a new crack tip. F_{ix},F_{iy} and F_{iz} are the forces in x, y and z direction. Δu_{ix}, Δu_{iy} and Δu_{iz} are the displacement in x, y and z directions. S represent the surface area of crack surface. Stated that, the range of cement thickness are from 0.05 to 0.271 mm. Therefore, three cement thickness used in this study are 0.271, 0.2 and 0.05 mm as thick, medium and thin cement layer models. Crack length started from 0.2 until 0.26 were tested for each cement layers. Total ERR for enamel-cement and cement-bracket interface obtained and discussed by using Griffith's energy balance as stated by Solderholm [7].

3. Simulation result

During bracket debonding, high stress exerted at the edge of enamel labial will caused the present of flaws or cracks to the enamel region [4]. In this study, comparison of thin, medium and thick cement thickness was conducted according to range of cement thickness obtained from previous study. In order to validate the result obtained from VCCT method, the results were compared with ERR obtained from J – integral method [5]. Validation was conducted for cement thickness 0.2 at enamel-cement interface. As shown in Table 2, percentage error of ERR between VCCT and J - integral method was less than 1 percent for all crack length. As shown in Figure 2, lowest ERR were obtained when the thin cement thickness applied. This behaviour explained less load needed to cause cement or adhesive failure when thin adhesive layer applied to attach the bracket. This result confirmed the lack of bonding ability of thin cement layer. Therefore, cement thickness applied must be thicker than 0.05mm to avoid bracket debonding due to adhesive failure. Comparing with thick and medium cement thickness, thin adhesive shows a relatively low energy value across all crack length tested and low peak energy value.

IOP Conf. Series: Journal of Physics: Conf. Series **908** (2017) 012016 doi:10.1088/1742-6596/908/1/012016

Table 2. Comparison of percentage error of ERR for VCCT and *J* – Integral

Crack length	ERR (VCCT)	Contour 1	Contour 2	Contour 3	ERR (*J* – Integral)	% error
0.2	6.37E-05	6.36E-05	8.19E-05	7.08E-05	7.21E-05	0.1272
0.4	8.81E-05	8.79E-05	8.83E-05	8.82E-05	8.81E-05	0.0359
0.6	1.02E-04	1.01E-04	1.02E-04	1.02E-04	1.02E-04	0.2329
0.8	1.20E-04	1.19E-04	1.20E-04	1.20E-04	1.19E-04	0.3156
1	1.47E-04	1.46E-04	1.47E-04	1.47E-04	1.47E-04	0.2268
1.2	1.84E-04	1.84E-04	1.85E-04	1.85E-04	1.84E-04	0.0886
1.4	2.31E-04	2.31E-04	2.32E-04	2.32E-04	2.31E-04	0.0706
1.6	2.96E-04	2.95E-04	2.96E-04	2.96E-04	2.96E-04	0.0720
1.8	3.96E-04	3.94E-04	3.96E-04	3.96E-04	3.95E-04	0.1079
2	5.73E-04	5.71E-04	5.73E-04	5.73E-04	5.72E-04	0.1764
2.2	9.55E-04	9.51E-04	9.55E-04	9.55E-04	9.53E-04	0.1475
2.4	1.99E-03	0.001986	0.001994	0.001994	1.99E-03	0.0050
2.6	6.27E-03	6.29E-03	6.31E-03	6.31E-03	6.30E-03	0.4691

For medium and thick cement thickness, both having rapid increasing in ERR value at crack length 2.2 mm which is about 2/3 of the cement length. Explaining using Griffith's energy balance, the rapid increasing of energy value possibly because of the increasing in surface energy as the crack length increased. Due to the increasing of crack length, higher mechanical energy required to withstand the high energy produced from the surface energy.

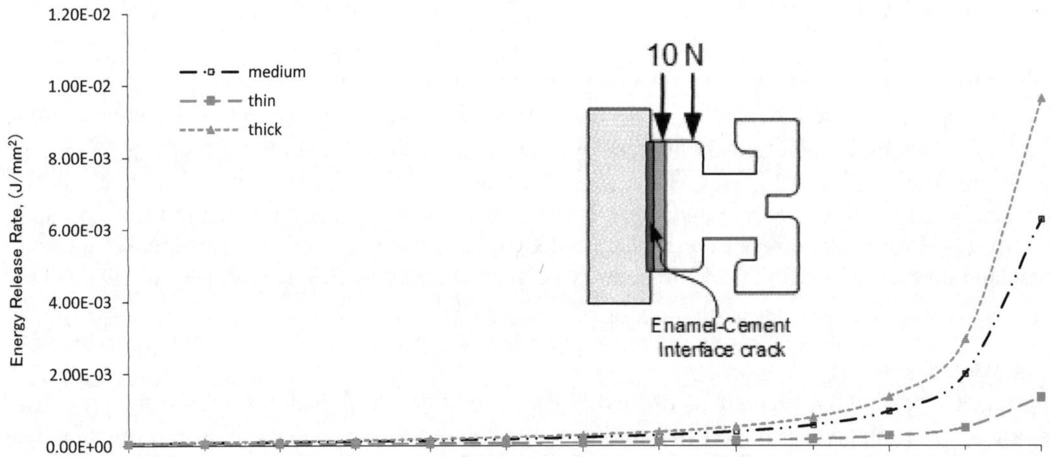

Figure 2. ERR for Enamel - Cement

Therefore, total ERR value should increase with increasing crack length value because total ERR is the summation of mechanical and surface energy occurred at the crack tip. From crack length 0.2 mm to 2.0 mm, medium and thick thickness showing a low energy value. The value of ERR increasing with a relatively small increment at the range of 0.00001 J/mm^2 to 0.0001 J/mm^2 for both medium and thick cement thickness. Therefore, it can be concluded that both thick and medium thickness can withstand debonding load because of the increasing in mechanical energy as the crack increased. However, medium thickness possibly be a better cement thickness to avoid enamel fracture due to high energy produced by thick thickness layer.

Figure 3 shows that low energy release rate value were obtained when thin cement thickness applied. However, increasing in energy value shown at crack length 2 mm. This result showing a different trend for thin cement thickness at cement-bracket interface crack compared to enamel-cement interface crack. This behaviour displayed higher energy when cement-bracket interface fracture occurred compared to enamel-cement interface crack. Therefore, when high possibility of bonding failure to occur when crack are at enamel-cement layer due to less energy required for debonding. For medium and thick thickness, increasing in energy release rate value occurred about half of the cement length. Peak value at crack length 2.6 mm for cement-bracket interface showing a higher value compared to enamel-cement interface for all cement thickness. This behaviour occurred should be because of the shear loading applied near the cement-bracket interface. Therefore, when fracture occurred near cement-bracket interface, bonding failure will easily happen compared to enamel-cement interface.

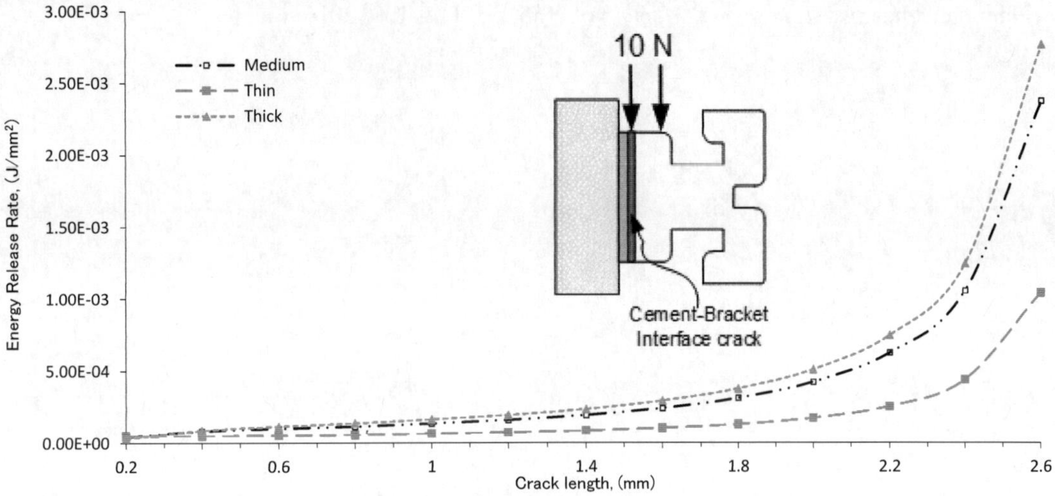

Figure 3. ERR for Cement – Bracket

4. Conclusion

As a conclusion, the validation model of to compare VCCT and J – Integral method result in less than 1 percent error. Therefore, validation considered as success. For comparison between cement thicknesses for enamel-cement interface fracture, medium thickness considered as the most suitable cement thickness because the ERRs value obtained are high enough to withstand loading and low enough to avoid enamel fracture. The ERR values obtained for cement-bracket interface crack showing a different trend compared to enamel-cement interface. Increasing in ERR occurred earlier at cement-bracket interface. Therefore, it can be concluded higher possibility of debonding occurred when crack or failure started from cement-bracket interface compared to enamel-cement interface.

5. Acknowledgement

The authors are grateful for the research support of Ministry of Higher Education (MOHE) Malaysia and University Malaysia Perlis through awarded Fundamental Research Grants (FRGS 9003-00491).

6. References

[1] C. Chen, M. Hsu, K. Chang, S. Kuang, and P. Chen, "Failure Analysis : Enamel Fracture after Debonding Orthodontic Brackets," vol. 78, no. 6, 2008.

[2] M. Hajizadeh, F. Ghalichi, and B. Mirzakouchaki, "Optimization of orthodontic bracket base geometry for planar enamel surface teeth: A finite element study," *Orthod. Waves*, vol. 75, no.

3, pp. 64–75, 2016.

[3] C. Holberg, P. Winterhalder, N. Holberg, A. Wichelhaus, and I. Rudzki-janson, "Orthodontic bracket debonding : risk of enamel fracture," pp. 327–334, 2014.

[4] M. Hajizadeh, F. Ghalichi, B. Mirzakouchaki, and S. Shahrbaf, "ScienceDirect Pattern of stress distribution in different bracket – adhesive – tooth systems due to debonding load application," *Orthod. Waves*, vol. 73, no. 1, pp. 8–16, 2014.

[5] Daud, R. , Ariffin, A.K., Abdullah, S., "Validation of crack interaction limit model for parallel edge cracks using two-dimensional finite element analysis," *International Journal of Automotive and Mechanical Engineering*, vol. 7, issue 1, pp. 993-1004, 2016

[6] S. E. Elsaka, S. M. Hammad, and N. F. Ibrahim, "Evaluation of stresses developed in different bracket-cement-enamel systems using finite element analysis with in vitro bond strength tests Evaluation of stresses developed in different bracket-cement-enamel systems using finite element analysis with in vitro bond strength tests," no. December 2015, pp. 1–8, 2014.

[7] K. Soderholm, "Review of the fracture toughness approach," vol. 6, pp. 63–77, 2009.

[8] P. S. Valvo, "On the calculation of energy release rate and mode mixity in delaminated laminated beams," *Eng. Fract. Mech.*, vol. 165, pp. 114–139, 2016.

Effect of elevated temperatures on flexural strength of hybrid Napier/glass reinforced epoxy composites

W M A W Ramli[1], M J M Ridzuan[1], M S Abdul Majid[1], M N Rahman Y[1] and K Azduwin[2]

[1]Mechanical Engineering Programme, School of Mechatronics Engineering, Universiti Malaysia Perlis, Pauh Putra Campus, 02600 Arau, Perlis, Malaysia
[2]Department of Chemical Engineering Technology, Faculty of Engineering Technology, Universiti Malaysia Perlis, 02600 Arau, Perlis, Malaysia

ridzuanjamir@unimap.edu.my

Abstract. The effects of elevated temperatures on the flexural strength of hybrid Napier/glass reinforced epoxy composites were investigated. Hybrid composites laminates were fabricated using untreated, 5%, or 10% alkali-treated Napier fibres with woven E-glass fibres and epoxy resin. The composites were manufactured using a vacuum infusion process; the volume fraction of the Napier, glass fibres and epoxy resin were 24%, 6% and 70% respectively. When tested at room temperature (RT), the maximum flexural strength was recorded for the hybrid composites with the 5% alkali-treated Napier fibres. When the test temperature greater than 60°C, the matrix cracking and delamination were occurred due to the temperature that approached the glass transition temperature (Tg) of the composites, which resulted in a reduction of the flexural strength. The fracture surface morphologies indicated that the 5% alkali-treated Napier fibres improved the fibre-matrix interfacial bonding of the hybrid Napier/glass reinforced epoxy composites.

Keywords: Hybrid composites; natural fibres; flexural strength; elevated temperatures; epoxy.

1. Introduction

Lately, the practice of natural fibres for reinforced composites has become a promising idea to emerge "green" based economy in industrial application. Natural fibres are completely renewable resources, environmentally friendly, low-cost, high biodegradability and possessed comparable mechanical properties as synthetic fibres[1–4]. Their weight also light with preeminent specific modulus and required little energy during processing; less wear and tear, which all these advantages have brought natural fibres as a solid reinforcing agent in polymer composites for various applications both in engineering and technology fields [5]. Plant-based natural fibre reinforced composites are mostly from plant fibres like jute, sisal, kenaf, hemp, Napier, ramie, pineapple leaf, abaca leaf, sugarcane, bamboo and so on [3-5]. Due to severe environmental problems and a high concern for the world plus easy availability and eco-friendly of natural fibres, many researchers are compelled to investigate the use of these continuous resources as an alternative to traditional polymer composites manufactured from synthetic fibres [9]. Therefore, the exploitation of natural fibres is improved remarkably and has substantially generated the interest in the study for the development of plant-based polymer composites [7–9].

However, the elevated temperature on the composites brought a serious impact towards the mechanical properties of the composites[13]. For example, increasing the temperature of surrounding can impair the structure of a composite which will create various damages such as intra-ply damage, matrix cracking, interlaminar cracks, fibre breakage and subsurface delamination [14]. The fact is natural fibres have less thermal stability. Thus, they tend to degrade at lower temperature [15]. An interesting technique to limit this problem is through hybridization rather than using chemical treatment since it has the disadvantage of being expensive. Concerning the hybrid composites which consist two kinds of fibres, the most virtue relies on the fact that the benefits of a fibre could surpass the lack of another fibre [15].

In this study, experimental analysis was implemented to analyze the flexural strength of hybrid Napier/glass reinforced composites. The tests were carried out in room temperature (RT) and also at elevated temperatures. By thoroughly elucidating and analyzing the mechanical properties of the composites which are cost-effective and environmentally friendly, this study is expected to produce supportive manifestation for further application and development.

2. Materials and methods

2.1. Materials

The hybrid composites consisted of Napier's fibres, woven E-glass fibres and epoxy resin. Napier grass was gained from a local plantation located at Bukit Kayu Hitam, Kedah, in Northern Peninsular Malaysia. Water retting process was practiced to facilitate the extraction of fibres from the Napier's stem internodes. By using a mallet, the stems were chopped into smaller portion at the early step. After that, the short stems were left in water for 2-4 weeks to ease the separation of the fibres from the skin and the cellulose. The woven E-glass fibres were ordered from a local supplier and prepared in the form of mats with dimensions of 300 × 300 × 15 mm per layer. EpoxAmite 100 series resin was used as a matrix, and mixed with a hardener at a ratio of 3:1.

2.2. Composite fabrication

The hybrid Napier/glass-reinforced epoxy composites were fabricated using a vacuum infusion method. The volume fractions of the Napier fibres, glass fibres, and epoxy resin were 24%, 6%, and 70%, respectively. The hybrid composite samples consist of either untreated, 5%, or 10% alkali-treated Napier fibres. The reinforcement plies were laminated above a glass mould. By using a high vacuum pump machine, the matrix resin was infused into the lamination plies. The in-mould pressure was ensured to be below than 2,000 Pa in preparation for the matrix resin to be infused into the mould. Then the composites were left at room temperature for 12 hours to cure. In the end, rectangular plates with dimensions of 300 × 300 × (3.2 ± 0.4) mm^3 were produced. Afterwards, post-curing was carried-out in an oven at 80 °C for 2 hours under air circulation [17]. Lastly, by using a Dremel 4000 tool, the plates were cut to prepare specimens for flexural testing according to the related standards.

2.3. Flexural testing

Flexural tests via three-point bending tests, was carried-out on specimens with dimensions of 125 × 13 × (3.2 ± 0.4) mm^3, in accordance with ASTM D790-1 [18]. The specimens were loaded with the recommended span-to-depth ratio of 16:1. Flexural tests were implemented in the Instron environmental chamber at RT (25°C) and at elevated temperature of 40, 60 and 80°C. The tests on the specimens were conducted using a universal micro tester with a load cell of 2 kN at a crosshead speed of 2.5 mm/min [20,21]. Five replicated specimens of each composite were tested for every different temperature. For those stated temperatures, the flexural stress–strain curves were plotted and recorded.

2.4. Morphological observation

The fractured surface morphologies of the hybrid Napier/glass-reinforced epoxy composites were observed using a field-emission scanning electron microscope (FESEM). For FESEM observation,

IOP Conf. Series: Journal of Physics: Conf. Series **908** (2017) 012017 doi:10.1088/1742-6596/908/1/012017

section of the fractured test specimens was cut. Platinum was then coated over the cutting section before scanning. The scanned images were acquired with accelerating voltage of 3–5 kV.

3. Results and discussion

Figure 1 shows the maximum stress curves for the different states of the hybrid Napier/glass reinforced epoxy composites for the untreated, 5% and 10% of alkali-treated that were subjected to the flexural test in RT and under certain elevated temperatures. Based on Figure 1, it can be simply concluded that the stress curves become increasing non-linear while the temperature is increased. The figure apparently showed that the hybrid composites when tested at RT and 40°C, the recorded stress was at the high for all composite states and significantly differed to those results obtained and recorded at 60 and 80°C which much lower. This situation is due to the increased temperature that softens the epoxy matrix which caused the structure of the specimens to be ductile and eventually failed at higher strain due to higher temperature.

The maximum stress recorded for hybrid composites with untreated Napier's fibres were at 69, 67, 33 and 15 MPa for RT, 40, 60 and 80°C respectively. From these values, it can be observed that the flexural strength at RT and 40°C were very similar due to the fibre reorientation which happened during the softening of the matrix. On the other hand, at greater temperatures, the characteristics of the matrix resin crucially effected the total potential of the hybrid Napier/glass reinforced epoxy composites. This is substantially proved by the decrement of the flexural strength at the testing temperature of 60°C, showing there was 50% reduction in strength than the result gained at room temperature.

From the data collected, a bit greater flexural strength and stiffness readings were recorded for the hybrid composites with 5% of alkali-treated Napier fibres at testing temperatures of RT and 40°C compared to the results acquired for the untreated hybrid composites. This is due to the excellent interfacial adhesion among the fibre and the matrix resin. The fact is alkali treatment caused the removal of the hemicellulose, waxes, lignin and other impurities. In the other word, this condition would offer a better fibre/matrix adhesion instead of causing a rough surface on the Napier fibres. Besides, the alkali treatment also led to fibre fibrillations which increased the effective surface area between the fibres and the matrix. As a consequence, the flexural strength of the composites improved indirectly. This has been verified by some researchers that apparently validated the increment in the composites' flexural strength of using alkali-treated natural fibres is because of the greater interface between the fibres and the matrix after the treatment [22–26]. However, a lower flexural strength was recorded for hybrid composites with 10% of alkali-treated Napier fibres. This decrement of flexural strength can be clarified because of the loss of hemicellulose and lignin at the interfibrillar region, thereby causing the fibres to be less dense and rigid. Das et al. [26] claimed that the decrement of long-chain cellulose molecules at the natural fibre interfaces in right concentration of alkali treatment will reduce fibre breakage and lead to optimal flexural strength of the composites. Though, when the composites were tested at elevated temperatures of 60 and 80°C, the flexural strength values recorded for the composites with the untreated Napier fibres were higher compared to the hybrid composites with 5% and 10% of alkali-treated Napier fibres.

Following test at RT, it was observed that brittle failure occurred at the centre of the specimens, causing in the formation of rough fracture surfaces. Similar fracture surfaces were indicated at elevated temperatures for both untreated and treated Napier fibres hybrid composites. At testing temperature of 40°C, the fracture surfaces were rough with noticeable fibres breakage. Though, part of the fracture surfaces exhibited a smoother profile if comparing to those of the specimens tested at RT. On the other hand, the resulting fracture surfaces of the specimens were rougher when tested at temperature of 60°C compared to the specimens tested at RT and 40°C, showing that the specimens failed because of the softening of the polymer matrix. This could be observed from the specimens that failed more gradually with progressive failure and more pull-out of the Napier fibres happened [27]. Eventually, when the specimens were tested at 80°C, high levels of defectiveness were noticed due to the crack propagation that hindered by the fibres, propagating at different points of the specimens, thereby numerous cracks were visible. This is specified by a gradual decrement of stress of the composites with the increasing temperature as showed by the figure 1.

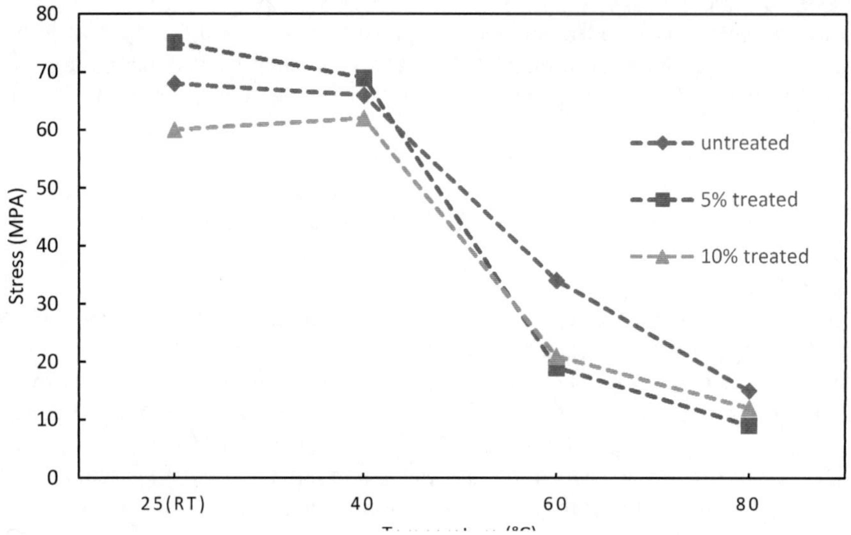

Figure 1. Maximum flexural stress behaviour of the hybrid Napier/glass reinforced epoxy composites subjected to RT and elevated temperatures for different state of composites.

4. Fracture surface morphology

The fracture surface morphologies for the different state of hybrid Napier/glass reinforced epoxy composites which subjected to flexural tests at RT and elevated temperatures shown by Figure 2. The surface morphologies of the hybrid composites with the untreated Napier fibres, when tested at RT, was rough and demonstrated typical long fibres pull-out as shown in Figure 2 (a). When tested at 40°C, matrix cracking has been observed as illustrated in Figure 2(b). When the specimens were tested at 60°C, voids because of to the fibres pull-out were visible as well as the porous structure like in the Figure 2(c), indicating that the testing temperature was approaching the T_g value of the composites. Lastly, delamination of glass fibres was observed in the fracture surface image of the specimens that were tested at 80°C as figured in Figure 2 (d).

The effects of elevated temperature on the flexural strength of hybrid Napier/glass reinforced epoxy composites were investigated. Untreated, 5% and 10% alkali-treated Napier fibres were used to fabricate the hybrid composites. The highest maximum flexural strength recorded following test at RT was given by the hybrid composites with 5% alkali-treated Napier fibres. All hybrid composites with untreated and treated Napier fibres showed maximum flexural strength when tested at RT and 40°C. However when the samples tested at elevated temperatures of 60°C and 80°C, the voids due to fibres pull out and delamination occurred. Consequently, the damages reduced the flexural strength of the composites since the temperatures approached and exceeded the glass transition temperature (T_g) of each composite. This situation was due to the softening of the matrix, leading to poorer interfacial bonding between the Napier fibres and the matrix.

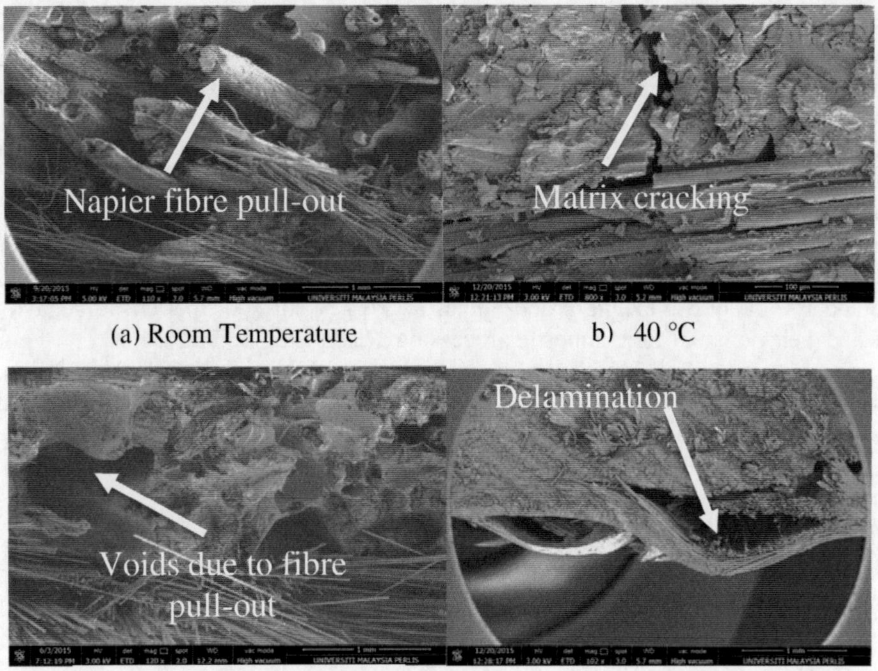

(a) Room Temperature b) 40 °C

(c) 60 °C (d) 80 °C

Figure 2. Fracture surface morphologies of hybrid Napier/glass reinforced epoxy composites subjected to flexural test following tests at RT and elevated temperatures.

5. Acknowledgements

The authors would like to express gratitude to School of Mechatronic Engineering, Universiti Malaysia Perlis for as providing the equipment and financial assistance for this project.

6. References

[1] Ridzuan M J M, Abdul Majid M S, Afendi M, Azduwin K, Amin N A M, Zahri J M and Gibson A G 2016 Moisture absorption and mechanical degradation of hybrid Pennisetum purpureum/glass–epoxy composites *Compos. Struct.* **141** 110–6

[2] Ridzuan M J M, Abdul Majid M S, Afendi M, Azduwin K, Kanafiah S N A and Dan-mallam Y 2015 The Effects of the Alkaline Treatment's Soaking Exposure on the Tensile Strength of Napier Fibre *Procedia Manuf.* **2** 353–8

[3] Ridzuan M J M, Abdul Majid M S, Afendi M, Kanafiah S N A and Nuriman M B M 2015 Effects of alkaline concentrations on the tensile properties of Napier grass fibre *Appl. Mech. Mater.* **786** 23–7

[4] Ridzuan M J M, Abdul Majid M S, Afendi M, Kanafiah S N A, Zahri J M and Gibson A G 2016 Characterisation of natural cellulosic fibre from Pennisetum purpureum stem as potential reinforcement of polymer composites *Mater. Des.* **89** 839–47

[5] Saba N, Jawaid M, Alothman O Y and Paridah M T 2016 A review on dynamic mechanical properties of natural fibre reinforced polymer composites *Constr. Build. Mater.* **106** 149–59

[6] Lee B H, Kim H J and Yu W R 2009 Fabrication of long and discontinuous natural fiber reinforced polypropylene biocomposites and their mechanical properties *Fibers Polym.* **10** 83–90

[7] Lu N and Oza S 2013 Thermal stability and thermo-mechanical properties of hemp-high density polyethylene composites: Effect of two different chemical modifications *Compos. Part B Eng.* **44** 484–90

[8] Alamri H, Low I M and Alothman Z 2012 Mechanical, thermal and microstructural

characteristics of cellulose fibre reinforced epoxy/organoclay nanocomposites *Compos. Part B Eng.* **43** 2762–71

[9] La Mantia F P and Morreale M 2011 Green composites: A brief review *Compos. Part A Appl. Sci. Manuf.* **42** 579–88

[10] Lu N and Oza S 2013 A comparative study of the mechanical properties of hemp fiber with virgin and recycled high density polyethylene matrix *Compos. Part B Eng.* **45** 1651–6

[11] Norul Izani M A, Paridah M T, Anwar U M K, Mohd Nor M Y and H'Ng P S 2013 Effects of fiber treatment on morphology, tensile and thermogravimetric analysis of oil palm empty fruit bunches fibers *Compos. Part B Eng.* **45** 1251–7

[12] Moothoo J, Allaoui S, Ouagne P and Soulat D 2014 A study of the tensile behaviour of flax tows and their potential for composite processing *Mater. Des.* **55** 764–72

[13] Ridzuan, M.J.M.;Abdul Majid, M.S.: Afendi, M.; Rahman, Y.M.N. ; Azduwin K 2016 Thermal, composition and morphological properties of untreated and alkali-treated Napier fibres *J. Eng. Appl. Sci.* **11** 2485–9

[14] Vieille B, Casado V M and Bouvet C 2013 About the impact behavior of woven-ply carbon fiber-reinforced thermoplastic- and thermosetting-composites: A comparative study *Compos. Struct.* **101** 9–21

[15] Araújo J R, Waldman W R and De Paoli M A 2008 Thermal properties of high density polyethylene composites with natural fibres: Coupling agent effect *Polym. Degrad. Stab.* **93** 1770–5

[16] Costa C S M F, Fonseca A C, Serra A C, Coelho J F J, Fonseca A C, Serra A C and Dynamic J F J C 2016 Dynamic Mechanical Thermal Analysis of Polymer Composites Reinforced with Natural Fibers Dynamic Mechanical Thermal Analysis of Polymer Composites **3724**

[17] Júnior J H S A, Júnior H L O, Amico S C and Amado F D R 2012 Study of hybrid intralaminate curaua/glass composites *Mater. Des.* **42** 111–7

[18] Specimens P and Materials E I 2011 Standard Test Methods for Flexural Properties of Unreinforced and Reinforced Plastics and Electrical Insulating Materials 1 *Annual Book of ASTM Standards* pp 1–11

[19] Sathishkumar T P, Navaneethakrishnan P and Shankar S 2012 Tensile and flexural properties of snake grass natural fiber reinforced isophthallic polyester composites *Compos. Sci. Technol.* **72** 1183–90

[20] Ochi S 2008 Mechanical properties of kenaf fibers and kenaf/PLA composites *Mech. Mater.* **40** 446–52

[21] Aziz S H and Ansell M P 2004 The effect of alkalization and fibre alignment on the mechanical and thermal properties of kenaf and hemp bast fibre composites: Part 1 – polyester resin matrix *Compos. Sci. Technol.* **64** 1219–30

[22] Goud G and Rao R N 2011 Effect of fibre content and alkali treatment on mechanical properties of Roystonea regia -reinforced epoxy partially *Bull. Mater. Sci.* **34** 1575–81

[23] Sinha E and Rout S K 2009 Influence of fibre-surface treatment on structural, thermal and mechanical properties of jute fibre and its composite *Bull. Mater. Sci.* **32** 65–76

[24] Rokbi M, Osmani H, Imad A and Benseddiq N 2011 Effect of Chemical treatment on Flexure Properties of Natural Fiber-reinforced Polyester Composite *Procedia Eng.* **10** 2092–7

[25] Boopathi L, Sampath P S and Mylsamy K 2012 Investigation of physical, chemical and mechanical properties of raw and alkali treated Borassus fruit fiber *Compos. Part B Eng.* **43** 3044–52

[26] Das M, Pal A and Chakraborty D 2006 Effects of mercerization of bamboo strips on mechanical properties of unidirectional bamboo–novolac composites *J. Appl. Polym. Sci.* **100** 238–44

[27] Manalo A C, Wani E, Zukarnain N A, Karunasena W and Lau K 2015 Effects of alkali treatment and elevated temperature on the mechanical properties of bamboo fibre–polyester composites *Compos. Part B Eng.* **80** 73–83

IOP Conf. Series: Journal of Physics: Conf. Series **908** (2017) 012018 doi:10.1088/1742-6596/908/1/012018

Influence of ceramic dental crown coating substrate thickness ratio on strain energy release rate

K Khasnulhadi[1a], R Daud[1b], F Mat[1], S N F M Noor[2], K S Basaruddin[1] and M H Sulaiman[1]

[1] Fracture and Damage Mechanics SIG, School of Mechatronic Engineering, University Malaysia Perlis, 02600 Arau, Malaysia.
[2]Craniofacial and Biomaterials Science Cluster, Advanced Medical and Dental Institute, Universiti Sains Malaysia, Bertam, 13200 Kepala Batas, Pulau Pinang

[a] khalilahkhasnulhadi@gmail.com, [b]ruslizam@unimap.edu.my

Abstract. This paper presents the analysis of coating substrate thickness ratio effect on the crown coating fracture behaviour. The bi-layer material is examined under four point bending with pre-crack at the bottom of the core material by using finite element. Three different coating thickness of core/substrate was tested which is 1:1, 1:2 and 2:1. The fracture parameters are analysed based on bilayer and homogenous elastic interaction. The result shows that the ratio thickness of core/veneer provided a significant effect on energy release rate.

1. Introduction

All-ceramic dental crown has been used extensively for clinical restorations because of their superior aesthetic, inertness and biocompatibility. They are usually fabricated into layered structures with aesthetic somewhat weak veneer porcelain on stiff and strong ceramic supporting-cores. The relationship between different layers of material that involves in production of dental crown, especially the ratio of thickness between veneer and core is not clearly understood. Such an understanding is necessary for the restoration to be fully functional and reliable. Wakabayashi [1] investigated on the influence of the alumina core thickness and the supporting substrate stiffness on the strength and failure mode of bilayered ceramic disks. They found that, the core/veneer thickness ratio is the dominant factor that controls the failure initiation site in the bilayered ceramic disks. Then, when the crack is starting at the lower center of the core surface, the crack will penetrating the veneer surface and leading to bulk fracture. According to Pross et al., [2], the maximum principal stress was located on the core region, and the value of the principal stress are greatly sensitive to thickness of the core. However, their research was only focusing on the maximum stresses that arise in a first pre-molar crown. Therefore, it just shows the critical region that will lead to the initial crack. Lima et al., [3] also study the influence of veneer thickness on the flexural strength of a bilayer ceramic system. The results revealed the thickness of the veneer ceramic has an influence on the mechanical strength of the bilayer ceramic. Considerable work has been devoted to assessing the strength of dental crown material. This is generally done by subjecting the material to some form of external force and examining the critical region that will be possibly be the initial location for crack to occur. A simple means for evaluation energy release rate for dental crown in this study is the four point bending bilayer configuration. The objective of this paper is to study the behaviour of energy release rate on different coating/substrate thickness ratios by using finite element.

2. Finite element modelling

Finite element model is developed based on four point bending configuration. The beam were modelled using 8 node quadratic elements (plane183) in the finite element code ANSYS version 14.5 specified to plane strain condition. The interface between layers was assumed to remain perfectly bonded at all stages of computation. After performing detailed convergence analysis, the element were meshed to be 0.2 mm in size for both layers. The loading was simulated as pressure applied on point load location as shown in Figure 1.

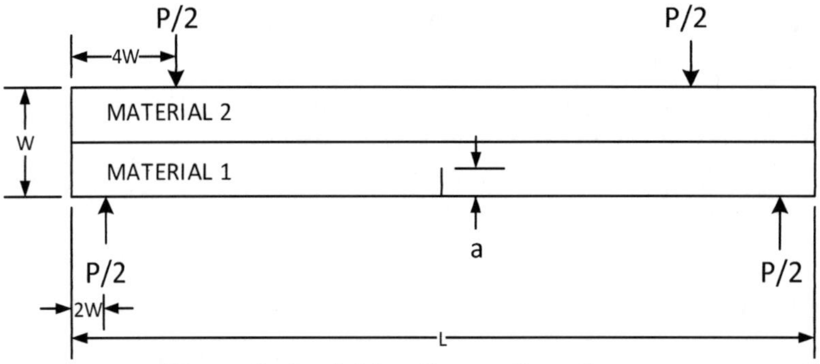

Figure 1: 4 point-bending configuration.

The following elastic properties were used in these study; (E_1, E_2) = (315, 80) GPa, (υ_1 , υ_2) = (0.26, 0.30). Three different types of In-Ceram Alumina/porcelain bilayered beam were studied with In-Ceram Alumina to porcelain thickness ratios of 1:1, 1:2 and 2:1, respectively. All beams were modelled in the desired form; total thickness, W= 2 mm, length, L = 16mm and crack length, a start from 0.2 to 1.4 mm. Details of geometry dimension were model according to standard reported from previous literature [4]. The evaluation of stress distribution is based on von-Misses stress σ_e and 1st Principle Stress σ_1, von-Misses stress can be expressed as

$$\sigma_e = [\frac{(\sigma_1-\sigma_2)^2+(\sigma_2-\sigma_3)^2+(\sigma_3-\sigma_1)^2}{2}]^{1/2} \quad (1)$$

where

$$\sigma_1 = \frac{1}{2}(\sigma_x+\sigma_y) + [\frac{1}{2}(\sigma_x-\sigma_y)]^{1/2} + \tau_{xy}^2 \quad (2)$$

$$\sigma_2 = \frac{1}{2}(\sigma_x+\sigma_y) - [\frac{1}{2}(\sigma_x-\sigma_y)]^{1/2} + \tau_{xy}^2 \quad (3)$$

$$\sigma_3 = \sigma_z = 0 \quad (4)$$

The energy release rate G equal to J integral value. The J-integral was calculated by the domain integral method for six contours. The J-integral evaluation is based on the domain integral method by Shih & Asaro [5] with assumption of linear isotropic elasticity material behaviors [6]. The domain integration formulation applies area integration for 2-D problems. For a 2-D problem, and in the absence of thermal strain, path dependent plastic strains, body forces within the integration of area, and pressure on the crack surface, the domain integral representation of the J-integral is given by:

$$J = \int_A \left[\sigma_{ij}\frac{\partial u_j}{\partial x_1} - w\delta_{1i}\right]\frac{\partial q}{\partial x_i}dA \quad (5)$$

where σ_{ij} is the stress tensor, u_j is the displacement vector, w is the strain energy density,δ_{ij} is the Kronecker delta, x_i is the coordinate axis, and q is referred to as the crack-extension vector. The direction of q is the simple x-axis of the local coordinate system ahead of the crack tip. The q vector is chosen as zero at nodes along the contour Γ and is a unit vector for all nodes inside Γ except the midside nodes, if there are any, that are directly connected to Γ. The first contour for the area integration of the J-integral is evaluated over the elements associated with the crack-tip node component. The second contour for the area integration of the *J*-integral is evaluated over the elements adjacent to the first contour of elements. This procedure is repeated for all contours.

3. Result and discussion

Figure 2 shows the stress distribution of In-Ceram Alumina/porcelain bilayered beam.

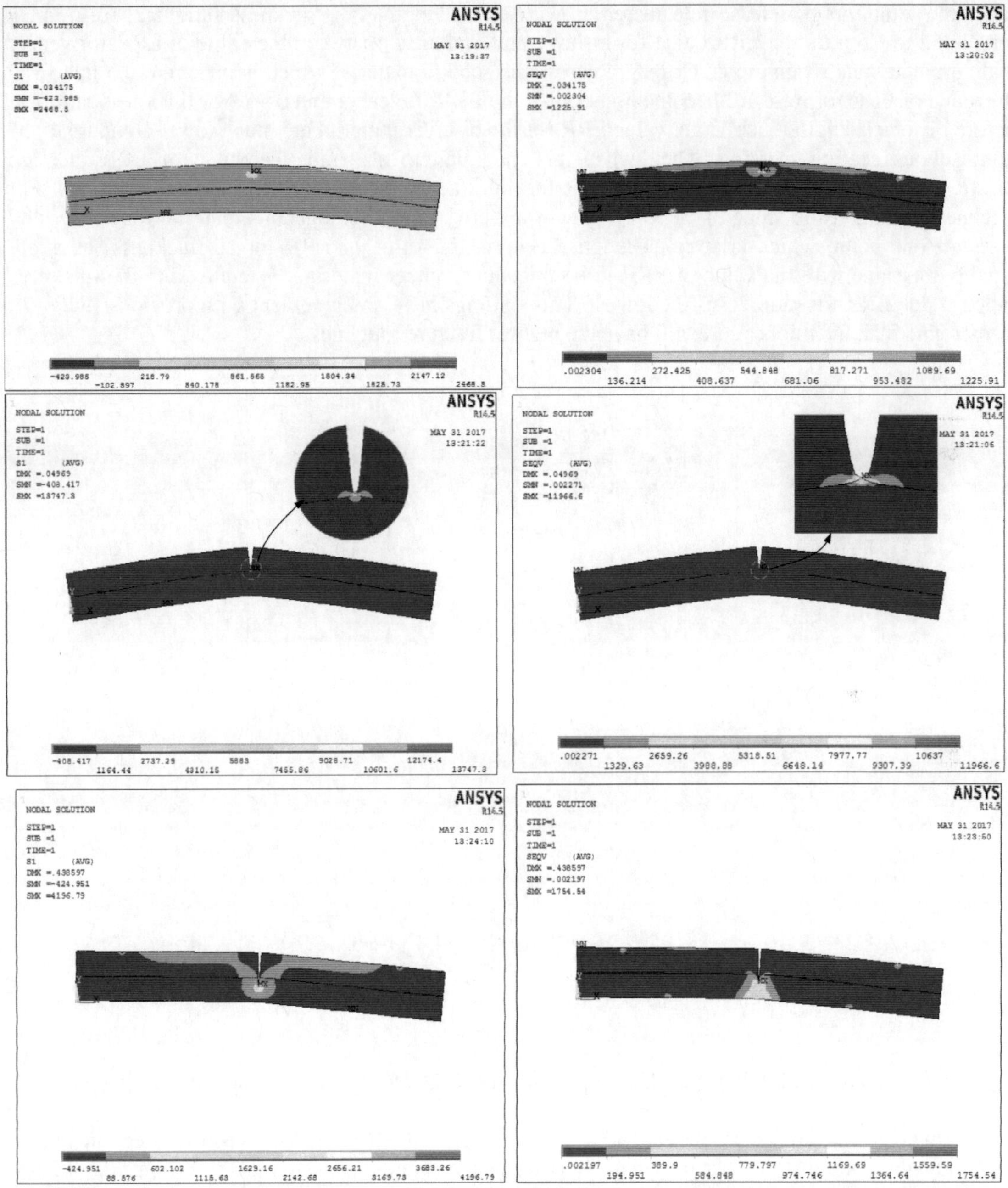

Figure 2 (a) 1st principal stress when crack length at core region. b) von-Mises stress when crack length at core region(c) 1st principal stress when crack length at interface (d) von-Mises stress when crack length at interface (e) 1st principal stress when crack length at veneer region.

A crack oriented perpendicularly to the interface is localized in the In-Ceram alumina (core).The effect is analyzed according to the veneer and core thickness ratio. Under the effect of the applied loading, the crack propagates to the interface. According to the behavior of the materials, the crack can penetrate the interface or deviate from it. In this case only the penetration effect is studied. The obtained result are presented in Figure 3. It illustrates the variation of the energy release rate, G according to crack length, a, for various thickness ratio. It can be seen that whatever be the thickness ratio, the total energy release rate increases with crack length. For thickness ratio 1:2, from crack length 0.2 to 0.5 mm, the ERR value for bi-layer material are in between the value of ERR for veneer and core material. When the crack length across the second material which is penetrate the interface, the value of ERR for ratio 1:2 become higher than the ERR in veneer and core. For thickness ratio 1:1, before the crack length reach 1mm, value ERR for the bi-layer material are much lower compare to the homogenous coating material. Then when the crack length at 1mm, the value of ERR increase drastically because there is an interaction at the interface between core and veneer materials. For thickness ratio 2:1, the value of ERR are remain in between the homogenous material for veneer and core. At one point, which is at crack length 1.33mm, the value of ERR for all thickness ratios are equally the same with the value of ERR in homogenous veneer material. By comparing the thickness ratios, value ERR for ratio 2:1 is the lowest. This findings have an agreement with previous study [7]. They found that, the thicker core will have the highest fracture toughness.

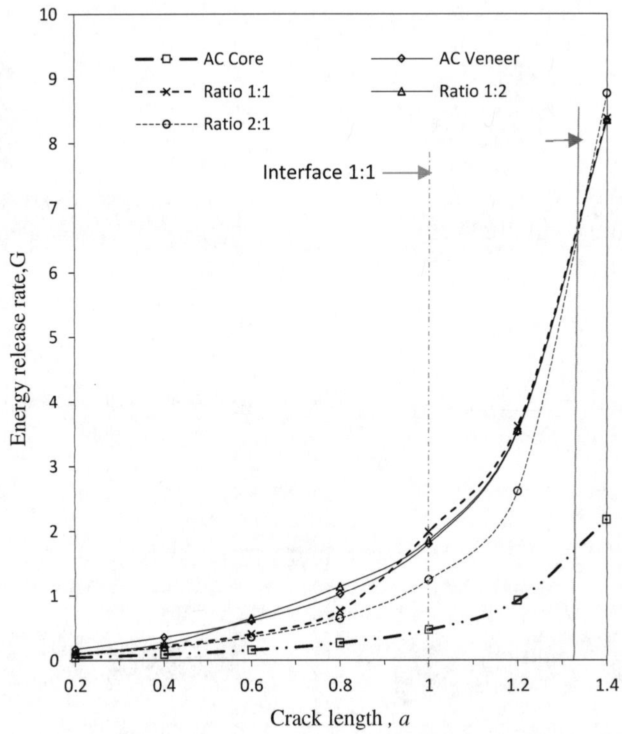

Figure 3. Influence of ratio thickness on energy release rate vs crack length.

Once the crack penetrate the second layer material, the ERR value for bilayer are become higher than the monolithic material because there is no influence of the first material. Thus the value of ERR for second material will act as the homogenous veneer material. Overall observation, the monolithic In-Ceram alumina have the lowest value of ERR. While, monolithic porcelain have the highest value of ERR and it depends on which thickness ratio was compared. The energy of penetration for core/veneer layer represents the energy release rate of a homogenous.

4. Conclusion
Core material have the lowest value of ERR due to its excellent material properties among the others. Ratio thickness give influence on the value of ERR. However, when the crack start to penetrate the second layer material (veneer), the value of ERR for bilayer material will act as a homogenous veneer material.

5. Acknowledgments
The authors are grateful for the research support of Ministry of Higher Education (MOHE) Malaysia and University Malaysia Perlis through awarded Fundamental Research Grants (FRGS 9003-00491).

6. References
[1] N. Wakabayashi and T. Medical, "Crack Initiation Modes in Bilayered Alumina / Porcelain Disks as a Function of Core / Veneer Thickness Ratio and Supporting Dental Research," no. April 2016, 2000.
[2] K. A. Pross, M. V. Swain, J. Ironside, and G. P. Steven, "Investigation of stress analysis on coating-substrate interfacial with pre-cracked model," *Int. J. Prosthodont.*, vol. 16, no. 5, 2003.
[3] J. M. C. Lima, A. C. O. Souza, L. C. Anami, M. A. Bottino, R. M. Melo, and R. O. A. Souza, "Effects of thickness , processing technique , and cooling rate protocol on the flexural strength of a bilayer ceramic system," vol. 9, pp. 1063–1072, 2013.
[4] D. P. Clausing, "Crack Stability in Linear Elastic Fracture Mechanics," *Int. J. Fract. Mech.*, vol. 5, no. 3, pp. 211–227, 1968.
[5] C. F. Shih and R. J. Asaro, "Elastic-Plastic Analysis of Cracks on Bimaterial Interfaces : Part 1 - Small Scale Yielding," *J. Appl. Mech.*, vol. 55, 1988.
[6] Daud, R. , Ariffin, A.K., Abdullah, S., "Validation of crack interaction limit model for parallel edge cracks using two-dimensional finite element analysis," *International Journal of Automotive and Mechanical Engineering*, vol. 7, issue 1, pp. 993-1004, 2016
[7] J. H. Kim, H. Park, Y. B. Park, and H. S. Moon, "Fracture load of zirconia crowns according to the thickness and marginal design of coping," 2011.

IOP Conf. Series: Journal of Physics: Conf. Series **908** (2017) 012019 doi:10.1088/1742-6596/908/1/012019

Stress analysis of implant-bone fixation at different fracture angle

B Izzawati[1, a], R Daud[1], M Afendi[1], MS Abdul Majid[1], N A M Zain[2], Y Bajuri[3]

[1] Fracture and Damage Mechanics (FDM) SIG, School of Mechatronic Engineering Universiti Malaysia Perlis (UniMAP), 02600 Arau, Perlis, Malaysia
[2] Institute of Mathematics, Universiti Malaysia Perlis, Perlis, Malaysia
[3] UKM Medical Centre, Department of Orthopedics and Traumatology, Universiti Kebangsaan Malaysia, Jalan Yaacob Latif, Bandar Tun Razak, Cheras, Kuala Lumpur, Malaysia

[a]izzawati.basirom@gmail.com

Abstract. Internal fixation is a mechanism purposed to maintain and protect the reduction of a fracture. Understanding of the fixation stability is necessary to determine parameters influence the mechanical stability and the risk of implant failure. A static structural analysis on a bone fracture fixation was developed to simulate and analyse the biomechanics of a diaphysis shaft fracture with a compression plate and conventional screws. This study aims to determine a critical area of the implant to be fractured based on different implant material and angle of fracture (i.e. 0°, 30° and 45°). Several factors were shown to influence stability to implant after surgical. The stainless steel, (S. S) and Titanium, (Ti) screws experienced the highest stress at 30° fracture angle. The fracture angle had a most significant effect on the conventional screw as compared to the compression plate. The stress was significantly higher in S.S material as compared to Ti material, with concentrated on the 4th screw for all range of fracture angle. It was also noted that the screws closest to the intense concentration stress areas on the compression plate experienced increasing amounts of stress. The highest was observed at the screw thread-head junction.

1. Introduction

Bone fractures occur throughout the population; causes can be the result of high load impact or minimal trauma injury. Statistical analysis has been carried out whereas the significantly, majority 81% sustained the fracture due to road traffic accidents. The rest is equally divided between low-energy trauma such as fall from height and assault (9.5%). The report stated the percentage of union (19.04%) achieved by 4 month, (66.66%) by 6 months and the rest complete union by 9 months [1]. However, the result may vary due to absolute the stability of bone fracture fixator and early range of motion to minimize the risk of reoperation.

Femur bone, subjected to a combination of axial, bending and torsion load that provides the variety type of fractures when the applied load exceeds its strength limit. Axial compression load can cause oblique fracture while the torsion delivers to the transverse fractures [2]. The different between these are the fracture patterns, which can be classified by the angle of fractures. According to the Müller AO Classification of fractures— long bones, it was classified as category 32-D. Sub-categories complete transverse with an obliquity of 30° or less and complete oblique or spiral more than 30° are simple fractures and the complex fractures [3]. However, the treatment of bone fracture continues to be a challenging problem for a surgeon particularly involving the complication of surgical approaches [4]. It is necessary to provide the orientation of the bone fragments and promotes the tissue regeneration. Thus, the precise implant device is needed to support the fractured bone and restore compromised functional.

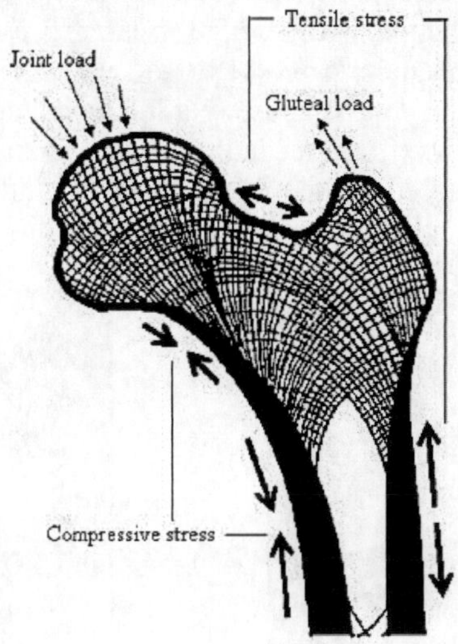

Figure 1: Load stress distribution in femur bone [5]

Besides that, mechanical loading applied to the bone structure allows the subsequent calculation of stress induced, for example, in femur bone structure as shown in Figure 1. The fracture pattern generated involving various soft tissues and the bone quality at the time of injury. This is further compounded by the fact that in these types of fractures, the articular surface may be involved which, if inappropriately treated, may lead to future complications such as osteoarthritis. The treatment of these fractures requires complete immobilization of patients who, ultimately result in prolonged recovery periods [6]. Thus, the study of proper internal fixation was necessary whereas it offers anatomical reduction of the bone fragment, stability and rigid of fixation for better formation of soft tissue and bone healing process.

Some researchers studied the stress analysis by using the finite element to determine the fracture mechanism of the long bone. They found that fracture angle 30° degree causes the highest stress on the screw and plate [7]. The bending stiffness and contact condition effects were considered [8]. It is expected to reduce contact with other living tissues may help diminish the foreign body reaction and thus increase curing efficiency. Similar to the investigation the plate failure reported takes place at the screw hole near the oblique 45° fracture site [2]. These studies reported that screw and plate have a significant effect on the implant failure. However, these studies were limited to the implant design geometry and materials used whereas should highlight the optimal parameters granting the implantation success.

Progressive losing of bone fixation screws by stress shielding and causing adaptive bone remodeling due to bone loss around the screw [9]. Bone adaptation caused losing implant due to the stress-shielding pattern which dependent on material and geometrical characteristics of the screw [10]. While the primary stability depends on the immediate mechanical engagement between the implant and the bone seat, the secondary stability requires the establishment of a resistant bone-implant adhesion. These research efforts pursue a clear goal: improve the implants' efficiency in standing conditions, i.e. as bone fracture fixator subjected to stress due to the mechanical loading. These are the conditions in which the bone-implant 'assembly' must grant stability and reliability. However, the high-stress concentration due to load can cause the loosening of the screw, and results implant failure and infection occurs on the fracture site. These study aims to determine a critical area of the implant to be fractured based on different implant material and angle of fracture (i.e. 0°, 30° and 45°).

2. Finite Element Modeling

The three-dimensional geometrical model of the bone fracture fixation and femur bone structure was created using Solidwork software 2013. The geometry of femur bone took the shape created from CT database through image segmentation with IGES format. All assembly processes of the screw into femur bone model were performed in Ansys 18v software by using Boolean operation in design modular as shown in Figure 2. The bone fracture fixation model comprised compression plate and four internal conventional cortex screws (full threaded) anchored along the diaphysis shaft as shown Figure 3. The fixation or implant was placed in the middle of the fracture line. Three type of the fracture line developed (i.e. 0°, 30° and 45°) and the screws orientation also used based on the fracture angle.

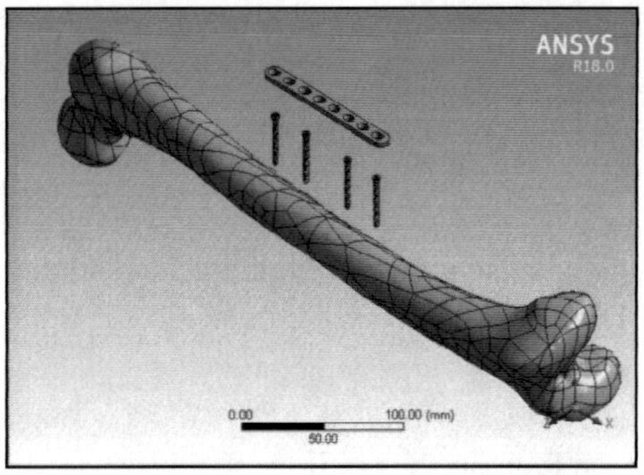

Figure 2: Component of bone fracture fixation and femur bone structure

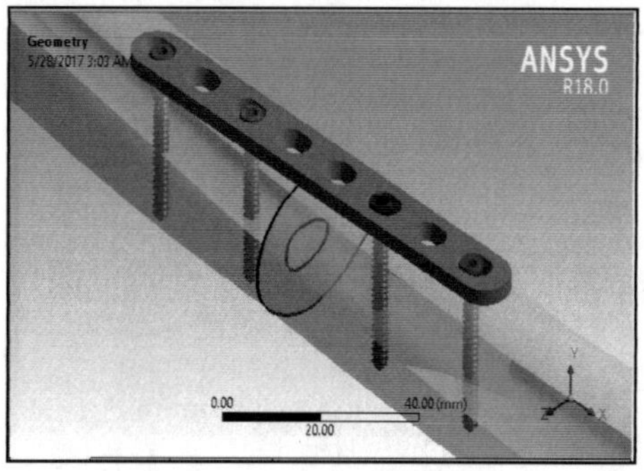

Figure 3: Assembled femur bone structure and implant

Table 1 describes the details of bone fracture fixation for conventional cortex screw and compression plate based on Synthes product. In this study, two type screw lengths (i.e. 36 mm and 40 mm) and two types of material (i.e. Stainless Steel (no product : 204.8xx) and Titanium (no product : 404.8xx)) were used where "xx" refer as screw length. These types of bone fracture fixation chose based upon the fracture type suggested based on AO Foundation.

Table 1: Dimensional parameters for bone fracture fixation

Conventional Cortex Screw		8-holes compression plate	
Diameter of thread (mm)	3.5	Width (mm)	11.0
Thread pitch (mm)	1.25	Thickness (mm)	3.3
Diameter of Core (mm)	2.4	Distance between center holes (mm)	13.0
Diameter of head (mm)	6.0		

Table 2 presents the information regarding the material properties of bone; cortical bones were assumed to be orthotropic, homogeneous, and linearly elastics as were the others' materials used in this study.

IOP Conf. Series: Journal of Physics: Conf. Series **908** (2017) 012019 doi:10.1088/1742-6596/908/1/012019

Table 2: Mechanical properties of bone structure and implant materials in FE analysis

Material	Young's modulus (GPa)	Poisson's ratio (υ)	Shear modulus (GPa)
Cortical Bone	$E_3 = 20.0$	$\upsilon_{12} = 0.376$	$G_{12} = 4.53$
(Longitudinal transverse)	$E_1 = 12.0$	$\upsilon_{23} = 0.235$	$G_{23} = 4.53$
	$E_2 = 12.0$	$\upsilon_{23} = 0.376$	$G_{13} = 4.53$
Stainless steel (S.S)	$E_{s.s} = 200$	$\upsilon_{S.S} = 0.3$	-
Titanium (Ti)	$E_{Ti} = 113.8$	$\upsilon_{Ti} = 0.34$	-

Femur bone model was categorized into five parts, which separated by using virtual topology to combine the vertical edge. It can be done by inserting a virtual topology branch. Thus, the model was separated into four parts, which are a femoral head, greater trochanter, diaphysis shaft, and condyle as shown in Figure 4. The surface that separates for each part has been considered as a bonded region in order to connect the part together with the elements of the upper part overlap with the elements of the lower part. The fracture section represents the fracture line whereas divided the diaphysis shaft into two section (i.e. upper and lower part). The static load based on standing conditions was applied by tension load 400N (body weight) and compression load 300N (abdument muscle) as shown in Figure 4, to produce significant influence instability.

An optimized convergence meshing has been developed using ANSYS Workbench an appropriate setting and values have been executed in order to use smaller elements on proximities and curvatures for the model. The numbers of elements used for the bone fracture fixation model are 1058958, 1078497 and 1075567 for fracture angle of 0°, 30° and 45° respectively.

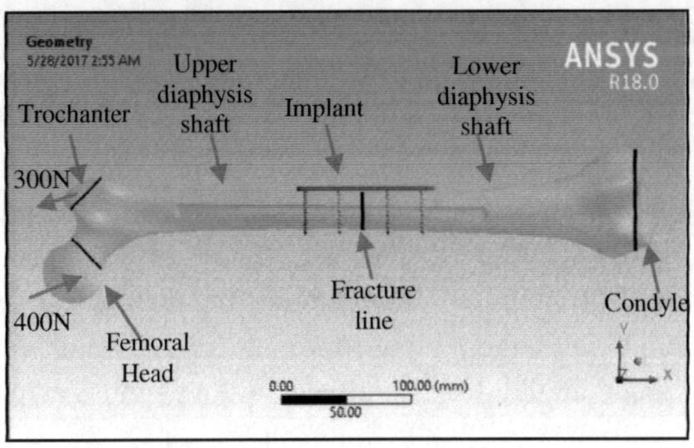

Figure 4: Boundary condition of bone fracture

3. Result

The primary stability of implant is a reference of the mechanical loading attained between screw surface and bone structural designed. This tension and compression loads is analyzed in terms of the implant bending, mainly in situations where a non-Osseo-integration device response is planned for supportive purposes. In this work, the minimal static loads based on standing conditions was applied by compression load 400N (body weight) and pulling load 300N (abdument muscle), at the different of fracture angle (0°, 30° and 45°). The magnitude of the applied forces was in accordance with the orthopedic demand for clinical needs.

All the different fracture angle and configurations of fixation were subjected to the same load, each one evaluated separately regarding the critical stress of implant. The stress results, even in the bone, screws and compression plate are presented in Table 3.

Table 3: Stress of bone and implant at different fracture angles and implant material

		Maximum von-Mises stress (MPa)		
Materials	Fracture type	Bone stress	Conventional screws	Compression plate
Stainless steel (S.S)	Transverse (θ=0°)	152.10	215.03	274.21
	Transverse (θ=30°)	175.12	335.59	234.26
	Oblique (θ=45°)	185.96	332.06	232.61
Titanium (Ti)	Transverse (θ=0°)	154.19	163.46	195.34
	Transverse (θ=30°)	177.25	222.5	154.64
	Oblique (θ=45°)	182.25	221.47	153.73

Table 3 shows the maximum von-Mises stress for bone, screws and compression plate el at the different fracture angle, and implant materials. The bone stress showed inversely proportional to the stress of compression plate. Dissimilar to the conventional screws it was challenging to correlate with bone stress due to unevenly force transfer to each screw. Nevertheless, the early assumption could be made which the higher screw stress will tend to optimize the stability of the fixator. This assumption meets the statement from some researchers that verified the higher the screw tension will reduce the displacement of bone segment [11]. Although, the stress of bone shows an increment as increasing the fracture angle for both implanted materials. The bone stress value of Ti implanted shows slightly higher than S. S implant. It was compared to the effect of increasing the fracture angle to the implant stress (screws and compression plate), its shows independent of each other. Thus, a comparison was made to analyze the affected parts that influence by the fracture angle and implant materials. The details of implant stress could describe in a data as shown in Figure 5.

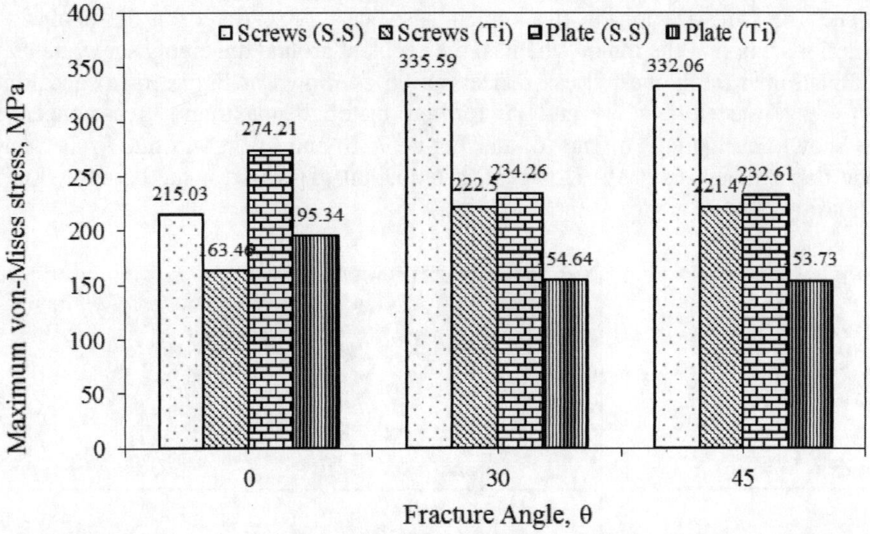

Figure 5: Stress of implant (screws and compression plate) at different angle fractures

The maximum equivalent von-Mises stress implants were shown in Figure 5. The Ti material for both the compression plate and screws materials experienced a lower amount of stress as compared to the S.S at fracture angle range (0°, 30° and 45°). This difference in stress was also observed on the screws of all range of fracture angle. The both materials of screws that anchored through diaphysis shaft on the femur bone experienced higher stress at 30° fracture angle as compared to the 0° and 45°, with mean differences of 335.59MPa and 222.5MPa for S.S and Ti material respectively. Dissimilar with the stress of compression plate whereas, it indicates decrement as increasing the fracture angles for both materials.

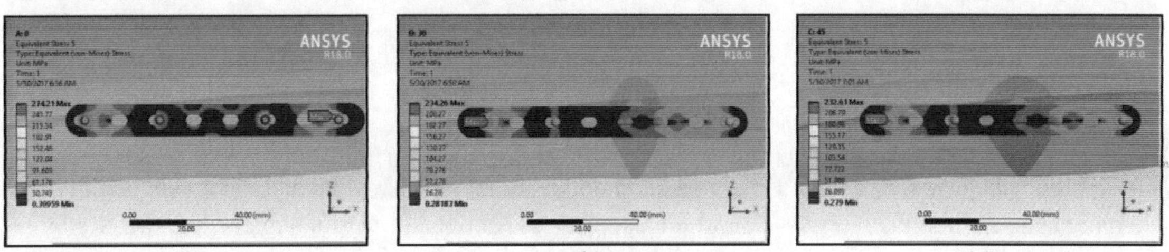

(a) (b) (c)

Figure 6: Stress contour of S.S compression plate (a) at 0°, (b) at 30°, and (c) 45° fracture angle

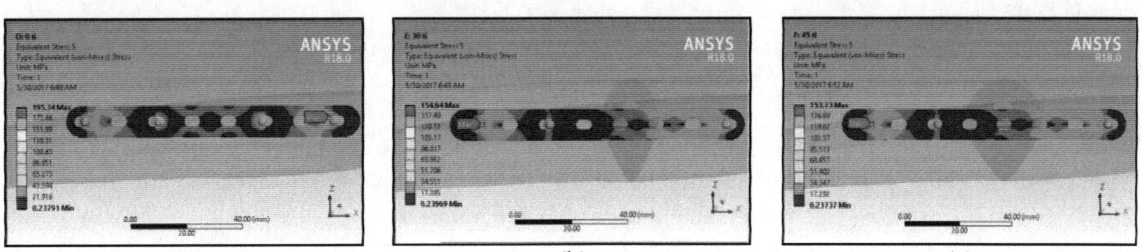

(a) (b) (c)

Figure 7: Stress contour of Ti compression plate (a) at 0°, (b) at 30°, and (c) 45° fracture angle

In the stress simulations, green color demonstrates less stress, and the red color is related to higher stress values. Figure 6 and Figure 7 shows the stress contour of compression plate for S. S and

Ti material. The stress distribution on the implant also was concentrated on the compression plate supporting anterior section of the femur whereas, the segment around the empty screw holes 4th on the section of the plate had the highest stress concentration as shown in Figure 6 (a) and Figure 7 (a). Comparing to stress distribution at 30° and 45° for both materials, maximum stress concentrated at 1st screw hole as shown in Figure 7 (b) and (c) and Figure 7 (b) and (c). It was due to the inclination of bending of the upper diaphysis shaft. However, a reduction of stress for the compression plate was shown by increasing fracture angle.

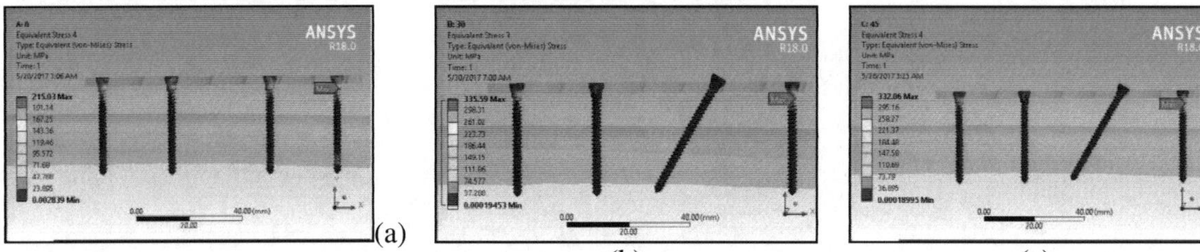

(a) (b) (c)

Figure 8: Stress contour of S.S conventional screw (a) at 0°, (b) at 30°, and (c) 45° fracture angle

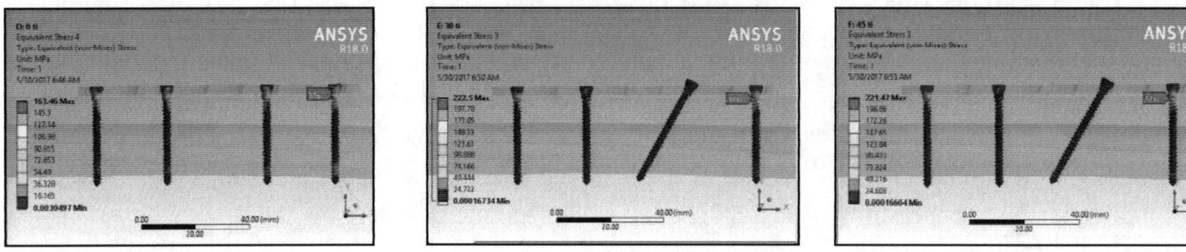

(a) (b) (c)

Figure 9: Stress contour of Ti conventional screw (a) at 0°, (b) at 30°, and (c) 45° fracture angle

Figure 8 and Figure 9 shows the stress contour of conventional screws for S. S and Ti material. It was also noted that in both materials, the screws closest to high up concentration stress areas on the compression plate experienced a higher amount of stress. Both materials observed that the highest stress concentration at the screw neck and the upper thread for each fracture angle, as shown in Figure 8 (a), (b) and (c) and Figure 9 (a), (b) and (c). However, the stresses' distribution appears to decrease along thread as it approaches the tip of the screws. Thus, screws shows had a major influence of fracture for all range of fracture angle compare to the compression plate. Hence, further studies using biomechanical testing on the screws models will need to establish the details of information in order to provide better accuracy of the screw failure.

The present study also established that conventional screw provides better stability for the different type of fracture regarding diaphysis shaft. Figure 10 indicates the results of the interfaces point of these two sections (bone-implant) undergo high stresses as a result of pull-out strength on the 4th screw created by the deformation of bone fragments. The present of screw thread, allow the implant to hold the bone fragmented and maintain the stability of bone-implant interaction [12]. However, the conventional screw implanted in each bone fragment will provide a critical section to fracture that will affect the surrounding tissue.

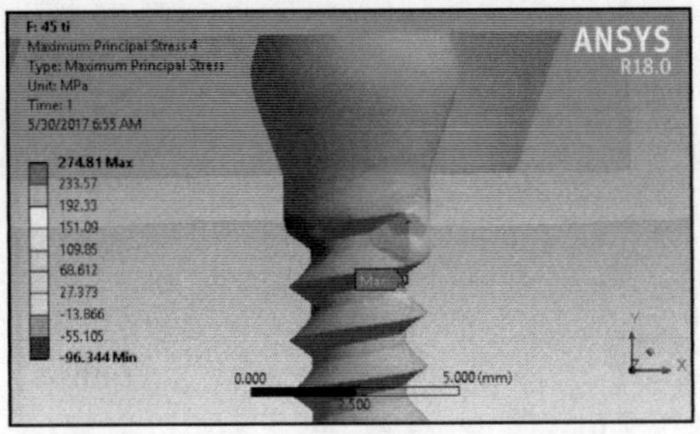

Figure 10: Stress concentration on the screw thread

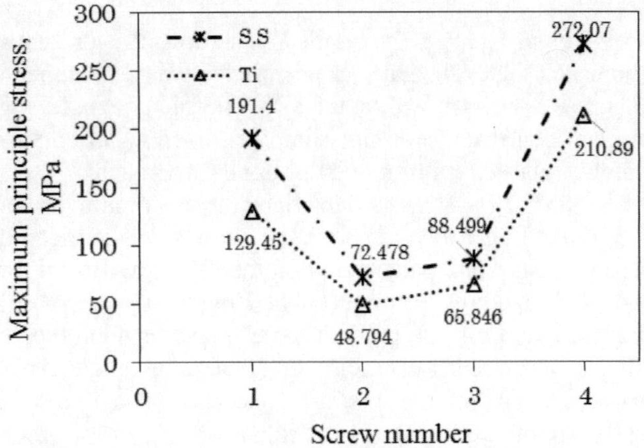

Figure 12 (a): Screws stress of S.S and Ti materials at 0° fracture angle

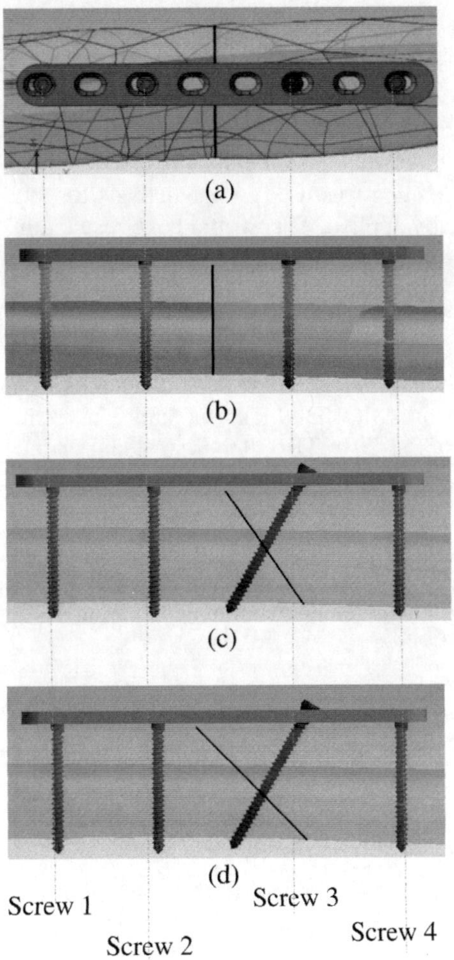

Figure 11: Screws orientation at different fracture angle.

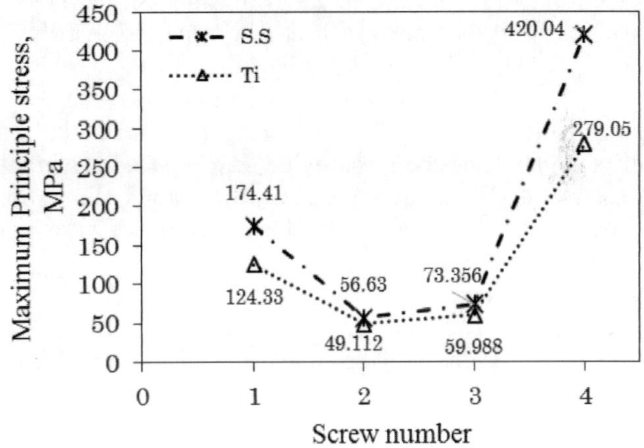

Figure 12 (b): Screws stress of S.S and Ti materials at 30° fracture angle

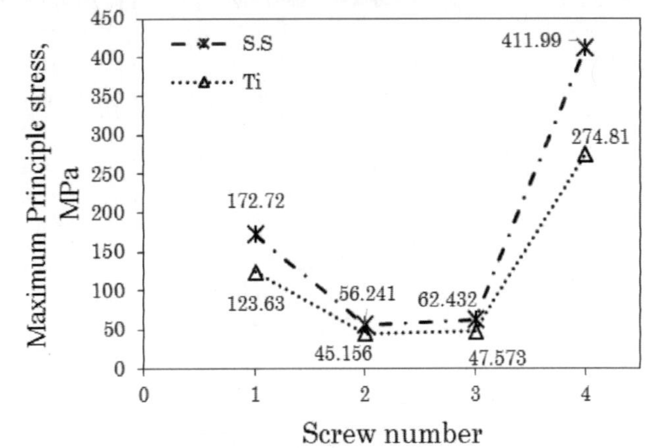

Figure 12 (c): Screws stress of S.S and Ti materials at 45° fracture angle

Figure 11 (a) shows the orientation of four screws along the compression plate through the diaphysis shaft. The stress concentration was determined on the individual screws (i.e. screw 1, screw 2, screw 3, and screw 4) at the different fracture angle. Figure 11 (b) shows the screws implant straight through the diaphysis shaft compare to screws orientation in Figure 11 (c) and Figure 11 (d) as the screw 3 placed tilt 30° based on surgical approach.

Based on screw orientation, the maximum principal stress of both implanted material was determined after load applied. The stress was significantly higher in S. S material as compared to the Ti material, with concentrated on the 4th screw for all range of fracture angle as shown in Figure 12 (a),(b) and (c). It was due on bending of the upper diaphysis, whereas the intensities of bending transferred. Thus, the pull-out force generates upon the bending on the 4th screw, and cause the higher stress concentration especially on 30° fracture angle. This would lead the likelihood of implant failure to occur. A reduction in stress concentration points on 2nd and 3rd screws for all fracture angles as shown in Figure 12 (a), (b) and (c).

In addition, the present study analyzes the stresses of a static load, which if cyclical loading were applied would result in the increased likelihood of implant loosening. The reason for determined stress concentration in the screws remains unclear since the various types of thread design, and the material used for implants. It can be expected that the critical region of implant would fail for the internal implant fixation of the diaphysis shaft.

Several limitations were observed in this study. Static loading was used in the analysis for this model whereas, in the real-life situation, cyclic loading is usually applied. This would have resulted in different outcomes being observed. Instead, findings in this study may be demonstrative to this type of implant fracture pattern. Moreover, fracture configuration is also shown to be one of the criteria to determine the screw orientation to be used for fracture fixation.

4. Acknowledgment

This study was funded by the Ministry of Higher Education (MOHE) Malaysia and Universiti Malaysia Perlis through the Fundamental Research Grants (FRGS/1/2016/TK03/UNIMAP/03/4).

5. Conclusion

The stainless steel (S.S) and Titanium (Ti) screws experienced the highest stress at 30° fracture angle. The fracture angle had the most significant effect on the conventional screws as compared to the compression plate. The stress was significantly concentrated on the 4th screw for all range of fracture angle. The highest was observed at the screw thread-head junction**.**

6. Reference

[1] S. ; Dr Biju and M. Dr Sabarisree, "Functional Outcome of Supracondylar Fracture Femur Treated By Locked Compression Plating – A Prospective Study," *J. Med. Sci. Clin. Reseach*, vol. 5, no. 5, pp. 21229–21240, 2017.

[2] K. R. Krishna, I. Sridhar, and D. N. Ghista, "Analysis of the helical plate for bone fracture fixation," *Intern. J. Care Inj.*, vol. 39, no. 12, pp. 1421–1436, 2008.

[3] J. F. Kellam and L. Audigé, "'A classification is useful only if it considers the severity of the bone lesion and serves as a basis for treatment and for evaluation of the results.' Maurice E Müller," pp. 1–18, 2016.

[4] B. French and P. Tornetta, "High-energy tibial shaft fractures.," *Orthop. Clin. North Am.*, vol. 33, no. 1, p. 211–30, ix, 2002.

[5] N. Passi and A. Gefen, "Trabecular Bone Contributes to Strength of the Proximal Femur Under Mediolateral Impact in the Avian," vol. 127, no. February 2005, 2016.

[6] D. J. Redfern, S. U. Syed, and S. J. M. Davies, "Fractures of the distal tibia: Minimally invasive plate osteosynthesis," *Injury*, vol. 35, no. 6, pp. 615–620, 2004.

[7] M. O. Kaman, N. Celik, and S. Karakuzu, "Numerical Stress Analysis of the Plates Used to Treat the Tibia Bone Fracture," *J. Appl. Math. Phys.*, no. May, pp. 304–309, 2014.

[8] S. H. Kim, S. H. Chang, and D. S. Son, "Finite element analysis of the effect of bending stiffness and contact condition of composite bone plates with simple rectangular cross-section on the bio-mechanical behaviour of fractured long bones," *Compos. Part B Eng.*, vol. 42, no. 6, pp. 1731–1738, 2011.

[9] A. Gefen, "Dynamic Simulations Of Cancellous Bone Resorption Around Orthopaedic Fixative Implants," *Int. Conf. IEEE Eng. Med. Biol. Soc.*, pp. 23–26, 2001.

[10] A. Gefen, "Optimizing the biomechanical compatibility of orthopedic screws for bone fracture fixation," *Med. Eng. Phys.*, vol. 24, no. 5, pp. 337–347, 2002.

[11] D. J. Stringhini, R. Sommerfeld, L. C. Uetanabaro, D. P. Leonardi, M. R. Araujo, N. L. B. Rebellato, D. J. da Costa, and R. Scariot, "Resistance and stress finite element analysis of different types of fixation for mandibular orthognathic surgery," *Braz. Dent. J.*, vol. 27, no. 3, pp. 284–291, 2016.

[12] R. M. Aizat, M. R. A. Kadir, S. Ab Rahman, T. Shihabudin, N. Robson, and T. Kamarul, "Biomechanical Comparative Analyses Between the Anterolateral and Medial Distal Tibia Locking Plates in Treating Complex Distal Tibial Fracture: A Finite Element Study," *J. Med. Imaging Heal. Informatics*, vol. 3, no. 4, pp. 532–537, 2013.

Effect of graphite loading on the electrical and mechanical properties of Poly (Ethylene Oxide)/Poly (Vinyl Chloride) polymer films

M D S Hajar[1], A G Supri[2], M P M Hanif[1], M I M Yazid[1]

[1] School of Materials Engineering, Kompleks Pusat Pengajian Jejawi 2, Taman Muhibbah, Universiti Malaysia Perlis (UniMAP), Jejawi, 02600, Arau, Perlis, Malaysia.
[2] Faculty of Engineering Technology, Universiti Malaysia Perlis (UniMAP), P.O Box 77, D/A Pejabat Pos Besar, 01000, Kangar, Perlis, Malaysia.

sitihajarmohddin@gmail.com

Abstract: In this study, films consisting of a blend of poly (ethylene oxide)/ poly (vinyl chloride) (PEO/PVC) and a conductive filler, graphite were prepared and characterized for their mechanical and electrical properties. Solid polymer blend films based on PEO/PVC (50/50 wt%/wt%) with different graphite loading were prepared by using solution casting technique. Electrical conductivity results discovered the conductivity increased with increasing of filler loading. However, increasing amount of graphite loading led to a decreased in tensile strength and young's modulus of PEO/PVC/Graphite polymer films. The dispersion of graphite and mechanism of conductive path in the polymer films were also investigated by scanning electron microscopy (SEM). The morphology of the PEO/PVC/Graphite polymer films shows that agglomeration occurred to complete the connection of conductive path, thus improving the conductivity behavior of the polymer films.

Keywords: Poly (vinyl chloride), Poly (ethylene oxide), graphite, electrical conductivity, mechanical testing

1. Introduction

Conductive polymer composites can be fabricated to achieve the desired properties for specific applications. It has been extensively used in various engineering applications including conductive adhesives, electromagnetic interference shields and electronic devices [1-3]. The choice of conductive fillers is the important criteria to produce highly conductive polymer composites since different conductive fillers have their own unique properties. Numerous types of carbon based fillers have been studied including graphite, carbon black and carbon fiber [4].

PEO is crystalline polymer which has solvating characteristics and commonly applied in polymer electrolytes system. The major problem of using PEO is the high degree of crystallization when used as single matrix in conductive polymer composites. This problem can be overcome by blending or mixing with another polymer such as PVC and PMMA. PVC has chlorine atom which containing lone pair electrons and inorganic salts able to be dissolved in electrolyte systems [5]. The PVC are inexpensive, able to form homogenous hybrid film and marketable. In this study, PVC forms miscible blend together with PEO due to its unique characteristic which is easy processability and compatible with various plasticizers. The PEO/PVC blends have been broadly used as a matrix in polymer electrolytes.

Graphite is crystal form of a pure carbon much resembling of mica-sheets of strongly linked atoms, with very weak bonds between the sheets. Graphite generally used as electronically conducting filler because it is economical and abundant naturally [6]. In this study, the electrical properties, morphology and mechanical properties of PEO/PVC blends containing graphite filler were investigated.

2. Experimental

2.1 Materials

PVC powder with molecular weight of 220 000 g/mol and PEO powder with molecular weight of 100 000 g/mol as the matrix for conductive polymer film was supplied by AR Alatan Sdn. Bhd., Alor Star, Malaysia. Graphite with molecular weight of 12.02 g/mol and particle size of <50 μm was supplied by R&M Chemicals. Plasticizer used in this research was Dioctyl terepthalate with molecular weight of 390 g/mol and tetrahydrofuran with molecular weight of 72.11 g/mol was provided from AR Alatan Sdn. Bhd.

Figure 1. Morphology of graphite particle

2.2 Composites Preparation

The PEO/PVC/Graphite polymer films were processed by solution casting method. Initially, PEO and PVC are dissolved in tetrahydrofuran solution (THF) and were blended together to get homogeneous solution. The DOTP as was incorporated at the same amount for all composition. After that, conductive filler were added in the solution. The PEO/PVC/Graphite polymer films solutions were poured and casted on a glass mould after 4 hours and let it to evaporate in a fume cupboard at room temperature. The PEO/PVC/Graphite polymer films were formed when THF are fully dried. The same process was repeated to produce films which consist of different graphite content.

2.3 Tensile Test

This testing was conducted using Universal Testing Machine Instron 5569 by following to ASTM D638. Five rectangular samples were used for each film composition and were tested with crosshead speed of 10 mm/min. The tensile strength and young's modulus of each of the films were obtained.

2.4 Scanning Electron Microscopy (SEM)

The morphology of PEO/PVC/Graphite polymer films was examined by scanning electron microscope, JEOL JSM-6460 LA. The sample surfaces were coated by auto fine coater, JEOL JFC 1600.

2.5 Electrical Conductivity Testing

The conductivity measurement of PEO/PVC/Graphite polymer films was conducted using I-V measurement systems. The conductivity was tested by Keithley Model 4200 Semiconductor Characterization System. The samples were conducted with voltage varied from 0 V to 10 V. The electrical conductivity was determined using its relationship with resistivity and conductivity as shown in Equation 1and Equation 2 below:

$$\rho = R.\left(\frac{w \times t}{l}\right) \quad (1)$$

Where R = resistance of the films, w = width, t = thickness, l = length between the probe contact.

$$\sigma = \frac{1}{\rho} \quad (2)$$

3. Discussion

3.1 Electrical Conductivity

The dependency of the electrical conductivity of PEO/PVC filled with graphite polymer films on the filler content is shown in Figure 2. The results illustrate that the electrical conductivity of PEO/PVC/Graphite polymer films increased with increasing in filler loading. When the graphite content in polymer films increased, the electrical conductivity of PEO/PVC/Graphite polymer films also increased. The graphite fillers act as mediums for the electron or electrical charge to flow. This allows the electrons free to flow through the graphite fillers and matrix. Electrical conducting paths and networks are formed in films with increasing filler concentration. It can be observed that about 15 wt% of the graphite content, the increase of electrical conductivity was observed. In this condition, infinite cluster of graphite particles is built within the polymer matrix. These infinite clusters go through to the sample and represent a conductive path for electrons to move throughout the sample [7]. Huang et al. (2015) have investigated the effect of graphene loading on the properties of polypropylene (PP). The result shows that an increasing amount of graphene in the PP results in a more complete conductive path thus contributed higher conductivity [8].

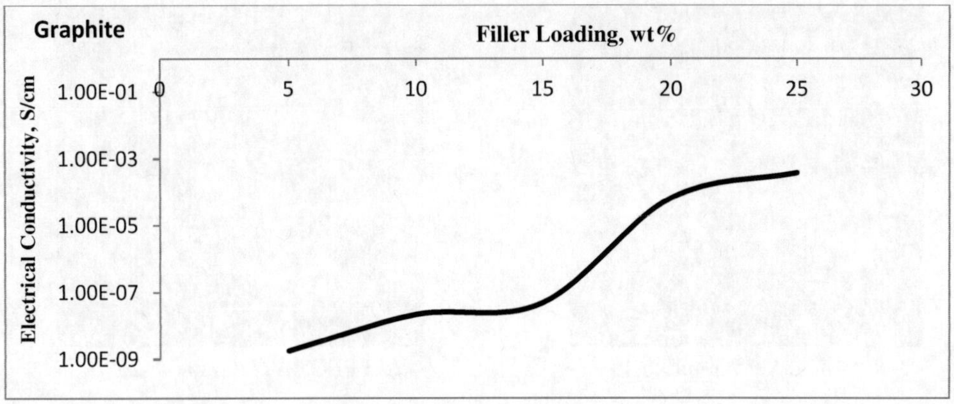

Figure 2. Electrical conductivity of PVC/PEO/Graphite polymer films at different graphite loading.

3.2 Morphology analysis

Figure 3 shows the surface morphology of pure PEO/PVC and PEO/PVC/Graphite polymer films with different filler loading varied from 5 wt% to 25 wt%. Figure 3(a) shows the surface morphology of the pure PEO/PVC polymer films without addition of graphite filler at x200 magnification. A flat surface was observed on the surface of pure PEO/PVC polymer films due to the miscibility between PVC and PEO blends.

Figure 3(b) shows the surface morphology of PEO/PVC polymer films with incorporation of 5 wt% of graphite loading. It can be observed that the graphite particles are distributed far because only a small amount was added into the blend. Figure 3(c) and Figure 3(d) shows the surface morphology of PEO/PVC polymer films with a graphite filler of 15 wt% and 25 wt% respectively. The distributions of filler particle are clearly uniform, but there is agglomeration of graphite on the surface of the sample at high filler loading. At 25 wt% of graphite filler, the filler are adequate to have intimate contact, thus conductive path is formed. This was convinced that by increasing amount of graphite in PEO/PVC polymer films has led to the high electrical conductivity behavior. The filler matrix interaction at higher filler content is substituted by the domination of filler-filler interaction.

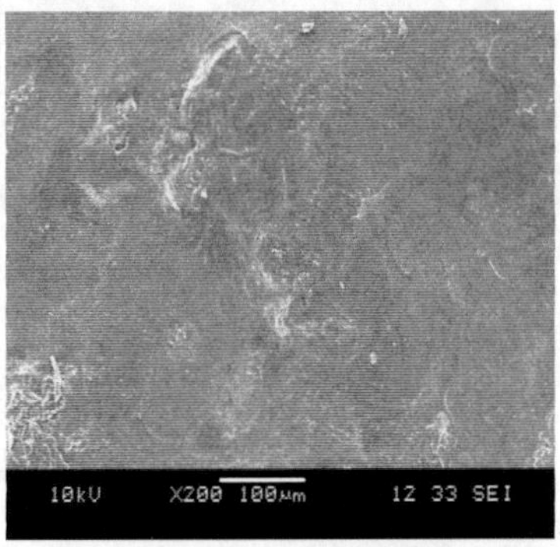

(a) Pure PEO/PVC

(b) PEO/PVC /Graphite-5

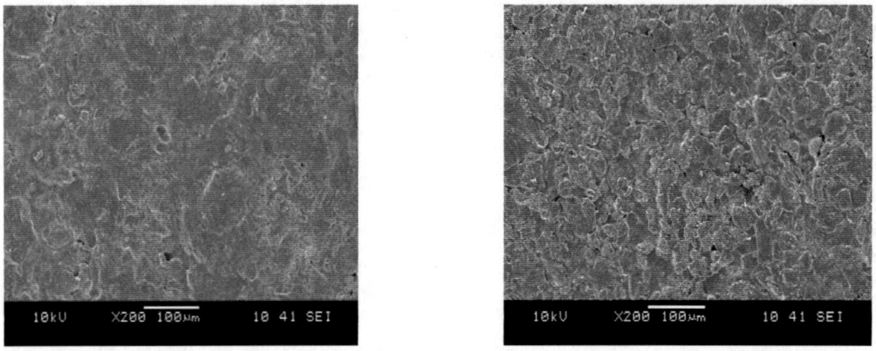

(c) PEO/PVC /Graphite-15 (d) PEO/PVC /Graphite-25

Figure 3. Surface morphology of PEO/PVC/Graphite polymer films at different graphite loading.

3.3 Tensile properties

The mechanical properties of polymer composites are particularly corresponded with the filler content, dispersion of fillers in the matrix, and the properties of the polymer matrix and the interaction between polymer matrix and filler. The results of tensile strength and young's modulus for pure PEO/PVC polymer films and PEO/PVC/Graphite polymer films containing various graphite loading are presented in Figure 4 and 5 respectively. By comparing to the tensile properties of pure PEO/PVC polymer films, the PEO/PVC/Graphite polymer films containing 5 wt% of graphite have 23 % lower tensile strength and 39 % lower young's modulus and the tensile properties of the polymer films are decreased continuously with addition of loading level of graphite. The addition of graphite fillers attributes a restrictive effect over the PEO/PVC polymer films by deformation of their molecular chains and restricting the motion [8].

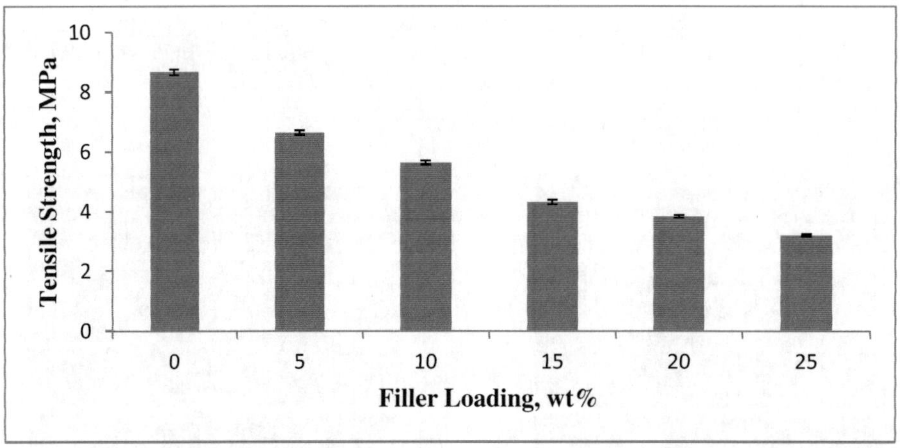

Figure 4. Tensile strength of PVC/PEO/Graphite polymer films at different graphite loading.

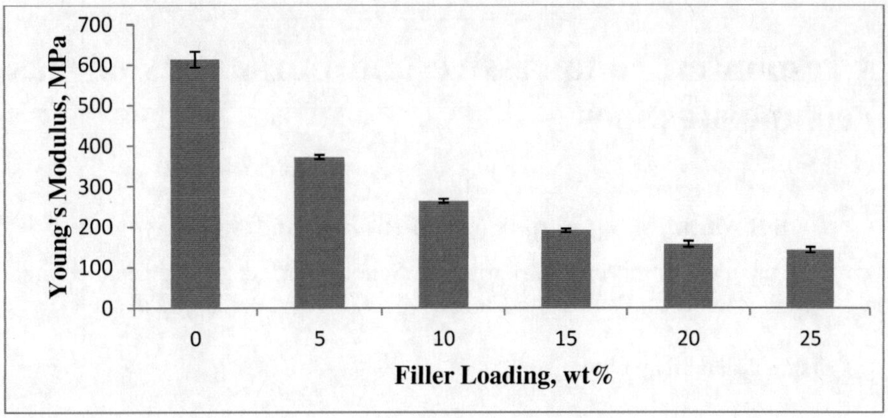

Figure 5. Young's Modulus of PEO/PVC/Graphite polymer films at different graphite loading.

4. Conclusion

In this study, the electrical and mechanical properties of PEO/PVC/Graphite polymer films were investigated. The addition of graphite filler to PEO/PVC blends induced an increased in electrical conductivity. It was observed that increasing amount of graphite filler has led to a decreased in tensile strength and young's modulus of the PEO/PVC/Graphite polymer films.

5. Acknowledgment

The authors acknowledge the support from the Ministry of Higher Education of Malaysia for offering scholarship which is MyBrain15 to further the PhD program at the Universiti Malaysia Perlis.

6. References

[1] Bouchet, J., Carrot, C., & Guillet, J. (2000). Conductive Composites of UHMWPE and Ceramics Based on the Segregated Network Concept. *Polymer Engineering and Science 40*, 36-45.

[2] Barba, A. A., Geatano, L., Matteo, D.A., & Domenico, A. (2006). Carbon Black/Silicone Rubber Blends as Absorbing Materials to Reduce Electro Magnetic Interferences (EMI). *Polymer Bulletin (Berlin) 57*, 587.

[3] Strumpler, R., & Glatz-Reichenbach, J. (1999). Conducting Polymer Composites. *Journal of Electroceramics 3(4)*, 329-346.

[4] Rupprecht, L. (1999). *Conductive Polymers and Plastics in Industrial Applications*. Norwich, New York: Plastics Design Library.

[5] Kulasekarapandian, K., Jayanthi, S., Muthukumari, A., Arulsankar, A., & Sundaresan, B. (2003). Preparation and Characterization of PVC–PEO Based Polymer BlendElectrolytes Complexed With Lithium Perchlorate. *International Journal of Engineering Research and Development*. 5 (11), 30-39.

[6] Ahlatcioılu Özerol, E., Şenkal, B. F., & Okutan, M. (2015). Preparation and characterization of graphite composites of polyaniline. *Microelectronic Engineering*, *146*, 76–80.

[7] Teh, P. L., Ng, H. T. & Yeoh, C. K. (2011). Recycled copper as the conductive filler in Polyester composites. *Malaysian Polymer Journal*, 6(1), 98-108.

[8] Huang, C.-L., Lou, C.-W., Liu, C.-F., Huang, C.-H., Song, X.-M., & Lin, J.-H. (2015). Polypropylene/Graphene and Polypropylene/Carbon Fiber Conductive Composites: Mechanical, Crystallization and Electromagnetic Properties. *Applied Sciences*, *5*(4), 1196–1210.

IOP Conf. Series: Journal of Physics: Conf. Series **908** (2017) 012021 doi:10.1088/1742-6596/908/1/012021

Impact responses, compressive and burst tests of glass/epoxy (GRE) composite pipes

F S B Abrar, M S A Majid[1], M J M Ridzuan and A R A Syayuthi

Mechanical Engineering Programme, School of Mechatronic Engineering, Universiti Malaysia Perlis, Pauh Putra Campus, 02600 Arau, Perlis, Malaysia

shukry@unimap.edu.my

Abstract. The paper presents the impact responses, compression and burst tests of glass reinforced epoxy (GRE) composites pipes. Impact loadings of three different energy levels (5 J, 7.5 J, and 10 J) were applied, followed by monotonic burst tests. Uniaxial compressive tests were conducted GRE samples using a universal testing machine in accordance with ASTM D695-10. In addition, the tests were also repeated with samples of different winding angles of ±45 ° and ±55 ° and tested at room temperature, and elevated temperatures of 45 °C and 65 °C. The result shows that the higher the impact energy applied to the pipes, the lower the burst strength of the pipes. The maximum burst strength found decreased with an increase in the impact energy level. The results also indicate that the strength of the GRE pipes significantly decreases with increase in temperature though, they are also found to increase as the winding angles decrease.

Keywords: Glass/epoxy composite pipes; composites materials; crushing behaviour; elevated temperatures; compression test

1. Introduction

In the current years, composite pipes are widely utilized in marine division and also as underground passage of fluid elements such as oil, firewater, natural gas, wastewater, and also drinking water. This is because of their high heat resistance, corrosion competent mechanical properties and easy to handle. They are more economical compared to steel and can be obtained at a lower price. The investigation on mechanical properties of glass fibre reinforced epoxy (GRE) pipes which including compression, tensile, and impact test have been conducted from recent studies [1,2].

The burst strength of a GRE pipes is a crucial aspect to be considered in their life cycle of service since they can be subjected to different combination of loadings. The effect of impact test and quasi-static on burst pressure have been studied by Matemilola et al. They discovered that the decline in the pressure bearing capacity is caused by fibre micro-buckling of the exterior layer at the variety of the impact point [3]. Before testing the composite tubes under internal pressure test, Curtis et al. implemented lateral indentation and low-speed impact test. They found that strain measurements confirmed compelling redistribution of strains when loading and the damage (micro-cracking) only weaken overall stiffness marginally but did result in altered strain distribution [4].

Other than that, there is also work reported on the effects of winding angles and elevated temperatures on the efficiency of the GRE pipes. The effect of winding angles on biaxial ultimate elastic wall stress (UEWS) has been studied by Abdul Majid et. al. [5], and they concluded that the different winding angles of GRE pipes have significant effect on the mechanical properties of GRE pipes. Abdul-Majid et. al [6] also looked into the performance of GRE pipes under hydrostatic and biaxial load conditioned at elevated temperatures. Similarly, the results also indicates that the failure of the GRE pipes greatly affected by the test temperatures.

2. Materials and experimental procedure

2.1 Composite fabrication

The filament-wound E-glass fibre epoxy pipes were fabricated at Advanced Materials Research Centre (AMREC), SIRIM Kulim High Tech Park. The process involved glass fibre roving wetting by passing through a resin bath. The wet fibre then wound onto the mandrel at angles of ±45° angle and ±55° to form 6 layers of composite shells. The dimension of the pipes is approximately ±1000 mm in length and ±2.5 mm in thickness. The pipes were then cured for 3 hours at 130°C. Then, the tubes were left to cool at room temperature before withdrawal from the mandrel.

2.2 Drop weight impact test

To study the impact response, a model IM10 drop weight-impact tester supplied by IMATEK was used. The impactor surface was hemispherical with a radius of 12.7 mm and the total mass including the impactor and carriage was 9.6209 kg in this experiment. Three different impact energy levels, which are 5 J, 7.5 J and 10 J were chosen and the test was performed at room temperature. A high-speed camera was used to record the movement of the impact load on the samples during the impact tests.

2.3 Burst test

As shown in Figure 1, the burst tests were performed using a in house-built pressure test rig in accordance to ASTM D1599 standard. The pipe was carefully filled with tap water from the inlet of one of the end fittings. It is of high importance to ensure that the entire system is 'bleed-out' for approximately 3-4 min before the test began to assure that no air bubbles were trapped. The strain gauge was installed near the impacted area on the pipe to record the axial strain experienced by the pipe. The pipes wthen hung in a free-free condition by using lubricated rubber rings to assure accuracy of the strain measurements. Once all checked out and the instrumentations functioning, the pipe then gradually pressurised until the it ultimately burst.

Figure 1. Test rig for conducting hydrostatic burst test.

2.4 Compression test

The universal uniaxial compression tests at increased temperatures were performed using a universal testing machine (UTM, Shimadzu) in accordance with ASTM D695-10. The test temperatures were ranged from room temperature to 45°C and 65°C. These selected temperatures were based on the determined T_g value obtained through differential scanning calorimetry (DSC) tests. To examine the failure of the pipes at different temperatures, the test temperatures were set at; below the T_g, and close to T_g. The tests were performed at a crosshead speed of 1.0 ± 0.3 mm/min as elucidated in ASTM D695-10 standard.

3. Results and discussion

3.1 *Impact responses*

All through the drop impact tests, the force–displacement curves of the impacted samples were documented and evaluated. Figure 2 shows the force–displacement response of the impacted pipes in the neutral settings with impact energies of 5 J, 7.5 J and 10 J. A rebounding-type curve was examined for the pipe. The closed curve obtained suggested that no severe damage that would cause functional failure was applied to the pipe [7]. At the beginning, the slope represents the elastic response, where the pipe underwent initial deformation. The intermediate slope illustrates the plastic response, suggesting the earliest permanent damages such as matrix cracking and delamination to take place. From physical assessment, it unveiled that matrix cracking occurred on 5 J impacted pipes and both matrix cracking and delamination damage were suspected for the impact energies of 7.5 J and 10 J. The final slope indicates the rebound of the striker from the pipe surface. The neutral samples force-displacement curve ended faster, yielding greater ultimate force after the onset of initial damage. The test was conducted for the chosen impact energy levels such that no fibre breakage and penetration nor perforation of the impactor was expected during the tests.

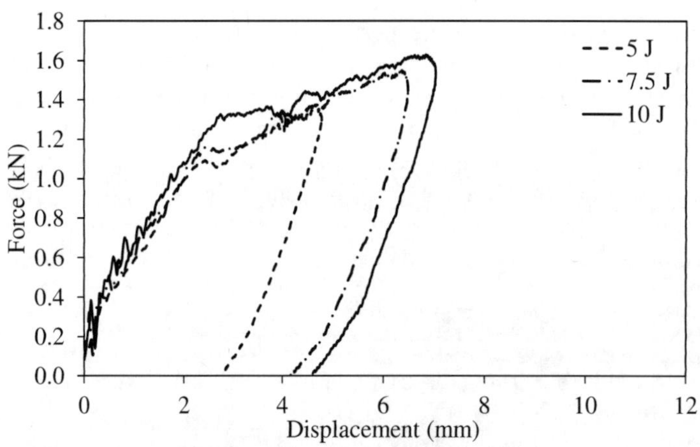

Figure 2. Force–displacement responses of the pipe after drop impact test.

The area under the force–displacement plot gives that the energy dissipated for plastic deformation as the energy absorbed by the specimen was consumed for failure. The absorbed energy by the specimen was computed mathematically from the force–displacement response curve by using the following Eq. (1):

$$W = \int F.ds = F_m(S_f - S_i) \quad (1)$$

where W is the sum of energy absorbed by the impacted pipe and F_m is the applied average force. S_f is the maximum displacement, and S_i is the initial displacement under impact loading [7].

Figure 3 shows an image of the damages that occurred near the impact point on the pipes. As illustrated in Fig. 4, the size of damage area enlarged as the impact energy increased. The lower slope illustrates, with an increase in the impact energy the evolution of matrix cracking damage and delamination will occur [4]. The size of the damaged area produced by the higher-energy impact event is an indicator of the material's capability to absorb energy. Visual analysis of the impact-damaged specimens explained that with an increase in the impact energy will cause the damaged area tended to become more localised. Identical information were also recorded by Deniz and Karakuzu [8].

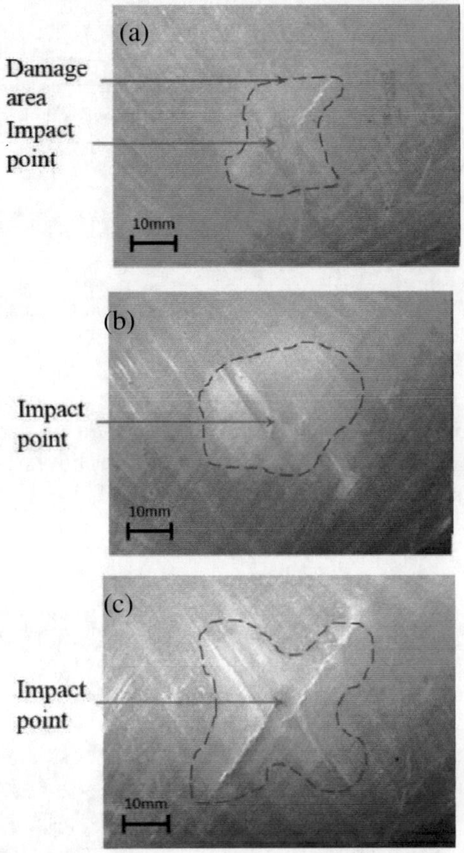

Figure 3. Physical damage to the pipe specimen after (a) 5 J, (b) 7.5 J, and (c) 10 J impact tests.

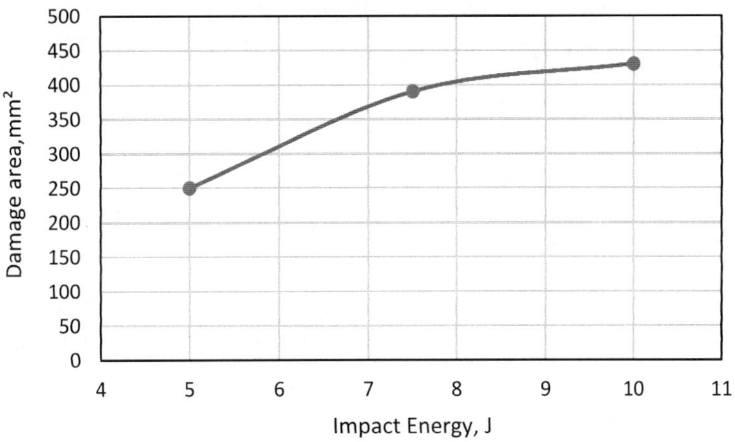

Figure 4. Damage area against impact energy of the impacted pipes

3.2 Burst strength

Monotonic burst tests were performed on the impacted GRE pipes in closed-ended conditions. The purpose these tests is to determine the burst strength of the pipes after they were subjected to loadings of different impact energies. Table 1 presents result analysis of the impacted specimens that underwent burst pressure tests. It shows that the increase in impact energy will decrease the burst strength This can be associated with the decrement in the mechanical strength of the pipes due to the plasticization of the matrix, which weakens the interfacial fibre–matrix interface.

Table 1. Results for monotonic burst tests on pipes impacted at various energy levels.

Impact energy (J)	**Max. burst pressure (MPa)**	**Axial stress (MPa)**	**Hoop stress (MPa)**	**Strain (%)**	**Failure type**
5	10.50	113.80	227.60	0.32	Weepage
7.5	6.57	75.78	151.55	0.18	Weepage
10	6.04	69.92	139.84	0.17	Eruption

The results shows that the pipe samples impacted with the energy level of 5 J persistently displayed greater burst pressure compared to specimens impacted by energy levels of 7.5 J and 10 J. The 5-J-impacted sample yielded a burst strength of 113.8 MPa, but reduced to 75.78 MPa and 69.92 MPa for the 7.5-J- and 10-J-impacted specimens, respectively. This signifies a strength degradation of more than 30% in both these cases. The authors recorded the same decrement in burst strength on impacted pipes in their report [4], [9]–[12].

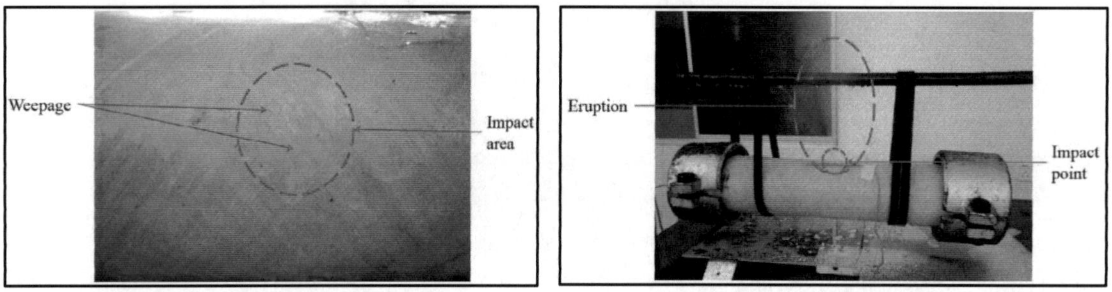

(a) (b)

Figure 5. The impacted pipe sample during monotonic burst test with (a) weepage and (b) eruption damage.

3.3 Compressive properties

Figure 6 displays the stress-strain response of a GRE pipe sample during the uniaxial compression test. Thee sample was taken from a ±45 ° angled pipe crushed at 65 °C. The uniaxial load increased continuously until it come to the failure point during the compression process. Abrupt cracks appeared when the specimens experience failure, and this caused the upper segment to lose its structural integrity. As a result, the strength drops swiftly before stabilising at a constant strength. Nonetheless, the load is still exerted on the specimen causing it to keep on crushing as the strain increases.

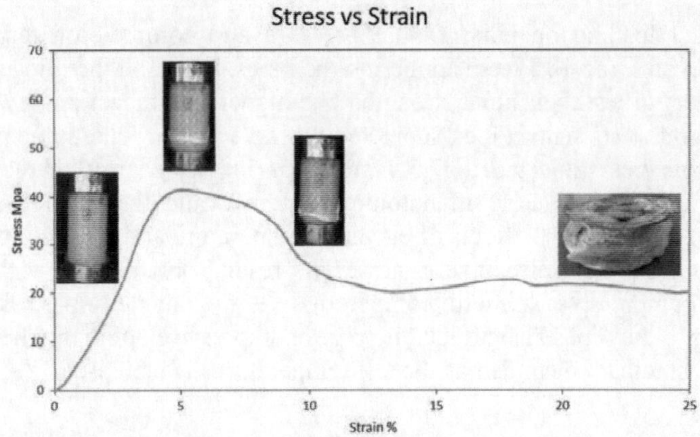

Figure 6. Stress-strain response of GRE pipe during uniaxial compression testing

As shown in Fig. 6, the stress increased linearly with the strain until it reached the highest peak. During this phase, the pipe is still in an elastic state, which indicates that the pipe can still turn back to its original state. As soon as the stress reached the peak, the stress reduced speedily. Now, the pipe is in a plastic state, which implies that the pipe can no longer turn back to its original state. It is permanently deformed, and buckling usually begins at this point. As the crushing process continues, the stress retained at a low but steady value following the initial failure. Following the initial failure, the stress decreases, and the load starts to propagate progressively. As the pipes reach ultimate failure, the stress begins to increase speedily. The rapid surge is because of the cumulative effect of the specimen materials, which causes to exhibit a greater value of stress compared with that of the initial failure. At this point, the pipes cannot deform anymore.

At the onset of failure, the pipe displays local buckling and the unsteady response. As the crushing progresses, more buckles formed which eventually lead to progressive buckling. As for low temperatures (RT, 45 °C), the stress-strain behaviour exhibited lower strain responses. It is suspected that, at low temperatures, the material is rigid thus after initial deformation, the brittle-like material would quickly fracture thus no further deformation could be observed.

Fig. 7a, in the case of the ±45° angled pipes, the maximum compressive stresses of the pipes compressed at RT, 45 °C, and 65 °C were 118.9 MPa, 57.3 MPa, and 23.9 MPa respectively. Figure 7b shows the compressive strength of the ±55 ° angled pipes. These outcomes show an noticable pattern, where the ±45 ° angle pipe has a higher compressive stress at RT, 45 °C and 65 °C compared to pipes at an angle of ±55 in the respective temperatures.

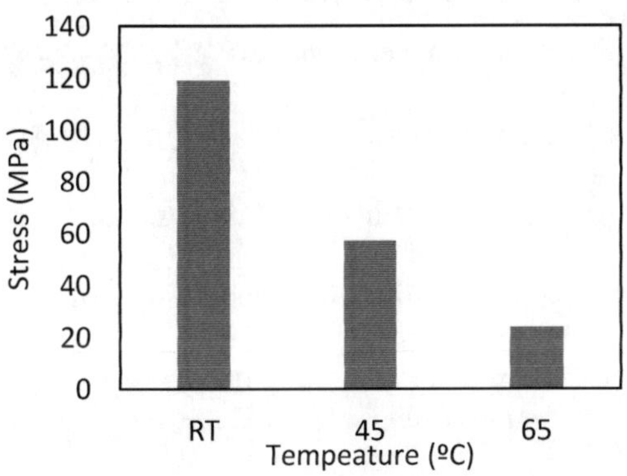

(a)

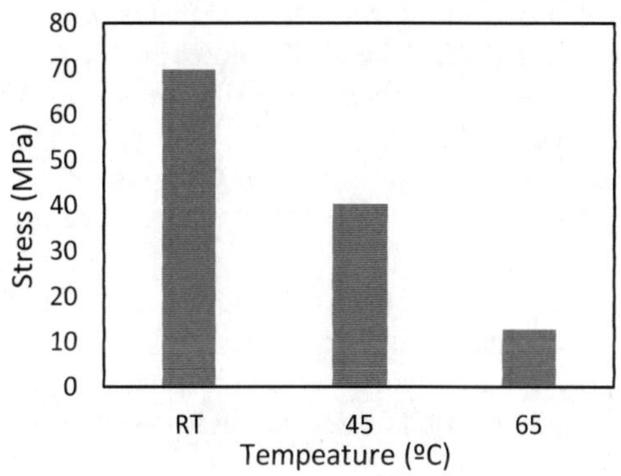

(b)

Figure 7. The compressive strength of (a) ±45 °, (b) ±55 ° wound GRE pipes under RT and elevated temperatures

4. Conclusion

The effects of impact loading on burst failure pressure and compression tests were investigated. Impact energy of 5 J, 7.5 J and 10 J were applied to the pipes. Later, all the impacted pipes were tested with burst pressure test to test their burst strength. The higher the impact energy applied to the pipes, the lower the pipe's burst strength. Pipes subjected with 5 J impact energy acquire 30% more burst strength compared to pipes subjected to 7.5 J and 10 J impact energy. Three failure modes were detected; whitening spot in the early monotonic burst test, and then will cause manifestation of weepage and eruption failures. The effects of elevated temperature and different winding angles on the pipe compressive strength were also investigated. The results obtained show that pipes with ±45 ° angle have higher compressive strength compared to ±55 ° angle pipes. Besides, the elevated temperatures applied to the pipes also affects the pipes compressive strength where the strength of the GRE pipes found substantially degraded as the test temperatures is increased.

5. Acknowledgements

The authors are thanking Universiti Malaysia Perlis and the Ministry of Education, Malaysia, for providing financial support (FRGS No: 9003-00519) and the access for using the facilities during the course of the research conducted. Many thank you also to the laboratory technician for the assistance throughout the project.

References

[1] A. Hawa, M.S. Abdul Majid, M. Afendi, H.F.A. Marzuki, N.A.M. Amin, F. Mat, and A.G. Gibson, "Burst strength and impact behaviour of hydrothermally aged glass fibre/epoxy composite pipes," *Mater. Des.*, vol. 89, pp. 455–464, Jan. 2016.

[2] Fitriah S N, Majid M S A, Ridzuan M J M, Daud R, Gibson A G and Assaleh T A 2017 Influence of hydrothermal ageing on the compressive behaviour of glass fibre / epoxy composite pipes *Compos. Struct.* **159** 350–60

[3] Stronge W J 2016 Low-Speed Impact Damage in Filament-Wound CFRP **1**

[4] Curtis J, Hinton M ., Li S, Reid S . and Soden P . 2000 Damage, deformation and residual burst strength of filament-wound composite tubes subjected to impact or quasi-static indentation *Compos. Part B Eng.* **31** 419–33

[5] Majid M S A, Afendi M, Daud R, Gibson A G and Hekman M 2013 Effects of Winding Angles in Biaxial Ultimate Elastic Wall Stress (UEWS) Tests of Glass Gibre Reinforced Epoxy (GRE) Composite Pipes, *Adv. Mater. Res.*, vol. **795** 424–8

[6] Majid M S A, Assaleh T A, Gibson A G, Hale J M, Fahrer A, Rookus C A P and Hekman M 2011 Composites : Part A Ultimate elastic wall stress (UEWS) test of glass fibre reinforced epoxy (GRE) pipe *Compos. Part A Appl. Sci. Manuf,* **42** 1500–8

[7] Belingardi G and Vadori R 2002 Low velocity impact tests of laminate glass-fiber-epoxy matrix composite material plates *Int. J. Impact Eng.* **27** 213–29

[8] Deniz M E and Karakuzu R 2012 Seawater effect on impact behavior of glass–epoxy composite pipes *Compos. Part B Eng.* **43** 1130–8

[9] Gning P B, Tarfaoui M, Collombet F and Davies P 2005 Prediction of damage in composite cylinders after impact *J. Compos. Mater.* **39** 917–28

[10] Chang D J 2003 Burst tests of filament-wound graphite-epoxy tubes *J. Compos. Mater.* **37** 811–29

[11] Uyaner M, Kara M and Şahin A 2014 Fatigue behavior of filament wound E-glass/epoxy composite tubes damaged by low velocity impact *Compos. Part B Eng.* **61** 358–64

[12] Kara M, Uyaner M, Avci A and Akdemir A 2014 Effect of non-penetrating impact damages of pre-stressed GRP tubes at low velocities on the burst strength *Compos. Part B Eng.* **60** 507–14

Prediction on fracture risk of femur with Osteogenesis Imperfecta using finite element models: Preliminary study

S B C Wanna[1], K S Basaruddin[1], M H Mat Som[1], M S Mohamad Hashim[1], R Daud[1], M S Abdul Majid[1] and A R Sulaiman[2]

[1]School of Mechatronic Engineering, Universiti Malaysia Perlis, Pauh Putra Campus, 02600, Arau, Perlis.
[2]Department of Orthopaedics, School of Medical Science, Universiti Sains Malaysia, 16150 Kota Bharu, Kelantan.

khsalleh@unimap.edu.my

Abstract. Osteogenesis imperfecta (OI) is a genetic disease which affecting the bone geometry. In a severe case, this disease can cause death to patients. The main issue of this disease is the prediction on bone fracture by the orthopaedic surgeons. The resistance of the bone to withstand the force before the bones fracture often become the main concern. Therefore, the objective of the present preliminary study was to investigate the fracture risk associated with OI bone, particularly in femur, when subjected to the self-weight. Finite element (FEA) was employed to reconstruct the OI bone model and analyse the mechanical stress response of femur before it fractures. Ten deformed models with different severity of OI bones were developed and the force that represents patient self-weight was applied to the reconstructed models in static analysis. Stress and fracture risk were observed and analysed throughout the simulation. None of the deformed model were observed experienced fracture. The fracture risk increased with increased severity of the deformed bone. The results showed that all deformed femur models were able to bear the force without experienced fracture when subjected to only the self-weight.

1. Introduction

Osteogenesis imperfect (OI) is a genetic disease associated with a brittle bone due to less of collagen count. To date, it was reported that osteogenesis imperfecta could affect approximately 6 or 7 per 100,000 people and occur in approximately 1 in 20,000 births with no gender dependence or race dependence [1]. Bone mainly made up of collagen where type I collagen is the most abundant which cover up almost 95% of the entire bone format [2] . Type I collagen is the result of cross-link between two main gene named collagen type 1 alpha 1 (COL1A1) and collagen type 1 alpha 2 (COL1A2) where the alteration in one of these two genes will lead to abnormal structure collagen produced in bone where it causes low bone mass and bone fragility which leads to health condition such as OI [3], [4]. Generally, OI can be classified into four main types namely Type I, Type II, Type III and Type IV which basically based on clinical and radiology feature [5]. However, up to now, there is no specific cure for OI patients. OI patients are undergoing several treatments based on the severity of the disease recommended by their surgeons. Those treatments included fracture care, physical therapy, bracing, and medication. Thus, in this situation, the surgeons are having a limitation in order to predict the fracture risk of OI bone before the bone break. In general, the capability of bone to resist fracture risk depend on the integrity of all levels in the bone principle which include the entire amount of bone material/mass, the size and shape of the whole bone, the microarchitecture of cortical and trabecular bone, trabecular bone volume per tissue volume, trabecular number and thickness, the intrinsic properties of the bone material, the constitution of the organic matrix and the interaction between mineral and organic phases [6]. Despite the medical treatment, the surgeon needs to perform the surgical procedure on OI patient as the last option to correct the bone deformity such bowing. For the past ten years, intramedullary rod implant had been become bone assist to ensure the long bone straight and prevent fracture [7], [8]. However, the drawback from this option is the rod itself might proximally and distally [9], [10] .Thus, it is very important to determine and to know the fracture risk of the long bone to help the surgeon to predict the occurrence. In doing so, the bone fracture risk of OI can only be predicted through the simulation which only can be done in finite element (FE) study. Finite element study had been extensively used in orthopedics biomechanics field for over almost a decade to study and investigate the behavior, process, mechanism, and properties biological system under the mechanical forces or also known as biomechanical phenomena [11]–[13]. Its ability to cope with biological structure especially non-linear material properties, irregular geometry, and complicated boundary condition are the main reason for its widely used among researchers [14].

Therefore, in this study, the fracture risk of femur associated with OI bone was investigated through FE analysis. Ten reconstructed femur models with variation of deformed angles were observed when directional force was applied vertically in the femoral head in static condition.

2. FE modelling of OI Bone

FE modelling of OI bone is the core process in this study in order to investigate the fracture risk associated with femur bone deformity. The load was applied vertically to the femoral head in sagittal plane of the femur model

2.1 Standard Femur (SF)

The standard femur (SF) was obtained from Biomedtown website which was first proposed by Viceconti et al [15] to make it available for download. The length of SF is 409.4mm. The SF then imported into finite element software, ANSYS for simulation process. The SF was first modified in the design modeler to separate the femoral head and femur shaft part as shown in figure 1. The next was to develop ten different forms of bending angle where the middle of the femur shaft act as reference point to resemble the real OI bone.

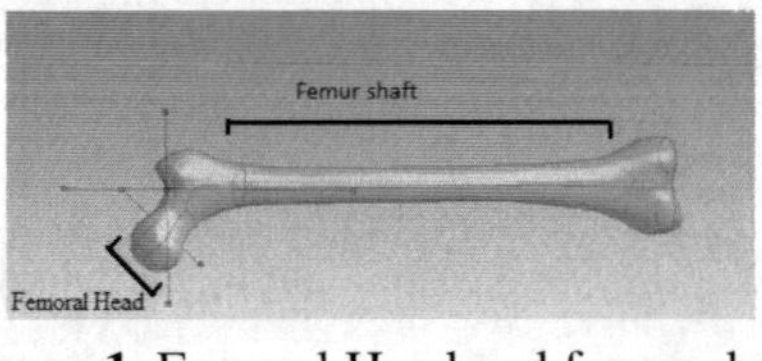

Figure 1. Femoral Head and femur shaft.

2.2 Mechanical properties

Mechanical properties was assigned based on the work conducted by Fan et al [16], [17] using nanoindentation test performed on femur bone sample. Thus, the same mechanical properties were applied in this simulation which the SF was set as isotropic linear elastic with the Young's Modulus of 19 GPa and Poisson ratio of 0.3.

2.3 Convergence Test

Convergence test was conducted to observe whenever the stress reached its consistency in response to the number of element providing those number of elements are the optimum element size. Ten convergence tests performed by varying the element size ranging from 10 mm to 1 mm. Figure 2(a) showed the number of element against Von Misses stress for convergence test graph. Figure 2(b)(c) and (d) showed the meshing result when the element size was set to 10 mm, 4 mm and 3 mm respectively. Compared to figure 2(b) and figure 2(c), figure 2(d) meshing seems more fine and able to properly mesh around the curve and edge region of the geometry with the element size of 3 mm. The meshing number started to show consistency when stress reached the value of 13.845MPa when the element size was set to 3mm. The bigger the element size will produce yield less complexity, however it will reduce the efficiency and effectiveness of the analysis as the distribution of stress among the bigger cell which made up of nodes and element is less accurate rather than smaller cell. Thus it is very important to choose the suitable element size as the meshing defines the model for the analysis. In this study, the suitable mesh for the SF geometry was consisted of 28105 number of nodes and 12745 number of elements of tetrahedron and hexahedron method when the element size was set to 3 mm.

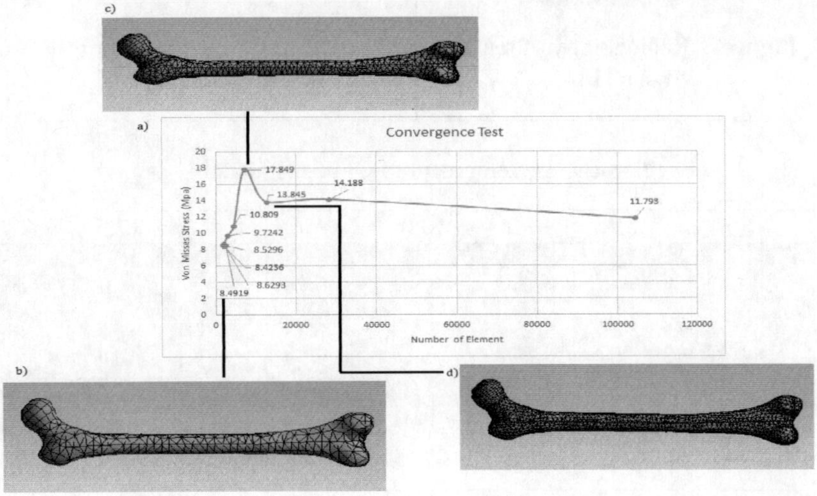

Figure 2. Convergence Test. (a) Von Misses stress against number of element. (b) Meshing with 10 mm element size. (c) Meshing with 4 mm element size. (d) Meshing with 3 mm element size.

2.4 OI Bone modelling

The SF was then undergoing deformation process using ANSYS software to obtain the deformed shape of bone to resemble real OI bone. Equation (1) was applied to deform the SF model up to ten different angles of bending.

$$\tan\phi = \frac{dx}{\left(\frac{L}{2}\right)} \qquad (1)$$

where $\emptyset$ denotes as deformed angle, dx as deformed length and L is length of femur. The bone length was measured and divided by half, where this middle point was used as the reference point to deform the bone to represent the OI bone geometry. Hence all the deformed models have the same deformed point of reference. Figure 3 showed the radiography of OI bone geometry and figure 4 demonstrated deformed model of SF bone geometry to resemble the radiography image as shown on figure 3. Ten deformed angles ranging from 7.5 ° to 30.0° with different of 2.5° each was simulated to represent the severity of the deformed bone shown on Table 1. Each of the deformed SF geometry was then applied with force in vertical direction at the femoral head which act as upper body weight and fixed support at the end of the femur shaft to sustain the force applied.

IOP Conf. Series: Journal of Physics: Conf. Series **908** (2017) 012022 doi:10.1088/1742-6596/908/1/012022

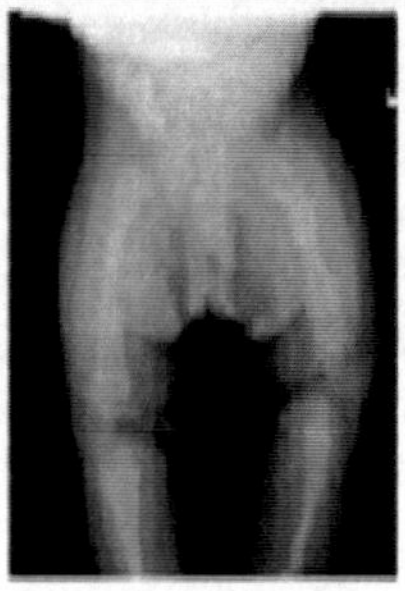

Figure 3. Radiography of real OI bone [18]

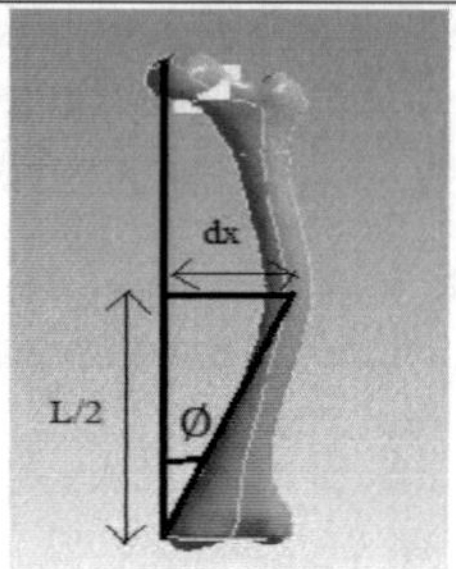

Figure 4. Deformed model of SF bone geometry

Table 1. Development of OI bones model.

7.5°	10.0°	12.5°
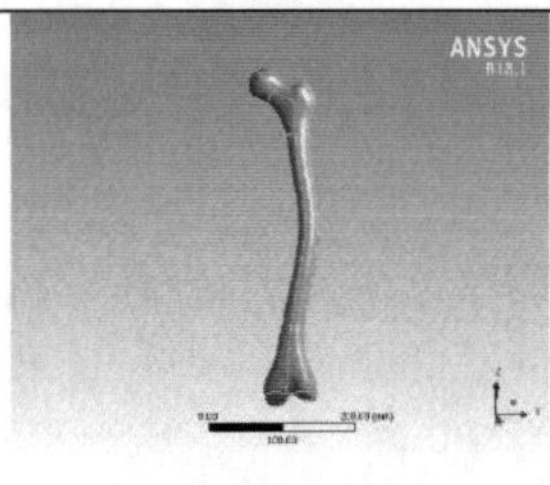	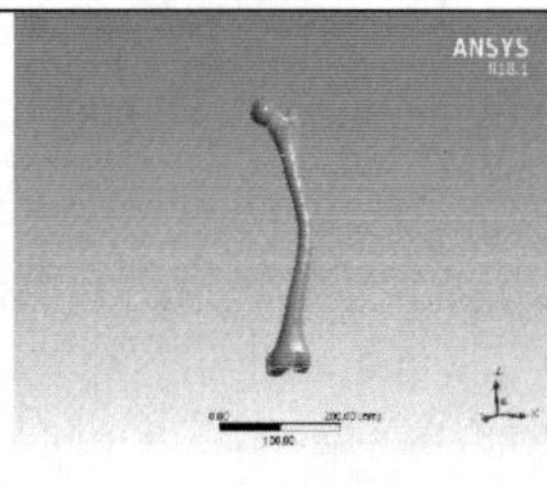	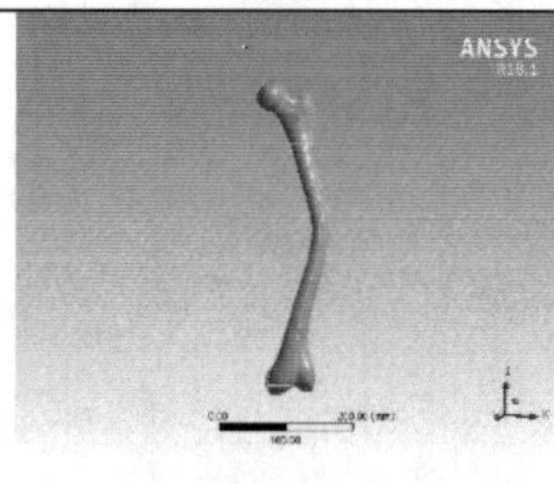
15.0°	17.5°	20.0°
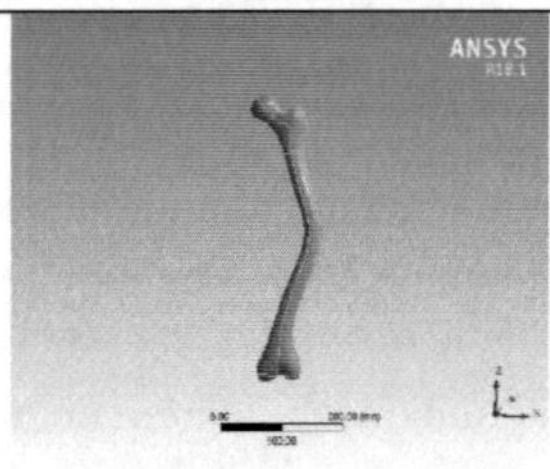	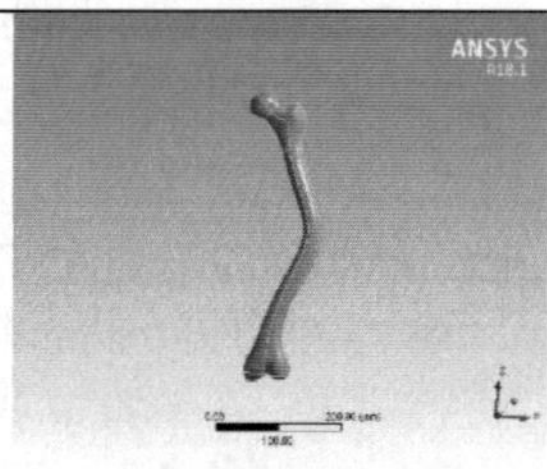	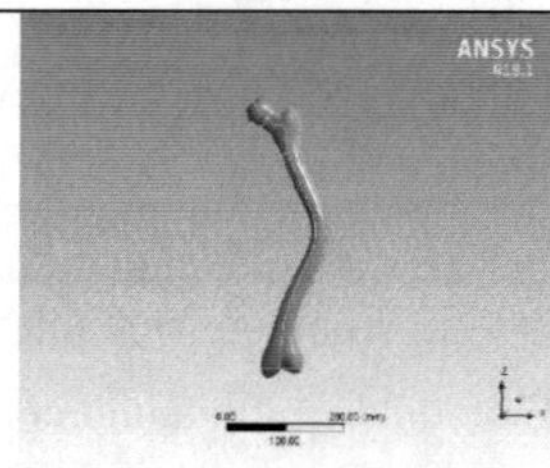
22.5°	25.0°	27.5°
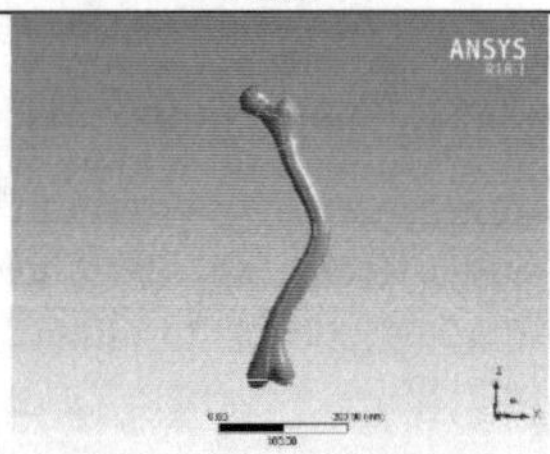	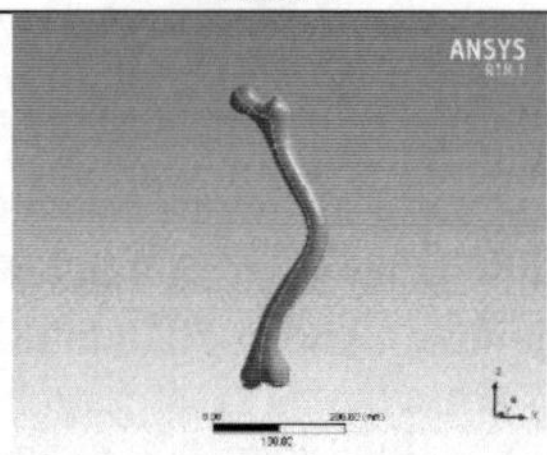	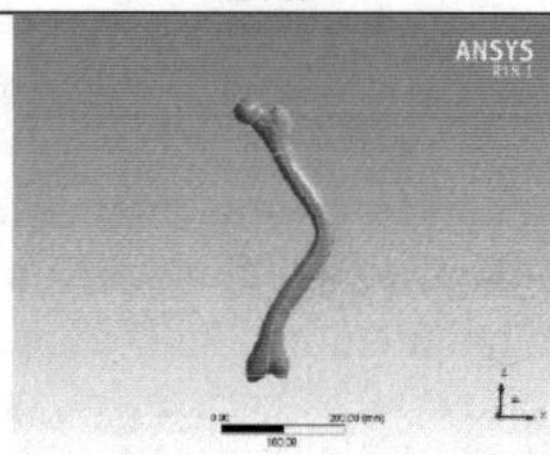
30.0°		
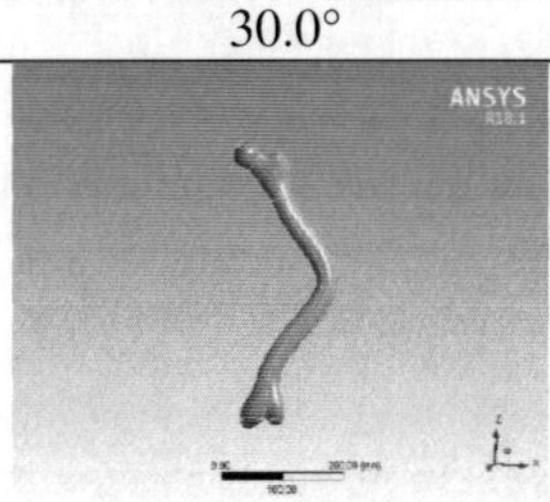		

2.5 Boundary condition
The boundary conditions for the SF were set to be in static condition. The static condition is considered one of the crucial standing position for OI patients as this condition provide the greatest support specifically in femur to balance the whole body as the forces act mainly on hip. This force influenced the weight in static condition of OI patients. Thus to observed this behaviour, normal weight of 512 N was assumed to represent the average normal weight for normal person. In order to included only for upper body weight, based on de Leva's Segment Weight data [19], the percentage of upper body weight was calculated then divided by half to represent femur on both legs. The weight of 151.705 N was considered after calculation to represent as force acting on SF during simulation. The force was applied vertically at the femoral head in sagittal plane for all deformed models and the end of femur shaft act as fixed support as shown in figure 5.

Figure 5. Anterior view of the SF with force acting on femoral head.

2.6 Fracture risk criteria
Fracture risk was calculated based on calculation develop by Keyak et al [20] as shown in Equation 2. Fracture strength of 115 MPa was assumed in this study based on the nanoindentation test done by previous study [21]–[23]. The femur bone considers undergo fracture if the value of fracture risk is equal or less than 1 (FR $\leq$1).

$$\text{Fracture risk} = \frac{\text{Fracture strength}}{\text{maximum principle stress}} \quad (2)$$

3. Results and Discussion
Maximum principal stress against severity of the deformed angle was presented in Figure 6. However only the upper body weight was included into this study to observe does the OI femur bone is able to support the upper body weight in standing position without involved any body movement or others activities. Figure 6 showed that for the least deformed angle, which is 7.5°, generate maximum principle stress of 8.64 MPa. For the most deformed angle of 30.0°, the maximum principle stress was found to be 10.37 MPa. The maximum principle stress increased linearly with the severity of the deformed angle. The fracture risk was then calculated based on Equation (2) and the correspond value of fracture risk against deformed angle was presented in Figure 7. From the graph, the fracture risk recorded 11.09 for the most deformed angle (30°). None of the model experienced fracture throughout the whole simulation process. Table 2 presented the stress contour for all developed OI models. The maximum principle stress was seen to be accumulated at the deformed region, in the middle of the femur shaft for all deformed models, thus making this deformed region were able to withstand the most stress compared to other region when force was applied vertically to the femoral head. As the

severity of deformed angle increase, the stress was seen appeared getting larger at the middle point of femur shaft as indicated by red region in Table 2 for deformed bowing angles of 10°, 20° and 30°.

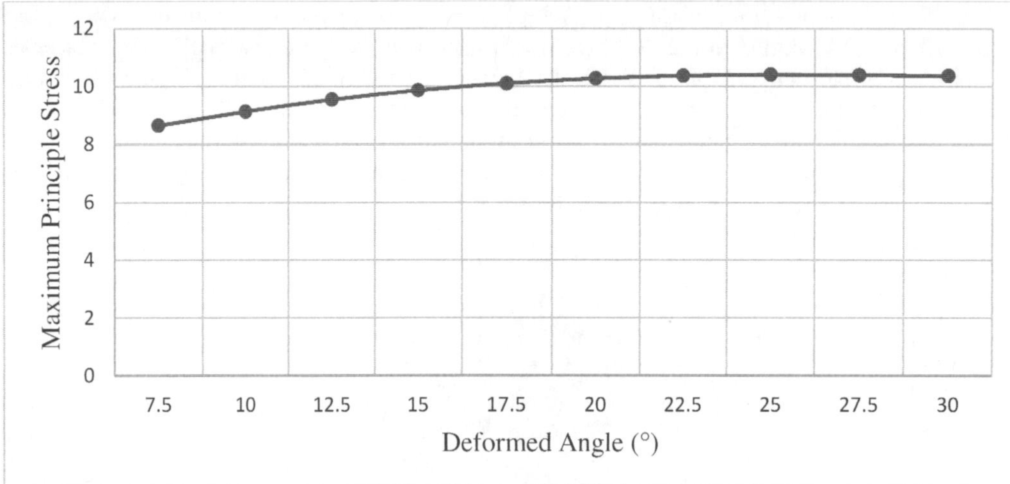

Figure 6. Effect on Deformed Angle on maximum principal stress.

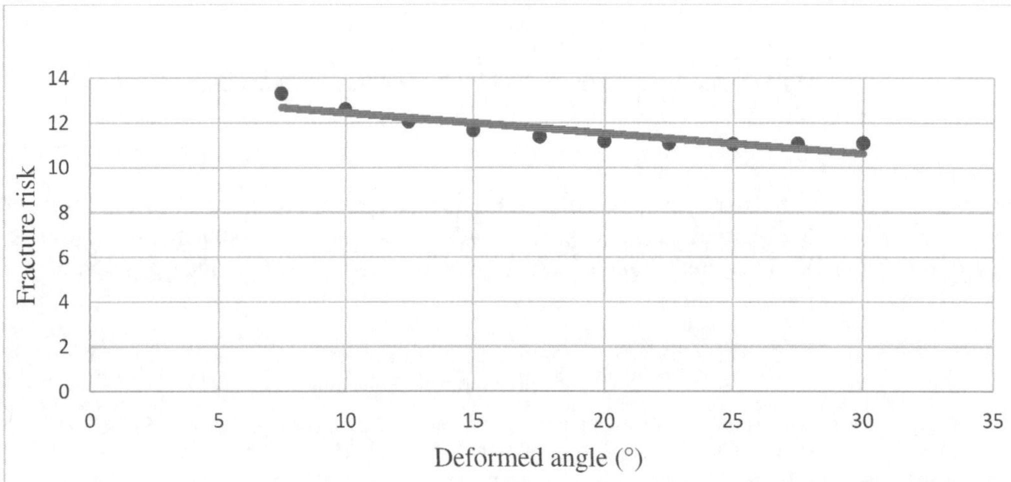

Figure 7. Effect of deformed angle on Fracture Risk.

IOP Conf. Series: Journal of Physics: Conf. Series **908** (2017) 012022 doi:10.1088/1742-6596/908/1/012022

Table 2. Stress contour for several deformed models.

<table>
<tr><td>10.0°</td><td>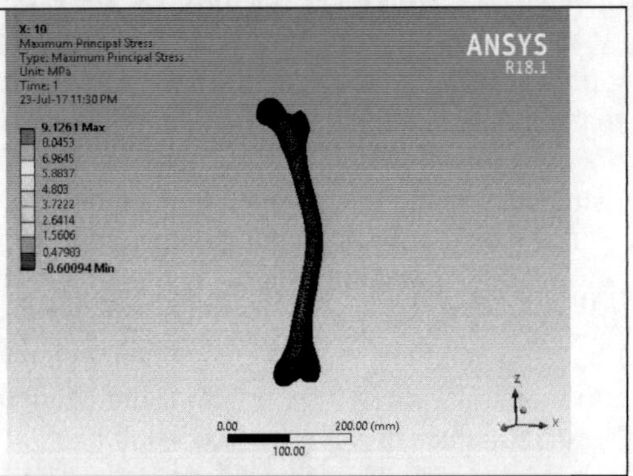
</td></tr>
<tr><td>20.0°</td><td>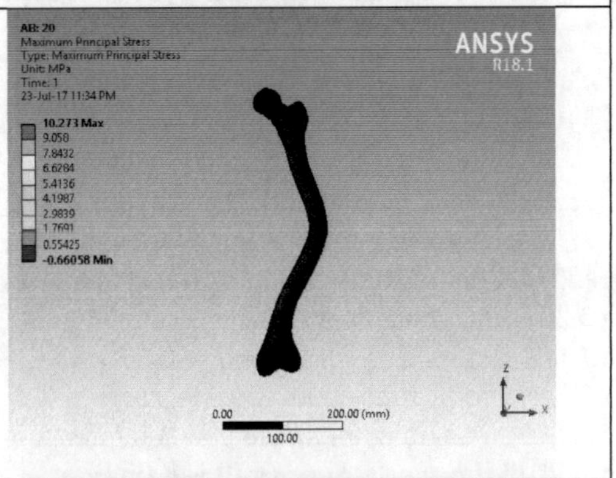
</td></tr>
<tr><td>30.0°</td><td>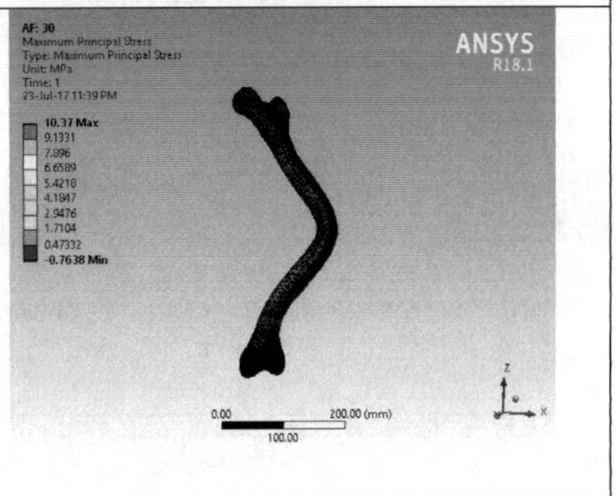
</td></tr>
</table>

This situation might lead to the possibility that static condition could bear the upper body weight by itself without fracture for all the severity level of OI. The fracture risk value was noticeable still high (11.089) even for the most deformed angle, 30°, however as mentioned early, the middle point of the femur shaft was used as reference point in this study to act as the bending point to estimate the fracture risk, meanwhile in real case, the bending angle might be slightly on different anatomical position which will produce different angle of bending and fracture risk. In addition, self-body weight was used as the force acting on femoral head in this study, thus, this result reflects that by the normal body weight, the OI bone was able to support the upper body on their own without fracture in standing position which involved no movement. Any external movement might actually cause fracture to occur. Nonetheless, this study was to provide preliminary result to show that the same approach could be applied when dealing with the similar issue in the real case.

4. Conclusion

The different OI bone models consisted of different severity of deformed bone have been developed and analysed in this FE study. The stress and fracture risk were observed against different deformed angle. Fracture risk associated with OI patients in femur bone could be fatal if there is no early detection. Thus, it very important to estimate and predict the fracture in femur in OI patient. In this study, it was noticed that in static condition, all the severity of the deformed bones were able to withstand the load in standing position. No fracture risk occurs for all the deformed angle in this position. On the other hand, this study also proves that modelling and simulation approach through finite element analysis could benefit to both the OI patients and medical support in order to predict the fracture in OI patients before surgical intervention.

5. Acknowledgments

The authors would like to thanks financial support from the Ministry of Education Malaysia through Fundamental Research Grant Scheme (FRGS) No. 9003-00606.

6. References

[1] I. Oakley and L. P. Reece 2010 *AANA Journal.* **78-1** 47.
[2] N. Fratzl-Zelman, B. M. Misof, K. Klaushofer, and P. Roschger 2015 *Wien. Med. Wochenschr* **165-13–14** 271.
[3] Primorac D, Anticević D, Barisić I, Hudetz D and Ivković A 2014 *Coll. Antropol.* **165-13** 271.
[4] M. JC 2011 *Nelson.Textbook of Pediatrics* S B Kliegman RM, Behrma RE, Jenson HB, p 19.
[5] D. D. M. Sillence D O, Senn A 1979 *J Med of Genet.* **16** 101.
[6] S and R. M. Boyd 2006 *J. Biomech.* **39** 1287.
[7] A. Clarke, S. Thomas, and F. Monsell 2013 *Orthop. Trauma.* **27-2** 101.
[8] D. Jones, H. Hosalkar, and S. Jones 2002 *Curr. Orthop.* **16-5** 374.
[9] B. M. Birke O, Davies N, Latimer M, Little D 2011 *J Pediatr Orthop.* **31** 458.
[10] L. J. Cho T, Kim J 2011 *J Bone Jt. Surg Br.* **93–B** 634
[11] G. F. H. J.M. Fritz, Y. Guan, M. Wang, P.A. Smith 2009 *Biomed Sci Instrument* 45 316.
[12] S. V. et Al 2007 *J. Biomech.* **40** 26.
[13] J. Jameson, C. Albert, P. Smith, R. Molthen, and G. Harris 2011 *Proc SPIE 7965 Med. Imaging 2011 Biomed Appl Mol Struct Funct Imaging.* **7965** 79650I.
[14] J. R. Jameson, C. I. Albert, B. Busse, P. a. Smith, and G. F. Harris 2013 *Proc SPIE 8672 Medical Imaging 2013: Biomedical Applications in Molecular, Structural, and Functional Imaging.* **8672**.
[15] T. A. Viceconti M, Casali M, Massari B, Cristofolini L, Bassini S 1996 *J. Biomech.* **29** 1241.
[16] Z. F. Fan, P. Smith, F. Rauch, and G. F. Harris 2007 *Compos Part B Eng.* **38** 411.
[17] Z. Fan, P. A. Smith, E. C. Eckstein, and G. F. Harris 2006 *J. Biomed Mater Res - Part A.*
[18] L. Zeitlin, F. Fassier, and F. H. Glorieux 2003 *J Pediatr Orthop B* **.12-2** 77.
[19] M. Damavandi, F. Barbier, J. Leboucher, N. Farahpour, and P. Allard 2009 *Med Eng Phys.* **31-7** 880.
[20] S. H. Keyak JH, Rossi SA, Jones KA 1998 *J. Biomech.* **31** 125.
[21] J. M. Fritz, Y. Guan, M. Wang, P. A. Smith, and G. F. Harris 2009 *Med Eng Phys.* **31** 1043.
[22] Z. Fan, P. a Smith, G. F. Harris, and F. Rauch 2007 *Connective Tissue Research.* **48** 70.
[23] M. C. H. Van der Meulen, K. J. Jepsen, and B. Mikic 2011 *Bone.* **29-2** 101.

Effect of physical properties of natural fibre on the sound absorption coefficient

N H Zunaidi[1], W H Tan[1], M S A Majid[1] and E A Lim[2]

[1] Mechanical Engineering Programme, School of Mechatronic Engineering, Universiti Malaysia Perlis (UniMAP), Pauh Putra Campus, 02600 Arau, Perlis, Malaysia[1,2,3].
[2] Institute of Engineering Mathematics, Universiti Malaysia Perlis, Kampus Pauh Putra, 02600 Arau, Perlis, Malaysia[4].

whtan@unimap.edu.my[2]

The aim of the study was to analyse the effect of physical properties of natural fibre to the sound absorption coefficient. It was also aimed to study the relation of the physical properties of natural fibre to wider the frequency range and increase the sound absorption coefficient. There were two types of natural fibre will be studied such as kenaf fibre and rice straw fibre. The performance of sound absorption coefficient was measured using fabricated impedance tube according to ISO 10534-2. The physical properties of the natural fibre may affect the sound absorption coefficient either shifted to low frequency or shifted to high frequency range. All the physical properties that influence the sound absorption coefficient of natural fibre will be explained. The physical properties of natural fibre that will be studied were fibre thickness, fibre diameter and compressed fibre. Results have shown that the fibre thickness and diameter have a big effect to sound absorption coefficient. From the results, the utilization of the sound absorption coefficient can be achieved by choosing the suitable natural fibre according to its physical properties for a suitable application.

1. Introduction.

Agriculture sector now days have growth rapidly due the development of technologies that help the agriculture process. It has contributed a lot of number to the growth of Malaysia's economy. As the number of agriculture products increase, the agriculture by-products will also increase. Thus, the agriculture by-products need to be managed properly to avoid opened burning. Opened burning may worsen the effect of global warning due to the release of carbon dioxide and heat to the environment [1-2]. Therefore, to help reduce the effect of global warming was by transforming the agriculture by-products into sound absorbent material.

Apparently, there were lot of sound absorbent materials made from synthetic fibre [3]. Prolong exposure to synthetic fibre may harm human health [4-5]. Introducing the natural fibre as sound absorbent material may reduce the sound intensity level and not harm human hearing. Thus, absorbent material from agriculture by-products will also benefit human health. Agriculture by-products were also known as natural fibre. The natural fibre is obtained from the stem, leaf, fruit and pulp of the plant. Besides that, natural fibre has been applied in many industries such as automotive, building,

IOP Conf. Series: Journal of Physics: Conf. Series **908** (2017) 012023 doi:10.1088/1742-6596/908/1/012023

factories and others. Natural fibre can be found in abundance in Malaysia. It was also cheap, renewable and bio-degradable [6-7]. Nevertheless, it also has its disadvantages such as weak in strength of mechanical properties, flammable, lead to fungi growth due to humidity [4].

Natural fibre was considered as promising sound absorbent materials [2,8]. Many researchers conducted the researches on the natural fibre for sound absorption purpose. Rice straw fibre was considered as a good sound absorbent material in frequency range of 2000Hz and above [1]. There was also study on composite of rice straw fibre for sound absorption purpose [9]. The result was also considered good. Moreover, tea waste fibre from tea industry was also tested for its sound absorption [5]. The result was great for the frequency range 1500-6300Hz of sound absorption coefficient with 0.5. Besides that, fibre from oil palm fruit has shown an average result of sound absorption coefficient of 0.75 for frequency 2500Hz and above [10].

In this study, two types of natural fibre will be studied for their physical properties that influence their sound absorption coefficient. The frequency range to be considered is 60-1800Hz. The sound absorption performance will be measured and analysed. There was abundance of natural fibre waste from agriculture waste every year. This small effort was an alternative to make use of the agriculture waste in order to reduce the global warming effect. Moreover, natural fibre might be promoted for the sound absorbent materials in future time.

2. Methodology

In this study, two types of natural fibre will be used; which were kenaf fibre and rice straw fibre. The sound absorption coefficient measurement was conducted using impedance tube. The fabrication of impedance tube was according to ISO standard 10534-2 Transfer Function Method. The impedance tube was considered as one direction flow of sound wave for the frequency range of 60-1800Hz. Below and beyond the frequency range was not valid. The value of sound absorption coefficient was range from 0 to 1. The value 0 was assumed no sound energy wave was absorbed and vice versa. Figure 1 is the setup of sound absorption coefficient measurement.

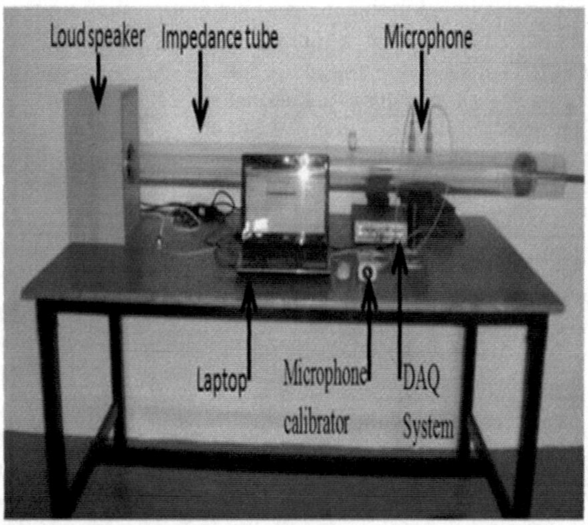

Figure 1. Setup for the sound absorption measurement setup.

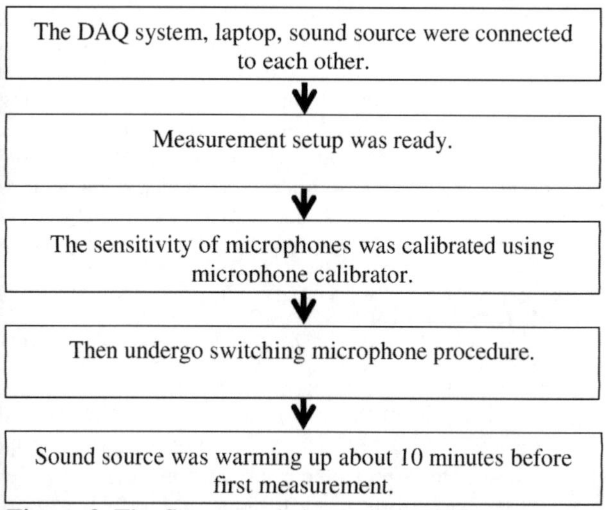

Figure 2. The flow sound absorption coefficient coefficient measurement.

Basically the measurement setup consists of laptop, sound source, microphones, microphone calibrator and DAQ system. The equipments were connected to each other before starting the measurement. Firstly, the microphones sensitivity were calibrated. Secondly, the microphones were tested its phase equalisation using switching method. It was important to equalize the phase of both microphones. Lastly, the setup was ready for measurement after 10 minute sound source was on. Figure 2 is the flow of sound absorption coefficient measurement setup.

IOP Conf. Series: Journal of Physics: Conf. Series **908** (2017) 012023 doi:10.1088/1742-6596/908/1/012023

The kenaf fibre and rice straw fibre were common fibre in Malaysia, in which both natural fibre can be found in abundance. Figure 3 and Figure 4 show the natural fibre in loose form and compressed form. The natural fibre was shaped of circular in order to be inserted into the impedance tube for sound absorption coefficient measurement. The fibre in loose form were inserted into the mould and compressed with an average force of 45kg/cm^2 using Universal Testing Machine (UTM) which shown in Figure 5 [11] .The experiment was conducted using impedance tube. The impedance tube was used for small scale of sample size, up to the maximum diameter of 10cm sample size.

Figure 3. Rice straw fibre; (a) rice straw fibre in loose form (b) rice straw fibre in compressed form

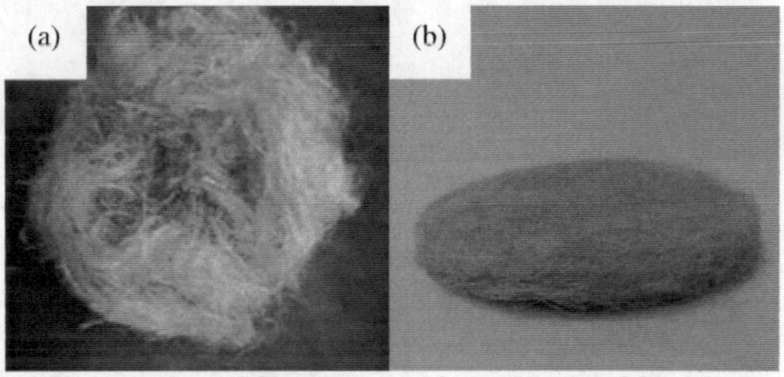

Figure 4. Kenaf fibre; (a) kenaf fibre in loose form (b) kenaf fibre in compressed form.

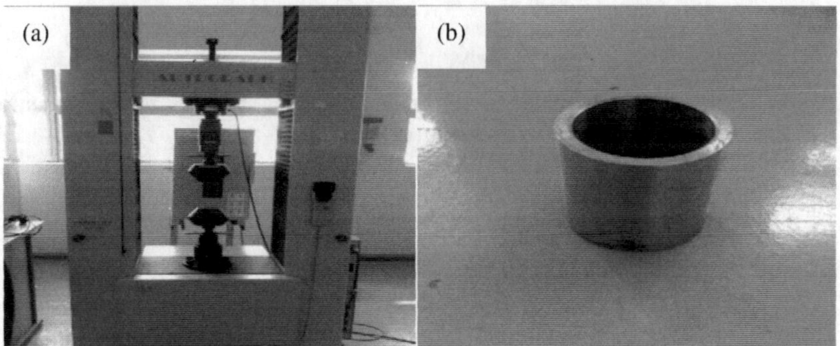

Figure 5. (a) Shimazu Universal Testing Machine, (b) mould used for the sample preparation.

3. Results And Discussions

There were three physical properties will be discussed in this study. There were fibre thickness, fibre diameter and compressed fibre. All the factors have a great influence on the sound absorption coefficient performance. Both kenaf fibre and rice straw fibre were able to absorb sound energy but with different capability.

3.1 Fibre Thickness

Figure 6 and Figure 7 show the results of sound absorption coefficient for 40mm and 60mm thickness of natural fibre respectively. For 40mm thickness, both natural fibres have different sound absorption performance according to their physical. The maximum sound absorption coefficient for kenaf fibre and rice straw fibre were 0.59 and 0.38 respectively. As referred in Figure 6, the average sound absorption coefficient of kenaf fibre was 0.56 for the frequency more than 1122Hz and rice straw has a maximum sound absorption coefficient of 0.5 at frequency 1778Hz.

Based on Figure 6 and Figure 7, it indicated that, as the thickness of fibre increases, the sound absorption coefficient also affected [11-12]. The sound absorption shows an increment and shifted to the left hand side of the low frequency range [13]. This observation shows that, there was an enhancement of sound absorption coefficient at low frequency. Even though, it was only shows a small increment but it has a positive motivation for future study. In additional, as the thickness of fibre increases, the fibre will have more porous structure and more sound energy wave produced by the sound source in the impedance tube will be absorbed by the sample. It can be seen in Figure 8, the difference of size and number of pore between kenaf fibre and rice straw fibre. It was the reason for the increasing of the sound absorption coefficient for both fibres although both have the same thickness. Compared to kenaf fibre and rice straw fibre, it was shown that kenaf fibre has a better sound absorption coefficient performance. Both fibres have a different diameter of fibre size. This physical factor has a significant effect on sound absorption coefficient performance.

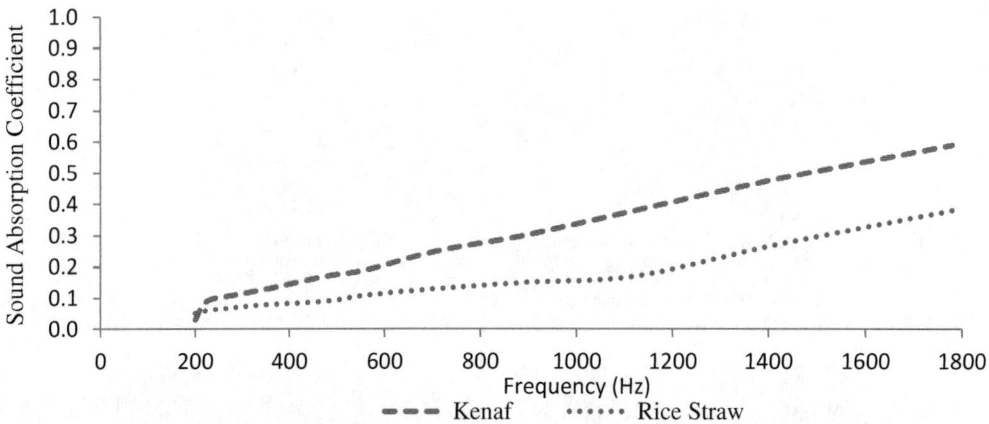

Figure 6. Sound absorption coefficient of 40mm thickness of kenaf fibre and rice straw fibre respectively.

3.2 Fibre Diameter

The effect of fibre diameter will be discussed in this section. Previous researcher has found that, diameter of fibre has a significant effect to the sound absorption coefficient performance. Both kenaf fibre and rice straw fibre has a significant different of diameter size. According to Figure 6 and Figure 7, it can be observed that, fibre with small diameter has an increase on sound absorption coefficient. As shown in Figure 8, it can be seen that kenaf fibre diameter is smaller than rice straw fibre diameter. In this study, kenaf fibre has an increase of sound absorption coefficient at low frequency compared to rice straw. Both have an increment towards low frequency as the thickness of fibre increase.

As the diameter of fibre was decreased, the number of porous structure will also increase when the thickness of fibre increase. Thus, compared to kenaf fibre and rice straw fibre, kenaf fibre will have more porous structure. When the porous structure increase the frictional between the sound source and natural fibre will increase. This will decrease the sound wave energy and been absorbed by the natural fibre. Moreover, reduction of fibre diameter will required more fibre for the equality of the fibre thickness between kenaf fibre and rice straw fibre. Both have the same thickness, but in term of volume fibre, kenaf has more volume of fibre. Thus, it created more porous structure for kenaf fibre sample.

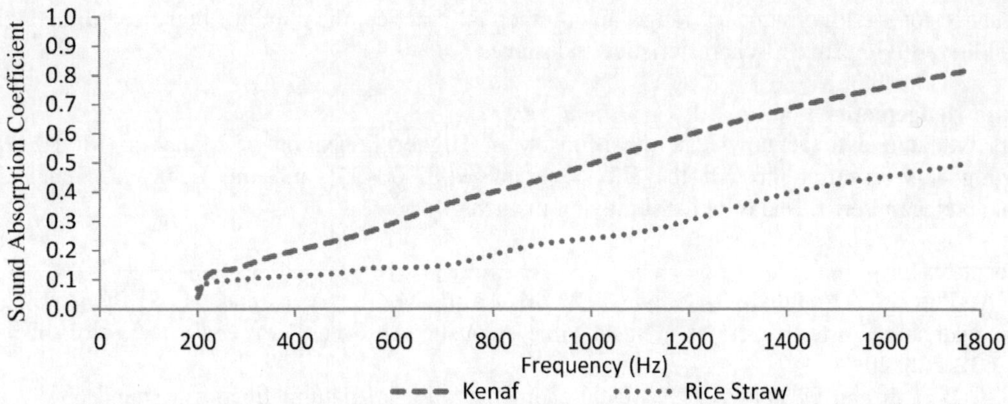

Figure 7. Sound absorption coefficient of 60mm thickness of kenaf fibre and rice straw fibre respectively.

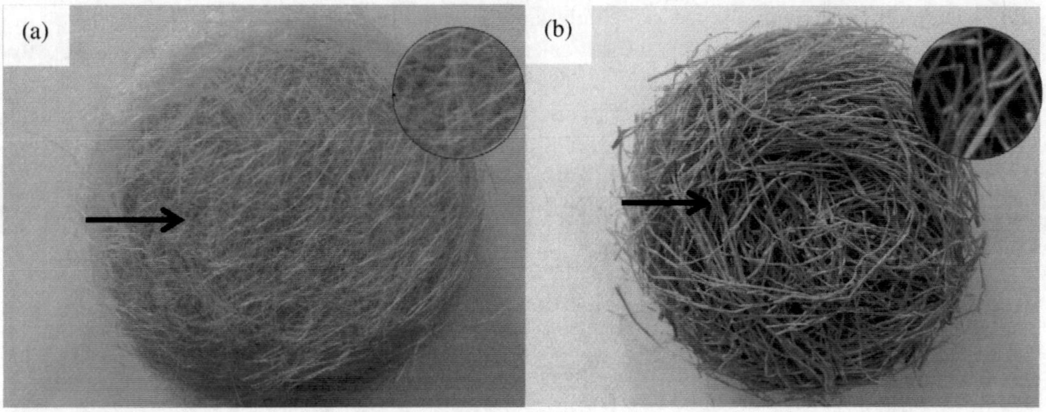

Figure8. Pores that resulted from the increases of fibre thickness of (a) kenaf and (b) rice straw fibre from normal vision.

3.3 Compressed Fibre

The impedance tube was in circular shape. Thus, natural fibre samples were needed to be shaped into circular form in order to be fitted into the impedance tube for measurement. In order to shape the natural fibre, it was compressed using the mould and UTM machine in Figure 5. As it been shape according to the size of the impedance tube, it will be fitted nicely into the impedance tube. The compression of the natural has a significant effect to sound absorption coefficient [7,14]. As the thickness fibre increase will provide more porous structure to the natural fibre sample. But at the same, as the force exerted onto the natural fibre during the shape process of the natural, the porous structure will become smaller and packed. This has contributed to the increase of frictional force and increase the sound absorption coefficient.

4. Conclusion

In this study, there were two types of natural fibre; kenaf fibre and rice straw fibre were investigated. The sound absorption coefficient of two types natural fibres were determined and measured successfully using impedance tube for the frequency range of 60-1800Hz. As a conclusion, both natural fibres were able to absorb sound wave energy effectively. It can be concluded that the fibre mass and diameter have a big effect to sound absorption coefficient. Based on the results, the utilization of the sound absorption coefficient can be achieved by choosing the suitable natural fibre according to its physical properties and other features for a suitable application. In future work, the natural fibres will be hybrid with Micro-perforated panel (MPP) to enhance its sound absorption performance for the low frequency range. In commercial practice, the natural fibres need to be added some additives to improve its characteristics and shape.

5. Acknowledgements
Authors would like to acknowledge the Ministry of Higher Education of Malaysia (MOHE) for supporting this research through the FRGS Grant (9003-00435) and the Lembaga Kenaf dan Tembakau Negeri Perlis, Malaysia for supplying the kenaf fibre.

6. References
[1] A. Putra, Y. Abdullah, H. Efendy, W. M. Farid, M. R. Ayob, and M. S. Py, "Utilizing Sugarcane Wasted Fibers as a Sustainable Acoustic Absorber," *Procedia Eng.*, vol. 53, pp. 632–638, 2013.
[2] U. Berardi and G. Iannace, "Acoustic characterization of natural fibers for sound absorption applications," *Build. Environ.*, vol. 94, no. July, pp. 840–852, 2015.
[3] M. Hosseini Fouladi, S. Y. Peng, S. Y. Wen, P. Z. Xin, M. J. Mohd Nor, M. A. Mohd Saleh, and T. C. Seng, "Replacement of Synthetic Acoustic Absorbers With Natural Fibers," *Vol. 12 Vib. Acoust. Wave Propag.*, vol. 12, no. February 2016, p. 83, 2012.
[4] S. Fatima and a. R. Mohanty, "Acoustical and fire-retardant properties of jute composite materials," *Appl. Acoust.*, vol. 72, no. 2–3, pp. 108–114, 2011.
[5] S. Ersoy and H. Küçük, "Investigation of industrial tea-leaf-fibre waste material for its sound absorption properties," *Appl. Acoust.*, vol. 70, no. 1, pp. 215–220, 2009.
[6] M. Hosseini Fouladi, M. J. M. Nor, M. Ayub, and M. Ghassem, "Enhancement of Coir Fiber Normal Incidence Sound Absorption Coefficient," *J. Comput. Acoust.*, vol. 20, no. 1, p. 1250003, 2012.
[7] N. Amin, M. H. Fouladi, and S. Jaya, "Effect of Compression on the Acoustic Absorption of Coir Fiber," vol. 7, no. 9, pp. 1285–1290, 2010.
[8] F. Asdrubali, F. D'Alessandro, and S. Schiavoni, "A review of unconventional sustainable building insulation materials," *Sustain. Mater. Technol.*, vol. 4, pp. 1–17, 2015.
[9] H. S. Yang, D. J. Kim, and H. J. Kim, "Rice straw-wood particle composite for sound absorbing wooden construction materials," *Bioresour. Technol.*, vol. 86, no. 2, pp. 117–121, 2003.
[10] M. R. Ishak, S. M. Sapuan, Z. Leman, M. Z. A. Rahman, U. M. K. Anwar, and J. P. Siregar, "Sugar palm (Arenga pinnata): Its fibres, polymers and composites," *Carbohydr. Polym.*, vol. 91, no. 2, pp. 699–710, 2013.
[11] M. Hosseini Fouladi, M. J. M. Nor, M. Ayub, and Z. A. Leman, "Utilization of coir fiber in multilayer acoustic absorption panel," *Appl. Acoust.*, vol. 71, no. 3, pp. 241–249, 2010.
[12] M. Nor, N. Jamaludin, and F. Tamiri, "A preliminary study of sound absorption using multi-layer coconut coir fibers," *Elect. J. Tech. Acoust*, 2004.
[13] L. Peng, B. Song, J. Wang, and D. Wang, "Mechanic and acoustic properties of the sound-absorbing material made from natural fiber and polyester," *Adv. Mater. Sci. Eng.*, vol. 2015, 2015.
[14] C. N. Wang, Y. M. Kuo, and S. K. Chen, "Effects of compression on the sound absorption of porous materials with an elastic frame," *Appl. Acoust.*, vol. 69, no. 1, pp. 31–39, 2008.

IOP Conf. Series: Journal of Physics: Conf. Series **908** (2017) 012024 doi:10.1088/1742-6596/908/1/012024

A review on mode-I interlaminar fracture toughness of fibre reinforced composites

N Nasuha, A I Azmi, C L Tan

School of Manufacturing Engineering, Universiti Malaysia Perlis (UniMAP), Pauh Putra Campus, 02000 Arau, Perlis, Malaysia
Faculty of Engineering Technology, Universiti Malaysia Perlis (UniMAP), UniCITI Campus, 02600 Pdg Besar, Perlis, Malaysia

azwaniskandar@unimap.edu.my

Abstract. Composite material has been growing rapidly throughout the year for its unique properties in comparisons with metal. Recently, there has been a growth on studying the way to reduce the delamination failure, which is the primary challenge on laminated fibre composite. This failure can degrade the strength of composite materials, hence loses its function. In this review, database search was performed using the keywords search on "interlaminar fracture toughness", "double cantilever beam", "delamination resistance" and "Mode-I G_{IC}". The searches were performed on Google Scholar, Scopus and Web of Science with further cross-referencing with other databases. Most relevant studies were selected for review and referencing by the author. This review paper gives a brief explanation on Mode-I interlaminar fracture toughness of composite material. This fracture mode is the most common modes on studying the delamination failure.

1. Introduction

By definition, composite material is a material comprises of two or more material; reinforcing agent (fibre, particles or sheets) and the matrix phase (polymers, metals or ceramics) [1-2]. Composite materials are also introduced with the aim to enhance the properties of the material that is being bonded together; matrix and reinforcement. The matrix functions as a support for the reinforcement, whereas; the reinforcement provides unique mechanical and physical properties that improve the matrix properties. The application of composite material has been growing rapidly in the field of automobile, aerospace, marine, nuclear, biomedical and other engineering field. For example, there is about 30% of the external surface of the Boeing 767 consists of composite material [3]. Moreover, in 2000, polyurethane reinforced with a mixed of flax/sisal material door trim panels were introduced by Audi during the launched of its A2 midrange car [4]. According to the market research published by Lucintel on 2016, the global market is expected to reach an estimated of $37.3 billion by 2021 and rise at a CAGR (Compound Annual Growth Rate) of 5.1% from 2016 to 2021 [5].

Nowadays, fibre reinforced polymer composite (FRP) is the most commonly used among other composite materials due to its unique properties; high specific strength and lightweight. However, FRP is unable to resist defect initiation and propagation when compared to metals [6]. These laminated fibre composites are most likely to form delamination damage. The characterization of the delamination resistance has gained a lot of interest among researchers. The Mode-I double cantilever beam (DCB) theory is commonly used for studying the delamination resistance [7-8]. In the DCB testing, a resistance-type fracture develops as the delamination grows from the insert. In the beginning of the delamination growth, the calculated G_{IC} will increase monotonically, and subsequently stabilize as the delamination growth. The reason to observe the resistance during delamination is due to the development of fibre bridging [9].

2. Interlaminar Fracture Toughness

Interlaminar fracture or delamination damage is the primary challenge for the laminated fibre composite. This is because these factors can degrade the strength of the laminate and subsequently leads to failure. The primary factors which affect the fracture of FRP composites are speed of loading, stress concentration, temperature, and thermal shock [10]. It is also known that the woven fabric composites provide more damage tolerant in comparisons to unwoven fabrics. This is because of the non-planar interply structure of woven-fabric composites, the interaction between the delamination crack, the matrix regions and the weave structure during propagation, will result in crack growth [11]. The interlaminar performance are characterized by their weakness towards the sheer stress and tensile stress [12]. According to Zulkifli R et al. [1], when the number of layers increases, the stiffness and interlaminar fracture toughness also found to be increasing. Hence, if the material has a high capability to dissipate elastic energy without a catastrophic failure during the crack propagation, the fracture toughness of the material is high.

Interlaminar fracture comprises energy losses in term of matrix cracking, fibre breakage, fibre pull-out, fibre slipping and fibre de-bonding, which differ from metals where fracture toughness depends on the energy dissipates at the crack tip [14]. There are three modes of fracture by which a load can operate a crack, that is, Mode-I (tensile), Mode-II (shear), Mode-III (tearing shear), or by the combination of these Modes as shown in figure 1 [15], [16]. Delamination damage, which is the most predominant problem caused, can be study using these fracture mechanics approach with the critical strain energy as a standard [1].

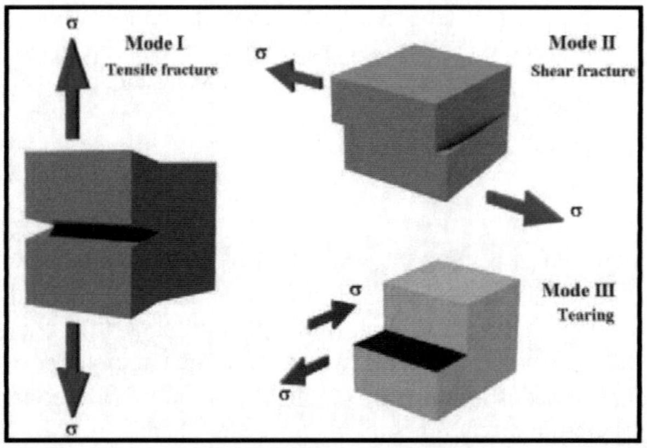

Figure 1. Fracture Modes [16]

Delamination damage is the result from low velocity impact caused by interlaminar shear stress and occur when the interlaminar stress exceed the interlaminar strength level. Interlaminar stress is caused by the geometry and the loading parameter, while the interlaminar strength depends on the material properties. Hence, to achieve the optimum delamination resistance, it is necessary to match the optimum parameter to achieve the low interlaminar stress level with high interlaminar strength level. Delamination resistance is increased when there is a decreasing of mismatch between the lamina, while there is a possibility of debonding decreasing with the decreasing mismatch between the material constituents. The difference between debonding and delamination is that debonding occurs when two materials stop adhering to each other. On the contrary, delamination is a failure of laminated material, which leads to separation of the plies. When a composite material has a large difference in fibre and the matrix stiffness, the mismatching coefficient will also be higher [13]. For example, carbon/epoxy composites have a higher mismatching coefficient compare to jute/epoxy, thus larger delamination area is expected to occur.

3. Mode-I Double Cantilever Beam Testing

The Mode-I testing follows the standard specimen of ASTM D5528-13 to measure the Mode-I fracture toughness, G_{IC}, of continuous fibre reinforced composite material. The specimens used in this testing are the double cantilever beam (DCB) specimen, figure 2.

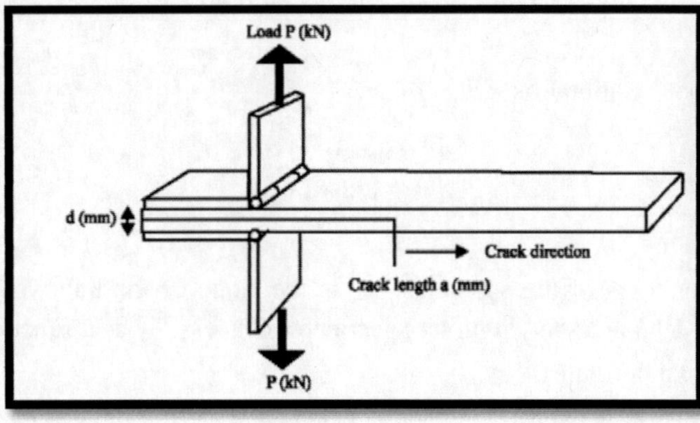

Figure 2. Double Cantilever Beam Specimen [1]

The position for delamination opening should be inserted between the plies using a non-adhesive film which approximately 13 μm thickness, and the length of the film is 50 mm, plus the extra length of the hinges. The length of the specimen should be at least 125 mm and 20 to 25 mm wide. The specimen used a constant crosshead rate between 1 to 5 mm/min to promote delamination. Delamination lengths are determined visually, but for more accurate readings, a travelling microscope is recommended.

The calculation for the analysis of the G_{IC} can be made through three different theories; namely the modified beam theory (MBT), a compliance calibration method (CC) and a modified compliance calibration method (MCC), as given in the equation respectively [9], [17]. It has been reported that each is being differ not more than 3.1%. However, MBT data reduction method yields the most conservative value of G_{IC} for 80% of the specimen tested and are the most recommended data reduction method [18].

- Modified Beam Theory (MBT)

$$G_I = \frac{3P\delta}{2b(a+\Delta)} \tag{1}$$

where, P is the load applied and δ is the load point displacement. Whereas, b specimen width and a is the delamination length. While lastly, $\Delta$is the correction for the delamination length. The way of correcting this theory is by treating the DCB as if it has a slightly longer delamination, $(a + \Delta)$, as the Δ can be determined experimentally by generate at least a square plot of the cube root of the $C^{1/3}$ as shown in the graph in figure 3 (a) [9]. C is the ratio of the load point displacement to the applied load, $\frac{\delta}{P}$.

- Compliance Calibration (CC)

$$G_I = \frac{nP\delta}{2ba} \qquad (2)$$

where, n is the exponent from the straight line drawn on the graph result. The value of n can be calculated when $\frac{\Delta y}{\Delta x}$,where Δ_y and Δ_x can be defined as shown in the graph below; figure 3 (b) [9].

- Modified Compliance Calibration (MCC)

$$G_I = \frac{3P^2C^{2/3}}{2A_1bh} \qquad (3)$$

where, *h* is the thickness of the specimen, A_1 is the slope of the line which generate a least square plot of the delamination length by specimen thickness, $\frac{a}{h}$, as a function of the cube root compliance as shown in figure 3 (c) [9].

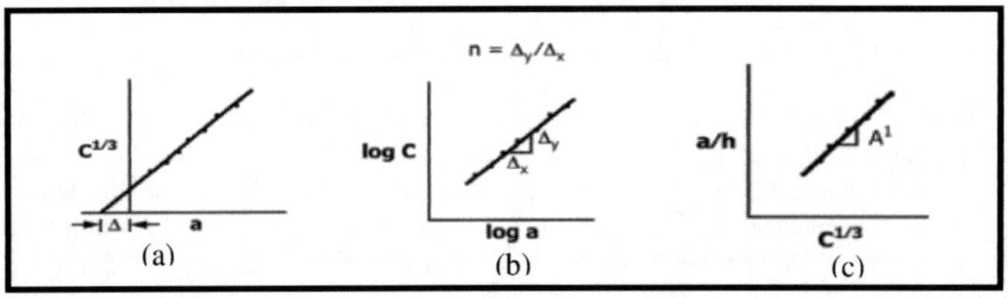

Figure 3. Graph of Fracture Modes Theory, (a) Modified Beam Theory, (b) Compliance Calibration and (c) Modified Compliance Calibration

As reported by F.A. Almansour et al. [17], the MBT gives the lowest and more conservative value for both initiation and propagation toughness for all specimens as depicted in Figure 4. Furthermore, the G_{IC} for a wet sample is low compared to dry because of the high moisture absorption by the flax fibre, which leads to a weak fibre matrix interface. Moreover, the presence of water can create surface degradation and eventually cause micro-cracking, reduce in the fracture toughness and de-bonding in the composites. The fracture mechanics showed energy dissipation through matrix deformation, fibre de-bonding, fibre breakage and fibre pull-out Furthermore, a high stress concentration around stitch fibres can also reduce the fracture toughness. This is because a higher amount stitched fibres can cause an in-plane fibre misalignment [17, 19].

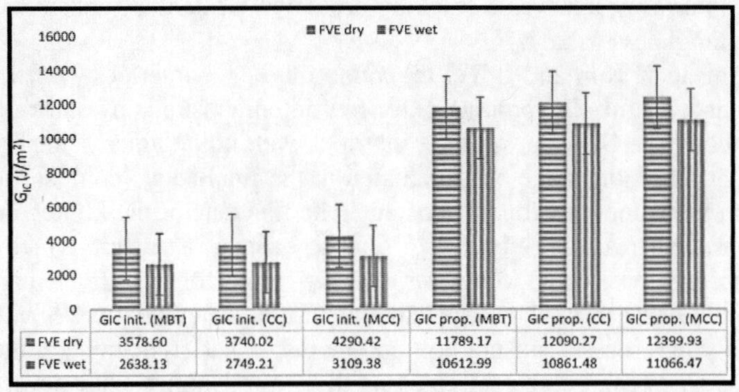

Figure 4. Graph of initiation G_{IC} and propagation G_{IC} against wet and dry flax fibre[17]

According to Zulkifli R et al., the number of layers of fibre gives high impact on the interlaminar fracture toughness. Hence, this results in the increase of stiffness of the fibre and as well as the interlaminar fracture toughness of the composites. When the interlaminar fracture toughness increases, the delamination damage grows slowly and stable. Besides, the propagation values away from the crack initiation were higher than the initiation values. The toughness of the initiation values is required. This is because most woven DCB specimens, fibre bridging occurs as the delamination progress further [1].

According to P. Suppakul and S. Bandyopadhyay [11], the mode-I of strain energy release rate (G_I) were generated to determine the initiation and propagation of delamination on a specimen. The twill weave shows the highest G_{IC} value, while 8 harness satin weave gives the highest value of G_{Ii} value. However, there is a greater number of fractured fibre occurred on the twill weave laminates compared to that of the plain weave even though the plain weave gives the lowest value for both the G_{IC} value and the G_{Ii} value. Fibre bridging is one of the mechanism which contributing to the fracture resistance on the twill weave, quadrant 4-harness satin weave and 8-harness satin weave laminates, even though twill weave has the highest values of fracture energies. Nevertheless, there is no indication of fibre bridging occur on the plain weave glass/vinyl ester laminate, due to its lowest values of fracture energies. This effect is associated with the overall weave pattern and the relative contribution of the individual microdamage on the interface [20].

Moreover, as reported by M. Pinto et el [21] , a woven fabrics showed 28% greater initiation toughness than the unidirectional because of the fibre bridging and increased in the interaction of the intra-ply. The author also stated that a plain wave experienced a high degree of nesting, which subsequently exhibited more interaction between the laminates. M. Ravandi et al. [22] also stated that the G_{IC} values for unidirectional glass fibre composite shown to be less than the unidirectional flax fibre, which is due to the large difference in the bridging across the delamination plane. However, the G_{IC} values for woven flax fibre shows three times higher than the unidirectional flax fibre.

The difference between flax fibre and glass fibre is that the flax fibre possesses superior mechanical properties such as low tensile properties and elongation at break but the density of natural fibre is half the difference between glass fibre. So, as reported by Y Zhang et el [23], by hybridizing the flax fibre and glass fibre, the new hybrid might possess such as lightweight and higher in strength and modulus. The result shows that the hybrid gives the highest interlaminar sheer strength compare to flax fibre reinforced polymer (FFRP) and glass fibre reinforced polymer (GFRP). Thus, the structure of the fibre led to a remarkable fibre bridging and improved the interlaminar properties.

A. Des Morias et al. [8] stated that there will be no fibre bridging occurred in the $[0°/90°]_{12}$ specimens. This is because the measured Mode I critical strain energy release rates G_{IC} were higher compared to $[0°]_{24}$ specimens. Transverse cracking often occurs when the specimens have a high interlaminar fracture toughness. Although the major problem for multidirectional laminates is the development of the interplay damage and it is often associated with fibre bridging [7], in the case of

IOP Conf. Series: Journal of Physics: Conf. Series **908** (2017) 012024 doi:10.1088/1742-6596/908/1/012024

Θ/Θ and +Θ/-Θ interfaces, transverse cracking promotes substantially fibre bridging, which is believed to result in high toughness values.

As referring to K Kim and L Ye, the value of G_{IC} for interlock weft knitted were a lot higher than uni-weave and hybrid. The production of hybrid composite is by stacking knitted and uni-weave composite together. The G_{IC} values for the interlock weft knitted increasing rapidly with the increased of crack length at the initial stage but fluctuated at the final stage, compare to hybrid and uni-weave composite which the values are almost constant. The fluctuate of the knitted composite means that the unstable fracture occur may cause by the difference of micro-structure which resulting from the non-uniform preform lay-up process. Moreover, the G_{IC} value for hybrid are higher than the uni-weave composite. This is because of the large amount of fibre bundles between the adjacent plies travel during manufacturing process, causing non-unique delamination planes. Furthermore, both the knitted and hybrid composites can be characterized as the wavy laminae with no distinct interlaminar layer between the plies. These wavy laminae structure can lead to a higher interlaminar fracture toughness in comparisons to uni-weave[24].

Table 1. Table of Past Research on Mode-1 Double Cantilever Beam

Author	**Material**		**Condition**	**G_{IC} Value (J/m^2)**
F. A. Almansour, H. N. Dhakal, and Z. Y. Zhang[17]	FVE ~ Flax fibre reinforced vinyl ester	MBT	Wet	2683.13
			Dry	3578.60
		MCC	Wet	3109.38
			Dry	4290.42
		CC	Wet	2749.21
			Dry	3740.02
	FBVEu ~ Flax fibre hybridised basalt reinforced vinyl ester unstitched	MBT	Wet	2821.70
			Dry	4236.64
		MCC	Wet	3781.00
			Dry	4640.97
		CC	Wet	3063.47
			Dry	4414.48
	FBVEs ~ Flax fibre hybridised basalt reinforced vinyl ester stitched)	MBT	Wet	2806.11
			Dry	2669.90
		MCC	Wet	3436.01
			Dry	3172.21
		CC	Wet	3096.85
			Dry	3113.49
M. Pinto et el [21]	Jute/epoxy (Silane Treated)	Weave	Flock	1400
			No flock	800
		UD	Flock	1000
			No Flock	600
A. De. Morais et el [8]	Carbon/epoxy	$[0°/90°]_{12}$		1295
		$[0°]_{24}$		369
Zulkifli R. et el [1]	Silk/polyester	8 plies		847
		10 plies		1530
		12 plies		1638
		14 plies		1872
P. Suppakul and S. Bandyopadadyay[11]	E-glass/vinyl ester	Plain weave	MBT	450
			CC	500
		Twill weave	MBT	1400
			CC	1600
		Quadrant	MBT	850

		4-harness stain weave	CC	870
		8-harness satin weave	MBT	1350
			CC	1410
K. Kim and L. Ye [24]	Glass Fibre	Interlock weft knitted		4960
		Uni-weave		230
		Hybrid		680

4. Conclusion

In conclusion, the Mode-I double cantilever beam testing shows a promising insight on delamination resistance. Based on past research as shown in table 1, the glass and carbon fibre gives lower interlaminar fracture toughness in comparisons to natural fibre such as flax, jute and silk. This is because of the mismatch coefficient of the glass and carbon fibre. The structure of the fibre has a large effect on the interlaminar fracture toughness, which means the structure of the fibre can lead to an improvement of fibre bridging and consequently increase the interlaminar fracture toughness.

5. Acknowledgments

The authors gratefully acknowledge the financial support of the Ministry of Education Malaysia through the Fundamental Research Grant Scheme (FRGS); code number FRGS/1/2016/TK03/UNIMAP/02/7.

6. References

[1] Z R, Pei K S, and A C H 2008 Interlaminar Fracture Properties of Multi-Layer Woven Silk Fibre/Polyester Composites," *Asian J. Appl. Sci. 1*, **vol. 2** pp. 177-84

[2] Saxena M, Pappu A, Sharma A, Haque R, and Wankheda S 2011 Composite Materials from Natural Resources : Recent Trends and Future Potentials, in *Advances in Composite Materials - Analysis of Natural and Man-Made Materials.*

[3] H. Hocheng and C. C. Tsao, 2005 The path towards delamination-free drilling of composite materials, *J. Mater. Process. Technol.*, **vol. 167** no. 2–3 pp. 251–64

[4] Sanjay M R, Arpitha G R, Laxmana-Naik L, Gopalakrishna K, and Yogesha B 2016 Applications of Natural Fibers and Its Composites: An Overview, *Nat. Resour.*, **vol. 7** no. 7 pp. 108–14

[5] Lucintel 2016 The global composites industry is expected to reach $37.3 billion by 2021," *GlobeNewswire.* [Online]. Available: https://globenewswire.com/news-release/2016/11/04/886908/0/en/The-global-composites-industry-is-expected-to-reach-37-3-billion-by-2021.html. [Accessed: 20-May-2017].

[6] Jacob G C, Starbuck J M, Fellers J F, Simunovic S, and Boeman R. G May 2005 The effect of loading rate on the fracture toughness of fiber reinforced polymer composites, *J.Appl. Polym. Sci.*, **vol. 96** no. 3 pp. 899–904

[7] Pereira A B, De Morais A B, De Moura M F S F and Magalhães A G 2004 Mode I Interlaminar Fracture Of Carbon / Epoxy Multidirectional Laminates, *Compos. Sci. Technol.*

[8] De Morais A B, De Moura M F S F, Marques A T and De Castro P T 2002 Mode - I interlaminar fracture of carbon / epoxy cross - ply composites, *Compos. Sci. Technol.*, no. 62, pp. 679–86

[9] ASTM Standard 2014 D5528-01 2001 Standard Test Method for Mode I Interlaminar Fracture Toughness of Unidirectional Fiber-Reinforced Polymer Matrix Composites," *Am. Soc. Test. Mater.*, no. Mode I, pp. 1–13

[10] Maleque M A and Salit M S 2013 Mechanical Failure of Materials, in *Materials Selection and Design*, pp. 17–37

[11] Suppakul P and Bandyopadhyay S 2002 The effect of weave pattern on the mode-I

interlaminar fracture energy of E-glass/vinyl ester composites, *Compos. Sci. Technol.*, **vol. 5** pp. 709–17

[12] Borowski E, Soliman E, Kandil U, and Taha M 2015 Interlaminar Fracture Toughness of CFRP Laminates Incorporating Multi-Walled Carbon Nanotubes, *Polymers (Basel).*, **vol. 7** no. 6, pp. 1020–45

[13] Dransfield K, Baillie C and Mai Y W 1994 Improving the delamination resistance of CFRP by stitching-a review, *Compos. Sci. Technol.*, **vol. 50**, no. 3, pp. 305–17

[14] Shukla A, 2004 Fracture Toughness of Fiber Reinforced Composites," in *Fiber Reinforced Composite*, pp. 313–35

[15] Sham-Prasad M S, Venkatesha C S and Jayaraju T 2011 Experimental Methods of Determining Fracture Toughness of Fiber Reinforced Polymer Composites under Various Loading Conditions, *J. Miner. Mater. Charact. Eng.*, **vol. 10**, no. 13, pp. 1263–75

[16] Rocha-rangel E, 2011 Fracture Toughness Determinations by Means of Indentation Fracture, in *Fracture Toughness Determinations by Means of Indentation Fracture, Nanocomposites with Unique Properties and Applications in Medicine and Industry, Dr. John Cuppoletti (Ed.).*

[17] Almansour F A, Dhakal H N, and Zhang Z Y, 2017 Effect of water absorption on Mode I interlaminar fracture toughness of flax/basalt reinforced vinyl ester hybrid composites, *Compos. Struct.*, **vol. 168** pp. 813–25

[18] KimJ K and Mai Y W, 1998 *Engineered Interfaces in Fiber Reinforced Composites.* Elsevier Science.

[19] Göktaş D, Kennon W R, and Potluri P, 2017 Improvement of Mode I Interlaminar Fracture Toughness of Stitched Glass/Epoxy Composites, *Appl. Compos. Mater.*, **vol. 24** no. 2, pp. 351–75

[20] Oskouei A R, Zucchelli A, Ahmadi M, and Minak G, 2012 Using Acoustic Emission to Evaluate Fracture Toughness Energy Release Rate (GI) at Mode I Delamination of Composite Materials, *Acoustic Emission*, InTech, pp. 151–172

[21] Pinto M, Chalivendra V B, Kim Y K and Lewis A F 2016 Improving the strength and service life of jute / epoxy laminar composites for structural applications, *Compos. Struct.*, **vol. 156**, pp. 333–37

[22] Ravand M, Teo W S, Tran L Q N, Yong M S and Tay T E, 2016 The effects of through-the-thickness stitching on the Mode I interlaminar fracture toughness of flax/epoxy composite laminates, *Mater. Des.*, **vol. 109**, pp. 659–69

[23] Zhang Y, Li Y, Ma H and Yu T 2013 Tensile and interfacial properties of unidirectional flax / glass fiber reinforced hybrid composites, *Compos. Sci. Technol.*, **vol. 88** pp. 172–77

[24] Kim K Y and Ye L 2012 Interlaminar fracture properties of weft-knitted/woven fabric interply hybrid composite materials, *J. Mater. Sci.*, **vol. 47** no. 20, pp. 7280–90. DOI: 10.1007/s10853-012-6682-x

IOP Conf. Series: Journal of Physics: Conf. Series **908** (2017) 012025 doi:10.1088/1742-6596/908/1/012025

Vibration analysis on compact car shock absorber

W H Tan[1,], J X Cheah[1], C K Lam[1], E A Lim[2], H G Chuah[3] and C Y Khor[4]

[1] School of Mechatronic Engineering, Universiti Malaysia Perlis (UniMAP), Pauh Putra Campus, 02600 Arau, Perlis, Malaysia.
[2] Institute of Engineering Mathematics, Universiti Malaysia Perlis (UniMAP), Pauh Putra Campus, 02600 Arau, Perlis, Malaysia.
[3] Department of Engineering, Computing and Build Environment, KDU Penang College Univeristy, 32, Jalan Anson, George Town, 10400 George Town, Pulau Pinang, Malaysia.
[4] Faculty of Engineering Technology, Universiti Malaysia Perlis (UniMAP), Aras 1, Blok S2 Kampus UniCITI Alam, Sungai Chuchuh, Padang Besar, 02100 Perlis, Malaysia.

whtan@unimap.edu.my

Abstract. Shock absorber is a part of the suspension system which provides comfort experience while driving. Resonance, a phenomenon where forced frequency is coinciding with the natural frequency has significant effect on the shock absorber itself. Thus, in this study, natural frequencies of the shock absorber in a 2 degree-of-freedom system were investigated using Wolfram Mathematica 11, CATIA, and ANSYS. Both theoretical and simulation study how will the resonance affect the car shock absorber. The parametric study on the performance of shock absorber also had been conducted. It is found that the failure tends to occur on coil sprung of the shock absorber before the body of the shock absorber is fail. From mathematical modelling, it can also be seen that higher vibration level occurred on un-sprung mass compare to spring mass. This is due to the weight of sprung mass which could stabilize as compared with the weight of un-sprung mass. Besides that, two natural frequencies had been obtained which are 1.0 Hz and 9.1 Hz for sprung mass and un-sprung mass respectively where the acceleration is recorded as maximum. In conclusion, ANSYS can be used to validate with theoretical results with complete model in order to match with mathematical modelling.

1. Introduction

Car suspension system is known to absorb vibration from the vehicle itself. Suspension system consists of two main components, which are springs and shock absorber. This allows the vehicle to absorb vibration or to minimize the vibration level to provide a comfort driving experience.

In the existing suspension system, different types of spring were used, such as leaf spring and coil spring. Leaf spring is arc-shaped spring steels which cascade together into few layers. It is normally locate at the axle of the wheels. While coil spring, unlike leaf spring consists of a few layers of spring steel, it is a helical-shaped spring. Coil spring mounted on both rear wheels and front wheels [1]. It is normally installed in light vehicles unlike leaf spring which is commonly used in heavy type vehicles such as trucks [2]. Kong et al. mentioned that the ease of maintenance and low cost of leaf spring are the reasons why leaf spring is commonly used in heavy vehicles [3]. Car suspension system can be considered as spring-mass-damper system in other words the system consists of spring, mass and damper.

For a car suspension system, shock absorber can have a huge impact on the comfort level of driving experience. According ISO 2631-1, ride comfort can be affected between 0.5 Hz and 80 Hz [4]. In order to analyse the suspension system, mathematical model can be used such as quarter-car, half-car and full-car models. The models can be generally defined with the governing equation.

Mathematical modelling can be used to analyse the car suspension system. A car suspension system can be expressed as spring-damper-mass system which means the system consists of these three main components: spring, damper and mass. Different car model can be used for the analysis such as quarter-car model, half-car model and full-car model. Quarter-car model offers simplicity by defining the system solely in vertical motion. Bhargav et al. also stated that quarter-car model is easily understandable as it only described the suspension system in vertical motion [5]. Half-car model, unlike quarter-car model, it takes into account rotational motion which is one of the important element that could has direct impact on the ride comfort [6]. Another study, Gang Wang et al. stated that half-car model is normally used due to its symmetry which analysis of both front wheel and rear wheel can be viewed [7]. Full-car model is the most complex model among the other two models which showed the suspension system for both front wheels and rear wheels. Full car model can either be seen as four quarter-car models which could be considered an approach to analyse or direct analysis on the model [8]. Full-car model has better accuracy than the other two models which are quarter-car model and half-car model because full-car model can analyse if in a case where asymmetry exist such as uncertainties of road.

Vibration control can be classified into three types and they are passive, semi-active, and active. Passive suspension system has the advantages of low cost and zero energy consumption and is commonly used in suspension system. Passive suspension system can be understand as a system consists of energy storing element and dampers according to Syed et al [9]. Javad et al. supported the usage of passive suspension system as this is an approach to obtain appropriate spring stiffness and damping coefficient with the lowest possible cost [10]. However, passive suspension system's performance can be limited because of unable to dissipate or add energy in a control manner like active suspension system.

Syed et al. also mentioned that semi-active suspension system can dissipate energy in a control manner but energy cannot be build up to the system. In recent researches, Magneto-rheological (MR) damper is mostly used in developing semi-active suspension system [9]. Low power consumption and safety provided by MR encouraged the usage of it in the suspension system [11].

Active suspension system performs better than both passive and semi-active suspension system. It provides better ride comfort due to its ability to attenuate vibration amplitude with the use of actuator [12]. Active suspension system is still not commonly used as the technology on active suspension system is not that advance enough and costly compare to passive and semi-active suspension system although it has advantages over passive and semi-active suspension system.

In this study, it aims to investigate and conduct analysis on car's suspension system. Another purpose is to investigate the resonance frequency that occurs on the suspension system. Resonance is a phenomenon when natural frequency is same with the forcing frequency. Excessive deflection will happened if suspension system is exposed to resonance which will eventually leads to system failure.

2. Methodology

In this study, there are two phases of analysis which are mathematical modelling and vibration analysis. Mathematical modelling is used to define the suspension system. Mathematical modelling used in this study is quarter-car model which represent one-quarter of the suspension system. Derivations of the equations are developed in phase 1 to obtain the theoretical result. Phase 2, the simulation, by using finite element analysis (FEA) based software – ANSYS to conduct the modal analysis on the suspension system. To import 3D model to ANSYS, a model is created using CATIA.

In ANSYS, it is aimed to study the characteristic of the shock absorber when expose to the natural frequency.

2.1 Mathematical modelling

Mathematical modelling on suspension system is vastly used and quarter-car model is the model most commonly used in the activities of researching due to its simplicity. Based on the several studies from literature review [12 - 14], quarter-car model has been used to analyse the vehicle suspension system. In this study, quarter-car model is developed according to a passive suspension system which consists of a few important elements which are sprung mass, un-sprung mass, spring and damper. Figure 1 shows a drawn 2 degree-of-freedom (DOF) quarter-car model of suspension system.

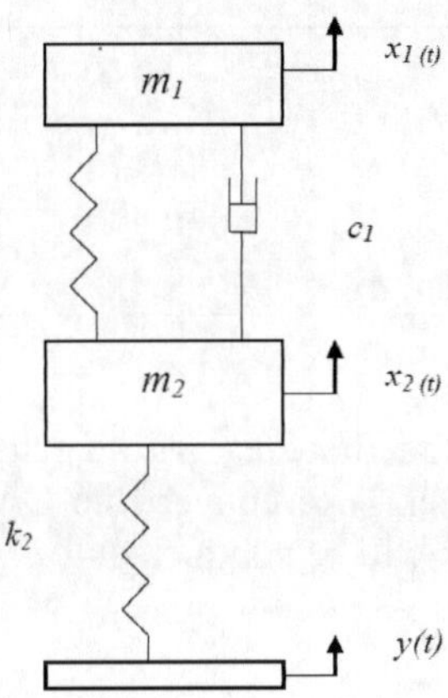

Figure 1. Two degree-of-freedom quarter-car model of shock absorber

By applying Newton's second law of motion to both sprung mass m_1 and un-sprung mass m_2, equations of motion can be represented the model in figure 1 as below:

$$m_1\ddot{x}_1 + c_1\dot{x}_1 + k_1x_1 - c_1\dot{x}_2 - k_1x_2 = 0 \quad (1)$$

$$m_2x_2 - c_1(\dot{x}_1 - \dot{x}_2) - k_1(x_1 - x_2) = k_2y \quad (2)$$

In figure 1, c_1 represents suspension damping coefficient while k_1 represents suspension stiffness and k_2 represents tire stiffness, m_1 and m_2 represent sprung mass and un-sprung mass respectively. Road, y acts as base excitation of this system and assuming it is sinusoidal. Equations (1) and (2) are equations of motion based on the quarter-car model.

$$x_1(t) = X_1e^{i\omega t} \quad (3)$$

$$x_2(t) = X_2e^{i\omega t} \quad (4)$$

$$y(t) = Ye^{i\omega t} \quad (5)$$

$x_1(t)$, $x_2(t)$ and $y(t)$ can be represent as in Equation (3) , Equation (4) and Equation (5) respectively and X_1 , X_2 and Y represent the amplitude for each equation. After that, Equation (3), Equation (4) and Equation (5) are to substitute into equation (1) and (2) and transform into matrix form

$$\begin{bmatrix} i\omega c_1 + k_1 - \omega^2 m_1 & -i\omega c_1 - k_1 \\ -i\omega c_1 - k_1 & i\omega c_1 + k_1 + k_2 \end{bmatrix} \begin{bmatrix} X_1 \\ X_2 \end{bmatrix} = \begin{bmatrix} 0 \\ k_2Y \end{bmatrix} \quad (6)$$

$$Z = \begin{bmatrix} i\omega c_1 + k_1 - \omega^2 m_1 & -i\omega c_1 - k_1 \\ -i\omega c_1 - k_1 & i\omega c_1 + k_1 + k_2 \end{bmatrix} \quad (7)$$

$$F_0 = \begin{bmatrix} 0 \\ k_2Y \end{bmatrix} \quad (8)$$

IOP Conf. Series: Journal of Physics: Conf. Series **908** (2017) 012025 doi:10.1088/1742-6596/908/1/012025

Equation (6) can be obtain after the substitution of the equations (3), (4) and (5) into equation (1) and equation (2). Then, Z represents the mechanical impedance which is also the first term of the matrix, Equation (6) while F_0 represents external force exerts on the system.

$$\begin{bmatrix} X_1 \\ X_2 \end{bmatrix} = Z^{-1} F_0 \tag{9}$$

X_1 and X_2 can be obtained by using Equation (9) where Z^1 is the inverse matrix of Z. Input of the parameters refer to previous study and is showed in table 1.

Table 1. Parameters of quarter-car model of shock absorber [15]

Parameters	Value
Sprung mass, m_1	290 kg
Un-sprung mass, m_2	59 kg
Shock absorber spring stiffness, k_1	16812 N/m
Tire stiffness, k_2	190000 N/m
Shock absorber damping coefficient, c_1	1000 N.s/m

2.2 Phase 2: Simulation

Phase 2 consists of two main steps. The first step is to create a 3D model in CATIA which will be imported to ANSYS. While second step is to convert the CAD file into a compatible file type and import to ANSYS, then modal analysis will be conducted.

3. Results and Discussions

In analytical analysis, acceleration against frequency graphs were plotted after solving by using Wolfram Mathematica 11. Equation of motions which represent the 2 DOF system were used for the calculation and further solving using Wolfram Mathematica.

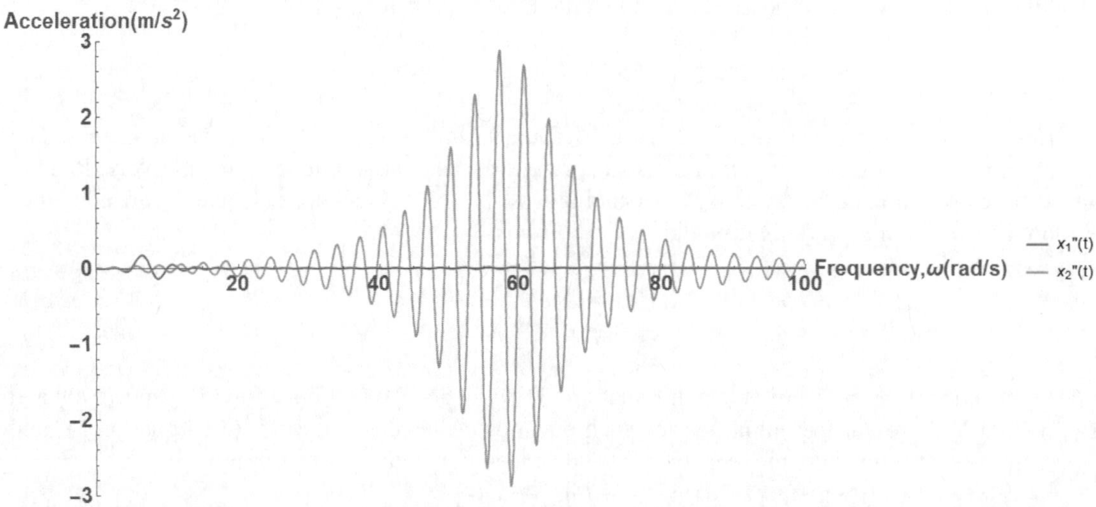

Figure 2. Graph of acceleration against frequency of both sprung mass and un-sprung mass at t=1s

In figure 2, it shows the acceleration against frequency for both sprung mass and un-sprung mass. At 6.3856 rad/s or 1.0 Hz the acceleration for sprung mass is 0.1779 m/s^2 while at 57.1314 rad/s, acceleration of un-sprung mass is 2.8562 m/s^2. The magnitude of acceleration of un-sprung mass is comparatively higher than that of sprung mass. The frequencies mentioned are the resonant frequency for sprung mass and un-sprung mass respectively where the maximum acceleration occurred. It could be explained that the resonance has greater effect on sprung mass than un-sprung mass.

3.1 Parametric Studies

There are three parametric study cases were investigated by changing one of the parameters by two times. The changing variables are k_1, k_2 and c_1 while the remaining variables are remained constant.

3.1.1 Case 1: k_1 = 33624 N/m

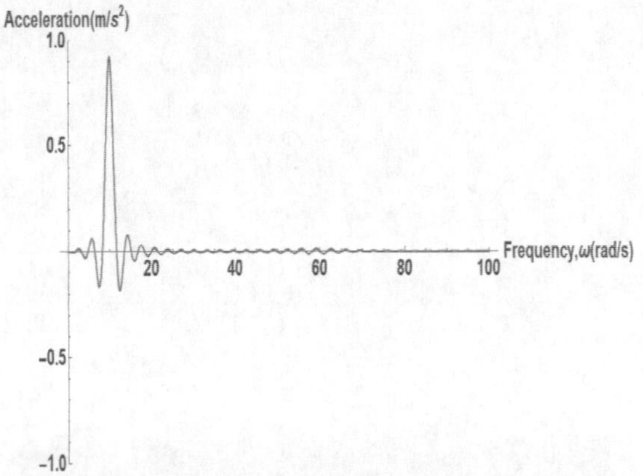

Figure 3. Graph of acceleration against frequency for sprung mass when k_1 = 33624 N/m

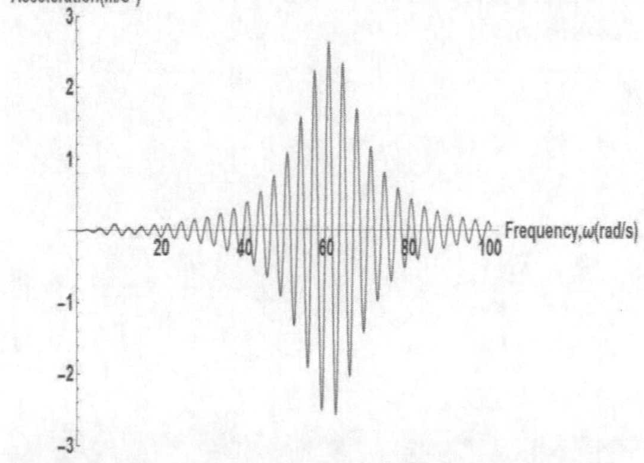

Figure 4. Graph of acceleration against frequency for un-sprung mass when k_1 = 33624 N/m

In figure 3, maximum acceleration 0.9133 m/s^2 occurs at frequency 10.0245 rad/s while in figure 4, maximum acceleration 2.6125 m/s^2 occurs at frequency 60.4339 rad/s. The natural frequency of sprung shifted to right compare to original due to the linear relationship of stiffness to natural frequency and the increase in magnitude of acceleration of sprung mass due to the greater stiffness greater potential energy which able to absorb as well as able to release more energy compare to original parameters. However, negligible change in un-sprung mass as no change in stiffness for un-sprung mass.

3.1.2 Case 2: k_2 = 380000 N/m

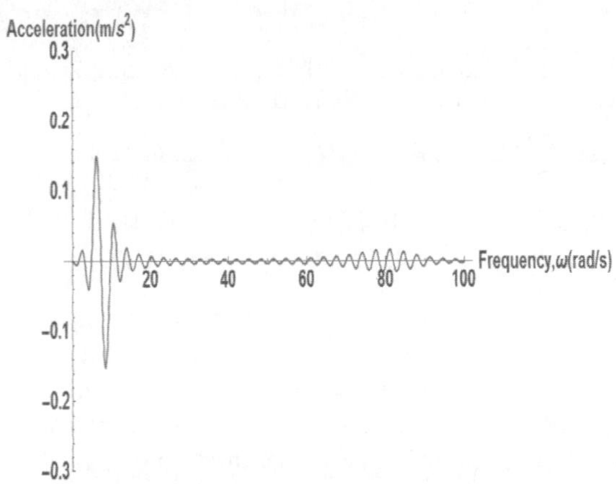

Figure 5. Graph of acceleration against frequency for sprung mass when k_2 = 380000 N/m

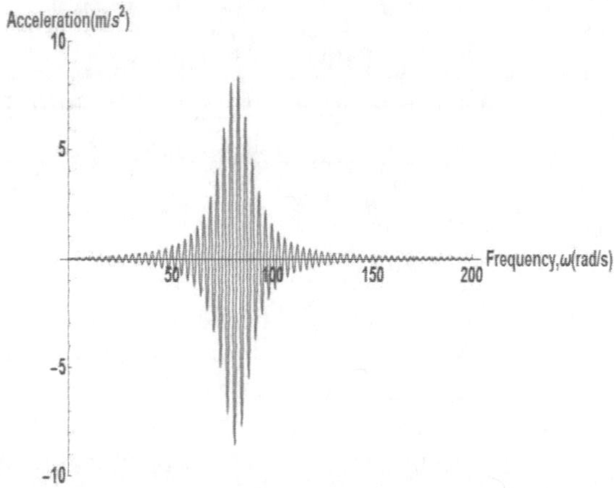

Figure 6. Graph of acceleration against frequency for un-sprung mass when k_2 = 380000 N/m

In figure 5, at maximum acceleration 0.2505 m/s^2 occurs at frequency 7.2777 rad/s. while in figure 6, at maximum acceleration -0.8.4645 m/s^2 occurs at frequency 80.9205 rad/s. In this case, negligible change happened in sprung mass while significant change as there is only increase of stiffness of un-sprung mass but not sprung mass.

3.1.3 Case 3: c_1 = 2000 N.s/m

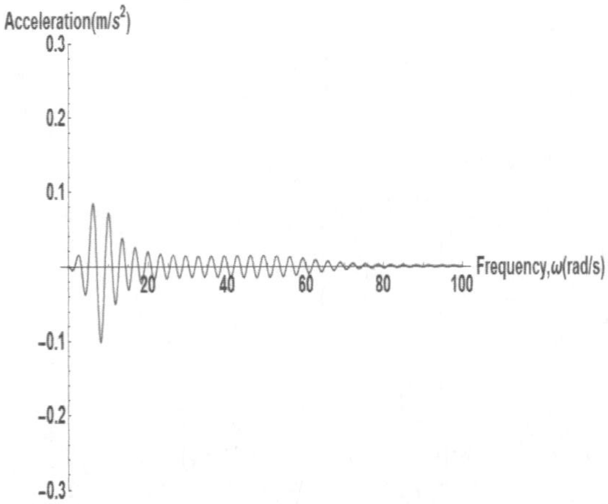

Figure 7. Graph of acceleration against frequency for sprung mass when c_1 = 2000 N.s/m

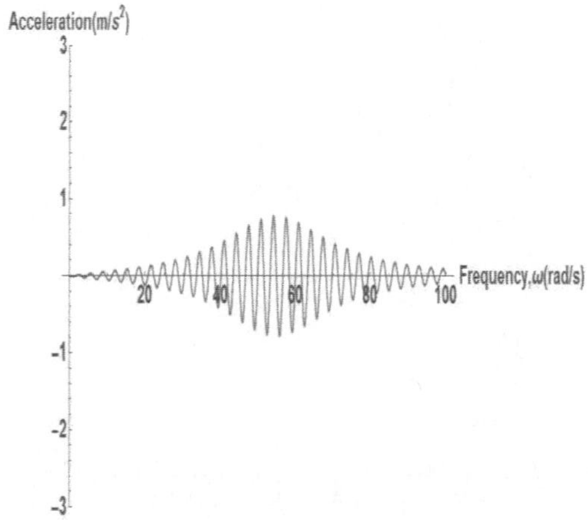

Figure 8. Graph of acceleration against frequency for un-sprung mass when c_1 = 2000 N.s/m

In figure 7, at maximum acceleration -0.0988 m/s^2 occurs at frequency 8.0861 rad/s. While in figure 8, at maximum acceleration -0.7686 m/s^2 occurs at frequency 55.7153 rad/s. The increase in damping coefficient also increase the energy loss when the car suspension system vibrate and affect both sprung mass and un-sprung mass with decrease in magnitude of acceleration.

3.1.4 Analysis

Table 2. Acceleration and frequency after changing parameters

Case no.	**Sprung mass**		**Un-sprung mass**	
	$\ddot{x}_1(t), m/s^2$	$\omega_1, rad/s$	$\ddot{x}_2(t), m/s^2$	$\omega_2, rad/s$
1.k_1 = 33624 N/m	0.9133	10.0245	2.6125	60.4339
2.k_2 = 380000 N/m	0.2505	7.2777	−8.4645	80.9205
3.c_1 = 2000 N.s/m	-0.0988	8.0861	-0.7686	55.7153

In table 2**,** it concludes the results from figure 3 to figure 8**,** shows that when increased the spring stiffness either of sprung or of un-sprung mass, the acceleration of sprung mass or un-sprung mass increased significantly, respectively, but with its natural frequency shifted away when compared to original parameters for both Case 1 and Case 2. While in Case 3, the change in damping coefficient has almost negligible effect on the natural frequency of both sprung mass and un-sprung mass but greatly reduced the magnitude of acceleration of both sprung mass and un-sprung mass. It can be explained that the natural frequency can be affected by the spring stiffness but not the damping coefficient and it is directly proportional to the spring stiffness.

3.2 Analysis Simulation

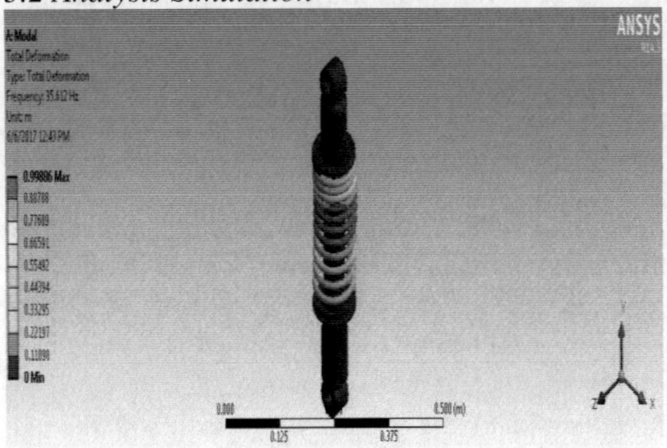

Figure 9. Total deformation at 35.612 Hz

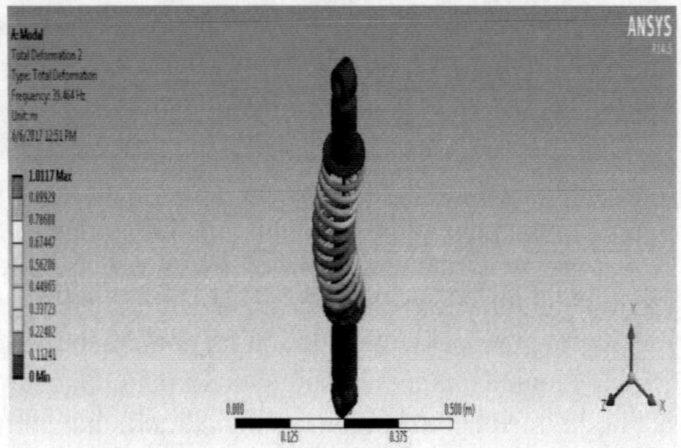

Figure 10. Total deformation at 39.464 Hz

In figure 9, it shows the first mode shape of the shock absorber at 35.612 Hz and in figure 10, shows the first mode shape of the shock absorber at 39.464 Hz. From the mode shapes, the coil spring vibrates the most at natural frequencies. It can be explained that the coil spring will most likely to fail first before the shock absorber itself failed. Once the coil spring failed, without the spring, the shock absorber can hardly return to original position since the function of coil spring is to absorb the potential energy and helps the shock absorber to return to original position so that the wheel can always keep intact with the road.

4. Conclusion

According to this study, natural frequencies of the car shock absorber can be obtained using both calculations and simulation. In calculation, it investigated the quarter-car model of the suspension system and obtained frequency response of the model to observe how the natural frequencies effect on the system. Based on the results, resonance has more significant effect on the un-sprung mass which is also known as the shock absorber compare to the sprung mass.

In simulation, it is further studies the effect of resonance on shock absorber only. The result explains that the coil spring of the shock absorber will fail first before the tubes of shock absorber fail. Thus, both theoretical and simulation results can be validated as the resonance has significant effect on the shock absorber.

5. Acknowledgement

Authors would like to acknowledge School of Mechatronic Engineering UniMAP for the financial support and assistance from colleagues, and Mechanical Design Lab 2 for providing the computer facilities.

6. References

[1] "The Role of Springs in Suspension Systems - Tire Review Magazine", *Tire Review Magazine*, 2017[Online]. Available: http://www.tirereview.com/the-role-of-springs-in-suspension-systems. [Accessed: 03- Dec- 2016]..

[2] Scuracchio B G, de Lima N B and Schön C G 2013 Role of residual stresses induced by double peening on fatigue durability of automotive leaf springs *Mater. Des.*, **vol. 47** pp. 672–76

[3] Kong Y S, Abdullah S, Omar M Z and Hari S M 2016 Failure assessment of a leaf spring eye design under various load cases *Eng. Fail. Anal.*, **vol. 63** pp. 146–59

[4] "ISO 2631-1:1997(en), Mechanical vibration and shock — Evaluation of human exposure to whole-body vibration — Part 1: General requirements", *Iso.org*, 2017. [Online]. Available: https://www.iso.org/obp/ui/#iso:std:iso:2631:-1:ed-2:v2:en. [Accessed: 03- Dec- 2017].

[5] Gadhvi B, Savsani V and Patel V 2016 Multi-Objective Optimization of Vehicle Passive Suspension System Using NSGA-II, SPEA2 and PESA-II *Procedia Technol.*, **vol. 23** pp. 361–68

[6] Gadhvi B, Savsani V and Patel V 2016 Multi-Objective Optimization of Vehicle Passive Suspension System Using NSGA-II, SPEA2 and PESA-II *Procedia Technol.*, **vol. 23** pp. 361–68

[7] Wang G, Chen C, and Yu S 2016 Optimization and static output-feedback control for half-car active suspensions with constrained information *J. Sound Vib.*, **vol. 378**, pp. 1–13

[8] Fang Z, Shu W, Du D, Xiang B, He Q, and He K 2011 Semi-active suspension of a full-vehicle model based on double-loop control *Procedia Eng.*, **vol. 16**, pp. 428–37

[9] Hur Rizvi S M, Abid M, Khan A Q, Satti S G and Latif J 2016 H 1 control of 8 degrees of freedom vehicle active suspension system.

[10] J. Marzbanrad J 2013 Optimization of a passive vehicle suspension system for ride comfort enhancement with different speeds based on design of experiment method (DOE) method *J. Mech. Eng. Res.*, **vol. 5** no. 3, pp. 50–59

[11] Federal D 2004, Journal of Applied Research and Technology *J. Appl. Res.*, **vol. 2** pp. 127–37

[12] Arias-Montiel M, Silva-Navarro G and Antonio-García A 2014 Active vibration control in a rotor system by an active suspension with linear actuators *J. Appl. Res. Technol.*, **vol. 12** no. 5, pp. 898–907

[13] Hassaan G A 2015, Car Dynamics using Quarter Model and Passive Suspension, Part VI: Sprung-mass Step Response," *IOSR J. Comput. Eng.*, **vol. 17** no. 1, pp. 2278–661

[14] Mitra A C, Desai G J, Patwardhan S R, Shirke P H, Kurn W M H and Banerjee N May 2008 Optimization of Passive Vehicle Suspension System by Genetic Algorithm *Procedia Eng.*, **vol. 144** pp. 1158–66

[15] Darus R, Modeling and Control of Active Suspension for a Full Car *Univ. Teknol. Malaysia*,

IOP Conf. Series: Journal of Physics: Conf. Series **908** (2017) 012026 doi:10.1088/1742-6596/908/1/012026

Fatigue life investigation of UIC 54 rail profile for high speed rail

P Gurubaran [1], M Afendi [1], M A Nur Fareisha[1], M S Abdul Majid [1], I Haftirman [2], M T A Rahman [1]

[1.] School of Mechatronic Engineering, Universiti Malaysia Perlis, Pauh Putra, Perlis, Malaysia.

[2.] University Mercu Buana, Kampus Menara Bakti Fakultas Teknik, Jakarta Barat, Indonesia.

gurubaranpanerselvan@gmail.com, afendirojan@unimap.edu.my

Abstract. This study is to investigate the fatigue life of high speed rail in Malaysia. This paper describes about the experimental and simulation analysis investigation on fatigue life of rail profile UIC 54 using bulk specimen according to ASTM E 466-15 standard. The Fatigue life testing was performed in the fatigue testing machine (Instron 8800) 100 kN. Meanwhile, the fatigue life analysis was performed in ANSYS Workbench 14.5. Furthermore, the stress levels for experimental testing were applied as 16.7%, 25%, 35%, 50%, 58.3%, 66.77% and 75% with machine frequency of 20 Hz. Apart from that, the total fatigue life cycles for rail profile UIC 54 were acquired from both experimental and simulation. The fatigue life S-N curves were plotted and validated with the results of the simulation analysis with experimental results.

1. Introduction

High speed railway is one of the primary advancement patterns to enhance the rail line transport capacity in recent years throughout the world [1]. Expanding requests for transportation items and travellers cause a quick progress in the transportation industry and railway vehicles are not avoided form this advance. The fatigue of high speed rail, it is important to find the endurance limit and the fatigue life cycle of rail steel. A lot of factors can contribute to fatigue life of railway material. Due to the long travelling distance and duration of high-speed trains running at high speed, structural materials of trains and track may change, and a wide range of shrouded issues be progressively uncovered, which may undermine security [2]. Fatigue failure is the failure of repeated or fluctuating stresses having high value, but below the ultimate strength of the material. The final fracture may have characteristics of brittle or ductile material, depends on the type of metal, stress apply and surrounding conditions. It is dangerous because it happened without any sign. The direction of applied stress is perpendicular to the surface of fatigue failure. Fatigue failure has three stages for rail steel that are initiation, propagation and final rupture as shown in figure 1. Initiation stage is the most complex stage and the irreversible changes in metal because of repetition of shear stresses. The initiation stage is quite small and not propagating more than two to five grains around the origin. Environmental condition, stress, metallurgical and strength conditions can be factors that affect crack on the surface. Site of stage 1 crack which is chosen for "fatal" crack is constantly competitive amid the distinctive times of loading [3]. Crack Propagation is stage 2 of fatigue failure. Micro structure start to change direction and develop perpendicular to the tensile stress. It is mostly prompt identifiable region of fatigue failure at this stage. The third stage is material rupture. The cross-sectional area of the part is decreasing when the propagation of fatigue crack continues, become weakened greatly and the fracture can occur completely with the application of only one more load. The fracture mode is whether ductile or brittle or the combination of both, depends on the metal, stress level and environment [4].

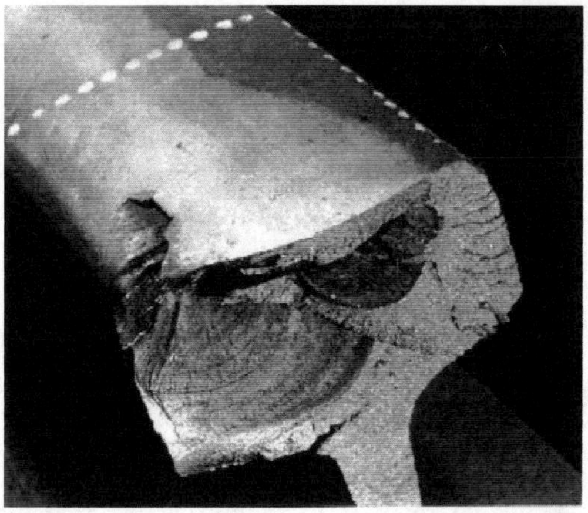

Figure 1: Fatigue fracture at rail profile head [5]

2. Rolling Contact Fatigue (RFC)

Rail defects because of rolling contact fatigue (RCF) undermine the activity security around the globe. That danger is more unmistakable on railways without sufficient maintenance technique [5]. It also will cause the increment of contact stress at the wheel/rail and wear decrement of rail steel. Rolling contact fatigue (RCF) is a phenomenon which occurs because of overemphasizing of material by a large number of contact cycles between wheel and rail. It is represented by initiation and propagation of cracks under Hertzian contact stress that will grow on surface amid contact of the system of rail and wheel. The complex stress state at growing crack's tip result from the contact stress may be overlaid by residual stresses. Moreover, Bending of the axle and thermal stress will lead to the normal stress between wheel/rail contact regions [6]. Rolling contact fatigue is one of the problems in high speed railway. Fatigue happens to rail steel when the stress is over a specific level and when the quantity of traffic loading cycles is sufficiently high [7]. Crack development of rolling contact fatigue is difficult to predict and depends on various aspects. Consequently, it is important to early recognize rolling contact fatigue defects [5]. Furthermore, excessive wear of the rail head and decrease of rail life is due to the speed increment of the railway. The RCF brought on mishaps may not just aim individual injuries and economical costs, they may likewise tend individuals to commute by car, which rise traffic jam, causes ecological issues and in the end may prompt an expansion in individual injuries since car traffic is essentially more risky than railroad transportation [8].

3. Materials and Methods

The materials which were tested are UIC 54 profile, which meets the requirements of the Codex standard. The technical condition of manufacture and delivery of railway rails was standardized in Codex UIC 54 of the International Railroad Union and which has been harmonized with the world trends [9]. Furthermore, the UIC 54 rail profile provided by Malaysian Railway Assets Corporation (RAC) and the chemical composition material range is given in table 1. Moreover, rail profile is made from steel, graded according to ultimate tensile strength (UTS) and designated according to their weight per unit length (kilogram per meter). The UIC 54 rail profile was made for Malaysia high speed train (ETS) with weight scale of 54 kg/m. The fatigue life tests were conducted with Instron 8800 fatigue machine at a laboratory room temperature. In addition, the surface of the bulk specimen was polished in order to avoid improvement of the fatigue performance by delaying crack initiation and remove possible machining imperfections. The fatigue tests were executed with machine frequency of 20 Hz with a constant stress ratio of $R = 0.1$ for all 7 sets of the specimen [10]. Along with this, the fatigue limit were set as the maximum stress the specimen can withstand after 10E5 cycles without failure. The fatigue S-N curve was determined with variable applied stress level for 7 sets of the specimen such as 16.7%, 25%, 35%, 50%, 58%, 66.7% and 75% (Figure 2). Moreover, the specimen was mounted onto Instron 8800 fatigue testing machine with clamping at both ends and similar procedure for all specimens (figure 3 and figure 4).

Table 1: Chemical composition and mechanical properties of UIC 54 profile [6]

Quality	**Chemical composition (%)**							**Tensile strength (MPa)**	**Yield strength (MPa)**
	C	Si	Mn	P (max)	S (max)	Cr	Ni		
Grade 900 A (UIC 54)	0.6-0.8	0.1-0.5	0.8-1.3	0.04	0.04	_	_	924	533

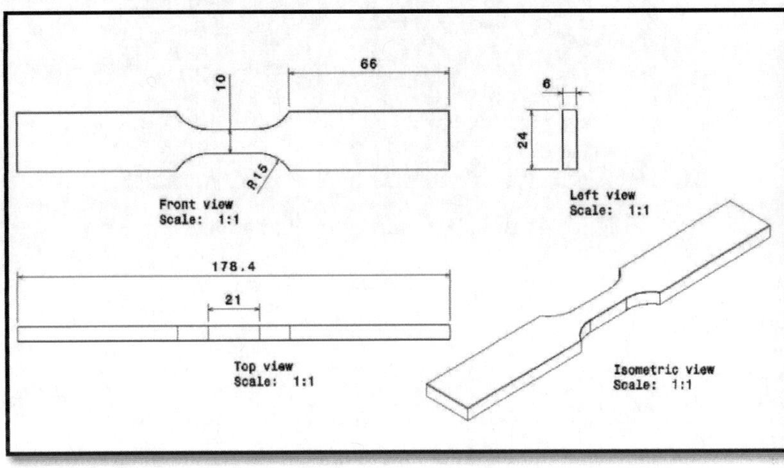

Figure 2: Bulk specimen technical drawing

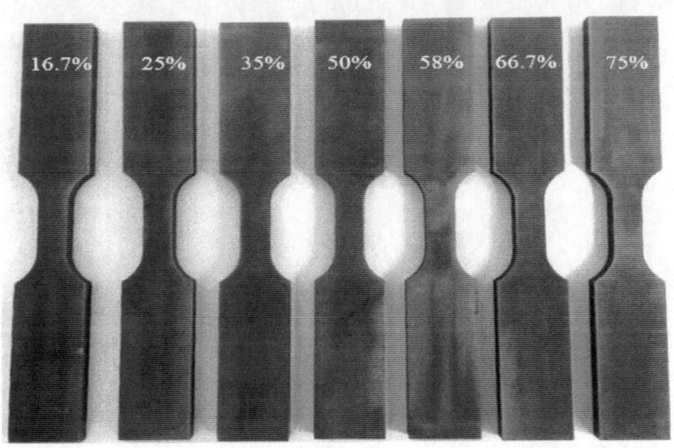

Figure 3: Bulk specimen extracted from UIC 54 profile head region

Figure 4: Bulk specimen mounted at 90°.

4. Results and Discussion

4.1 Fatigue life Experimental Testing

In railways service, the rail steel parts subjected to repetitive or cyclic stresses will fail due to the fatigue loading. The UIC 54 profile structural damages occur even when the experienced stress range is far below the static material strength. In most cases, the failure of railway structures experienced sudden failure from fatigue loading. Furthermore, the testing was conducted with the variable stress level and constant stress ratio 0.1 (table 2). In addition, the applied stress level for fatigue life testing vary 16.7 % to 75% and this is below the value of UTS, 924MPa of UIC 54 profile.

Table 2: Fatigue life cyclic data for testing

Specimen	Stress level %	Stress max (kN)	Stress min (kN)	Stress ratio	Ramp (kN)	Amplitude (kN)
1	16.7	9.258	0.926	0.1	5.092	8.333
2	25.0	13.860	1.386	0.1	7.623	12.474
3	35.0	19.404	1.940	0.1	10.672	17.464
4	50	27.720	2.772	0.1	15.246	24.948
5	58.3	32.322	3.232	0.1	17.777	29.089
6	66.7	36.978	3.698	0.1	20.338	33.281
7	75	41.580	4.158	0.1	22.869	37.422

Apart from that, the fatigue life S–N curve for UIC 54 rail profile was successfully acquired as shown in figure 5. The fatigue life S-N curve is in relation to stress amplitude, S against the number of cycles, N. From the result, the UIC 54 profile loading condition was determined. Therefore, the UIC 54 profile fail in loading condition exceeds above the fatigue S–N curve and any loading condition below the S–N curve will be safe. The fatigue life testing result shows, for the maximum applied stress 639MPa (75% stress level from UTS) the fatigue cycle number was smaller (8000 cycles). As expected the result of fatigue life testing for UIC 54 profile was shown in the trend of i.e. at the smaller number of fatigue cycle with high stress amplitude. In the fatigue life S–N curve result, the endurance limit was obtained for UIC 54 profile. Based on the result, if the loading condition is lower than fatigue endurance limit of UIC 54 profile, the rail part will never fail due to fatigue for an infinite number of cycles. Subsequently, for the loading condition intercept with the endurance limit at S–N curve the rail part able to survive the failure. Furthermore, the fatigue endurance limit was determined for rail steel at lowest stress level with non–failure occurs.

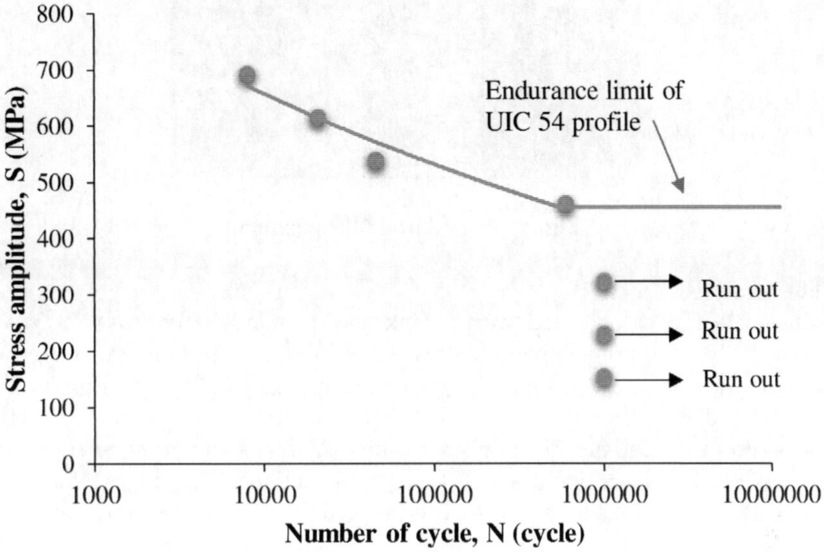

Figure 5: Fatigue life performance of UIC 54 profile

In general, there are two regimen of fatigue cycle which are high cycle fatigue (HCF), N > 10E5 and low cycle fatigue (LCF), N < 10E4 or 10E5. From the results plotted on figure 5, the stress level 75%, 66.7% and 58.3% LCF meanwhile for stress level 50%, 35%, 25% and 16.7% HCF region. Moreover, for HCF the stresses are low enough that the stress–strain relation can be considered elastic. The high nominal strength of the rail steel (924 MPa) acts in favor of the very good HCF performance assisting late crack initiation. The carbon content of 0.6–0.8 wt% seems to have a beneficial influence by moderately assisting the ductility and fatigue performance. Meanwhile, the carbon content of UIC 54 profile shows with strength of material but reduced fatigue strengths. From the observation on failure bulk specimen, the crack propagation zone was at the middle of the specimen and this because of the principle load applied normal to the crack plane to open the crack (figure 6).

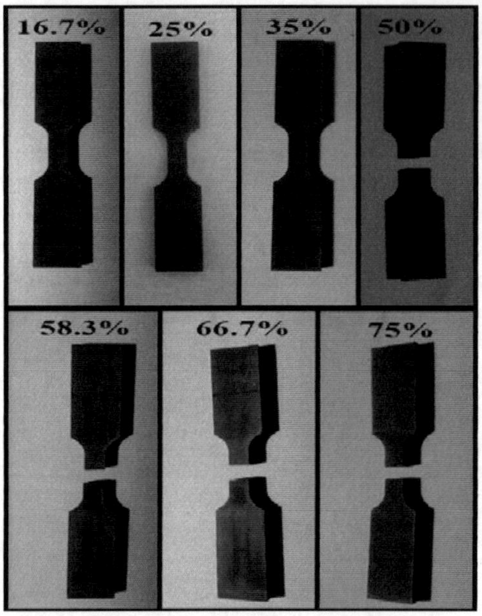

Figure 6: Failure bulk specimen

4.2 Fatigue life Analysis

The fatigue life analysis was executed with a bulk specimen model on ANSYS Workbench 14.5. Furthermore, this analysis was run with exact stress level value as used in fatigue experimental testing. The analysis result was used to validate the fatigue life experimental testing result and to observe the fatigue life gradient during testing. Table 4 tabulates the result of fatigue life analysis for bulk specimen model. From the fatigue life gradient result (figure 7), the affected stress region of bulk specimen was observed and predicts the UIC 54 profile fatigue life pattern whenever the load is applied. Based on the result, the maximum stress is subjected to the middle of the bulk specimen.

Table 3: Fatigue life analysis results above the endurance limit of UIC 54 profile (ANSYS Workbench 14.5)

Stress level (%)	Stress (MPa)	Total Number of cycles
50	462	16521
58.3	538.692	9786
66.77	616.9548	6455
75	693	4492

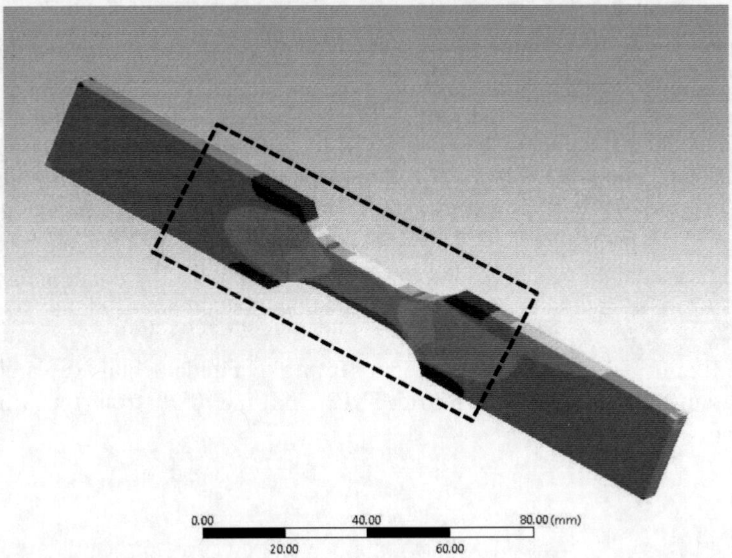

Figure 7: Fatigue life gradient for maximum stress level at 75 % (693 MPa)

The fatigue S-N curve for experimental and analysis testing was plotted in figure 8. From the result, the fatigue life performance for UIC 54 profile was validated with experimental and simulation testing. Furthermore, the curve pattern between experimental and simulation results was acquired in range and the material fatigue endurance limit was identified. The high cycle fatigue performance of this rail steel is consumed in a narrow range of 230 MPa between 693 MPa and 463 MPa. From the finding, the recommended applied stress level for the rail steel is below endurance limit, to have an infinite life cycle. The result illustrated, at the stress amplitude of 616.3MPa, the intercept between experimental and simulation results. Therefore, the number of fatigue cycle for UIC 54 profile was obtained similar at HCF regions.

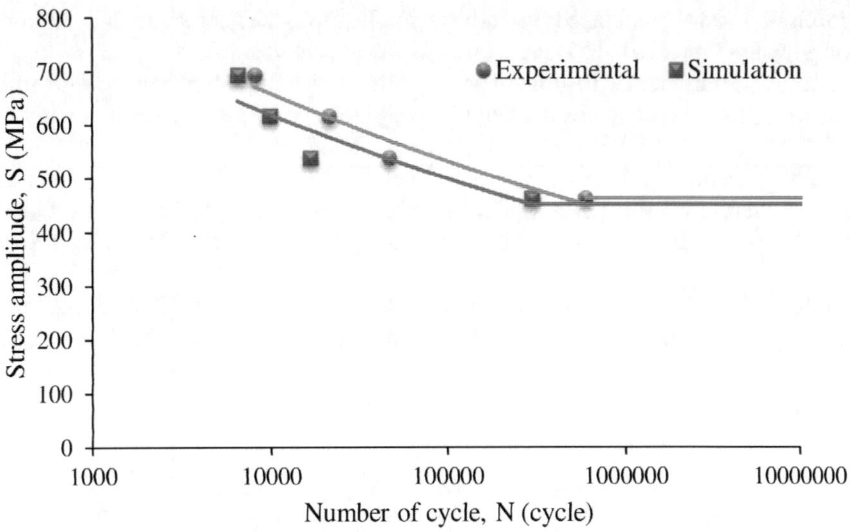

Figure 8: Fatigue life prediction for UIC 54 profile

IOP Conf. Series: Journal of Physics: Conf. Series **908** (2017) 012026 doi:10.1088/1742-6596/908/1/012026

5. Conclusion

From the fatigue life testing, the endurance limit for UIC 54 profile was obtained. From the finding, the lowest stress amplitude for infinite number of cycle was identified for UIC 54 profile head region. Furthermore, from fatigue S–N curve the fatigue life performance of Malaysia high speed train rail was acquired. This result will be significant for KTM for selecting the high fatigue life performance rail for ETS and to calculate the life span of UIC 54 profile with number of the fatigue cycle to failure. The carbon content of 0.6–0.8 wt% seems to have a beneficial influence by moderately assisting the ductility and fatigue performance for UIC 54 profile. Whereas, the higher carbon content steel of UIC 54 profile shows higher strength properties but lower fatigue strengths [6].

Acknowledgements

The author would like to acknowledge the support from the Fundamental Research Grant Scheme (FRGS) under a grant number of FRGS/1/2015/TK03/UNIMAP/02/8 from the Ministry of Higher Education Malaysia.

References

[1] C. He, J. Guo, Q. Liu and W. Wang, "Experimental investigation on the effect of operating speeds on wear and rolling contact fatigue damage of wheel materials", Wear, vol. 364-365, pp. (2016) 257-269.

[2] S. Tarafder, S. Sivaprasad And V. Ranganath, "Comparative assessment of fatigue and fracture behavior of cast and forged railway wheels", Fatigue & Fracture of Engineering Materials and Structures, vol. 30, no. 9, pp. (2007) 863-876.

[3] Z. Zhu, G. Li, G. Dai, J. Zhao, L. Xu and Q. Zhang, "Mechanisms and new parameter attribute reduction of high-speed railway wheel rim steel subjected to low temperature fatigue", Materials Science and Engineering: A, vol. 673, pp. (2016) 476-491.

[4] X. JIN, "Key problems faced in high-speed train operation", Journal of Zhejiang University Science A, vol. 15, pp. (2014) 936-945.

[5] U. Zerbst, R. Lunden, K. Edel and R. Smith, "Introduction to the damage tolerance behaviour of railway rails – a review", Engineering Fracture Mechanics, vol. 76, no. 17, pp. (2009) 2563-2601.

[6] P. Christodoulou, A. Kermanidis and G. Haidemenopoulos, "Fatigue and fracture behavior of pearlitic Grade 900A steel used in railway applications", Theoretical and Applied Fracture Mechanics, vol. 83, pp (2016) 51-59.

[7] Z. Popovic, L. Lazarevic, L. Brajovic and M. Vilotijevic, "The Importance of Rail Inspections in the Urban Area -Aspect of Head Checking Rail Defects", Procedia Engineering, vol. 117, pp. (2015) 596-608.

[8] Ekberg and E. Kabo, "Fatigue of railway wheels and rails under rolling contact and thermal loading—an overview", Wear, vol. 258, no. 7-8, pp. (2005) 1288-1300.

[9] I. Vitez, D. Krumes and B. Vitez, "UIC-Recommendations for the Use of Rail Steel Grades". (2017).

[10] ASTM E466-15, Standard Practice for Conducting Force Controlled Constant Amplitude Axial Fatigue Tests of Metallic Materials, ASTM International, West Conshohocken, PA.(2015).

IOP Conf. Series: Journal of Physics: Conf. Series **908** (2017) 012027 doi:10.1088/1742-6596/908/1/012027

Determination of effective elastic properties of metal matrix composites with damage particulates using homogenization method

S Z Halim, K S Basaruddin, I Ibrahim, M S Abdul Majid and M J M Ridzuan

School of Mechatronic Engineering, Universiti Malaysia Perlis, Pauh Putra Campus, 02600 Arau, Perlis, Malaysia.

khsalleh@unimap.edu.my

Abstract. The present study aims to investigate the effect of micro-damage in particulates metal matrix composite on the elastic properties. The micro damage that perhaps could occurs during manufacturing process or due to environmental effects was modelled in three different types, namely shattered, debonded and breakage particulates with variation of volume fraction. The modelling and analysis were conducted based on homogenization theory by utilizing multiscale finite element software (Voxelxon). The results suggest that the elastic properties of metal matrix composite was sensitive to the geometrical defects of its particle.

1. Introduction

Metal matrix composite (MMCs) is a combination of at least two or more materials. The materials consist of metal that combine with another metal or materials. Matrix is a monolithic material where the reinforcement of MMCs is embedded and completely continuous. The reinforcement is used to change the physical properties of the MMC such as the friction coefficient. The invention of MMC is to reinforce the existence metal particles by improved the mechanical properties of MMC such as the Shear modulus, Young's modulus, and Poisson's ratio. The process state of metal matrix composite is divided into three types that are solid, liquid and vapour state. For solid, the forming method can be powder blended and stir casting method is applied in liquid state of MMC. In addition, physical vapour deposition is the method in forming the MMC in vapour state. MMCs are normally acts as a structural usage material to support loading and also to sustain the environmental effect for a long time. The metal of the metal matrix composite can be zinc, magnesium, titanium, copper, nickel iron and cobalt. For example, the aluminium-silicon carbide is metal matrix composite where the silicon carbide been added into aluminium and change from metal to composite.

By adding the silicon carbide, the elastic modulus of the composite will with the same density. In means, the density would not change. The common particles reinforced MMCs such as *SiC* and *Al* are usually used in the field of automobile and aviation. Besides, MMCs are normally in the characteristics of low density but high in strength, specific modulus and fatigue resistance. These characteristics are influenced by the arrangement of the particles, shape and size of the particles. The mechanical properties of MMCs can be determined from some analysis for the micro structure of the material. Based on previous study [1] the potential of MMC in improving the mechanical properties of materials with monolithic alloys has been considered as an interest in MMC. Mostly, it was also being considered in the increases of the specific mechanical properties. The general degradation in the properties of fracture also have been compiled by these improvement. The damage of MMC can be affected during manufacturing process or due to environmental temperature. The common types of damage that exist were void, crack, debonding and shattered in particles of the metal matrix composites [2,3,4].These damage of particles in metal matrix composites also usually occurs when during mechanical testing on MMC. At different metal matrix composite being test, the different types of damage also will occurs [5, 6, 7]. For instance of metal matrix composites of aluminium and silicon carbide three tests which were fatigue test, creep test and monoatomic loading being conducted to the metal matrix composites shows voids, small crack nucleation, dominant cracks propagation and also fatigue damage. The fatigue damage was usually occurred when there was a formation around particles due to single slip bands.

In addition, damage also would appears in the bulk of material and it types would be depends on anisotropy of the material. The nature of material, size, length and type of draping plays an important role in anisotropy of the material. In composites, damage can occurs for the first 20% of fatigue life with more than 50% and can withstand the presence of crack [8, 9, 10]. In increasing the mechanical properties such as strength and stiffness usually depends on the improvement of the material properties. The relation in degradation on properties of fracture also can improve the mechanical properties such as the ductility and toughness[11]. Based on previous studies, the microstructural parameters were have effects on mechanical properties. The increasing of the volume fraction will increased the modulus, yield and the ultimate tensile stresses but the strain to failure would reduce in a shear-lag-type models. Besides, the increasing in particle size effects the fracture mechanism by changing the particles decohesion into particle cracking [11, 12].

Bian and Wong [4] have studied about the influenced of particle size and volume fraction on the micro damage composites. The bulk and shear modulus of the metal matrix composite have been modified with various volume fraction of particles. Besides, by introducing a new model and modified modulus, the effect of size particles on stress-strain relation under damage of debonding can be predicted. Fatigue was one of the factors in failure of metal matrix composites. It can occurs and lead to the failure of structures under certain loading conditions. Besides, fatigue fracture may occur by the environment condition, faulty design, types of stress and presence of manufacturing defects. A lower compressions resistance than tensile resistance of the composite material under static loading also lead to fatigue when the parameter of fatigue influenced by the viscosity and ductility of the matrix composite [8, 13, 14]. Gupta et al. [15] studied about the failure mode of the *SiC* that being reinforced with *Al* due to its interfacial characteristics. The pre-information was about the effect in different cycles of heat treatment. Heat treatment in a cast of 6061 *Al/SiC* were conducted based on the variation of MMCs.

The damage in particles may be occurs during the manufacturing process of metal matrix composite (MMC) or due to the environmental effects. Damage of particles also may occur when the metal matrix composite is being subjected to a large deformation such as plastic deformation. The types of damage in particles of MMCs that usually occurs during the manufacturing process are shattered, debonding and breakage particles. Thus, the purpose of this study is to investigate the effect in micro-damage of particulate of MMC on its elastic properties. Through homogenisation method, the distribution of the metal matrix can be determined. Thus, the mechanical properties of the MMCs with damage also can be determined.

2. Computational modelling

2.1 Periodic microstructure

Four models been developed through CATIA software with variation of damage and volume fraction to carry out the analysis of how damage occurs in MMC of Aluminium and Silicon Carbide. The periodic microstructure of metal matrix composite can be imagine as in figure 1.

IOP Conf. Series: Journal of Physics: Conf. Series **908** (2017) 012027 doi:10.1088/1742-6596/908/1/012027

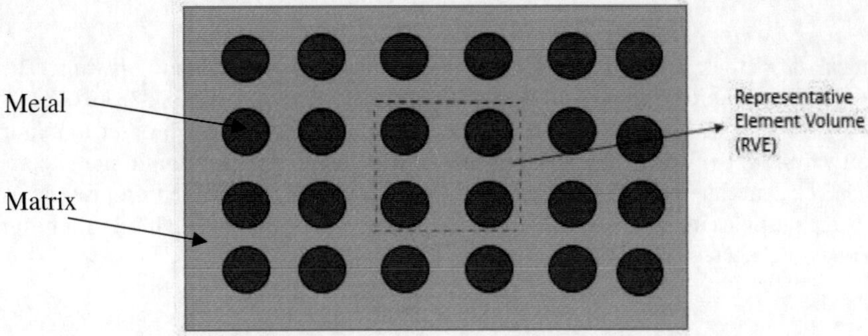

Figure 1: Periodic microstructure of MMC

As demonstrative computation, aluminium was chosen as constituent for matrix and silicon carbide (SiC) for the particles. Three mode of damages in particle were developed namely, breakage, debonded and shattered. Additional of one model of MMC with undamaged particles to compare the properties of damage and undamaged MMCs. For variation of volume fraction, four types of metal matrix composite with different condition of particles were modelled in three different diameters of particles as in figure 2.

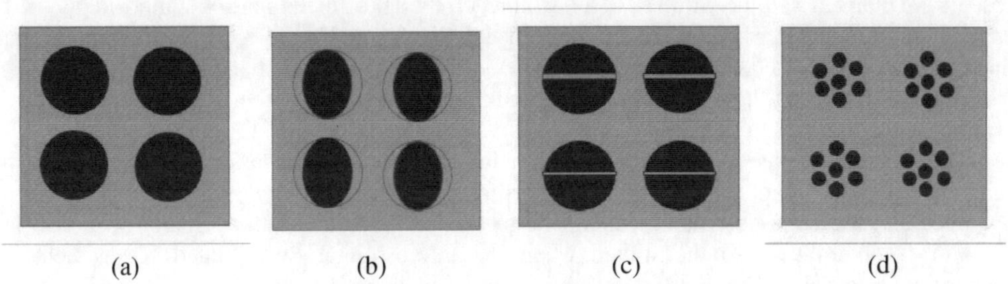

Figure 2: model of metal matrix composites for (a) Undamaged particles, (b) Debonding particles, (c) breakage particles, (d) Shattered particles

2.2 Convergence test

The convergence test was conducted by referring to the selected size of voxel element mesh to get the finest results. But the problem arise when the size element of 0.05 show error in received result due to the software limits. It was generally known that the smallest size of element tend to take longer time to be compute or need a powerful workstation in order to male it successfully done. Thus the voxel size was decided to the minimum size of 0.1 for all model with element type of cubical voxel to be analyse using Voxelcon software. Figure 3 shows the view of meshed model. The software generates the polyhedral mesh that can approximate the geometric domain of the model. The voxels, volume, voxel space and voxel size were shown out in the interface of the software.

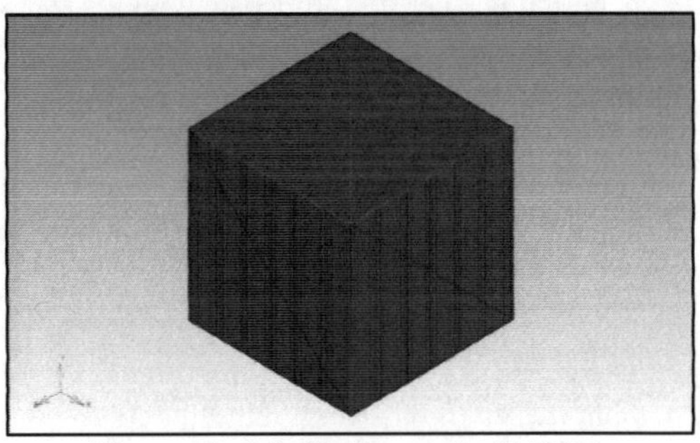

Figure 3 : Meshed model of MMCs

2.3 Mechanical properties of constituents

The mechanical properties of the material will be calculated by Voxelcon software. The analysis regarding the details of micro structure from the models that have been developed from CATIA software will be done by Voxelcon. The files of models need to save as .stl format in order to export into the Voxelcon software. Type of MMC that had chosen for this research is aluminium metal compose with silicon carbide. The mechanical properties in this research was decided based on previous study [16] about the damage particles in metal matrix composite. It was clearly stated in table 1 as a summarization for the properties of selected constituents to be analysed.

Table 1: Micro properties of MMCs [16]

Material	Young's modulus (MPa)	Poisson's Ratio
Aluminium	70	0.33
Silicon Carbide	430	0.22

2.4 Homogenization method

Homogenization analysis will be carry out in this research. A unit of cell of metal matrix composite will be taken as a representative element volume in applying the homogenization method. The representative element volume will be chosen from each model of metal matrix composite as the best since it can represents the whole micro structure of the composite. The selection of the representative element volume due to the behaviour of the metal matrix composite that have heterogeneous type of microstructure. By using homogenization method, the behaviour of the whole material can be determined and simplified to be considered as an idealised unit cell.

The microstructure size and local region being periodically array assumed to be smaller than global body [17]. The periodicity of the microstructure was remained in the local region after the load was applied and the large deformed occurs. The method of homogenization usually occurs in large deformation problem that can be solved when the microstructure was stated by element. The heterogeneous material can be replaced by equivalent homogenized model with periodically repeated element [18, 19].

The homogenization method being formulated by describing the theory of large deformation and followed by two coordinate system which was microscopic (x) and microscopic (y). Then, the theory of conventional linear homogenization can be derived. By using two scale singular perturbation theory, all equations were followed based on previous studies [17,20].

$$u_i(\boldsymbol{x},\boldsymbol{y}) = u_i^0(\mathrm{x}) + \varepsilon u_i^1(\boldsymbol{x},\boldsymbol{y}) \qquad (1)$$

In this equation, u_i^0 is the microscopic displacement and u_i^1 is a perturbed term due to the microscopic heterogeneity. We suppose applied the traction t_i on the boundary Γ and neglected the body force. An elastic tensor is denoted as X .By taking $\varepsilon \to 0$ in homogenization, the decoupled microscopic and macroscopic equations can be obtained.

$$u_i^1(x,y) = -\chi_i^{kh}(y)\frac{\partial u_k^0(x)}{\partial x_h} \qquad (2)$$

χ_i^{kh} is a characteristic displacement that is a periodic function with respect to the microscale. The microscopic equation to solve for the unit cell Y under the periodic boundary condition is as follows:

$$\int_Y (E_{ijkh} - E_{ijmn}\frac{\partial \chi_m^{kh}}{\partial y_n})\frac{\partial v_i^1}{\partial y_j}\mathrm{d}Y = 0 \quad \text{for} \quad \forall v_i^1 \qquad (3)$$

Due to the existence of the solution equation (3), the macroscopic equation can be derived as below:

$$\int_\Omega E_{ijkh}^H \frac{\partial u_k^0}{\partial x_h}\frac{\partial v_i^1}{\partial x_j}\mathrm{d}\Omega = \int_\Gamma t_i\, v_i^0 \mathrm{d}\Gamma \quad \text{for} \quad \forall v_i^0 \qquad (4)$$

Where E_{ijkh}^H is the homogenized elasticity tensor that is also symmetric defined by,

$$E_{ijkh}^H = \frac{1}{|Y|}\int_Y (E_{ijkh} - E_{ijmn}\frac{\partial \chi_m^{kh}}{\partial y_n})\mathrm{d}Y) \qquad (5)$$

Where $|Y|$ is the volume of the unit cell.

3. Results and discussion

There are three model have been developed in order to investigate the effect of the variation of volume fraction on mechanical properties behaviour of MMCs. The particles of MMCs were divided into four categories which are undamaged, shattered, breakage and debonding model. Besides, homogenisation analysis also has been done by using the same procedure and setting to each model. The results were shown in a linear graph of Young's modulus against volume fraction as in figure 4 while graph of Poisson's ratio versus volume fraction as in figure 5, and graph of Shear modulus against volume fraction in figure 6.

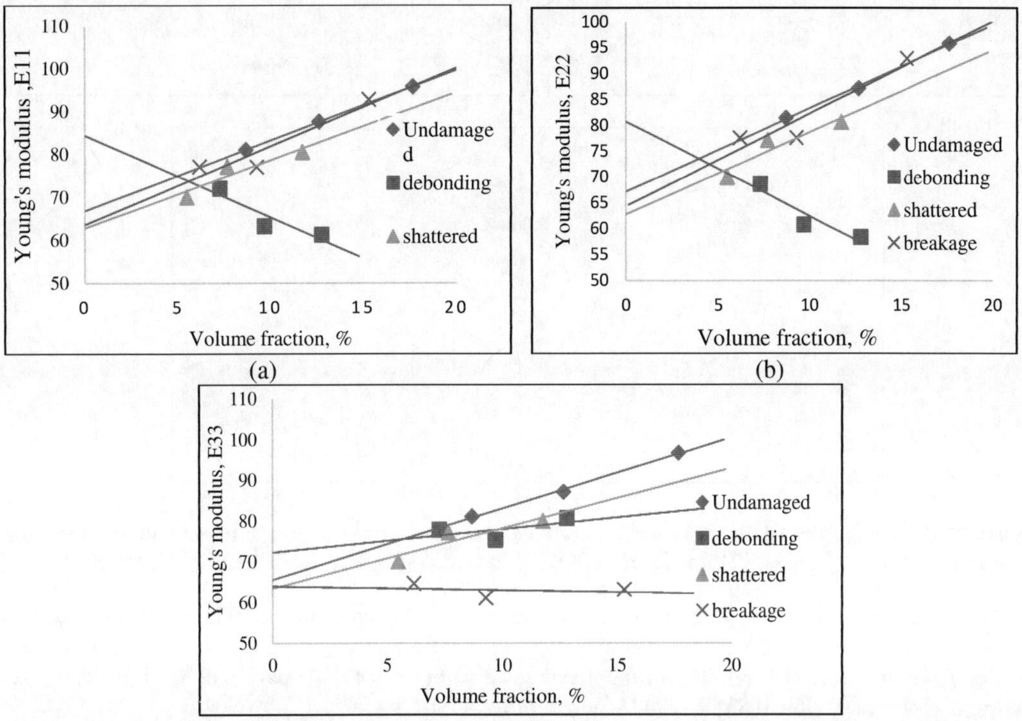

(a) (b) (c)

Figure 4: Young's modulus against volume fraction (a) : Young's modulus, E_{11} against volume fraction, (b) : Young's modulus, E_{22} against volume fraction, (c) : Young's modulus, E_{33} against volume fraction.

Figure 4(a) shows the graph of Young's modulus, E_{11} versus volume fraction of each model. Based on the graph, three types of metal matrix composite model show the incline value while model of debonding shows the decline value of the Young's modulus, E_{11}. The coefficient of determination, R^2 for the linear correlation plot of undamaged, debonding, shattered and breakage were 1.000, 0.8849, 0.8670, and 0.8357. From figure 4(b), it shows that three model of metal matrix composite that were undamaged, shattered and breakage which have an increased in Young's modulus, E_{22} value but decreased value for debonding model. The coefficient of determination, R^2 for the linear correlation plot of undamaged, debonding, shattered and breakage were 0.9975, 0.8894, 0.8670, and 0.8687. Based on figure 4(c), the graph shows that an increasing value of Young's modulus, E_{33} for undamaged and shattered model while the value for debonding and breakage model were fluctuated. The coefficient of determination, R^2 for the linear correlation plot of undamaged, debonding, shattered and breakage were 0.9963, 0.3668, 0.8397and 0.0649.

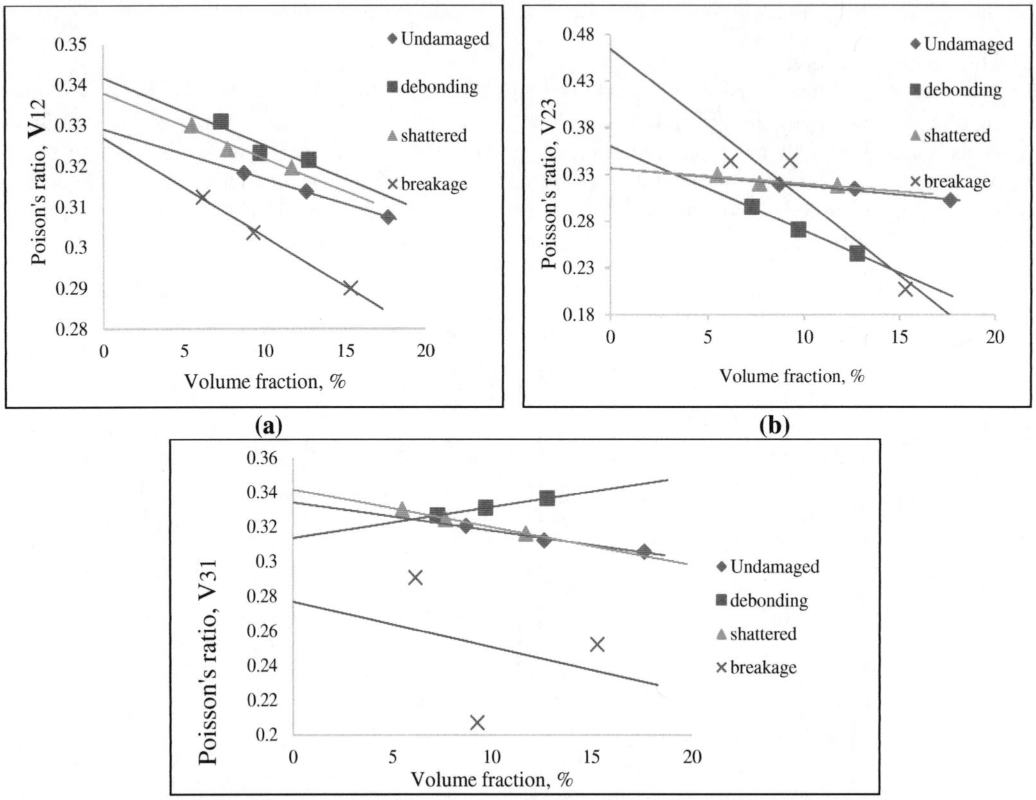

(c)

Figure 5: Poisson's ratio against volume fraction (a) : Poisson's ratio, v_{12} against volume fraction, (b) : Poisson's ratio, v_{23} against volume fraction, (c) : Poisson's ratio, v_{31} against volume fraction.

Figure 5(a) shows the graph of Poisson's ratio versus volume fraction. Based on the graph, it can be seen that there was a declined value of Poisson's ratio for each type of micro damage model. The coefficient of determination, R^2 for the linear correlation plot of undamaged, debonding, shattered and breakage were 0.9969, 0.8358, 0.9423, and 0.9969.

In Figure 5(b), the graph of Poisson's ratio against volume fraction shows that the value of Poisson's ratio, v_{23} for debonding and breakage model are declined while the value for undamaged and shattered model were constant. The coefficient of determination, R^2 for the linear correlation plot of undamaged, debonding, shattered and breakage were 0.9643, 0.9968, 0.7546 and 0.8868.

Figure 5(c) shows the graph of Poisson's ratio versus volume fraction. Three model of micro damage that are undamaged, shattered and breakage show that their Poisson's ratio, v_{31} were linearly decreased while debonding was linearly increase. The coefficient of determination, R^2 for the linear correlation plot of undamaged, debonding, shattered and breakage were 0.9784, 0.9991, 0.9919 and 0.0858.

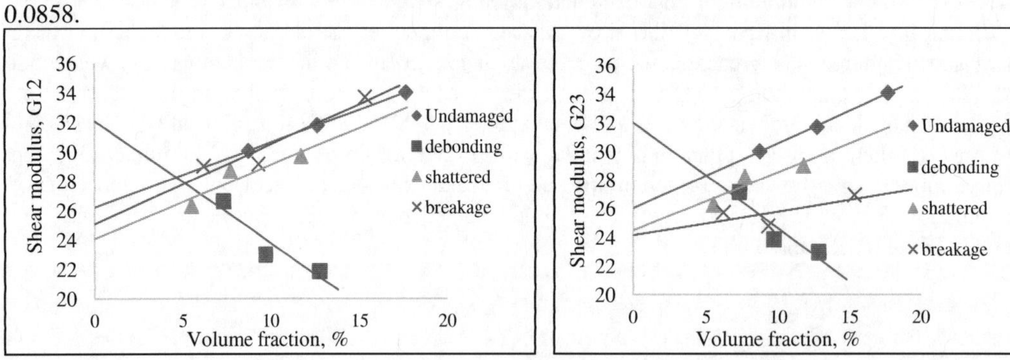

(a) (b)

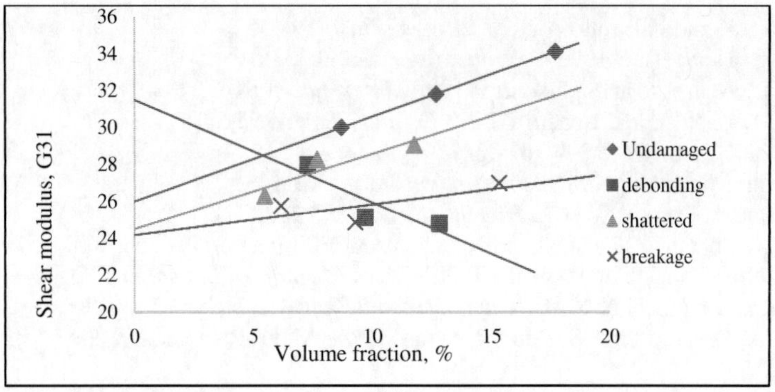

(c)

Figure 6: Shear modulus against volume fraction (a) : Shear modulus, G_{12} against volume fraction, (b) : Shear modulus, G_{23} against volume fraction, (c) : Shear modulus, G_{31} against volume fraction.

Based on figure 6(a), Shear modulus, G_{12} of three micro damage model of metal matrix composite that are undamaged, shattered and breakage were linearly inclined as the volume fraction was increased. The coefficient of determination, R^2 for the linear correlation plot of undamaged, debonding, shattered and breakage were 1.0000, 0.8729, 0.8476, and 0.9034.

Shear modulus, G_{23} of undamaged, shattered and breakage model of metal matrix composite that shown as in figure 6(b) were linearly increased as the value of volume fraction was increased while shear modulus of debonding model was directly decreased even as the volume fraction was increased. The coefficient of determination, R^2 for the linear correlation plot of undamaged, debonding, shattered and breakage were 0.9990, 0.8553, 0.8403 and 0.4956.

Shear modulus, G_{23} of undamaged, shattered and breakage model of metal matrix composite that shown as in figure 6(c) were linearly increased as the value of volume fraction was increased while shear modulus of debonding model was directly decreased even as the volume fraction was increased. The coefficient of determination, R^2 for the linear correlation plot of undamaged, debonding, shattered and breakage were 0.9990, 0.8553, 0.8403 and 0.4956.

4. Conclusion

Through this study, the results of analysis by using finite element method (FEM) software showing that the particles with regular arrangement and bigger size of damage particle obtain greater elastic modulus. Therefore, by knowing the variation of particles in terms of particles arrangement, size and shape, the mechanical properties of metal matrix composite can be predicted. It is important to know the mechanical properties of metal matrix composite because by knowing this metal matrix composite with optimized mechanical characteristics will able to be manufactured for certain application without any damage occurs.

5. Acknowledgement

This study was partially supported by University Malaysia Perlis under Short Term Research Grant (STG No. 9001-00483).

6. References

[1] Mummery P and Derby B 1991 *Mater. Sci. Eng. A* **135** 221
[2] Tong W and Ravichandran G 1994 *Compos. Sci. Technol.* **52** 247
[3] Rutecka A, Kowalewski Z L, Pietrzak K, Dietrich L, Makowska K, Wožniak J, Kostecki M, Bochniak W and Olszyna A 2011 *Procedia Eng.* **10** 1420
[4] Bian L and Wang Q 2013 *Arch. Appl. Mech.* **83** 445
[5] Kim E, Rim M, Lee I and Hwang T 2013 *Compos. Struct.* **95** 123
[6] Brechet Y, Embury D and Luo L 1991 *Acta Metallurgica Mater* **39** 1781
[7] Jun L, Guiqiong J, Bo W, ChengpengY and Gang W, 2014 *Chinese J. Aeronaut.* **27** 1586
[8] Jollivet T, Peyrac C and Lefebvre F 2013 *Procedia Eng.* **66** 746
[9] Agrawal S and Singh K K 2014 *J Compos Mater* **48** 317

[10] Naboulsi S K and Palazotto A N 2002 *Int. J. Plast.* **19** 435
[11] Fang D, Qi H and Tu S 1996 *Comput. Mater. Sci.* **6** 303
[12] Davis L C, Andres C and Allison J E 1998 *Mater. Sci. Eng. A* **249** 40
[13] Greco F, Leonetti L and Lonetti P 2013 *Compos. Struct.* **95** 582
[14] Xia Z H and Curtin W A 2001 **61** 2247
[15] Manoharan M and Gupta M 1997 *J Compos Mater* **31** 1431
[16] Park M S and Kwon Y W 2013 *Comput. Struct.* **123** 28
[17] Takano N, Ohnishi Y, Zako M and Nishiyabu K 2000 *Int J Solids Struct* **37** 6517
[18] Azizi R, Niordson C F and Nyvang B 2014 *Acta Mechanica Sinica* **30** 175
[19] Suzuki K and Kikuchi N 1991 *Comput Methods Appl Mech Eng* **93** 291
[20] Halim S A, Basaruddin K S, Daud R, Safar M J A, Abdul Rahman A S 2015 *ARPN J Eng Appl Sci.* **10** 9353

IOP Conf. Series: Journal of Physics: Conf. Series **908** (2017) 012028 doi:10.1088/1742-6596/908/1/012028

Characterisation of static and dynamic responses of natural rubber bio-composites

S Kamaruddin[1], A B Chai[2], C L Lim[2] and J-H Ho[2]

[1]Rubber Research Institute of Malaysia, Malaysian Rubber Board, Malaysia
[2]The University of Nottingham Malaysia Campus, Malaysia

kshamsul@lgm.gov.my

Abstract. Characterisation of static and dynamic behavioural are presented in this study of hybrid filled natural rubber. The hybrid filler consists of a combination of oil palm ash (OPA) and egg shell powder (ESP). Testing on the samples are conducted in three different aspects namely static compression testing, time dependent stress relaxation testing in compression and dynamic compression testing. Mechanical properties such as the static stiffnesses, hysteresis loss ratios, stress relaxation responses, dynamic stiffnesses, phase angles and damping ratios are reported. The static stiffness values of the hybrid filled rubber, although lower than rubber filled carbon black, presented potential if the ESP content were to be increased. The stress relaxation responses lacked consistency in the uploading stages but were fairly consistent in the unloading stages where the relaxation was close to that of rubber filled carbon black. The dynamic properties appear to be no significant improvement over that of rubber filled carbon black filled. All hybrid filled rubber samples conformed to general dynamic properties of viscoelastic material and did not display any unexpected behaviour. Collectively, there is a trend of improvement with increasing ESP content but no prominent improvement over rubber filled carbon black is observed.

1. Introduction

A majority of rubber is used in anti-vibrational and shock absorbance applications include tyres, rubber isolators and seismic bearings as their energy dissipating capacity and viscoelasticity generally outperform that of other materials [1]. Rubber's ability to conform to any surface it comes into contact with also enables it to be an effective seal in engine gaskets, window seals and waterproof seals [2]. Natural rubber on its own has limited adaptability, hence to improve its properties, fillers are usually mixed in during production. The most common type of rubber filler used today is carbon black. Carbon black is an unsustainable option for rubber fillers as it is manufactured by the intentional incomplete combustion of petroleum products [3]. This petroleum based filler is mainly used because it is has superior filler properties to almost all other types of fillers. This in itself presents two problems. Firstly, petroleum will face eventual depletion making it a non-sustainable source of rubber fillers. Secondly, carbon black filled rubber is difficult to recycle. The aftermath of this on the environment is catastrophic as the tyres are non-biodegradable. Efforts in recycling rubber tyres are possible but these are limited in terms of processing rates, cost, as well as technology [3]. Today, the main method of recycling rubber is by cryogenic crushing where the rubber is frozen using liquid nitrogen and crushed to a powder and melted to be reused in other industries [4]. Cryogenic crushing of carbon black rubber however is very energy intensive and costly.

Hence, one solution is to replace the type of fillers used in natural rubber. In this study, the filler used is a hybrid made up of both egg shell powder (ESP) and oil palm ash (OPA). The advantage of using this hybrid filler is the wide availability of the material, cost potential as well as sustainability prospects as they are made from natural sources. The main advantage of using bio-fillers is their reduced harm to the environment. Research has been conducted on the ability of OPA [6] and ESP [7] to be used as fillers individually. It can be proven that ESP has excellent toughness and strength on its own as a filler and, in addition to that, it can be observed that properties such as the stress softening and hysteresis loss are not diminished by the use of bio-fillers [8].

This paper will focus on the effect of varying the content of the ESP against static compression testing, stress relaxation testing and dynamic testing to analyse the properties of the hybrid filled rubber. As a whole, the results of this study can serve as a reference point for behavioural characteristics of OPA/ESP hybrid filled rubber under both static and dynamic compression testing.

2. Materials and Methods

2.1 Materials

Five different of rubber compounds were mixed at Rubber Research Institute of Malaysia. Each compound was mixed using two roll mill machine and cured using press machine at the temperature of 140°C in the shape of hardness button with a dimension of 8mm in thickness and 44mm in diameter. The composition of each compound is described in Table 1. Their compositions are described in table 1.

Table 1. A composition of rubber compounds.

Rubber Compound Id. /Ingredients (phr[a])	Natural Rubber (NR)	OPA/ESP10	OPA/ESP20	OPA/ESP30	Carbon Black
SMR L[b]	100	100	100	100	100
Zinc Oxide	5	5	5	5	5
Stearic Acid	1	1	1	1	1
CB N550	-	-	-	-	30
Eggshell Powder	-	10	20	30	-
Oil Palm Ash	-	1	1	1	-
Permanax TMQ[c]	1	1	1	1	1
Sulphur	2.5	2.5	2.5	2.5	2.5
CBS[d]	0.5	0.5	0.5	0.5	0.5

[a]parts per hundred of rubber
[b]Standard Malaysian Rubber Grade L
[c]1,2-dihydro.2,2,4-trimethyl quinolone
[d]N-cyclohexyl-2-benzothiazole sulphonamide

2.2 Methods

2.2.1 Static compression tests

The static compression test was carried out on an Instron 5582 Universal Testing Machine (UTM) with a 100kN capacity of load cell. Figure 1 shows the uniaxial compression set up. Emery paper of grit 120 was used to prevent the samples from sliding during the compression cycles. The machine software, Blue Hill, is used to create a compression testing procedures. Virgin samples were used to obtain a highest possible accuracy. The strain magnitudes were kept constant at 40% thickness for all samples to ensure all responses obtained were based on material properties rather than situational properties.

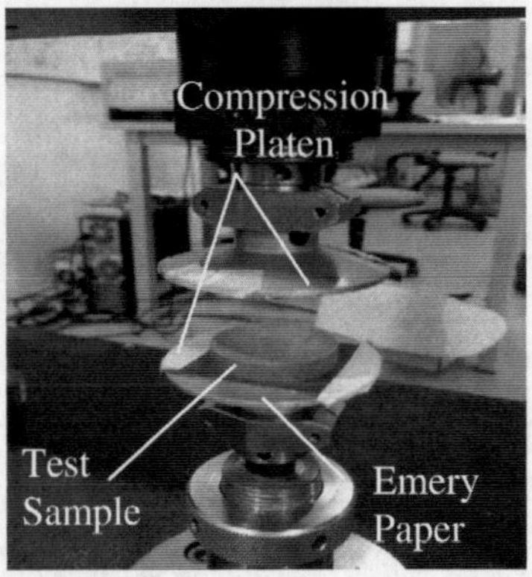

Figure 1. Compression test set up.

Before compressing, the virgin samples were measured with a vernier calliper to obtain their diameters, thicknesses and hence the depth of compression to achieve the required compression strain. All tests were carried out at a rate of 5mm/min up to 40% of the sample thickness and once reached, the compression reverses back to 0% strain. This cycle is repeated 5 times and all tests were conducted at room temperature. Although, temperature has a significant influence on the stiffness of rubberlike materials [9], all tests were conducted at room temperature to establish a more fundamental understanding on the effect of the hybrid filler first.

2.2.1.1 Maximum Compressive Stress at 40% Strain
The maximum stress is obtained by simply reading the maximum response at 40% strain. This is a simple yet good way to directly compare material properties and determine if the material being tested possesses any improved characteristics or has potential to improve upon conventional materials.

2.2.1.2 Static Stiffness
In order to calculate the static stiffness, two points are taken from the loading curves for each material on a force vs. displacement graph. Equation 1 below is then used to calculate the static stiffness of each individual material where F is the force and x is the displacement (strain) at points 1 and 2 on the loading curve.

$$\textit{Static stiffness, } K_s = \frac{F_2 - F_1}{X_2 - X_1} \quad (1)$$

2.2.1.3 Hysteresis Loss Ratio
The hysteresis loss ratio is defined using the energy provided to the rubber sample during the uploading phase and the energy loss during the unloading phase. Essentially the energy dissipated is represented by the difference in area of the uploading and unloading curves and the energy supplied is represented by the area under the uploading curve on a stress vs. strain graph [9]. The purpose of using fillers are to increase the hysteresis loss ability of rubber so that it may absorb more energy during vibrational damping or shock absorbance [10].

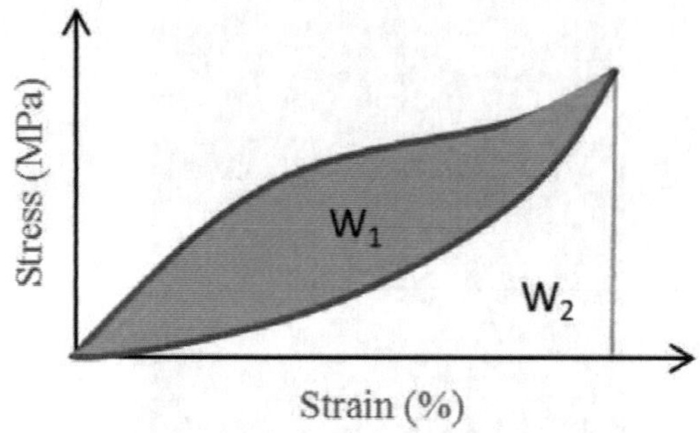

Figure 2. Compression test set up.

Figure 2 represents a generic hysteresis curve with the hysteresis loss (W_1) and area under the unloading curve (W_2). Hence, to calculate the hysteresis loss ratio equation 2 is used.

$$Hysteresis\ loss\ ratio = \frac{W_1}{W_1+W_2} \tag{2}$$

2.2.1.4 Stress relaxation

The stress relaxation test was also conducted using an Instron 5582 Universal Testing Machine by creating a testing program using a Blue Hill software. The physical set up of the test is identical to that of the compression test. In the case of a stress relaxation test, all samples used were pre-stressed from the compression tests to simulate real loading and relaxation conditions. Testing conditions used were similar to the compression tests with a strain rate of 5mm/min. The static strain regions profiled into the testing parameters are 20 minutes long and occurred at every 15%, 30% and 45% strain on both loading and unloading. The dimensions of the samples used in the stress relaxation test were recorded previously and the strain levels for each sample were calculated individually based on the sample thickness. The main purpose of the final static region (45% strain) is to obtain distinct results for the region before and after completion of the test. This is because the results for the final relaxation region is often difficult to read due to the values of the static region blending in with the values of the unloading region immediately after it. Hence, since results for strain levels up to 30% are desired, an additional strain level is tested at 45% strain. Figure 3 shows the strain profile used in this test.

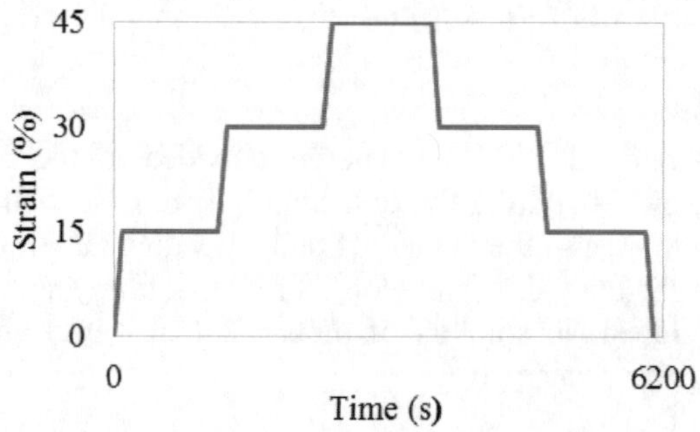

Figure 3. Strain profile for stress relaxation test.

The results of the stress relaxation test will indicate the stress relaxation ratios of each material. Stress relaxation is the residual elasticity present in the material up to the point where it achieves equilibrium stress. In order to compare the stress relaxation capacity of each material at each strain magnitude, they are converted into a relative form. The ratios of the relaxation stress with time at each strain level, σ_t, and the initial stress at each strain level, σ_o, gives the stress relaxation ratio or stress ratio (Equation 3). Hence, this conversion makes it easier to make direct comparisons between different rubber compositions as they are now relative.

$$Stress\ ratio = \frac{\sigma_t}{\sigma_o} \quad (3)$$

2.2.2 Dynamic test

Dynamic tests were carried out to determine the frequency dependent behaviour of the rubber materials. Virgin samples were utilised in this case because properties such as the dynamic stiffness are more accurate when there is minimal breakage of bonds from pre-stressing [11]. A servo-hydraulic testing machine Material Testing System (MTS) Multi-axis with a load cell of 25kN was used. The testing set up was similar with the previous tests by applying 120 grit emery paper to prevent sliding. The rubber samples were subjected to a strain amplitude of 10% at 4 different frequencies of 0.5 Hz, 1 Hz, 5 Hz and 10 Hz. The dynamic test was programmed using a MultiPurpose TestWare (MPT) integrated with a MTS Multi-axis testing machine.

2.2.2.1 Dynamic Stiffness

To calculate the dynamic stiffness of the rubber samples, the force amplitude applied, $\bar{F}$, is divided by the varying displacement amplitude, $\bar{\gamma}$, at every antinode in the sinusoidal stress and strain response with respect to time (Equation 4). These values are then averaged out within each tested frequency.

$$Dynamic\ stiffness,\ k_d = \frac{\bar{F}}{\bar{\gamma}} \quad (4)$$

2.2.2.2 Phase Angle

The definition of the phase angle, δ, is shown in figure 4 where dynamic strain response lags behind the dynamic stress input.

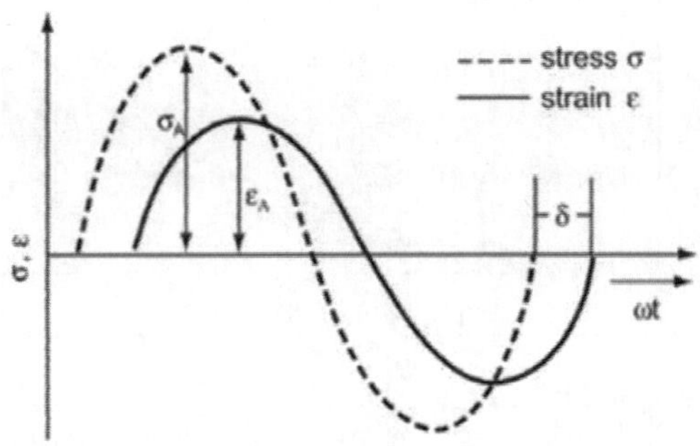

Figure 4. Dynamic stress input and strain response demonstrating phase lag [12].

According to the Kelvin-Voigt model, the tangent of the phase angle is a factor of the damping coefficient, c, elastic stiffness, k_s, and frequency of harmonic input, ω [13]. The elastic stiffness and input frequency are both known and the equation of motion is used to calculate the damping coefficient. Hence, we get the phase angle from equation 5 below:

$$Tan\ \delta = \frac{cw}{k_s} \quad (5)$$

2.2.2.3 Damping Ratio

The damping ratio, ξ, is calculated from equation 6 below [13]:

$$\xi = \frac{W_L}{2\pi k' \bar{\gamma}^2} \tag{6}$$

where,

$$W_L = \pi \bar{F} \bar{\gamma} \sin(\delta) \tag{7}$$

The symbol, k', represents the storage modulus of the sample which is similar to the stiffness of a rubberlike material. According to the ISO 6721-1 standard [14], it can be assumed to be equivalent to the dynamic stiffness for rapid stresses at low frequencies with reversible deformation. Hence, here it is assumed equal to the dynamic stiffness, k_d.

3. Results and Discussions

3.1 Static Compression Behaviour of Rubber Compounds

The stress strain responses of each rubber compound are compiled in figure 5. The uploading and unloading curves for all samples appear almost identical in behaviour and shape to each other. Carbon black is immediately observed to have a more significant response compared to the other rubber compounds and the hybrid filled rubber can be seen to already obey an incremental trend between themselves. The hybrid filled rubber samples show gradually increasing stress strain curves with respect to increasing ESP filler loading.

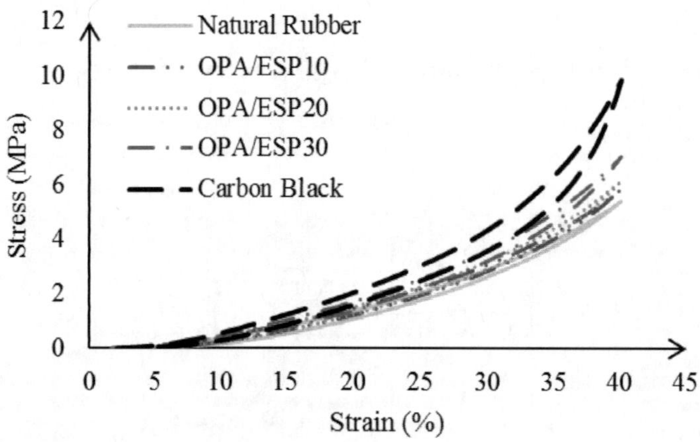

Figure 5. Stress vs. strain responses for each rubber compound

3.1.1 Maximum Compressive Stress at 40% Strain

The compressive stress experienced by the rubber compounds are an initial indication of the strength of samples. As shown in figure 6, it can be observed that the maximum compressive strength of the hybrid filled compounds are already higher than that of natural rubber. The hybrid filled rubber themselves follow an increasing trend in compressive stress as expected. There is a bigger gap in this trend between an ESP loading of 20 phr and 30 phr indicating the possibility of a larger increment in strength per content of filler used. However, carbon black is still observed to have the largest maximum compressive stress response as compared to the hybrid filled rubber.

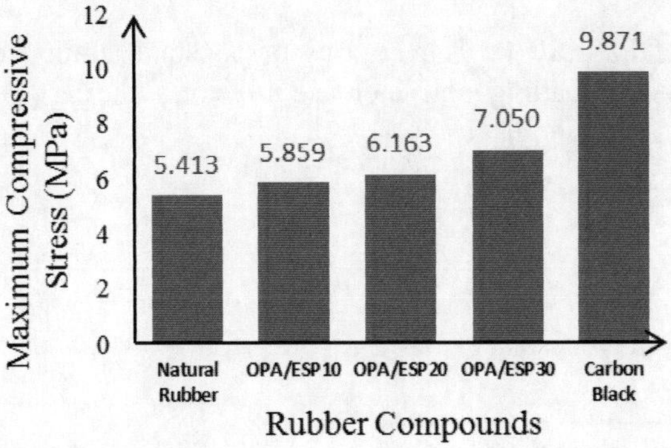

Figure 6. Maximum compressive stress for each rubber compound.

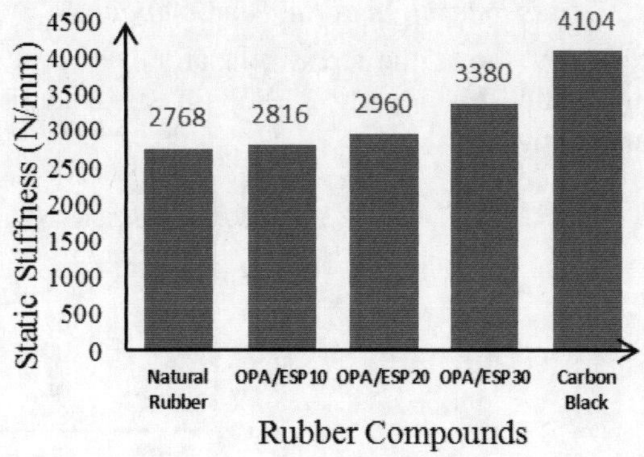

Figure 7. Static stiffness for each rubber compound at third cycle.

3.1.2 Static Compression Stiffness

The third cycle static stiffness values under uniaxial compression testing is shown in figure 7. Similar to the compressive stress, the results point towards an increasing pattern of static stiffness with increasing ESP content. Although the improvement in static stiffness of the rubber compound appears marginal at lower ESP content, the sudden increase between the results of OPA/ESP20 and OPA/ESP30 is also observed here. This could suggest an acceptable application of the hybrid rubber for lower magnitudes of force where the strength of carbon black is unnecessary and the bio-fillers present a more economical alternative. Additionally, increasing the loading of ESP could further improve the stiffness.

3.1.3 Hysteresis Loss Ratio

The effect of the hybrid fillers on the hysteresis ability of rubber is shown in figure 8 for a constant 40% of strain. The reason the hysteresis loss ratio was not analysed at other amplitudes of strain was due to each sample being approximately 7.6mm to 8mm in depth. Testing the amplitude at 10%, for example, although significant in percentage, was not significant in absolute value and hence would not produce any discernible difference in results. On a smaller scale, the effect of this property is less prominent but has a significant difference on a larger scale. The hysteresis loss ratio of the hybrid filled rubber is lower compared to that of both natural rubber and carbon black rubber although this deficit is somewhat negated at 30 phr of eggshell powder. This could mean that any suitable applications would be more effective in smaller scaled pieces.

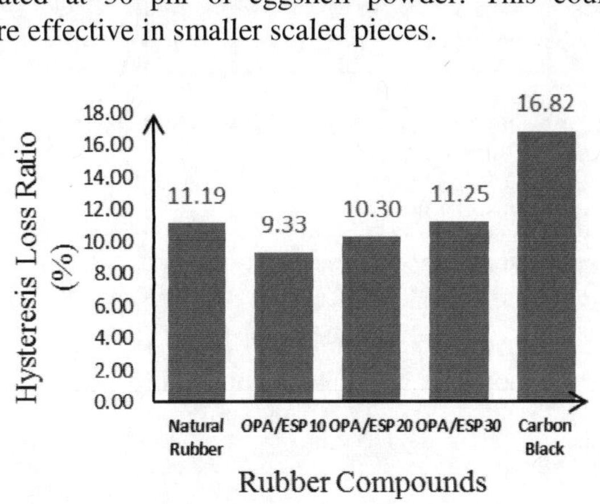

Figure 8. Hysteresis loss ratios for each rubber compound.

3.2 Stress Relaxation of Rubber Compounds

Figure 9 shows the stress relaxation response for OPA/ESP30. The regions of constant strain are discernible by a vertical drop in stress during the uploading and increase in stress during the unloading.

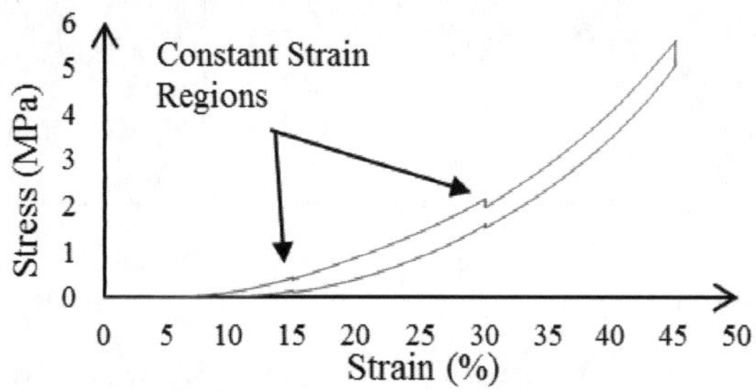

Figure 9. Stress strain response for OPA/ESP30 under the stress relaxation test.

The time dependent behaviour of the rubber compounds in terms of the stress relaxation are shown in figure 10 (15% strain, uploading), figure 11 (30% strain, uploading), figure 12 (15% strain, unloading) and figure 13 (30% strain, unloading). The trends and behaviour of these results are consistent with the findings of [15]. In figure 10, the stress relaxation responses for the hybrid filled rubber are not observed to obey a pattern or trend of any sort between the two graphs. This effect is more pronounced at lower strain levels and at higher strain levels the stress relaxation curves seem to be almost identical. This similarity could be due to the material being more compact at higher strain levels and hence due to the physical thickness of the sample the effect was reduced slightly. For the unloading curves, the samples appear to obey a trend in relaxation with OPA/ESP10 generally having the fastest increase in stress and OPA/ESP30 having the slowest among the three hybrid samples.

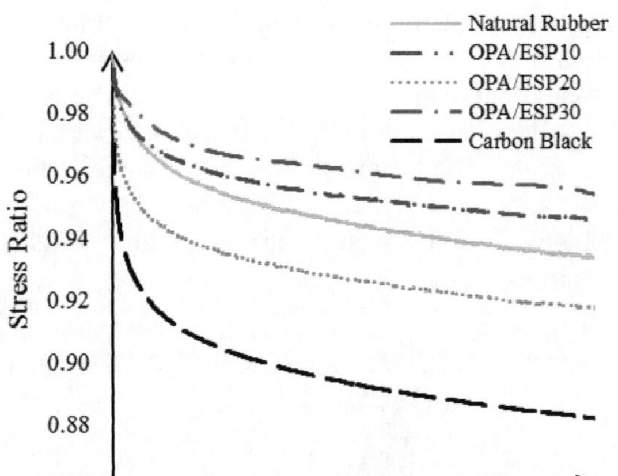

Figure 10. Stress relaxation response during uploading for each rubber compound at 15% strain.

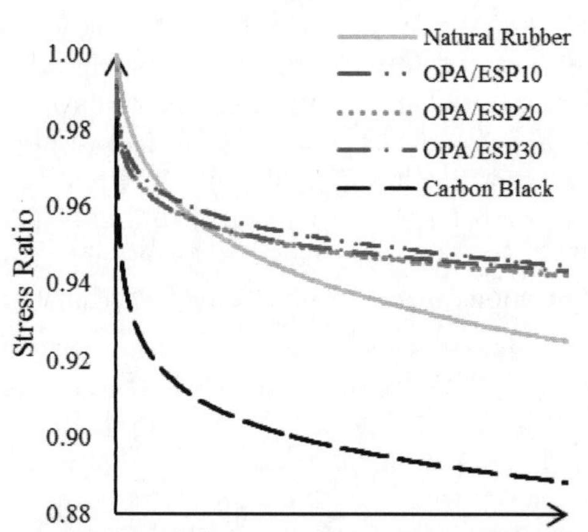

Figure 11. Stress relaxation response during uploading for each rubber compound at 30% strain.

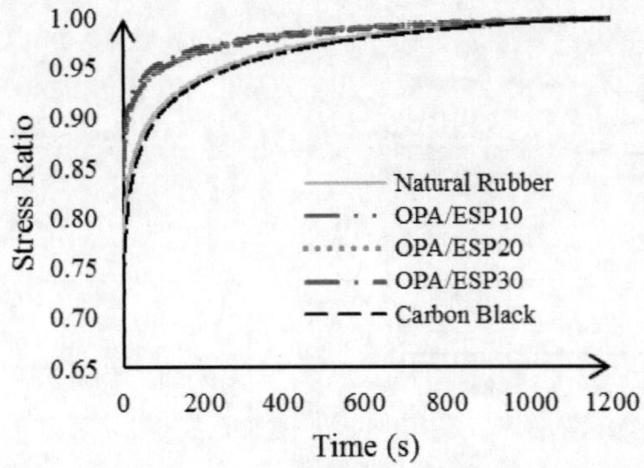

Figure 12. Stress relaxation reponse during unloading for each rubber compound at 15% strain.

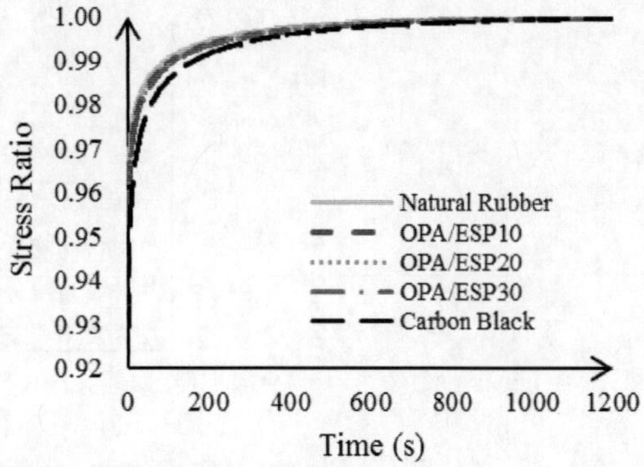

Figure 13. Stress relaxation response during unloading for each rubber compound at 30% strain.

All three hybrid samples appear to achieve equilibrium at almost the same time with each other for all cases with the exception of figure 10. The behaviours of both natural rubber and NR filled carbon black are consistent across all situations with NR filled carbon black achieving equilibrium the quickest in the uploading stage.

3.3 Dynamic Test

In general, the dynamic stiffness (figure 14), phase angle (figure 15) and damping ratio (figure 16) of the hybrid compounds do not appear to have a prominent difference between their magnitudes within each respective property. Increasing the loading of the hybrid filler does not seem to significantly affect the frequency dependency of the rubber. This can be seen where the trend lines for all compounds emerge very closely to each. It is not known if this is a phenomenon due to the type of filler used or if it is because the range of frequencies being tested is too small. Overall the dynamic analysis generally ranks the rubber samples in the following order carbon black > OPA/ESP30 > OPA/ESP20 > OPA/ESP10 > natural rubber with the carbon black being the most dynamically favourable.

Note there is a visible drop in properties at a frequency of 1 Hz across all five test samples in all three properties. It is a possibility that this is either a human or machine induced error as carbon black rubber does not undergo a transition phase at such a low frequency [17] and the drop is consistent for all compounds.

3.3.1 Dynamic Stiffness

The low gradient of the dynamic stiffness across the frequencies in figure 14 can be explained by rubberlike materials being expected to have minimal change in dynamic stiffness at "low" frequencies [18].

IOP Conf. Series: Journal of Physics: Conf. Series **908** (2017) 012028 doi:10.1088/1742-6596/908/1/012028

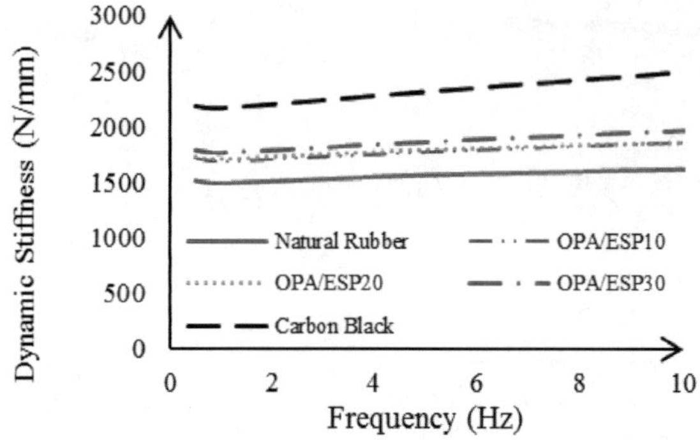

Figure 14. Dynamic stiffness against frequency for each rubber compound.

The observable dynamic stiffness of the hybrid samples lack the performance aptitude of carbon black as their change in magnitude is not as significant as that of carbon black. In contrast, rubber filled carbon black experiences a larger increment in dynamic stiffness per increment in frequency.

3.3.2 Phase Angle

In most cases, it is usually expected that the phase angle values across several frequencies will result in a flat line unlike the slope presented in figure 15. However, it is not unusual to obtain experimental results that appear to have a slight increment in phase angle values with frequency [16].

Trends in phase angle values are generally used to determine transition regions within a specific material, hence, their magnitude is usually of less importance. However, as mentioned earlier, the phase angle also represents a sort of delayed displacement reaction of rubber when subjected to a force. Therefore, the position of the carbon black trend line above that of the hybrid fillers is a further indication of carbon black rubber having a higher stiffness although the actual magnitude would be better referred from stiffness calculations directly.

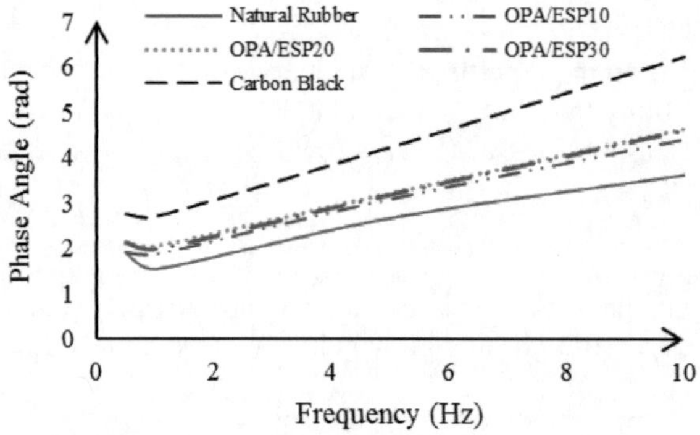

Figure 15. Phase angle against frequency for each rubber compound.

3.3.3 Damping Ratio

The results of the damping ratio are consistent with experimental works previously conducted [17]. The seemingly steep gradient is an expected occurrence even with small testing frequencies. The damping ratios of the hybrid fillers here are still compellingly lower than carbon black meaning either the damping ability of the hybrid samples are very limited or their critical damping coefficients are rather large. Rubber filled carbon black still maintains better overall damping with respect to the critical damping coefficient (damping ratio) and appears to actually have a damping ratio increment per increment in frequency, albeit a small one.

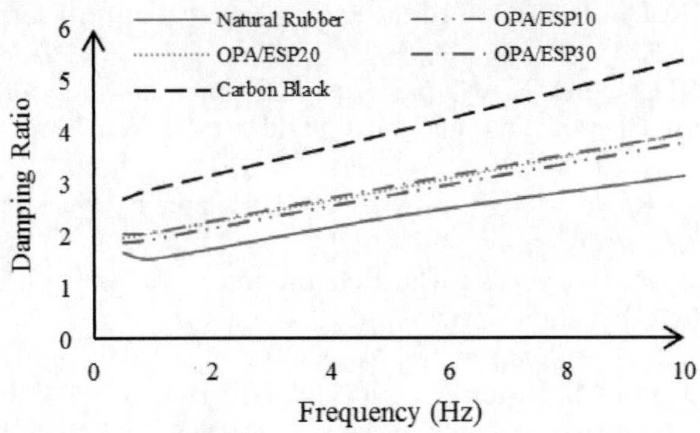

Figure 16. Damping ratio against frequency for each rubber compound.

4. Conclusion

Based on the results of static compression tests, it does not suggest that the hybrid bio-fillers have a significant advantage over rubber filled carbon black even though they do indeed have substantial mechanical and dynamic properties. Stress relaxation results do not contribute at any potential over rubber filled carbon black and there is actually a great deal of difference in this property. Across all dynamic applications, content of loading does not seem to significantly affect the dynamic properties. While the dynamic properties could be considered at a scalable magnitude with increasing frequency, the hybrid bio-fillers appear to be at a disadvantage again. The rubber with OPA/ESP content does not affect the dynamic properties of the hybrid filled rubber significantly, hence a lower content could be used for a range of vibrational scenarios.

5. Acknowledgments

The authors gratefully acknowledge the financial support from management of Malaysian Rubber Board and also their permission to publish this work.

6. References

[1] J. M. Kelly and D. Konstantinidis 2011 *Mechanics of Rubber Bearings for Seismic and Vibration Isolation. Somerset Wiley p 1*
[2] R. M. Gibbons 1986 Silicon rubber gasket and material *Google Patents p 2*
[3] International Carbon Black Association 2016 *Carbon Black User' s Guide* p 36
[4] T. Amari, N. J. Themelis, and I. K. Wernick 1999 Resource recovery from used rubber tires *Resour. Policy* vol 25 no. 3 pp 179–188
[5] Y. Fang, M. Zhan, and Y. Wang 2001 The status of recycling of waste rubber *Mater. Des.*, vol 22 no. 2 pp 123–128
[6] Z. X. Ooi, H. Ismail, and A. A. Bakar 2013 Optimisation of oil palm ash as reinforcement in natural rubber vulcanisation: A comparison between silica and carbon black fillers *Polym. Test*. vol. 32 no. 4 pp 625–630
[7] S. Lumlong, S. Wanapan, B. Khamsri, and P. Pungpo 2016 Effect of Eggshell as a Filler on Rubber Composite Properties
[8] A. S. M. Bashir, Y. Manusamy, L. C. Thiam, H. Ismail, and S. Ramasamy 2017 Mechanical, Thermal, and Morphological Properties of (Eggshell Powder)-Filled Natural Rubber Latex Foam *J Vinyl Addit. Technol*, no. 23 pp 3–12

IOP Conf. Series: Journal of Physics: Conf. Series **908** (2017) 012028 doi:10.1088/1742-6596/908/1/012028

[9] L. Kari, P. Eriksson, and B. Stenberg 2001 Dynamic stiffness of natural rubber cylinders in the audible frequency range using wave guides *Kautschuk Gummi Kunststoffe* vol. 54 no. 3 p 106

[10] R. Houwink and H. K. De Decker 1971 Elasticity, plasticity and structure of matter. Cambridge University Press p 1

[11] Y. Zhou, J. Wang, H. Chen, Q. Nie, Q. Sun, and Y. Wang 2009 The influence of pre-stressing on breakdown characteristics in liquid silicone rubber *J. Electrostat.* vol 67 no. 2–3 pp 422–425

[12] G. W. Ehrenstein, G. Riedel, and P. Trawiel 2004 Chapter 6 Dynamic Mechanical Analysis *Therm. Anal. Plast. Theory Pract*. pp 236–299

[13] E. J. Graesser and C. R. Wong 1991 The Relationship of Traditional Damping Measures for Materials with High Damping Capacity pp 1-3

[14] B. ISO,6721-1: 2011 Plastics *Determ. Dyn. Mech. Prop. Gen. Princ.*

[15] M. Cespi, G. Bonacucina, M. Misici-Falzi, R. Golzi, L. Boltri, and G. F. Palmieri 2007 Stress relaxation test for the characterization of the viscoelasticity of pellets *Eur. J. Pharm. Biopharm*. vol. 67 no. 2 pp 476–484

[16] M. Abdelmouleh, S. Boufi, M. N. Belgacem, and A. Dufresne 2007 Short natural-fibre reinforced polyethylene and natural rubber composites: Effect of silane coupling agents and fibres loading *Compos. Sci. Technol.* vol 67 no. 7–8 pp 1627–1639

[17] A. I. Medalia, 1978 Effect of Carbon Black on Dynamic Properties of Rubber Vulcanizates *Rubber Chem. Technol.* vol 51 no 3 pp 437–523

Biodegradation of PLA-*Pennisetum purpureum* based biocomposite scaffold

R Revati[1], M S Abdul Majid[2], M J M Ridzuan[3], M Normahira[4], N F Mohd Nasir[5], E M Cheng[6]

Universiti Malaysia Perlis (UniMAP), Kampus Tetap Pauh Putra, 02600, Pauh, Perlis

shukry@unimap.edu.my

Abstract. The *in vitro* degradation and mechanical properties of a 3D porous *Pennisetum purpureum* (PP)/polylactic acid (PLA) —based scaffold was investigated. In this study, composite scaffolds with PP to PLA ratio of 0%, 10%, 20%, and 30% were immersed in PBS solution at 37 °C for 40 days. Interestingly, the degradation rate was reduced for the PLA/PP_{20} scaffold, though insignificantly, this could be attributed to the improved mechanical properties and stronger fibre-matrix interface. The FESEM results indicated that a sound fibre-matrix interface was formed in the PLA/PP_{20} scaffold, which reflected the addition of *P. purpureum* into PLA decreasing the degradation rate compared to in pure PLA scaffolds. The results suggest that the *P. purpureum*/PLA scaffold degradation rate can be altered and controlled to meet the requirement imposed by a given tissue engineering application.

1. Introduction

Various biodegradable polymers have been utilised to produce scaffolds in tissue engineering. These microstructures maintain the extracellular matrix (ECM) production by cells and are presumed to evenly degrade, allowing the surrounding tissue to recover the supporting function of the scaffold [1]. Thus, biodegradable polymers are used to provide temporary support for cell growth, and the chemical properties of these polymers allow hydrolytic degradation through de-esterification [2]. Once degraded, the monomeric components of each polymer are removed by intrinsic pathways [3]. The biological performance of biomaterials highly depends on their degradation behaviour owing to the influence of degradation on cell performance and inflammatory response [4]–[6]. Therefore, it is crucial to investigate the degradation behaviour of biodegradable materials as the degradation rate is a major factor affecting cartilage tissue regeneration.

Utilisation of *P. purpureum* as a reinforcement filler in a PLA matrix offers many advantages; it results in the production of biocomposites with favourable mechanical properties and controllable biodegradability. Ridzuan et al. determined that *P. purpureum* fibres could potentially be used as reinforcement materials in polymer composites. They discovered that the application of alkaline treatment on fibres could effectively increase the surface roughness and minimises the hemicelluloses quantities of the fibre [22]. Hameem et al. studied the tensile and flexural properties of *P. purpureum* and concluded that when polyester composites were reinforced with *P. purpureum*, their mechanical properties improved [23]. Similarly, Ridzuan et al. reported that moisture absorption influences the mechanical properties of hybrid *P. purpureum*/glass–epoxy composites [24]. They noticed that as the glass fibre content of the samples increased, the degree of moisture absorption decreased while the tensile and flexural strength of the composites increased. Based on several recent studies, it can be suggested that *P. purpureum* fibres could be used effectively as reinforcement filler in polymer composites for cartilage tissue engineering applications [23]–[27].

In this review, the influence of fillers content on the *in vitro* degradation behaviours of the PLA and *P. purpureum*/PLA scaffolds will be described. Therefore, the composite scaffolds were immersed in PBS to study the scaffolds' *in vitro* degradation behaviour. The scaffold-immersed buffer solution was not changed until the experiment was ended to observe the effect of the PBS solution pH on the degradation behaviour of the scaffolds (pH was evenly lowered by the formation of acid due to the degradation of PLA). Finally, the surface morphology of *P. purpureum*/PLA scaffolds influenced by various fillers compositions will also be reviewed, implying the use of PBS medium to mimic conditions in the human body.

2. Materials and Methods

2.1. Materials

Pennisetum purpureum fillers were produced from Napier grass received from Bukit Kayu Hitam, Malaysia. Polylactic acid (PLA) was acquired from NatureWorks LLC. Organic solvents, chloroform and dichloromethane, were purchased from Fisher Chemical Co. and used as received. Sodium chloride (NaCl) with analytical grade was provided and used as a porogen.

2.2. Preparation of Pennisetum purpureum/PLA composite scaffolds

Composite scaffolds based on polylactic acid and *Pennisetum purpuruem* fillers were prepared by solvent casting and particulate leaching methods. The first step is to prepare 60.0g/L PLA solution by dissolving PLA pellets in (1:1) dichloromethane and chloroform solution with stirring at 70°C to get a clear solution. Then *Pennisetum purpureum* fillers were dispersed in PLA solution for 30 min by continuous stirring on a magnetic stirrer at medium speed. After the filler was evenly distributed, the sodium chloride particles were added to the composite with the ratio of 1 to 9. To eliminate sodium chloride, the obtained *Pennisetum purpureum*/PLA porous scaffolds were then soaked in distilled water for two days to leach out the sodium chloride by refreshing the medium every 4h. The filtered samples were dried overnight at room temperature and then kept in a desiccator for characterization.

2.3. In vitro degradation of P. purpureum/PLA composite scaffolds

2.3.1. Sample preparation

Before degradation, the *P. purpureum*/PLA scaffolds were pre-wetted to make sure that phosphate-buffered saline (PBS) solution penetrated through all the pores of the composite scaffolds [28]. The prepared scaffolds were pre-wetted by immersing the samples in ethanol for one h and then transferred into buffer (PBS) solution for one h; due to the hydrophobic character of PLA, the composite scaffolds cannot be directly immersed in the PBS solution [29]. Various *P. purpureum*/PLA composite scaffolds and pure PLA scaffold were prepared as a rectangle with a length of 13 mm, width of 13 mm and thickness of 25 mm. The samples were immersed in glass vials containing 20 ml of PBS solution and were placed in a 37 °C. The in vitro degradation of the scaffolds was further evaluated by pH changes, weight loss and water uptake test using PBS solution (0.1 M, pH 7.4) as degradation medium. All experiments were carried out in an incubator (37 °C) and the degradation period lasted for 40 days.

2.3.2. pH changes of degradation medium

The measurement of pH changes was performed by immersing the composite scaffolds in PBS solution (pH=7.4) for up to 40 days. The degradation medium was not refreshed in the whole degradation period. The pH value of the degradation medium was measured every five days using a pH meter (Model H1 2213 pH/ORD HANNA Instrument). Three samples were prepared for each composition, and the average pH of three samples was reported.

2.3.3. Water uptake and weight loss of P. purpureum/PLA scaffolds

The measurement of water uptake and weight loss was carried out by immersing the composite scaffolds in PBS solution at 37 °C for up to 40 days. At the end of each time point, the weight of the swollen samples was recorded after removing the excess surface water with filter paper. Water uptake percentage of the scaffolds was determined from the weight of the wet sample (w_w) and the initial weight of the dry sample (w_i) using the following equation (1). The reported water uptake was considered as the average value of three samples.

$$Water\ uptake\ (\%) = \frac{(W_w - W_i)}{W_i} \times 100 \quad (1)$$

The weight loss of the scaffolds was determined from the initial weight of dry sample (w_i) and weight of final dry sample (w_f) at the end of each time point using equation (2)

$$Weight\ loss\ (\%) = \frac{(W_i - W_f)}{W_i} \times 100 \qquad (2)$$

At the end of predetermined time intervals, all samples from each composition were removed and weighed for the calculation of the percentage of water uptake (equation 1). Samples then washed thoroughly with distilled water to eliminate any soluble inorganic salt and weighed after being completely vacuum dried at room temperature to constant weight for calculation of the percentage of weight loss (equation 2).

3. Result and Discussions

3.1. In vitro degradation of P. purpureum/PLA composite scaffolds

3.1.1. Effect of P. purpureum on the pH changes in degradation medium

The change in pH value over the degradation is presented in Figure 1. Acidic groups resulting from the PLA degradation may decrease the pH value of the degradation medium, while the dissolution of the *P. purpureum* fillers could alkalize the medium. Therefore, the pH of the PBS medium is highly dependent on both the degradation rate of PLA matrix and the dissolution of the *P. purpurem* fillers. As shown in Figure 1, the pH value of degradation medium for incubating pure PLA scaffold decreased with a sharp drop through the 0-10th day and gradually increased through the rest of the degradation period and reached around 7.43 at the end of 35th to 40th day of incubation. Also, the pH value of the medium for incubating PLA/PP$_{10}$ scaffold decreased at a fast rate through the 0-10th day and reached around 7.32 at the end of 10th day of incubation, but after the 10th day, the pH began to increase up to 7.40 at the end of 40th day. Then, owed to the high molecular weight (M_W=200, 000) of PLA, it degraded quite fast. The pH of degradation medium for incubating PLA/PP$_{20}$ and PLA/PP$_{30}$ scaffolds were down sharply through the 0-15th day of incubation, and both scaffolds reached around 7.28 at the end of the 15th day of incubation. However, after the 15th day, the pH began to stable and exhibited an upward trend from the 20th day and reached 7.33 and 7.34 at the end of 40th day of incubation, respectively.

The pH drops during the degradation time up to 10-15 days; releasing more acidic degradation products into the medium. After the initial 15 days of incubation, the pH value increases; owed to the *P. purpureum* degradation and OH^- release from the bio-composite scaffolds, which neutralise the acid degradation products of PLA. Hence, the results showed that the pH value of the degradation medium decreased at the beginning of the degradation and it has increased up to the end of incubation (figure 1). It could be concluded that the pH is decreased, by introducing *P. purpureum* fillers in PBS medium.

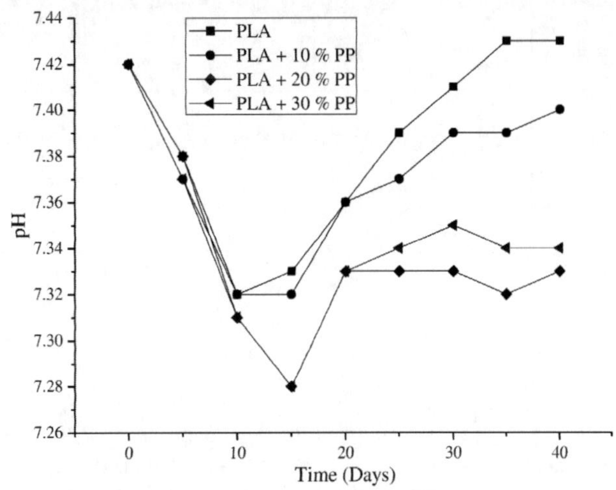

Figure 1: pH changes vs. incubation time in PBS for porous *P. purpureum*/PLA scaffolds: (a) PLA_C, (b) PLA/PP$_{10}$, (c) PLA/PP$_{20}$, and (d) PLA/PP$_{30}$

3.1.2. Water uptake and weight loss of P. purpureum/PLA scaffolds

The scaffolds based on *P. purpureum* and PLA were prepared by solvent casting and particulate leaching methods. The main key point in the in vitro degradation studies was to use enzymes present in human serum for stimulating physiological conditions, which are responsible for the degradation of *P.purpureum*/PLA scaffolds. Degradation with PBS medium was performed under incubation. The water permeability of a scaffold and its ability to absorb water influence the absorption of cell culture medium or physiologically compatible buffers and the transportation of cell nutrients and metabolites throughout the scaffold. The PBS uptake of the scaffold material and that of the structure, give various measures of the ability of the scaffold to bind to the fluid.

Figure 2 shows the water uptake of the *P. purpureum*/PLA scaffolds with different *P. purpureum* content. According to the results, the percentage of water absorption is decreased throughout the degradation period. It can be observed that after the immersion of scaffolds in medium, compared to PLA/PP_{20} scaffold, the water content of PLA_C, PLA/PP_{10}, and PLA/PP_{30} scaffolds decreased rapidly, though with the introduction of *P. purpureum* fillers which have a hydrophilic behaviour. However, the water uptake of the biocomposite scaffolds decreased remarkably in the presence of PBS medium (figure 2), which owed to the degradation of the scaffolds in the presence of these enzymes, which similar to the presence of the enzyme in human serum.

The PLA/PP_{20} scaffold shows that the percentage of water uptake decreased slowly and reached around 245%, compared to than that of the PLA_C, PLA/PP_{10}, and PLA/PP_{30} scaffolds, which decreased significantly and reached around 175, 165, and 155%, respectively. PLA/PP_{20} composite scaffold exhibits the highest water absorption percentage throughout the whole incubation period. The water uptake ability of the *P.purpureum*/PLA scaffold gradually increases with the further addition of *P. purpureum* content due to the presence of *P. purpureum* fillers on the surface of the composite scaffold. *P. purpureum* is naturally hydrophilic because they are cellulose fibres. Thus the presence of P. purpureum within the scaffold can significantly enhance the water absorption ability of scaffold. In contrast, PLA/PP_{30} scaffold shows that the percentage of water uptake decreased rapidly, compared to PLA/PP_{20} scaffold, owing to the weakness of the fillers to protrude from the surface of the scaffolds. This result agrees with the result of the weight loss analysis (Figure 3).

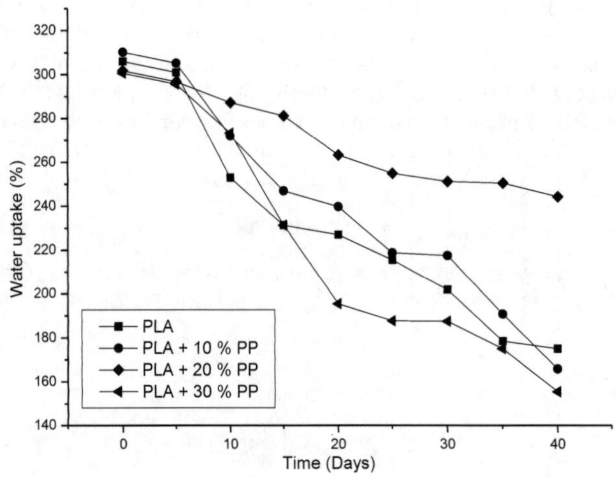

Figure 2: Water uptake vs. incubation time in PBS for porous *P. purpureum*/PLA scaffolds: (a) PLA_C, (b) PLA/PP_{10}, (c) PLA/PP_{20}, and (d) PLA/PP_{30}

In tissue engineering, it is expected that the degradation rate of *P. purpureum*/PLA scaffolds is controllable in the regeneration and repair process of tissue. This is highly achieved by modifying the composition of the polymer composites. However, it is compulsory to highlights the influence of *P. purpureum* on the *in vitro* degradation behaviour of scaffolds as degradation rate is a critical factor affecting the recovery of damaged cartilage. The degradation mechanism of biodegradable polymer is chemical degradation, and it was observed that the chain ends cleavage resulted in weight loss, while the random scission influences the reduction in weight loss. Therefore, the water absorption is considered to be particularly important for the degradation of the scaffold (as illustrated in figure 2). The weight loss for the different porous *P. purpureum*/PLA scaffolds after incubation in PBS is shown in figure 3. It can be found that the weight loss of the PLA/PP_{20} scaffold proceeds slowly throughout the whole degradation period. After 40 days of incubation, the scaffold loses about 7.4% of its initial weight. The weight loss trend of PLA_C, PLA/PP_{10}, and PLA/PP_{30} scaffolds were similar, characterised by a rapid weight loss phase at initial 15 days, followed by a dramatic weight loss at the remaining weeks. As a result, 20.1, 28.8, and 27.4% of the original weight are lost after 40 days of incubation, respectively.

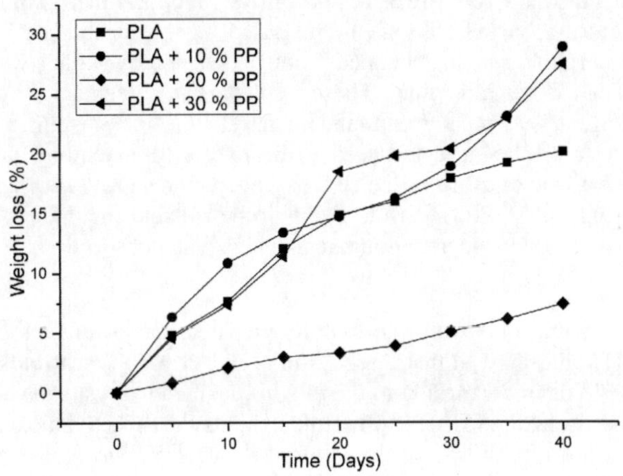

Figure 3: Weight loss vs. incubation time in PBS for porous *P. purpureum*/PLA scaffolds: (a) PLA_C, (b) PLA/PP_{10}, (c) PLA/PP_{20}, and (d) PLA/PP_{30}

In our study, the weight loss of PLA_C and PLA/P_{10} scaffolds increased faster from the early period than the PLA/PP_{20}. Based on FESEM results, we can conclude that the micro-structure of PLA/PP_{10} and PLA_C scaffold is similar. This increase in weight loss is possibly because the medium could diffuse more readily into the scaffolds than into PLA/PP_{20} as the pore size of the PLA_C and PLA/PP_{10} scaffolds are larger and wider compared to that of PLA/PP_{20}. Therefore, it is deduced that the PLA_C and PLA/PP_{10} scaffolds displayed a faster degradation than PLA/PP_{20} in good accordance with the literature [28]–[30]. Wu et al. reported that the scaffold with larger pore size tends to degrade more quickly (i.e., are less resistant to hydrolysis) than those with smaller pore size. The reason is that they possess thicker pore walls and therefore a smaller surface area to volume ratio, which delay the diffusion of small chains (acidic-by-products), resulting in a stronger acid-catalyzed (auto-catalyzed) hydrolysis [31]. Additionally, if the adhesion is weak, the PLA/PP_{10} scaffold has poor strength and fatigue properties. In this study, PBS medium can diffuse quickly along the interface of *p. purpureum*/PLA scaffold is owing to insufficient adhesion and disturb the interface which leads to rapid strength loss of the scaffolds.

The development of excellent adhesion between matrix and reinforcement is still a challenge in composite science. PLA/PP_{20} scaffold caused reduction of degradation when comparing to PLA_C, PLA/PP_{10}, and PLA/PP_{30} scaffolds. Moreover, the addition of 20 wt% of *P. purpureum* into PLA composite scaffold increased the water absorption from 175 to 245 % demonstrating the moisture effect of the *P. purpureum* additives. It is well known that fibre-matrix bonding generated on the scaffold surface can act as a load bearer and improve thermal stability. It has been described that the hydrolytic chain cleavage preceded preferentially in the semi-crystalline regions, and thus leading to the increase in composite crystallinity. Therefore, the well dispersed of *P. purpureum* particles helped in the easy breach of the medium into the inner of the PLA/PP_{20} scaffold from the interface between *P. purpreum* particles and PLA matrix. Afterwards, the degradation of semi-crystalline regions of PLA matrix in the PLA/PP_{20} scaffold happened before the crystalline regions. When the polymer chains in semi-crystalline regions degrade, the number of semi-crystalline regions decrease, the percentage of crystalline to semi-crystalline increased. Thus, the increased *P. purpureum* is expected to reduce the degradation of the PLA/PP_{20} scaffold. This might be owing to small molecules produced by the first degradation of grafted PLA molecules onto *P. purpureum* particles played a role of plasticiser and thus improved the toughness of the PLA/PP_{20} scaffold. Small molecules produced by the slower degradation of PLA/PP_{20} scaffold also played a role of plasticiser and enhanced the toughness of the PLA/PP_{20} scaffold to some extent.

However, in this study, after 40 days of incubation, it was found that overloading *P. purpureum* fillers could increase the degradation of PLA/PP_{30} scaffold. The weight loss of the PLA/PP_{30} scaffold gradually increases with the further addition of *P. purpureum* content due to the tendency of fillers to agglomerate; some of the fillers would not have been fully enclosed within the composite scaffolds, hence, some of will have protruded from the surface of the scaffolds. With the *in vitro* degradation results, we can conclude that the interface between *P. purpureum* and PLA matrix in PLA/PP_{20} scaffold was more susceptible to erosion by the PBS medium. Based on the mechanical, water absorption, weight loss and FESEM morphology data obtained from *in vitro* study which was performed in a controlled environment using PBS as a medium. These results are in agreement with the behaviour of the same *in vivo* in cartilage tissue repair, as found in few studies suggesting a low regenerative capacity of cartilage. Therefore, chondrocytes need to be expanded in culture and seeded onto a scaffold that can degrade slowly and resorb as the new tissue structures grow *in vitro* and *in vivo* [32]. According to the present study, PLA/PP_{20} scaffold seems to be suitable for the treatment of artificial cartilage where the fixation needs slow degradation rate and high water absorption.

4. Conclusions

P. purpureum/PLA composite scaffolds were successfully prepared via solvent casting and particulate leaching technique. The pH value showed that the PLA/PP_{20} and PLA/PP_{30} scaffolds were kept nearly 7.33 and 7.34 at the end of 40 days. According to the *in vitro* degradation test showed that of the three composites and pure PLA scaffold, the PLA/PP_{20} scaffold exhibits the aspected bioactivity for cartilage regeneration. Biodegradation results show that the reinforcement of *P. purpureum* could increase the water absorption of PLA scaffolds and significantly influence the degradation rate of the PLA matrix. The results suggest that the *in vitro* degradation of PLA/PP_{20} scaffold more suitable for culture of chondrocyte-like cells and facilitates their application in cartilage tissue engineering applications.

Acknowledgment

The authors sincerely appreciate the financial support from the Ministry of Education, Malaysia through the Research Acculturation Grant Scheme (RAGS) 2014 under the project number 9018-00073. Also thanks to School of Mechatronic Engineering, Universiti Malaysia Perlis for the use of its facilities.

References

[1] G. Chen, T. Ushida, and T. Tateishi, "Scaffold design for tissue engineering," *Macromol. Biosci.*, vol. 2, pp. 67–77, 2002.
[2] T. Cordonnier, J. Sohier, P. Rosset, and P. Layrolle, "Biomimetic materials for bone tissue engineering - State of the art and future trends," *Adv. Eng. Mater.*, vol. 13, no. 5, pp. B135–B150, 2011.
[3] H. Tian, Z. Tang, X. Zhuang, X. Chen, and X. Jing, "Biodegradable synthetic polymers: Preparation, functionalization and biomedical application," *Prog. Polym. Sci.*, vol. 37, pp. 237–280, 2012.
[4] T. Fakhrul and M. A. Islam, "Degradation Behavior of Natural Fiber Reinforced Polymer Matrix Composites," *Procedia Eng.*, vol. 56, pp. 795–800, 2013.
[5] G. S. D. Hetal Patel*, Minal Bonde, "Biodegradable polymer scaffolds for tissue engineering," *Trends Biomater. Artif. Organs*, vol. 25, no. 1, pp. 20–29, 2011.
[6] S. I. A. Razak, N. F. A. Sharif, and W. A. A. Rahman, "Biodegradable polymers and their bone applications : A review," *Int. J. Basic Appl. Sci.*, vol. 12, pp. 31 – 49, 2012.
[7] C. Tu, Q. Cai, J. Yang, Y. Wan, J. Bei, and S. Wang, "The fabrication and characterization of poly(lactic acid) scaffolds for tissue engineering by improved solid-liquid phase separation," *Polym. Adv. Technol.*, vol. 14, pp. 565–573, 2003.
[8] S. Vogt, Y. Larcher, B. Beer, I. Wilke, and M. Schnabelrauch, "Fabrication of highly porous scaffold materials based on functionalized oligolactides and preliminary results on their use in bone tissue engineering.," *Eur. Cells Mater.*, vol. 4, pp. 30–38, 2002.
[9] L. Zhang, J. Hu, and K. A. Athanasiou, "The role of tissue engineering in articular cartilage repair and regeneration," *Crit. Rev. Biomed. Eng.*, vol. 37, pp. 1–57, 2009.
[10] Z. Zhang and H. Cui, "Biodegradability and biocompatibility study of poly(chitosan-g-lactic acid) scaffolds," *Molecules*, vol. 17, pp. 3243–3258, 2012.

[11] N. Zhu, D. Cooper, X.-B. Chen, and C. H. Niu, "A study on the in vitro degradation of poly(l-lactide)/chitosan microspheres scaffolds," *Front. Mater. Sci.*, vol. 7, no. 1, pp. 76–82, Mar. 2013.

[12] Q. Chen, A. Bruyneel, K. Clarke, C. Carr, and J. Czernuszka, "Collagen-based scaffolds for potential application of heart valve tissue engineering," *J. Tissue Sci. Eng.*, vol. S11, pp. 1–6, 2012.

[13] S. I. Jeong, M. D. Krebs, C. A. Bonino, S. A. Khan, and E. Alsberg, "Electrospun alginate nanofibers with controlled cell adhesion for tissue engineering," *Macromol. Biosci.*, vol. 10, no. 8, pp. 934–943, 2010.

[14] A. R. Shah, S. R. Shah, S. Oh, J. L. Ong, J. C. Wenke, and C. M. Agrawal, "Migration of co-cultured endothelial cells and osteoblasts in composite hydroxyapatite/polylactic acid scaffolds," *Ann. Biomed. Eng.*, vol. 39, no. 10, pp. 2501–2509, 2011.

[15] I. Olivas-armendariz, E. Santos-Rodríguez, L. E. Alvarado-Gutiérrez, M.L. Meléndez-Molina, Z.A. Márquez-Chávez, L. A. Valencia-Gómez, C. L. Vargas-Requena, S. A. Martel-Estrada, and 1In, "Biocomposites scaffolds for bone tissue engineering," *Int. J. Compos. Mater.*, vol. 5, pp. 167–176, 2015.

[16] B. Gupta, N. Revagade, and J. Hilborn, "Poly(lactic acid) fiber: An overview," *Prog. Polym. Sci.*, vol. 32, pp. 1–84, 2007.

[17] E. Hoveizi, M. Nabiuni, K. Parivar, S. Rajabi-Zeleti, and S. Tavakol, "Functionalisation and surface modification of electrospun polylactic acid scaffold for tissue engineering," *Cell Biol. Int.*, vol. 38, no. 1, pp. 41–49, 2014.

[18] M. S. Lopes, A. L. Jardini, and R. M. Filho, "Poly (lactic acid) production for tissue engineering applications," *Procedia Eng.*, vol. 42, pp. 1402–1413, 2012.

[19] F. M. AL-Oqla, S. M. Sapuan, T. Anwer, M. Jawaid, and M. E. Hoque, "Natural fiber reinforced conductive polymer composites as functional materials: A review," *Synth. Met.*, vol. 206, pp. 42–54, Aug. 2015.

[20] G. Bogoeva-Gaceva, M. Avella, M. Malinconico, A. Buzarovska, A. Grozdanov, G. Gentile, and M. E. Errico, "Natural fiber eco-composites," *Polym. Compos.*, vol. 28, no. 1, pp. 98–107, Feb. 2007.

[21] Chandramohan and Marimuthu, "Characterization of natural fibers and their application in bone grafting substitutes," *Acta Bioeng. Biomech.*, vol. 13, no. 1, pp. 77–84, 2011.

[22] M. J. M. Ridzuan, M. S. Abdul Majid, M. Afendi, S. N. Aqmariah Kanafiah, J. M. Zahri, and A. G. Gibson, "Characterisation of natural cellulosic fibre from Pennisetum purpureum stem as potential reinforcement of polymer composites," *Mater. Des.*, vol. 89, pp. 839–847, 2016.

[23] M. Haameem J.A., M. S. Abdul Majid, M. Afendi, H. F. A. Marzuki, I. Fahmi, and A. G. Gibson, "Mechanical properties of Napier grass fibre/polyester composites," *Compos. Struct.*, vol. 136, no. February, pp. 1–10, 2016.

[24] M. J. M. Ridzuan, M. S. Abdul Majid, M. Afendi, K. Azduwin, N. A. M. Amin, J. M. Zahri, and A. G. Gibson, "Moisture absorption and mechanical degradation of hybrid Pennisetum purpureum/glass-epoxy composites," *Compos. Struct.*, vol. 141, no. January, pp. 110–116, 2016.

[25] M. J. M. Ridzuan, M. S. Abdul Majid, M. Afendi, S. N. Aqmariah Kanafiah, J. M. Zahri, and A. G. Gibson, "Characterisation of natural cellulosic fibre from Pennisetum purpureum stem as potential reinforcement of polymer composites," *Mater. Des.*, vol. 89, no. January, pp. 839–847, 2016.

[26] M. J. M. Ridzuan, M. S. A. Majid, M. Afendi, M. N. Mazlee, and A. G. Gibson, "Thermal behaviour and dynamic mechanical analysis of Pennisetum purpureum/glass-reinforced epoxy hybrid composites," *Compos. Struct.*, vol. 152, no. July, pp. 850–859, 2016.

[27] M. Haameem, M. S. Abdul Majid, M. Afendi, H. F. A. Marzuki, E. A. Hilmi, I. Fahmi, and A. G. Gibson, "Effects of water absorption on Napier grass fibre/polyester composites," *Compos. Struct.*, vol. 144, no. September, pp. 138–146, 2016.

[28] X. Niu, Q. Feng, M. Wang, X. Guo, and Q. Zheng, "In vitro degradation and release behavior of porous poly(lactic acid) scaffolds containing chitosan microspheres as a carrier for BMP-2-derived synthetic peptide," *Polym. Degrad. Stab.*, vol. 94, no. 2, pp. 176–182, 2009.

[29] Y. You, B.-M. Min, S. J. Lee, T. S. Lee, and W. H. Park, "In vitro degradation behavior of electrospun polyglycolide, polylactide, and poly(lactide-co-glycolide)," *J. Appl. Polym. Sci.*, vol. 95, no. 2, pp. 193–200, Jan. 2005.

[30] F. Mansourizadeh, V. Mirir, F. Shahabi, F. Mansourizadeh, E. Pasban, and A. Asadi, "In vitro degradation of electrospinning PLLA/HA nanofiber scaffolds compared with pure PLLA," *Int. J. Plant, Anim. Environ. Sci.*, vol. 4, pp. 330–334, 2014.

[31] L. Wu and J. Ding, "In vitro degradation of three-dimensional porous poly(D,L-lactide-co-glycolide) scaffolds for tissue engineering," *Biomaterials*, vol. 25, pp. 5821–5830, 2004.

[32] D. W. Hutmacher, "Scaffolds in tissue engineering bone and cartilage.," *Biomaterials*, vol. 21, pp. 2529–2543, 2000.

Effect of stress ratio on the fatigue behaviour of glass/epoxy composite

A R A Syayuthi , M S Abdul Majid, M J M Ridzuan, K S Basaruddin, Peng, T L

School of Mechatronic Engineering, Universiti Malaysia Perlis, Pauh Putra Campus, 02600 Arau, Perlis, Malaysia.

syayuthi@unimap.edu.my,

Abstract. The effect of stress ratio on the fatigue behaviour of the GFRE composite has been investigated. The glass fibre reinforced epoxy (GFRE) composite plates were fabricated using vacuum infusion method. Static tensile was performed in accordance with the ASTM D5766 standard, and the cyclic test was conducted according to ASTM D3479 with three different stress ratio, R= 0, 0.5, -1. Static tensile tests were carried out to determine the ultimate strength of this composite. Subsequently, fatigue tests loads ranging from 30% to 90% of the ultimate load were applied to each specimen. The S–N curve of different stress ratio loading of fibreglass/epoxy composites was then established. The results show that the number of cycles to failure increases as the loading is decreased. The specimens for fatigue tests loads 30% at R= 0 and -1 recorded the highest number of cycles at 2 million cycles. The results obtained from this test indicated a significant life reduction for R = -1 compared with the tension-tension loading, with the life reduction for R = - 1 being greatest. The fatigue behaviour of the GFRE composite materials is not only influenced by the percentage of fatigue tests load but with different of stress ratio.

Keywords: Glass fibre reinforced composites; tensile properties; fatigue strength; stress ratio.

1. Introduction

Glass fibre reinforced epoxy (GFRE) composites exhibit desirable physical and chemical properties that include light weight coupled with high stiffness and strength along the direction of the reinforcing fibre, dimensional stability, temperature and chemical resistance, flex performance, and relatively easy processing [1]. The stiffness and strength properties of the GFRE composite can be controlled efficiently by laying various layers of glass fibre on top of each another with oriented (stacking) in different preferred directions [2]. GFRE exhibits desirable specific mechanical and physical properties such as light weight, good in strength properties and stiffness properties, excellent with temperature and chemical resistance and relatively easy processing. Also, GFRE laminated composite is also popularly used in various applications such as aircraft manufacturing, automotive engineering, structure engineering and also sport equipments. However, composite materials exhibit complex failure mechanism under static and fatigue loading and therefore, it is a necessity to investigate the fatigue behaviour of GFRE composite.

Sureshkumar et al. [3] studied the mechanical response of glass fibre reinforced polyurethane resin composite in three different volume ratios. They found that as the fibre content increases, the strength of the composite also increases, though, at some point, it will decline due to lack of resin (Polyurethane). Moreover, Arifin et al. [4] studied the failure characterization in polymer matrix composite for unnotched and notched (open-hole) specimens under tension condition. It was founded that unnotched samples for both types of epoxy and polyester composites had high values concerning strength and stress.

Besides that, composite materials also very well know with the fatigue resistance properties. However, the fatigue behaviour of the composite can influence the notch size or the stress concentration factors. Sonsino and Moosbrugger [5] studied the effect of notches on axial fatigue behaviour of short glass fibre reinforced polyamide-6.6. The difference stress ratio used in the test can affect the fatigue life of a specimen. Furthermore, Jason et al. [6] explored the effect of R-ratio on fatigue life and residual strength of glass-reinforced polymer composite materials. The composite materials were undergoing fatigue test with three different stress ratio (R=0.1, R=-1, and R=10). It was shown that the smaller the stress ratio, the longer the fatigue life, the lower the maximum stress.

2. Materials and Methods

2.1. Materials
Woven E-glass fibre and Epoxy Amite 100 series resin were supply by a local supplier. The woven glass fibre was prepared with dimension 600 x 600 mm per layer, and the diameter of the single fibre has been measured approximate to 12.8 µm. The Epoxy Amite 100 series resin was mixed with a hardener to a ratio of 4:1.

2.2. Composite fabrication
The GFRE composites were laminated with 14 layers; 12 layers in the $0^0/90^0$ orientation and two layers at the mid of the laminates with 45^0 orientation. This GFRE composite was fabricated using vacuum infusion method at room temperature (25°C). The resin was infused into the lamination plies, and the curing process took about 12 h. Finally, rectangular $600 \times 600 \times (2.6 \pm 0.1)$ mm^3 plates were obtained and the specimens were cut using a Dremel 4000 tool, in accordance with ASTM D5766 standard [7].

2.3 Tensile Test
The tensile specimens were prepared as illustrated in figure 1. The tensile tests were performed using a universal testing machine with a load cell of 100 kN at a crosshead speed of 1mm/min. Five specimens were tested, and the average result of the tensile strength and tensile elastic modulus were determined. From the load-displacement tensile test results, the stress and strain curves were plotted by using the formula as below:

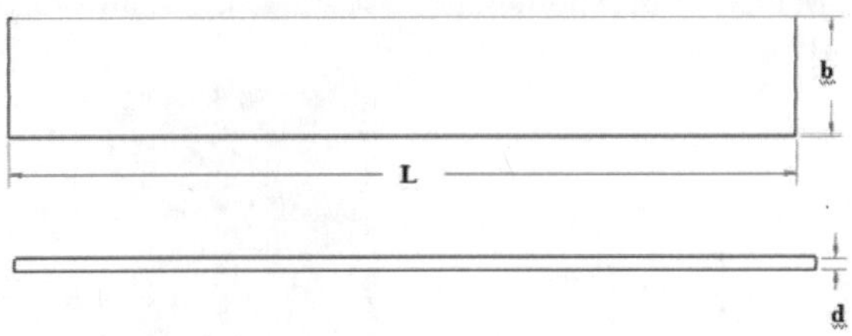

Figure 1. GFRE composite specimens.

Eq.1 show the formula Area of specimen,

$$A = b \times d \tag{1}$$

where, b = width of specimen;
d = thickness of specimen;

The stress was calculated from the ratio of force to the area of specimen shown in Eq. 2

$$\sigma = \frac{F}{A} \tag{2}$$

where, F = Force;
A = Area of specimen;

Strain of specimen on tensile test,

$$\varepsilon = \frac{\Delta d}{L_o} \tag{3}$$

where Δd = elongation of the specimen from tensile test;
L_o = Original length of specimen;

Elastic Young's Modulus of each specimen, $E = \frac{\Delta\sigma}{\Delta\varepsilon}$ (slope of stress-strain curve)

(a) **(b)** **(c)**

Figure 2. Experimental set-up for the tensile test. (a) Universal Testing Machine (b) Tensile test of GFRE specimen using Universal testing machine (c) Schematic diagram of the experimental set-up

2.4 Fatigue test

The experimental set up of the fatigues tests was shown in figure 3. The fatigue tests were performed using Instron 8801 Servo hydraulic Fatigue Machine 100kN at three different stress ratio, R= 0, 0.5, -1, respectively. Fatigue load was applied in a sinusoidal waveform with a frequency of 20 Hz, and five displacement levels were obtained corresponding to 30, 45, 60, 75 and 90% of the maximum load of Tensile test.

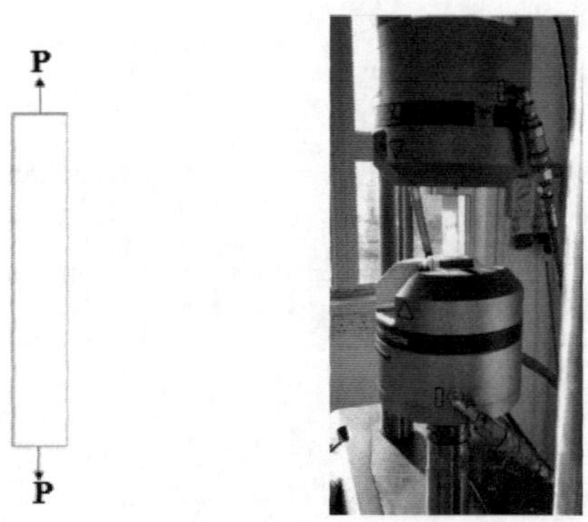

(a) (b)

Figure 3: Experimental set-up for fatigue test (a) Schematic diagram of the experimental set-up (b) Fatigue test of GFRE composite using fatigue machine

2.5 Fracture observation
The fracture side surface of the composite was observed using Optical Microscope CV-M50IR.

3. Results and Discussion

3.1 Tensile properties
Figure 4 shows a typical stress-strain curve GFRE composites obtained from the static tensile test. The figure shows the stress linearly increases with increasing strain prior reaching the maximum stress of the specimen failure. The average maximum tensile strength and Young modulus of GFRE composites were 221.79 MPa and 7.03 GPa, respectively. The fibre exhibited a brittle behaviour with a sudden load drop following a slight softening when fibre failure occurred. Figure 5 shows the sample of failure specimen of the tensile test to observed the bonding between the resin and single fibre are non-uniform.

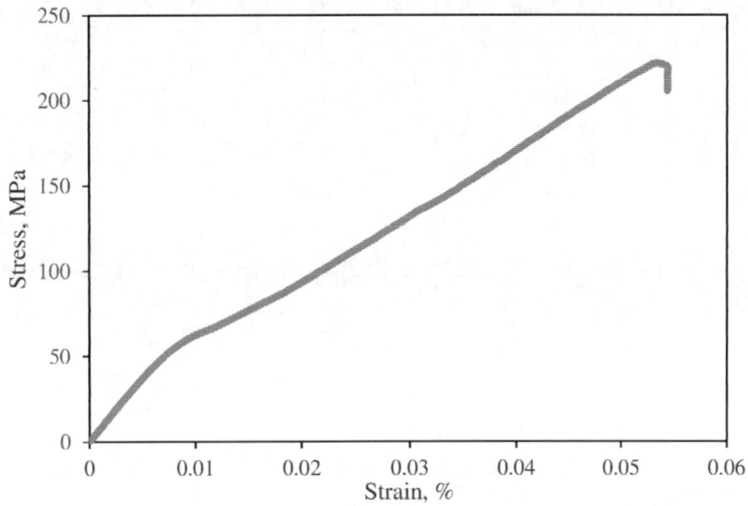

Figure 4. The typical stress-strain curve for none hole specimen.

Figure 5. Side view of failure specimen of fatigue test (a) for R $\geq$ 0 and (b) for R< 0

3.2 S-N curves with different stress ratio
Figure 6 show comparisons between the S–N relationships at different stress ratios R = 0, -1, 0.5 for representative GFRE composites respectively. These results clearly show that the number of cycles to

failure increases as the loading is decreased. The specimens 90% loading from ultimate stress at R= 0.5 recorded the lowest number of cycles, 70N respectively compared to others stress ratio. The results show at R=0.5 the number of cycles for specimen 30% (lower)loading from ultimate stress are 780000N. The fatigue strength of the GFRE composites under R=0 and 0.5 were almost similar exhibit stress loading higher than R=-1. The R ratio equal to -1 when minimum stress would be negative of the maximum stress also known as fully reversed (R=-1) testing. The compressive loads in fatigue produce a significant reduction in fatigue life when compared with results for tension tension loading. The tensile and compressive parts(R=-1) of the stress do not contribute equally to the damage[8][9]. The compressive loads in fatigue produce a significant reduction in fatigue life when compared with results for tension tension loading. For R=-1, the energy associated with the tensile loading causes damage. Overall, the fatigue behaviour of the GFRE composite materials is not only influence by the percentage of fatigue load from ultimate stress, but also of stress ratio applied.

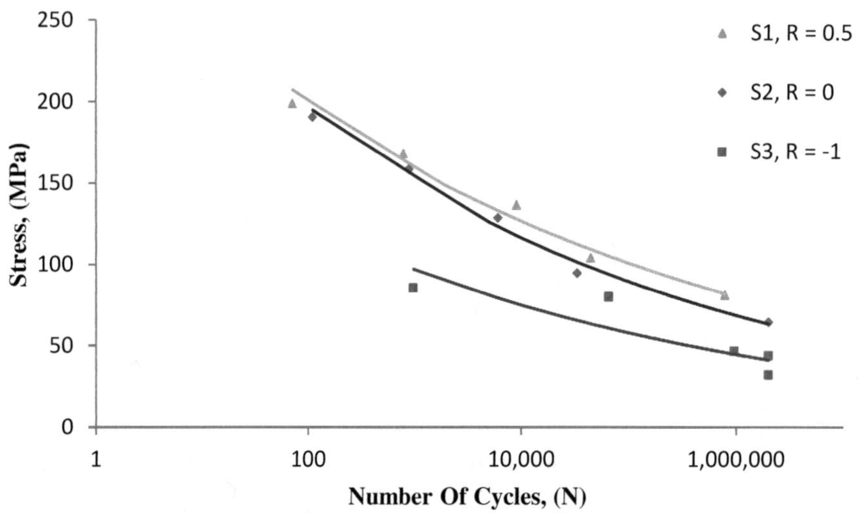

Figure 6: S-N curve under different stress ratio (R) 0,-1 and 0.5 of FGRE specimen.

Figure 7 shows the definition of the strain energy for various values of the stress ratio for negative stress ratios, R= -1 if it is assumed that only the energy associated with the tensile loading causes damage, then the damaging energy would be represented by the shaded area

4. Conclusions

In this study, the GFRE composite has been fabricated using the vacuum infusion method and the effect of stress ratio on the fatigue behaviour of an un-notched unidirectional GFRE was examined at room temperature. A phenomenological fatigue that considered the effect of stress ratio on the fatigue behaviour of composites was evaluated through observed fatigue behaviours at different stress ratios.
A fatigue failure criterion for GFRE based on the input strain energy is presented. This criterion takes into account the effect of stress ratio. The number of cycles of GFRE composites is found to highly influenced by stress ratio. The results clearly show that the number of cycles to failure increases as the loading is decreased. The specimens for fatigue tests loads 30% at R= 0 and -1 recorded the highest number of cycles. The results obtained from this work that the fatigue behaviour of the GFRE composite materials are not only influence by the percentage of fatigue tests load, but with different of stress ratio. The strain energy may be used as a fatigue failure criterion for fibre-reinforced materials[10]. Since this parameter does not rely on the different failure modes obtained in

composites, it gives equally good results independent of the failure mechanism. It is shown that, in general, the tensile and compressive parts(R=-1) of the stress do not contribute equally to the damage. The compressive loads in fatigue produce a significant reduction in fatigue life when compared with results for tension tension loading.

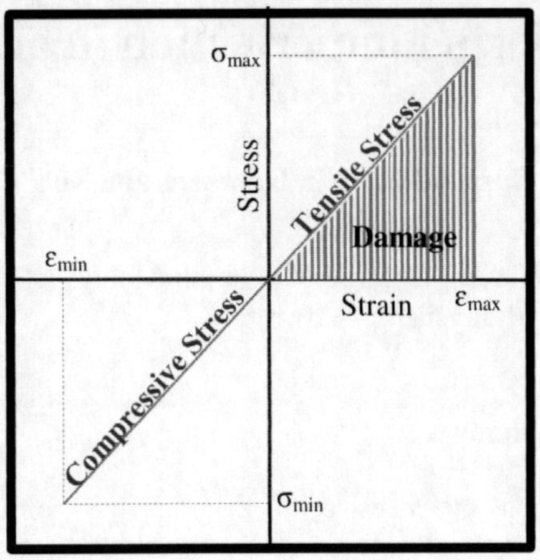

Figure 7: Definition of the strain energy for different values of the stress ratio, R < 0

5. References

[1] A. R. Syayuthi, Haftirman, K. S. Basaruddin, and M. S. Abdul Manan, "The Effect of Thicknesses on Impact Damage Advanced Composite Material for Aircraft Application," *Appl. Mech. Mater.*, vol. 754–755, pp. 802–806, 2015.

[2] Y. C. Keshavamurthy, N. V Nanjundaradhya, R. S. Sharma, and R. S. Kulkarni, "Investigation of Tensile Properties of Fiber Reinforced Angle Ply laminated composites," vol. 2, no. 4, pp. 700–703, 2012.

[3] P. Sureshkumar, B. Karthick, S. Dinakaran, and R. Rajapradeepan, "Experimental Investigation of Mechanical Behavior of Glass- Fiber Reinforced Polyurethane ResinComposite in three Different ratios," vol. 4, no. 3, pp. 36–41, 2014.

[4] A. M. T. Arifin, S. Abdullah, R. Zulkifli, and D. A. Wahab, "Failure Characterisation in Polymer Matrix Composite for Un-notched and Notched (open-hole) Specimens under Tension Condition," vol. 15, no. 8, pp. 1729–1738, 2014.

[5] E. M. C.M. Sonsinoa, "Fatigue design of highly loaded short-glass-fibre reinforced polyamide parts in engine compartments," *Int. J. Fatigue*, vol. 30, no. 7, pp. 1279–1288.

[6] J. J. Cain, N. L. Post, S. W. Case, J. J. Lesko, and J. J. Cain, "R-Ratio Effects on Glass-Reinforced Polymer Composite Life and Remaining Strength," In *16th International Conference On Composite Materials*.

[7] ASTM, "D5766/ D5766M Filled-Hole Tension and Compression Testing of Polymer Matrix Composite Laminates," *ASTM Stand.*, no. May, pp. 1–7, 2012.

[8] A. Rotem and H. G. Nelson, "Failure of a laminated composite under tension-compression fatigue loading," *Compos. Sci. Technol.*, vol. 36, no. 1, pp. 45–62, 1989.

[9] A. Rotem, "The fatigue behavior of composite laminates under various mean stresses," *Compos. Struct.*, vol. 17, no. 2, pp. 113–126, 1991.

[10] M. Kawai and K. Kato, "Effects of R-ratio on the off-axis fatigue behavior of unidirectional hybrid GFRP/Al laminates at room temperature," *Int. J. Fatigue*, vol. 28, no. 10 Spec. Iss., pp. 1226–1238, 2006.

Effective elastic constants of corrugated core sandwich plate microstructure considering imperfection in adhesive bonding

M F Hazman, K S Basaruddin, N S Kamarrudin, M S Abdul Majid and M Afendi

School of Mechatronic Engineering, Universiti Malaysia Perlis, Pauh Putra Campus, 02600 Arau, Perlis, Malaysia.

khsalleh@unimap.edu.my

Abstract. Imperfection of adhesive bonding in the corrugated core sandwich plate microstructure is commonly occured due to inaccuracies in fabrication process or environmental effect. Considering the geometrical changed due to the adhesive imperfection, it could influence the mechanical properties of sandwich plate structure. Hence, this paper was caried out to predict the effective elastic constants of corrugated core sandwich plate microstructure by considering the effect of adhesive imperfecction. Unit cell of corrugated core microstructure with variation of adhesive imperfection was developed using multiscale finite element software named Voxelcon. Homogenization method was integrated with probability function to predict the effective elastic constants of corrugated core sandwich plate structure. The proposed method could potentially be extended to other types of periodic microstrostructure in predicting the reliable homogenized properties of heterogeneous materials.

1. Introduction

Corrugated core sandwich plate microstructures are commonly used as structural part in various applications. Sandwich plate structure [1], [2] is consist of two thin stiff and high performance composite faces that bonded (adhesive) to a thick lightweight core. As for the result its create structures that high tensile strength to density ratio and corrosion resistance. Adhesive bonding is one of famous technique to join two different materials. It high tensile strength to weight ratio and other advantages has attracted numerous interest from researchers [3]–[5] and also demand in industries. While the design and analysis of non-defect structures in adhesive bonding has rapidly grown today, there is still less progress has been made in understanding the defect structures. Adhesives come in forms of film, paste, and liquid. Sandwich structures may contain a number of various defects induced during the manufacturing process such as debond, insufficient adhesive, etc. or inherent imperfections in core material like flaws and voids.

The application of adhesive is critical in the overall structure of sandwich plate because the mechanical properties of sandwich plate is strongly dependant on bonding capability. The defects on adhesive can cause stress concentration that could lead to reducing the mechanical strength of the sandwich plate structure. Grove et al. [6] mentioned that any sandwich plate structural construction (particularly tensile strength) is relies on upon the nature of the adhesive bond quality in the middle of core and skin. The structural strength or quality is depends on the arrangement of filet in the adhesive between cell divider and skin at the interface [7].

Moreover, the defects in adhesive of sandwich plate structure are commonly occurs due to inaccuracies of fabrication process or even environmental effects. One of the adhesive imperfection is inconsistency in amount of adhesive bonding between the sandwich plate structures. Hence, this study was undertaken to predict the effective elastic properties of sandwich plate microstructure considering inconsistency of adhesive volume using integrated homogenization method and modified probability function.

2. Homogenization Theory

Homogenization method was employed in the present study to calculate the effective properties of corrugated core sandwich plate microstructure [8], [9]. Since the homogenization method has been established and used extensively in numerous studies [10], [11], only the outline of this theory is highlighted in this paper. Equation (1) was analysed for sandwich plate microstructure model with periodic boundary condition, where D is the elastic tensor.

$$\int_Y D_{ijmn} \frac{\partial \chi_m^{kl}}{\partial y_n} \frac{\partial \delta u_i^1}{\partial y_j} dY = \int_Y D_{ijkl} \frac{\partial \delta u_i^1}{\partial y_j} dY \quad \forall \delta u_i^1 \tag{1}$$

Here, χ is a periodic function with respect to the microstructure, known as the characteristic displacement to represents the microscopic perturbation of displacement due to the heterogeneity. Once the characteristic displacement is obtained, the macroscopic homogenized elastic tensor is computed using eq. (2).

$$D_{ijkl}^H = \frac{1}{|Y|} \int_Y \left(D_{ijkl} - D_{ijmn} \frac{\partial \chi_m^{kl}}{\partial y_n} \right) dY \tag{2}$$

where D^H is homogenized elastic properties, *Y* is the region of the microstructure model and $|Y|$ is the volume of microstructure model. Then, eq. (3) which is the macroscopic equation, coincides with the classical micromechanics theory is written as,

$$\int_\Omega D_{ijkl}^H \frac{\partial u_k^0}{\partial x_l} \frac{\partial \delta u_i^0}{\partial x_j} d\Omega = \int_\Gamma t_i \delta u_i^0 d\Gamma \quad \forall \delta u_i^0 \tag{3}$$

t denotes the traction applied on the surface Γ of domain Ω. Hence, macroscopic response such as displacement can be obtained based on eq. (3).

3. Geometrical model

Figure 1 shows the dimension of unit cell and the periodic microstructure of sandwich plate. Imperfection on adhesive bonding was modelled based on inaccuracy in amount of adhesive. During fabrication process, the amount of adhesive used commonly is not consistent throughout the microstructure. Hence, five unit cell models were developed with variation of adhesive imperfection, which is categorised as insufficient bonding (80% and 90% of adhesive), perfect bonding (100% of adhesive) and excessive bonding (110% and 120%).

IOP Conf. Series: Journal of Physics: Conf. Series **908** (2017) 012031 doi:10.1088/1742-6596/908/1/012031

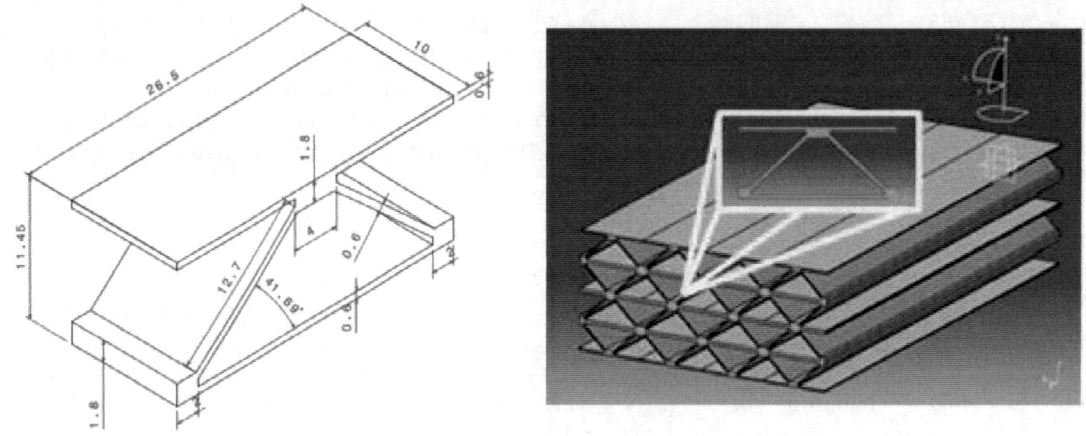

(a) (b)

Figure 1. Unit cell of sandwich plate microstructure. (a) Dimension of unit cell (in mm). (b) Periodic microstructure.

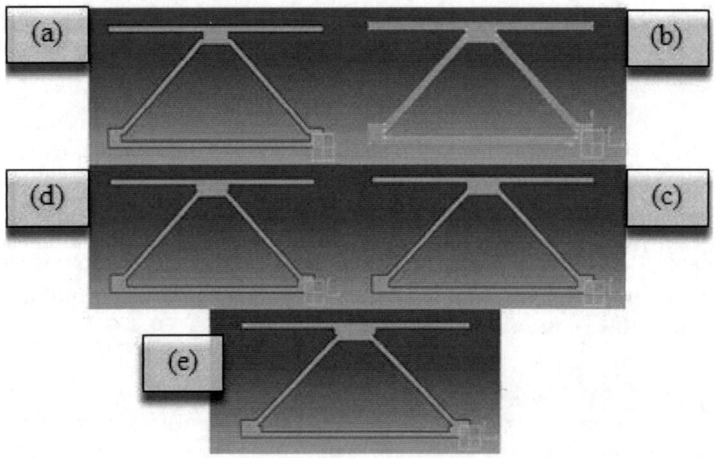

Figure 2. Geometrical model of sandwich plate with variation of adhesive imperfection. (a) 80% of adhesive, (b) 90% of adhesive, (c) 100% of adhesive, (d) 110% of adhesive, and (e) 120% of adhesive.

4. Computational Implementation

The use of sections to divide the text of the paper is optional and left as a decision for the author. Where the author wishes to divide the paper into sections the formatting shown in table 2 should be used. As demonstrative example of the present calculation, AISI type 304 stainless steel with Young's modulus of 210 GPa and Poisson's ratio of 0.3 was used as plate whereas epoxy with Young's modulus of 1.3 GPa and Poisson's ratio of 0.38 was assigned as adhesive. Both materials was set as linear isotropic.

Finite element models of sandwich plate's unit cell were created using Voxelcon (Quint Corp., Tokyo). The meshed size was determined based on convergence test that yield at 0.1 mm. The element type was cubical eight-node element.

In order to predict the effective elastic constants with considering variation of adhesive imperfection, the probability function based on normal distribution (as shown in figure 3) was applied. Each

IOP Conf. Series: Journal of Physics: Conf. Series **908** (2017) 012031 doi:10.1088/1742-6596/908/1/012031

category of adhesive bonding was assigned with probability (Pr). Hence, the effective constant can be calculated as eq. (4)[12].

$$\left(D^H\right)_{effective} = \sum_{i}^{n} (\mathrm{Pr})_i \left(D^H\right)_i \quad (4)$$

where *i* represents the category of imperfection number which is five in total.

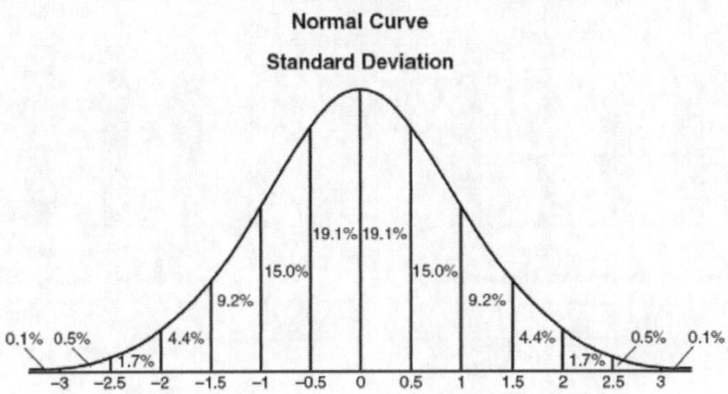

Figure 3. Normal distribution curve with percentage of probability for each interval.

5. Results and Discussion

Figure 4 shows the effect of adhesive imperfection on Young's modulus of unit cell in sandwich plate microstructure. Obviously, the increase of adhesive percentage from insufficient to excessive bonding increased the Young's modulus in all directions. Although the highest Young's modulus was obtained in E_{22}, however its increment rate was found lower compared to E_{11} and E_{33}. It seems that E_{22} not sensitive to fluctuations in adhesive volume. Increment rate between E_{11} and E_{33} was found similar.

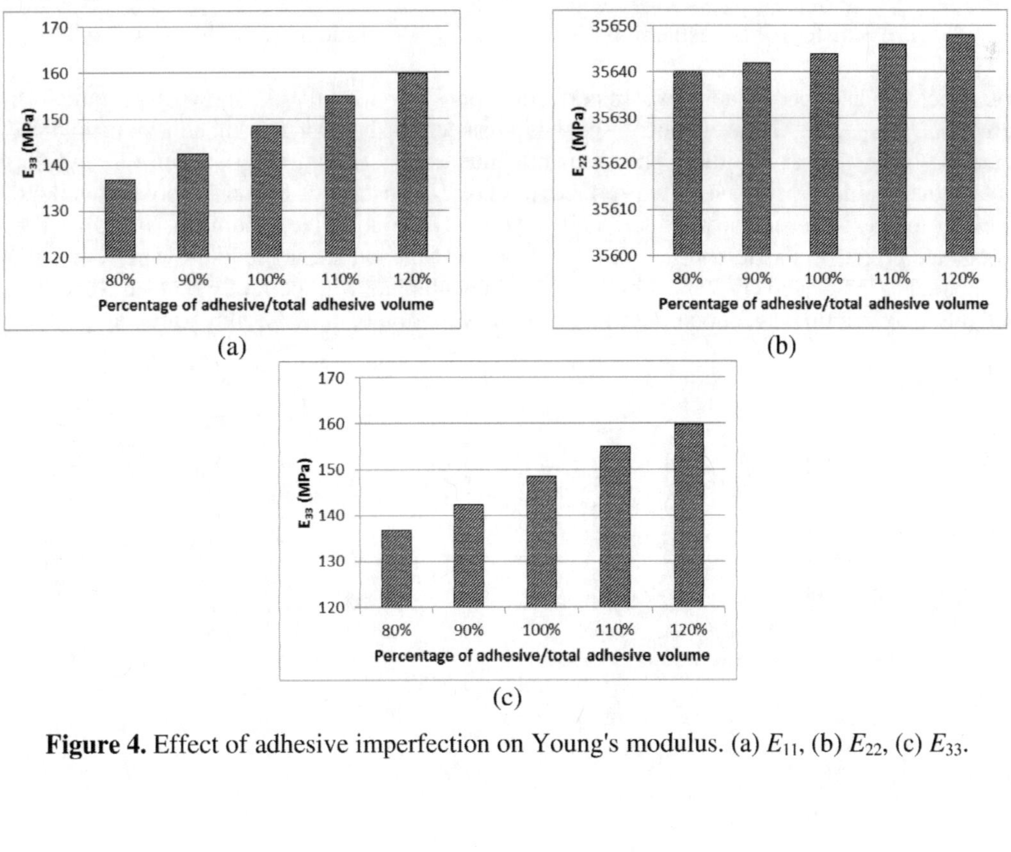

Figure 4. Effect of adhesive imperfection on Young's modulus. (a) E_{11}, (b) E_{22}, (c) E_{33}.

In contrary, Poisson's ratio of sandwich plate microstructure was found not significantly influenced by the amount of adhesive bonding as shown in figure 5. Only the slight different was obtained for ν_{12}, ν_{23} and ν_{31} with respect to increase of adhesive bonding. Hence, this result suggest that inconsistency in adhesive bonding on sandwich late microstructure did not affected the Poisson's ratio, at least for the present case which is ± 20% of adhesive imperfection.

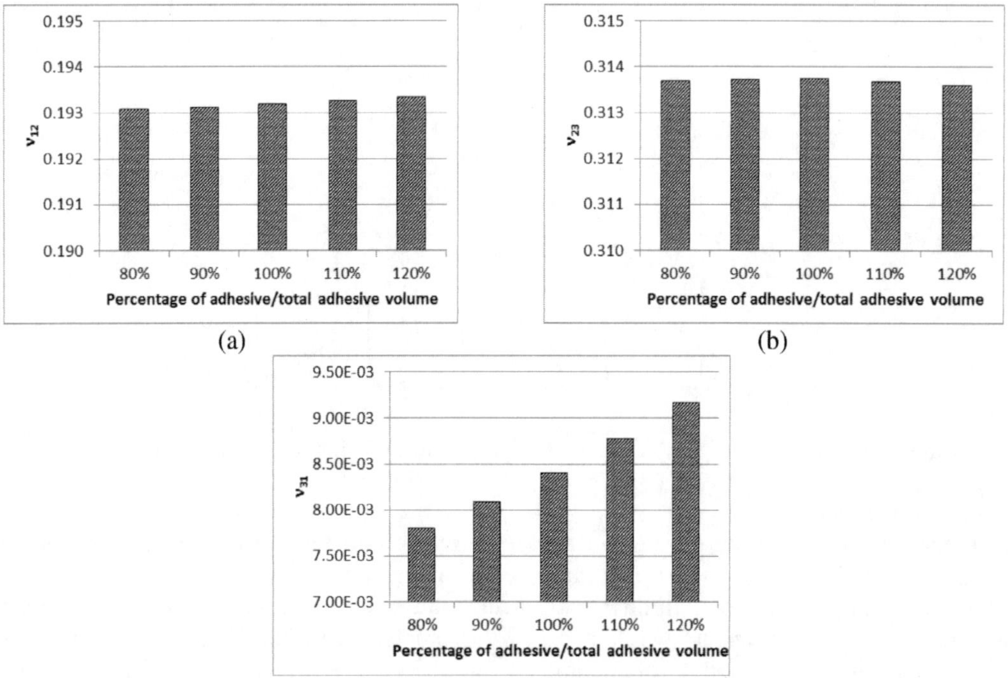

(a) (b)

(c)

Figure 5. Effect of adhesive imperfection on Poisson's ratio. (a) ν_{12}, (b) ν_{23}, (c) ν_{31}.

On the other hand, effect of adhesive imperfection on shear modulus is shown in figure 5. Shear moduli of G_{12}, G_{23} and G_{31} were increased with respect to the increase of adhesive volume. G_{12} obtained the highest shear modulus. The shear modulus and increment rate between G_{23} and G_{31} was found almost similar. Moreover, the effective elastic constants of the proposed method and comparison to elastic constants with perfectly bonded (100% adhesive volume) is listed in Table 1. The predicted effective elastic constants in Young's modulus and shear modulus were found slightly low compared to the elastic constants obtained in sandwich plate with perfectly bonded. However, the predicted Poisson's ratio was found not sensitive to the variation of adhesive imperfection.

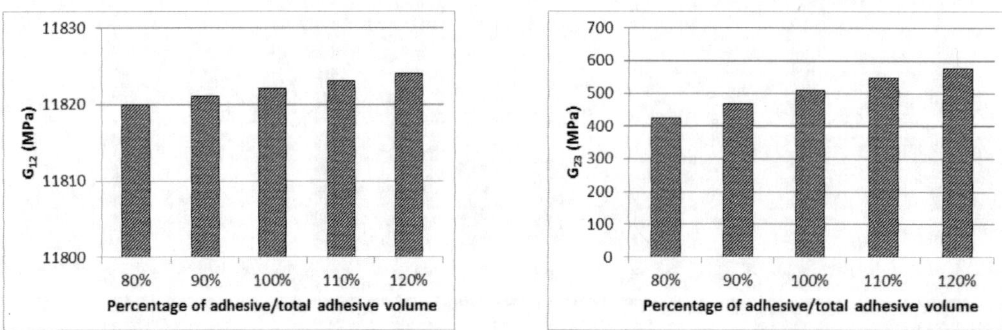

(a) (b)

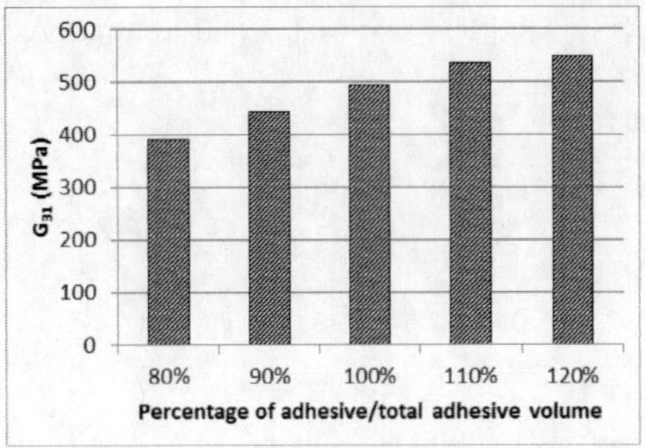

(c)

Figure 6. Effect of adhesive imperfection on shear modulus. (a) G_{12}, (b) G_{23}, (c) G_{31}.

Table 1. Comparison between effective elastic constants and elastic constants with 100% adhesive.

	Effective Elastic Constants	Elastic Constants with 100% adhesive
E_{11} (MPa)	22677.080	22953.000
E_{22} (MPa)	35216.270	35644.000
E_{33} (MPa)	146.736	148.490
ν_{12}	0.191	0.193
ν_{23}	0.310	0.314
ν_{31}	0.008	0.008
G_{12} (MPa)	11680.140	11822.000
G_{23} (MPa)	501.834	509.670
G_{31} (MPa)	482.636	494.110

6. Conclusion

The presents study was carried out to predict the effective elastic constants by considering the variation of adhesive imperfection. The increase of adhesive volume contributed to the increase of elastic constants in Young's modulus and shear modulus. The effective elastic constants was predicted well since the value obtained slightly lower compared to the value of sandwich plate with perfectly bonded. The predicted value much more realistic because the fluctuation of adhesive volume that commonly occurred due to inaccuracies in manufacturing process was considered in the calculation. The method proposed in this paper could potentially be extended to other periodic microstructures with variation of uncertainties in heterogeneous materials.

7. References

[1] Rejab M R M and Cantwell W J 2013 *Compos. Part B* **47** 267.
[2] Foo C C, Seah L K and Chai G B 2008 *Compos. Struct.* **85** 20.
[3] Kazemahvazi S and Zenkert D 2009 *Compos. Sci. Technol.* **69** 913.
[4] Tumino D, Ingrassia T, Nigrelli V, Pitarresi G and Miano V U 2014 *Fract. Struct. Integr.* **30** 317.
[5] Deshpande V S and Fleck N A 2014 *Int. J. Mech. Sci.* 1.
[6] Grove S M, Popham E and Miles M E 2006 *Compos. Part A* **37** 804.
[7] Kooistra G W and Wadley H N G 2007 *J. Appl. Mech.* **74** 259.
[8] Basaruddin K S, Kamarrudin N S, Ibrahim I, 2014 *Lat Am J Solids Struct.* **11** 755.
[9] Carlsson L A, Nordstrand T and Westerlind B *J. Sandw. Struct. Mater.* **3** 253.
[10] Guedes J and Kikuchi N *Comput. Methods Appl. Mech. Eng.* **83** 143.
[11] Takano N, Fukasawa K and Nishiyabu K *Int. J. Mech. Sci.* **52** 229.

[12] Halim S A, Basaruddin K S, Daud R, Safar M J A and Rahman A S A 2015 *ARPN J. Eng. Appl. Sci.* **10** 9353.

IOP Conf. Series: Journal of Physics: Conf. Series **908** (2017) 012032 doi:10.1088/1742-6596/908/1/012032

Homogenized Properties of Porous Microstructure: Effect of Void Shape and Arrangement

M Thomas, K S Basaruddin, M J A Safar, S F Khan and I Ibrahim

School of Mechatronic Engineering, Universiti Malaysia Perlis (UniMAP), Pauh Putra Campus, 02600 Arau, Perlis, Malaysia.

khsalleh@unimap.edu.my

Abstract. This paper aims to investigate the effect of void shape and arrangement on the effective elastic properties of porous microstructure. The characteristics of the voids are in different shapes, sizes and arrangement. The porous microstructure models were developed using CATIA. Then, Voxelcon was employed to analyse the multiscale finite element model and determine the homogenized properties. Based on the results, void shape, size, and arrangement of porous microstructure were found sensitive to the elastic (homogenized) properties. Ellipsoidal shape having the highest Young's modulus, whereas the spherical shape has the highest Poisson's ratio and shear modulus. Cubical shape was the lowest for all the elastic properties. Moreover, the formation arrangement in void cubical shape produced the highest Young's modulus and shear modulus.

1. Introduction

Porous material is a material containing pores or it can be called as voids. The pores have porosity that might contained air, and other particles that can stick on the wall as a pores. The characteristics of a porous material may vary depending on its size, arrangement and shape of the pores, as well as the porosity. Porous materials has been used in many applications. For instance, wood contain pores and it may not always can be seen by naked eye, but it has strong material properties that can serve as the structural foundation of a home. Porous materials with open pores are mainly used for industrial applications such as filters and carriers for catalysts and bioreactors [1]. Closed porous materials are used for thermal insulators and low specific gravity structural components and biomedical field.

Chan and Francis [2] had developed a multi scale method that allows one to predict the mechanical performance, where it includes the elastic modulus and yield strength of the 3D scaffold of a given pore geometry on the basis of properties of constituents at smaller length scales and higher structural hierarchies. From their researched, a first-principles computational method had been used to compare and analyse the mechanical properties of nano-scaled particles. On the other hand, Wang and Hu [3] had studied on the mechanical properties and pores structure deformation behaviour for porous titanium. Porous titanium has been exhibited into desirable properties as biomedical materials. Porosity of porous titanium is varying from 36% to 66% and pores size has average of 230μm, then, were fabricated by powder sintering. Its microstructure features were using scanning electron microscopy as its characteristics. Porous titanium has mechanical properties that found to be close to those of human bone. Its stiffness value is ranging from 1.86GPa to 14.7GPa and compressive strength values are 85.16MPa to 461.94MPa.

According to Roberts and Garboczi [4], the elastic properties of 2D phase (solid phase) porous materials depend on the geometrical nature of the pore space and solid phase and also the porosity ratio. The relevant aspects of porous materials include pore shape, size and type of the interconnections between solid areas. Besides that, they used finite element method to derive simple formulae that relate with Young's modulus and Poisson's ratio to porosity and microstructure for three different models of microstructure. They have created or test the three different models by overlapping solid spheres, overlapping spherical pores and overlapping ellipsoidal pores. Hence, the main objective of this study was to develop the finite element model of porous microstructure with variation of voids characteristics in order to analyse the effect of voids shape and formation on the effective elastic properties of porous microstructure.

2. Computational Analysis

2.1 Homogenization Method

Homogenization Method is macroscopic structure supposed to be made of a material that has the microscopic heterogeneity. If we can define a microscopic unit cell structure Y that can represent the global heterogeneity, the macroscopic properties can be defined as the volumetric average of the microscopic properties in the unit cell. The heterogeneous material can be replaced by the equivalent homogenized model. Then, it assumed that the unit cell is repeated periodically. By using the two scale singular perturbation theory, all the equation was followed based on previous studies [5-7].

$$u_i(\boldsymbol{x},\boldsymbol{y}) = u_i^0(\mathrm{x}) + \varepsilon u_i^1(\boldsymbol{x},\boldsymbol{y}) \tag{1}$$

In this equation, u_i^0 microscopic or homogenized displacement and u_i^1 is a perturbed term due to the microscopic heterogeneity. Suppose that the traction t_i applied on the boundary Γ and the body force is neglected. An elasticity tensor is denoted as E_{ijkh}. By taking the lim of $\varepsilon \to 0$ for homogenization, the decoupled microscopic and macroscopic equations are finally obtained. In this derivation, u_i^1 in Equation (1) is assumed to be written as below :

$$u_i^1(x,y) = -\chi_i^{kh}(y)\frac{\partial u_k^0(x)}{\partial x_h} \tag{2}$$

Where χ_i^{kh} is a characteristic displacement that is a periodic function with respect to the microscale. The microscopic equation to solve for the unit cell Y under the periodic boundary condition is as follows:

$$\int_Y (E_{ijkh} - E_{ijmn}\frac{\partial \chi_m^{kh}}{\partial y_n})\frac{\partial v_i^1}{\partial y_j}\mathrm{d}Y = 0 \qquad \text{for} \quad \forall v_i^1 \tag{3}$$

Due to the existence of the solution in Equation (3), the macroscopic equation can be derived as below:

$$\int_\Omega E_{ijkh}^H \frac{\partial u_k^0}{\partial x_h}\frac{\partial v_i^1}{\partial x_j}\mathrm{d}\Omega = \int_\Gamma t_i\, v_i^0 \mathrm{d}\Gamma \qquad \text{for} \quad \forall v_i^0 \tag{4}$$

where E_{ijkh}^H is the homogenized elasticity tensor that is also symmetric defined by

$$E_{ijkh}^H = \frac{1}{|Y|}\int_Y (E_{ijkh} - E_{ijmn}\frac{\partial \chi_m^{kh}}{\partial y_n})\mathrm{d}Y \tag{5}$$

where $|Y|$ is the volume of the unit cell.

2.2 Geometrical Model

The development of porous model for void regular and void formation arrangement in this study is using CATIA Software. The model that decided to be investigated for one of the porous characteristic are cubical, spherical and ellipsoidal, where there are 18 models. Each of the arrangement has 9 models. There are 3 sizes; small, medium and large. The dimensions for the shapes were obtained by calculating the same volume for each shape. The number of voids is set to be 4 x 4 x 4 in 3D, where there are 64 numbers of voids. Total volume is 1000 mm³, where 10 mm x 10 mm x 10 mm of an outer box dimensions. Porosity or volume fraction can be calculated by number of voids divided by total volume. Porosity that decided to be determined is 15%, 20% and 25% for cubical and spherical void while ellipsoidal void is 8.03%, 10.03% and 12.78%.

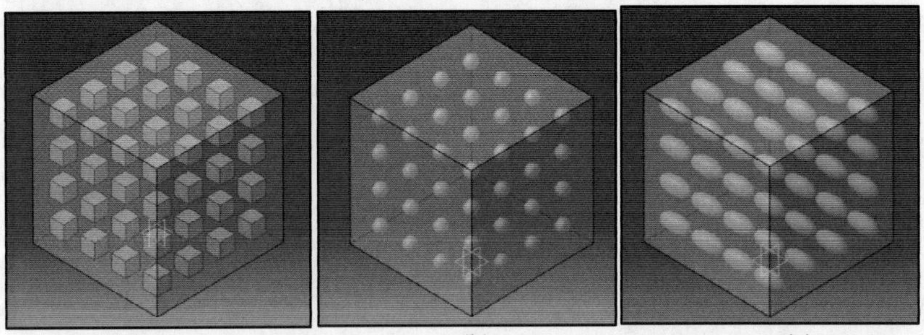

(a) (b) (c)

Figure 1: Void regular arrangement porous model; (a) Cubical, (b) Spherical, (c) Ellipsoidal.

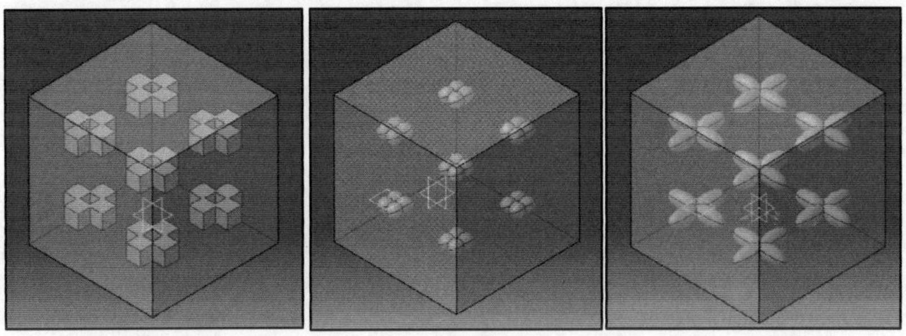

(a) (b) (c)

Figure 2: Void formation arrangement porous model; (a) Cubical, (b) Spherical, (c) Ellipsoidal.

2.3 Meshing

Voxel (cubical) was used as element type for all models in this study. Since validation of the present analysis using experiment work is very difficult due to the small scale size imposed in the modelling, the element size becomes very crucial in order to ensure the accuracy of numerical result. Hence, convergence test was performed in order to obtain the optimum element size. Table 1 shows the results of convergence test for all models with different void shape. The test was conducted for 0.1^3, 0.2^3, 0.3^3. 0.4^3 and 0.5^3 mm^3 of element size. Based on the result, element size of 0.2 was selected for all analysis in the present study.

Table 1: Convergence test of unit cell models with variation of void shape.

No.	Element Type	Cubical
1	Convergence Test	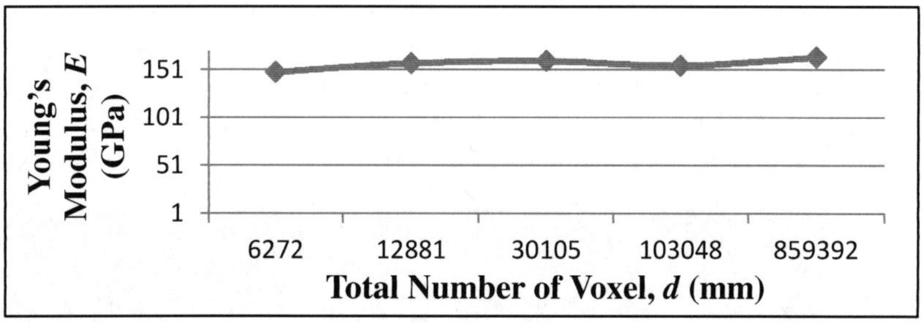
	No. of element	6272, 12881, 30105, 103048, 859392
2	Element Type	Spherical
	Convergence Test	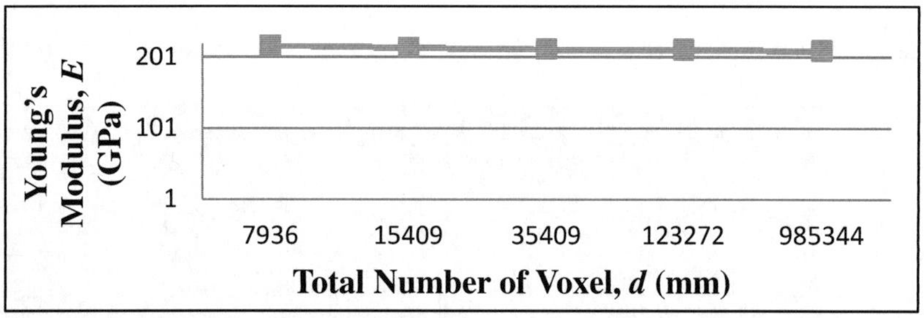
	No. of element	7936, 15409, 35409, 123272, 985344
3	Element Type	Ellipsoidal
	Convergence Test	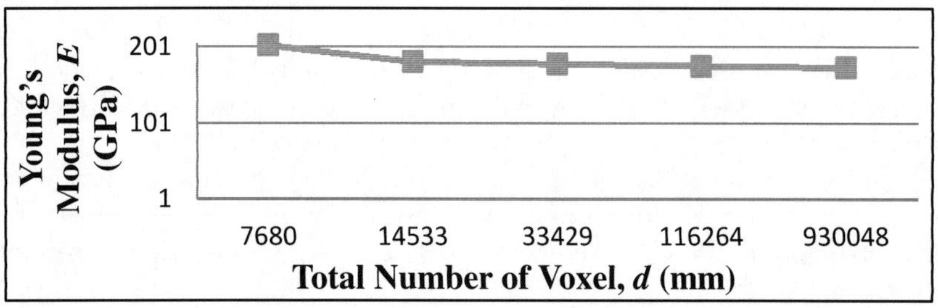
	No. of element	7680, 14533, 33429, 116264, 930048

2.4 Mechanical Properties of Constituent

The constituent material used is porous alumina. It has been widely used in industrial such as nanostructure template in various nano-applications. The porous structure consists of numerous hexagonal cells perpendicular to the aluminium substrate and each cell has several tens or hundreds of nano-scale pores at its centre. Because the nano-morphology of anodic porous alumina is limited by the electrolyte during anodizing, the discovery of additional electrolytes would expand the applicability of porous alumina. Porous alumina has Young's modulus (E) of 404 GPa and Poisson's ratio (v) of 0.239.

3. Results & Discussion

Based on all the obtain result, the project is more clarified in term of analysis of model and theoretical of project. Convergence test was conducted in order to get the finest result for the analysis. The size of voxel element mesh that chosen to analyse all the model is 0.1mm. It is generally known that, small size of element tends to take longer time to compute the result analysis.

The result for void shape that on Young's modulus (E); E_{11}, E_{22}, E_{33} is shown in figure 3. Ellipsoidal shape has the highest value than spherical shape, while cubical shape is the lowest value. Next, the result on Poisson's ratio (ν); ν_{12}, ν_{23}, ν_{31} is depicted in figure 4. The ν_{12} result shows that spherical shape has the highest value than ellipsoidal shape, while for cubical shape is the lowest. Result for ν_{23}, and ν_{31} is the same where ellipsoidal is the highest that spherical, while cubical shape is still the lowest. Then, figure 5 shows the effect of void shape on shear modulus (G). G_{12}, G_{23}, G_{31} was found has the same pattern as Young's modulus (E); E_{11}, E_{22}, E_{33}.

The result for void formation effect on Young's modulus and shear modulus is shown in figure 6. The result shows that the elastic properties was not sensitive to the void arrangement. The line graph for both arrangement (regular and formation) almost in the same line. Based on the past research, some research papers found that the elastic properties of the porous materials are not sensitive to the shape of the voids but depends on the void's number and size. But, some papers are said the opposite. Therefore, this project is conducting for investigate and compare the problem whether the shape of the voids really affect the elastic properties and mechanical properties of the porous microstructure. The result for void shape and void formation are affect the effective elastic properties of porous microstructure.

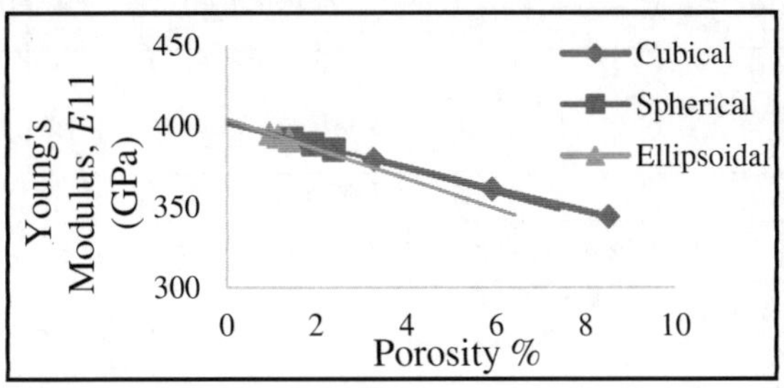

(a)

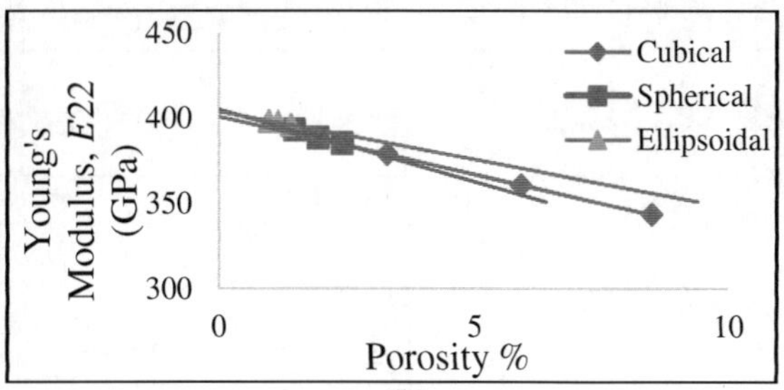

(b)

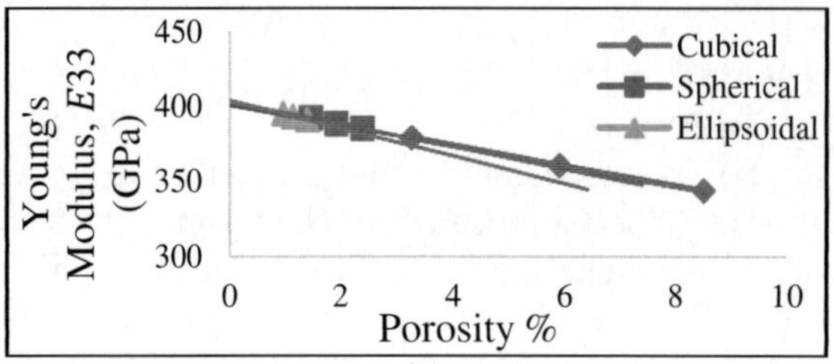

(c)

Figure 3: Effect of shape on; (a) E_{11}, (b) E_{22}, (c) E_{33}.

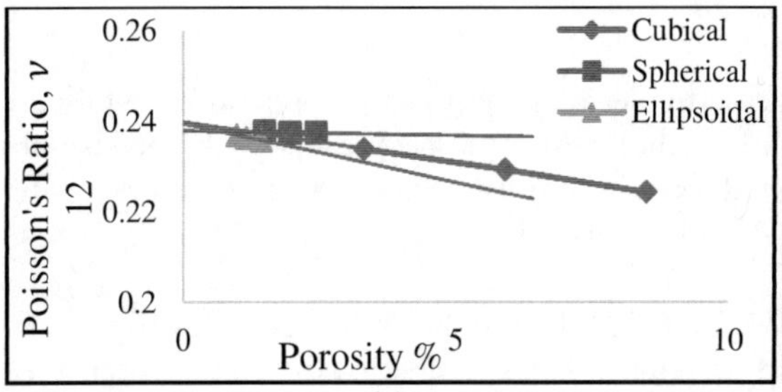

(a)

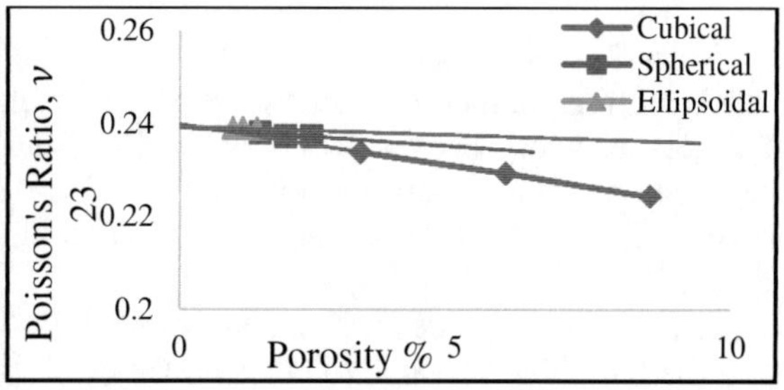

(b)

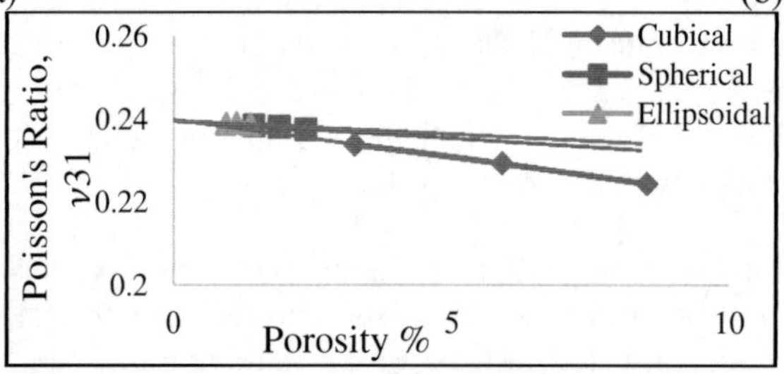

(c)

Figure 4: Effect of shape on ; (a) ν_{12}, (b) ν_{23}, (c) ν_{31}.

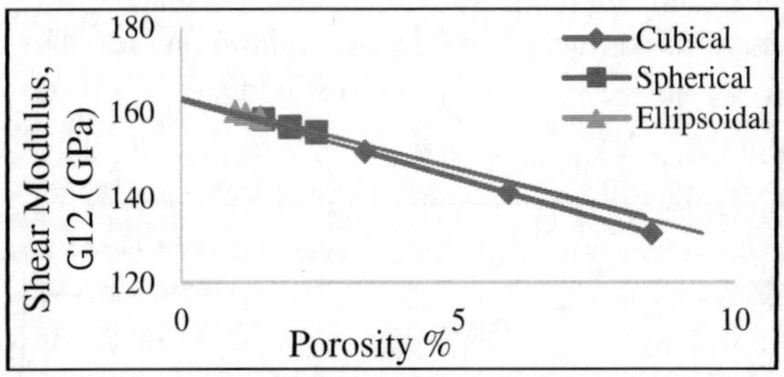

(a)

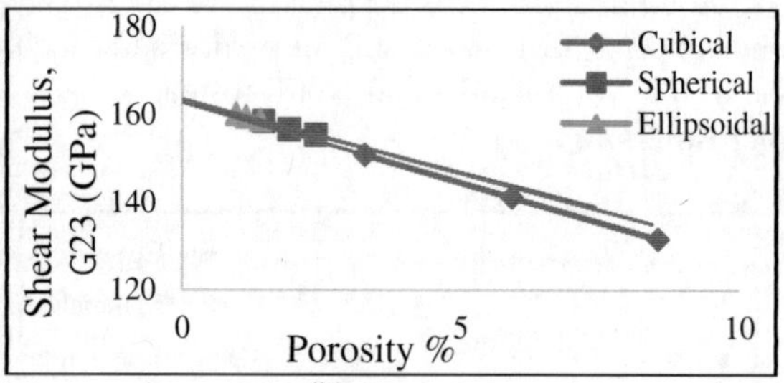

(b)

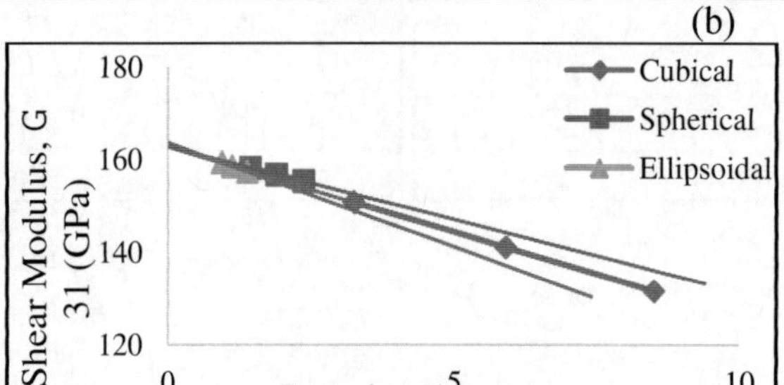

(c)

Figure 5: Effect of shape on; (a) G_{12}, (b) G_{23}, (c) G_{31}.

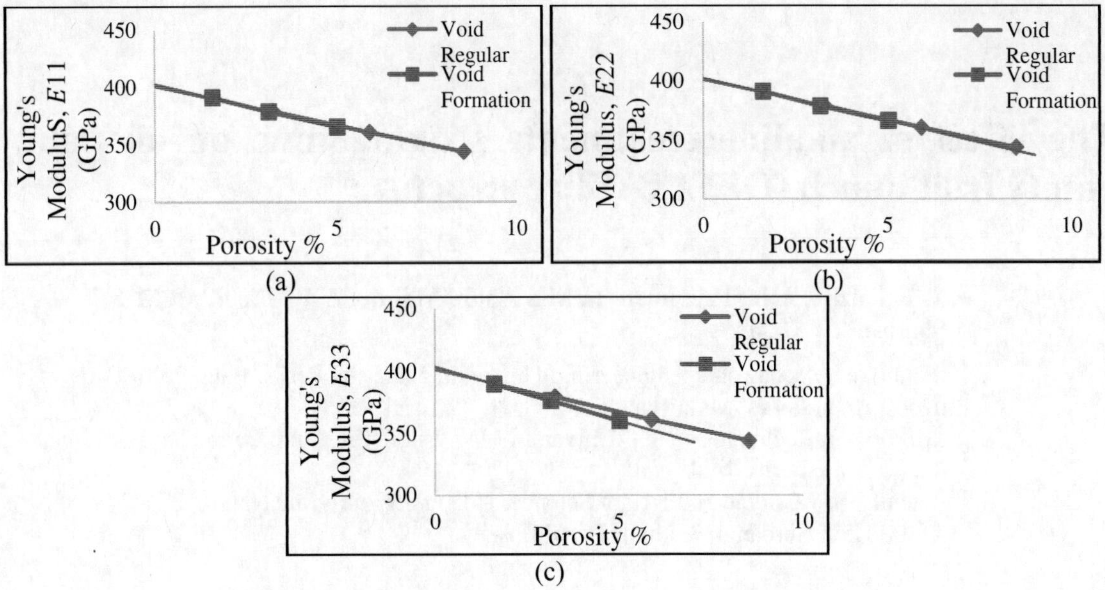

Figure 6: Effect of arrangement on; (a) *E*11, (b) *E*22, (c) *E*33.

4. Conclusion

As the conclusion, the objective of the project had been achieved. The project successfully developed the finite element model of porous microstructure with variation of voids characteristics. The project also managed to analyse the effect of voids shape on the effective elastic properties of porous microstructure. Besides that, the project had also investigated the effect of voids formation on the effective elastic properties of porous microstructure. Based on the results and analysis, ellipsoidal shape having the highest Young's modulus. Next, spherical shape has the highest Poisson's ratio and shear modulus. Whereas, cubical shape is the lowest for all the elastic properties. Besides that, the effect of void arrangement on cubical shape, void formation has the highest Young's modulus and shear modulus. While for spherical shape void arrangement, void formation has the highest value for both Young's modulus and shear modulus. Moreover, ellipsoidal shape has the highest Young's modulus and shear modulus on void regular arrangement. Hence, all of the factors does affect the elastic properties of porous microstructure.

The result obtained based on the study of previous research papers and significant information also knowledge on the effect of void in porous microstructure. It is important to study and know whether the void affect or not on the structural element of the porous microstructure. This is especially for those in manufacturing or building industry.

5. References

[1] Liu P S and Chen G F 2014 *Porous Materials* 221.
[2] Chan K S, Liang W, Francis W L and Nicolella D P 2010 *J Mech Behaviour Biomed Mater* **3** 584.
[3] Wang X, Li J, Hu R and Kou H 2015 *Transac. Nonferrous Metals Soc China* **25** 1543.
[4] Roberts P and Garboczi E J 2000 *J Am Ceram Soc* **83** 3014.
[5] Takano N, Zako M, Kubo F and Kimura K 2003 *Inter J Solids Struct* **40** 1225.
[6] Halim S A, Basaruddin K S, Daud R, Safar M J A, Abd Rahman A S, 2015 *ARPN J Eng Appl Sci* **10** 9353.

IOP Conf. Series: Journal of Physics: Conf. Series **908** (2017) 012033 doi:10.1088/1742-6596/908/1/012033

The effect of alkaline treatments soaking time on oil palm empty fruit bunch (OPEFB) fibre structure

M K Faizi[1,2], A B Shahriman [1,2], M S Abdul Majid[1], Z A Ahmad[1], B M T Shamsul[3], Y G Ng [3]

[1]School of Mechatronic Engineering, Universiti Malaysia Perlis, Pauh Putra Campus 02600 Arau, Perlis, Malaysia.
[2]Motorsport Technology Unit, Universiti Malaysia Perlis, Pauh Putra Campus 02600 Arau, Perlis, Malaysia.
[3]Faculty of Medicine and Health Science, Universiti Putra Malaysia, 43400 UPM Serdang, Selangor Darul Ehsan.

mkhairulfaizi@unimap.edu.my

Abstract. The primary objective of this work was to study the effect of different soaking time in an alkaline solution for oil palm empty fruit bunch (OPEFB) fibre. For this purpose, the OPEFB fibres were treated with the sodium hydroxide (NaOH) at 5% concentration for ½ hour, 1-hour, 3-hour, 5-hour, 7-hour and 24-hours. The single tests of treated and untreated fibre were then performed in accordance with the ASTM D3822-07 standard. Next, the surfaces of the fibres prior and after the treatment were observed with a scanning electron microscope (SEM) TM3000. The result shows that at 7 hours of soaking time exhibits the highest tensile strength compared to the other soaking time and untreated fibre.

1. Introduction

The enthusiasm for utilising natural fibre, for example, another plant and wood fibre as support material in plastics have expanded significantly in a recent couple of years. Due of the light weight, high quality to weight proportion and decomposition resistance natural fibre based composites are getting to be necessary composite materials in building and designing areas [1]. Oil palm trees can develop well in a Malaysian atmosphere. Without a doubt, it has turned into the most critical rural product in Malaysia and has been the way to the national financial development [2]. The oil palm empty fruit bunch (OPEFB) is another potential natural fibre to be explored [3]. Just 10% of the aggregate biomass delivered in the estate comprises of oil, the staying 90% of the generation consists of large amounts of lignocellulosic material, for example, OPEFB, trunks and oil palm fronds [4]. In the process produce bunch remains as compost; OPEFB are mostly incinerated, which exposes to an ecological issue. These agriculture waste, particularly of OPEFB, can be changed to useful by-products utilising fibre based (natural fibre) composite as opposed to blazing the OPEFB. The structure of OPEFB comprises of three primary constituents; celluloses, hemicellulose and lignin. Steady holding among matrix and fibre can increase the composites mechanical properties. In order to achieve good consolidation between fibre and matrix, surface treatment on fibre is the common approach adopted via chemical treatments [5]-[8]. The common chemicals that used for surface treatments were sodium hydroxide, acetyl acid, and silane [9]-[14]. This research focuses on the effect of soaking time during alkali treatment of the OPEFB fibre. The elastic properties; tensile, modulus of elasticity and morphological structure of the fibres are discussed.

2. Materials and methods

The oil palm empty fruit bunch (OPEFB) fibres are obtained from United Oil Palm Industries, Nibong Tebal, Pulau Pinang. The fibres were manually extracted hair by hair and continue to water retting process. The water retting process takes about one week. After a week, the fibre was cleaned using distilled water and dried under the sun to ensure removal of the moisture content of the fibres. The processed oil palm empty fruit bunch (OEFB) fibre was then treated with 5% aqueous solutions of NaOH at room temperature for 1/2, 1, 3, 5, 7 and 24 hours, maintaining the liquor ratio 40:1 to remove the hemicelluloses and surface impurities of the fibre. Finally, the fibres were cleaned using distilled water and dried at room temperature. The physical properties of the treated and untreated fibres are presented in Table 1. The measurement of the length and diameter were taken before the tensile test to ensure the properties of the specimens were maintained for analysis purpose.

Table 1. Physical properties of OPEFB fibre. (20 samples for each soaking time).

Soaking time	Average diameter (mm)	Standard deviation	Average length (mm)
Untreated	0.52	0.09	70
½ hour	0.34	0.05	70
1 hour	0.30	0.04	70
3 hours	0.32	0.04	70
5 hours	0.33	0.04	70
7 hours	0.34	0.04	70
24 hours	0.34	0.03	70

The tensile strength of the fibre was determined through a single fibre tensile testing in accordance with the ASTMD3822-07 standard. The treated and untreated fibres were weighed using an analytical balance device, and the length of the fibre strand was measured microscopically. The surface and the diameter of the fibres were observed using Scanning Electron Microscope (SEM) TM3000. Prior to the test, the fibres were dried by using oven to the temperature of 60°C to 100°C about 30 minutes to an hour. The fibre was mounted onto a tab-shaped piece of paper with the gauge length greater than 50 mm. The tensile strengths of the treated and untreated fibres were then determined using an INSTRON 5848 Micro universal testing machine with a load cell of 2kN.

3. Result and discussion

The SEM micrographs of the surfaces of untreated and treated of OPEFB fibre before performing the single test are shown in Figure 1. The surface morphology of the fibre is important to observe the changes that happen on the surface of the OPEFB fibre because of the alkali treatment. The surface of the fibre after the treatment is found rougher than the untreated fibre. This is because the alkali treatment causes reduce in hemicellulose and lignin contents within the fibre thus make it more compact. The compact structure of the treated fibre will lead to making the treated fibre increase in tensile strength.

Figure 1 shows the stress–strain responses for selected of untreated and alkali-treated OPEFB fibres. The treated fibres exhibit a higher ultimate tensile stress because of the changes in the cellulose crystallinity. The graph shows that the 7 hours of soaking time have the highest ultimate tensile test.

IOP Conf. Series: Journal of Physics: Conf. Series **908** (2017) 012033 doi:10.1088/1742-6596/908/1/012033

The removal enables the fibril to rearrange them more compactly manner, thus enhancing the tensile strength of the fibre. Following the 3 hours, 24 hours, 1 hour, 5 hours, ½ hour of soaking time, there is an increase in the stiffness and this result in the increase in the tensile stress but a decrease in strain compared to untreated fibres.

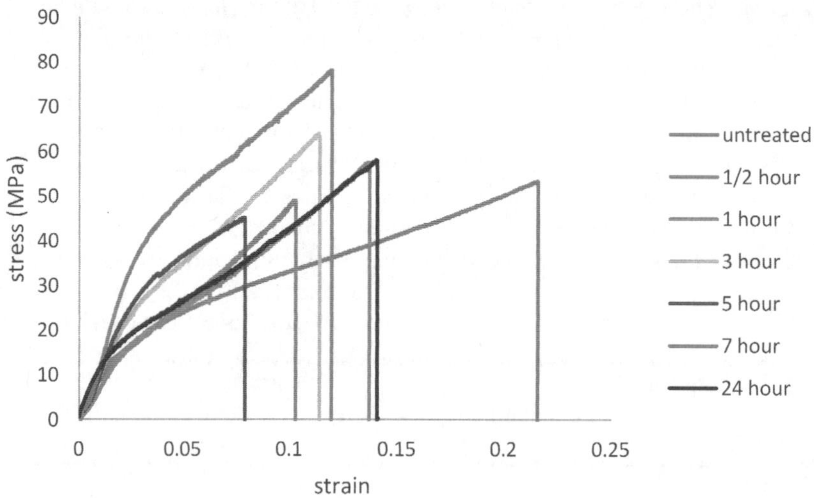

Figure 1. Stress–strain responses for selected untreated and alkali-treated OPEFB fibre.

As shown in figure 2, it is seen that the highest strength obtained is at 7 hours of soaking time in 5% concentration of NaOH for OPEFB fibre with its 77.93 MPa. The second highest is for 24 hours, followed by 3 hours, 1 hour, untreated, ½ hour and 5 hours. The lowest strength is 43.42 MPa recorded for 5 hours of soaking time in alkali treatment. This indicates that the 7 hours of soaking time had increased almost 49 % from the untreated fibre. The alkali treatment causes fibrillation, which is a process that causes the fibre bundle to break into smaller bundles. Smaller bundles help to distribute the load applied throughout the fibres.

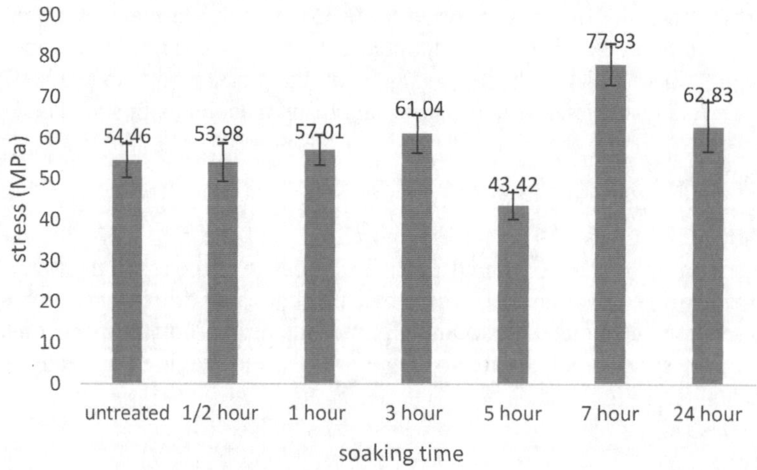

Figure 2. The average tensile stress of OPEFB fibre (20 samples for each soaking time).

Based on figure 3, the highest average Young's Modulus of OPEFB fibre recorded is 1.4 GPa for 5 hours of soaking time in 5% of NaOH. The lowest average modulus of elasticity, on the other hand, is; after 1-hour treatment fibre with 0.67 GPa. Seeing these results, the alkali treatment indicates changes of the fibre elasticity concerning the soaking time applied during the treatment. Also, the elastic modulus remains below 1 GPa between specimens under 3 hours of treatment, registering between 0.67 GPa to 0.96 GPa. However, the fibre showed improved elasticity after being treated for more than 5 hours, recording at 1.4GPa, an increase of almost 65%. The modulus then indicates a slight reduction in elasticity with prolonged soaking times though higher than untreated fibres. Hence, the different modulus elasticity may result in the different strain to failure of specific fibre.

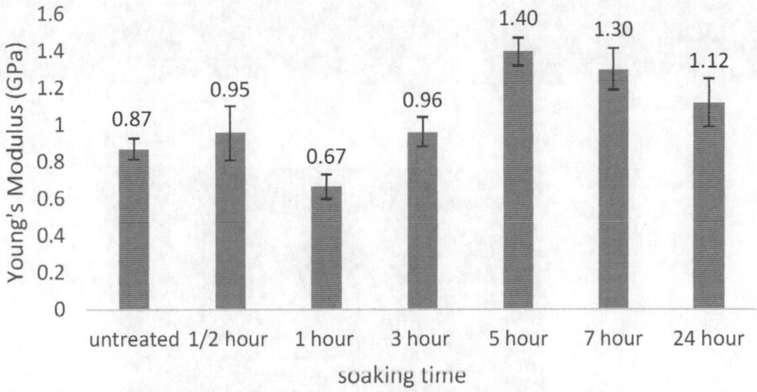

Figure 3. Young's Modulus of OPEFB fibre

The SEM micrographs of the surface of the untreated and alkali treated OPEFB fibres are shown in figure 4. Figure 4 (a) shows the images of the untreated fibre. The white spots represent the presence of impurities on the surface of the fibre. Figure 4(b), (c), (d), (e), (f), and (g) represents the images of ½ hours, 1 hour, 3 hours, 5 hours, 7 hours, and 24 hours of soaking time respectively. Based on the observation, these alkalis treated OPEFB fibres showed cleaned surfaces due to the removal of impurities.

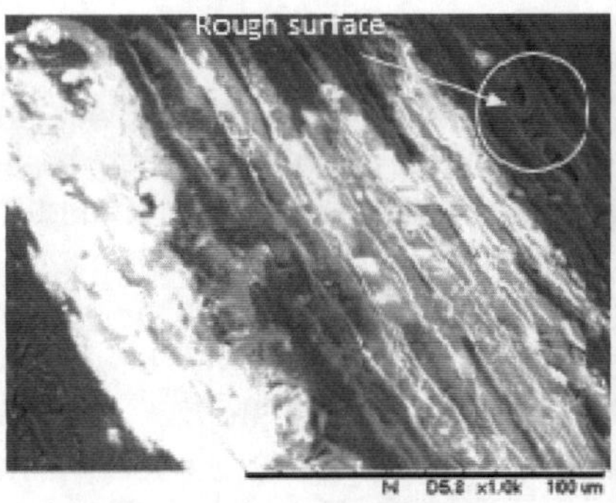

(a) Untreated

(b) ½ hour

(c) 1 hour

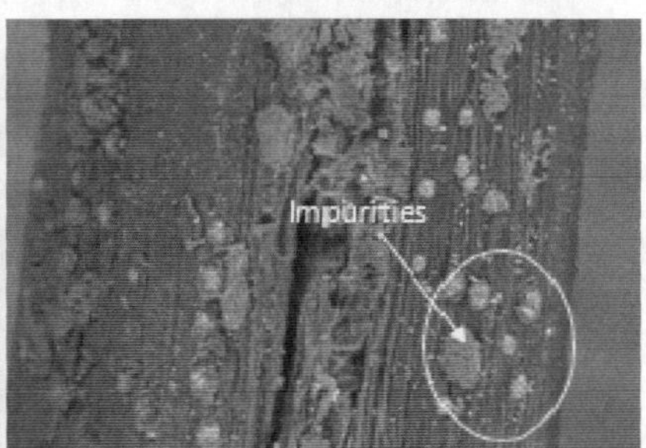

(d) 3 hour

IOP Conf. Series: Journal of Physics: Conf. Series **908** (2017) 012033 doi:10.1088/1742-6596/908/1/012033

(e) 5 hour (f) 7 hour

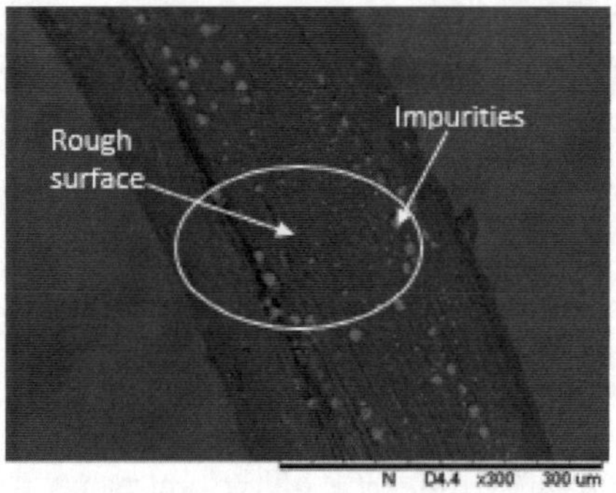

(g) 24 hour

Figure 4. Surface morphologies of untreated and alkali treated OPEFB fibre.

4. Conclusions

The mechanical properties of the treated and untreated OPEFB fibre with 5% NaOH aqueous solutions accompanied by different soaking times were investigated in this paper. It is shown that the OPEFB fibre is increased in their mechanical properties when treated in alkaline solution. The structure of the fibre also differences between the treated and untreated fibre. The result shows that the treated fibre exhibited better tensile strength whereas the 7 hours of soaking time is the highest compared to the other soaking time and untreated fibre. The surface observation through Scanning Electron Machine (SEM) shows the texture of the fibres achieved surface roughness after treating the OPEFB fibres. For future work, the surface roughness of fibre is an important factor to form it into composite since it promotes the best fibre-matrix bonding compared to untreated. The study further supported the feasibility of utilising OPEFB fibres as reinforcing materials in polymer composites.

5. Acknowledgement

The authors would like to be obliged to Universiti Malaysia Perlis (UniMAP) for providing laboratory facilities, Motorsport Technology Unit (MOTECH) UniMAP for providing financial assistance, and Universiti Putra Malaysia (UPM) as collaboration partnership.

6. References

[1] R. K. and R. S. Sandeep Bhardwaj, "Natural Fibre Composites and Its Potential," vol. 1, no. 4, pp. 37–40, 2013.

[2] M. F. Awalludin, O. Sulaiman, R. Hashim, and W. N. A. W. Nadhari, "An overview of the oil palm industry in Malaysia and its waste utilization through thermochemical conversion, specifically via liquefaction," Renew. Sustain. Energy Rev., vol. 50, no. October, pp. 1469–1484, 2015.

[3] M. K. Faizi, Shahriman A.B., M.S. Abdul Majid, Shamsul B. M. T., S. N. Basah, E. M. Cheng, M. Afendi, Zuradzman M. R., Khairunizam WAN, and D. Hazry, "An overview of the oil palm empty fruit bunch (OPEFB) potential as reinforcing fibre in polymer composite for energy absorption applications," MATEC Web of Conferences 90, 2017.

[4] M. F. Awalludin, O. Sulaiman, R. Hashim, and W. N. A. W. Nadhari, "An overview of the oil palm industry in Malaysia and its waste utilization through thermochemical conversion, specifically via liquefaction," *Renew. Sustain. Energy Rev.*, vol. 50, pp. 1469–1484, 2015.

[5] M. M. Kabir, H. Wang, K. T. Lau, and F. Cardona, "Effects of chemical treatments on hemp fibre structure," Appl. Surf. Sci., vol. 276, pp. 13–23, 2013.

[6] W. Fatra, R. Sanjaya, H. Rionaldo, and Z. Helwani, "Alkaline Treatment of Oil Palm Frond Fibers by Using Extract of Oil Palm Empty Fruit Bunch Ash for Use in Natural Fiber Reinforced Composite," vol. 47, no. October, pp. 1–10, 2013.

[7] W. Fatra, H. Rouhillahi, Z. Helwani, Z. Zulfansyah, and J. Asmura, "Effect of Alkaline Treatment on the Properties of Oil Palm Empty Fruit Bunch Fiber-reinforced Polypropylene Composite," Int. J. Technol., vol. 7, no. 6, p. 1026, 2016.

[8] M. J. M. Ridzuan, M. S. Abdul Majid, M. Afendi, S. N. A. Kanafiah, and M. B. M. Nuriman, "Effects of alkaline concentrations on the tensile properties of Napier grass fibre," Appl. Mech. Mater., vol. 786, no. June, pp. 23–27, 2015.

[9] Li Xue, Tabil Lope G., Panigrahi Satyanarayan, "Chemical treatments of natural fibre for use in natural fibre reinforced composites: A review," Journal of Polymers and the Environment., vol. 15, February, pp. 25-33, 2017.

[10] Islam M.S., Pickering K.L., Foreman N.L., "Influence of alkali treatment on the interfacial and physic-mechanical properties of industrial hemp fibre reinforced polylactid acid composites," Composites Part A: Applied Science and Manufacturing., vol. 41, August, pp. 596-603, 2009.

[11] Islam M.S., Pickering K.L., Foreman N.L., "Influence of accelerated ageing on the physic-mechanical properties of alkali-treated industrial hemp fibre reinforced poly (lactic acid) (PLA) composites," Polymer Degradation and Stability., vol. 95, October, pp. 59-65, 2009.

[12] Ming Cai, Hitoshi Takagi, Antonio N. Nakagito, Mahasiro Katoh, Tomoyuki Ueki, Geoffrey I.N. Waterhouse, Yan Li, "Influence of alkali treatment on internal microstructure and tensile properties of abaca fibres," Industrial Crops and Products., vol. 65, November, pp. 27-35, 2015.

[13] Mohammad Asim, Mohammad Jawaid, Khalina Abdan, Mohamad Ridzwan Ishak, "Effect of alkali and silane treatments on mechanical and fibre matrix bond strength of kenaf and pineapple leaf fibre," Journal of Bionic Engineering., vol. 13, pp. 426-435, 2016.

[14] A. Orue, A. Jauregi, U. Unsuain, J. Labidi, A. Eceiza, A. Arbelaiz, "The effect of alkaline and silane treatments on mechanical properties and breakage of sisal fibers and poly (lactic acid)/sisal fiber composites," Composites Part A: Applied Science and Manufacturing., vol. 84, January, pp. 186-195, 2016.

Evaluation of the effectiveness of elastomeric mount using vibration power flow and transmissibility methods

M N Arib Rejab[1], S A Abdul Shukor[1], M R Mohd Sofian[1], J I Inayat-Hussain[2], A Nazirah[3], I Asyraf[3]

[1]School of Mechatronic Engineering, Universiti Malaysia Perlis, Pauh Putra Campus, 02600 Arau Perlis Malaysia.
[2]College of Graduate Studies, Universiti Tenaga Nasional.
[3]Malaysian Rubber Board (MRB), Engineering & Product Design Unit, Technology & Engineering Division.

mnarib@unimap.edu.my

Abstract. This paper presents the results of an experimental work to determine the dynamic stiffness and loss factor of elastomeric mounts. It also presents the results of theoretical analysis to determine the transmissibility and vibration power flow of these mounts, which are associated with their contribution to structure-borne noise. Four types of elastomeric mounts were considered, where three of them were made from green natural rubber material (SMR CV60, Ekoprena and Pureprena) and one made from petroleum based synthetic rubber (EPDM). In order to determine the dynamic stiffness and loss factor of these elastomeric mounts, dynamic tests were conducted using MTS 830 Elastomer Test System. Dynamic stiffness and loss factor of these mounts were measured for a range of frequency between 5 Hz and 150 Hz, and with a dynamic amplitude of 0.2 mm (p-p). The transmissibility and vibration power flow were determined based on a simple 2-Degree-of-Freedom model representing a vibration isolation system with a flexible receiver. This model reprsents the three main parts of a vehicle, which are the powertrain and engine mounting, the flexible structure and the floor of the vehicle. The results revealed that synthetic rubber (EPDM) was only effective at high frequency region. Natural rubber (Ekoprena), on the other hand, was found to be effective at both low and high frequency regions due to its low transmissibility at resonant frequency and its ability to damp the resonance. The estimated structure-borne noise emission showed that Ekoprena has a lower contribution to structure-borne noise as compared to the other types of elastomeric mounts.

1. Introduction

One of the research and development efforts in automotive area is for controlling noise and vibration problems in order to achieve improvements in ride comfort. This can be done by improving the isolation system of its powertrain. Engine and mounting systems play critical roles in noise, vibration and harshness (NVH) of the vehicle. The main causes of vibration are the engine excitation force generated by gas pressure of fuel explosion in the cylinder, and the inertia force of the piston and connecting rod. The vibration is transferred through the mounting system to the seat track structure causing discomfort to passengers.

Various types of isolators have been proposed to attenuate the unwanted vibration of powertrain that is transferred to the body structure. Elastomers have been used as engine mounts to reduce vibration from the powertrain to the structure since 1930s. Compactness, cost-effectiveness and low maintenance are the main advantages of these elastomeric mounts.

Currently, most of the elastomeric mounts used for this application are made from petroleum-based synthetic rubber (SR). This is mainly due to their better resistance to heat, oxidation and oil as compared to natural rubber (NR). The use of petroleum-based raw material not only depletes the earth's non-renewable natural resources but also causes environmental hazards. Elastomeric mounts, which are disposed everyday, usually ends up in the landfill. Elastomers buried in landfill sites release high toxic chemicals into the groundwater, and carcinogens are also released to the environment. The increasing awareness of environmental sustainability in recent years has motivated researchers to explore the use of environmental-friendly non-petroleum-based raw materials in the development of eco-friendly elastomeric mounts.

NR is considered as an alternative to SR due to its advantages as a way to conserve land, that also acts as a sink for CO_2 generated by automobiles [1-6]. From the perspective of energy consumption in the preparation of these elastomers, NR also has an advantage over SR, as shown in table 1.

Table 1. Energy consumption for preparation of various elastomers [2]

Material	Approximately Energy Comsumption (GJ/t)
Natural Rubber (NR)	15-16
Polybutadiene Rubber (BR)	108
Polypropylene (PP)	110
Styrene-Butadiene-Rubber (SBR)	130-156
Ethylene-Propylen-Dien-Modify (EPDM)	142-179
Butyl Rubber (IIR)	174-209
Chloroprene (CR)	120-144

NR is categorized as 'green' material because its production uses natural material or renewable resources and it is produced with minimal waste. The use of these eco-friendly and low-energy products exerts less stress on the environment and improves carbon life cycle.

Ekoprena and Pureprena are advanced natural rubber products resulting from research and development activities conducted by Malaysian Rubber Board (MRB) in order to improve the properties of natural rubber depending on a conventional grade used such as Standard Malaysia Rubber (SMR). These advanced natural rubber products have been tested on curing characteristics and physical mechanical properties. The properties of Ekoprena and Pureprena were found to be quite similar to the properties of synthetic rubber. Ekoprena has properties such as oil resistant, low water absorption, high damping, high wet grip, low rolling resistance and gas permeability, while Pureprena has properties such as low creep, stress relaxation and low compression set, low water absorption, good dynamic properties, low protein and low ash. Unfortunately, there is no data or information available on the vibration control characteristics of these advanced natural rubber. Therefore the main aim of this work is to investigate the potential and suitability of these advanced natural rubber mounts as an alternative to the more widely used synthetic rubber mount in controlling vibration and noise. This paper presents the comparison between natural rubber and synthetic rubber mounts' effectiveness in controlling vibration and noise using transmissibility and vibration power flow methods. Vibration power flow is used as a tool, in addition to vibration transmissibility, in quantifying the effectiveness of the elastomeric mount, and to identify the elastomeric mount that contributes the most to structure-borne noise. The natural rubber considered in this work is a conventional grade, SMR CV60, and the advanced natural rubber are Ekoprena and Pureprena. The syntetic rubber used in this work is EPDM.

IOP Conf. Series: Journal of Physics: Conf. Series **908** (2017) 012034 doi:10.1088/1742-6596/908/1/012034

2. Methodology

This section describes the experimental set-up and procedures for the dynamic test used to determine the dynamic stiffness and loss factor of the elastomeric mounts, as well as the analytical model to determine the characteristic of elastomeric mount associated with its contribution to structure-borne noise, namely the transmissibility and vibration power flow.

2.1. Dynamic Test

Dynamic test was conducted according to Japan Industry Standard (JIS) K6385-1977 [7] using the MTS 830 Elastomer Test System. The elastomeric mount was placed on a jig plate which was then fixed between two plates, figure 1. Preload considered in the dynamic test was 313.92 N. This preload which is about ¼ of the weight of the powertrain was also applied for the testing of the original elastomeric mounts. The frequency range was between 5 Hz and 150 Hz, and the dynamic amplitude applied in the radial direction was set to 0.2 mm (p-p). LVDT was used to measure the displacement and load cell was used to measure the force. Figure 2 shows the schematic drawing of the dynamic testing machine. The elastomeric mount was subjected to a single frequency sinusoidal signal at increment of 5 Hz. The time domain signals were analysed, and the dynamic stiffness and loss factor of these elastomeric mounts were derived from the measured force, displacement and the phase angle between the force and the displacement.

Figure 1. Elastomeric mount placed on a jig plate and fixed between two plates of the MTS 830 machine for static and dynamic tests.

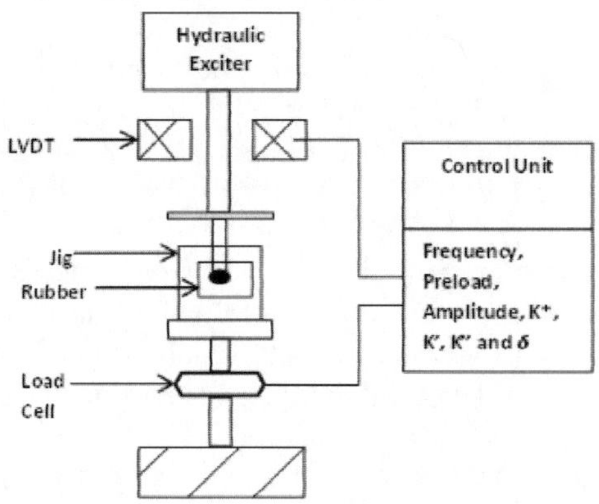

Figure 2. Schematic drawing of the dynamic testing machine.

2.2. Analytical Model to Evaluate the Effectiveness of Elastomeric Mount

In the analytical work, a simple 2 Degree-of-Freedom (DOF) vibration isolation system with a flexible receiver was modelled. The model is based on three main parts of a vehicle which are the powertrain and engine mounting, the flexible structure and the floor of the vehicle. The model is shown in Figure 3. M_1 represents the block mass, and K_1 represents the elastomeric mount's stiffness in complex form. M_2 represent the mass of the flexible structure, which consists of the mass of the beam and the mass of the plate that is attached to it. K_2 represents the beam stiffness. A harmonic force **F** of frequency ω acts upon M_1, in the line of motion (vertical direction).

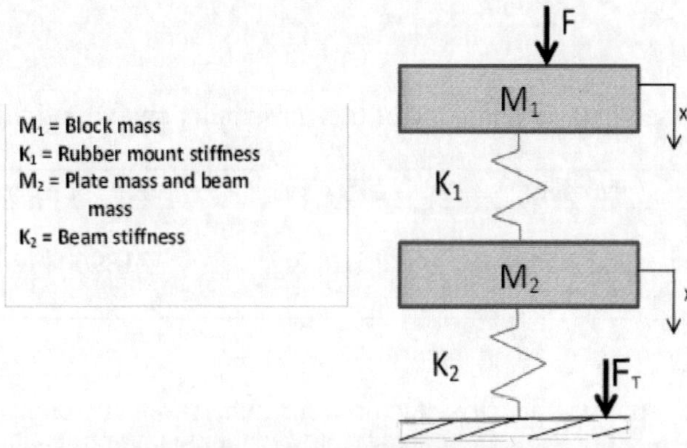

Figure 3. The 2- DOF model.

The transmissibility (*Tr*) for the elastomeric mounts was evaluated using measured values of dynamic stiffness and loss factor of elastomeric mount for a range of frequencies from 10 Hz to 150 Hz. The beam stiffness (K_2) has a constant value. The general equation of transmissibility (Equation 1) was derived from the 2-DOF model shown in figure 3. *Kd* is the dynamic stiffness of the elastomeric mount, δ is the loss factor of elastomeric mount, and *Kb* is the beam stiffness. *M* is the mass of the block and *Mb* is the mass of the beam and the plate.

$$Tr = \frac{F_T}{F} = \sqrt{\frac{(KdKb)^2 + (\delta KdKb)^2}{(KdKb - Mb\omega^2 Kd - M\omega^2 Kd - M\omega^2 Kb + M\omega^4 Mb)^2 + (\delta KdKb - \delta KdMb\omega^2)^2}} \quad (1)$$

The source acceleration (Equation 2) and receiver acceleration (Equation 3) were also derived from the 2-DOF model shown in Figure 3.

$$\tilde{a}_s = \omega^2 \left(\frac{F(Kd + Kb - Mb\omega^2 + j\delta Kd)}{KdKb - Mb\omega^2 Kd - M\omega^2 Kd - M\omega^2 Kb + M\omega^4 Mb + j\delta KdKb - j\delta KdMb\omega^2} \right) \quad (2)$$

$$\tilde{a}_r = \omega^2 \left(\frac{F(Kd + j\delta Kd)}{KdKb - Mb\omega^2 Kd - M\omega^2 Kd - M\omega^2 Kb + M\omega^4 Mb + j\delta KdKb - j\delta KdMb\omega^2} \right) \quad (3)$$

The vibration power flow through the elastomeric mount was evaluated using Equation 4 and Equation 5 [8-9], where M_{rs} is the apparent mass, which is obtained from the dynamic stiffness (K') and loss factor (η) of the elastomeric mount. The other components, a_s and a_r, represent the source and receiver acceleration, respectively. Both accelerations are in complex forms. The dynamic stiffness and loss factor from the measurement were required to determine vibration power flow through the elastomeric mount. The total vibration power flow was determined by the summation of the vibration

IOP Conf. Series: Journal of Physics: Conf. Series **908** (2017) 012034 doi:10.1088/1742-6596/908/1/012034

power flow at each frequency from 30 Hz to 150 Hz. The values used to calculate transmissibility and vibration power flow are given in Table 2.

$$P = \frac{1}{2\omega} Im(\widetilde{M}_{rs}\tilde{a}_s\tilde{a}_r) \tag{4}$$

$$\widetilde{M}_{rs} = \frac{K\prime}{\omega^2} + j\delta\frac{K\prime}{\omega^2} \tag{5}$$

Table 2. Values used in the calculation of transmissibility and vibration power flow.

F (N)	M (Kg)	Mb (Kg)	Kb (N/m)
2.5 (average)	10 Kg	15.26 Kg	2189604 (constant)

3. Results and Discussion

Measurements of dynamic stiffness and loss factor were undertaken for eight different elastomeric mounts. These elastomeric mounts were SMR CV60 with carbon black (CB) content of 45% and 20%, EPDM with CB content of 45% and 20%, Ekoprena with CB content of 45% and 20% and Pureprena with CB content of 45% and 20%. Figures 4(a) and 4(b) show the dynamic stiffness and loss factor of these elastomeric mounts for a range of frequency between 5 Hz and 150 Hz, and for dynamic amplitude of 0.2 mm (p-p) and preload of 313.92 N.

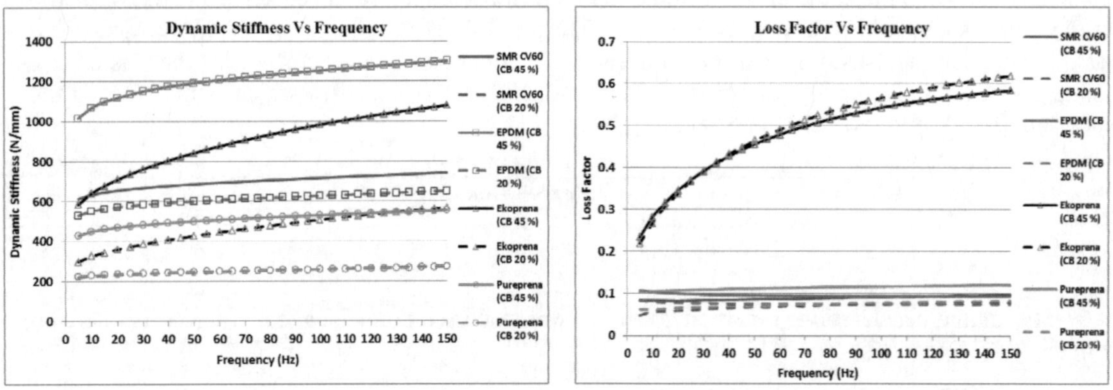

(a) (b)

Figure 4. (a) Dynamic stiffness of the elastomeric mount and (b) Loss factor of the elastomeric mount.

Figures 4 (a) shows that the dynamic stiffness increases with the increase in the frequency excitation from 5 Hz to 150 Hz for the natural rubber and synthetic rubber with 45% and 20% CB content. The dynamic stiffness also increases with the increase in CB content for both the natural rubber and synthetic rubber. Figure 4(a) shows that the synthetic rubber (EPDM) has the highest dynamic stiffness followed by natural rubber (SMR CV60, Ekoprena and Pureprena) for both cases of CB 45% and 20%. Ekoprena is however seen to have a relatively high dynamic stiffness although its hardness is close to SMR CV60 and Pureprena. Figure 4(b) shows that the loss factor increases with the increase in the frequency excitation from 5 Hz to 150 Hz for both natural rubber and synthetic rubber. The loss factor also increases with the increase in CB for both natural rubber and synthetic rubber. Figures 4(b) shows that Ekoprena has the highest loss factor followed by Pureprena, EPDM and SMR CV60, for both CB content of 45% and 20%.

Figure 5(a) represents transmissibility of elastomeric mounts with 45% carbon black (CB), while Figure 5(b) represents transmissibility of elastomeric mounts with 20% CB.

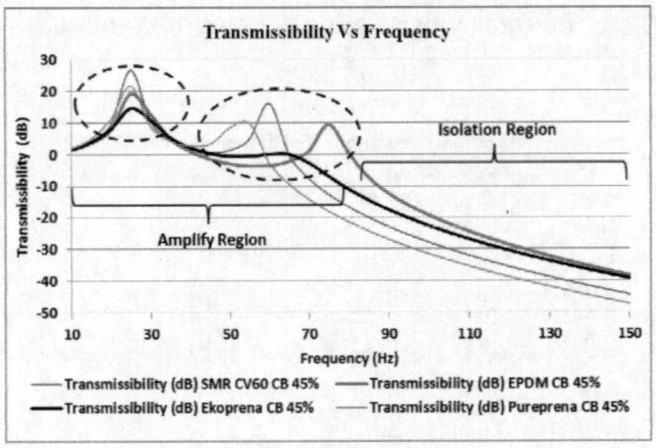

(a)

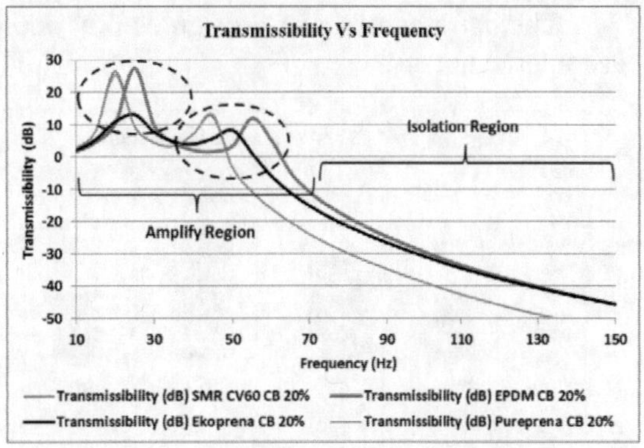

(b)

Figure 5. (a) Transmissibility of elastomeric mounts with 45% CB and (b) Transmissibility of elastomeric mounts with 20% CB.

Two regions can be observed in Figures 5(a) and 5(b), namely the amplification and isolation regions. There are two resonance peaks (dot circle) in amplification region which are primary and secondary resonances. Based on the observation of the natural frequencies of the system in Figure 5(a), the natural frequency is about 25 Hz for primary resonance and 55 Hz to 75 Hz for secondary resonance. The natural frequencies of the system in figure 5(b) is about 23 Hz for primary resonance and 45 Hz to 55 Hz for secondary resonance. These natural frequencies are based on the dynamic stiffness of the elastomeric mounts. The dynamic stiffness of elastomeric mounts are in turn dependent on the CB content. The dynamic stiffness increases with the increase in CB content for both natural rubber (NR) and synthetic rubber (SR). The synthetic rubber (EPDM) has the highest dynamic stiffness followed by natural rubber (SMR CV60, Ekoprena and Pureprena) for both cases of CB content of 45% and 20%. The characteristics of high dynamic stiffness produced natural frequency (primary and secondary) at relatively higher frequencies for both cases of CB coontent of 45% and 20%. However natural rubbers (SMR CV60 and Pureprena) showed that the secondary resonance occurred at a lower frequency compared to EPDM and Ekoprena for both cases of CB contnet of 45% and 20%. It was also observed that natural rubber, SMR CV60 and Pureprena, are soft materials that produce low dynamic stiffness and low loss factor even for the same CB content as the EPDM and Ekoprena. Natural rubber, SMR CV60 and Pureprena, also showed low transmissibility at higher frequency as compared to EPDM and Ekoprena for both cases of CB content of 45% and 20%. Pureprena has low loss factor which makes it suitable to be used to reduce vibration at high frequency.

It is observed from Figures 5(a) and 5(b) that the elastomeric mounts based on synthetic rubber (EPDM) and natural rubber (SMR CV60 and Pureprena) produced higher transmissibility at resonant frequency (dot circles), about 25 Hz for primary resonance and 55 Hz to 75 Hz for secondary resonance for CB content of 45% and about 23 Hz for primary resonance and 45 Hz to 55 Hz for secondary resonance for CB content of 20%. Ekoprena, on the other hand, produced lower transmissibility at the same resonant frequencies for both cases of CB 45% and 20%. Ekoprena appeared to have a higher ability to reduce or damped resonance compared to the other elastomeric mounts. This is attributed to the higher loss factor of Ekoprena. By having a high loss factor, the magnitude of the transmissibility at the amplification region is reduced. On the other hand, the higher loss factor, also caused the value of the transmissibility to increase in the isolation region. The high loss factor is beneficial in the amplification region as it reduces the transmissibility at the resonance frequency. The results show that the natural rubber from Ekoprena and Pureprena are suitable to be used as an alternative to synthetic rubber (EPDM) for vibration isolation application. This is because Ekoprena has characteristic of high loss factor that makes it suitable to be used at low frequency to reduce vibration amplitude at resonance, while Pureprena has characteristic of low loss factor that makes it suitable for vibration reduction at high frequency.

The bar graphs shown in figures 6(a) and 6(b) are the total vibration power flow through each elastomeric mounts.

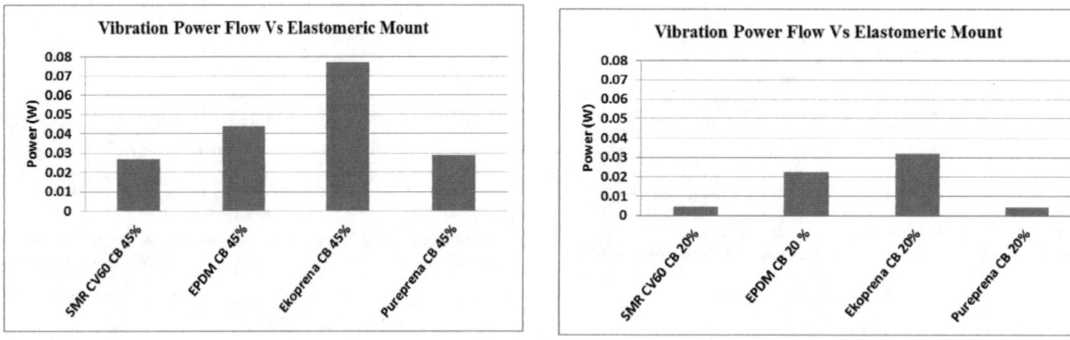

(a) (b)

Figure 6. (a) Total vibration power flow through elastomeric mounts with 45% CB and (b) Total vibration power flow through elastomeric mounts with 20% CB.

Figure 6(a) shows the level of total vibration power flow through each elastomeric mount based on synthetic rubber (EPDM) and natural rubber (SMR CV60, Ekoprena and Pureprena) for the case of CB content of 45%. The results showed that Ekoprena was the highest contributor to the transmission of energy to the structure, followed by EPDM. SMR CV60 and Pureprena showed a relatively lower total vibration power flow. The higher total vibration power flow for Ekoprena is due to its high dynamic stiffness and loss factor as compared to the other natural rubbers (SMR CV60 and Pureprena) and synthetic rubber (EPDM). Referring to the vibration power flow formula, $P = \frac{1}{2\omega} Im\left(\widetilde{M}_{rs}\tilde{a}_s\tilde{a}_r\right)$, the apparent mass, $\widetilde{M}_{rs} = (\frac{K\prime}{\omega^2} + j\delta\frac{K\prime}{\omega^2})$ is a combination of dynamic stiffness and loss factor. Therefore, the characteristics of elastomeric mount that have high dynamic stiffness and high loss factor will produce high vibration power flow, while elastomeric mount that have low dynamic stiffness and low loss factor will produce low vibration power flow. If the CB content of Ekoprena is reduced from 45% to 20%, the total vibration power flow is reduced from 0.077W to 0.032W; approximately 58% reduction of the energy transmission to the structure is reduced as shown in figure 6(b). This is because the dynamic stiffness and loss factor of Ekoprena is reduced with the decrease in the CB content.

4. Conclusions

The results for the transmissibility and vibration power flow, presented in this work, indicate that vibration power flow can be used as a tool to augment vibration transmissibility in assessing the effectiveness of rubber isolators in controlling vibration and noise. Ekoprena and Pureprena were found to be suitable for this application, as an alternative to the more widely used synthetic rubber (EPDM). This is because Ekoprena has high loss factor, while Pureprena has low loss factor. Ekoprena is therefore suitable to be used at low frequency to reduce amplitude at resonance while Pureprena is suitable to be used to reduce vibration at high frequency.

IOP Conf. Series: Journal of Physics: Conf. Series **908** (2017) 012034 doi:10.1088/1742-6596/908/1/012034

5. References

[1] Eleanor W T, Paul M D and David P E 2015 Increasing Demand for Natural Rubber Necessitates a Robust Sustainability Initiative to Mitigate Impacts on Tropical Biodiversity *Conservation Letters*. 8(4): 230–241.

[2] Mukhopadhyay R, Sengupta R and Gupta S D 2008 Recent Advances in Eco-Friendly Elastomer Technology In: Current Topic in Elastomer Research. United State. Taylor & Francis Group Publisher. 1021-1041.

[3] Zairossani M N 2010 Sustainability of the Natural Rubber Industry. *Malaysian Rubber Board: Rubber Technology Developments*. 10(2): 3-11.

[4] Cook, S 2010 Sustainability as a Global Issue. *Malaysian Rubber Board: Rubber Technology Developments*. 10(2): 12-15.

[5] Wei T, Lei L, Kang, H., Qiao B, Wang Z, Zhang L, Coates P, Hua K C and Kulig J 2012 Tough Bio-Based Elastomer Nanocomposites with High Performance for Engineering Applications. Advanced Engineering Materials. 14, 112-118.

[6] Swetha D V and Rani K D 2014 Effect of Natural Rubber on the Properties of Bitumen and Bituminious Mixes. *International Journal of Civil Engineering and Technology*. 5(10) 9-21.

[7] Japanese Standard Association. *Testing Methods for Rubber Vibration Isolators*. Japan, JIS K6385. 1977.

[8] Lee S K 2004 Identification of a Vibration Transmission Path in a Vehicle by Measuring Vibrational Power Flow. *Proceedings of the IMECHE Part D. Journal of Automobile Engineering* 218 (2), 167-175.

[9] Lee S K 2000 Application of Vibration Power Flow to a Passenger Car for Reduction of Interior Noise. *Shock and Vibration*. 7: 277-285.

Finite element modelling of Plantar Fascia response during running on different surface types

A H A Razak, K S Basaruddin, A F Salleh, W M R Rusli, M S M Hashim and R Daud

School of Mechatronic Engineering, Universiti Malaysia Perlis, Pauh Putra Campus, 02600 Arau, Perlis, Malaysia.

khsalleh@unimap.edu.my

Abstract. Plantar fascia is a ligament found in human foot structure located beneath the skin of human foot that functioning to stabilize longitudinal arch of human foot during standing and normal gait. To perform direct experiment on plantar fascia seems very difficult since the structure located underneath the soft tissue. The aim of this study is to develop a finite element (FE) model of foot with plantar fascia and investigate the effect of the surface hardness on biomechanical response of plantar fascia during running. The plantar fascia model was developed using Solidworks 2015 according to the bone structure of foot model that was obtained from Turbosquid database. Boundary conditions were set out based on the data obtained from experiment of ground reaction force response during running on different surface hardness. The finite element analysis was performed using Ansys 14. The results found that the peak of stress and strain distribution were occur on the insertion of plantar fascia to bone especially on calcaneal area. Plantar fascia became stiffer with increment of Young's modulus value and was able to resist more loads. Strain of plantar fascia was decreased when Young's modulus increased with the same amount of loading.

1. Introduction

Plantar fascia is one of the major structures in human foot that acting as a major ligaments connected between foot bones where located underneath the foot bones which it is function as the major stabilizing structures of the longitudinal arch of human foot, sustains a high tensions during weight-bearing[1], [2]. Plantar fascia plays a significant role in the movement of human either in normal walk or running. The injury namely plantar fasciitis occurs when large amount of tension or forces acting toward it [3]. Several studies have been conducted in investigating the mechanical strain of plantar fascia through experiment [4]–[6] and finite element simulation [7]–[9]. Welk et al [10] conducted a study on changes in plantar fascia thickness from tissue creep in runners and walkers by using high resolution ultrasound. The study was conducted in two groups where subject from group 1 need to walk on treadmill for 10 minutes while subject from group 2 need to run on treadmill for 30 minutes. The subjects were imaged by using high resolution ultra sound with 12-MHz linear transducer probe immediately before and after session on treadmill. The changed in mean of plantar fascia thickness where it is reduced about 0.03 mm ± from its original thickness for group 1. The result for group 2 shows that the changed of 0.06 mm ± of the plantar fascia thickness. Moreover, Clark et al [11] investigated about the mechanical strain in the intact of plantar fascia. The frozen intact limb cadaveric was used as experiment's subject. The frozen intact limb was thawed before begin the experiment.

Three-way rosette gauges were used to measure the strain in plantar fascia. The foot of intact limb was dissected to attach six strain gauge. This study resulted the mean of oscillation frequency of the limbs was similar to the data for live subject. It is shows that the plasticity of plantar fascia was similar between live subject and cadaver intact limb.

On the other hand, Cheung et al [3, 4] developed the 3D geometrical of foot bone based on reconstruction of MR images obtained from CT scan. The model comprises of 28 bony segments. A total of 72 ligaments and plantar fascia were included and defined as the attachment point between the bones. Similarly, Cheng et al [7] developed the 3D model of foot bone from images. The DICOM image from CT scan was taken and then reconstructed by using image processing software AMIRA 3.1.1 (mercury Computer System, Germany). The profile of the bones were edited to make it smooth in term of shape and gathered by threshold adjustment. The 3D bones model then imported to the ANSYS software for assembly and further analysis. Cheung et al [3,4] set the mesh for foot bones 3D solid model and soft tissue with total mesh of 50 775 tetrahedral elements (included bony structures, cartilage and soft tissue) and 103 tension-only truss element for ligamentous structure (included plantar fascia). All tissue were set as linearly elastic, homogeneous and isotropic. The meshed for the 3D solid foot bone model with total of 50 482 elements and 72 930 nodes was generated. Cheung et al [12] reported that Achilles tendon forces with 75% of the total weight on one foot was the closest match of the measured centre of pressure during balanced standing of the subject. An increment in tension of the plantar fascia was obtained due to the weight of the foot and Achilles tendon loading. In the previous study by Cheung et al [13], there was a significant increment in the long and short plantar and spring ligament when the value of Young's modulus decreased. The maximum stress was found concentrated near to the medial calcaneal tubercle. Three dorsiflexion angle are selected in the study which is 15, 30, and 45 degree. The stresses of plantar fascia was increased when the dorsiflexion angle is increasing. Hence, kinematics movement of foot was found contributed the biomechanical response of plantar fascia. Since running on different surface could affect the kinematics of foot [14], this study was undertaken to investigate the effect of surface types on plantar fascia response during running.

2. Materials and Method

2.1. Geometrical Modelling

The foot bone model was obtained from Turbosquid database. The foot model was then imported to Solidworks 2015. The foot bones contains separated 26 bony structures. Between each bone structure, there is a small gap that was filled up to connect all the bone into one single body. The effect of articular cartilages in this study was neglected. The plantar fascia of 1.5 mm thickness was constructed to connect medial calcaneal tubercle to metatarsal head. Five stripes of plantar fascia were connected to metatarsal head as shown in Fig. 1. Each stripe was attached to each metatarsal bone and four connecting bodies were constructed between the stripes. The stripes were then joined to become a single stripe in the middle of foot before connected to calcaneus. Figure 2 shows the ground support with dimension of 150 mm × 300 mm × 20mm that was constructed to act as a base to the foot bone model where it is attached at the top surface of the ground support body.

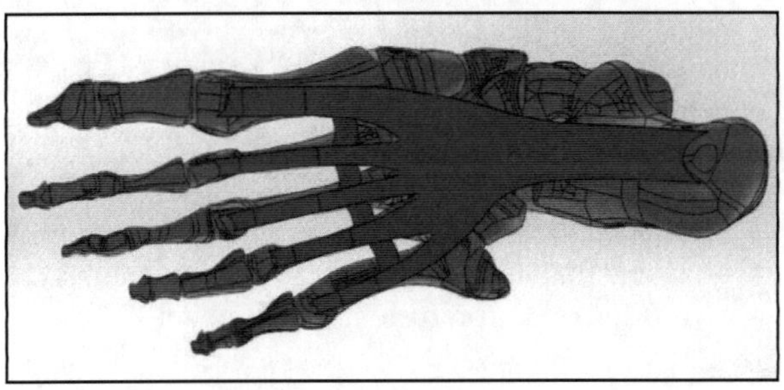

Figure 1. CAD model of plantar fascia with foot bone.

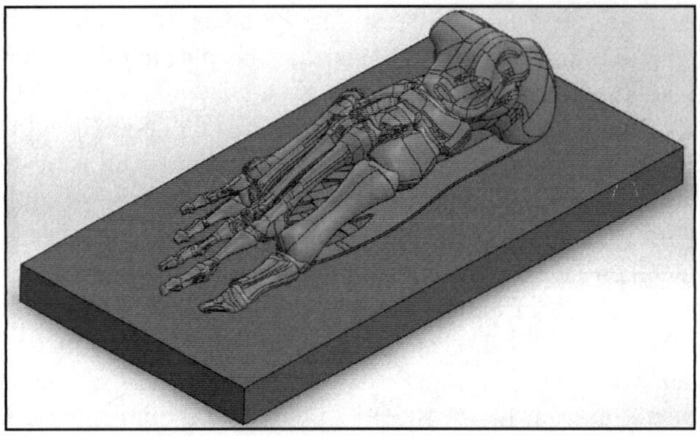

Figure 2. Foot bone model with ground support plate.

2.2. Finite Element Modelling

All materials were considered as isotropic and linearly elastic. The mechanical properties of bone, plantar fascia and ground support plate are listed in Table 2. Three Young's modulus were set on plantar fascia in order to analyse the effect of Young's modulus variation on plantar fascia response. The contacts between bones, plantar fascia and ground support were defined as ‘bonded type’ for the present finite element analysis (FEA). Figure 3 shows a complete meshed model. The element size of 5 mm tetrahedral type was used for foot bones and plantar fascia, whereas 6 mm of 3D-brick element was used for ground support plate. The optimum element size of 5 mm for tetrahedral element were obtained from the convergence test.

Table 1. Material properties of foot bone, plantar fascia and ground support model.

Component	Young's modulus (MPa)	Poisson’s Ratio
Foot bone	7300	0.3
Plantar fascia	350, 500 & 700	0.4
Ground support plate	210 000	0.4

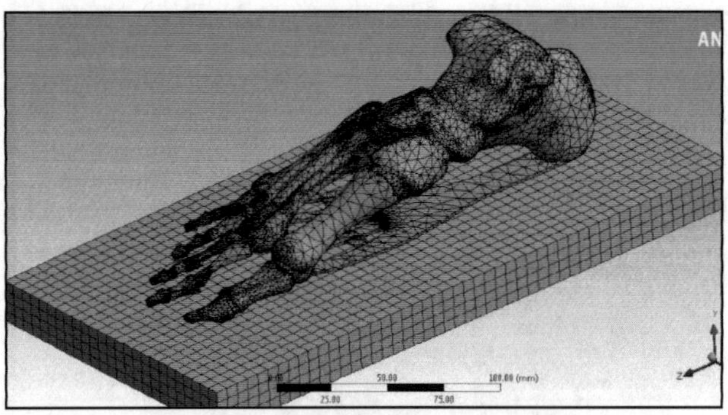

Figure 3. Complete meshed model.

Boundary conditions were set out to represents the foot condition during running by applying the peak ground reaction force (GRF) vertically from the bottom of the ground support plate. The peak

GRF for running on three different surfaces; concrete, rubber and artificial grass, was obtained from the previous experimental study [15]. The top surface of the foot was fixed. The peak GRF load was applied with 1373 N for concrete surface, 1490N for artificial grass surface and 1620N for rubber surface.

3. Results and Discussion

Figure 4 shows the result of equivalent (von-Mises) stress for plantar fascia under running on three different surfaces with Young's modulus of 350 MPa. The highest value for maximum equivalent stress is 0.92285 MPa which occurred during running on rubber surface. Whereas running on concrete and artificial grass have the maximum equivalent stress of 15.2% and 8% less than the value from running on rubber tile respectively. Figure 5 shows the result of equivalent elastic strain under the same boundary conditions. Similarly, running on rubber surface contributes to the highest value for maximum equivalent elastic strain which is 0.0026772 mm/mm.

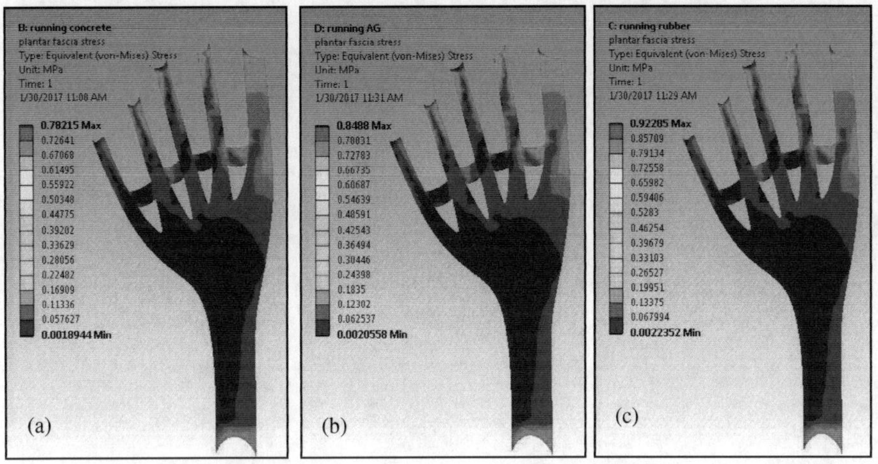

Figure 4. Equivalent (von-Mises) stress for plantar fascia when running on three different surfaces with Young's modulus of 350 MPa: (a) concrete, (b) artificial grass, (c) rubber.

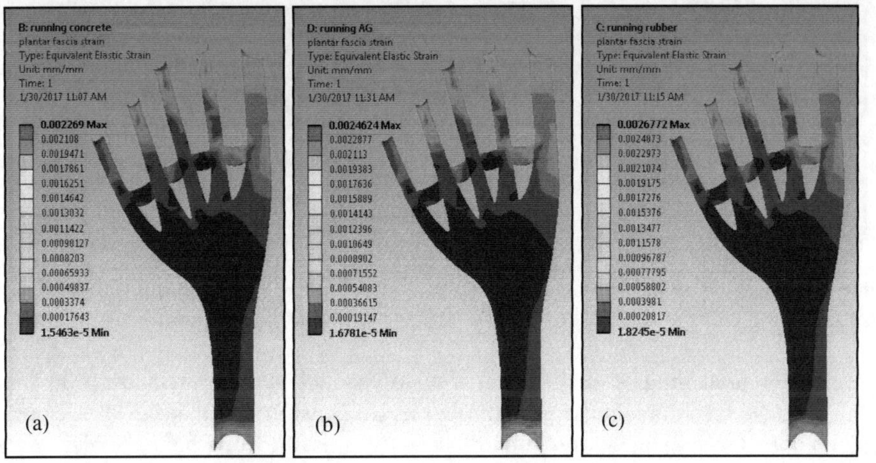

Figure 5. Equivalent (von-Mises) stress for plantar fascia when running on three different surfaces with Young's modulus of 350 MPa: (a) concrete, (b) artificial grass, (c) rubber.

Moreover, the stress and strain contour of plantar fascia with Young's modulus of 500 MPa and 700 MPa was found similar to the contour obtained when Young's modulus is 350 MPa as shown in Figs. 6 to 9. However, the increase of Young's modulus of plantar fascia reduced the equivalent elastic strain and increased von-misses stress as shown in Fig. 10 (a) and (b), respectively. Obviously, the plantar fascia became stiffer with the increase of Young's modulus and was able to resist load and prevent large deformation.

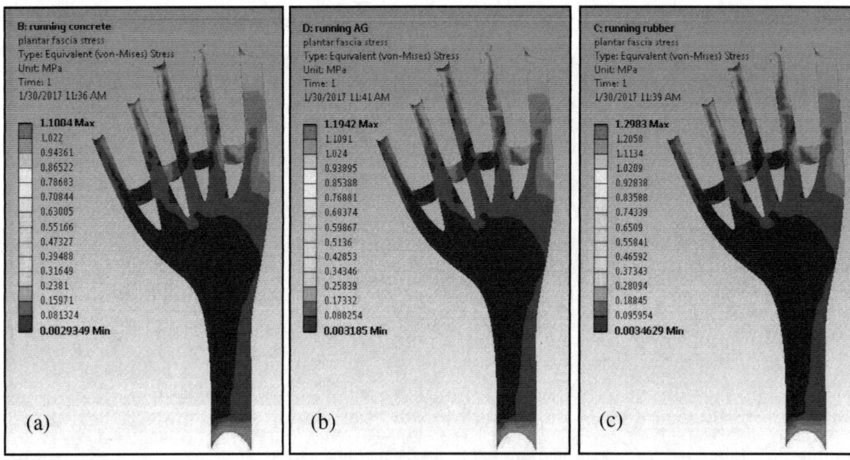

Figure 6. Equivalent (von-Mises) stress for plantar fascia when running on three different surfaces with Young's modulus of 500 MPa: (a) concrete, (b) artificial grass, (c) rubber.

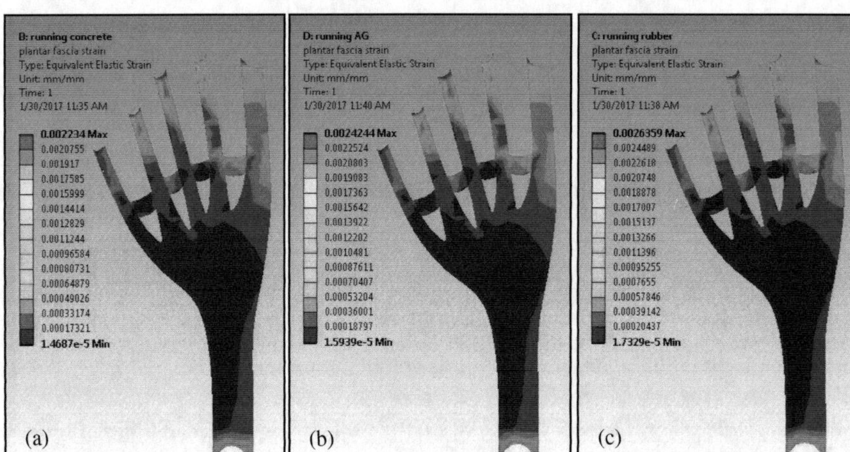

Figure 7. Equivalent (von-Mises) stress for plantar fascia when running on three different surfaces with Young's modulus of 500 MPa: (a) concrete, (b) artificial grass, (c) rubber.

Finite element modelling is very useful tool to enhance the understanding on biomechanics response. Based on the result obtained in the present analysis, we found that the stress and strain were concentrated on the insertion area of plantar fascia towards the foot bones which is near to the calcaneus and metatarsal head regardless of the value load applied to the model. This result in correlation to the common injury of plantar fascia namely, plantar fasciitis which is always occurred on the heel area around the insertion area of plantar fascia to calcaneus. The contour of these result were same because the boundary conditions applied for all three model are same. The only difference for these models are Young's modulus values. It contributes to the difference in minimum and maximum value of equivalent stress for all models. The stress and strain were concentrated on the insertion area of plantar fascia which is on calcaneal and metatarsal area. The stress and strain concentration were very low in the medial area of plantar fascia.

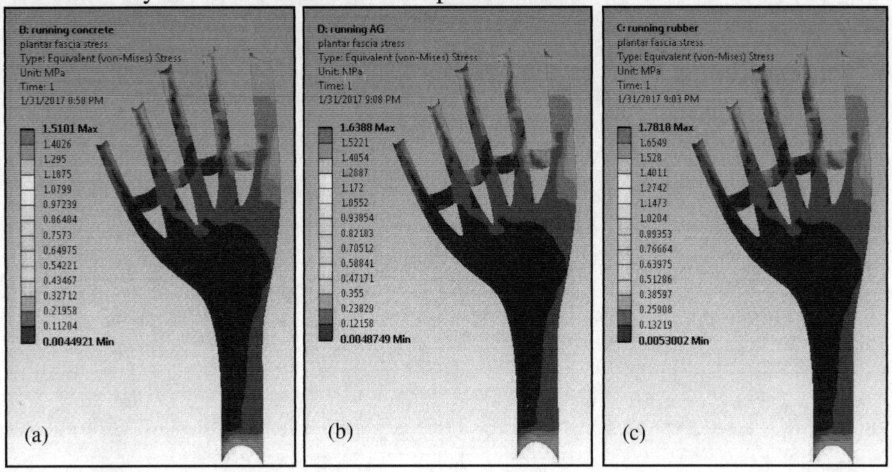

Figure 8. Equivalent (von-Mises) stress for plantar fascia when running on three different surfaces with Young's modulus of 700 MPa: (a) concrete, (b) artificial grass, (c) rubber.

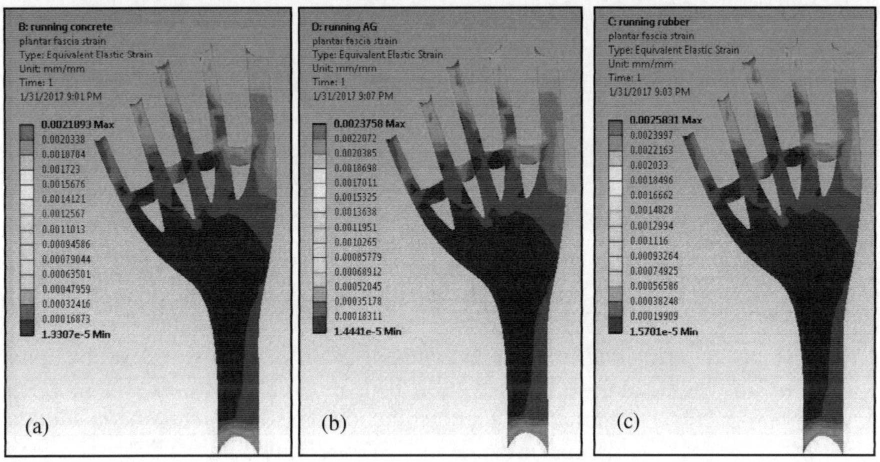

Figure 9. Equivalent (von-Mises) stress for plantar fascia when running on three different surfaces with Young's modulus of 700 MPa: (a) concrete, (b) artificial grass, (c) rubber.

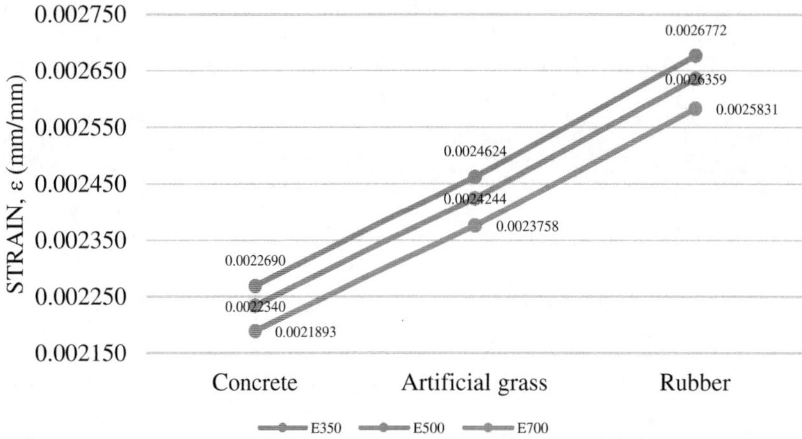

(a)

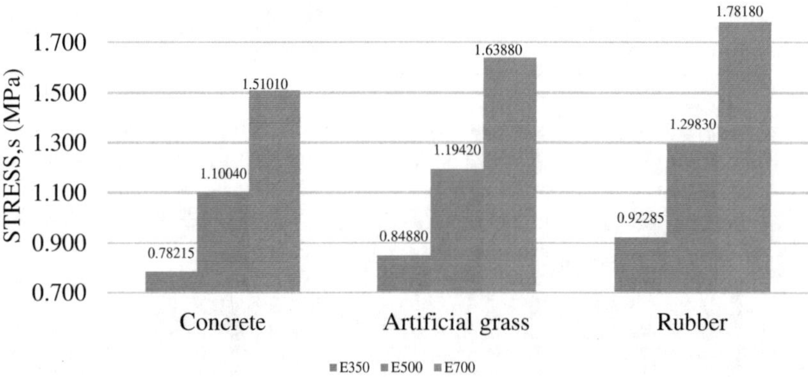

(b)

Figure 10. Effect of surface types on plantar fascia response with Young's modulus variation. (a) Equivalent elastic strain. (b) Von-misses stress.

Since there is no experimental study conducted to validate the result of the present finite element study, the result was compared to other published finite element studies[16]. The equivalent stress obtained in this study was found in agreement with other published results when Young's modulus of 350 MPa was used on plantar fascia. Hence, Young's modulus of 350 MPa predicted well on the stress response of plantar fascia. There are some limitations associated to the present study. All materials of the analyzed model were assigned to linearly elastic. The soft tissue and other connective tissue such as other ligaments and tendons other than plantar fascia were not included. However, these limitations did not compromise the general findings of the present study. The predicted stress and strain response of the plantar fascia was found similar to other published results.

4. Conclusion

The plantar fascia was observed to have high concentrated stress and strain at the insertion area that connected to calcaneus and metatarsal. The highest stress and lowest strain response of plantar fascia was found when running on rubber surface. The increase of plantar fascia's modulus contributed to the increase of equivalent stress and reduce the elastic strain. The present study could provide additional insight on study of plantar fascia response due to forces reacted during running with different surfaces that could lead to plantar fasciitis.

5. Acknowledgments
This work was supported by Ministry of Higher Education, Malaysia under Fundamental Research Grant Scheme (FRGS) No. 9003-00489.

6. References
[1] L. A. Bolgla and T. R. Malone, "Plantar fasciitis and the windlass mechanism: a biomechanical link to clinical practice," J. Ahletic Train., vol. 39, no. 1, pp. 77–82, Jan. 2004.

[2] S. J. Bartold, "The plantar fascia as a source of pain—biomechanics, presentation and treatment," J. Bodyw. Mov. Ther., vol. 8, no. 3, pp. 214–226, 2004.

[3] A. Gefen, "Stress analysis of the standing foot following surgical plantar fascia release," J. Biomech., vol. 35, no. 5, pp. 629–637, 2002.

[4] M. A. Mohseni-Bandpei, M. Nakhaee, M. E. Mousavi, A. Shakourirad, M. R. Safari, and R. Vahab Kashani, "Application of ultrasound in the assessment of plantar fascia in patients with plantar fasciitis: a systematic review," Ultrasound Med. Biol., vol. 40, no. 8, pp. 1737– 1754, 2014.

[5] N. Karabay, T. Toros, and C. Hurel, "Ultrasonographic Evaluation in Plantar Fasciitis," J. Foot Ankle Surg., vol. 46, no. 6, pp. 442–446, 2007.

[6] Y. Gu and Z. Li, "Mechanical Information of Plantar Fascia during Normal Gait," Phys. Procedia, vol. 33, pp. 63–66, 2012.

[7] H.-Y. K. Cheng, C.-L. Lin, H.-W. Wang, and S.-W. Chou, "Finite element analysis of plantar fascia under stretch-the relative contribution of windlass mechanism and Achilles tendon force.," J. Biomech., vol. 41, no. 9, pp. 1937–44, Jan. 2008.

[8] J. Yu, J. T. M. Cheung, Y. Fan, Y. Zhang, A. K. L. Leung, and M. Zhang, "Development of a finite element model of female foot for high-heeled shoe design," Clin. Biomech., vol. 23, no. Supp.1, pp. 31–38, 2008.

[9] Y. Wang, Z. Li, and M. Zhang, "Biomechanical study of tarsometatarsal joint fusion using finite element analysis.," Med. Eng. Phys., vol. 36, no. 11, pp. 1394–1400, 2014.

[10] A. B. Welk, D. W. Haun, T. B. Clark, and N. W. Kettner, "Use of High-Resolution Ultrasound to Measure Changes in Plantar Fascia Thickness Resulting From Tissue Creep in Runners and Walkers," J. Manipulative Physiol. Ther., vol. 38, no. 1, pp. 81–85, 2015.

[11] R. A. Clark, A. Franklyn-Miller, E. Falvey, A. L. Bryant, S. Bartold, and P. McCrory, "Assessment of mechanical strain in the intact plantar fascia," Foot, vol. 19, no. 3, pp. 161–164, 2009.

[12] J. T. M. Cheung, M. Zhang, and K. N. An, "Effect of Achilles tendon loading on plantar fascia tension in the standing foot," Clin. Biomech., vol. 21, no. 2, pp. 194–203, 2006.

[13] J. T.-M. Cheung, M. Zhang, and K.-N. An, "Effects of plantar fascia stiffness on the biomechanical responses of the ankle–foot complex," Clin. Biomech., vol. 19, no. 8, pp. 839–846, 2004.

[14] N. A. A. A. Yamin, K. S. Basaruddin, W. M. R. Rusli, A. F. Salleh, N. A. Razak, and W. Z. A. W. Muhamad, "Effect of Surface Hardness on Three-Segment Foot Kinematics during Barefoot Running," Int. J. Mech. Mechatronics Eng., vol. 16, no. 6, pp. 18–26, 2016.

[15] N. A. A. A. Yamin, M. N. A. Amran, K. S. Basaruddin, A. F. Salleh, and W. M. R. Rusli, "Ground Reaction Force Response during Running on Different Surface Hardness," ARPN J. Eng. Appl. Sci., vol. 12, no. 7, pp. 2313–2318, 2017.

[16] T.-X. Qiu, E.-C. Teo, Y.-B. Yan, and W. Lei, "Finite element modeling of a 3D coupled foot–boot model," Med. Eng. Phys., vol. 33, no. 10, pp. 1228–1233, 2011.

A study on contraction of pneumatic artificial muscle (PAM) for load-lifting

W S W A Najmuddin[1], M T Mustaffa[2]

School of Manufacturing Engineering, Universiti Malaysia Perlis, Kampus Pauh Putra, 02600 Arau, Perlis, Malaysia.

wahmad@studentmail.unimap.edu.my[1], mohammadtaufiq@unimap.edu.my[2]

Abstract. Pneumatic Artificial Muscles (PAMs) have been known for its wide application in various aspects of industrial automation and robotic equipments. Many advantages in terms of high power-to-volume ratio, high power-to-weight ratio, stick-slip-free operation and high degree of safety offer by PAM compare to traditional actuators. However, behind this benefits lie a limitation of significant compatibility of PAM mechanism which have to be considered so as to fully understand how the PAM works during load-lifting. In this study, the mesh suitability experiment and the effect of force load on PAM contraction experiment have been carried out. PAM is constructed and compatibility of bladder and the braided mesh to produce uniform expansion is investigated. Moreover, the first experimental result of finding compatibility is used to verify the contraction value under various loads.

1. Introduction

Robots become significant to serve and assist human, especially for older people and disabled people. However, high rigidity of robot joints will be hazardous to a human because a human has soft arms and compliant joints activated by numerous muscles. Necessity for human-like actuators system, many researchers have motivated to study in this field and created the Pneumatic Artificial Muscle (PAM). The PAM is used for converting pneumatic power to pulling force and eventually causing movements of the mechanism. These are the systems that have performance characteristic alike to and stronger than that of human muscle. Besides PAM, it has another name which are Fluidic Muscle [1], Pneumatic Muscle Actuator [2], Air Muscle [3], Axially Contractible Actuator [4], Tension Actuator [5-6], Fluid Actuator [7] and Fluid-Driven Tension Actuator [8].

Muscles of human involve biceps and triceps which help to move the human hand, are roughly classified as flexor and extensor muscles. Biceps included in the flexor group and it will bend the arm by decreasing the angle between the forearm and upper arm whereas, triceps muscle included in the extensor muscle. In flexion condition, biceps muscle will contract while triceps muscle will relax. While in extension condition, the biceps will relax and triceps contracted. The length of muscle shorter and the contracting force becomes smaller [9]. The PAM uses a layer that demonstrations a muscle and pneumatics as its source of activation force. The membrane ought to have the capacity to either radially stretch or alter its radial section.

The PAM has a long history with few accomplishments in different applications. In the 1950s, the rubber-tube type of PAM was initially developed by the physician, Joseph L. McKibben for an artificial limb of the handicapped [10]. McKibben muscle basically made from an inner rubber tube or called bladder and a non-extendable double- helix woven sheath or called braided mesh shell. The PAM was upgraded in the 1980s and turn out to be all the more capable. It had been presented by Bridgestone Company and was utilized for painting applications as a part of the industry and a few applications to help debilitated people and administration mechanical autonomy [11].

The PAM used to various types of humanoid applications and propelled robots. The PAM is basically utilized as an automated actuator where consistence and low power to weight proportion are essential, e.g. strolling or running machines or even humanoid robots. There are few companies manufactured PAMs for robotic and modern applications such as Case in point, Shadow Group and FESTO [3,10-11]. PAM can be classified into four types which are embedded muscle, netted muscle, pleated muscle and braided muscle [11]. PAM has numerous natural focal points for human-machine interface. For example, PAM is low weight, high force to weight proportion, working principle same as human, safe energy and can change the stiffness of PAM.

It is important to understand the correlation of main components includes braided mesh and bladder PAM's design. In this study, the mesh suitability and the effect of force load on PAM contraction experiment have been carried out. PAM is constructed and compatibility of bladder and the braided mesh to produce uniform expansion is investigated. The first experimental result of finding compatibility is used and the performance of PAM is accessed through testings using various loads and the contraction value is verified.

2. Experimental

2.1 Materials

The PAM consists of bladder and braided mesh. Material for the bladder is rubber while for braided mesh is polyethylene. The PAM experiment setup is illustrated in figure 1a where one side is the air inlet and the other side is the closed end and connected with the load. Figure 1b shows the equipment involved and assembled for PAM's load lifting test. The material and experiment apparatus includes bladder or inner tube, braided mesh shell or expandable braided cable sleeve, pressure regulator, connector and loads.

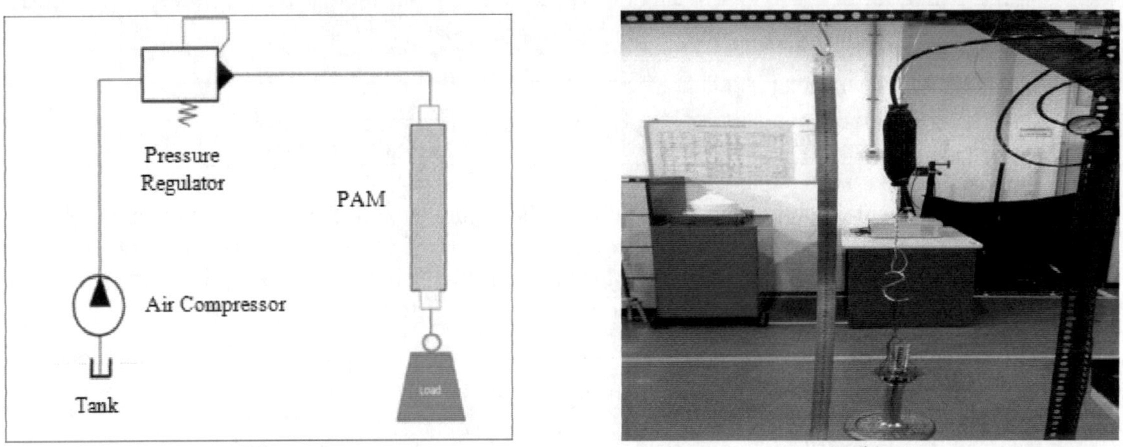

(a) (b)

Figure 1. Pneumatic diagram (a) and assembled PAM's load lifting test (b)

IOP Conf. Series: Journal of Physics: Conf. Series **908** (2017) 012036 doi:10.1088/1742-6596/908/1/012036

2.2 Mesh Suitability

The purpose of this test is to prove the significance of choosing the suitable braided mesh and how using the optimum size braided mesh can lead to maximum contraction. Contraction percentage can be achieved by using formula as shown in Eq. 1.

$$Contraction = \frac{Displacement}{Initial\ Length} \times 100\% \quad (1)$$

The parameter used for this experiment is specified in table 1. Three samples are used with each different bladder diameter and constant length. The experiment is set up according to the figure 1. For the first sample includes 2.5cm braided mesh with 3kg of selected load is attached to PAM.

Table 1. Experimental parameters.

No. of Sample	1	2	3
Bladder Diameter (cm)	2.0	2.5	3.0
Bladder Length (cm)		17	
Braided Mesh Diameter (cm)		2.5, 3.0, 3.5	
Load (kg)		3	

Next, PAM is pressurized manually and the displacement result of the PAM is recorded. The experiment is then repeated using different diameter of braided mesh and samples accordingly.

2.3 Effect of Force Load on PAM

The primary objective of this experiment is to study how the contraction percentages of the bladder changes as the different amount of loads are applied to the bladder. Contraction percentages can be calculated by using Eq. 1. The parameter used for this experiment is specified in Table 2.

Table 2. Experimental parameters.

No. of Sample	1	2	3
Bladder Length (cm)	12	17	22
Bladder Diameter (cm)		2.0	
Braided Mesh Diameter (cm)		3.5	
Load (kg)		3, 4, 5, 6, 7, 8, 9, 10	

The experiment is conducted according to the figure 1. For the first sample includes 12cm of bladder length with fix bladder and braided mesh diameter. Minimum of the 3kg load is used until maximum of 10kg loads is reached. The PAM is pressurized manually and the displacement of the PAM is recorded. Each data taken includes different loads with three different samples accordingly.

3. Results and Discussion

3.1 Mesh Suitability Experiment

Figure 2a, figure 2b and figure 2c show contraction percentage versus pressure of the 2 cm, 2.5 cm and 3 cm diameter of bladder respectively. Each diameter using three different diameters of braided mesh. The compressed air pressure of 1, 2, 3 and 4 bar have been tested. The experiment was performed by using load 3 kg. It can be seen that if air pressure increase, the contraction will also increase.

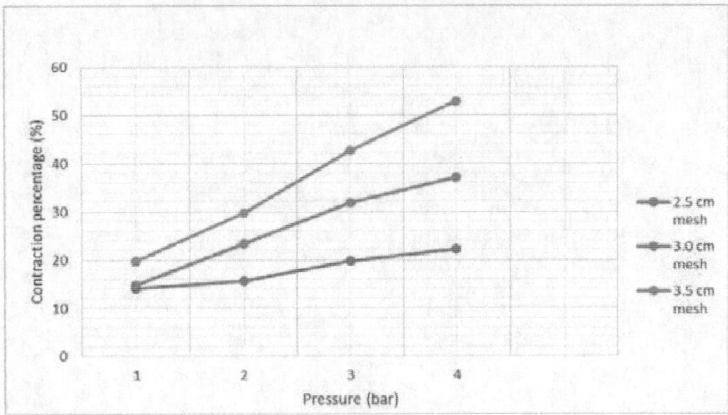

(a)

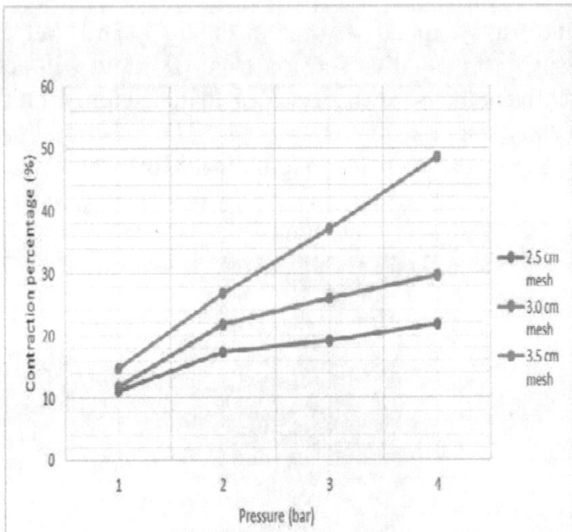

(b)

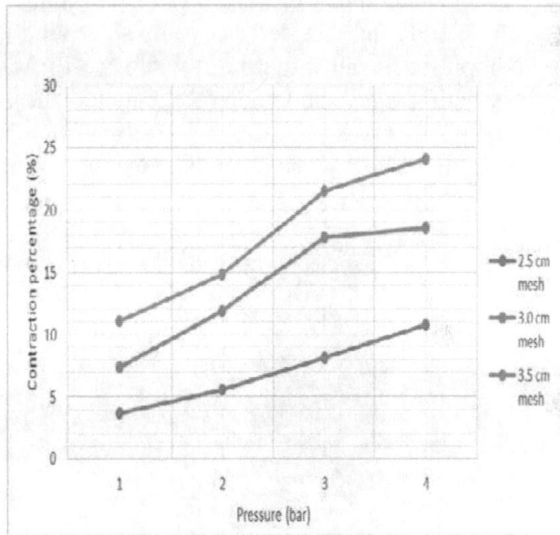

(c)

Figure 2. Contraction percentage versus pressure of diameter bladder (a) 2 cm (b) 2.5 cm (c) 3 cm

Figure 3 shows the contraction percentage versus pressure of 3.5 cm braided mesh using three different diameters of bladder. Value of 2.0 cm bladder (green line) gives the best result in contraction percentage which is 51.85%. It shows, the mesh gives fully cooperating with the expansion of the bladder. The mesh can be categorized as the mesh that has optimum size when it can contract half from its original length.

IOP Conf. Series: Journal of Physics: Conf. Series **908** (2017) 012036 doi:10.1088/1742-6596/908/1/012036

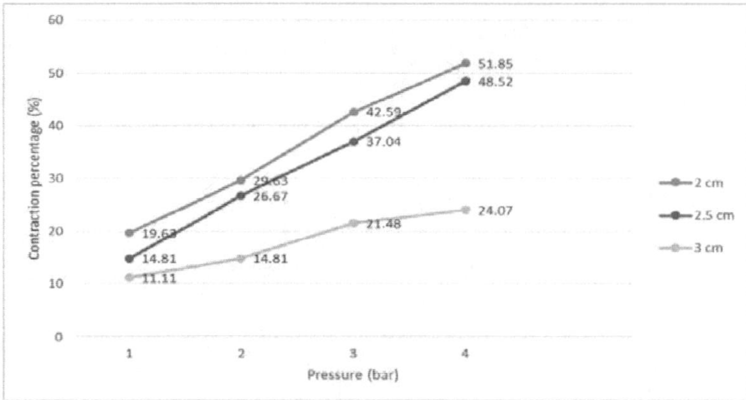

Figure 3. Contraction percentage versus pressure for 3.5 cm braided mesh

3.1.1 Compatibility braided mesh with bladder.

The bladder has braided sleeve mesh around it and uniformly expands as seen in Figure 4. The bladder is compressed against the mesh when the PAM is pressurized manually. Tension from the braid will act as restricting agent and therefore balance the pressure from the bladder. Besides that, tension from the material is mingled at the end points of braided mesh to balance external load [11]. The mesh is suitable with the bladder when it can balance the constrictive force of mesh and expanding force from the bladder.

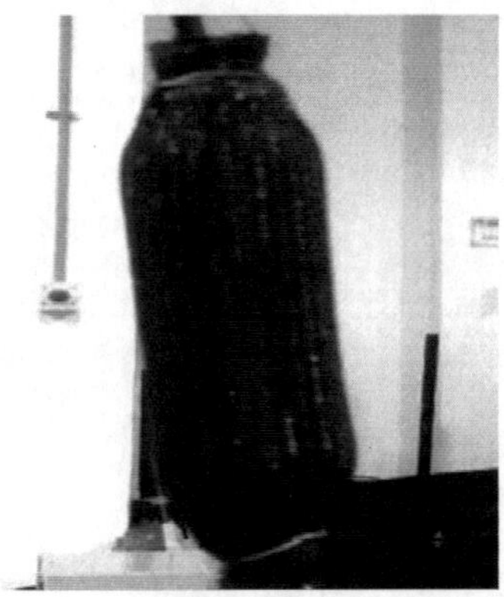

Figure 4. Uniform expansion of PAM

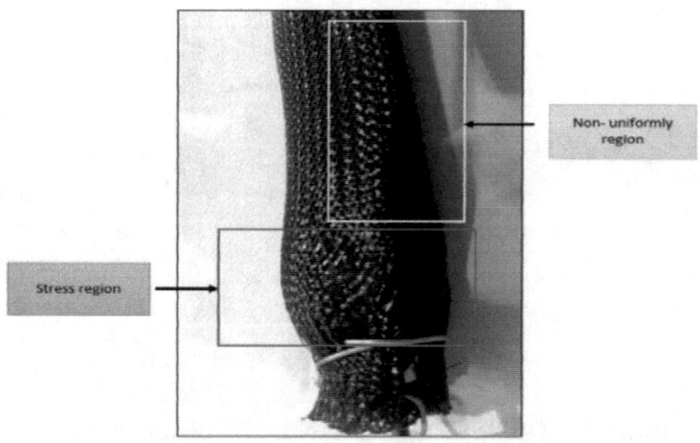

Figure 5. Non-uniform expansion of PAM

3.1.2 Incompatibility braided mesh with bladder.

Figure 5 shows the stress regions where the non-uniformly of bladder expanded. Other diameter bladder is incompatible with the braided mesh due to the mesh that unable to support and balance when the air pressure installed. When bladder fails to expand uniformly, the energy is loss and contraction percentage is reduced. The condition of PAM at the stress concentration region in red line and non- uniform expansion region of the bladder in yellow line with low concentration percentage is seen in Figure 5. From observation, the red region will be torn and the yellow region indicates the flat surface of the tube due to non-uniform expansion. The reading of displacement and contraction percentage can be taken only from 1 bar to 4 bar due to the capability of the rubber tube.

3.2. Effect of Load on PAM Contraction

The contraction percentage versus load for 2 cm diameter bladder using 3.5 cm braided mesh is illustrated in figure 6. From this experiment, when bladder is stress-free which is no air pressure is applied it will elongate twice of its original length. When air is supplied to the PAM, it will expand horizontally instead of vertical direction which causes the PAM to contract.

The 22 cm length of bladder presents the higher contraction percentage compared to 12 cm and 17 cm of bladder length. This means that the length increment will increase the value of PAM contraction. The result of contraction between 22 cm long and 17 cm are slightly similar since both have a limitation contraction. The bladder will reach the maximum potential to contract at certain point of length. The contraction of PAM has a limited value which is PAM unable to develop any forces and its volume has reached a maximum value.

The contraction percentage is decreased when a large amount of load is applied. This is because the force transmitted by the muscle remains constant even though the amount of load is increasing. The force remains the same due to the diameter of PAM which is constant. Thus, the experiment shows that diameter and length of bladder are the main factors to increase the performance of PAM.

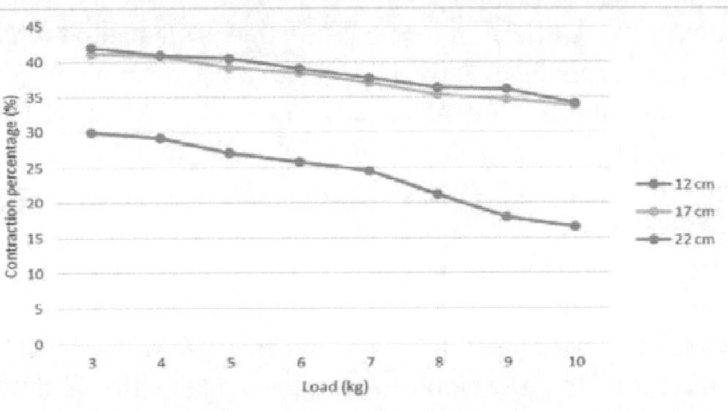

Figure 6. Contraction percentage versus load for 2 cm diameter bladder using 3.5 cm braided mesh

4. Conclusion

In the present study, the compatibility of braided mesh shell with bladder has been identified in uniform and non uniform condition. Research efforts were important on the new developed of PAM to provide an option for load-lifting mechanisms, while possessing the safety, portability, reliability and low cost. The experimental results showed the effectiveness value of mesh suitability. The suitable result indicates the value of 2.0 cm bladder in contraction percentage which is 51.85%. Under various loads, the PAM contraction indicates that the value of 22 cm length of bladder present the higher contraction percentage compared to others. Obviously, these experiments contribute knowledge on correlation of PAM mechanisms as well as assisting by avoiding the failure of PAM during PAM contraction process.

Future recommendation for this project is to increase the thickness of bladder's wall. If the tube thickness is high, it can withstand more pressure applied. Besides that, the type of braided mesh can be recommended for better efficiency of PAM.

5. Acknowledgment

The authors would like to thank University Malaysia Perlis (UniMAP) and the Ministry of Higher Education (MOHE) for funding support through Research Acculturation Grant Scheme (RAGS).

6. References

[1] Andrikopoulos G, Nikolakopoulos G and Manesis S 2011 A survey on applications of pneumatic artificial muscles *Control & Automation (MED)* 19th Mediterranean Conf. on 2011 Jun 20 pp 1439-46 IEEE.

[2] Daerden F and Lefeber D 2002 Pneumatic artificial muscles: actuators for robotics and automation *European J. of Mech and Env. Eng. 47*(1) pp 11-21.

[3] Daerden F, Lefeber D, Verrelst B and Van Ham R 2001 Pleated pneumatic artificial muscles: compliant robotic actuators *Intelligent Robots and Systems Proc. IEEE/RSJ Int. Conf.* Vol 4 pp 1958-63 IEEE.

[4] Gonzalez C 2015 What's the Difference Between Pneumatic, Hydraulic, and Electrical Actuators? *Machine Design.*

[5] Jobbágy B, Šimšík D, Karch J and Onofrejová D 2014 Robotic arm with artificial muscles in rehabilitation *Procedia Engineering* 96 pp 195-202.

[6] Kang B S, Kothera C S, Woods B K S and Wereley N M 2009 Dynamic Modeling of Mckibben Pneumatic Artificial Muscles for Antagonistic Actuation. *Proc. - IEEE Int. Conf. on Robotics and Automation* Kobe Japan pp 182–87.

[7] Laksanacharoen S 2004 Artificial muscle construction using natural rubber latex in Thailand *The 3rd Thailand and Material Science and Technology Conf.* pp. 10-11.

[8] Lee Y K and Shimoyama I 1999 A Skeletal Framework Artificial Hand Actuated by Pneumatic Artificial Muscles *Proc. - IEEE Int. Conf. on Robotics and Automation* pp 926–31.

[9] Lightner S and Lincoln R 2002 The fluidic muscle: a 'new'development *The Int. J. of Modern Engineering* 2(2) pp1-8.

[10] Minh T V, Kamers B, Ramon H and Brussel H Van 2012 Modeling and Control of a Pneumatic Artificial Muscle Manipulator Joint – Part I: Modeling of a Pneumatic Artificial Muscle Manipulator Joint with Accounting for Creep Effect *Mechatronics* 22(7) pp 923–33.

[11] Saga N and Saikawa T 2008 Development of a pneumatic artificial muscle based on biomechanical characteristics *Advanced Robotics 22*(6-7) pp 761-70.

Numerical simulation of two-phase slug flow with liquid carryover in different diameter ratio T-junction

W Pao, L Hon, A Saieed and S Ban

Universiti Teknologi PETRONAS, 32610 Seri Iskandar, Perak, MALAYSIA

william.pao@utp.edu.my

Abstract. A smaller diameter conduit pointing at 12 o'clock position is typically hot-tapped to a horizontal laying production header in offshore platform to tap produced gas for downstream process train. This geometric feature is commonly known as T-junction. The nature of multiphase fluid splitting at the T-junction is a major operational challenge due to unpredictable production environment. Often, excessive liquid carryover occurs in the T-junction, leading to complete platform trip and halt production. This is because the downstream process train is not designed to handle excessive liquid. The objective of this research is to quantify the effect of different diameter ratio on phase separation efficiency in T-junction. The liquid carryover is modelled as two-phase air-water flow using Eulerian Mixture Model coupled with Volume of Fluid Method to mimic the slug flow in the main pipe. The focus in this paper is 0.0254 m (1 inch) diameter horizontal main arm and vertical branch arm with diameter ratio of 1.0, 0.5 and 0.3. The present research narrowed the investigation to only slug flow regime using Baker's map as reference. The investigation found that, contrary to common believe, smaller diameter ratio T-junction perform worse than larger diameter ratio T-junction.

1. Introduction

T-junction is an appendage of small diameter pipe, which is attached to the main pipe to tap the fluid source from the main stream. A T-junction consists of a main arm, run arm and side arm as shown in figure 1.

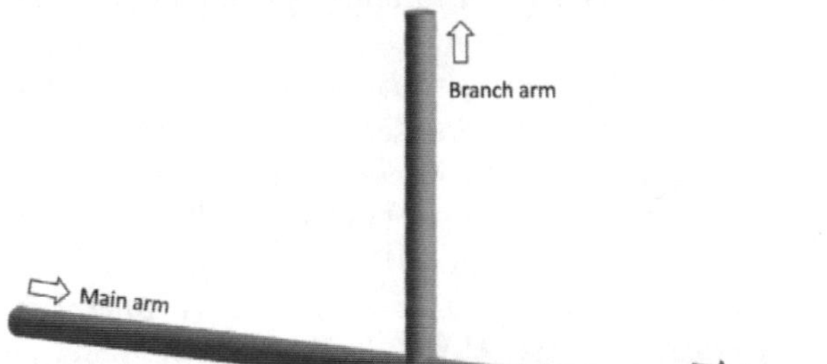

Figure 1. T-junction and its components.

T-junction configuration is most commonly deployed in offshore platform where it is used to tap gas directly from the production header. The tapped gas is channel downstream as fuel gas for power generation plant or as source gas for other purposes. When a two-phase mixture flows pass a T-junction, lighter phase will preferentially enter the branch arm, while the heavier phase tends to continue flowing into the run arm [1]. Deployment of T-junction as a mini 'phase separator' has substantial advantages over conventional separators due to its simple installation, saving in terms of cost, space and minimal maintenance [2]. However, liquid carryover is a frequent problem that occurs in T-junction whereby excessive amount of liquid is take off into the gas stream and channeled into downstream equipment which is not designed to handle liquid. Consequently, platform trip and production must be stopped to allow offshore crews to manually drain out the excessive liquid.

2. Literature review

The initial focus for studying the phase separation at T-junctions with a reduced side-arm diameter was for industrial situations, where the branch arm of a piping system could have a smaller diameter than the main pipe ($D_3 < D_1$, $D_1 = D_2$). Reduced diameter T-junction referred to the diameter ratio of the branch arm D_3 to the main arm D_1. Diameter of main and run arms are usually the same. A reduction in the branch arm diameter will have two distinct effects; the pressure drops and the entrance area available for liquid take off. It is well known that the distribution of the phases at a T-junction depends not only on the geometry and approaching flow pattern but also rely on the two downstream pressures and the pressure drop across the T-junction. Pressure drop measurements around T-junction showed a loss between the inlet and side-arm but a pressure recovery in the run arm due to Bernoulli's effect [3]. The pressure drop in the branch arm can be attributed to the sudden momentum change for the fluids moving from the inlet into the branch arm [4]. The effect caused by a reduction in side-arm diameter, as first suggested by Azzopardi [5], would be to reduce the axial distance available for phase take-off to occur. This reduces the liquid travel time, the time available for the liquid to flow into the side-arm instead of flowing straight into the run arm. As such there is less chance of the liquid to be dragged towards the side-arm. By the same token, smaller diameter side arm will have lesser liquid carryover issue than a regular T-junction. Pandey et al [6] carried out experiment to verify the effect of side arm diameter ratio on phase split for stratified flow and plug flow. Experiments have been performed with kerosene and water in a reduced diameter T-junction whose branch arm is 0.0127 m in diameter while the main and run arms are 0.0254 m (diameter ratio: 0.50) and compared with phase split result obtained from a regular T-junction. There is an increased in the fractional liquid take off with a decrease in the branch arm diameter. Azzopardi [7] studied the effect of branch arm to main arm diameter ratio on phase split of liquid-gas flow at T-junction for annular and stratified flow. The data were taken at 2 scales: small scale T-junction with main arm diameter of 0.038 m and side arm bore of 0.038 m, 0.025 m and 0.0127 m; large scale T-junction with main arm diameter of 0.125 m and side arm bore of 0.125 m and 0.076 m. The medium was air and water. For small scale T-junction, decrease in diameter ratio of a junction leads to lower liquid take off for both stratified an annular flow. For the larger scale T-junction, data supports that decrease in side arm diameter results in lower liquid take off. The effect of diameter ratio is most significant at lower gas rates while least at lower liquid and high gas flow rate conditions. Marti and Shoham [8] carry out experiment to compare the reduced tee data and regular tee data for horizontal branch arm. Main arm diameter of 0.051m was paired with side arm of 0.051m (regular T) and 0.0255m (reduced Tee). At very low gas fraction intakes, less liquid was diverted for reduced T-junctions than for regular T-junctions as the pressure drop at the T-junction, due to increased gas velocity in the reduced diameter side arm, was not yet significant enough to compensate for the dominating axial inertia forces within the liquid phase. Furthermore, effect of reduced tee is more pronounce in lower superficial liquid velocity. This was due to the accelerated gas velocities in the reduced T-junction forcing a greater proportion of the liquid phase to be drawn off its axial course into branch arm. More recently, Pao et al. [9,10,11] investigated methane and water separation in a T-junction numerically by varying the diameter ratio, density and inlet saturation of gas phase. They found that the optimum range of diameter ratio is between 0.5 and 0.75, the differential density has an adverse effect on the separation efficiency and inlet gas saturation is inversely proportional to the gas extraction.

The objective of this paper is to use commercial CFD Software FLUENT to simulate a two-phase liquid carryover to the branch arm of T-junction with different diameter ratio. Specifically, the research aims to correlate T-junction diameter ratio on liquid/gas separation and to predict the fraction of liquid taken off in a T-junction at different air and water inlet velocity.

3. Methodology

This research aims to correlate the effect of T-junction diameter ratio and different superficial velocity of air and water at the inlet on the fraction of liquid carryover in the branch arm. The T-junction with desired geometry is modelled in 3D using ANSYS FLUENT software. In the present paper, Eulerian Mixture Model coupled with Volume of Fluid Method is used to model the air-water separation in T-junction. The schematic model of T-junction is shown in figure 2.

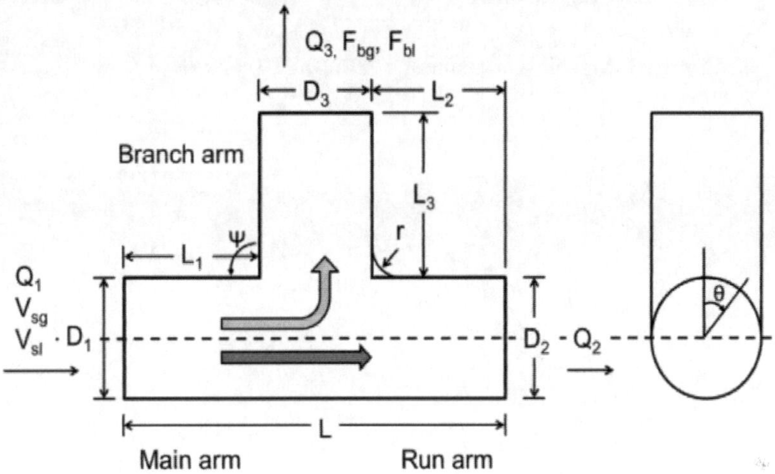

Figure 2. Schematic model of T-junction and its parameters.

Table 1 listed the input data for the benchmark and the variation of parameters for present study. The volume fraction of the air-water is assumed 50-50. Phasic velocities are prescribed at the pipe inlet while the atmospheric pressure is assumed at both the outlet.

Table 1. Input parameters and phasic velocities for benchmarking and parametric studies

Input parameters	Wren [12]	Present studies
Diameter, D_1 & D_2 (mm)	127	25.4
Diameter D_3 (mm)	76, 127	127, 25.4, 7.62
Length, L_1 (mm)	4000	1500
Length, L_2 (mm)	4000	500
Length L_3 (mm)	500	300
Water density, ρ_l (kg/m^3)	998.2 (water)	998.2(water)
Air density, ρ_g (kg/m^3)	1.225 (air)	1.225 (air)
Water velocity, V_l (m/s)	0.0283, 0.0401, 0.0535, 0.404, 0.558	1, 2
Air velocity, V_g (m/s)	4, 12, 24	1, 2

Before the full study was conducted, mesh dependency test is carried out to ensure the produced solution is mesh independent and to determine the optimum mesh density for simulation. Figure 3 show two cases of different mesh densities used for testing. The averaged pressure in the branch arm is used as convergent criteria for mesh independent solution study.

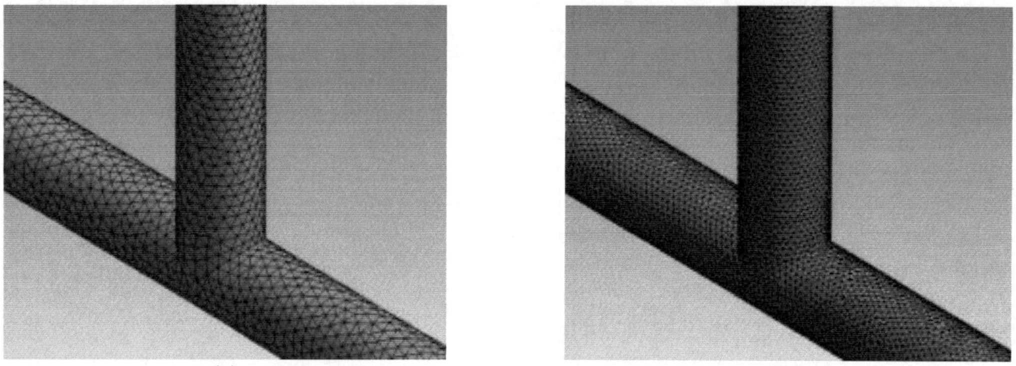

(a) (b)

Figure 3. Meshing at element size of (a) 2.0 and (b) 1.0 mm.

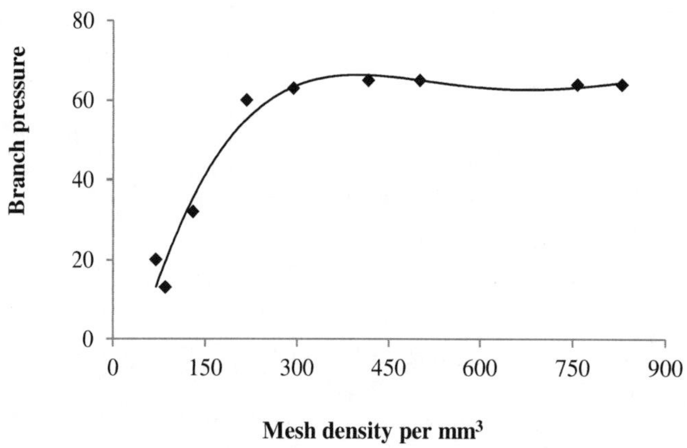

Figure 4. Branch pressure convergence against mesh density.

Figure 4 shows the convergence behavior of averaged branch pressure versus mesh density per mm^3. When the mesh density increases, the averaged branch pressure increases until approximately 300 elements per mm^3, whereby the averaged branch pressure fluctuated within a very narrow margin. Figure 4 also showed that the averaged branch pressure converges from below. Apparently, the smaller the element mesh size, the bigger the number of element division, thus, the longer CPU hour is needed to run the simulation. It was decided that the lowest bound for mesh density of 450 elements per unit of mm^3 is acceptable in terms of accuracy and computational time. This translated into approximately 120 hours per simulation.

There are seven possible types of flow regimes for two-phase flow. The flow regime is dependent on mass flow rate of each phase, pipe diameter and fluid properties. Some distance, which was tested by trial and error in the present study, is required for a flow regime to fully develop. Inlet mass flow rate for each phase are manipulated to reproduce different flow regime. Flow regime maps are used to identify types of flow regime based on the fluids inlet parameters. Different flow regime map introduced by different author refers only to specific pipe diameter and type of mediums. The Baker (1954) map [13] for two-phase flow in horizontal pipes is the most commonly used flow regime map for air-water flow because it covers pipe diameter up to 10 cm, as shown in figurer 5. To use the map, mass flux, G, of gas and liquid phase, gas parameter, λ, and liquid parameter, ψ, need to be determined using equation (1) – (3).

$$G = \frac{\dot{m}}{A} = \rho v \tag{1}$$

$$\lambda = \left(\frac{\rho_G}{\rho_a} \frac{\rho_L}{\rho_w} \right)^{\frac{1}{2}} \tag{2}$$

$$\psi = \left(\frac{\sigma_w}{\sigma} \right) \left(\frac{\mu_L}{\mu_w} \left(\frac{\mu_w}{\mu_L} \right)^2 \right)^{\frac{1}{3}} \tag{3}$$

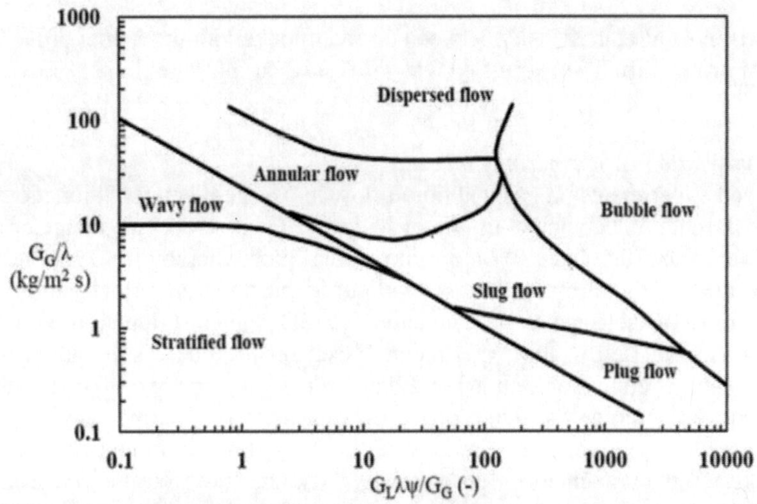

Figure 5. Baker's flow regime map [13].

The value of the *x*- and *y*-axis are then determined to locate the interested flow regime in Figure 5.

4. Results and discussion

4.1. Qualitative validation of flow regime based on Baker's map

The contours obtained from current simulation were compared with results by de Schepper et.al [14] in figure 6, that carried out VOF method simulation on a 7 m long horizontal tube with 0.08 m diameter. Five flow regimes were compared which are stratified, stratified wavy, slug, plug and bubbly flow. It is clear from the comparison that the current model of slug flow pattern is much more realistic. It was found that without interface perturbation at the air-water contact, it is impossible for slug and plug flow to be generated. The perturbation may be achieved by either starting the flow as stratified wavy flow or artificially disturbed the air-water interface by FLUENT's User Defined Function, e.g. using a simple sinusoidal function. This perturbation at the interface allows the high velocity incoming air from the inlet to pushes the waves to form liquid slug. The crest slowly increases until it reaches the top of the pipe to formed plug flow.

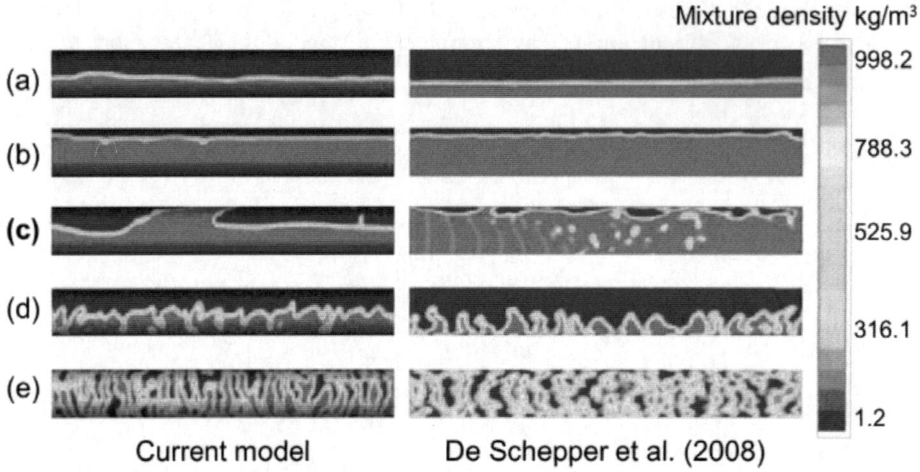

Figure 6. Comparison of mixture density between current model with the model of De Schepper et al. [14]: (a) stratified, (b) stratified wavy, (c) slug, (d) plug, (e) bubbly flow.

4.2. Quantitative validation with experimental data

Wren [12] preformed experiments on gas and liquid flow in a large diameter T-junction and correlated the flow split with T-junction geometry. In one of his experiment setup, two diameter ratio of branch arm to main arm were used (0.5 and 1.0) on 8 m horizontal pipe whereby the 1.5 m height T-junction is in the middle of the horizontal main pipe. For validation purpose, air velocity of 12 m/s and water velocity of 0.0283 m/s was selected to be simulated in a 3D regular T-junction with vertical branch arm with same dimension used in the experiment. Based on the phase split graph in figure 7, the current simulation result is comparable to Wren's data. For liquid carryover to occur, the gas velocity need to be sufficient to overcome the liquid phase flowing inertia force and gravity forces that settle down the water. By reducing branch arm diameter, air velocity flowing in the branch will increase and cause a pressure drop in the branch arm. Subsequently, back pressure from the run arm will act on the pressure drop in the branch arm to suck up liquid in to the vertical branch arm. Therefore, the amount of liquid carryover is negligible unless a gas takes off reach its critical value.

4.3. Parametric studies

A parametric study was conducted to further investigate the behavior of phase splitting for different diameter ratio T-junction. The horizontal arm was 1.5 m long with vertical branch located 1 m away from the inlet. The main and run arms were 0.0254 m or 1 inch in diameter. Three different branches to main diameter ratios were investigated, namely 0.3, 0.5 and 1.0. The air and water inlet velocity was both prescribed as 1 m/s and 2 m/s, respectively, for each phase which made up of four combinations of velocities for each diameter ratio. Figure 8 shows phase split graph for slug flow regime that pass through a T-junction with diameter ratio of 1.0, 0.5 and 0.3. The data points were collected by varying the water inlet velocity from 1 m/s to 2 m/s while air inlet velocity was kept constant at 1 m/s for case figure 8(a), 8(c), 8(e) while for figure 8(b), 8(d) and 8(f), the air velocity was kept constant at 2 m/s. The results show that at higher water inlet velocity, the data points shifted towards liquid dominant region which indicates more liquid carryover into the branch arm.

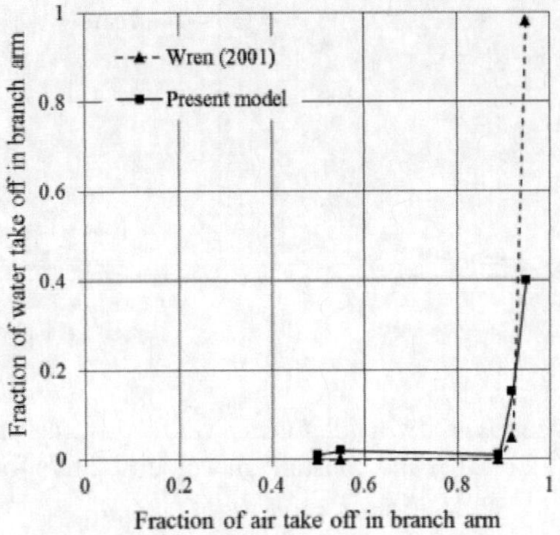

Figure 7. Validation phase split graph.

Figure 9 shows phase split graph for slug flow regime that pass through a T-junction with diameter of 1.0, 0.5 and 0.3. The data points were collected by changing air inlet velocity from 1 m/s to 2 m/s while water inlet velocity was kept constant for case figure 9(a), 9(c), 9(e) at 1 m/s and case figure 9(b), 9(d), 9(f) at 2 m/s. At higher air inlet velocity, the data points shifted towards liquid dominance region, indicating more liquid carryover into the branch arm. This is mainly due to Bernoulli's principle, whereby higher air velocity at branch arm creates a pressure drop, sucking up more liquid.

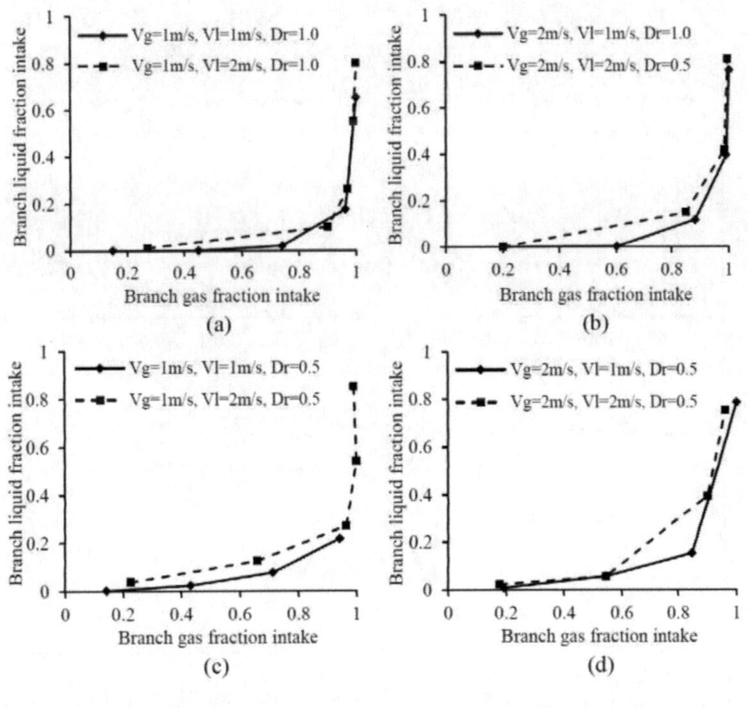

IOP Conf. Series: Journal of Physics: Conf. Series **908** (2017) 012037 doi:10.1088/1742-6596/908/1/012037

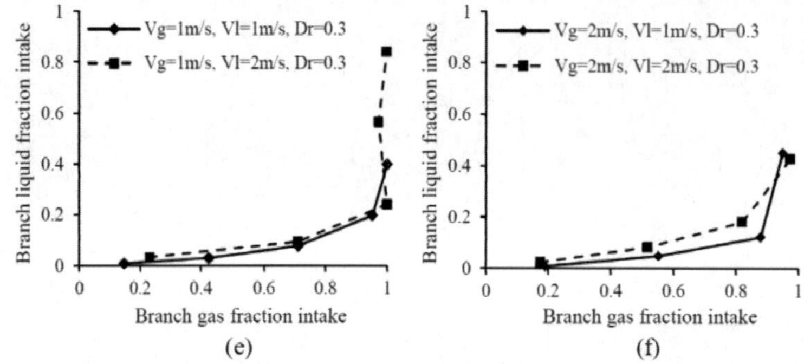

Figure 8. Phase split graph when *increasing inlet water velocity* from 1 m/s to 2 m/s for constant gas velocity 1 m/s in (a), (c), (e) and constant air velocity 2 m/s for (b), (d) and (f).

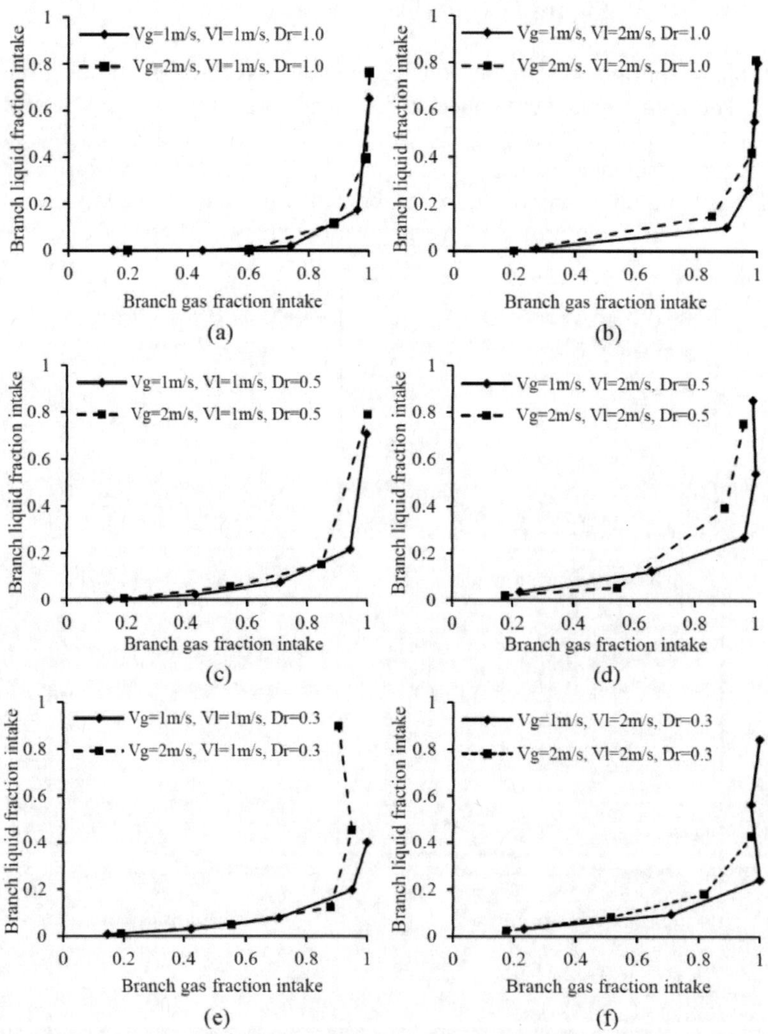

Figure 9. Phase split graph when *increasing inlet air velocity* from 1 m/s to 2 m/s with constant gas velocity 1 m/s for (a), (c) and (e) and constant gas velocity 2 m/s for (b), (d) and (f).

5. Conclusions

In petroleum industry, T-junctions are very common in piping networks as they are used to transport liquid within the industry. It is important to understand the correlation of diameter ratio of T-junction towards phase split on the two-phase flow. In this project, VOF formulation is selected as it provides excellent surface tracking ability specialize for stratified and slug flow whereby their interface is distinct. Air and water was used as the medium.

The current flow regime simulation using Eulerian Mixture Model coupled with VOF Method is comparable with published data based on Baker's map flow regime. In fact, the current simulation achieved a better slug flow phenomenon in a horizontal pipe. Contrary to common believe that the smaller the diameter ratio, the better is the separation, it is found that 0.3 diameter ratio of branch to main arm lead to earlier liquid carryover when compared to 1.0 and 0.5 diameter ratio T-junction. This is because smaller branch diameter will increase air velocity within it, leading to greater pressure drop in the branch pipe which eventually causes earlier liquid carryover. Furthermore, liquid carryover into the branch arm also increases when both inlet air and water velocity increase.

6. Acknowledgments

The authors would like to acknowledge that this research is funded by Malaysia Ministry of Higher Education via FRGS Grant 0153AB-L03, Petroliam Nasional Berhad PETRONAS via YUTP 0153AA-E03 and Universiti Teknologi PETRONAS via Graduate Assistantship.

7. References

[1] Wren E and Azzopardi B J 2004 *Lecture Notes in Statistics* vol 61 ed J Breckling (Germany: Springer) pp 364-37
[2] Mohamed M A, Soliman H M and Sims G E 2007 *Int. J. Multiphase Flow* **47** pp 66-72
[3] Okto D, Yosephus A K P, Akmal I M, Akhmad Z D, Yusnanda A N, Adhika W and Deendarlianto 2016 *Experimental Thermal and Fluid Science* **81** pp 93 – 108
[4] Penmatcha V R, Ashton P J and Shoham O 1996 *Int. J. Multiphase Flow* **22** pp 1105 – 1122
[5] Azzopardi B J 1984 *Int. J. Multiphase Flow* **10** pp 509-512
[6] Pandey S, Gupta A, Chakrabarti D P, Gas G and Ray S 2006 *Chemical Engineering Research and Design* **84** pp 895 - 904
[7] Azzopardi B J 1999 *SPE Annual Technical Conference and Exhibition* Houston
[8] Marti S and Shoham O 1997 *Int. J. Multiphase Flow* **23** pp 725 – 748
[9] Pao W, Hashim F M and Low H M 2014 MATEC Web of Conference **13** 03009.
[10] Saieed A, Ban S, Pao W, Hashim F M and Norpiah R 2016 ARPN J. Eng. and App. Sci. **11** pp 14233-14238
[11] Ban S, Pao W, Nasif M S and Norpiah R 2016 ARPN J. Eng and App. Sci. **11** pp 12011-12016
[12] Wren E M K 2001 *PhD thesis* University of Nottingham, United Kingdom
[13] Baker O 1954 Oil and Gas Journal **53** pp 385-389
[14] De Schepper S C K, Heynderickx G J and Marin G B 2008 Chemical Engineering Journal **138** pp 349-357

IOP Conf. Series: Journal of Physics: Conf. Series **908** (2017) 012038 doi:10.1088/1742-6596/908/1/012038

Numerical investigation on the effect of balcony open and solid upstand on smoke contamination in atrium balconies

M A Mutafi[1] , M S Nasif[1] , W Pao[1] and F B Ismail[2]

[1] Mechanical Engineering Department, Universiti Teknologi PETRONAS, 32610 Bandar Seri Islander, Perak, Malaysia.
[2] Power Generation Research Centre, College of Engineering, Universiti Tenaga Nasional, Selangor, Malaysia.

mohammad.nasif@utp.edu.my

Abstract. Atrium is a modern architectural design of buildings that has an open space at the middle and contains several floors. In atrium structure the smoke due to fire can easily flow through the open spaces and causes smoke contamination. CFD simulation using fire dynamic simulator (FDS) software is conducted in this work, to investigate the effect of balcony open upstand on smoke contamination in atrium's balconies and compare it with balconies with solid upstand. It was found that in case of open upstand, the smoke contamination occurrence decreased inside the balcony comparing with balconies with solid upstand.

1. Introduction

In a fire incident, significant amount of smoke is released which may cause hazards such as toxicity, visibility and thermal hazards. Past research shows that inhalation of the toxic gases released from the smoke is the main cause of death in a fire accident. In United states, it was reported that most fire victims died because of smoke inhalation rather than burns [1].

Atrium is a modern structural design of buildings that is used in airports, hospital and shopping malls. It has large open spaces in the middle and contains multiple floors. In a fire incident inside an atrium, the smoke can easily spread due to the open space which may endanger occupant's life. When fire occurs in an atrium compartment, It is expected that the smoke plume leaving the fire compartment rises vertically towards the atrium ceiling. However, past research found that sometimes smoke may curl into the atrium upper balconies instead of completely rising upward, resulting in smoke contamination. This phenomena may causes risk to the occupants.

Cox [2] investigated the reason of the smoke curling and found that the static pressure at the balcony is affected by the rising plume that cause it to drop in that region. This pressure drop inside the balcony creates a low-pressure region that cause the spill plume to curl inward the balcony and result in a smoke contamination in the particular balcony.

Previous researches [3-4] found that changing the atrium structure has a great effect on the occurrence of smoke contamination. Balcony breadth, fire compartment opening and down stand depth, and width effects on smoke contamination were well investigated. However, balconies upstand effect on smoke contamination occurrence was not investigated. Some balcony upstand (railing) design has openings, whilst other designs are solid with no opening. . These openings may affect the pressure distribution inside the balcony and hence the smoke contamination occurrence, past researches investigated only the effect of balcony upstand as a solid obstacle attached to the end of a balcony.

The objective of this paper is to compare the behavior of smoke and smoke contamination occurrence when open balcony and solid upstand are used.

2. Model development

A full scale three storey atrium was modelled using Fire Dynamic Simulator (FDS) as it is shown in Figure 1(a). FDS is a Computational Fluid Dynamic (CFD) software that models fluid flow driven by fire. It solves numerically a form of the Naiver-Stokes equations of a low speed fluid flow, thermally – driven flow with an emphasis on smoke and heat transport from fires [7].

FDS software is used to numerically study the effect of balcony open-upstand on smoke contamination in atrium's balconies. The modelled geometry has parameters that varies in every numerical simulation including balcony breadth, b, and fire compartment opening width, w. 1.6 MW, 3.2 MW and 4.7 MW heat release rates were selected for this study to be consistent with the recommended design fire range in an atrium [8] and corresponded to those used by [5-6]. The dimensions of the atrium used in this research are like the atrium dimensions used in Tan [9] experiment but for full scale Upstand structure modelled as shown in figure 1(c).

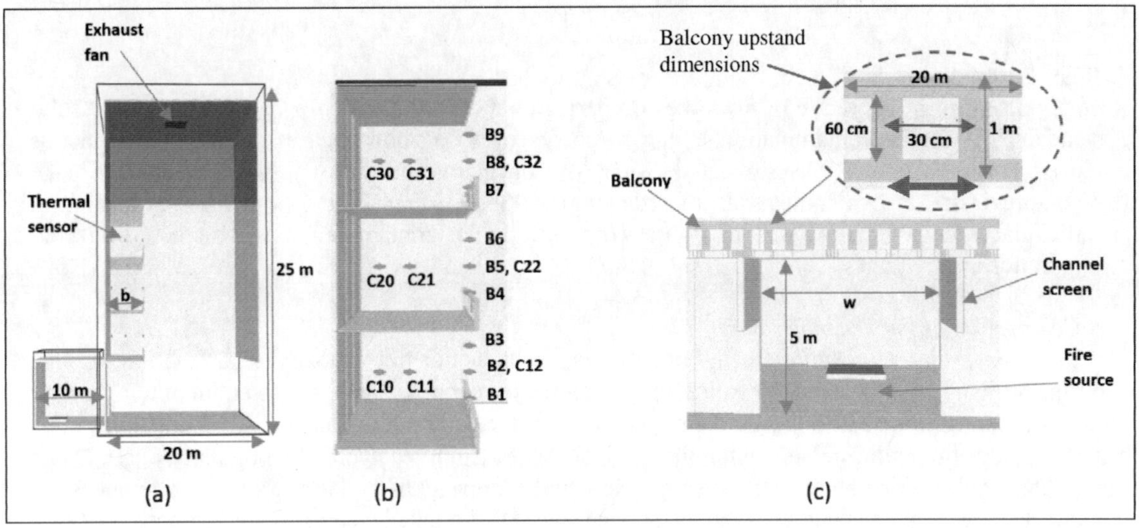

Figure 1. Schematic drawing of atrium and fire compartment (a) side view of atrium, (b) Location of temperature sensor, (c) Front view of fire compartment with the balcony upstand.

FDS utilized $\frac{D^*_{spill}}{\Delta x}$ ratio grid size determination in modelling [7]. The criterion that was utilized in this study for choosing an appropriate mesh size was set by Harrison [3]. The criterion for the size of grid cells is shown in equation (1) and (2).

$$n^*_{spill} = \frac{D^*_{spill}}{\Delta x} \geq 0.9 \quad (1)$$

$$D^*_{spill} = \left(\frac{Q_c/_w}{\rho C_p T\sqrt{g}}\right)^{\frac{2}{3}} \quad (2)$$

Where D^*_{spill} is characteristic length of plume for determining the grid size (m), n^*_{spill} is coefficient for determining the grid size, ΔX is grid dimension in x-axis, Qc is convective heat flow layer below spill edge, ρ is air density at ambient condition, Cp is a specific heat, T is absolute temperature and g is gravity acceleration. Hence, a grid size of 200 mm met the criteria mentioned above and it was selected for all simulations. The total number of mesh elements in the atrium space and fire compartment is 2332800 and 81000 respectively. Values adopted to the simulation like the balcony breadth (b), fire compartment opening width (w) and heat release rate (HRR) were varied for each case. However, down stand depth was kept constant as 1m for all cases. The maximum heat release rate used in this study is 4.8MW which is considered acceptable according to past studies performed for shopping mall fires with sprinklers fire protection system. The parameters used in the simulation are listed in Table 1.

Table 1. Values adopted to simulation.

Case	Balcony Breadth b (m)	Fire compartment opening width w (m)	Heat Release Rate (kW)
1	5	10	1581
2	3	8	1581
3	3	6	3160
4	2	6	3160
5	1.5	4	3160
6	1.5	2	4740

3. Results and discussion

Results presented in this section shows the comparison between the effect of balcony open and solid upstand on the smoke contamination in the balconies of an atrium. 6 cases were selected in this comparison and each case was simulated twice with open upstand and with solid upstand. Smoke temperature inside the balconies was recorded via FDS software sensors, which were located at specific places as it is illustrated in figure 1(b) and smoke contamination was assessed in both configurations.

3.1. Comparison of smoke temperature

Tan [3] found that temperature can be used as an indicator for smoke contamination. In his experiment, Tan recognized that when the smoke temperature was ≥ 10 °C above ambient temperature which is assumed to be **20** °C the smoke contamination appeared in the balcony. [5] and [6] used this criterion to determine the smoke contamination inside the atrium's balconies. In this study, 30 °C was highlighted in the temperature scale in smokeview and it appears as a black color in the temperature contours to determine the height as well as the location of the smoke layer within the balcony as it was very hard to determine them from FDS smoke contours directly.

Figure 2 clearly illustrates that the temperature along the balcony edges are higher for the balconies with solid upstand in comparison with the balconies with open upstand. This indicates that the smoke plume is closer to the atrium structure in the case of solid upstand than with open upstand. Figures 3 and 4 show the temperature inside the balconies for case 1. As seen in Figure 3 that the temperature at balconies 2 and 3 was higher than 30°C indicating that smoke contamination has occurred. However, when open upstand was modelled; Figure 4 shows that all temperatures were lower than 30°C which indicates that no smoke contamination occurred.

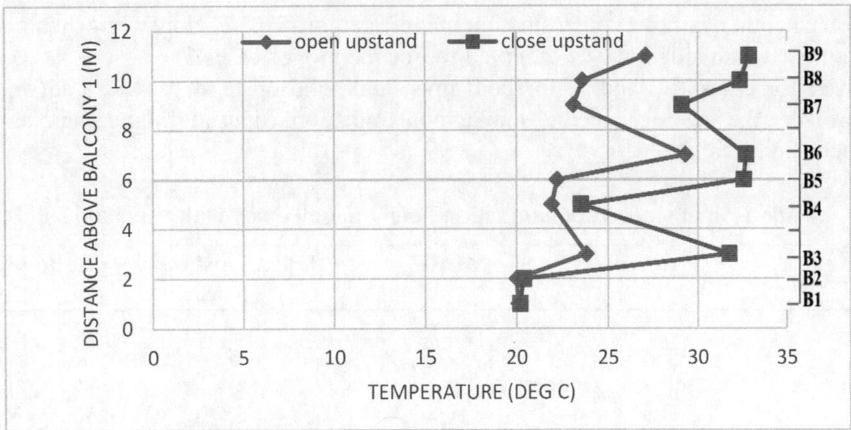

Figure 2. Temperature profile across balcony edge.

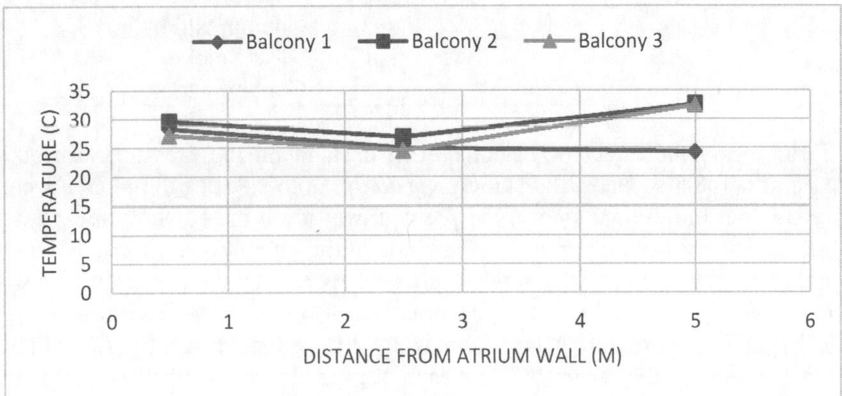

Figure 3. Temperature profiles along balcony breadth with solid upstand.

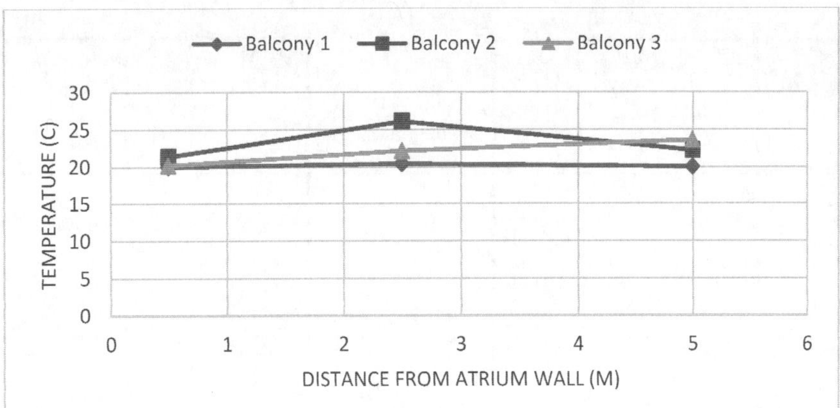

Figure 4. Temperature profiles along balcony breadth with open upstand.

3.2. Comparison of smoke contamination height

As mentioned in Table 1 that the atrium dimensioned varied for each case. Table 2 shows the simulation results including the height of smoke in the balcony (above balcony 1). As seen, that although in cases 3, 5 and 6, the smoke contamination occurred at the balconies when solid upstand was used and the height of smoke contamination is indicated in Table 2. However, when open upstand was used, no smoke contamination was observed for the mentioned cases.

Moreover, for cases 1,2, and 4, for both open and solid upstands, smoke contamination has occurred, however, the severity of the smoke contamination reduced (higher smoke height) as compared with solid upstand.

Table 2. A comparison of smoke contamination height in balconies with open and solid upstand.

Case	b(m)	w(m)	Open Upstand H(m)	Solid Upstand H(m)
1	5	10	7	2.7
2	3	8	8.7	1
3	3	6	No Smoke Contamination	12.4
4	2	6	13.9	8.3
5	1.5	4	No Smoke Contamination	7
6	1.5	2	No Smoke Contamination	13.4

Figure 5 (a) shows the effect of balcony open upstand on the pressure distribution. As the simulation progress to 158.5 seconds, the static pressure on both sides of the plume decreases below the ambient pressure but in the right side of the plume (away from the balcony) the pressure dropped to 0.78 Pa which is lower than the pressure on left side of the plume (near or inside the balcony) that is 0.52 Pa. Hence, the smoke was pushed outward the balcony and smoke contamination will not occur inside the balcony. Figure 5 (b) illustrates the pressure distribution of case 1 with balcony solid upstand at 158.5 second. The pressure inside the balcony dropped more than the side of the plume that is facing the open space of the atrium. The reason behind the static pressure drop in the region between the plume and the atrium structure is the continuous air entrainment and less air replacing the entrained air in that region. Hence, the smoke will curl inward and the smoke contamination will occur at that balcony.

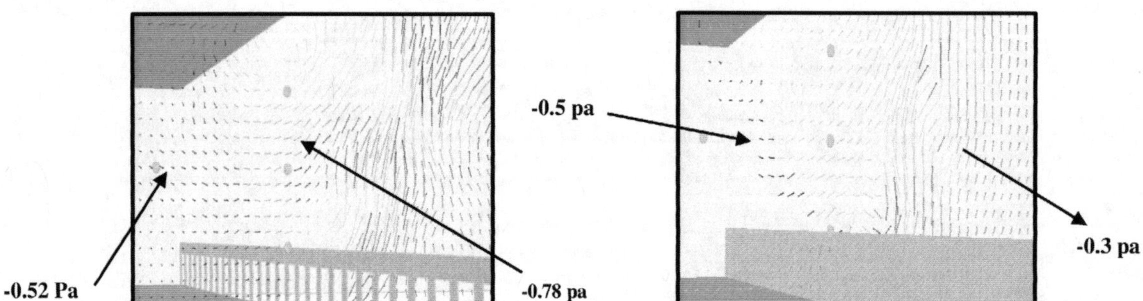

(a) (b)

Figure 5. Pressure distribution of case number 1 at simulation time = 158.5 sec
(a) Open balcony upstand (b) Solid balcony upstand.

Cases 1 and 5 were selected to present the effect of upstand type on smoke contamination. For case 1, at time 158.5 sec and 198.5 sec and for case 5 at time 162.5 sec and 246.5 sec. The smoke contamination is highlighted in a red dotted circle, in the contours. As seen in figure 6 that for both cases with solid upstand, the temperatures at the balconies were higher than 30°C indicating smoke contamination has occurred. However, when open upstand is modelled, the temperature inside the

IOP Conf. Series: Journal of Physics: Conf. Series **908** (2017) 012038 doi:10.1088/1742-6596/908/1/012038

balcony were lower than 30°C indicating no smoke contamination occurrence. As seen in Figure 6 (a) he smoke contamination decreased in the case of open upstand. (b) The smoke contamination was deep in balcony 1 and 2 and it becomes shallow in the presence of open upstand. Similarly, in figure 6 (c) and (d) balconies with open upstand are clear after it had been contaminated in the case of solid upstand. That reflects the effect of balcony open upstand in reducing the smoke contamination occurrence inside the balcony.

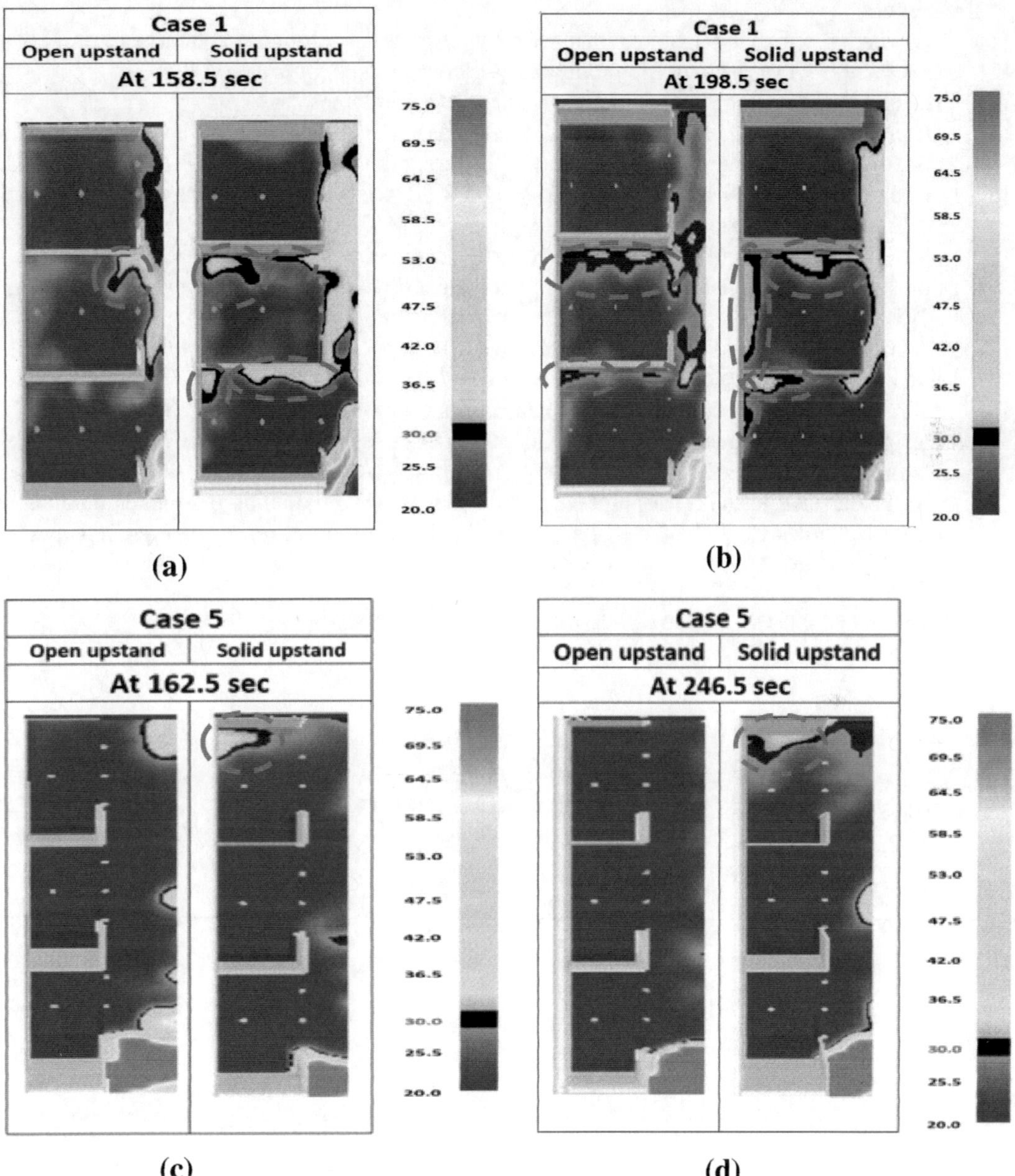

Figure 6. Comparison of smoke contamination occurrence with open and solid upstand (a) and (b).

Case 1 at 158.5 sec and 198.5 sec, (c) and (d) Case 5 at 162.5 sec and 246.5 sec

It can be concluded that in case of fire, using open upstand provide safer occupant environment than solid upstand.

4. Conclusions

CFD simulation using fire dynamic simulator is carried out by varying structural parameters of an atrium such as balcony breadth, fire compartment width and the heat release rate. The effect of balcony open upstand on smoke contamination in atrium's balconies is investigated in this study and compared to the effect of balconies with solid upstand. It was found that in case of open upstand, the smoke contamination is decreased inside the balcony comparing with balconies with solid upstand.

5. References

[1] Association N.F.P. 1991 Smoke management systems in malls, atria and large areas 92B.
[2] Cox G 1995 Compartment fire modelling *J. Combust. fund. of fire* 329-404
[3] Harrison R 2009 Entrainment of air into thermal spill plumes
[4] Hasnain SA, Nasif MS, Pao W and Al-Waked R 2017 Numerical investigation of smoke contamination in atrium upper balconies at different down stand depths *J. Building Simulation*
[5] Hasnain SA, Nasif MS, Pao W and Al-Waked R 2014 Effect of Down Stand Depth Variation on Atriums Upper Balconies Smoke Contamination *J. MATEC Web of Conferences*
[6] Ho CL, Li KY and Spearpoint MJ 2013 Numerical simulation of scale-model smoke contamination of upper atrium levels by a channelled balcony spill plume *J. Fire and Materials* **37(8)** 581-96
[7] McGrattan K, Hostikka S, Floyd J, Baum H, Rehm R, Mell W and McDermott R 2010 Fire dynamics simulator (version 5), user's guide **1019(5)** 1-186.
[8] Morgan HP, Ghosh BK, Garrad G, Pamlitschka R, De Smedt JC and Schoonbaert LR 1999 Design Methodologies for Smoke and Heat Exhaust Ventilation. Construction Research Communication Ltd, Londres, R, U.
[9] Tan F, Harrison R and Spearpoint MJ 2010 Physical scale modeling of smoke contamination in upper balconies by a channeled balcony spill plume in an atrium *J. of Fire Sciences* **1;28(4)** 313-36.

Numerical investigation on effect of aortic root geometry on flow induced structural stresses developed in a bileaflet mechanical heart valve

S S Abbas[1], M S Nasif [1], M A M Said [1] and S K Kadhim[2]

[1] Mechanical Engineering Department, Universiti Teknologi Petronas, Bandar Seri Iskander, Perak, Malaysia.
[2] Control and System Engineering Department, University of Technology Baghdad, Iraq.

mohammad.nasif@utp.edu.my

Abstract. Structural stresses developed in an artificial bileaflet mechanical heart valve (BMHV) due to pulsed blood flow may cause valve failure due to yielding. In this paper, von-Mises stresses are computed and compared for BMHV placed in two types of aortic root geometries that are aortic root with axisymmetric sinuses and with axisymmetric bulb, at different physiological blood flow rates. With BMHV placed in an aortic root with axisymmetric sinuses, the von-Mises stresses developed in the valve were found to be up to 47% higher than BMHV placed in aortic root with axisymmetric bulb under similar physiological conditions. High velocity vectors and therefore high von-Mises stresses have been observed for BMHV placed in aortic root with axisymmetric sinuses, that can lead to valve failure.

1. Introduction

The natural aortic valve allows unidimensional blood flow from the left ventricle into the great aorta, through which blood is transferred to all other parts of the body. The native valve may be diseased due to inborn heart defects, rheumatic fever, bacterial infection etc. and thus can cause stenosis or valvular insufficiency [1]. Such valve can be replaced by a mechanical prosthetic Bileaflet Mechanical Heart Valve (BMHV), which is as shown in Figure 1. BMHV is known to be the most durable and bio-compatible among other types of prosthetics available [2].

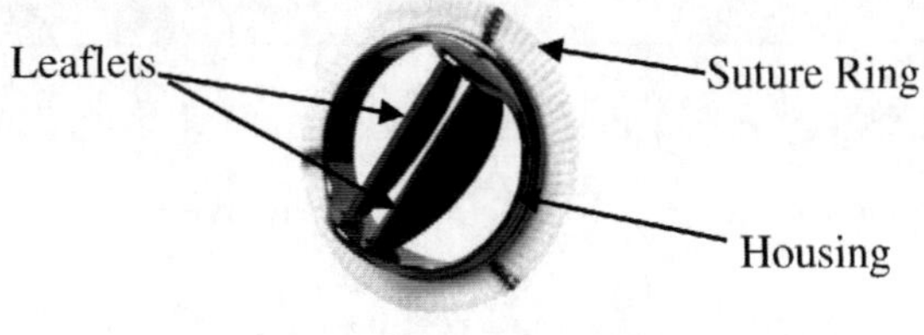

Figure 1. A typical BMHV [2].

Past research investigations on the dynamics of BMHV can be broadly classified into experimental (in-vitro and in-vivo) and numerical ones. Originating with the pioneer work of [3], the numerical methods have evolved as a promising tool to give deep insight into general dynamics of BMHV. However, numerical simulation of BMHV is a complex fluid-structure interaction (FSI) case as the interaction between fluid and the moving leaflets of valve is strong. Based on kinematical description of domain, the FSI approaches can be classified either as moving or fixed grid methods. The moving grid technique is based on Arbitrary Lagrangian Eulerian (ALE) approach, and is more accurate compared to fixed grid methods [4-5]. With ALE, greater distortions of the continuum can be handled better than would be allowed by a purely Lagrangian method, with more resolution than that afforded by a purely Eulerian approach. Having applied ALE to 2-D geometry of BMHV, [6] obtained velocity profiles for different cardiac phases. [7] and [8] developed a 3-D model of the valve, and employed ALE to study haemolytic properties of blood flow during different phases of cardiac cycle. Their results were found in good agreement with experimental measures. It's important to note that due to large structural displacements involved while simulating BMHV, subsequent remeshing and smoothing is required to maintain a good quality mesh [9].

The BMHV is located at the downstream of the left ventricle, and corresponds to the beginning of aorta. The aortic root, which is the initial tract of aorta, is characterized by three sinuses of Valsalva, from two of which, the coronary arteries originate as shown in Figure 2.

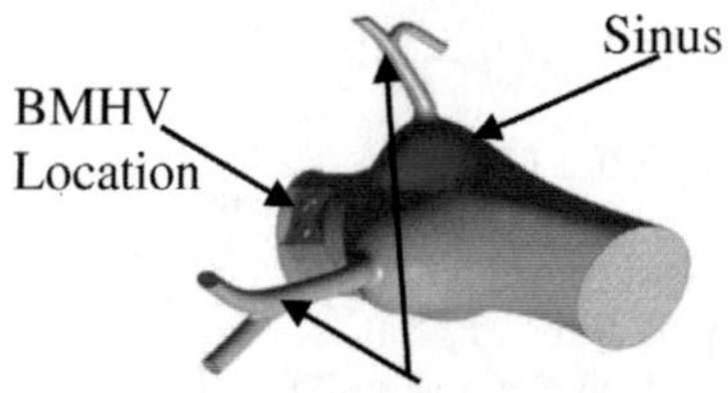

Figure 2. Native aortic root [10].

During the valve replacement process, the aortic root needs to be re-constructed or replaced with an artificial aortic root. Two of the available configurations for prosthetic aortic root are the aortic root with axisymmetric sinuses (mimics to natural root) or the aortic root with axisymmetric bulb [11], as can be seen in Figure 3. [12] compared the velocity profiles and blood flow shear stress contours of two types of aortic root models. High velocity vectors and therefore, high shear stresses were reported for the root with axisymmetric sinuses. Therefore, it is important to investigate the effect of sinuses configuration on blood flow hydrodynamics.

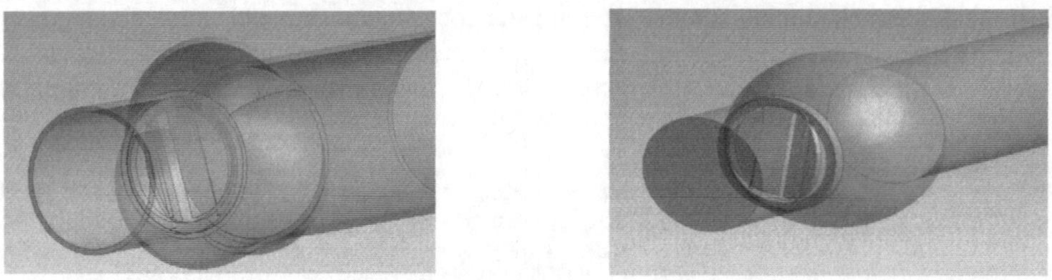

a- Aortic root with axisymmetric sinuses b- Aortic root with axisymmetric bulb

Figure 3. Different aortic root geometries.

The structural strength of BMHVs is an important matter while designing the valve as structural stresses are developed in the valve that may cause valve failure. During its lifetime operation, the valve is exposed to blood pressure force [13], force due to regurgitant flow at end of systole [14] and due to vortices formed in the wake region downstream the BMHV [15]. High equivalent stresses may cause yielding and thus failure of the structure.

In this study, we have placed the BMHV in two types of aortic root geometries, one with axisymmetric sinuses and the other with axisymmetric bulb and an algorithm based on ALE has been applied to account for the mesh motion. The von-Mises stresses developed in the structure have been computed and compared for the two aortic root geometries. The velocity profiles for the fluid domain are also reported at different physiological flow rates

IOP Conf. Series: Journal of Physics: Conf. Series **908** (2017) 012039 doi:10.1088/1742-6596/908/1/012039

2. Model development and boundary conditions

The aortic root model characterized by three axisymmetric sinuses has been developed based on the geometric parameters of sinuses presented by [16]. While the geometric model for the aortic root with an axisymmetric bulb has been designed following the study of [17]. The valve consists of two leaflets made up of pyrolytic carbon, which is fully closed at 25° and fully opens at 85°. The leaflets rotate about struts attached to their housing, which is made up of titanium. Simulations have been carried out using finite volume code ANSYS FLUENT coupled with finite element code Static Structure. The number of cells used in Static Structure and Fluent are 178,426 and 599,771, respectively (Figure 4) and maximum mesh skewness of 0.76 was maintained.

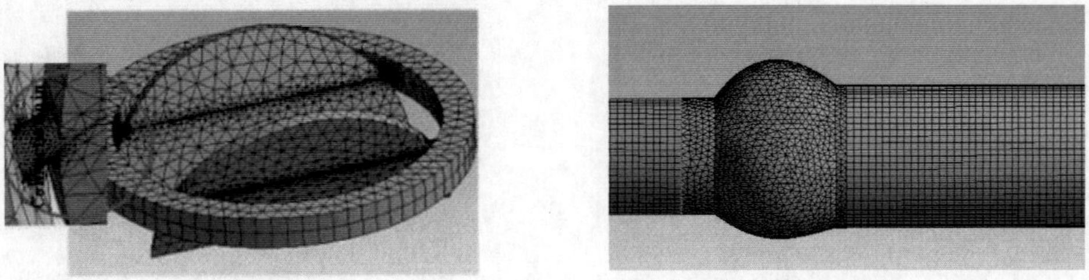

(a) Static Structural Model (b) Aortic root model

Figure 4. Meshing of aortic root model.

To account for leaflet motion, a user defined function was developed and incorporated in ANSYS. The set of Navier Stokes equations solved by the software are written as:

$$\nabla . v = 0 \quad (1)$$

$$p\left(\frac{\partial v}{\partial t} + v.\, \nabla v\right) = \nabla . \sigma + p.f \quad (2)$$

Where, p is the fluid density, v is the fluid velocity, σ is the Cauchy Stress tensor and f the body force.

A no-slip boundary condition has been applied to wall boundary. Blood has been modelled as laminar, Newtonian fluid, since the Newtonian behaviour of blood in large vessels like aorta has been widely agreed [18]. The viscosity of blood becomes asymptotically constant at high shear rates, which is the case for aorta. The blood's dynamic viscosity and blood density used in the model are 3.5 cP, and 1060 kg/m^3 respectively.

A physiological pressure waveform, which was approximated using the Fourier series, has been incorporated as the inlet and outlet boundary condition, as mentioned in equations (3) and (4) respectively [8]. The governing equations for inlet and outlet pressures are listed below:

$$p(inlet) = 44.24 - 50.76\cos wt + 44.4 sinwt - 0.4108 cos2wt - 19.45 sin2wt - 0.3481 cos3wt + 6.66 sin3wt + 8.032 cos4wt + 2.597 sin4wt + 0.3551 cos5wt + 2.903 sin5wt \quad (3)$$

where w represents a factor in the pulsed wave equation and is equalled to 8.72.

$$p(outlet) = 96.71 - 18.66 coswt + 2.59 sinwt + 0.182 cos2wt - 8.67 sin2wt + 4.41 cos3wt + 0.46 sin3wt + 0.059 cos4wt + 1.39 sin4wt - 2.101 cos5wt + 1.19 sin5wt \quad (4)$$

where w is equal to 8.438.

Three physiological flow rates were simulated which are 5 L/min, 5.5 L/min and 6 L/min corresponding to inlet Reynolds number of 1350, 1500 and 1700 respectively. It is known that blood flow rate through the human heart varies between the flow rates that have been mentioned [19-20].

3. Results and Discussion

The velocity profiles and the von-Mises stress contours at Re = 1350 for both the geometries are presented in subsequent figures, while results for other simulated Reynolds numbers have been tabulated in Table 1 and 2.

It has been observed that high velocity vectors are found in aortic root with axisymmetric sinuses as compared to aortic root with axisymmetric bulb. At Re=1350, the maximum velocity vectors in aortic root with axisymmetric sinuses at fully opened angle of the valve leaflets (85^0, t= 0.057sec) are 1.96m/s, which are 28% higher than those in aortic root with axisymmetric bulb at same time and angle (Figure 5).

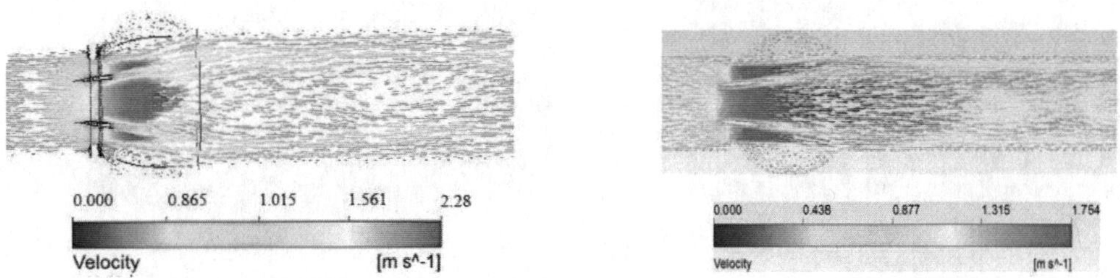

(a) Axisymmetric sinuses (b) Axisymmetric bulb

Figure 5. Velocity vectors at t=0.057sec (Re=1350).

Similarly, at the same Reynold's number, high velocity vectors are observed at leaflet's start to close angle (63^0, t=0.237sec) and fully closed angle (25^0, t= 0.8sec) in aortic root with axisymmetric sinuses as compared to aortic root with axisymmetric bulb. The velocity profiles for both geometries at time t=0.237sec and t=0.8sec are illustrated in Figures 6 and 7 respectively.

The same trend, that is high velocity vectors in aortic root with axisymmetric sinuses compared to root with an axisymmetric bulb was observed for the other two physiological flow rates corresponding to inlet Reynold number of 1500 and 1700 (Table 1 and 2).

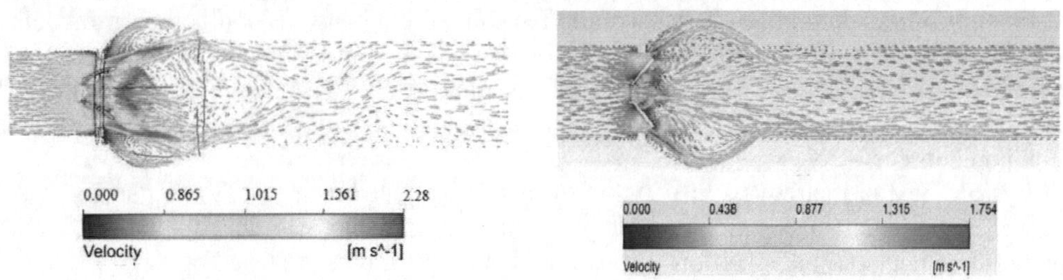

(a) Axisymmetric sinuses (b) Axisymmetric bulb

Figure 6. Velocity vectors at t=0.237sec (Re=1350).

Its also important to note that the flow recirculation and vorticity is higher in aortic root with axisymmetric sinuses, as compared to aortic root with axisymmetric bulb.

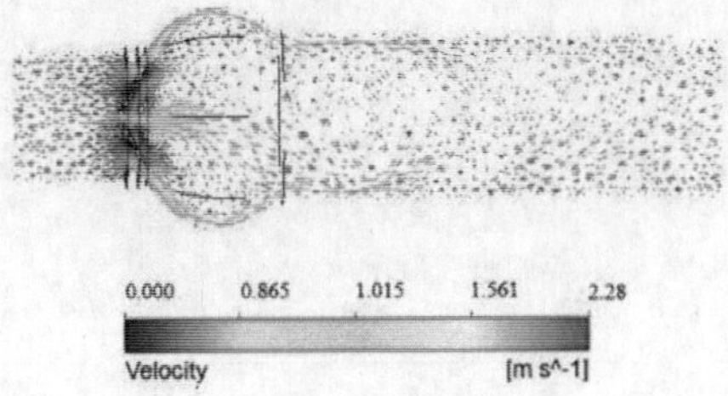

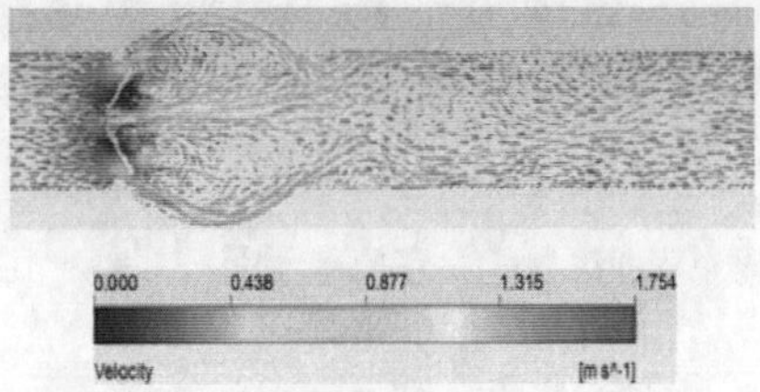

(a) Axisymmetric sinuses (b) Axisymmetric bulb

Figure 7. Velocity vectors at t= 0.8sec (Re=1350).

Based on above results, it can be well said that significantly high velocity vectors are found in aortic root with axisymmetric sinuses as compared to aortic root with axisymmetric bulb. Consequently, on coupling the fluid and structure domains, it has been observed that high von-Mises stresses are developed in BMHV placed in axisymmetric sinuses compared to axisymmetric bulb. The von-Mises stresses contours at different physiological phases for the two geometries can be seen in Figure 8 at Re= 1350. At fully closed phase of leaflets (angle 25^0, t= 0.8sec), the maximum von-Mises stress developed in BMHV placed in aortic root with axisymmetric sinuses is 56 MPa, nearly 48% higher than that in BMHV placed in aortic root with axisymmetric bulb, which is 35.5 MPa. These stresses occurred at the leaflet pin in a short time, however it may cause valve failure in the long run due to yielding as the allowable stress is 32 MPa [21]. At leaflet angle 63^0 and 85^0, the maximum von-Mises stresses in BMHV placed in axisymmetric sinuses are respectively 37% and 22% higher than those in BMHV placed in axisymmetric bulb at Re=1350.

It is worth mentioning that for all the Reynold's numbers and both the aortic root geometries, the von-Mises stresses developed in BMHV at fully open phase (angle 85^0, t=0.057sec) are lower than the allowable stresses. Other than that, higher von-Mises stresses than allowable are developed in BMHV placed in aortic root with axisymmetric sinuses. Therefore, the valve structure is more likely to fail if it has been implanted in such aortic root geometry, that closely mimics to natural aortic root.

IOP Conf. Series: Journal of Physics: Conf. Series **908** (2017) 012039 doi:10.1088/1742-6596/908/1/012039

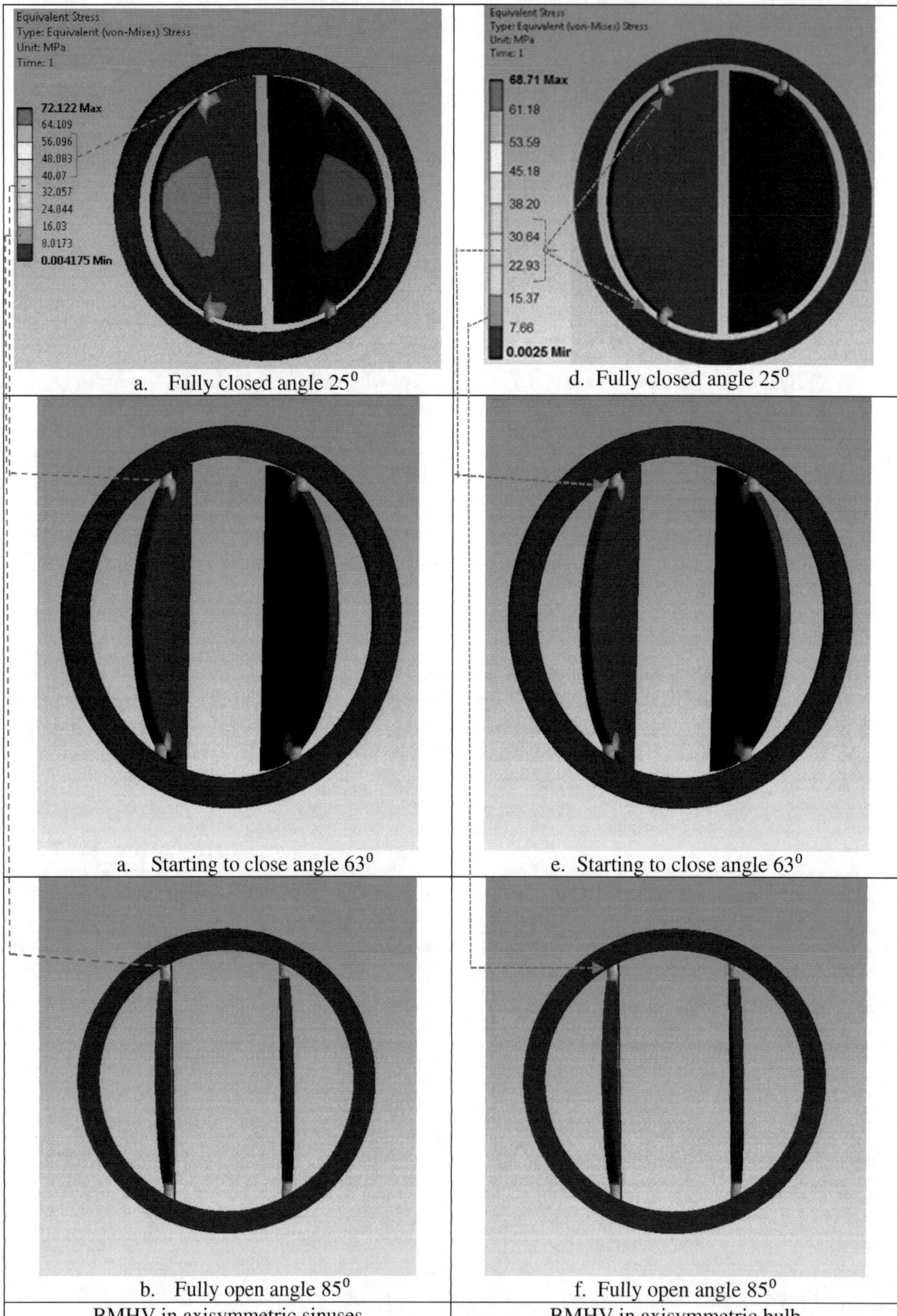

a. Fully closed angle 25^0 d. Fully closed angle 25^0

a. Starting to close angle 63^0 e. Starting to close angle 63^0

b. Fully open angle 85^0 f. Fully open angle 85^0

BMHV in axisymmetric sinuses | BMHV in axisymmetric bulb

Figure 8. von-Mises stresses contours at Re= 1350.

Table 1. Velocity and von-Mises stresses for axisymmetric sinuses.

	Re = 1350		Re = 1500		Re = 1700	
Valve Leaflet position (θ°)	von-Mises Stress (MPa)	V_{max} (m/s)	von-Mises Stress (MPa)	V_{max} (m/s)	von-Mises Stress (MPa)	V_{max} (m/s)
25	56	1.60	65	1.67	71.5	1.68
63	35	1.88	58.5	1.96	63.5	2.01
85	14	1.96	31	2.10	32	2.34

Table 2. Velocity and von-Mises stresses for axisymmetric bulb.

	Re = 1350		Re = 1500		Re = 1700	
Valve Leaflet position (θ°)	von-Mises Stress (MPa)	V_{max} (m/s)	von-Mises Stress (MPa)	V_{max} (m/s)	von-Mises Stress (MPa)	V_{max} (m/s)
25	35.5	0.96	40.5	0.88	47.5	0.74
63	25.5	1.35	40	1.45	43.5	1.18
85	11.5	1.54	20.5	1.75	29	1.97

4. Conclusion

Velocity profiles and von-Mises stresses developed in bileaflet mechanical heart valves placed in two types of aortic root geometries have been determined and compared at different physiological flow rates. It has been found that high velocity vectors and high von-Mises stresses, especially at the leaflet pins are developed in valves placed in aortic root with axisymmetric sinuses as compared to that placed in aortic root with axisymmetric bulb. Since the structural stresses developed in BMHV for both the geometries are higher than allowable stresses for most of the physiological conditions, the valve structure is expected to fail due to yielding and this could be one of the unknown reasons of valve failure.

5. References

[1] Hong T H, Choi C R and Kim C N 2009 *World Academy of Science, Engineering and Technology* **25** 679-684
[2] Zhou F, Cui Y Y, Wu L L, Yang J, Liu L, Maitz M F, Brown I G and Huang N 2016 *Journal of Medical and Biological Engineering* **36** 110-120
[3] Peskin C S 1977 *Journal of Computational Physics* **25** 220-252
[4] Kuhl E, Askes H and Steinmann P 2004 *Comput. Methods Appl. Mech. Engrg.* **193** 4207–4222
[5] Michler C, Hulshoff S J, Brummelen E H and Borst R 2004 *Computers & Fluids* **33(5)** 839-848
[6] Dumont K, Stijnen J M, Vierendeels J, van de Vosse F N and Verdonck P R 2004 *Comput Methods Biomech Biomed Engin.* **3** 139-46
[7] Nobili M, Morbiducci U, Ponzinic R, Gaudiod C D, Balduccid A, Grigionid M, Montevecchib F M and Redaelli A 2008 *Journal of Biomechanics* **41** 2539–2550
[8] Yeh H H, Grecov D and Karri S 2014 *Journal of Medical and Biological Engineering,* **34(5)** 482-486
[9] Nguyen V T, Kuan Y T, Chen P Y, Ge L, Sotiropoulos F, Yoganathan A P and Leo H H 2012 *Cardiovascular Engineering and Technology* **3** 88-100
[10] Hanafizadeh P, Mirkhani N, Davoudi M R, Masouminia M and Sadeghy K 2016 *Artificial Organs* **40** 179-191
[11] de Tullio M D, Pedrizzetti G, and Verzicco R 2010 *Acta Mech* **216** 146-163

[12] Grigioni M, Daniele C, Gaudio C D, Morbiducci U, Balducci A, D'Avenio G and Barbaro V 2005 *ASAIO Journal* **51(3)** 176-183
[13] Kwon Y J 2008 *Journal of Mechanical Science and Technology* **22** 2038-2047
[14] Mazumdar J 1989 *An introduction to Mathematical physiology and biology* (Adelaide: Cambridge University Press)
[15] Nakamura T and Kaneko S 2008 *Flow Induced Vibrations* (Tokyo: Elsevier Science)
[16] Reul H, Vahlbruch A, Giersiepen M, Schmitz-Rode T, Hirtz V, and Effert S 1990 *J Biomech.* **23** 181- 191
[17] De Paulis R, De Matteis G M, Nardi P, Scaffa R, Colella D F and Chiarello L 2000 *Ital Heart J.* **7** 457-463
[18] Annerel S, Claessens T, Van Ransbeeck P, Segers P, Verdonck P and Vierendeels J 2011 *Aortic Valve* (Croatia: Intech) p 29
[19] Guyton A C and Hall J E 1994 *Textbook of medical physiology* (Philadelphia: SAUNDERS ELSEVIER)
[20] Schlosser T, Pagonidis K, Herborn C U, Hunold P, Waltering K U, Lauenstein T C and Barkhausen J 2005 *AJR Am J Roentgenol* **184** 765-773
[21] Nasif M S, Kadhim S K, Al-Kayiem H H and Al Waked R 2016 *ARPN Journal of Engineering and Applied Sciences* **11** 11971-11974

Formation of ultra Si/Ti nano thin film for enhancing silicon solar cell efficiency

T Adam[1], T S Dhahi[2], M Mohammed[3], A M Al-Hajj[4], U Hashim[4]

[1]Faculty of Technology, Universiti Malaysia Perlis (UniMAP), Kampus Uniciti Alam Sg. Chuchuh, 02100 Padang Besar (U), Perlis, Malaysia.
[2]Department of physics, University of Basra, Basra, Iraq.
[3]Center of Excellence Geopolymer and Green Technology, School of Materials Engineering, Universiti Malaysia Perlis (UniMAP), 01007, P.O Box 77, D/A PejabatPosBesar, Kangar, Perlis, Malaysia.
[4]School of Electrical system Engineering, Universiti Malaysia Perlis (UniMAP)Malaysia.
[5]Institute of Nano Electronic Engineering, University Malaysia Perlis, Perlis, 01000 Kangar, Malaysia.

tijjani@unimap.edu.my

Abstract. An alternative electrical source has l has become the major quest of every researchers due to it numerous advantages and applications of power supply and as electronic devices are becoming more and more portable. A highly efficient power supply is become inevitable. Thus. in this study, present ultrasonic based assisted fabrication of electrochemical silicon-Titanium nano thin film by in-house simple technique, uniformly silicon Nano film was fabricated and etched with HF (40%): C2H5OH (99%):1:1, < 20 nm pore diameter of silicon was fabricated. The surface and morphology reveal that the method produce uniform nano silicon porous layer with smaller silicon pores with high etching efficiency. The silicon-Titanium integrated nano porous exhibited excellent observation properties with low reflection index ~ 1.1 compared to silicon alone thin film.

1. Introduction

Solar cell is one of the most recent development that has attracted [1-5]], due to it potential, the quest several material has keep up growing especially silicon nano porous has a potential to play an important role in visible light emission related applications, moreover, silicon nano porous (PS) devices constitute emerging platforms for including selective molecule separation and sensing, with great potential for high throughput and economy in manufacturing and operation. However, fabrication of this device poses some challenges that researcher find it necessary to solve [6-11]. The most commonly used method of fabricating PS is direct current (DC) anodic electrochemical etching and during the DC anodic etching process, the reaction results in silicon fluoride. This product tends to cover the pore tips during deposition process, hydrogen gas bubbles are adsorbed at the surface of silicon pillars because of interfacial constraint, blocking the silicon pores and leading to a reduction of HF concentration inside the pores [12-17]. This will lead to the etching process slowing down and dissolved species will increase the resistance of silicon wafer and hence decrease the current density [1].

The previous study presented using an ultrasonically enhanced anodic electrochemical Etching which is developed by [1, 18-20] to fabricate light-emitting PS material. They study took the advantage of the ultrasonic press effect and the diffusion of the dissolved species and H2 bubbles from silicon pores was accelerated [1, 21-24. They results has improved qualities in surface morphology, layer interface smoothness etching efficiency and optical characteristic compared with the sample prepared by DC etching. However, the current study presented and improve results smoothness etching efficiency, improved qualities in surface morphology and pore size that could increase it application in various field including biomedical, environment and life sciences such as water filtration etc[25, 26]. The study demonstrates a simple in-house fabrication of enhanced silicon solar cell using ultrasonic vibration etching. Well distributed photovoltaic cell can be produced in just 5 simple steps. The Experimental study results show that the silicon based fabricated photovoltaic devices using titanium as anti-reflector exhibit high photo energy absorption with high voltage when exposed to the spectrum of the solar radiation. The high efficiency obtained as a result of structures of Ti in which it shows the surface layers energy greatly enhanced the light-absorption capability of cell is greatly increased. Hence the methods are simple and at no cost, a stable solar cell can be produced.

2. Methods

The silicon wafer was cleaned using Standard Cleaning (SC) or RCA and Buffered Oxide Etch (BOE) solution. There are four (4) steps of procedure for wafer cleaning process. First, the wafer was dipped into RCA 1 at 70oC to 80oC for 10 minutes.Then, the wafer was dipped into BOE solution for 10 to 15 seconds to remove native oxide.After that, the wafer was dipped into RCA 2 solution for 10 to 15 minutes while the solution was heated to 75oC to 80oC., Then the wafer was rinsed in DI water or distilled water. Finally, the wafer was dried by blowing with spin dry. A single crystal p-type with (100)-oriented highly doped silicon substrate is used an it was placed in a Teflon etching cell and etched in the dark with a HF (40%): C2H5OH (99%):1:1 (by volume) electrolyte solutions were prepared for ultrasonic anodic etching using the ultrasonic wave frequency of the ultrasonic generator was 33 ^ 3 kHz. After the etching process, all the samples was immediately rinsed by deionized water and dried, for drying, a vacuum dryer was used. The etching current density was 55 (mA/cm2) for 1 minute and duty cycle of 0.7.

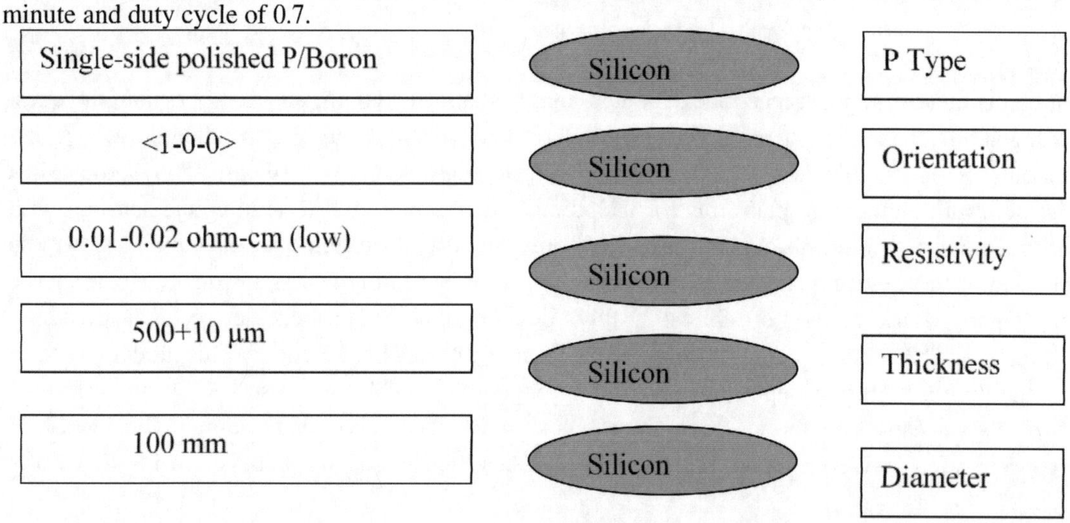

3. Results and Discussion

The study has successfully demonstrated fabrication and testing of the device , the figure 1, the sem image of the fabricated silicon based solar cell, as it can be observed, there is uniform distribution of size and approximately, 1um is obtained, thus it demonstrated a simple in-house fabrication of enhanced silicon solar cell using ultrasonic vibration etching. Well distributed photovoltaic cell can be produced in just 5 simple steps. The Experimental study results show that the silicon based fabricated photovoltaic devices using titanium as anti-reflector exhibit high photo energy absorption with high voltage when exposed to the spectrum of the solar radiation. The high efficiency obtained as a result of structures of Ti in which it shows the surface layers energy greatly enhanced the light-absorption capability of cell is greatly increased. Hence the methods are simple and at no cost, a stable solar cell can be produced.

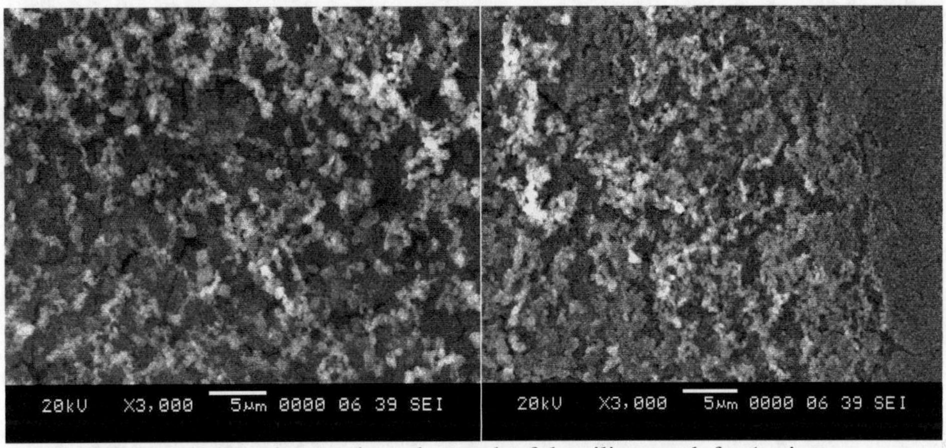

Figure 1. SEM image show the result of the silicon etch for 1 minute.

The figure 1, show the etch profile and the behavior of charged molecules present in the sensor surface, from the series of the modification step, the device electrical profile are continue to altered as at any time when device is subjected surface treatment. This behavior is resultant from the presence of two distinct polarized charges groups that are promoted to come closer with each other as results of minimizing distance of each electrode, thus the band energy gap is reduced there by higher electrons exchanges occurred, the exchange is further becomes higher with modification elements are in aqueous form, this claim can be observed form figure where the silicon application, is normally in aqueous and upon drop in, it was subject to test before fully dried and this has allowed a higher current to be recorded with Si/Ti compared with si, the reason is simple be at the aqueous the charge accumulation increase becomes eligible to jump from one electrode to another resulting in higher negative charge on the Ti/Si surface. The increase in ion charge enhance the charge accumulation on each electrode and this behavior consistent with many results of a similar experiment performed. The study has successfully demonstrated fabrication and testing of the device , the figure 1, the sem image of the fabricated silicon based solar cell, as it can be observed, there is uniform distribution of size and approximately, 1um is obtained, thus it demonstrated a simple in-house fabrication of enhanced silicon solar cell using ultrasonic vibration etching. Well distributed photovoltaic cell can be produced in just 5 simple steps. The Experimental study results show that the silicon based fabricated photovoltaic devices using titanium as anti-reflector exhibit high photo energy absorption with high voltage when exposed to the spectrum of the solar radiation. The high efficiency obtained as a result of structures of Ti in which it shows the surface layers energy greatly enhanced the light-absorption capability of cell is greatly increased. Hence the methods are simple and at no cost, a stable solar cell can be produced.

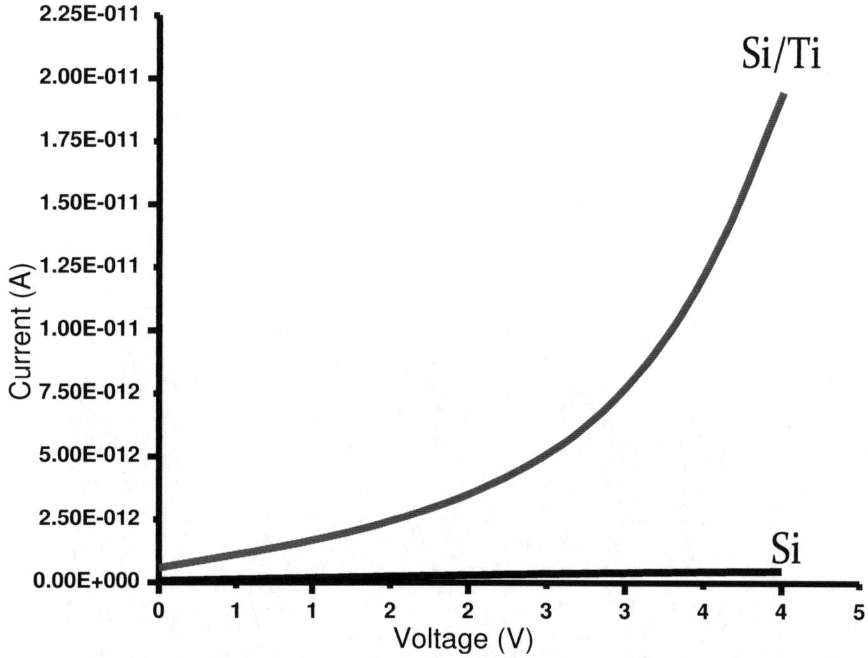

Figure 2. Photo absorbed current-voltage (I–V) density characteristics of photovoltaic devices based on the Si structures and the Si/Ti structure.

The voltages were swept from 0V to 5V using semiconductor analyzer with Volt internal 40mV input signal (0 V D.C Offset). The reference point probe is calibrated with reference gauge of step 0.04V and probing stations are set up fully to smooth and reliable measurements with vacuums environment. The measurement of current–voltage curve for two devices Si and Si/T layers have been conducted with a four-terminal micro-probing Keithley, 6487. Thus, current response of the structures has been measured. In order the to test the device and validate, the two devices were exposed to high bright of tungsten lamp as the energy source because the bright lamp have similar radiation as solar energy do have from the sun photocurrent, after the exposure and upon measuring of the two devices, the resultant voltage obtained from Si/Ti is much higher compared to the one obtained with Si alone and this light-absorption ability coming from its improved photo absorption as a result of the anti-reflection activity of the Ti figure 2. To examine response of the devices silicon configuration was prepared, the traced curved was recorded prior to surface prior deposition, the depostion could allow and enhance the silicon electrical properties by allowing the polarized charged molecules attached to the sensor surface, the modified silicon device was tested in UV with putting gold electrode , the curve in figure above indicated the range of the electrical response of Si/Ti, it exhibited increases in conductance in turn allow more electrons to jump the ide barriers electrode , this is repressed in red curves with ~1nA. The highest curve with indicating highest current values recorded, this wonderful behavior was seen with first light radiation dropped on treated Si/Ti, The curve shown the enhancement achieved as a result of the Ti application to the surface hence the study has successfully demonstrated fabrication and testing of the device , the figure 1, the sem image of the fabricated silicon based solar cell, as it can be observed, there is uniform distribution of size and approximately, 1um is obtained, thus it demonstrated a simple in-house fabrication of enhanced silicon solar cell using ultrasonic vibration etching. Well distributed photovoltaic cell can be produced in just 5 simple steps. The Experimental study results show that the silicon based fabricated photovoltaic devices using titanium as anti-reflector exhibit high photo energy absorption with high voltage when exposed to the spectrum of the solar radiation. The high efficiency obtained as a result of structures of Ti in which it shows the surface layers energy greatly enhanced the light-absorption capability of cell is greatly increased. Hence the methods are simple and at no cost, a stable solar cell can be produced.The increase in the photo voltage indicates higher conversion efficiency will be obtained by application of The Ti structure. The photons ray incident which is projected to on the surface of silicon will be generally reflected from the top surface due silicon low absorption capabilities of the bulk silicon substrate.

4. Conclusions

The study demonstrates a simple in-house fabrication of enhanced silicon solar cell using ultrasonic vibration etching. Well distributed photovoltaic cell can be produced in just 5 simple steps. The Experimental study results show that the silicon based fabricated photovoltaic devices using titanium as anti-reflector exhibit high photo energy absorption with high voltage when exposed to the spectrum of the solar radiation. The high efficiency obtained as a result of structures of Ti in which it shows the surface layers energy greatly enhanced the light-absorption capability of cell is greatly increased. Hence the methods are simple and at no cost, a stable solar cell can be produced.

References

[1] Y. Liu, Z.H. Xiong, Y. Liu, S.H. Xu, X.B. Liu, X.M. Ding, X.Y. Hou “A novel method of fabricating porous silicon material: ultrasonically enhanced anodic electrochemical etching” Solid State Communications 127 (2003) 583–588.

[2] Jun-Wei Huang Chien-Kuo Chang , Lu, K.C.-C. Jung-Tang Huang, Chii-Ruey Lin novel approach for nanoporous gas sensor fabrication using anodic aluminum oxidation and MEMS process, Nanotechnology, 2009. IEEE-NANO 2009. 9th IEEE Conference, 26-30 July 2009, Genoa

[3] T.S. Dhahi, U. Hashim, M.E. Ali, Nazwa Taib, "Polysilicon nanogap fabrication using a thermal oxidation process", Microelectronics International, 2012, 29(1), pp.40 – 46.

[4] U. Hashim, Shahrul A. B Ariffin, Tijjani Adam “Fabrication of Polysilicon Nanowires using Trimming Technique” Journal of Applied Sciences Research, 20128(4): 2175-2186,

[5] Tijjani.Adam, U.Hashim, Highly sensitive silicon nanowire biosensor with novel liquid gate Control for detection of specific single-stranded DNA molecules Biosensors and Bioelectronics67 (2015) 656–661.

[6] Md. Eaqub Ali, Th.S. Dhahi, Rasel Das, U. Hashim, DNA hybridization detection using less than 10-nm gap silicon nanogap structure, Sensors and Actuators A 199 (2013) 304–309.

[7] Chen, X., Guo, Z., Yang, G.-M., Li, J., Li, M.-Q., Liu, J.-H., & Huang, X.-J. (2010). Electrical nanogap devices for biosensing. Materials Today, 13(11), 28-41. doi

[8] M Wesam Al-Mufti, U Hashim, M Rahman, Tijjani Adam, M Arshad, Studying Effect Dimensions of Design and Simulation Silicon Nanowire Filed Effect Biosensor, Applied Mechanics and Materials 754, 854-858, 2015.

[9] Yi, M., Jeong, K.-H., & Lee, L. P. (2005). Theoretical and experimental study towards a nanogap dielectric biosensor. Biosensors and Bioelectronics, 20(7), 1320-1326

[10] [6] Uda Hashim, N. T., Thikra S. Dhahi, Azizullah Saifullah. (2011). Polysilicon nanogap structure development using size expansion technique. Microelectronics International, 28(3), 24-30

[11] Wang, C., Huang, J., Wang, J., Gu, C., Wang, J., Zhang, B., & Liu, J. (2009). Fabrication of the nanogapped gold nanoparticles film for direct electrical detection of DNA and EcoRI endonuclease. Colloids and Surfaces B: Biointerfaces, 69(1), 99-104

[12] Yexian Wu, T. A., Sebastian Gautsch, Nico De Rooij. (2011). Development of passivated Heteregeneous metal nanogaps using E-Beam Overlay Technique. Paper presented at the Proc. Eurosensors XXV, Anthens, Greece.

[13] Francois Beguin, E. F. (Ed.). (2013). Supercapacitors, Materials,Systems, and Applications (Vol. 4). Germany: Wiley-VCH Verlag GmbH & Co.

[14] Dimitrios P.Nikolelis, T. V., Arzum Erdem, Georgia-Paraskevi Nikoleli (Ed.). (2014). Portable biosensing of food toxicants and environmental pollutants (Vol. 1). New York: CRC Press.

[15] Tijjani Adam U. Hashim and Th S. Dhahi, Silicon nanowire formed via shallow anisotropic etching, Si-ash-trimming for specific DNA and electrochemical detection, Chinese Physics B Vol. 24, No. 6 (2015) 06810

[16] Tijjani Adam, U Hashim, Design and fabrication of micro-mixer with short turns angles for self-generated turbulent structures, Microsystem Technologies, 1-8, 2015

[17] T Adam, U Hashim, TS Dhahi, KN Khor, PS Chee, PL Leow, ELECTROCHEMICAL ETCHING: An Ultrasonic Enhance Method of Silicon Nano Porous Fabrication, Wulfenia Journal 20 (1), 45-55, 2013

[18] Tijjani Adam, H Uda, M Eaqub, PL Leow, The electroosmosis mechanism for fluid delivery in PDMS multi-layer microchannels, American Scientific Publishers All rights reserved.2013

[19] U Hashim, T Adam, J Lung, P Ling, Fabrication of Microchannel and Micro Chamber for Microfluidic Lab-on-Chip.Australian Journal of Basic & Applied Sciences 7 (1)2013

[20] T Adam and U. Hashim Light Observation in Polymer: A Study of Silicon-Based Organic Polymer Using Visible Spectroscopy, Australian Journal of Basic and Applied Sciences 7 (1), 76-80, 2013.

[21] U Hashim, MW Al-Mufti, T Adam, Silicon Nanowire Geometry: Investigation of Interaction Site Potential in Semiconductor-DNA Interaction, Australian Journal of Basic and Applied Sciences 7 (5), 242-245, 2013

[22] MW Al-Mufti, U Hashim, T Adam, Simulation of Nano lab on chip devices by using COMSOL MultiphysicsJournal of Applied Sciences Research 9 (2), 1056-1061, 2013

[23] T Adam, U Hashim, KL Foo, TS Dhahi, T Nazwa, Technology development for nano structure formation: Fabrication and characterization, Advanced Science Letters 19 (1), 132-137, 2013

[24] T Adam, U Hashim, PL Leow, KL Foo, PS Chee, Selection of optimal parameters in fabrication of poly (dimethylsiloxane) microfluidics using taguchi method, Advanced Science Letters 19 (1), 32-36,2013

[25] T Adam, U Hashim, KL Foo, Microfluidics design and fabrication for life sciences application, Advanced Science Letters 19 (1), 48-53, 2013

[26] T Adam, U Hashim, ME Ali, PL Leow, The electroosmosis mechanism for fluid delivery in PDMS multi-layer microchannel, Advanced Science Letters 19 (1), 12-15, 2013

IOP Conf. Series: Journal of Physics: Conf. Series **908** (2017) 012041 doi:10.1088/1742-6596/908/1/012041

Investigating the effects of PDC cutters geometry on ROP using the Taguchi technique

A A Jamaludin[1], N M Mehat[2] and S Kamaruddin[1]

[1] Mechanical Engineering Department, Universiti Teknologi PETRONAS, 32610 Bandar Seri Iskandar, Perak, Malaysia.

[2] Mechanical Engineering Technology Department, Universiti Malaysia Perlis, 02600 Arau, Perlis, Malaysia.

amirulaliff87@yahoo.com[1], nikmizamzul@unimap.edu.my[2], shahrul.k@utp.edu.my[1]

Abstract. At times, the polycrystalline diamond compact (PDC) bit's performance dropped and affects the rate of penetration (ROP). The objective of this project is to investigate the effect of PDC cutter geometry and optimize them. An intensive study in cutter geometry would further enhance the ROP performance. The relatively extended analysis was carried out and four significant geometry factors have been identified that directly improved the ROP. Cutter size, back rake angle, side rake angle and chamfer angle are the stated geometry factors. An appropriate optimization technique that effectively controls all influential geometry factors during cutters manufacturing is introduced and adopted in this project. By adopting L_9 Taguchi OA, simulation experiment is conducted by using explicit dynamics finite element analysis. Through a structure Taguchi analysis, ANOVA confirms that the most significant geometry to improve ROP is cutter size (99.16% percentage contribution). The optimized cutter is expected to drill with high ROP that can reduce the rig time, which in its turn, may reduce the total drilling cost.

1. Introduction

PDC cutters have gained popularity in petroleum drilling due to its durability together with its ability to maintain a high ROP. Nevertheless, due to abrasive wear, impact and abrasive damage, the extension of the cutters application into hard rock drilling has been the greatest challenges that faced by today manufacturer. A lot of studies have been conducted for a better understanding of cutters geometry with an intention to enhance drilling performance.

Designing cutters with higher ROP efficiency and longer cutter's life are quite often the objective. There is a various investigation conducted discussing the influence of cutter geometry towards the drilling performance. Knowlton [1] proposed a modified geometry of cutter with positive rake angle which overcame the difficulties encountered in drilling shale formations. While, in another study, Gerbaud et al. [2] adding more complexities to cutter geometry and incorporating the chamfer made on some cutters in their established model for better drilling efficiencies. Generally speaking, cutters optimization refers to designing a cutter with the highest ROP efficiency and longest drilling. Thus, in order to minimize the critical condition during drilling, optimization technique was quite often applied to simplify the manufacturer's work in determining the optimal design features during their manufacturing process. Taguchi technique is known as one of best and effective optimization techniques [3].

Developed by Dr. Genichi Taguchi, the Taguchi technique [4] is widely applied to optimize the manufacturing process through the design parameters setting. In order to plan a good framework and deciding the ideal controllable parameter settings, which are invulnerable to noise, the Taguchi technique has the solution [5]. Taguchi technique presents an integrated approach that is very straightforward in order to find the best range of design parameters for the best performance and

IOP Conf. Series: Journal of Physics: Conf. Series **908** (2017) 012041 doi:10.1088/1742-6596/908/1/012041

quality [6]. The (1) parameter design, (2) system design and (3) tolerance design are the most three common phases adapted in this technique [7]. Out of all the three, parameter design is recognized as the most vital for process optimization [8]. Orthogonal Array (OA), signal-to-noise (S/N) ratios, and ANOVA are the special features that have been introduced in Taguchi technique. Taguchi OA is a fractional orthogonal design used to appraise the main effects by running only a few experiments. As a consequence, using OA can decrease the costs and at the same time examining a huge number of parameters rather than studying one parameter at a time as done in conventional techniques [9]. The Taguchi concept was further discussed by Roy [10].

The primary objective of this paper is to study the impacts of cutter geometry on the ROP. By incorporating with L_9 Taguchi OA, the explicit dynamics finite element analysis (FEA) were run to obtain the ROP results. Using the S/N ratio and ANOVA, The optimal geometry factors or levels and the impacts of cutter geometry on ROP performance are further evaluated.

2. Methodology

For the finite element analysis, a single cutter with the dimension of 8mm, 13mm and 16mm diameter, 8mm height and 2mm diamond table thickness is used as a model and is shown in Figure 1. In this study, the ROP is simulated by using ANSYS explicit dynamic finite element analysis model. The initial conditions file consist of a pre-stress object to control the exchange of information from an implicit static or transient structural analysis to the explicit dynamics analysis. Transferable data consists of the displacements or a complete material state while the analysis settings incorporate boundary condition, body interaction and erosion are defined under explicit dynamics.

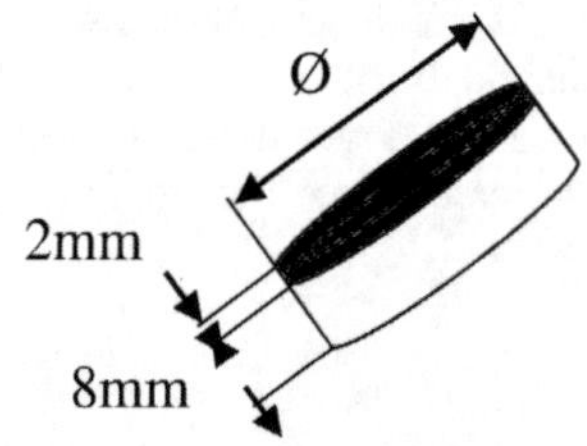

Figure 1. Geometry and dimension of PDC cutter.

With 300 rpm, the angular velocity is defined as the initial condition. Other parameters like the formation fixed support, the bit constant angular velocity and the bit force are embedded under analysis settings. In the real situation, weight on bit is the weight of the drill pipe up to the surface. This weight acting as a force to push the bit into the rock formation. In this simulation, the value of weight on bit was set as 100,000 Newton.

2.1. Geometrical factors and simulation

The objective of this work is to find out the influences of geometry factors on drilling performance. ROP is chosen as the quality objective to be considered. Based on the previous study, four geometries are selected to be run in FEA simulation. Table 1 shows the geometry factors and level values. L_9 (3^7) OA is chosen as the experimental layout for simulation runs. Table 2 presented the L_9 OA.

Table 1. Geometry factors and levels studied.

Column	Factors	Level 1	Level 2	Level 3
A	Cutter size (mm)	8	13	16
B	Back rake angle (°)	0	5	25
C	Side rake angle (°)	0	15	30
D	Chamfer angle (°)	15	30	45

Table 2. L_9 OA used in simulation.

Factors/ Run No.	A Cutter size (mm)	B Back rake angle (°)	C Side rake angle (°)	D Chamfer angle (°)	Results ROP (m/hr)
1	8	0	0	15	85.62
2	8	5	15	30	75.60
3	8	25	30	45	68.32
4	13	0	15	45	11.24
5	13	5	30	15	11.78
6	13	25	0	30	11.21
7	16	0	30	30	10.36
8	16	5	0	45	8.54
9	16	25	15	15	8.75

The simulation run shows the number of conducted simulations (Table 2). Nine simulation runs are conducted in this study and allocated as trial 1 to trial 9. Factor A to factor D represent the geometry factors: A for cutter size, B for back rake angle, C for side rake angle and D for chamfer size (Table 1). Each factor is allocated to column 1 to column 4. While, the ROP is recorded in the last column. The simulations are run as displayed in Table 2. In Trial 1, the simulation is conducted by a set of geometry factors consist of cutter size (A) at 8mm, back rake angle (B) at 0°, side rake angle (C) of 0° and chamfer angle (D) of 15° respectively.

3. Analysis method

3.1. Signal-to-noise (S/N) ratio

In order to identify the quality of the output, Taguchi presented the utilization of signal-to- noise (S/N) ratio. The S/N ratio is a measurement of performance focusing on developing processes and products that invulnerable to uncontrollable factors [11]. Depending on the quality characteristic, there are differences of S/N available that can be categorized to smaller-the-better, nominal-the-best and higher-the-better where each type of them are represented by different formula calculation and depend on experiment objective [12]. The larger value of S/N ratio indicates the robustness of the product against the noise and this principle is similarly applicable for all three smaller-the-better, nominal-the-best and higher-the-better characteristics [13]. In this study, the equation higher-the-better can be used for analysis of ROP properties. In this study, the equation for "higher-the-better", as shown in equation (1) can be used for analysis of ROP.

$$S/N = -10\log_{10}\left[{(y_1^2 + y_2^2 + y_3^2 + \cdots + y_n^2)}/{n} \right] \tag{1}$$

where the value of the quality characteristics is denoted as y and the number of repetitions is denoted as n.

4. Results and discussion

The following steps were conducted to decide the optimal cutter geometry; (1) S/N ratios for the simulation results were computed, (2) Main effect graph was evaluated (3) Geometry factor analysis of variance (ANOVA) were calculated.

4.1. S/N ratio computation

In this study, the simulation results of ROP performance for all nine runs are analyzed by using S/N ratios based on the larger-the-better quality characteristic (Equation 1). Table 3 summarized the S/N ratios for all nine runs.

Table 3. S/N response for L_9 OA.

Run No.	ROP (dB)
1	38.65
2	37.57
3	36.10
4	21.02
5	21.42
6	20.99
7	20.31
8	18.63
9	18.84

4.2. Main effect analysis

The main effects for each factor are calculated from the S/N ratio results shown in Table 3. Through the main effect graph, the changes in ROP due to the variations in cutter geometry can be analyzed. The optimal sets of geometry factors and levels can be predicted further. Figure 2 shows the main effects of Taguchi technique.

Basically, the larger value of S/N ratio will produce the better ROP performance. Referring to Figure 2, A_1, B_1, C_1, and D_2 show the largest value of the S/N ratio for geometries A, B, C, and D respectively. As a result, the optimal geometry which statistically results in the maximum ROP are predicted to be the cutter size of 8mm, back rake angle of 0°, side rake angle of 0° and chamfer angle of 30°.

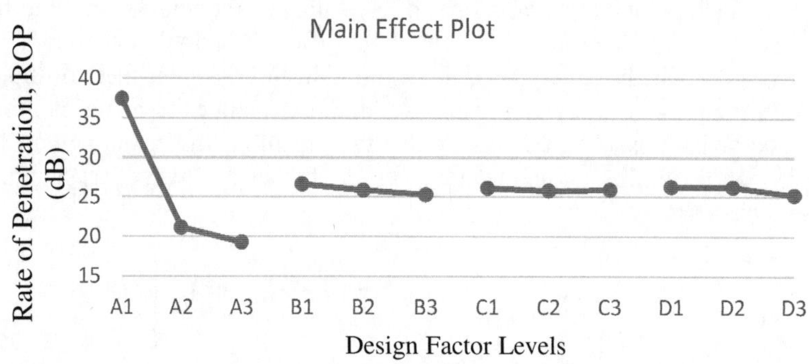

Figure 2. Main effect plot for ROP.

4.3. Analysis of variance (ANOVA)

Analysis of variance (ANOVA) was conducted on the S/N ratio results (Table 3) to investigate the influence of individual factor on ROP performance. Table 4 present the degrees of freedom (DOF), the sum of square (S), variance (V), and the percentage contribution (P).

Table 4. ANOVA for ROP S/N ratio.

Column	Factors	DOF	S	V	P (%)
A	Cutter size	2	599.64	299.82	99.16
B	Back rake angle	2	2.76	1.38	0.46
C	Side rake angle	2	0.12	0.06	0.02
D	Chamfer angle	2	2.19	10.95	0.36
	Other/error	0	0	0	0
	Total	8	604.71		100

Referring to Table 4, as a level of confidence of 95% is used in this study, the significance of the geometry factors to the ROP was cutter size, back rake angle, side rake angle and chamfer angle accordingly. The most decisive factor will be given the maximum value of the percentage of contribution for each geometry. As shown in Table 4, it can be observed that the cutter size is the most influential geometry which demonstrates the highest P of 99.16%. The analysis revealed that the size had the strongest correlation to the ROP performance in cutter operation and in this case the predicted optimum melt temperature for the part is 8mm.

5. Conclusion

The optimal geometry conditions of PDC cutter geometry were investigated by using the Taguchi technique. The ability of the Taguchi's technique to improve the ROP in response with a few numbers of runs with an acceptable accuracy level has been determined. The findings demonstrated that the Taguchi technique can maximize the ROP and can, along this line, be applied in future studies for an assortment fields. In addition, the contribution of each geometry to ROP was also decided by using the procedure explained. Through a series of analysis and optimization, the outcomes displayed that the optimal cutter geometry to results in maximum ROP was decided as A_1, B_1, C_1, and D_2. The cutter size demonstrates the strongest influence among the four vital geometries investigated in this study followed by back rake angle, chamfer size and side rake angle. The information gained in this work should help engineers to make better decisions in maximizing the ROP.

6. Acknowledgement

The author acknowledges the FRGS grant (No. 0153ABK92) provided by MOHE for defraying the work that resulted in this manuscript.

References

[1] Knowlton R H 1990 January PDC bits using positive rake cutters. In *SPE/IADC Drilling Conference*. Society of Petroleum Engineers.

[2] Gerbaud L, Menand S and Sellami H 2006 February PDC bits: all comes from the cutter rock interaction. In *IADC/SPE Drilling Conference*. Society of Petroleum Engineers.

[3] Tsai K M, Hsieh CY and Lo WC 2009 A study of the effects of process parameters for injection molding on surface quality of optical lenses. *J. Mater. Process. Technol.* 209 **7** 3469-3477.

[4] Nellian S 1996 Integration of Taguchi design of experiments and finite element method for robust design *A Masters Project directed by Prof. Ian R* http://www. ecs. umass. edu/mie/labs/mda/fea/sankar/chapter.

[5] Gaitonde V N, Karnik S R, Mata F and Davim J P 2008 Taguchi approach for achieving better machinability in unreinforced and reinforced polyamides. *J. Reinf. Plast. Compos*27 **9** 909-924.

[6] Chang, T.C. and Faison, E., 2001. Shrinkage behavior and optimization of injection molded parts studied by the Taguchi method. *Polym. Eng. Sci.* 41 **5** 703-710.

[7] Yang W P and Tarng Y S 1998 Design optimization of cutting parameters for turning operations based on the Taguchi method. *J. Mater. Process. Technol.* 84 **1** 122-129.

[8] Antony J 2001 Simultaneous optimisation of multiple quality characteristics in manufacturing processes using Taguchi's quality loss function. *Int. J. Adv. Manuf. Technol.* 17 **2** 134-138.

[9] Ozcelik B and Sonat I 2009 Warpage and structural analysis of thin shell plastic in the plastic injection molding. *Mater. Des.* 30 **2** 367-375.

[10] Roy R K 1990 *Handbook on A primer on the Taguchi method.* Van Nostrand Reinhold: Hanser.

[11] Oktem H, Erzurumlu T and Uzman I 2007 Application of Taguchi optimization technique in determining plastic injection molding process parameters for a thin-shell part. *Mater. Des.* 28 **4** 1271-1278.

[12] Ganapathy T, Murugesan K, Gakkhar R P 2009 Performance optimization of Jatropha biodiesel engine model using Taguchi approach. *Applied Energy* **86** 2476–2486.

[13] Maghsoodloo S, Ozdemir G, Jordan V, Huang C H 2004 Strengths and Limitations of Taguchi's Contributions to Quality, Manufacturing, and Process Engineering. *Journal of Manufacturing Systems* **23** 2 73-126.

Design and static structural analysis of a race car chassis for Formula Society of Automotive Engineers (FSAE) event

M L Mohamad, M T A Rahman, S F Khan, M H Basha, A H Adom and M S M Hashim

Mechanical Engineering Programme, School of Mechatronic Engineering, Universiti Malaysia Perlis (UniMAP) Pauh Putra Campus, 02600 Arau, Perlis, Malaysia.

tasyrif@unimap.edu.my

Abstract. The main purpose of this study is to make improvement for the UniMAP Automotive Racing Team car chassis which has several problems associated with the chassis must be fixed and some changes are needed to be made in order to perform well. This study involves the process of designing three chassis that are created based on the rules stated by FSAE rules book (2017/2018). The three chassis will undergo analysis test that consists of five tests which are main roll hoop test, front roll hoop test, static shear, side impact, static torsional loading and finally one of them will be selected as the best design in term of Von Mises Stress and torsional displacement. From the results obtained, the new selected chassis design which also declared as the new improved design poses the weight of 27.66 kg which was decreased by 16.7% from the existing chassis (32.77 kg). The torsional rigidity of the improved chassis increased by 37.74%.

1. Introduction

In automobile, chassis is one of the important parts of car because chassis is a skeletal edge that holds much type of different mechanical parts such as the suspension system, braking and handling, power train, engine, body and tires [1]. Poor chassis design and strength may lead to failure for other mechanical parts to function well. Therefore, chassis can be called as backbone to all the car system [2]. Just like human body which the backbone plays an important role in holding other body organ, muscle and skin. The other main function of a car chassis is to deal with the dynamic and static load which is applied to the chassis with the hope that the chassis will have less failure such as distortion and deflection due to the load.

The basic foundation of the FSAE chassis car is that the supporting chassis to which the engine and other mechanical parts are fastened [3]. The new design must undergo several testing before it will be develop. This is important to guarantee the best performance of the new chassis with less failure due to applied load during racing [4]. In order to design new chassis, the method should follow the guidelines set by FSAE (Formula Society Automotive Engineering) International. All the guidelines are already mentioned in the 2017-2018 Formula SAE Rules [5]. The regulation does not make the burden because it involves the driver's safety apart to teach engineering students to apply what they have learn in lecture and make practice on their skills [4],[6].

IOP Conf. Series: Journal of Physics: Conf. Series **908** (2017) 012042 doi:10.1088/1742-6596/908/1/012042

The FSAE car chassis must be rigid enough because it has to withstand vibration, twist, buckling, shock and many other stresses. Other than strength issue, the chassis design must be good enough to allow adequate bending and torsion stiffness in order to improve the handling system [5]. To make sure the good chassis are created, the design of chassis, ability of the chassis to handle load with less failure, ergonomics issue, weight [7] and ability to withstand several stresses must be consider while making the chassis.

In the existing FSAE car chassis, the chassis is too heavy due to the design that is too dense and many unimportant structures also cause in increasing the weight. Second problem is the engine bay and cockpit is too narrow causing problem when installing engine and the narrow cockpit affecting the driver's comforts. The existing design also is inappropriate due to loads that are not properly distributed along the chassis.

Many researches develop their own methods in order to make an analysis test of their chassis design. In designing process, there are many type of software that are used while designing the chassis. Some of the researchers use CATIA V5 software [1, 6, 8-11], AutoCAD [4], Solidwork [6], Pro-E [12]. Every CAD software used to design chassis have their own advantages and disadvantages, but most of the researches prefer to use CATIA V5 software because CATIA is better at complex bended surfaces and is utilized widely and maybe only as a part of the aviation and automotive enterprises [8]. For analysis test, there are many software that can be used for gaining the precise analysis data. Familiar analysis software that always be used by researches are ANSYS [1-4, 7, 10], and LS-DYNA [7, 9].

2. Materials and methods

2.1. Materials

The selection of material must not against the rules that stated in FSAE 2017-2018 Rules Book [5]. In the previous study, steel that always been used was Steel Grade ASTM A36 due to its suitability in many factors. This material is always be the best choice because of its availability, relatively low cost steel compare to others such as stainless steel and aluminium, and what is more importance than other factors is the ability of the material to handle certain load and impact if sudden load is applied to the material [14]. For this study, the same kind of material is used because it is easy to be welded with any type of welding such as Shielded Metal Arc Welding (SMAW), Metal Inert Gas (MIG) and Gas Tungsten Arc Welding (GTAW) and will produce high quality join [12]. In machining process, this type of steel is good for machining process because the rate is calculated to be more than 72%. The average of surface cutting feed is around 120 feet per minute [12].

2.2. Method Used

The study of chassis consists of several processes before full results can be obtained. Proper method must be well planned and considered before and during the study process in order to achieve the objective and solve all the problems as stated. The process started by designing the chassis by considering the specification in the FSAE 2017/2018 rules book [6]. For this study, three different designs were created. The chassis design can be referred in the Figure 1.

IOP Conf. Series: Journal of Physics: Conf. Series **908** (2017) 012042 doi:10.1088/1742-6596/908/1/012042

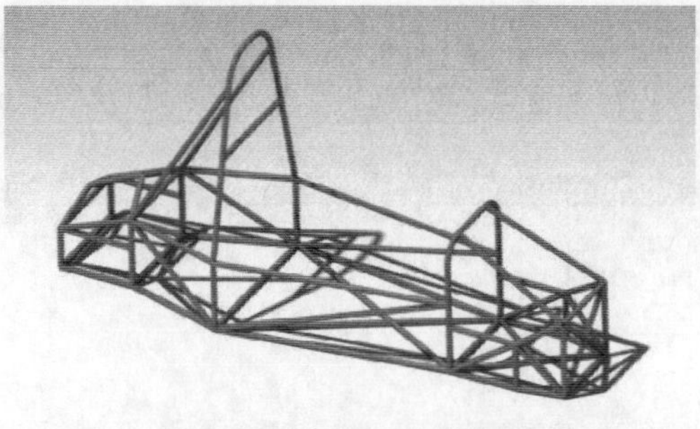

(a): Existing design

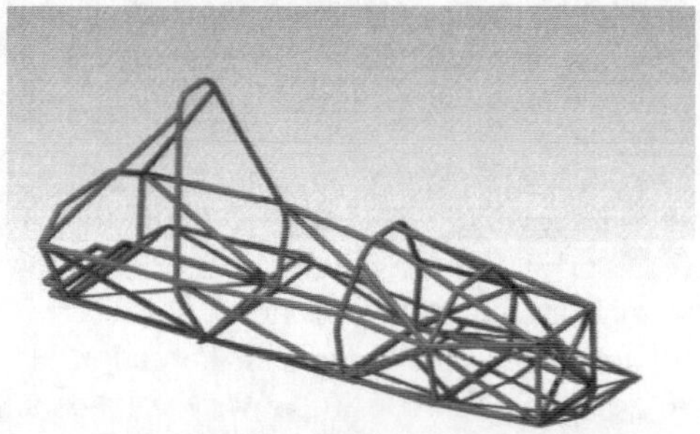

(b): Design 1

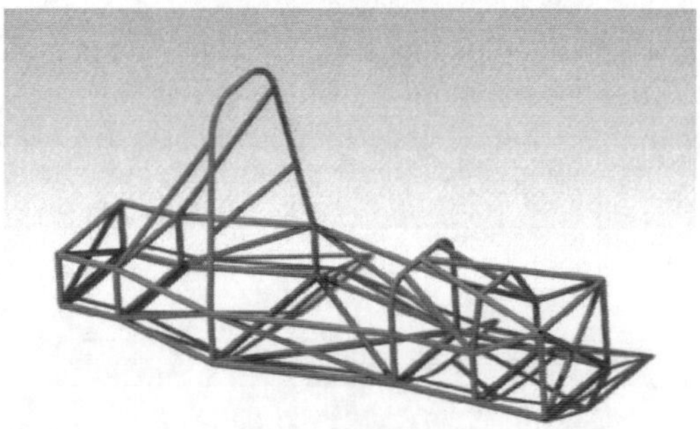

(c): Design 2

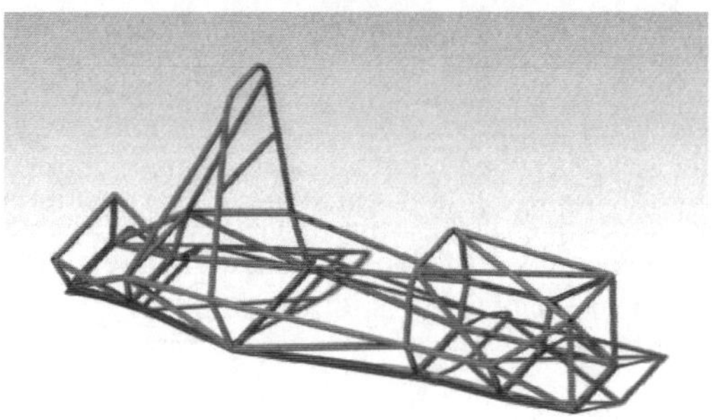

(d): Design 3

Figure 1. Chassis designs of UniART Racing Car.

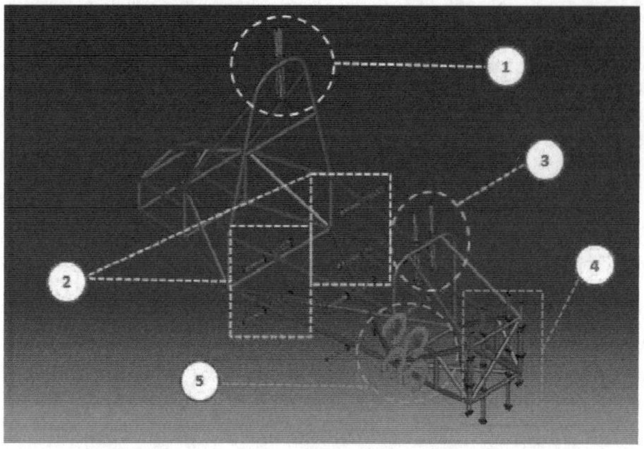

Figure 2. Load applied to different parts of chassis designs.

Based on Figure 2, Number 1 shows the load applied at the top of main roll hoop with the load in x, y and z-axis of 6.0kN, 5.0kN and -9.0kN respectively. The boundary condition for this test is by fixing the both bottom structure of front and main roll hoop. By referring to Number 2, the loads are applied towards the driver with loads of 7.0kN in y-axis direction. The boundary condition is by clamping both front and rear suspensions. Number 3 shows the loads applied at the top of front roll hoop with the same loads value and boundary condition as Number 1, while Number 4 shows the loads applied in downwards direction at the front bulkhead with value of -12.0kN. The boundary condition before this analysis test can be performed is by clamping the rear suspensions. The load applied in Number 5 is for static torsional loading test. The loads are applied at the right of side impact with moment value of 3.16kN.m. Initial condition involve in this test is by clamping rear suspensions while one side of front suspension is left unclamped. The entire analysis test, their boundary condition and load applied are summarised in Table 1 below.

Table 1. Summary of Analysis Test on Chassis Designs.

No	Analysis Test	Boundary Condition	Load Applied
1	Main Roll Hoop	Clamp the bottom structure of both main and front roll hoop.	Fx = 6.0kN Fy = 5.0kN Fz = -9.0kN
2	Side Impact	Both rear and front suspensions are clamped.	Fy = 7.0kN
3	Front Roll Hoop	Clamp the bottom structure of both main and front roll hoop	Fx = 6.0kN Fy = 5.0kN Fz = -9.0kN
4	Static Shear	Rear suspensions are clamped	Fz = -12.0kN
5	Static Torsional Loading	Rear suspensions are clamped and left side of front suspension is left unclamped.	T = 3.16kN.m

3. Results and Discussion

After an analysis has been made, all the results in term of Von Mises Stress and torsional displacement are compared in order to determine which one is the best. Results show that Design 1 is the best results in term of Von Mises Stress and torsional displacement. Below in Figure 3 show the analysis results that had been obtained during analysis test using CATIA V5 static structural analysis. The maximum value of Von Mises Stress and torsional displacement is represented by red colour.

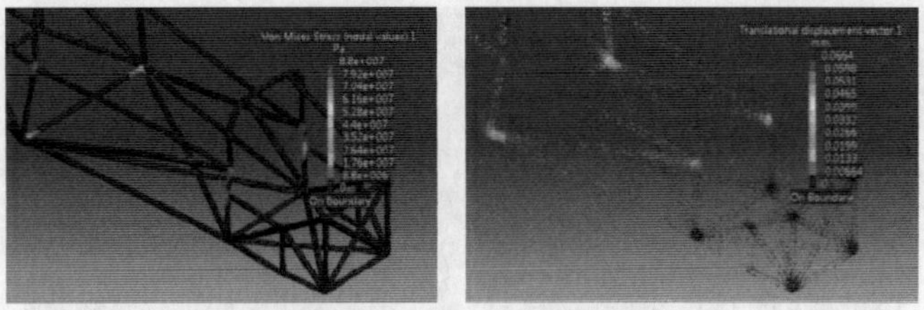

(a): Main roll hoop test

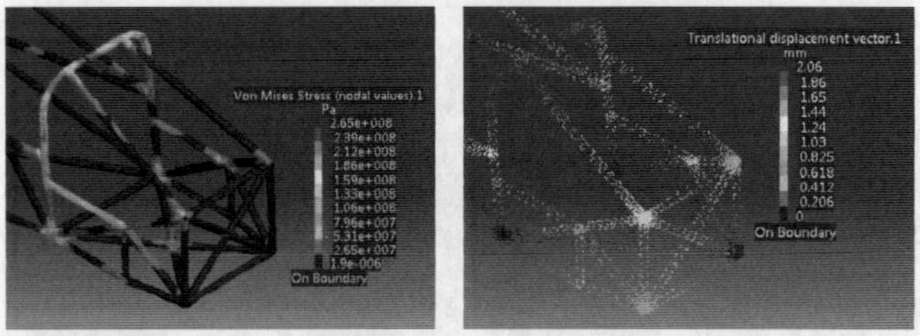

(b): Side impact test

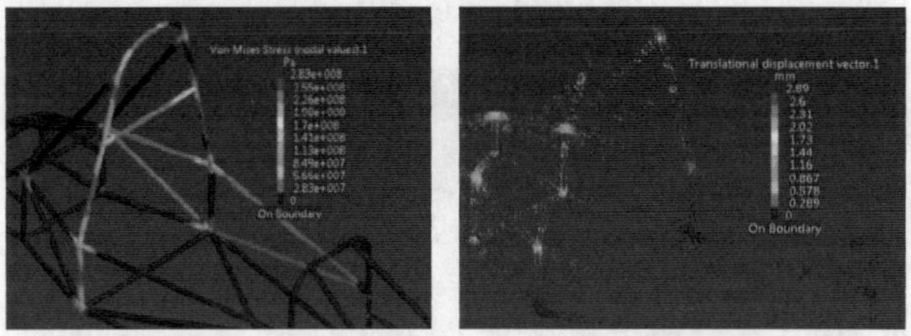

(c): Front roll hoop test

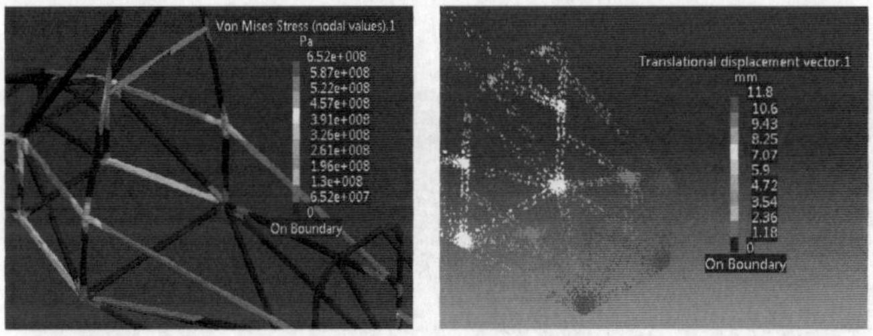

(d): Static shear test

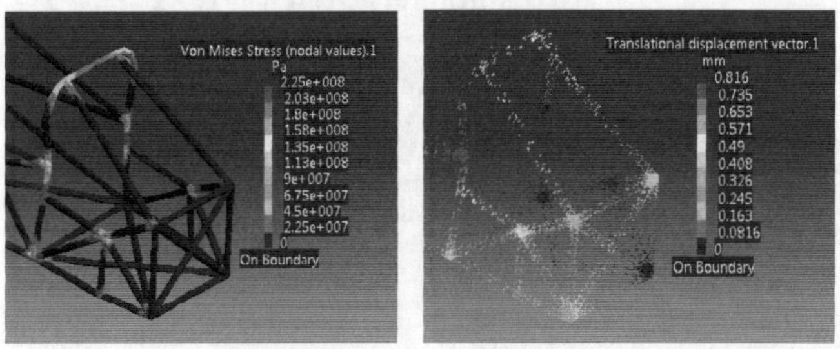

(e): Static torsional loading test

Figure 3. Von Mises stress and torsional displacement results for existing chassis.

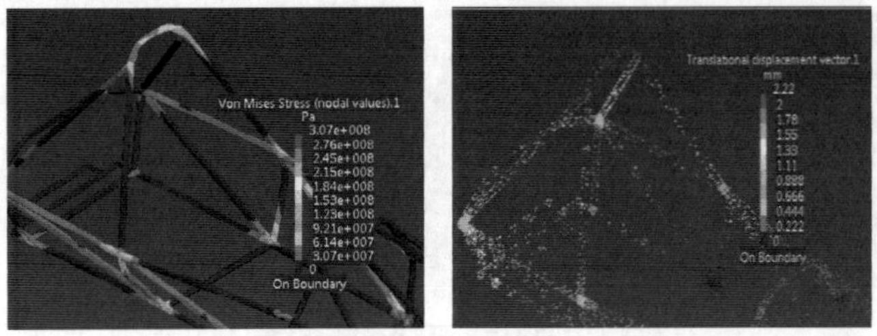

(a): Main roll hoop test

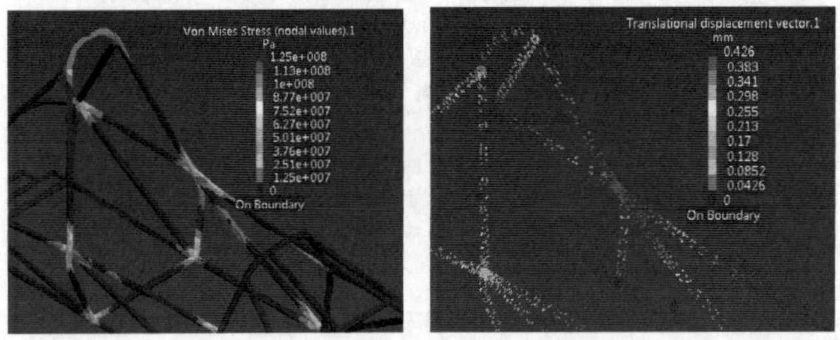

(b): Side impact test

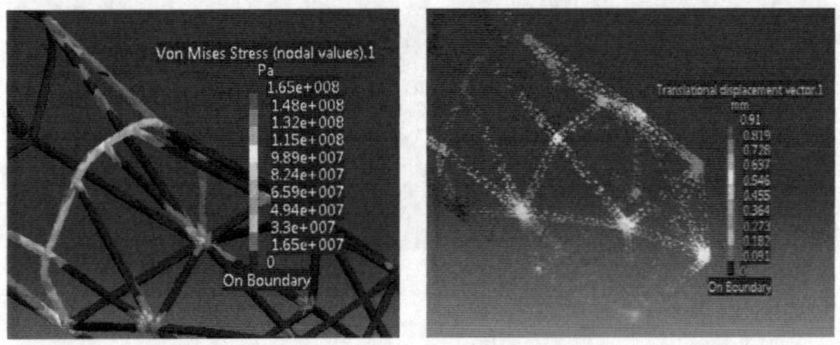

(c): Front roll hoop test

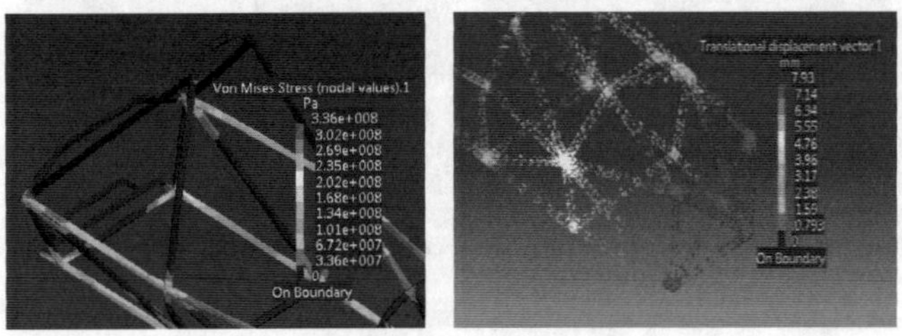

(d): Static shear test

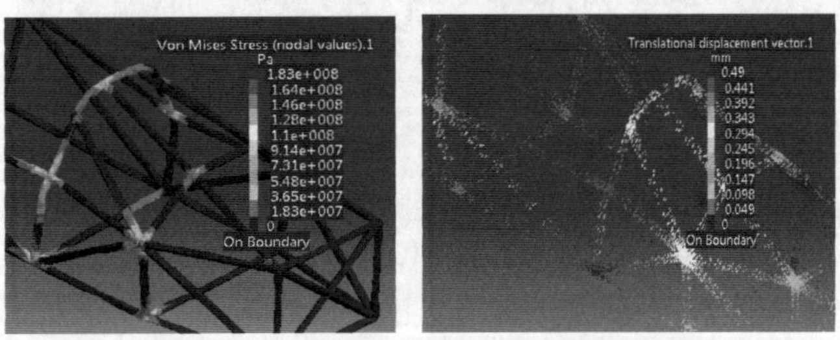

(e): Static torsional loading test

Figure 4. Von Mises Stress and torsional displacement results for improved chassis.

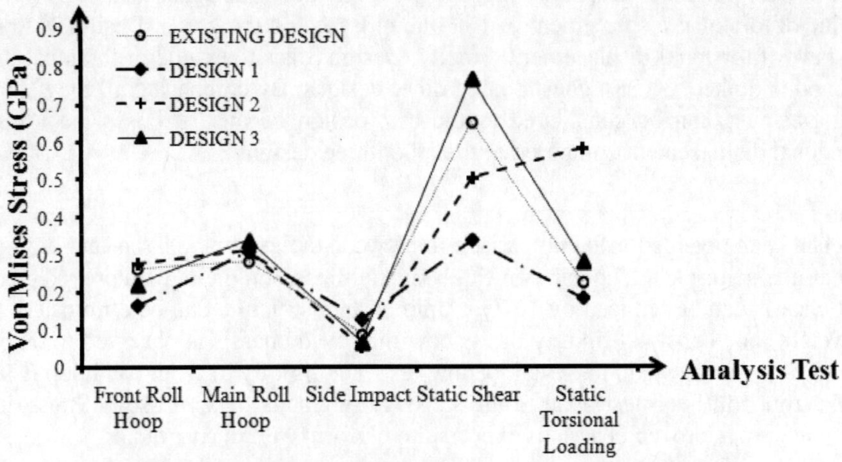

Figure 5. Graph of Analysis Test versus Von Mises Stress.

Figure 5 shows the graph obtained by plotting the type of analysis test versus the value of Von Mises Stress. By referring to the graph in Figure 4, the maximum value of Von Mises Stress for the existing design is 0.652GPa which is the second highest among all the designs. Design 2 poses the Von Mises value of 0.583GPa which is the third highest value comparing to the other three designs.For Design 1, the maximum value of Von Mises Stress occured during static shear too which is 0.336GPa and can be considered as the lowest value among other designs based on the graph. Finally, for Design 3, the maximum value of Von Mises Stress which occured suring static shear is 0.769GPa which is the first highest value of Von Mises Stress among other designs. By comparing to all the maximum value of Von Mises Stress for each test in each design, Design 1 can be declared as the best design due to its lowest value of Von Mises Stress compared to other designs.

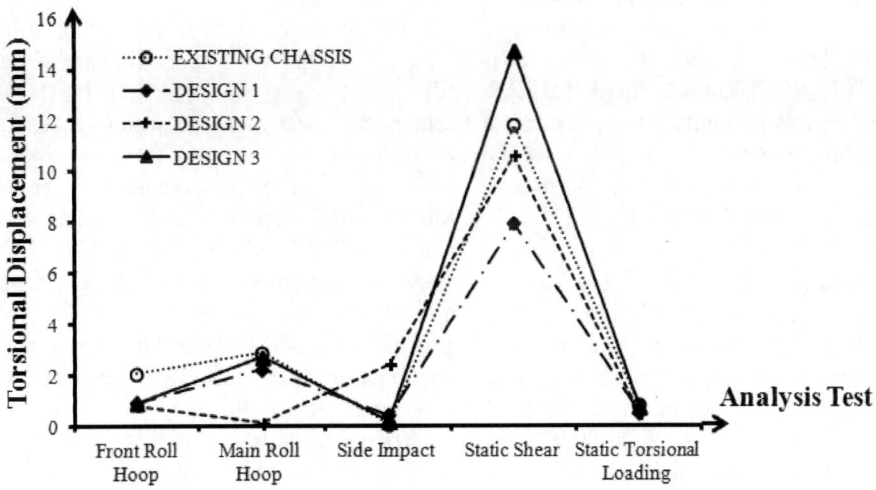

Figure 6. Graph of Analysis Test versus Torsional Displacement.

Figure 6 presents the whole graph results by plotting the type of analysis test versus torsional displacement which obtained during static structural analysis. By comparing to the maximum value of torsional displacement in each analysis test, the lowest value is considered as the best design. Existing design poses the maximum torsional displacement of 11.80 mm and the value is the second highest among the other three designs. Design 1 has the maximum displacement value of 7.930 mm that take places as the lowest torsional displacement compared to other designs, while Design 2 has the maximum value of torsional displacement of 10.60 mm. Based on the graph, Design 2 is ranked as the third highest rank of torsional displacement. Finally, Design 3 has the value of torsional displacement of 14.70 mm and is ranked as the highest among other designs. By comparing all the data of torsional displacement, Design 1 can be considered as the best design because it poses the lowest value of maximum torsional displacement compared to the other three designs.

4. Conclusion

After all data had been obtained, and comparison in between the existing chassis and improved chassis had achieved a satisfactory level, conclusion can be made. Based on the new improved chassis design, the weight of chassis can be reduced by 16.7%. Unimportant structure can be eliminated thus helping in reducing weight too. Torsional rigidity of the new improved chassis is observed to be better which increased by 18.66% during static torsional loading test. The rigidity of front roll hoop is increased by 37.74% during front roll hoop test. Due to the satisfactory results of Von Mises Stress and torsional displacement, the new improved chassis is expected to perform well during racing.

5. References

[1]. Rakesh Kumar Sahu, Suman Kumar Sahu, Srikant Behera, & V. Santosh Kumar (2016). Static Load Analysis of a Ladder Type Chassis Frame, Imperial Journal of Interdisciplinary Research, 2(5), 2454–1362.

[2]. Juvvi Siva Nagaraju, & U. Hari Babu (2012). Design And Structural Analysis of Heavy Vehicle Chassis Frame Made of Composite Material By Varyin Reinforcement Angles of Layers, International Journal of Advanced Engineering Research and Studies, 1(2), 70–75.

[3]. Naveen Ala, K. Tejdeep Reddy, & BVSS Bharadwaja (2016). Static Analysis of Truck Chassis using different materials, International Journal of Innovative Research in Science,Engineering and Technology , 5(1), 326–337.

[4]. Prajwal Kumar M. P, Vivek Muralidharan, & G. Madhusudhana (2014). Design and Analysis of Tubular Space Frame Chassis of A High Performance Race Car, International Journal of Research in Engineering and Technology, 3(2), 497–501.

[5]. 2016 SAE International. (2016, April 11). Retrieved November 28, 2016, from 2017-18 Formula SAE® Rules, http://www.fsaeonline.com/content/2017-18/20FSAE/20Rules/20PRELIMINARY.pdf

[6]. Ravinder Pal Singh (2010). Structural Performance Analysis of Formula SAE Car. Chassis Loading 46–61.

[7]. Kiran Ghodvinde, & S. R. Wankhade (2014). Structural Stress Analysis of an Automotive Vehicle Chassis, International Journal on Mechanical Engineering and Robotics 44–49.

[8]. John Burrill, Ankit Dahiya, Ankit Sharma, & Syed Zillay Ali. (2015). What are the advantages of using CATIA? Retrieved November 28, 2016, from Quora, https://www.quora.com/What-are-the-advantages-of-using-CATIA

[9]. Venkatesh S P, Mayur D Pawar, & Maruthi B H (2014). Design, Optimisation and Impact Analysis of a Chassis System of Race Car. International Journal of Enhanced Research in Science Technology & Engineering, 3(7), 377–386.

[10]. Suraj B. Patil, & Dinesh G. Joshi (2015). Structural Analysis of Chassis : A Review. International Journal of Research in Engineering and Technology, 4(4), 293–295.

[11]. A. Hari Kumar, & V. Deepanjali (2016). Design and Analysis of Automobile Chassis. International Journal of Engineering Science and Innovative Technology, 5(1), 187–195..

[12]. Vijayan, Sendhilkumar, & Kiran Babu (2015). Design and Analysis of Automotive Chassis Considering Cross-Section and Material. International Journal of Existing Research, 7(5), 15697–15701.

Improvement of cam performance curve using B-Spline curve

A B Shahriman [1,2,3], A K M Syafiq[1,2,3], M S M Hashim[1,2], D Hazry [3], Z M Razlan[2,3], K Wan[2,3], R Daud[2], E M Cheng[2], S K Zaaba[2] and Azizi Azizan[3]

[1]Unit Penyelidikan Teknologi Sukan Permotoran, MoTECH, Universiti Malaysia Perlis (UniMAP), Pauh Putra Campus, 02600 Arau, Perlis, Malaysia.
[2]School of Mechatronic Engineering, Universiti Malaysia Perlis (UniMAP), Pauh Putra Campus, 02600 Arau, Perlis, Malaysia.
[3]Centre of Excellence for Unmanned System (UAS), Universiti Malaysia Perlis (UniMAP), Perlis, Malaysia.

shahriman@unimap.edu.my, syafiq.akbar89@gmail.com

Abstract. The mathematical modeling approach has been applied in order to increase the cam profile curve of Modenas CT115s performance by using MATLAB software as a programmed to calculate the mechanism of the cam profile. Cam is used inside the engine to push the rocker and consequently open and close the engine valve that allows the fuel-air mixture to be entered during the combustion process. The B-Spline curve was implemented in order to enhance the current performance of the cam profile. The calculation had been done by using manual and MATLAB software. The results obtained has been analyzed and interpreted in plotting the graphs. From the analysis, the profile that had the highest displacement factor, *sk* produced higher cam curve performance of the engine. Thus, it can be concluded that the increase of the displacement factor, *sk* can increase the engine performance as the valve displace further in which allow higher fuel-air mixture entrance during the combustion process.

1. Introduction

Cam is a component that works to transmit a power by the motion of mechanical from a speed constant to a variety of speeds and widely used in the automotive industry due to the superior properties for high speed in operation, accuracy in motion, the rigidity of the structure and low cost production. The location and shape of the lobe will determine the timing of the valve opening-closing motion, its lifting height and valve movement rate of one complete cycle which could affect the performances of the MODENAS motorcycles.

In an automotive engine, the cam does not work individually because it attaches with follower, camshaft, valve and others. The follower is the element in which direct contact with the cam [1]. Cam also can be classified into two types which are eccentric and concentric based cam. An eccentric cam is a cam that has symmetrical lift location and in contrast, concentric cam has offset lift location in which can be adjusted in order to achieve a greater lift of the valves during internal combustion. Generally, profiles of the cam can affect the engine performances. It is because the profiles it can manipulate the intake duration, intake opening and closing angle, exhaust opening and closing angle, lobe separation angle and the valve lift of the cam. Thus, by using proper mathematical modeling all of the combustion process parameters can be manipulated successfully. The B-Spline method is more suitable and advantage because of its effectiveness and has no problem with the stability of numerical in the parameterization of cam shape. By then, the performance of MODENAS Motorcycle engine can be enhanced by several percentages than before.

2. Methodology

The main objective of this study to increase the cam performance by enhancing the cam profile curve by implementing the B-Spline method. Cam curve performance analysis will be conducted based on its several displacement factors, sk from 6th orders of the B - spline method and all possible engineering parameters. This is necessary to determine the best cam design that will produce high performances throughout. The analysis of the cam will be done by using computerized engineering software such as CATIA and MATLAB. This analysis is designed so that optimization of the cam profile can be done and thus, the factor affecting the optimization of the cam can be determined [2]. The collected data will be analyzed properly so that the pattern graph of displacement, velocity, acceleration and jerk versus angle of cam rotation can be established. Then, the results from the analysis could contribute in enhancing the engine performance for the MODENAS motorcycles.

The dimension and parameters of the cam that impact the motion of cam are measured by using 3D Scanner and CMM machine. This is necessary to calculate cam profile curve. Due to shining properties of cam, 3D Scanner facing a problem in cam measurement. Apart from that, the camshaft does not have the assembly part and cannot measure the cam lobe only. 3D Scanner more suitable to do the improvement design.

Therefore, CMM machine was used in order to obtain the dimension of the existing cam lobe profile due to the function that can give all dimensions of assembly parts. The dimension is in 3 axis, x, y, and z-axis, but the dimension that had been obtained are not very precise and accurate due to the obtained reading is in a wide range. There is also some error when taking the dimension because of condition changes of the cam while cam measurement. Besides, sometimes the condition of the cam is not correct or unstable. As a solution, CATIA software had been used to get any existing parameter that's needed in calculating the displacement, velocity, acceleration and jerk.

The fundamental principle of B-Spline is the number of splines can be mingled together in knots to get the desired cam function and it has been confirmed with the classical spline of 6th orders [3]. Figure 1 shows that the displacement spline and knot curve. In producing a smoother curve, four polynomials are put together at the three knot intermediate. This is because the intermediate knots are specifications consider for a single dwell cam displacement by using classical spline with 6th orders. By constructing the curve of the classical spline, the polynomial pieces and knots can be defined.

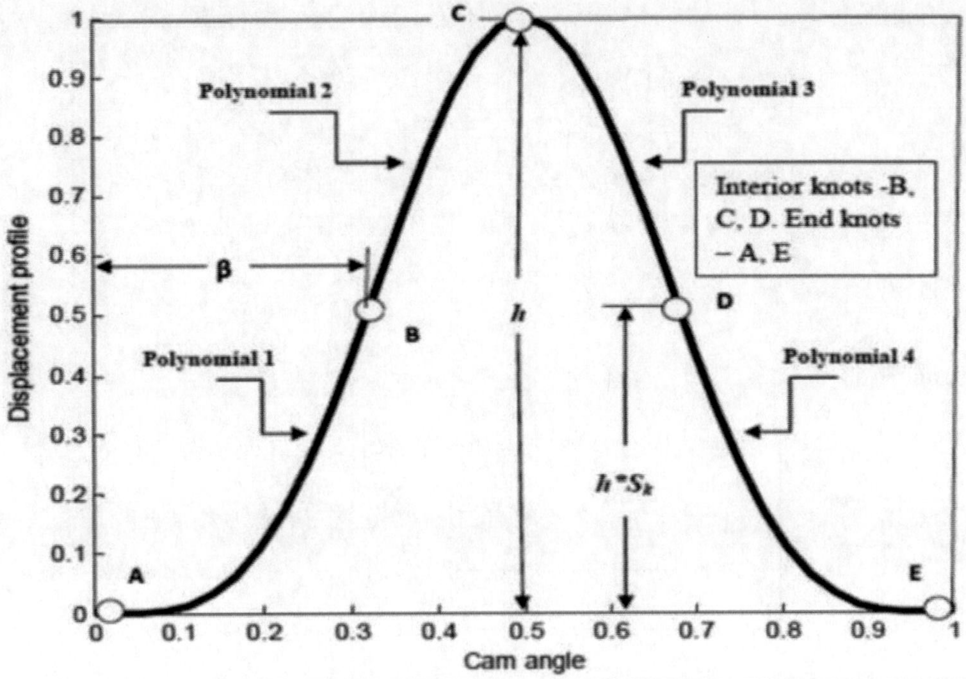

Figure 1. Displacement spline and knot curve [4].

The calculation involved in this research is the maximum rise of the cam is h, the rise of the cam rotation angle (2ß) and fall cam rotation angle h is in next 2ß. The balance of cam dwells rotation angle is (2π -4ß) and the angular velocity of the camshaft is taken as constant due to not enough parameter obtain. The starts and closure angle of cam rotation is 0 and 4ß, while the intermediate angle is ß, 2ß, and 3ß. A formulation that involved in this research is [5]:

Displacement Profile:

$$s(\theta)=\begin{cases}\left[Cnj\left(\frac{\theta}{\beta}\right)^{\bar{m}-1}+Cnj\left(\frac{\theta}{\beta}\right)^{\bar{m}-2}+Cnj\left(\frac{\theta}{\beta}\right)^{\bar{m}-3}+Cnj\left(\frac{\theta}{\beta}\right)^{\bar{m}-4}+Cnj\left(\frac{\theta}{\beta}\right)^{\bar{m}-5}+Cnj\right], & 0\le\theta\le\beta\\ \left[Cnj\left(\frac{\theta-\beta}{\beta}\right)^{\bar{m}-1}+Cnj\left(\frac{\theta-\beta}{\beta}\right)^{\bar{m}-2}+Cnj\left(\frac{\theta-\beta}{\beta}\right)^{\bar{m}-3}+Cnj\left(\frac{\theta-\beta}{\beta}\right)^{\bar{m}-4}+Cnj\left(\frac{\theta-\beta}{\beta}\right)^{\bar{m}-5}+Cnj\right], & \beta\le\theta\le2\beta\\ \left[Cnj\left(\frac{\theta-2\beta}{\beta}\right)^{\bar{m}-1}+Cnj\left(\frac{\theta-2\beta}{\beta}\right)^{\bar{m}-2}+Cnj\left(\frac{\theta-2\beta}{\beta}\right)^{\bar{m}-3}+Cnj\left(\frac{\theta-2\beta}{\beta}\right)^{\bar{m}-4}+Cnj\left(\frac{\theta-2\beta}{\beta}\right)^{\bar{m}-5}+Cnj\right], & 2\beta\le\theta\le3\beta\\ \left[Cnj\left(\frac{\theta-3\beta}{\beta}\right)^{\bar{m}-1}+Cnj\left(\frac{\theta-3\beta}{\beta}\right)^{\bar{m}-2}+Cnj\left(\frac{\theta-3\beta}{\beta}\right)^{\bar{m}-3}+Cnj\left(\frac{\theta-3\beta}{\beta}\right)^{\bar{m}-4}+Cnj\left(\frac{\theta-3\beta}{\beta}\right)^{\bar{m}-5}+Cnj\right], & 3\beta\le\theta\le4\beta\end{cases}\quad\text{a)}$$

Velocity Profile:

$$s'(\theta)=\begin{cases} \frac{1}{\beta}\left[5Cnj\left(\frac{\theta}{\beta}\right)^4+4Cnj\left(\frac{\theta}{\beta}\right)^3+3Cnj\left(\frac{\theta}{\beta}\right)^2+2Cnj\left(\frac{\theta}{\beta}\right)^1+Cnj\right], 0\le\theta\le\beta \\ \frac{1}{\beta}\left[5Cnj\left(\frac{\theta-\beta}{\beta}\right)^4+4Cnj\left(\frac{\theta-\beta}{\beta}\right)^3+3Cnj\left(\frac{\theta-\beta}{\beta}\right)^2+2Cnj\left(\frac{\theta-\beta}{\beta}\right)^1+Cnj\right], \beta\le\theta\le2\beta \\ \frac{1}{\beta}\left[5Cnj\left(\frac{\theta-2\beta}{\beta}\right)^4+4Cnj\left(\frac{\theta-2\beta}{\beta}\right)^3+3Cnj\left(\frac{\theta-2\beta}{\beta}\right)^2+2Cnj\left(\frac{\theta-2\beta}{\beta}\right)^1+Cnj\right], 2\beta\le\theta\le3\beta \\ \frac{1}{\beta}\left[5Cnj\left(\frac{\theta-3}{\beta}\right)^4+4Cnj\left(\frac{\theta-3}{\beta}\right)^3+3Cnj\left(\frac{\theta-3}{\beta}\right)^2+2Cnj\left(\frac{\theta-3}{\beta}\right)^1+Cnj\right], 3\beta\le\theta\le4\beta \end{cases} \quad \text{b)}$$

Acceleration Profile:

$$s(\theta)=\begin{cases} \frac{1}{\beta^2}\left[20Cnj\left(\frac{\theta}{\beta}\right)^3+12Cnj\left(\frac{\theta}{\beta}\right)^2+6Cnj\left(\frac{\theta}{\beta}\right)^1+2Cnj\right], 0\le\theta\le\beta \\ \frac{1}{\beta^2}\left[20Cnj\left(\frac{\theta-\beta}{\beta}\right)^3+12Cnj\left(\frac{\theta-\beta}{\beta}\right)^2+6Cnj\left(\frac{\theta-\beta}{\beta}\right)^1+2Cnj\right], \beta\le\theta\le2\beta \\ \frac{1}{\beta^2}\left[20Cnj\left(\frac{\theta-2\beta}{\beta}\right)^3+12Cnj\left(\frac{\theta-2\beta}{\beta}\right)^2+6Cnj\left(\frac{\theta-2\beta}{\beta}\right)^1+2Cnj\right], 2\beta\le\theta\le3\beta \\ \frac{1}{\beta^2}\left[20Cnj\left(\frac{\theta-3\beta}{\beta}\right)^3+12Cnj\left(\frac{\theta-3\beta}{\beta}\right)^2+6Cnj\left(\frac{\theta-3\beta}{\beta}\right)^1+2Cnj\right], 3\beta\le\theta\le4\beta \end{cases} \quad \text{c)}$$

Jerk Profile:

$$s(\theta)=\begin{cases} \frac{1}{\beta^3}\left[60Cnj\left(\frac{\theta}{\beta}\right)^2+24Cnj\left(\frac{\theta}{\beta}\right)^1+6Cnj\right], 0\le\theta\le\beta \\ \frac{1}{\beta^3}\left[60Cnj\left(\frac{\theta-\beta}{\beta}\right)^2+24Cnj\left(\frac{\theta-\beta}{\beta}\right)^1+6Cnj\right], \beta\le\theta\le2\beta \\ \frac{1}{\beta^3}\left[60Cnj\left(\frac{\theta-2\beta}{\beta}\right)^2+24Cnj\left(\frac{\theta-2\beta}{\beta}\right)^1+6Cnj\right], 2\beta\le\theta\le3\beta \\ \frac{1}{\beta^3}\left[60Cnj\left(\frac{\theta-3\beta}{\beta}\right)^2+24Cnj\left(\frac{\theta-3\beta}{\beta}\right)^1+6Cnj\right], 3\beta\le\theta\le4\beta \end{cases} \quad \text{d)}$$

3. Results And Discussions

3.1 Cam Curve Performance

In this research, cam specification used data the maximum lift h = 0.009m, the rise and fall of cam angle 2ß x2 = 146°, roller radius rr = 0.0065m and base circle radius is rb= 0.025m. By using all the information, the displacement, velocity, acceleration, and jerk graph are obtained after calculation. In improves the cam curve performance part, the displacement factor, sk was used to optimize the current graph. The displacement factor, sk use in this improvement part is sk=0.25, sk=0.35, sk0.45, sk=0.5 and sk=0.6.

Figure 2 shows that the significant effect on the displacement curve performance. The area under the displacement curve graph increase with the increment of the displacement factor, sk. Figure 2 shows the existing cam curve performance is not in the best condition yet compare to the improvement cam curve performance using displacement factor,sk=0.6.

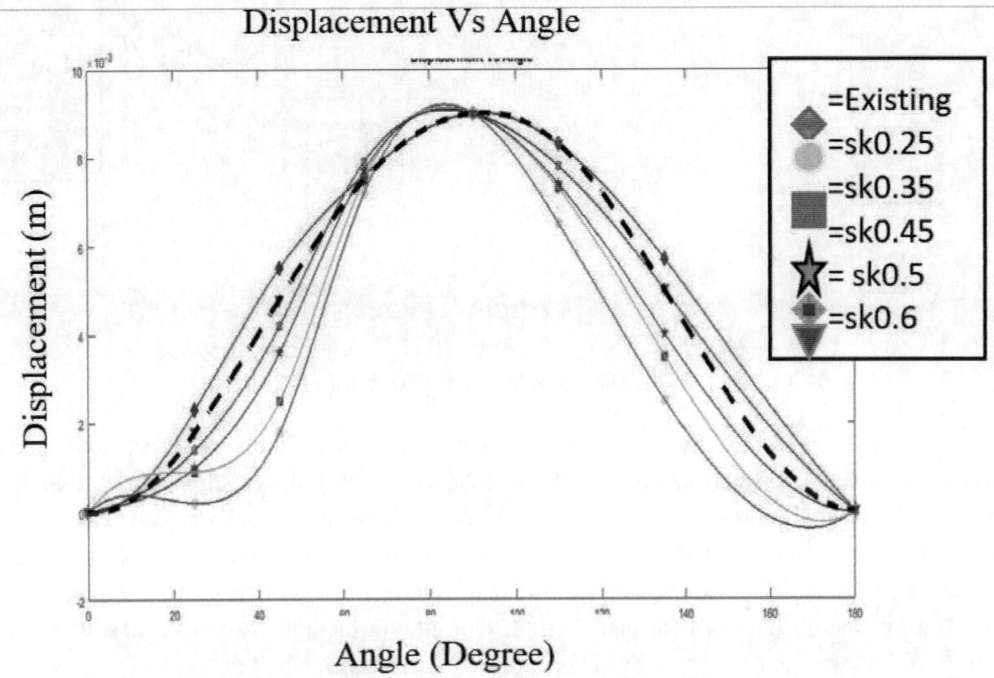

Figure 2. Displacement curve for all designs.

Comparison between existing and improvement cam curve performance for velocity can be seen in Figure 3. In the graph, it is shown that the velocity curve lowers with an increment displacement factor, sk and it remains constant at intermediate knots. Sk=0.6 is the best cam curve performance due to almost the same with the ideal velocity cam curve. The existing velocity cam curve performance for MODENAS CT115s did not optimize yet due to more different with ideal and improvement cam curve performance.

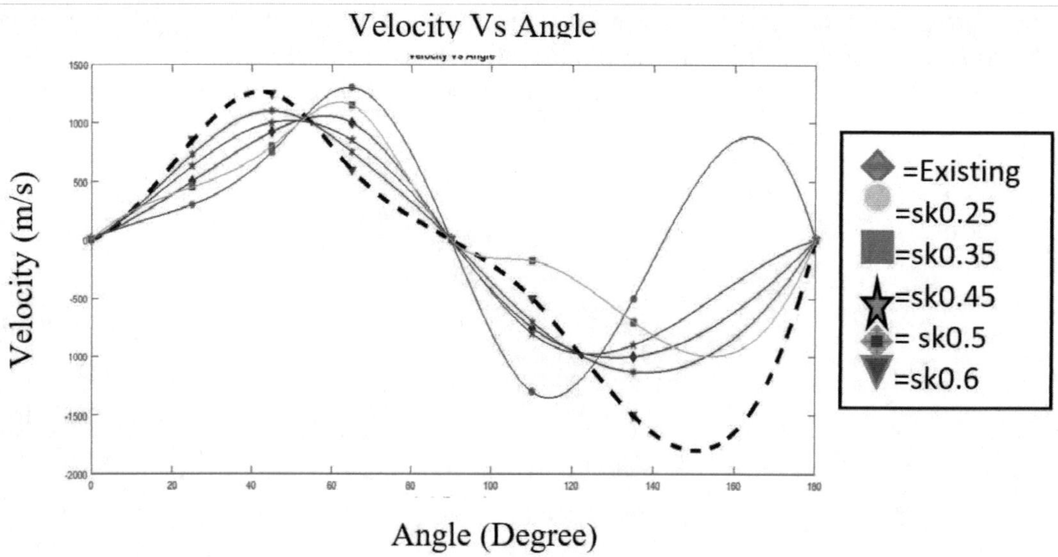

Figure 3. Velocity curve for all designs.

Figure 4 shows the effect of having different displacement factor, sk on the acceleration curve. The graph shows drastic changes in each displacement factor also to the existing acceleration curve. By using different knot the important observation can be clearly seen that negative acceleration decrease with the increment of displacement factor, sk. This is good for improvement due to less negative acceleration will reduce the tendency follower to jump. In order to improve the acceleration curve performance, negative acceleration should be less than the existing. One of the methods to reduce negative acceleration and minimize jump of the follower is by choosing the higher displacement factor, sk but it cannot be too much higher because it will tend to have more negative acceleration. Increasing displacement factor reduces negative peak acceleration.

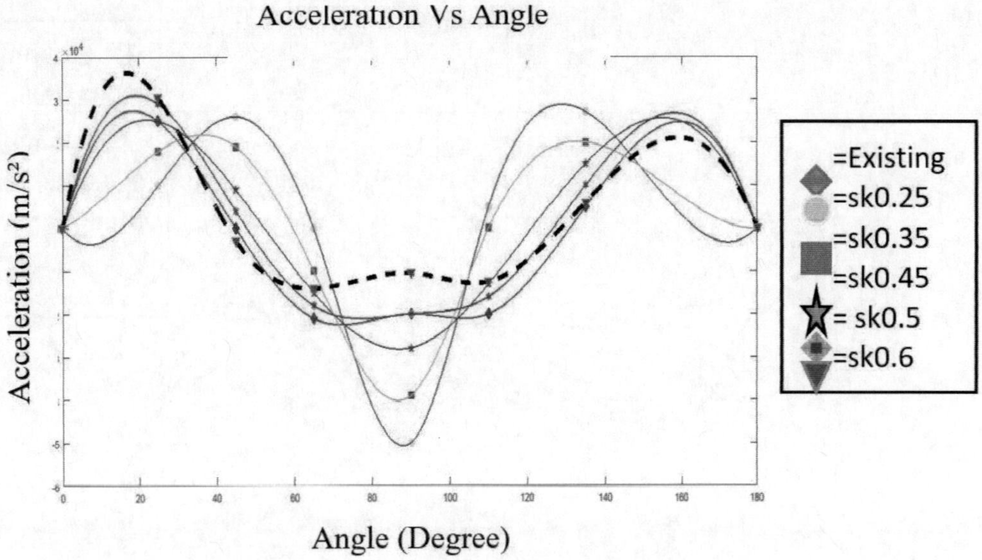

Figure 4. Acceleration curve for all designs.

Smoother jerk profiles will reduce the residual vibrations in cam system. In any profile with continuity, the curve will begin and ends at zero amount in each cycle but not for jerk profile. This is because jerk profile does not disregard the principal law of cam plan and rather desirable for smooth operation of cam driven mechanisms. In Figure 5, the graph showed the statement is correct which is jerk profile does not disregard the principal law of the cam. Thus, there is an exchange off between reducing the peak of acceleration and increasing the peak of a jerk. Therefore, in order to optimize the cam curve performance, cam systems with excessive jerk need to be avoided.

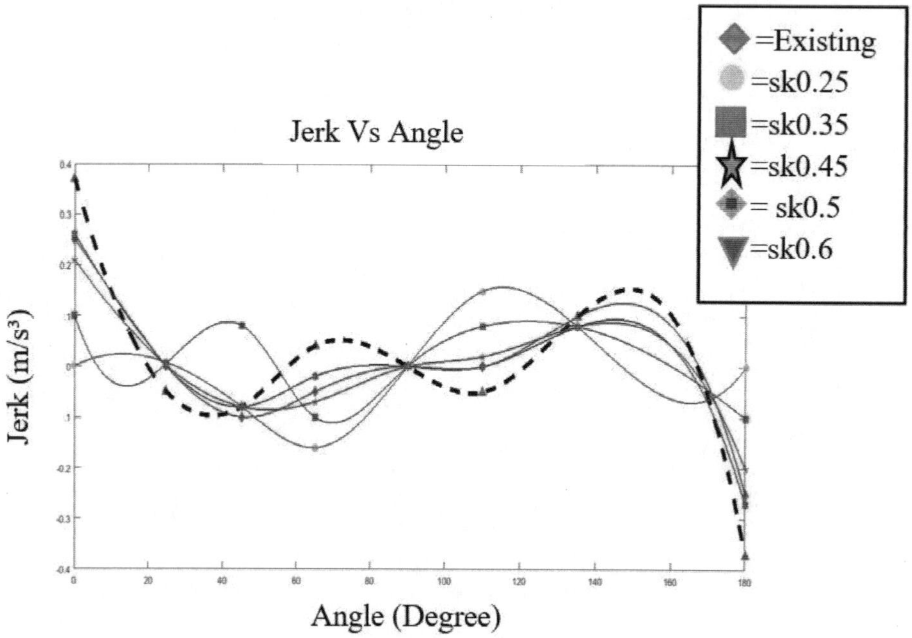

Figure 5. Jerk curve for all designs.

In order to optimize cam curve performance of MODENAS CT115s, there is few methods need to follow such as a B-Spline method. This is because, in the B-Spline method, knot specification and displacement factor, sk need to choose wisely to produce the great cam curve performance. To optimize the cam curve performance using B-Spline is important to know the meaning of negative acceleration, the peak of a jerk, and how it affects the performance of the cam.

The higher displacement factor, sk, was used in order to optimize the acceleration profile. This is to reduce the percentage of the follower to jump during high speed cam rotation condition. This is because higher negative acceleration will tend the follower to jump in the operation. It will give a bad impact on the cam curve performance. Apart from that, by increasing the peak of jerks it will give smooth operation in cam system and reduced the high vibration [6].

Large positive acceleration against load will make the cam follower interfaces forces that can bring about inordinate wear. Negative acceleration has a tendency to lessen the cam adherent interface constraints and if the negative acceleration is adequately worn bounce between the cam and follower can happen. A single dwell cam displacement function is characterized by a B-Spline method that is comprised of four polynomial sorts that tied it out at their closures and called as a knot. The particulars of these branches are considered for combination and investigation of Cam supporter followers. The mathematical connection between interface force and knot area is displayed as wear and jump models. These models are valuable to decrease wear [4].

4. Conclusions

The B-Spline method was implemented in calculating the overall cam performance. By the aid of MATLAB software, the performance curve that consists of displacement, velocity, acceleration, and jerk were plotted. The increasing displacement factor, sk produced the better cam curve performance in terms of displacement, velocity, acceleration and jerk profile. The selection of displacement factors is decided in investigating the effect of displacement factor, sk to the cam performance curve. The value of displacement factor must not exceed 0.60. This is because the value over 0.60 will reduce the performance of cam. This is because 0.60 was the ideal value in plotting the cam performance. Therefore, the increasing value of displacement factor, sk will improve the cam performance but the value of displacement factor must not exceed 0.60.

5. Acknowledgments

The authors acknowledge MoTECH, MODENAS, School of Manufacturing Engineering, Universiti Malaysia Perlis for the lab facilities and technical support. Special thanks to those who contributed to this project directly or indirectly.

6. References

[1] Hansong Xiao, Jean W. Zu, Cam profile optimization for a new cam drive, Journal of Mechanical Science and Technology 23, 2009, pp 2592-2602.

[2] A.K.Jamkhande, S.S.Tikar, S.S.Ramdasi, N.V. Marathe, Design high speed engine's cam profile using B-Spline functions for controlled dynamic, SAE Technical Paper 2012-28-0006

[3] Vu-Thinh Nguyen, Do-Joong Kim, Flexible cam profile synthesis method using smoothing spline curves, Mechanism and Machine Theory 42, 2007, pp 825-838

[4] Jose De Miguel, How to Build a Bobber on a Budget, http://xorl.wordpress.com, March 2011.

[5] Cobalt327, Crashfarmer, Crosley, Curtis73, Jon, The inspector 1, The right curve, Valkyrie 57, How to Choose a Camshaft, www.crankshaftcoalition.com, 2013.

[6] Harold k. Rothbart, Cam Design Handbook, McGraw-Hill Companies, 2004, pp 90-150

IOP Conf. Series: Journal of Physics: Conf. Series **908** (2017) 012044 doi:10.1088/1742-6596/908/1/012044

Numerical investigations of internal stresses on carbon steel based on ultrasonic LCR waves

R Ramasamy[1], Z Ibrahim[1] and H K Chai[2]

[1] Faculty of Engineering, University of Malaya, 50603 Kuala Lumpur, Malaysia.
[2] School of Engineering, University of Edinburgh, EH89YL United Kingdom.

ramasamy.ramesh@gmail.com; ramesh.ramasamy@siswa.um.edu.my

Abstract. Internal stresses or residual stresses in the structural elements are very crucial in carrying out in-service evaluations and fitness-for-purpose assessments. The generation of these internal stresses can occur as result of the fabrication of the steel members, installation sequence or other ad-hoc events such as accidents or impact. The accurate prediction of the internal stresses will contribute towards estimating the integrity state of the structural elements, with respect to their material allowable stresses. This paper investigates the explicit FE based numerical modelling of the ultrasonic based non-destructive technique, utilising the measurable longitudinal critical refracted wave (LCR) and relating these to the internal stresses within the structural elements by the evaluation of the material dependent acoustoelastic factors. The subsurface travel path of the LCR wave inside the structural elements makes it a sub-surface stress measurement technique and the linearised relationship with corresponding internal stresses can be systematically applied repeatedly. The numerical results are compared against laboratory tests data to correlate the findings and to establish modelling feasibility for future proof-of-concepts. It can be concluded from this numerical investigation, that the subsurface ultrasonic LCR wave has great potential to be implemented for in-situ structural residual stress measurements, as compared to other available surface measurements such as strain gauges or x-ray diffraction.

1. Introduction

1.1. Background

Residual stresses or internal stresses are major contributing factors in mechanical structures, especially when carrying out integrity assessments and subsequent rejuvenation plans. These residual stresses are possibly induced at various stages, from fabrication to installation sequences, and in some cases, are imparted from ad-hoc scenarios such as accidents or impacts. When stress analyses are carried out, almost always, only the operational design loads are considered in assessing the in-place integrity state, whilst these residual stresses are ignored completely in some cases, or analytical and numerical evaluations are carried out to provide some approximations. In most situations, these approximations are over-conservative, and results in unnecessary and costly repairs on structures with adequate resistance, whilst on the other extreme may result in underestimated stresses which may result in catastrophic failures without any warning.

1.2. Available Measurement Method

Currently, there are several prominent methods being used in measurement of residual stresses. One of the most elegant non-destructive method is the x-ray diffraction method [1][2] which involves the use of x-ray to measure the strain in the material microstructural crystal lattice arrangements, and calculation of the residual stress that causes this pre-existing strain assuming a linear elastic distortion of the lattice. This technique utilises Bragg's law in determining the diffraction angle, and any shift in the lattice spacing due to residual strain results in shift in the diffraction angle. The problem associated with this method is the depth of penetration of the monochromatic x-rays are limited to about 5μm beneath the surface [1]. This can be unfeasible as the bulk of the aged structural steel may be covered with rust or coatings and therefore thorough chemical based surface preparation is required, such that this does not induce any unnecessary stresses on the surface (which may be the case when sanding or grinding is involved). The presence of radioactive material within this system is also a problem with the in-place guidelines for explosive environments such as ATEX [3] and overall bulky configuration of the system makes it less portable for field operations. Another method for residual stress measurement is the hole drilling method as specified in ASTM [4], which involves the precision drilling of a hole through a residual strain gauge. The residual strain gauge is in fact a strain rosette with a hole in the middle to be able to accommodate the hole drilling on the structure, thus picking up the relieved strain around the hole, and the residual stress can be calculated from the principal stress equations, as provided in ASTM [4]. This method is classified as semi-destructive, since a small hole (approximately 3mm to 10mm) will be drilled on the structure, and may be very resource consuming since it involves drilling, strain gauging and data logger operations. The excessive heat and metal chips resulting from the drilling can also potentially damage the strain gauge, and the sparks from drilling can pose to be an ATEX [3] noncompliance issue, in addition to being a fire hazard especially if hydrocarbon is present, for instance in a petrochemical plant. To carry out reliable measurements of the residual stresses in critical structures, a non-destructive and robust method is therefore needed. In recent years, the vast use of piezoelectric ultrasonic transducers to measure structural dimensions and detect internal flaws have been very successful in various engineering sectors. This work will investigate and present the theoretical foundation, and numerical validation of using the established ultrasonic technology and the principles of acoustoelasticity in accurate prediction of the internal stresses in structural members.

2. Ultrasonic Stress Measurement Method

2.1. The Concept

Ultrasonic wave propagation in a solid structure is affected by the internal stress field [5][6][7], and this makes the non-destructive internal stress measurement to be feasible. This fundamental behaviour allows the measurement of time of flight (TOF) of the wave from the transmitting probe to the receiving probe. The change in measured speed depends on the internal stress magnitude along the propagation direction, in a linear relationship hence forming the foundation for the acoustoelastic theory. When an ultrasonic transmitter probe emits the wave into the solid medium, there are three components of wave generated at different speeds and angles in the medium, as shown in **Figure *1***. The incident wave at an angle of θ_1 with the vertical axis entering the medium will be refracted at angle θ_2 into the shear component, longitudinal component and the surface ripple or Rayleigh wave. The particles riding the shear wave will oscillate perpendicular to the propagation path, whereas the particles riding the longitudinal wave will be oscillating in parallel to the propagation path. The Rayleigh component, however, oscillates the particles in an elliptic motion near the surface of the medium causing the weak ripple moving away from the source. In the situation where residual stress measurement is present in the solid medium, the wave propagation path should ideally be parallel to the principal stress axis, i.e. the wave components travel along the stress direction. In this work, the longitudinal beam is utilised to measure the residual stress in structures due to the more significant changes in measurable propagation speed, as compared to the other components. Furthermore, the longitudinal wave which travels subsurface along the principal stress axis will vary much more significantly in the presence of stress in the medium, in addition to the material properties. This longitudinal component which is refracted at $\theta_2 = 90°$ is termed the longitudinal critical refracted (LCR) wave, and will be the focus of this work.

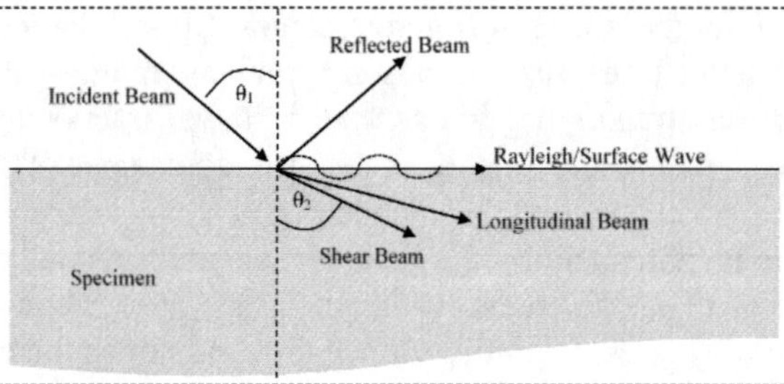

Figure 1. Schematics of Wave Components.

2.2. Theoretical Overview

For an unstressed elastic solid, the propagation speed terms of the ultrasonic wave components can be written as follows [8][9]:

$$V_L = \sqrt{\frac{\lambda + 2\mu}{\rho}} \tag{1}$$

$$V_S = \sqrt{\frac{\mu}{\rho}} \tag{2}$$

$$V_R \approx \frac{0.862 + 1.14\nu}{1+\nu} . V_S \tag{3}$$

where the terms V_L, V_S and V_R are the longitudinal, shear and Rayleigh wave speeds respectively, ν and ρ are the Poisson's ratio and density of the medium. The terms λ and μ are the second order elastic constants for the materials. Snell's law is used to orientate the signal source such that to produce the LCR wave, by introducing an intermediate medium with the critical incident angle of 90°. Polymethylmethacrylate (PMMA), or more commonly known as plexiglass or acrylic is found to be ideal for this purpose, and a machined wedge with the correct incident angle can create the LCR waves close to the surface to facilitate the residual stress measurements. The relationship between the LCR wave speed (V) and stress (σ) is as shown in Equation (4) [10], and reduces to Equation (1) in the absence of any residual stresses.

$$V^2 = \lambda + 2\mu + \frac{\sigma}{3\lambda + 2\mu}\left[\frac{\lambda + \mu}{\mu}(4\lambda + 10\mu + 4m) + \lambda + 2l\right] \tag{4}$$

Here, the parameters *m* and *l* are the part of the third order elastic constants of the material. Further manipulations and rearranging will lead to the more useful form in Equation (5)

$$\sigma = \frac{1}{k}\left(\frac{V^2}{V_0^2} - 1\right) \tag{5}$$

where

$$k = \frac{\frac{4\lambda + 10\mu + 4m}{\mu} + \frac{2l - 3\lambda - 10\mu - 4m}{\lambda + 2\mu}}{3\lambda + 2\mu} \tag{6}$$

The relative change in LCR wave is now directly proportional to the stress along the same direction with by a factor *k*. Experimental procedures can therefore be setup to calibrate the relative LCR speed to the stress.

2.3. Materials

The specimen selected for this investigation is Grade-B carbon steel (with yield strength of 240MPa), which is commonly used in the constructions of various structural systems decades ago ranging from offshore platforms to infrastructural buildings, and now can be found on majority aged structures worldwide where the residual stress measurements are ever more crucial. PMMA wedges cut to the correct angle are used to generate the LCR waves. The material specifications for the specimen and wedge are presented in Table 1, followed by the evaluation of the second order elastic constants and the respective wave speeds in Table 2.

Table 1. Material Specifications.

Material	**ρ (kg/m³)**	**E (GPa)**	**ν (-)**
Steel	7950	207	0.3
PMMA	1178	4.3	0.4

Table 2. Material Elastic & Acoustic Parameters.

Material	**λ (GPa)**	**μ (GPa)**	**V_L (m/s)**	**V_S (m/s)**	**V_R (m/s)**
Steel	119.42	79.62	5920.37	3164.57	2930.88
PMMA	6.15	1.54	2798.00	1142.28	1077.98

To generate the LCR which refracts into the steel from PMMA medium at 0° from vertical, the wedges need to be cut at about 28.2° based on Equation (5). To account for some small variations in material properties for repeatability of the tests, the base of the wedge is slightly curved to enable adjustments of up to ±15° on the beam incident angle such that it produces the maximum LCR wave amplitudes during tests as well as usable on most types of metal specimens, as shown in **Figure** *2*. The ultrasonic transducer is placed tightly into the hole on the wedge, with water-based ultrasound gel applied into the hole to create an acoustical contact to propagate the longitudinal waves through into the specimen. The resulting LCR wave at right angle from the vertical axis will be accompanied by a shear wave component at about 32° as calculated from Snell's Law. Placing a similar arrangement of receiving transducer at a specified distance away from the transmitting transducer along the specimen will permit the acquisition of the LCR signals during the test.

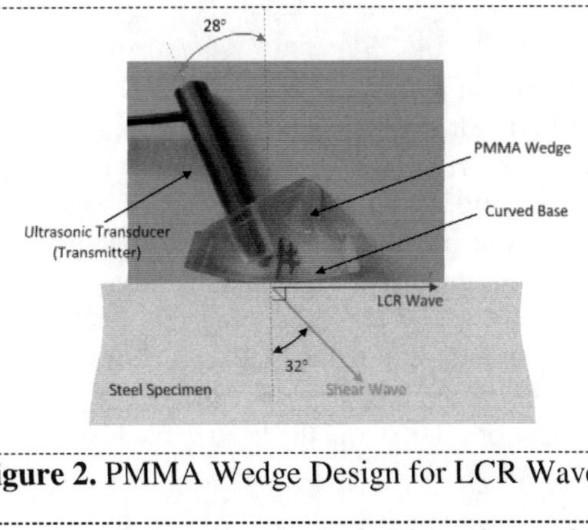

Figure 2. PMMA Wedge Design for LCR Wave on Steel.

The steel test specimens selected for the experimental work consist of hollow sections of various shapes and sizes cut at 500mm length. It should be noted that the objective of the study is to test the LCR's feasibility on stress measurement on the steel material, and the shapes/sizes are merely introduced to verify that the derived properties are independent of shapes/sizes and exclusively material dependant. Their limiting load limits are also calculated to ensure loadings remain within elastic limits and to ensure overall safety.

3. Numerical Analysis

3.1. Model Description

The numerical model, utilising the finite element (FE) method is carried out in the general purpose nonlinear code ABAQUS [13] to correlate to the theoretical procedures described in the preceding sections of this paper. The model is setup to analyse the case as outlined in **Figure *3***, highlighting the specimen and wedges modelled in a two-dimensional plane strain assembly. The specimen is modelled at 10mm depth and 100mm long, while the wedge is about 5mm depth and 20mm long. The wedges are modelled with straight base, as compared to the curved ones used in the tests since the specific material properties are used and the distinctive critical incident angle can be modelled directly to generate the LCR waves. The specimen is restrained along the vertical by boundary conditions at its base, and compressive residual stresses at the specimen edges, applied as shown over a range of 10MPa to 1000MPa.

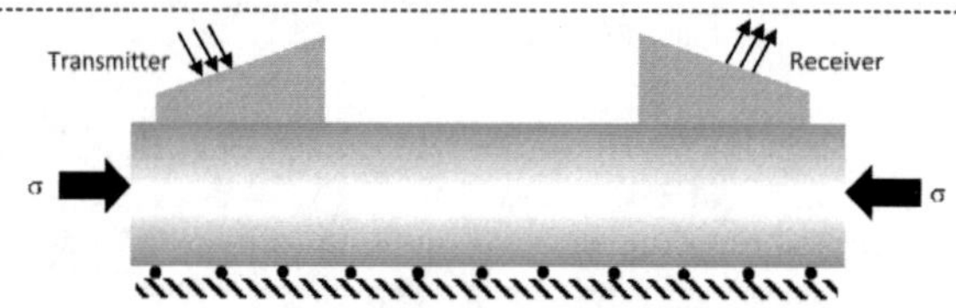

Figure 3. Outline of the FE Model for LCR Analysis.

The ultrasonic transducer signals are modelled by force amplitude F(t), and depends on the frequency of the transducer (f) used, and can be expressed over a range of time (t) in a cosine function as follows [7]:

$$F(t) = \left[1 - \cos\left(\frac{2\pi f t}{3}\right)\right] . \cos(2\pi f t) \tag{7}$$

The model is constructed in a plane strain condition, using the quadrilateral plane strain elements with reduced integration or CPE4R [13]. The interface between the wedge and specimen is kinematically coupled for all degrees of freedom, creating a continuity to enable the ultrasonic perturbation to travel across these boundaries.

3.2. Results

The LCR model will now be analysed and results extracted to investigate the acoustoelastic constant from a numerical perspective. The pressure contours are extracted and are presented for the wave propagation leaving the wedge in **Figure *4***, to identify the various wave components. The LCR wave components, being the more prominent and travelling at higher speed can be seen travelling parallel to the specimen surface, with its wave front propagating across the specimen depth, whereas the shear wave is propagating at an angle of approximately 30° from the vertical. The Rayleigh surface ripple is propagation outward away from the source and lagging before the much faster LCR wave. The head wave component which is observed as adjoining the LCR and shear components, i.e. interactions between the surface and inclined planes of the propagation path. The LCR wave can be expressed also as a net effect of the longitudinal, shear, Rayleigh and the head wave [14].

Figure 4. FE Pressure Output Showing All Wave Components.

The series of LCR wave propagation sequence can also be extracted and are presented in **Figure** *5* (i) to (vi) showing the progression of the LCR wave from the transmitting wedge until it reaches the receiving wedge. The longitudinal wave front is also observed to be reflected by the specimen backwall, and must be identified and ignored when post-processing the signal acquisition data. A quick check on the depth of the LCR wave on the FE output also shows an approximately 2mm depth for a 5MHz transducer which is being used, and is within acceptable order of magnitude from the measurements made earlier in [10][5] of same order of magnitude.

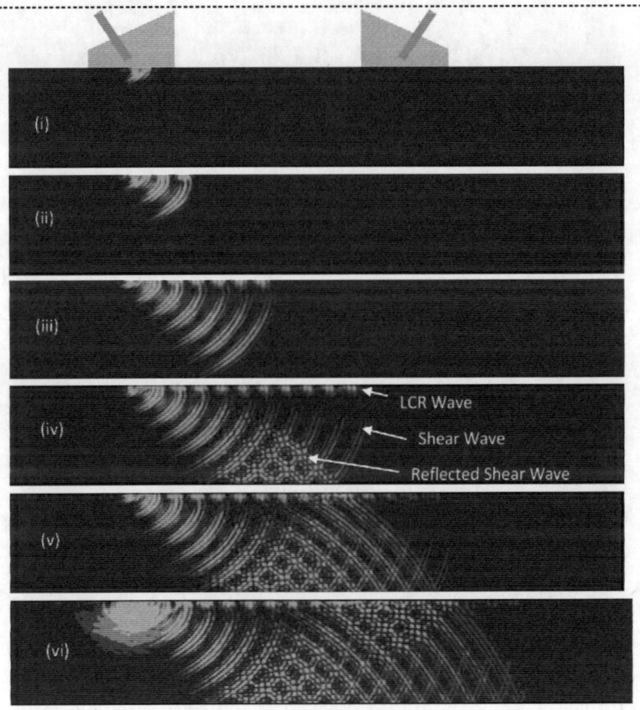

Figure 5. FE Output Showing Only the LCR Wave Propagation.

A quick comparison is made to check the signals received at the receiving probing location on the numerical model and the test data, as shown in **Figure** *6*. The plot shows an acceptable correlation with the test data [15], in terms of the LCR TOF arriving at the receiving probe point, and the approximate size of its amplitude.

The signals at the receiver probing point is further extracted and plotted in **Figure** *7*, with the exaggerated view on total TOF varying against residual stress. This is subsequently used to generate the plot for relative speed against stress shown in **Figure** *8*, resulting in the acoustoelastic constant value of 9.13 x 10-4 MPa-1, and lies within 5% limit in comparison of the experimentally measured average value due to the numerical idealisations.

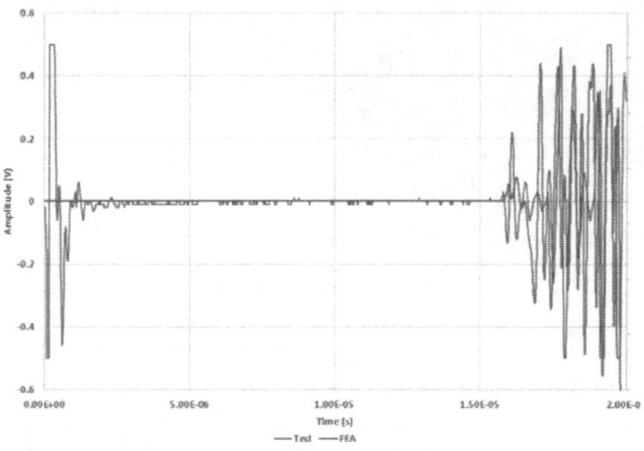

Figure 6. Comparison of FE and Test Signals.

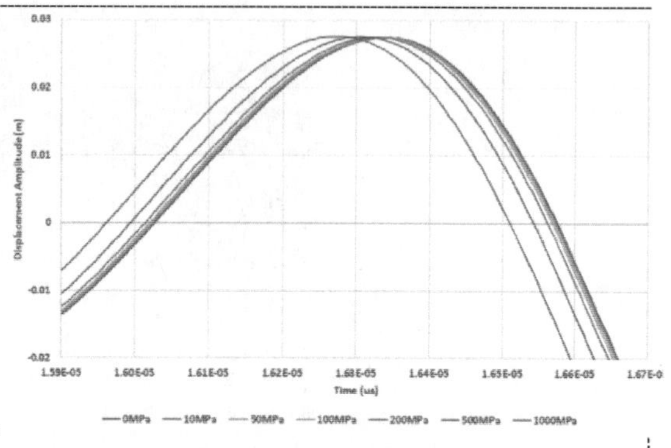

Figure 7. Numerical TOF Measurement at Receiver Wedge Location.

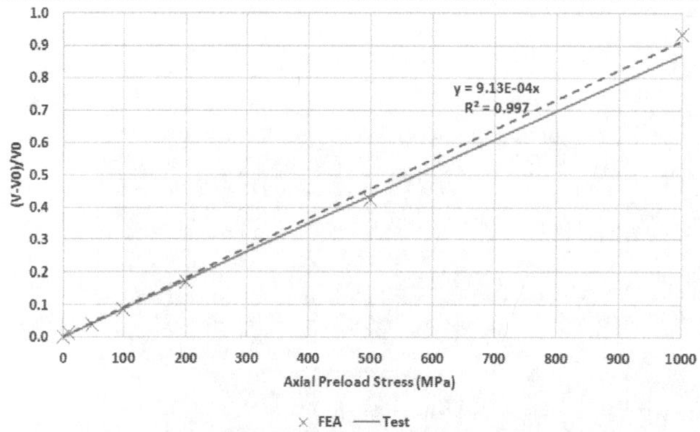

Figure 8. Comparison of Experimentally and Numerically Obtained LCR Relative Speed and Stress Relationship.

4. Conclusion

The primary objective of this study is to investigate the feasibility of using explicit FE based numerical simulations to effectively predict the internal stresses on carbon steel structural components based on the ultrasonic LCR wave propagations. LCR wave is generated by use of approximately 28° PMMA wedge which produces critically refracted wave components from a piezoelectric ultrasonic transducer. Compressive stresses were introduced into the specimen at controlled increments whilst the relative speed change is evaluated from the measured TOF of the LCR wave. The numerical model constructed was used to validate the experimental procedures carried out in a similar work, and to study the wave behaviours and extending the loadings beyond material allowable which are limited during the tests and found to correlate well with the experimentally measured values within 5% deviation.

In a field inspection of internal stresses, the sample material is required to evaluate the acoustoelastic constant prior to site measurements. This accounts for material properties and compositions used in the actual fabrication and construction. Once the constant is determined, it can be directly input into a customised computer routine to automatically produce internal stress magnitude along the ultrasonic wave propagations or transducer orientations during in-situ measurement activities. The transmitting and receiving wedges can also be integrated to be more easily handled, and the ultrasonic box interfaced with a tablet for mobility. The numerical model presented in this study is also deemed suitable for quick estimate of the acoustoelastic constant of idealised materials prior to setup of costly experiments to assess the feasibility and practicality of the ultrasonic LCR based residual stress measurements and the acquisition system resolutions and sampling rates required to successfully undertake any specific field measurements. Overall the LCR method is regarded as very portable and safe to operate within any conditions and adheres to stringent requirements as compared to other available techniques and is deemed to have great potential as a robust and reliable method in determining internal stresses on structures.

5. Acknowledgments

Authors wishing to acknowledge funding from the research grants RP004A-13AET and FP004-2014B, and academic support from the University of Malaya to undertake this research.

References

[1] P. S. Prevey, X-Ray Diffraction Residual Stress Techniques, www.lambdatechs.com (1986).
[2] Pulstec Industrial Co. Ltd., U-X360s Portable X-Ray Residual Stress Analyser, www.pulstec.co.jp (2013).
[3] European Commission, ATEX 2014/34/EU Guidelines, ec.europa.eu (2014).
[4] American Society for Testing and Materials, ASTM E837 Standard Test Method for Determining Residual Stresses by the Hole-Drilling Strain-Gage Method (2001).
[5] D. Crecraft, Ultrasonic Measurement of Stresses, Ultrasonics, 6(2), 117-121. doi:10.1016/0041-624x(68)90205-9 (1968).
[6] Acoustoelasticity using longitudinal waves for residual stress evaluation, NDT & E International, 28(2), 118. doi:10.1016/0963-8695(95)93327-3 (1995).
[7] M. N. Srinivasan, S. Joshi, J. Zhou, Development of a Numerical Model to Predict Residual Stress Based on Acoustoelastic Effect Using Critically Refracted Ultrasonic Longitudinal (LCR) Waves. Volume 12: Vibration, Acoustics and Wave Propagation. doi:10.1115/imece2012-89146 (2012)
[8] K.F. Graff, Wave motion in elastic solids. New York: Dover (1991).
[9] S. Timoshenko, Theory of elasticity. Auckland: McGraw-Hill (1970).
[10] D.E. Bray, W. Tang, Evaluation of Stress Gradients in Steel Plates and Bars with the LCR Ultrasonic Wave, Nuclear Engineering and Design, 207,pp. 231-240. DOI: 10.1016/S0029-5493(01)00334-X (2001).
[11] R. Marklein, K. Langenberg, K. Mayer, EFIT Simulations for Ultrasonic NDE, NDT.net., vol. 8., No. 3, 1-6 (2003).
[12] S. Cochran, Ultrasonics Part 12, Fundamentals of ultrasonic phased arrays. Insight - Non-Destructive Testing and Condition Monitoring, 48(4), 212-217. doi:10.1784/insi.2006.48.4.212 (2006).
[13] Abaqus Theory Manual, Version 6.11 (2011).
[14] W. Yaacoubi, S. Yaacoubi, Investigations on the critically refracted longitudinal wave for optimized use in the NDT/E field, 10th International Conference on NDE in Relation to Structural Integrity for Nuclear and Pressurized Components, Cannes, France (2013).
[15] R. Ramasamy, Z. Ibrahim, H.K Chai, T.F Chau, Measurement of platform conductor preload in streamlining the life extensions of ageing offshore wells, Applied Ocean Research, Volume 65, April 2017, Pages 12-22, ISSN 0141-1187, doi.org/10.1016/j.apor.2017.03.004.

IOP Conf. Series: Journal of Physics: Conf. Series **908** (2017) 012045 doi:10.1088/1742-6596/908/1/012045

Extraction of gold (Au) particles from sea water by *Delftia Acidovorans* microbes

A H M Yusoff, M E Nading and M N Salimi

School of Bioprocess Engineering, Universiti Malaysia Perlis, 02600 Arau, Malaysia.

nabil@unimap.edu.my

Abstract. Gold-mining activities have been an issue, especially when it involves in contamination of chemicals, for example arsenic and mercury. However, despite of these hazards, gold-mining activities are still being conducted. This is because the gold is worth, regardless of the problems. Gold-mining, as known needs a very large area of land, or site plant. Vast amount of labor force, mechanical force and fund are a must in order for the mining process to be continued. High demand of gold, made gold-mining industries as ones of the most profitable industries in the world. Thus, this has encouraged another alternative way to extract gold. At the mining site, researchers found that biomineralization of gold by *Delftia acidovorans* can be conducted. How it is done still cannot be understood. It is said that the bacteria secretes secondary metabolites, Delftibactin as a defensive mechanism against the toxicity of the soluble gold. Researchers try to find another source of elemental gold besides of the earth's core. The options are either lava of a volcano or ocean. Here, the focus is seawater. The problem of seawater is that its composition still not yet to be proved. Dissolve gold existed as gold chloride in seawater, but in a very small amount. So, the gold separation should be focused, in order to make this process to be a successful one. Factors such as depth, climate, region, temperature need to be considered. If this difference affecting the separating process, standardized seawater composition have to be proposed.

1. Introduction

Seawater is a complex mixture of 96.5% of water and 2.5% of salts and other smaller substances, which are either organic, inorganic materials or a few atmospheric gases [1]. 70% of Earth's surface is covered with seawater, and by this enormous composition of the foundation, this automatically marks seawater as the biggest source for gold metal [2]. However, in reality, there is only a small amount of gold existed in seawater. In fact, gold, Au is the least inorganic compound that presents in seawater, other than its periodic table neighbors such as Ag, Pd and Pt.

In seawater, gold, Au exists in its most stabilized form that is gold chloride, $AuCl_3^-$ [3]. This soluble form of gold is very toxic, and it harms human being. However, in spite of its toxicity, scientists have found that there is a superman-strength bacteria can survive in that toxicity condition. This bacteria is a prokaryote bacteria named *Delftia acidovorans. D. Acidovorans* is said to be resisting the toxic effects of this heavy metal. It deposited around the gold nuggets as a bacterial film, which make it available for redepositing in purer form somewhere else in the mine [4]. Therefore, researchers started to focus on this bacteria in order to develop other ways of extracting gold that are much cheaper and safe compare to the gold-mining process [5].

IOP Conf. Series: Journal of Physics: Conf. Series **908** (2017) 012045 doi:10.1088/1742-6596/908/1/012045

2. Methodology

2.1 Materials

2.1.1 Raw materials preparation
There will be eight types of raw materials used in this research. The materials are *Delftia acidovorans* strains, Acidovorax complex medium (ACM), Delftibactin, gold chloride ($AuCl_3$), seawater, 0.5% Agarose solution, double distilled water, and calcium carbonate.

2.1.2 Bacteria cultures
Delftia acidovorans strains will be cultured on Acidovorax complex medium (ACM) plates at 30°C.

2.2 Methods

2.2.1 Gold biomineralization
Gold biomineralization will be done by precipitation of gold on agar plate. *Delftia acidovorans* will be streaked onto a chelex-treated (deferrated) Acidovorax Complex Medium (ACM) agar plate. The bacteria will be grown for three days at 30°C. The culture will be centrifuged and resuspended in 5 mL sterilized ddH_2O. During the research, gold biomineralization step will be the main focus. There will be two factors involves; A (first factor), the concentration of gold chloride, and B (second factor), the concentration of Delftibactin, and C (third factor), the volume of seawater.

2.2.2 Factor A: Concentration of Gold Chloride
The assays will be set up as in 50 μL of ddH_2O containing 0.2 mM, 0.4 mM, 0.6 mM, 0.8 mM, 1.0 mM of Gold Chloride. Each sample will be added with 50 μL of 1.0 mM Delftibactin. Controls containing only water, which is No- Delftibactin and No- $AuCl_3$ will be added as a sample. The mixture will be incubated for 30 minutes in room temperature. Then, 100 μL of concentrated bacteria culture will be added to each sample. Incubate again for 30 minutes in room temperature. The mixture will be then incubated for 48 hours at 30°C. However, the absorbance reading will be recorded at first 120 minutes.

2.2.3 Factor B: Concentration of Delftibactin
The assays will be set up as in 50 μL of ddH_2O containing 0.2 mM, 0.4 mM, 0.6 mM, 0.8 mM, 1.0 mM of Delftibactin. Controls containing only water, which is No- Delftibactin and No- $AuCl_3$ will be added as a sample. Then, 100 μL of concentrated bacteria culture will be added to each sample. Incubate again for 30 minutes in room temperature. Seawater with volume of 5 liters will be introduced with 500ml of 5.0 M of Calcium Carbonate. The mixture will be allowed to settle, and sludges should be formed after several hour. The times normally vary based on the varying contents of said seawater. But, in this project, it will be limited to 48 hours. After that the water will be siphoned out, and sludges will be collected. Weight the wet weight of the sludge. The sludges are then introduced to the assays samples with different concentration of Delftibactin. The mixture is then incubated for 48 hours. However, the absorbance reading will be recorded at first 120 minutes.

2.2.4 Factor C: Volume of Seawater
The assays will be set up as in 50 μL of ddH_2O 1.0 mM of Delftibactin. Controls containing only water, which is No- Delftibactin and No- $AuCl_3$ will be added as a sample. Then, 100 μL of concentrated bacteria culture will be added to each sample. Incubate again for 30 minutes in room temperature. Seawater with volume of 1, 5, 10, and 15 liters will be introduced with 500ml, 1000ml, 1500ml, and 2000ml respectively of 5.0 M of Calcium Carbonate. The mixture will be allowed to settle, and sludges

should be formed after several hour. The time normally varies based on the varying contents of said seawater. But, in this project, the will be limited to 48 hours. After that the water will be siphoned out, and sludges will be collected. Weight the wet weight of the sludge. The sludges is then introduced to the assays samples with different concentration of Delftibactin. The mixture is then incubated for 48 hours. However, the absorbance reading will be recorded at first 120 minutes.

2.2.5 Extraction method for gold nanoparticles
The suspension will be centrifuged and washed once with water to concentrate the precipitate. The precipitate will be resuspended in ddH_2O. Au^{3+} in aqueous solution will be monitored by periodic sampling of aliquots (0.2ml) of the suspension. Each time the sample will be diluted with deionized water. UV-VIS spectra will be recorded by the use of Spectraphotometer at absorbance value of 600nm. UV-VIS will be used to examine the size and shape controlled nanoparticles in aqueous suspensions.

3. Results and Discussion

3.1 Wet weight of sludges
The sludges were obtained from the settlement of the mixture of the seawater and calcium carbonate. From Table 1.0, we can see that the wet weight of the sludges was varied. Eventhough, there was an increment of volume on both concentration of gold chloride, volume of seawater and volume of calcium carbonate added, the weight of the sludges diverged. Adding calcium carbonate in seawater was one of the assimilation of formation of calcrete on soils (as it occurs in nature). The vast different in wet weight was mainly due to a different consistency in the seawater composition. This different composition would react differently, and thus amount of sludges formed showed a discrepancy.

Table 1. Wet weight of sludges.

Concentration of Delftibatin, mM	**Wet weight of sludges,g**
0.2	84.6
0.4	72.5
0.6	80.3
0.8	116.2
1.0	94.1
Volume of Seawater, liters	**Wet weight of sludges,g**
1.0	486.9
5.0	725.3
10.0	1068.4
15.0	981.5

3.2 Factor A: Concentration of Gold Chloride

Controlled Variable: 1.0 mM Delftibactin

During the incubation process, gold element was separated from chloride and forming a colloidal. From Table 2.0, which shows weight of gold particles extracted, we can see that the number of particles existed in the sample affecting the reading of absorbance. Thus, we can

IOP Conf. Series: Journal of Physics: Conf. Series **908** (2017) 012045 doi:10.1088/1742-6596/908/1/012045

concluded that at 0.2 mM concentration, it shows the lowest weight, which is proven on Table 2.0

Table 2. Factor A.

Concentration of Gold Chloride, mM	Incubation Time, min	UV-VIS Reading(600nm), nm	Final Weight, g
0.2	0	0.038	0.183
	5	0.033	
	15	0.037	
	30	0.071	
	60	0.085	
	120	0.102	
0.4	0	0.044	0.267
	5	0.041	
	15	0.041	
	30	0.096	
	60	0.134	
	120	0.221	
0.6	0	0.047	0.445
	5	0.058	
	15	0.074	
	30	0.159	
	60	0.193	
	120	0.282	
0.8	0	0.050	0.522
	5	0.059	
	15	0.071	
	30	0.204	
	60	0.286	
	120	0.283	
1.0	0	0.078	0.764
	5	0.104	
	15	0.126	
	30	0.232	
	60	0.255	
	120	0.308	

3.3 Factor B: Concentration of Delftibactin

IOP Conf. Series: Journal of Physics: Conf. Series **908** (2017) 012045 doi:10.1088/1742-6596/908/1/012045

Controlled Variable: 5 liters Seawater

From Table 3.0, it is observed that the absorbance value increased as the concentration of Delftibactin increased. During the incubation process, gold element was separated from the sludges. At 1.0 mM
concentration, the absorbance value shows it highest value. Factor B showed to produced lesser amount of gold final weight as compared to Factor A. This show that data in Factor A show wider range of light scattered compare to Factor B since there is more number of gold particles existed in the samples.

Table 3. Factor B.

Concentration of Delftibactin, mM	Incubation Time, min	UV-VIS Reading(600nm), nm	Final Weight, g
0.2	0	0.152	0.000
	5	0.158	
	15	0.172	
	30	0.194	
	60	0.252	
	120	0.280	
0.4	0	0.182	0.0016
	5	0.186	
	15	0.204	
	30	0.211	
	60	0.282	
	120	0.307	
0.6	0	0.215	0.0000
	5	0.230	
	15	0.256	
	30	0.294	
	60	0.325	
	120	0.372	
0.8	0	0.240	0.0000
	5	0.247	
	15	0.290	
	30	0.342	
	60	0.385	
	120	0.403	
1.0	0	0.301	0.0045
	5	0.320	
	15	0.354	

30	0.372
60	0.394
120	0.399

3.4 Factor C: Volume of Seawater

Controlled Variable: 1.0 mM Concentration of Delftibactin

From Table 4.0, it is observed that the absorbance value increased as the volume of seawater increased. During the incubation process, gold element was separated from the sludges. At 20 liters of seawater, the absorbance values show its highest value. When comparing with Table 3.0, which shows no gold particles were extracted, we can see that the number of particles exists in the sample was not affecting the reading of absorbance. But instead, the composition or mostly the salinity of seawater affecting the reading of absorbance.

Table 4. Factor C.

Volume of Seawater, liters	Incubation Time, min	UV-VIS Reading(600nm), nm	Final Weight, g
1.0	0	0.1402	0.0000
	5	0.1486	
	15	0.1830	
	30	0.2361	
	60	0.2522	
	120	0.2818	
5.0	0	0.3005	0.0028
	5	0.3188	
	15	0.3561	
	30	0.3538	
	60	0.3684	
	120	0.3820	
10.0	0	0.3205	0.0004
	5	0.3372	
	15	0.3584	
	30	0.3804	
	60	0.3887	
	120	0.3951	
15.0	0	0.3258	0.0000
	5	0.3294	
	15	0.3406	
	30	0.3784	
	60	0.3902	
	120	0.4050	

20.0	0	0.3108	0.0000
	5	0.3237	
	15	0.3409	
	30	0.3552	
	60	0.3885	

3.5 Final Weight of Gold Particles

Table 5.0 shows the comparison of final weight of gold particles between gold chloride and seawater. As gold chloride is the stable form of gold particles in nature, we can directly assume that the extraction from gold chloride will definitely produce the highest number of gold particles. As mentioned previously, eventhough there is a constant amount of gold in seawater, but in different collection of seawater samples there is no guarantee there is gold present in the samples. This explains why most of the test extracted no gold despite the increment in seawater volume used.

Table 5. Final Weight of Gold Particles.

Gold Chloride	Seawater
0.183	0.0000
0.267	0.0028
0.445	0.0004
0.522	0.0000
0.764	0.0000

4. Conclusion

The reduction of Au^{3+} ions by *Delftia sp.* resulted in the formation of stable particles from gold chloride solution. However, the purpose of this main project is to be able to extract gold particles from seawater. The objectives were successfully accomplished eventhough the amount of gold particles extracted was almost to non-existent. In spite of this, the success in extracting this small amount of gold particles can proved that it is possible to obtain gold from the vast amount of seawater. Eventhough it is possible, the small amount of particles make this project impractical to be done in industries level. The expenditure to start the project was expensive, but the recovery of gold from seawater unable to compensate the start-up modal. Engineered microbes or improved equipment might help in the extraction in the future.

5. References

[1] F J Millero, R Feistel, D G Wright and T.J McDougall (2008) *The composition of Standard Seawater and the definition of the Reference-Composition Salinity Scale, Deep- Sea*, **50** 72

[2] G F Gadd (2010) *Metals, minerals and microbes: geomicrobiology and bioremediation, Microbiology*, **156** 609

[3] Goldberg, E. D. (1987) *Heavy metal analyses in the marine environment—approaches to quality control. Mar. Chem.*

[4] C. Ganesh Kumar, Y. Poornachandra, Suman Kumar Mamidyala (2014) *Green synthesis of bacterial gold nanoparticles conjugated to resveratrol as delivery vehicles, Colloids and Surface Biointerfaces*, 927

[5] F. Reith, L. Fairbrother, G. Nolze, O. Wilhelmi, P.L. Clode, A. Gregg, J.E. Parsons, S.A. Wakelin, A. Pring, R. Hough,G. Southam and J. Brugger (2010), *Nanoparticle factories: Biofilms hold the key to gold dispersion and nugget formation, Geology*, **38** 843

Functional design of heat exchange for pneumatic vehicles

Z G Xu[1], D Y Yang[2], W D Shen[2] and T T Liu[1]

[1] Shenzhen Research Institute of Shandong University, 518057, High-tech Zone A301, Virtual University Part, Shenzhen, China.
[2] School of Mechanical Eng, Shandong University, 250061, No 17923, Jingshilu, Lixiaqu, Jinan, Shandong, China.

zhgxu@sdu.edu.cn

Abstract. With the increasingly serious environmental problems, especially the impact of fog and haze, the development of air powered vehicles has become an important research direction of new energy vehicles. Quadrature test was done with different materials, i.e. stainless steel and aluminum alloy, at different inlet pressures, using different expansion gases, i.e. air, CO_2, for heat exchanging properties for pneumatic vehicles. The mathematics as well as simulation methods are used to analyze the different heat exchanging effects in the multistage cylinder. The research results showed that the stainless steel has better effects in heat exchanging than Aluminum Alloy; the intake pressure has little effect on CO_2 than the air in heat exchanging effect. CO_2 is better in heat exchanging than air.

1. Introduction

With the increasingly serious environmental problems, the transformation and upgrading of China's auto industry is becoming more and more urgent. In recent years, due to the deterioration of the natural environment, especially the impact of fog and haze, new energy vehicles are becoming more and more concerned.

French engineer Guy Negre [1] led the research team to develop a compressed air powered vehicle, more than 20 patents have been obtained in this field. There is no essential difference between the use of compressed air or compressed nitrogen for engines.

Researchers in University of Washington developed a liquid nitrogen powered vehicle prototype [2]. Their energy source is liquid nitrogen. Liquid nitrogen is heated to expand into the engine cylinder to drive the vehicle. There is no essential difference between the use of compressed air or compressed nitrogen for engines. Researchers in University of North Texas did the study of liquid nitrogen powered vehicles. A team lead by Dr. Carfos Ordonez, uses liquid nitrogen as power source, to drive vane type pneumatic motors [3].

Tsu-Chin Tsa from the University of California, Los Angeles, lead a R & D team, funded by Ford Motor Company, developed a hybrid power engine with no cam shaft [4]. High-pressure air is to start the car, when it is moving, normal gasoline combustion is to drive the car. When the car brakes slow down or downhill, pneumatic engine can also be used as air compressor energy recovery, the new compressed air can be used as the starting power for the next start use.

A Korean ENERGINE company studied a compressed air and electric hybrid vehicles, called PHEV research. The weight of the major part is 30 kg, cylinder working volume is 0.8L, gas tank volume is 100L, the intake pressure of the engine is 1MPa, gas pressure is 30 mPa [5].

In China, researchers in Zhejiang University studied the working principle of a pneumatic engine, the characteristics of energy release of high pressure air under different conditions are analyzed, a number of experiments have been carried out on the pneumatic engine, it provides the theoretical basis and experimental data for the design and manufacture of pneumatic engines [6].

2. Heat exchanging for air powered vehicles

Heat exchanger, as in figure 1, is a key component for air powered vehicles, the heat transferring efficiency determines the performance of an air powered vehicle. However there exist contradictions, the typical one is that, if the heat transferring efficiency is to increased, the heat exchanger will be very large. Heat exchange efficiency determines the performance of the air powered vehicle. The heat transferring efficiency is to be increased, but the volume is constrained to a certain space, it is a contradiction.

This contradiction is problem, and this problem is to solved by TRIZ contradiction matrix. Two contradiction parameters should be firstly recognized, according to the problems existing in the heat exchanger, the no.39 parameter, i.e. productivity, and the no.8 parameter, i.e. the static volume is identified. The solution principles in the conflict matrix can be found as the follows: no 2 principle, i.e. separation; no 35 principle, i.e. parameter changing; no 37 principle, i.e. thermal expansion.

The no.2 principle, i.e. separation is accepted. The "instance effect increased with the increase of temperature of heat exchanging can be used as an example, so the single cylinder is changed to a multi-stage cylinder, a two-cylinder structure is utilized, and the analyzing of the heat exchanging, and the two stage expansion diagram is shown in figure 6.

The high pressure gas in the first level cylinder, exchanges heat with low pressure gas discharged through a heat exchanger, and the first cylinder absorbs heat from the external environment, and then enters the second stage cylinder, this process can improve the thermodynamic properties, which was close to the isothermal process, and the aerodynamic efficiency of the engine is further improved.

The heat exchanging efficiency has a great influence on the performance of the engine, and the heat exchanger is a device for realizing the heat exchange between two or more than two fluids with different temperatures [6].

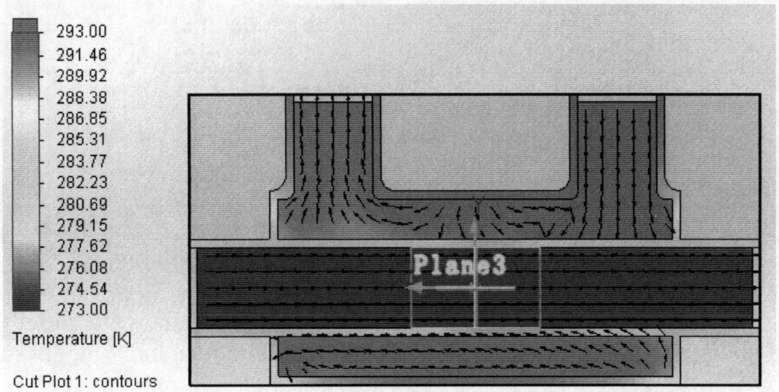

Figure 1. Heat transfer effect of casing heat exchanger.

According to its working principle, the heat exchanger can be divided into three types: the wall-type, the mixed type and the regenerative type. This paper chooses the wall-type heat exchanger.

In this paper, the casing heat exchanger is analyzed firstly, and the heat exchanger is made of several concentric tubes with different diameters. In this kind of heat exchanger, some fluid go through the inner tube, other fluid go through the gap between tubes, both can use a higher flow rate, so the heat transferring coefficient is relatively high. In addition, in this heat exchanger, two kinds of fluid can go oppositely, and the heat exchanging effect is getting higher.This heat exchanger has the advantages of simple structure, convenient operation, high pressure etc.

The low temperature gas enters from the gas entrance, and departs from the air outlet, the heated water comes into the liquid from the entrance, and exits from the outflow of liquid. The air velocity is 10m/s, the temperature is 273K, the liquid flow is 0.02kg/s, the temperature is 293K, and the heat exchanger material is stainless steel, the analyzing results are as follows.

It can be seen from the figure that the temperature of the gas outlet is very low, and the heat transfer efficiency of the heat exchanger is not high enough, which is not suitable for the heat exchange of the air powered vehicle. So it is necessary to improve the efficiency of the heat exchanger. Considering the "heat radiation" effect, i.e, solar energy is introduced to heat the gas, and the above method of heating the liquid gas is more convenient and better, so the design of a solar heat exchanger is as in figure.5. The low pressure low temperature gas discharged from the first stage cylinder, enters the heat exchanger, and the heat exchanging is done in the two stage cylinder to continue to expand and do works.

3. Orthogonal experiment of heat exchanger

The solar heat absorbing plate is equivalent to a constant power source, its power is influenced by many conditions, this paper set up the power of 50W, the gas entrance temperature is 273K, temperature 293K, under different conditions, the effect of heat transferring is different, corresponding to a number of different exchanging plan, for each program number with different numbers for different schemes, its meaning is shown in figure. Its specific meaning is shown in the following table:

IOP Conf. Series: Journal of Physics: Conf. Series **908** (2017) 012046 doi:10.1088/1742-6596/908/1/012046

Table 1. Different program details.

No	Material	Inlet pressure	Expansion gas
H141	Stainless steel	4atm	air
H142	Stainless steel	4atm	CO_2
H151	Stainless steel	5atm	air
H152	Stainless steel	5atm	CO_2
H241	Aluminum alloy	4atm	air
H242	Aluminum alloy	4atm	CO_2
H251	Aluminum alloy	5atm	air
H252	Aluminum alloy	5atm	CO_2

In order to compare the heat transferring effects of the influence of different materials, different inlet pressures and different gas. Eight schemes should be compared, in order to reduce the number of experiments, this paper used orthogonal test to select a representative research program.

The orthogonal test is very suitable for multi-factor analysis, through the orthogonal experiment to analyze the effect of each factor on the test results, the relationship can also be distinguished among the various factors and their interaction between primary and secondary factors etc, and the best combination is finally selected. More information could be obtained with less number of tests by orthogonal test [7]. The variable factor is the material, inlet pressure, expansion gas, each factor is set to two levels, this paper uses the three factor two level experiment, as shown in the following table2.

Table 2. Factors and different levels.

level	Factor A, material	Factor B, inlet pressure	Factor C, expansion air
1	Stainless steel	4atm	air
2	Aluminum alloy	5atm	CO_2

Take L4（2^3）as the orthogonal results as in table 3.

IOP Conf. Series: Journal of Physics: Conf. Series **908** (2017) 012046 doi:10.1088/1742-6596/908/1/012046

Table 3. Orthogonal test table.

no	A	B	C
1	1（Stainless steel）	1（4atm）	1（air）
2	1（Stainless steel）	2（5atm）	2（CO2）
3	2（Aluminum alloys）	1（4atm）	2（CO2）
4	2（Aluminum alloys）	2（5atm）	1（air）

Table 3 shows that the program are chosen as H141, H152, H242, H251, and the simulation results are as follows:

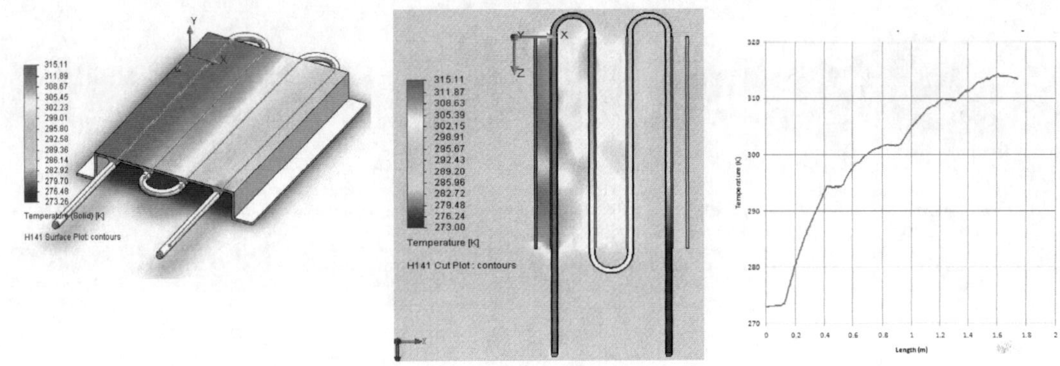

(a) Surface temperature. (b) Temperature of air duct section (c) Temperature change

Figure 2. Temperature change of H141 scheme.

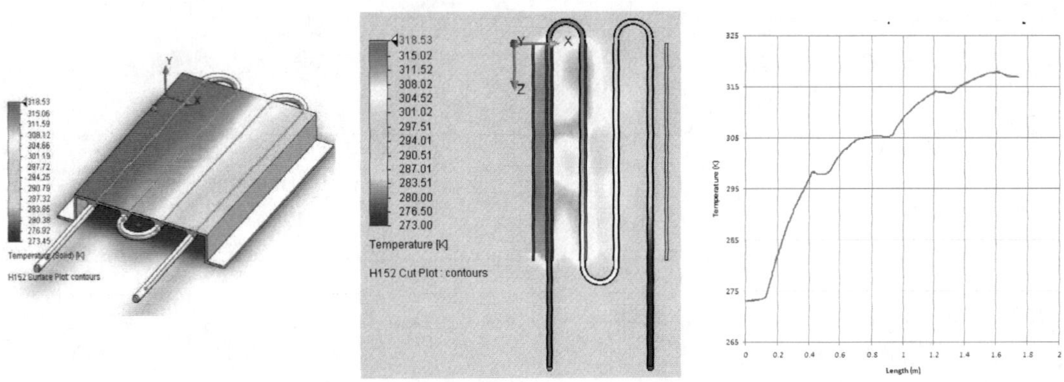

(a) Surface temperature. (b) Temperature of air duct section (c) Temperature change

Figure 3. Temperature change of H152 scheme.

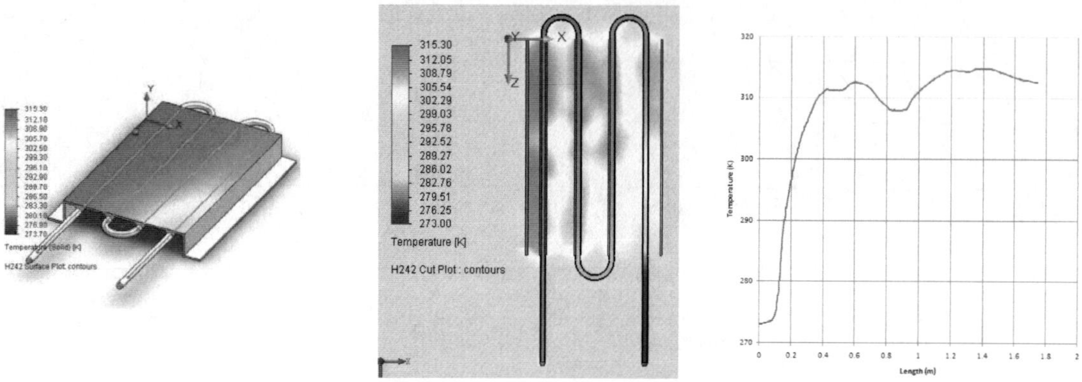

(a) Surface temperature. (b) Temperature of air duct section (c) Temperature change

Figure 4. Temperature change of H242 scheme.

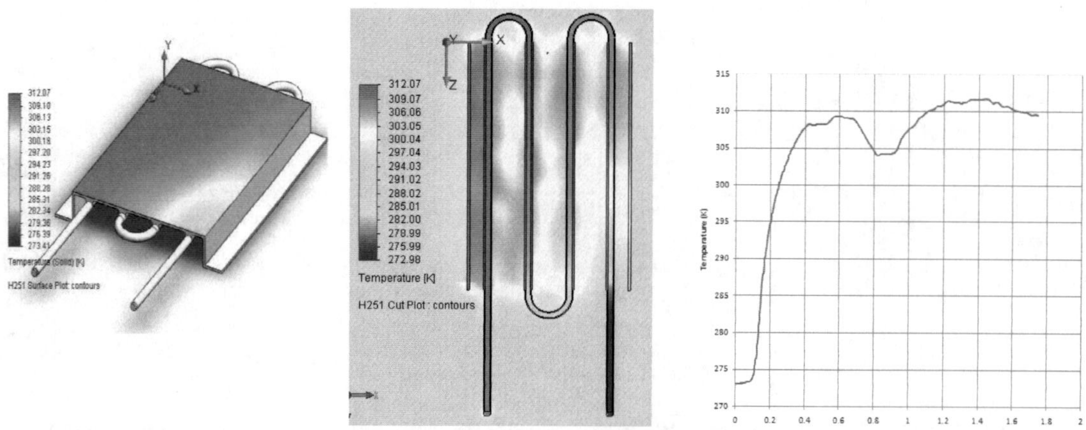

(a) Surface temperature. (b) Temperature of air duct section (c) Temperature change

Figure 5. Temperature change of H251 scheme.

Table 4 orthogonal test results

No	plan	A	B	C	Outlet temperature
1	H141	1	1	1	315.11
2	H152	1	2	2	318.53
3	H242	2	1	2	315.3
4	H251	2	2	1	312.07

From table4, the optimal solution is set as $A_1B_2C_2$, namely H152 scheme, the simulation results showed that the stainless steel material is better than Aluminum Alloy in heat exchanging effect. And the intake pressure has little effect on heat exchanging. CO_2 is better than air in heat exchanging. When the inlet temperature is 273K, through the absorption of solar energy, the outlet temperature is about 318K. And the thermodynamic process of the cylinder is approximately equal to a isothermal expansion process, thereby the output power of the pneumatic engine is increased.

4. Thermodynamic analysis of multistage expansion

The thermodynamic process of the multistage expansion is different from that of the single stage expansion. Take the two stage expansion as an example, the thermodynamic process consists of the following three stages, the first is the polymorphic process, experience and process, and finally the polymorphic process is shown in the end of the expansion.

Thick line in Figure 6 is the thermodynamic process of two stage expansion, 1-3-5 curve stands for the isothermal process, 1-2-5 curve stands for the 'polymorphic process.

Assuming that the environmental temperature is set in 293K, the detailed process is as follows, 1-2 curve is a polymorphic process, and the gas pressure decreased, the volume expanded and temperature reduced. Curve 2-3-4 is an isochoric process, gas absorbs heat from the outside environment, the temperature and the pressure rises with the constant volume; curve 4-5 is a two stage expansion polymorphic process.

The ordinary polymorphic multistage expansion processes consider the heat transferring with the environment, the temperature can not exceed the ambient temperature, when the solar energy is absorbed, the temperature will be higher than the ambient temperature, it can reach 318K, where more energy are released.As is shown in the figure 8, the energy released is the area of 3-4-5-3, and the energy of the two stage expansion is more than that of the single stage expansion, which is the area of 2-3-4-5-5 '-2.

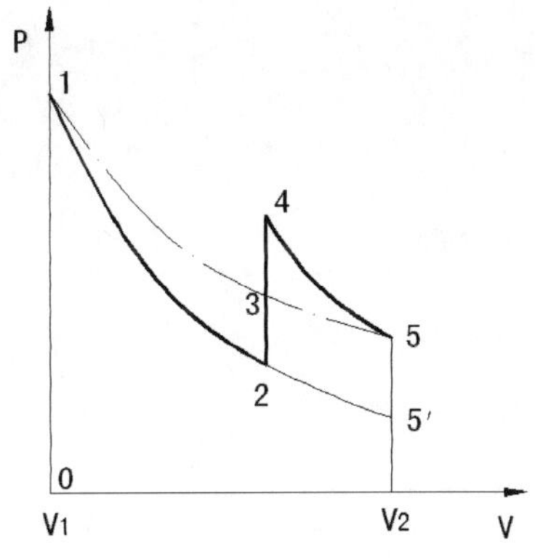

Figure 6. Two stage heat expansion thermodynamic processes.

5. Conclusions

The research results showed that the stainless steel has better effects in heat exchanging than Aluminum Alloy; the intake pressure has little effect on CO_2 than the air in heat exchanging effect. When the inlet temperature is 273K, through the absorption of solar energy, the outlet temperature is about 318K, make the thermodynamic process of cylinder approximately equal to an isothermal expansion process, thereby the output power of the pneumatic engine is increased.

6. Acknowledgment
The above-mentioned research work is supported by the Science, Technology and Innovation Commission of Shenzhen Municipality, JCYJ20160510165328965 and the Chinese NSFC, 61272017, P.R. China.

References

[1] Cheng Sen, Li Sanchao. Carbon fiber gas car tank. automotive and safety,2001,(9):44-45.
[2] He Cunxing, Zhang Tiehua. Hydraulic transmission and pneumatic drive, Huazhong University of Science and Technology press, Wuhan, China, 2000.
[3] Prabhakar G, Goel A.A functional modeling for adaptive design of devices in new environments, Artificial Intelligence in Engineering Journal(Special Issue), 1998,12(4):417-444.
[4] G. Altshuller. Creativity as an Exact Science. Gordon and Breach Science Publishers,1984.
[5] Tai, Chun, Tsao, Tsu-Chin, M. B. Levin, G. Barta,and M.M.Schechter,"Using Camless Valvetrain for Air Hybrid Optimization,"SAE Paper 2003-01-0038 presented in SAE World Congress, March 2003, Detroit, MI. Also in SAE World Congress 2003 Special Publications (SP-1752), ISBN: 0-7680-1175-2.
[6] Robert F V. Fuel of the future? In: Electric and Hybrid Vehicles Technology'98, 1998: 124-127
[7] Williams J, Knowlen C, Mattick A T, et al. Quqsi isothermal expansion engines for liquid nitrogen automotive propulsion. SAE 972649, 1997

IOP Conf. Series: Journal of Physics: Conf. Series **908** (2017) 012047 doi:10.1088/1742-6596/908/1/012047

Key issues in theoretical and functional pneumatic design

Z G Xu[1] , D Y Yang[2], W M Liu[1] and T T Liu[1]

[1] Shenzhen Research Institue of Shandong University, 518057, High-tech Zone A301, Virtual University Part, Shenzhen, China.
[2] School of Mechanical Eng, Shandong University, 250061, No 17923, Jingshilu, Lixiaqu, Jinan, Shandong, China.

zhgxu@sdu.edu.cn

Abstract. This paper studies the energy release of the pneumatic engine in different thermodynamic processes, the isothermal process is the highest power output process, while adiabatic process is the lowest energy output process, and the energy release of the pneumatic engine is a multi-state thermodynamic process between them. Therefore heat exchanging should be increased between the pneumatic engine and the outer space, the gas expansion process in the cylinder should be as close as possible to the isothermal process. Heat exchange should be increased between the cylinder and the external spaces. Secondly, the fin structure is studied to increase the heat exchanging between the cylinder body and the outside space. The upper part has fin structures and the lower cylinder has no fin structure, this structure improved the working efficiency of pneumatic engine. Finally the cam and the hydraulic bottle of pneumatic engines are designed. Simulation and theoretical calculation are used to the analysis of the whole structure, which lay the foundation for the manufacturing and design of the pneumatic engines.

1. Introduction
In recent years, due to the deterioration of the natural environment, especially the impact of fog and haze, new energy vehicles are becoming more and more concerned. Researchers in Hefei University of China, converted a R715 type diesel engine to a two-stroke pneumatic engine[2]. Also, H Wang, and J Zhou, in the Chinese University of science and technology did the research of heat exchange of the air in the pneumatic engine with the outer-space[3]. W, He in Beijing Industrial University, did the research of the performance of the single-screw pneumatic engine[4]. C, J, Marquand designed a experimental two-stroke eccentric vane-type pneumatic engine. Terry Millers, a French company called MDI etc, have also done many works on pneumatic engine.

2. Working principal for air powered vehicles
The working principle of an air powered engine is similar to that of an internal combustion engine, however, there are differences, the structure is shown in figure1. When compressed air enters the cylinder, the piston moves back and forth under the pushing force of the compressed air, which drives the crankshaft to rotate continuously, to move the car. High pressure gas in compressed air bottle 1

IOP Conf. Series: Journal of Physics: Conf. Series **908** (2017) 012047 doi:10.1088/1742-6596/908/1/012047

flows into the mechanical valve 2 through the air inlet of the mechanical reversing valve 2, then enter the first cylinder 3 though the first reversing interface. It pushes the piston 3 of the first cylinder 5 downward, drives the crankshaft 8 to rotate [5].

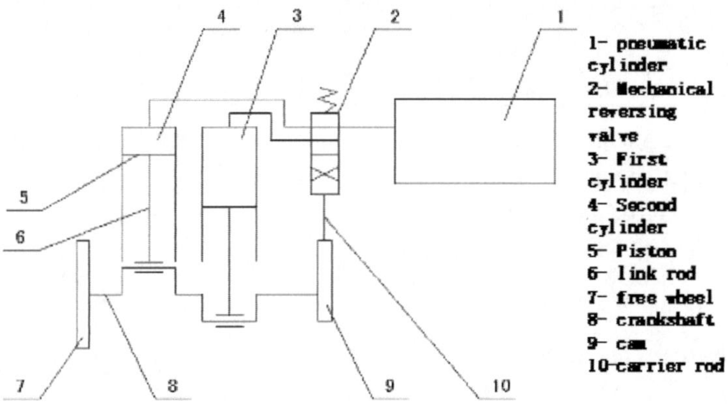

Figure 1. Working principal of air-powered engine.

Crankshaft 8 drives the second cylinder piston upward. Cam 9 reverses valve 2 through the ejector rod 10. The second piston in cylinder discharge the gas in into the atmosphere, through the second reversing interface of the mechanical reversing valve 2. When the first cylinder piston moves down to the lower stop point, the second cylinder piston moves up to the top stop point, cam 9 reverses the mechanical reversing valve 2 through the ejector rod 10.

High pressure gas in compressed air bottle 1 flows into the second cylinder 4, through the mechanical valve 2, pushes the second cylinder piston downwards, to rotate the crankshaft 8. Crankshaft 8 drives the first cylinder piston upwards, Cam 9 Make mechanical reversing valve 2 change direction, through the ejector rod 10.

The pressure/ volume change of the gas in the cylinder is shown in fig 4. During the intake stroke, high pressure gas first enters the cylinder with pressure P1, push the piston down, performs works. When the piston moves to position 3, the mechanical reversing valve stops the inlet air. The cylinder becomes a closed system, the high-pressure gas inside the cylinder expands, pressure energy releases, and pushes the piston downward to continue to work outside.

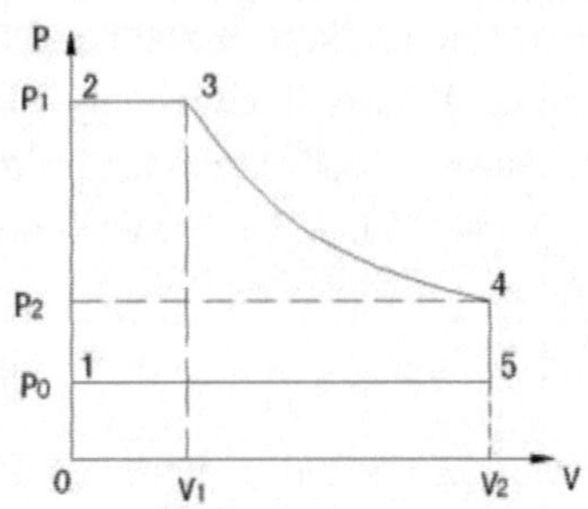

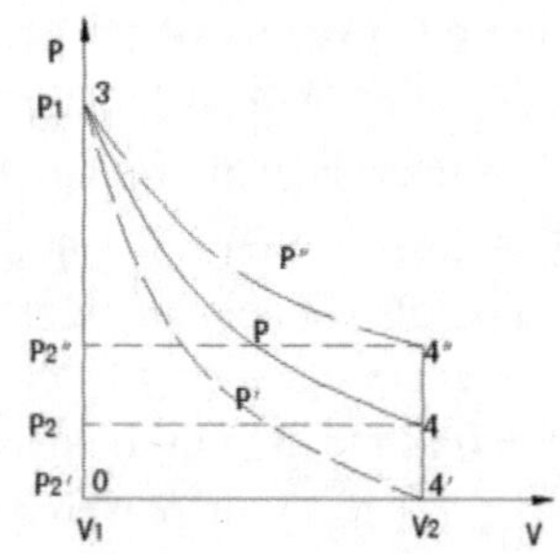

Figure 2. Gas expansion process P-V diagram. **Figure 3.** P-V curves of gas in cylinder.

When the piston reaches the bottom dead point, i.e. point4 in the picture, the mechanical reversing valve controls the cylinder to exhaust, reduced the inner pressure to atmospheric pressure P_0, in the meanwhile, the piston moves upward to discharge low-pressure gas in cylinder. The enclosed area by

IOP Conf. Series: Journal of Physics: Conf. Series **908** (2017) 012047 doi:10.1088/1742-6596/908/1/012047

1-2-3-4-5-1 in the chart is the theoretical work of high pressure gas V_0 , during the intake and exhaust processes, pneumatic engine exchanges the refrigerant with the outside world. In an ideal state, gas pressure has no change, the output function by the cylinder equals to gas technical work Ws [2]; i.e.

$$W_s = P_1V_1 - P_0V_2 \quad (1)$$

When the cylinder is a closed system, i.e. the mechanical reversing valve closed completely. Gas expansion performs the external work. The Cylinder output function is the expansion function Wv:

$$W_v = \int_{V_1}^{V_2} pdV \quad (2)$$

Power generated by the compressed air W equals the sum of the technical power and the expansion power.

$$\mathrm{W} = W_s + W_v \quad (3)$$

3. High pressure gas design

The power source of air powered vehicle is the high pressure tank. Because the air powered vehicle is a high-speed moving vehicle, so compared with the ordinary pneumatic equipment, the high pressure tank should be high in pressure, light in weight, good in safety, small in size etc. So steel tank is replaced with aluminum alloys, and is wrapped with glass fibers, which can increase 30% in capacity, reduce 30% in weights. 100% carbon fiber can also be used, so high in pressure and light in weight. This tank equipped with high-pressure gas has been tested under the condition of shooting test, fire test, falling down and explosion test etc.

The heat capacity molar ratio K of the pneumatic engine gas is, the ratio of Cp (the molar constant heat capacity of gas), to the Cv (molar constant heat capacity), i.e.

$$k = \frac{C_p}{C_v} \quad (4)$$

Molar heat capacity ratio of monatomic molecular gas:

$$k = \frac{5}{3} \quad (5)$$

Molar heat capacity ratio of diatomic molecular gas:

$$k = \frac{7}{5} \quad (6)$$

Molar heat capacity ratio of polyatomic molecular gas:

$$k = \frac{4}{3} \quad (7)$$

The external work of isothermal expansion has nothing to do with the molar heat capacity ratio of gas. However under the adiabatic expansion condition, the output expansion work is related to the molar heat capacity ratio of gas, i.e.

$$W_v = (\frac{p_1 V_1}{k-1})[1-(\frac{V_1}{V_2})^{k-1}] \tag{8}$$

For the convenience of calculation, the expansion ratio is expressed by e, i.e.

$$W_v = (\frac{p_1 V_1}{k-1})[1-e^{1-k}] \tag{9}$$

The molar heat capacity of a monatomic molecule is the largest, with the worst expansion effect. The polyatomic molecule is the best in expansion effect, while the diatomic molecule is between them.

4. Cam design

For a four stroke internal combustion engine, for every one work needs two strokes, while for a two stroke internal combustion engine, for every one work needs one stroke. The air powered engine can be re-manufactured by the existing internal combustion engine, the motion law of the cam follower is showing in table 1.

The cam contour for intake and exhaust of internal combustion engine is designed for four stroke gas distribution, cam contour should be redesigned for two stroke gas distribution in order to adapt to the set of air powered engine. Mechanical valve is used in the control of intake and exhaust of the air powered engine, which usually rely on cam, block or other mechanical external forces.

Table 1. Motion law of the cam follower.

movement	Shock property	v-max/(hω/Φ)	a-max/(hω2/Φ2)	j-max/(hω 3/Φ3)	application
Constant speed	Rigidity	1.00	∞	--	Low speed light load
Equal acceleration, deceleration	Flexibility	2.00	4.00	∞	Medium speed light load
Cosine acceleration	Flexibility	1.57	4.93	∞	Medium speed, medium load
Sinusoidal acceleration	NONE	2.00	6.28	39.5	High speed light load
3-4-5 degree polynomial	NONE	1.88	5.77	60.0	High speed, medium load

The cam mechanism has the advantages of simple in structure, compact in assembly, reliable in performance etc. The cam contour can be designed to drive the follower accurately to realize various anticipated movement. Roller type mechanical control valve has a roller at the top of the valve core, the roller has a direct contact to the end of the cam, the roller transfers forces to the core of the cam.

One of the advantages of this roller control is that, it can reduce the lateral force of the cam core, improved the loading conditions between the cam core and the valve body, the life and reliability of the valve is increased. The motion of the follower is decided by the cam contour, i.e. different motion laws of follower require that the cam has different shapes and contours:

Sinusoidal acceleration is selected for the follower motion to meet for the high speed, light load of the mechanical valve to control the air powered engine. The cam contour is designed according to the motion law of the engine follower. The cam contour equation is as the followings. Where, *e* is the offset , R_b is the basic radius , R_R is the radius of the roller, $s=s(\varphi)$, *s* is the motion law of the follower.

$$x_B = (s_0 + s)\sin\varphi + e\cos\varphi \tag{10}$$

$$y_B = (s_0 + s)\cos\varphi - e\sin\varphi \tag{11}$$

Where s_0 is:

$$s_0 = \sqrt{r_b^{\,2} - e^2} \tag{12}$$

The actual contour curve equation is:

$$x_A = x_B + r_r \frac{dy_B / d\varphi}{\sqrt{(\frac{dx_B}{d\varphi})^2 + (\frac{dy_B}{d\varphi})^2}} \tag{13}$$

$$y_A = y_B - r_r \frac{dx_B / d\varphi}{\sqrt{(\frac{dx_B}{d\varphi})^2 + (\frac{dy_B}{d\varphi})^2}} \tag{14}$$

Based on the above formula, the theoretical contour curve equation and the actual contour curve equation can be calculated by using the correlation function in **MATLAB**. The **MATLAB** graphical user interface (GUI for short) can be embedded in the written programs, and the running results can also be presented to the operator or customer in a man-machine interactive way. And the results can be data or graphics, and the code can also be invisible. MATLAB/GUI is composed of the cursor, window, menu, button, text and other objects (Object) consisting of a user interface. These graphical objects can be activated through mouse or keyboard operation, so as to control some parameters in the program, for the ultimate calculation, drawing and other functionaries.

5. Cylinder design

The different thermodynamic energy release properties have been studied for the pneumatic engine, the isothermal process is the highest power output expansion thermodynamic process, while the

IOP Conf. Series: Journal of Physics: Conf. Series **908** (2017) 012047 doi:10.1088/1742-6596/908/1/012047

adiabatic process is the lowest power output process, and the pneumatic engine is polymorphic process between them.

The heat exchanging should be increased for the pneumatic engine with the external space, and the gas expansion process in the cylinder should be as close as possible to the isothermal thermodynamic process, where the working efficiency of the pneumatic engine can be effectively improved. The outer part of the cylinder should be heated to increase the temperature or absorb energy from the outer space.

"Temperature increasing" effects have many examples, i.e. electromagnetic induction effect, dielectric effect, Joule-Lenz effect, arc discharge effect, "absorption" effect, emission-focusing effect, thermal radiation effect, Perle effect, hot electron emission effect, Thomson effect and thermo-electric phenomena etc.

From the above effects, "absorption" effect is specified. "Absorption" effect refers to the process of energy absorption of one material from the other objects or processes etc. There are many "absorption" effects, i.e. heat sink, heat exchanger, noise absorption, X-ray absorption etc, which can be combined with the actual situation of the cylinder. The fin structure is combined with the heat exchanger to complete the effect of "temperature increasing". The actual structure corresponding to the heat sink is the fin, which can appropriately increase the heat transfer effect between the cylinder body and the outside. The preliminary design of the cylinder is as in figure 4 (a).

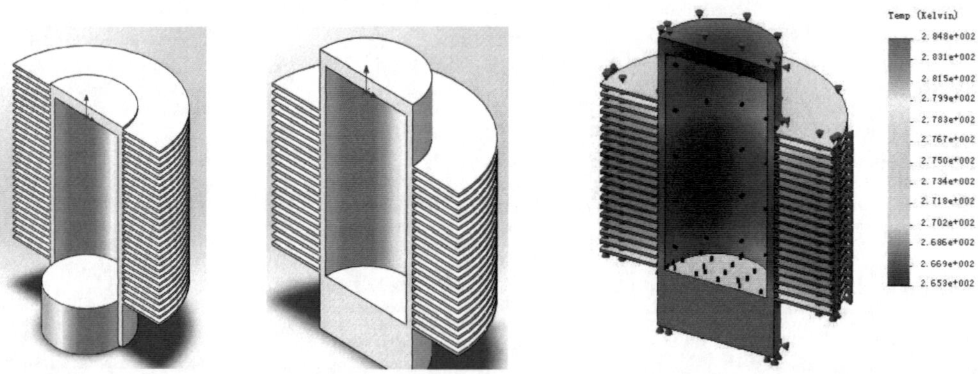

(a) primary design. (b) innovative design. (c) Thermodynamic effect of the fin structure.

Figure 4. Cylinder with fin.

When high pressured air enters the cylinder, the gas in the cylinder are compressed, and the piston did not immediately move down, which results in the gas temperature in the cylinder rises rapidly. If heat exchanging is increased, the energy in the cylinder will loss.

So when the cylinder fin need to absorb heat from the out space, it will also cause heat loss when the gas temperature is higher than the outside. So from the physical conflict in the TRIZ theory, the fins should be exit, and not be exit. The solution are as the follows: separation in space, separation in time, and conditional separation.

In this paper, the inventive principle of space separation is utilized. According to the innovative idea given by the above principle, a cylinder with no fin at the upper part of the outer wall of the cylinder is designed as in figure 4 (b). When the pneumatic engine works continuously, the gas in the cylinder experienced a poly-morphism process, the gas temperature is lower than the outside space, so extra heat needs to be absorbed from the outside mainly by convection and heat transferring. A heat exchanging simulation is carried out under the following conditions. The cylinder material is Aluminum Alloy, its thermal conductivity is λ=204W/(m·K)/ (M - K), and the heat transferring convection coefficient is 25W/ (M2 - K), the ambient temperature is 293K. The simulation result of heat transferring of the finned cylinder is shown in figure 4(c).

From the simulation result, it could be found that, the gas temperature is still very low, due to the reason that the heat transferring effect is poor, because the fin structure is small in size for the sake of compact structure. The compressed air consume too much energy in the work process due to heat transferring. In the practical work, the engine has a larger power output, the temperature of the environment is not high enough, so the fin structure has limited effects, so other heat exchanging methods must be considered.

6. Conclusions

The air powered vehicle can be used in the short distance transport to meet the urgent need of more serious air pollution in the city. In the first-order phase transition, compressed air is used as the power source to drive the vehicle. Compressed air is used to push the piston and the crankshaft to move the vehicle to achieve zero emissions. Some key components of air powered cars have been designed in this paper. The theoretical and actual cam profile have been drawn by MATLAB. Heat exchanging effect have been improved by using TRIZ theory.

7. Acknowledgment

The above-mentioned research work is supported by the Science, Technology and Innovation Commission of Shenzhen Municipality, JCYJ20160510165328965 and the Chinese NSFC, 61272017, P.R. China.

References

[1] Cheng Sen, Li Sanchao. Carbon fiber gas car tank. automotive and safety,2001,(9):44-45.

[2] Zuo Chengji, Qian, Ye Jian, an da, experimental research of pneumatic engine, Journal of mechanical engineering (China), 2007, (04), pp.93-97.

[3] Jia Guangzheng, Wang Xuanyin, and Liu Hao, Research on the energy distribution and dynamic characteristics of pneumatic car during the process of pressure reducing, Chinese mechanical engineering, 2004, (14), pp.74-78.

[4] W, He, Wu Yuting, Ma Chongfang, analysis of the energy efficiency of the single screw air engine power system. Journal of mechanical engineering(China), 2011, (06), pp.150-155.

[5] Wu Hongchao, Hu h L. (CBR)research on the model of product configuration unit. Computer engineering and design, 2009, : 5482-5484+5488.

[6] G. Altshuller. Creativity as an Exact Science. Gordon and Breach Science Publishers,1984.

[7] Tai, Chun, Tsao, Tsu-Chin, M. B. Levin, G. Barta,and M.M.Schechter,"Using Camless Valvetrain for Air Hybrid Optimization,"SAE Paper 2003-01-0038 presented in SAE World Congress, March 2003, Detroit, MI. Also in SAE World Congress 2003 Special Publications (SP-1752), ISBN: 0-7680-1175-2.

Conceptual functional-to-form mapping for green design

Z G Xu[1], W M Liu[1], W D Shen[2], D Y Yang[2] and T T Liu [2]

[1] Shenzhen Research Institue of Shandong University, 518057, High-tech Zone A301, Virtual University Part, Shenzhen, China.
[2] School of Mechanical Eng, Shandong University, 250061, No 17923, Jingshilu, Lixiaqu, Jinan, Shandong, China.

zhgxu@sdu.edu.cn

Abstract. Design for dis-assembly (DFD for short) is the key issue for green design automation. In this paper an assembly-level function-to-form mapping CAD system is reported for green design computing. The research work mainly includes: the assembly-level function definitions, product network model, two-step mapping mechanisms, dis-assembly sequencing based on graph theory, dis-assembly analysis etc. The function-to-form mapping is divided into two steps,i.e. mapping of function-to-behavior, called the first-step mapping, and the mapping of behavior-to-structure, called the second-step mapping. After the first step mapping, the three dimensional transmission chain (or 3D sketch) is established, and the product network model is created, on the basis of which the assembly/dis-assembly analysis and sequencing of the whole mechanism could be fulfilled. A mechanical hand is illustrated to verify the feasibility of the design methodologies.

1. Introduction

Green green is divided into the following several contents, 1) Computer-based representation of product specifications; 2) testing of product performance against the product specification. 3) design intent management with LCA inventories, where design intent management and the interface for acquisition, representation, analyzing, recording and retrieving of the user's intent information from the product life cycle design perspectives are outlined. 4) Product oriented assembly/disassembly (DFA/DFD) [1,2] analysis, optimal product assembly/disassembly sequence planning and design methods[3] and components maximum reusability design method [4,5]; 5) design method for recycling, recovery and recycling of waste products into the design of the assessment; Life Cycle Assessment (LCA for short) [6] etc. Traditionally, Suh.N.P [7], Yoshikawa.H [8], etc laid down the traditional design theories. So far, traditional CAD systems make use of the manifold entities, where, feature modeling, feature recognition and feature mapping, etc, based mainly on specific quantitative information [9], i.e. the design theory and methodologies coped with DFX (X means, A:Assembly, C:Cost, D:Disposal, F:Function, M:Manufacture, Q:Quality, R:Recycling, S:Services and T:Test) etc.

However, conceptual design systems should have the following functional features, i.e. the visualization of uncertain entity; growth and expansion of uncertain entity; adaptive and dynamic growth of the entity model within a specific design. Therefore, default logic reasoning [10] is applied to the study of geometric entity reconstruction under uncertain information, and a new theory of default geometric reasoning[11] is proposed. Automatic mapping among design domains could dramatically increase exploration proficients etc.

And reconstitution algorithms evolving default logic is to cope with ambiguous information occurred in conceptual design stage, which is also effective exploration tool for concept creation and visualization, called exploitation. The F-F mapping model cooperating axiomatic design theory is proved to be effective to integrate requirement, function and form domains for a creative design tool.

2. Function-to-form mapping for greem design

Let U be a functional domain (Function Requirements, referrd to as **FRs**), V is the domain (Design Properties, referred to as **DPs**), $P(U)$, $P(V)$ is the power set of U,V, $FRs \subset P(U)$, $DPs \subset P(V)$. $f_{FRs \to DPs}$, where, ***FRs*** is called the definition domian,

Definition 1, functional carrier, if there exist transitional domains, between functional domain and form domain, called ***CRs***, make $x \subset CRs$, if and only when x satisfies the following two terms:

- A simple, intuitive, deterministic correspondence that satisfies ***FRs*** to ***CRs***
- At the same time, mapping from ***CRs*** to ***DPs*** satisfies the 'deterministic' criterion, where 'deterministic' is referred to as reconfigurability.

Figure 1 is the generalized mapping process model of green products design. This model describes the demand analysis, functional design, function-to-form mapping, form reconstitution etc, in a "top-down" style. The green product requirement model similar to the LCA (Life Cycle Assessment) system, acted as the top constraints. These constraints are propagated to the function domain, functional carrier domain, form domain etc. The green design requirement LCA model support the design process automation.

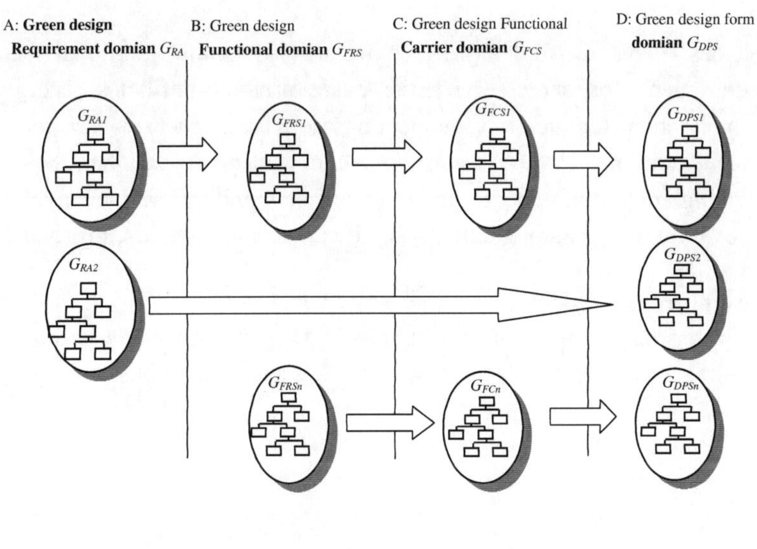

Figure 1. Function-to-form mapping for green design.

IOP Conf. Series: Journal of Physics: Conf. Series **908** (2017) 012048 doi:10.1088/1742-6596/908/1/012048

3. Management of green design system

In view of the special requirements of green product design, the functional requirements of easy manufacturing, the functional requirements for easy assembly/disassembly, the functional requirements for the maximum reuse value of of product parts, etc, are put forward, as in figure 2.

Green products involved many research fields, including: A. requirement domain GRA. B. green product function domain GFR; C. green product functional carrier domain GDPS. And d. green product functional domain GDPS. Green product design is to complete the mapping of generalized domains from A, B, C, and D. The mapping results are the D form domain, which is consisted of discrete feature/functional face sets etc,. The comprehensive reconstruction is to completed the structural innovative design of the product.

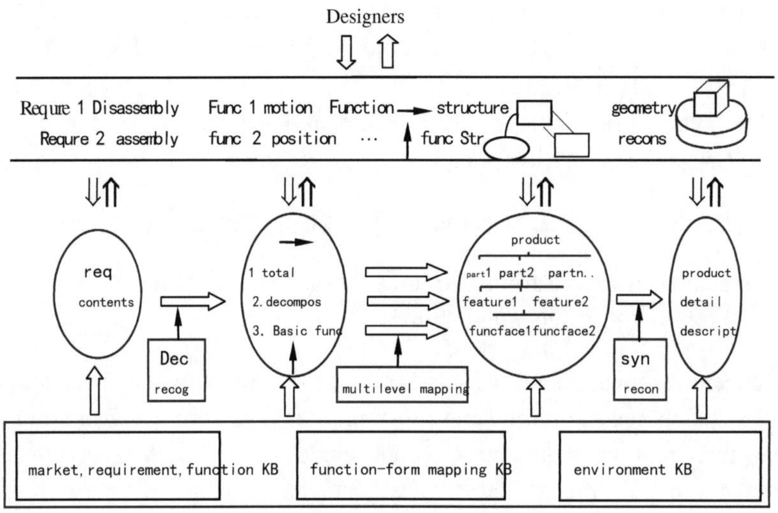

Figure 2. Green design automation system.

Green product design put forward higher requirements than the traditional mech-tronic design, these higher requirement are expressed in the basic product attributes, energy characteristics, manufacture/assembly characteristics, environment characteristics, energy characteristics and resource characteristics etc. Traditional CAD systems do not consider these higher requirements, so it is difficult to achieve high quality product design in an efficient and innovative way. So this research on green product innovation design automation has significant social and economic benefits.

4. The green design process model

The functional characteristics of the green product DFA/DFD analysis method based on automatic analysis structure to realize the product life cycle, and provide design recommendations, as in figure3.

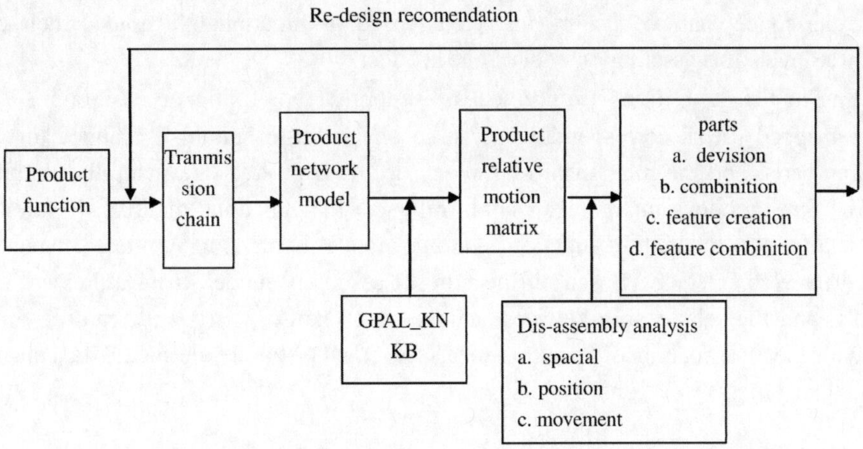

Figure 3. The dis-assembly analysis procedure.

Based on the "strong points map" theory of complex product structure assembly / disassembly sequence planning and design, and based on the concept of generalized double bonds, "green products" optimal "assembly / disassembly sequence planning analysis and design.

5. Product assembly model based on functional features

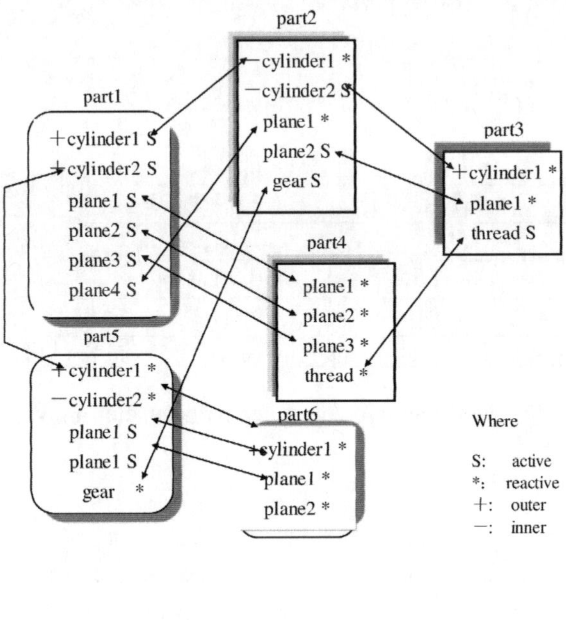

Figure 4. Product assembly model.

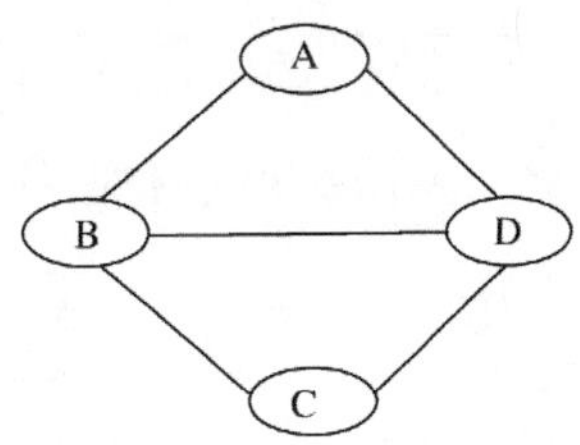

Figure 5. A four-parts assembly.

After function-to-form mapping (**F-F mapping** for short), the product conceptual model has been established, the conceptual product model is composed of functional faces and their correlations. There are three levels in a assembly model of one product.

According to the functional requirement of support, stop, fixing or positioning etc, different features are mapped to the corresponding parts, so one specific feature is mapped to two or more corresponding parts, and the relationship between parts in a product are naturally established, as in figure 3. The Sub-functions in figure 3 mainly refers to the functions of drive, location, clamping, sealing, blocking, etc. for motion function *motioni*, after **F-F** mapping, a meta_element is formed, defined as meta_elementArray. Based on the product assembly model, the features are expressed in the assembly, and the relationship between features and features, parts and features, and parts and parts etc, with the information of location, size, constraints etc, are defined in a data structure to describe the three types of relations, as in figure 4.

6. Dis-assembly analysis based on the relative motion matrix

After the function-to-form mapping, functional features are created, one functional feature is mapped to one part as the active feature, the other functional feature is mapped to another part as the re-active feature, and the correlation of the two parts are set up.The relative motion matrix is therefore establish based on the part-net and GPAL_KN.

Take the four parts assembly in figure 5 as one example, the connection in XY plane is established in table 1, where "1" means the part can be moved, "0" manes, the part can not move. There are six numbers to illustrate, -X, Y, -Y, Z, -Z directions respectively. For example, in table 1, column 2 and row 3, there is six numbers, i.e. 0,0,0,1,0,0, means part A can move relatively to part B in the -Y direction, and it can not move in the rest 5 directions. If two parts do not contact, the two parts can move in any direction relatively.

Table 1. the relative movement between the parts of the matrix.

	A	B	C	D
A	1,1,1,1,1,1	0,0,0,1,0,0	1,1,1,1,1,1	1,0,1,1,0,0
B	0,0,1,0,0,0	1,1,1,1,1,1	1,0,0,0,0,0	1,1,1,0,0,0
C	1,1,1,1,1,1	0,1,0,0,0,0	1,1,1,1,1,1	0,0,1,0,0,0
D	0,1,1,1,0,0	1,1,0,1,0,0	0,0,1,0,0,0	1,1,1,1,1,1

Note: 1. the two parts that are not connected in the connection diagram is set as: "111111111111".

2.The data of the relative motion matrix between any two connected parts is determined

by the mapped features.

The dis-assemblilty of any one part in an assembly can be analyzed by the relative motion matrix. The specific methods are as follows: take part B as the example, in table 1, find the corresponding partB in a row or column, look at the part B and other parts in the space with the same moving direction {x, -x, y, -y, Z, -z}, if the number is "1" in one direction, then it can move in this direction, i.e. it can be disassemblied in this direction.

If the number is "0", then it can not be dis-assembled in this direction. In the vector {0,0,0,1,0,0}, there is one "1" in the -Y direction, means part A in -Y direction can be removed.

7. Case study

Requirement analysis, there are many special considerations for green product design. Firstly, LCA oriented requirement domain need to be added to the generalized mapping process model and to increase the performance of the CAD systems. Secondly, product DFD/DFA requirements are also needed to add to the CAD systems; for the product assembly/disassembly analysis and path planning, etc. The higher requirement are for the product disassembly analysis and sequencing, for the parts reuse in the most economical way; and lastly, for product renovation design, re-manufacturing (de-manufacturing) design etc.

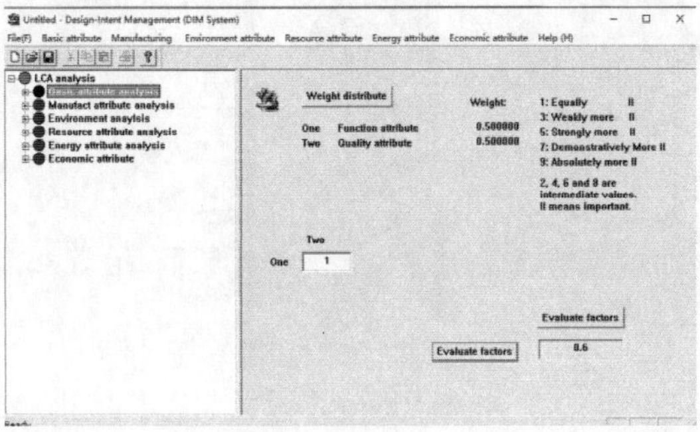

(a) requirement model

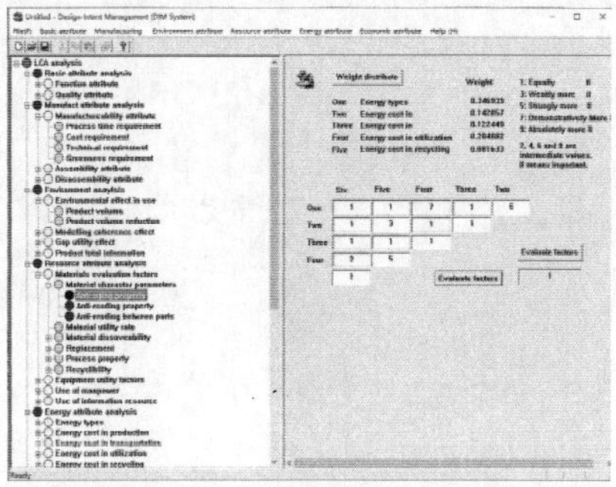

(b) constitution of this model

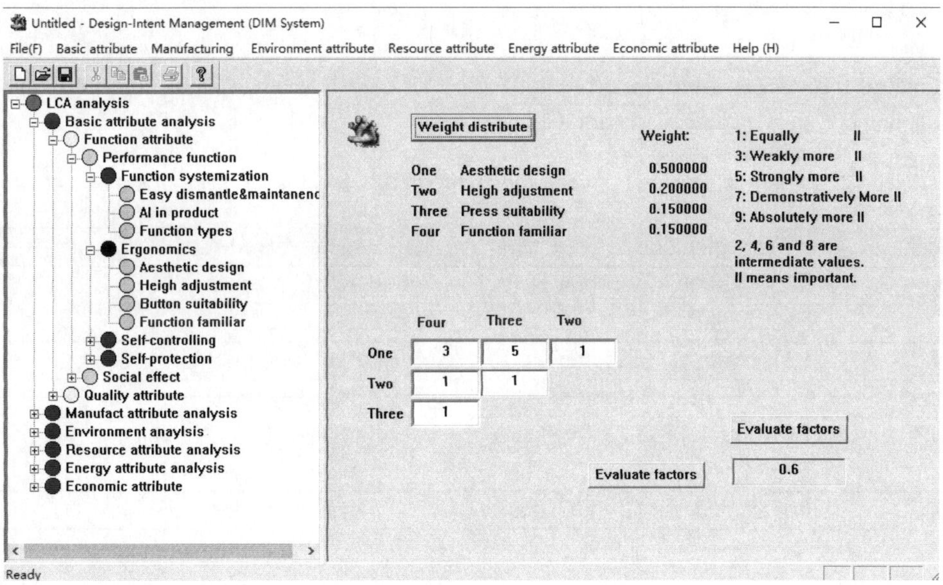

(c) the weight and evaluation result.

Figure 6. Green design requirement module.

To clamp one work-piece, the mechanical hand have the opening and closing movement, the Behavior1 can be further decomposed into a transmission principle diagram as shown in Figure 7(a).

Stepping motors is firstly selected interactively. The stepper motor is related to several parts as, drive parts, nuts, screw drive components. In combination with figure 7 (a), the 3-D transmission chain of the mechanical hand is deduced, as in figure 7(b), and the corresponding parts of the product are created at the same time.

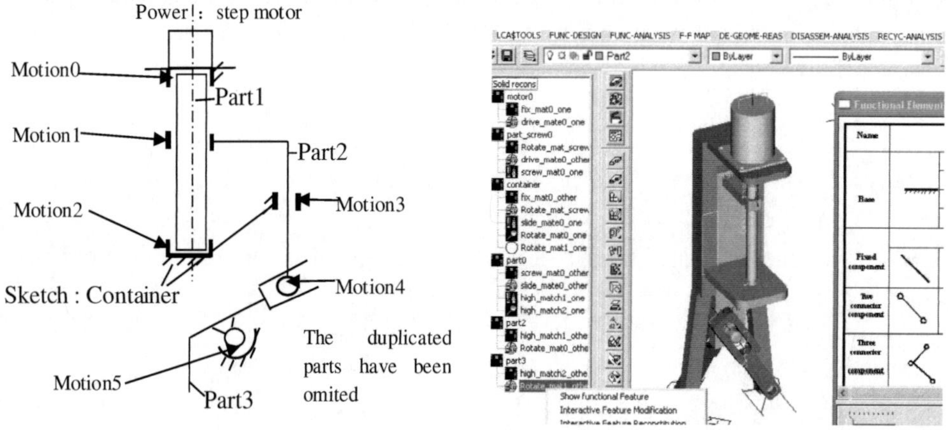

(a) conceptual model (b) model

Figure 7. Green design of the mechanical hand.

As shown in figure7 (a), from the driving principle of the mechanical hand, four correlated parts are created, namely, Part1, Part2, Part3 and Container.

Table 2. The relationship between the created parts and the resulting meta bodies

function	meta_element	Parts involved
Motion0	F01(Cylinder),F02 (plane)	Part1, Container
Motion1	F11(thread)	Part1, part2
Motion2	F21(Cylinder),F22(plane)	Part1, container
Motion3	F31,F32 F33,F34(plane)	Part2, container
Motion4	F41(double),F42、F43(Plane)	Part2(F411,F412(plane)) part3(F413(cylinder))
Motion5	F51(cylinder),F52、F53(plane)	Part3, container

Table 2 illustrates the five motion in the mechanical hand, meta_elements are firstly deducted. For example, the meta_element F41 is deduced by the motion4, F41 is mapped on Part2, and the corresponding feature, F411 and F412 are created, in the mean while F412 is created on Part3.

And the relationship between the two parts, i.e. Part2, Part3, and the corresponding features F411,F412 and F413 are established. The functional surface sets of Part1, part2, and part3 formed by the function structure mapping are shown in figure7(b).

8. Conclusion

An effective assembly level function-to-form mapping model is proposed in this paper. It could be concluded that, the **decomposition**, F-F **mapping** and **reconstitution** model could be a common exploration, transformation and exploitation procedure for management of computational design tools, therefore helpful for creative work, specifically the synthesizing of form domain is not simply re-arrangement of physical elements, rather it needs geometrical as well as algebraic reasoning on un-manifold polyhedral. Novel mathematical as well as AI technologies need to be introduced. Several packages of design tools have been developed to testify the effectiveness of the design methodology.

9. Acknowledgment

The above-mentioned research work is supported by the Science, Technology and Innovation Commission of Shenzhen Municipality, JCYJ20160510165328965 and the Chinese NSFC, 61272017, P.R. China.

References

[1] Brahim, S. B. Smith, A. Bidanda, E. (1993). Relating product specifications and performance data with a neural network model for design improvement. Journal of intelligent manufacturing, 4: 367-74.

[2] Cavallucci, D. and Lutz, P. (2000). Intuitive design method (IDM): a new framework for design method integration. In: M. Shpitalni, Edt, Proceedings of the International CIRP Design Seminar. Haifa, Israel. 447–452.

[3] Crocker, G. A. and Reinke, W. F. (1991). An editable non-manifold boundary representation. IEEE Computer Graphics and Applications, 11,(2), 39-51.

[4] Darrell L. M. (2003). Better technology forecasting using systematic innovation methods, Technological Forecasting and Social Change, 70,(8),779-795.

[5] Dietz, P. Ort, A. and Penschke, S. (1998). Classification of product knowledge — an approach to optimal feedback strategies for design. In: G. Jacucci, G.J. Olling, K. Preiss and M. Wozny, Eds, Globalization of manufacturing in the digital communications era of the 21st century: innovation, agility, and the virtual enterprise, Kluwer Academic Publishers, Dordrecht, 783–799.

[6] El-Mehalawi. M, and Miller, R. A, (2003). A database system of mechanical components based on geometric and topological similarity. Part II: indexing, retrieval, matching, and similarity assessment, Computer-Aided Design, 35,(1), 95-105.

[7] Suh N P. The principle of design, 1988, Oxford University Press, New York.

[8] Tomiyama T, Yoshikawa H. Extended general design theory. In: Yoshikawa H, and Warman E A, Ed Design Theory for CAD: 95-130, Amsterdam, NorthHoland , 1987.

[9] D.Spath. The Utilization of Hypermedia-Based Information Systems for Developing Recycling Products and for Disassembly Planning. Annals of the CIRP, 1994, 43(1):153~156.

[10] Diplng et al. Grouping of used products for cellular recycling systems, Annals of the CIRP, 1995, 44(1):11-14.

[11] SETAC A. Technical Framework for Life-cycle Assessments. published by society for environmental toxicology and chemistry, Washington D C. 1991.

Investigation on the effects of processing parameters on shrinkage behaviour and tensile properties of injection moulded plastic gear via the Taguchi method

N M Mehat[1], S Mohd Kassim[2] and S Kamaruddin[3]

[1]Department of Mechanical Engineering Technology, Faculty of Engineering Technology, Universiti Malaysia Perlis (UniMAP), Uniciti Alam Campus, Sungai Chuchuh, Padang Besar, Perlis, Malaysia.
[2]School of Mechanical Engineering, Universiti Sains Malaysia, Engineering Campus, 14300, Nibong Tebal, Penang, Malaysia
[2]Mechanical Engineering Department, Universiti Teknologi Petronas, 32610 Seri Iskandar, Perak Darul Ridzuan, Malaysia

nikmizamzul@unimap.edu.my

Abstract. Controlling numerous processing parameters and sustaining the optimal performance of multiple quality characteristics give a great challenge to manufacturer to produce a high-quality product with low operating cost. It was quite tricky to find the optimal combination parameters if used unsystematic techniques such as trial and error method. In this study, the Taguchi method was adopted to identify the significant processing parameters affected the shrinkage and tensile properties of plastic gears. The effectiveness of Taguchi's OA in reducing the number of experiment was studied and the results were analysed by ANOVA. Through the analysis, six parameters; melting temperature, mould temperature, packing pressure, packing time, cooling time and injection pressure, were carried out as significant parameters that influence dimensional stability and mechanical properties of plastic moulded gear. The findings showed that the Taguchi method is a systematic and straight forward approach that can be used in obtaining the significant processing parameters particularly in plastic gear manufacturing.

1. Introduction

Plastic gears can be manufactured by either machining or injection moulding. The machining of plastic gears involves most of the same processes used in the machining of metal gears, such as milling or hobbing. Injection moulding, on the other hand, is the most common, yet important process used to manufacture plastic gear. More than one third of all thermoplastic materials are injection moulded. With the advancement of plastic injection moulding, it is becoming increasingly more challenging for today's plastic production industry to manufacture precision gearing components within tight tolerance limits. Injection moulding can be used to mass produce an ample range of plastic parts with multifarious and intricate geometries [1-2]. In Addition, final products that exhibit good dimensional accuracy, excellent surface finish at a high rate of production, with reduced cost and within a shorter time period, has further proven the value of the injection moulding process [3-4].

Despite the numerous advantages, there are some drawbacks associated with plastic injection moulding. Due to the intricacy of the injection moulding process, keeping the final product quality under control requires an intense effort. The quality characteristics of the injection moulded products can be characterized in terms of dimensional stability, aesthetical features, such as weld lines and sink marks, and mechanical properties [5]. As with plastic gears, the practicability of injection moulding in producing low cost plastic moulded gears still restricted by the occurrence of shrinkage in the final steps of the process. Severe shrinkage can lead to deflection or warpage in the moulded part, as well as negatively influence the dimensional stability and accuracy of the involute profile, the concentricity, the roundness, the tooth spacing uniformity, and the size of the gear. Thus, it can have a great impact on the quality of the final moulded gear, due to noise and vibration. It also shortens the gear's service life due to different damage mechanism, such as tooth fatigue, creep, excessive wear, and plastic deformation [6].

Manufacturing an accurate plastic gear via the injection moulding process is more complicated than machining a similar gear, due to the interaction of many factors including material consideration, part and mould design, as well as variation in injection moulding processing parameters. With plastic materials exhibiting extremely convoluted properties, an intricacy of the moulding process, attaining the desired quality and accurate dimension of final moulded plastic gear is a very challenging task. Improper setting of processing parameters during the manufacturing process may lead to the degradation of dimensional accuracy, the shape of the final parts, the surface finish, and mechanical properties [7-8]. Therefore, in this work, the Taguchi method was adopted to evaluate the effects of injection moulding processing parameters on the quality of plastic gear produced in related with shrinkage behaviour and tensile properties. The findings will be contributed to an attempt in improving the quality of plastic injection moulded gear in particular from the manufacturing point of view, therefore the conventional trial and error method which would incurred high production cost and long set up times can be avoided.

2. Methodology

2.1. Material

In this study, High Density Polyethylene (HDPE) was used as raw material for injection moulding process. The HDPE HD5218AA grade was supplied by Etilinas Polyethylene Malaysia Sdn. Bhd. The general properties of virgin HDPE are listed in Table .1. Before running the experiment, all materials were dried in furnace at 80°C for 3 hours to remove moisture.

Table 1. General Properties of High Density of Polyethylene

Properties	Values
Density (g/cm^3)	0.95
Melt Flow Rate (190°C / 2.16kg) (g/10min)	4.0
Tensile Strength at break (N/mm^2)	26
Ultimate Elongation (%)	10
Izod Impact Strength at 23°C (kJ/m^2)	20

2.2. Gear description

The spur gear based on American Gears Manufacturers Association (AGMA) standard was used as a main product in this study. The detail specifications and geometry of spur gear are illustrated in Figure 1. The spur gear was designed with outer addendum diameter of 33mm, dedendum diameter of 27 mm, core diameter and thickness of gear are 10mm respectively. Additionally, the moulded gear has 20 teeth with 20° pressure angle for each.

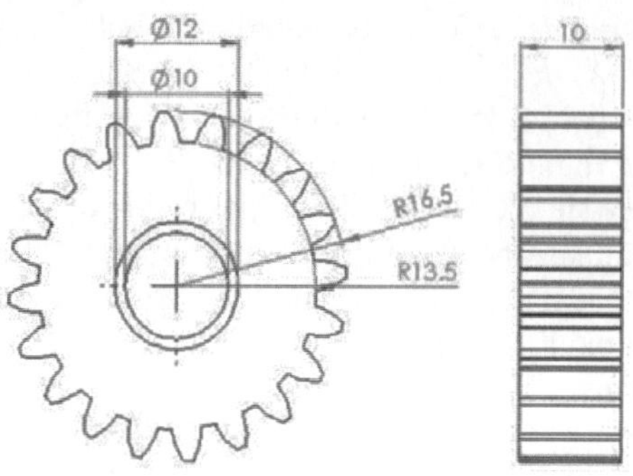

Figure 1. Geometry and Specification of Spur Gear

2.3. Design of experiment by Taguchi method

The objective of this study to identify the most significant processing parameter of injection moulding that influences the quality characteristic of moulded gear. Numerous processing parameters could affect the quality of moulded gear and the experiment runs will involve a large number of trials. Therefore, it is very difficult and requires a huge effort to obtain the optimal processing parameters setting. The procedures of designing the experiments by the Taguchi method are illustrated in Figure 2 and each stage is elaborated in the following section

2.3.1. Determining the Quality of Characteristics

The dimensional stability and mechanical properties are the most important features need to be focused for good quality of plastic gear. The shrinkage of plastic material will lead some defect on the dimensional accuracy of gear especially on tooth spacing uniformity and concentricity of roundness shape (Mehat et al., 2013). Senthivelan and Gnamoorthy (2008) have claimed in their study that shrinkage defect on injection moulded gear significantly influences the imperfection of geometric. Regarding to mechanical properties, the optimum performance of plastic material is required as characteristics of plastic gear for better power transmission to sustain high load capacity. Comprehending this fact, the quality measurement in this study is done in term of shrinkage and mechanical properties of the plastic gear.

2.3.2. Selection of Processing Parameters

In this study, seven processing parameters are selected to be analyzed as the influencing factor toward dimensional stability and mechanical properties for optimization study. The parameters involved are melting temperature, mould temperature, packing pressure, packing time, cooling time, injection pressure and injection time. Three levels of parameter were considered for better interpretation of nonlinear effect and the details was tabulated in Table 2.

Table 2. Processing Parameter and Levels

Label	Parameters	Unit	Level 1	Level 2	Level 3
A	Melt Temperature	˚C	170	190	210
B	Mould Temperature	˚C	40	50	60
C	Packing Pressure	%	60	80	100
D	Packing Time	second	5	10	15
E	Cooling Time	second	30	40	50
F	Injection Pressure	Bar	80	90	100
G	Injection Time	second	1	2	3

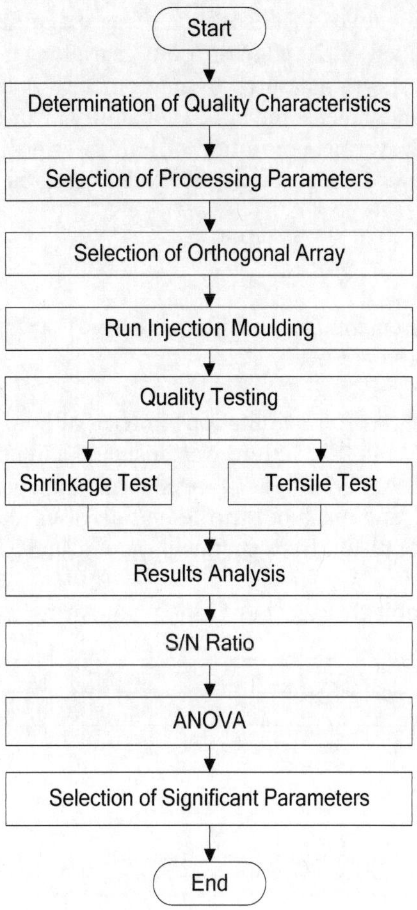

Figure 2. Flow chart of design of experiment by Taguchi method

2.3.3. Selection of Orthogonal Array (OA)
According to Taguchi rules, the selection of suitable OA is depending on total number degree of freedom (DOF) of processing parameters and levels. In this study, seven processing parameters with three levels indicates 14 DOF where 2 DOF for each parameter. DOF is obtained by number of levels minus with one (DOF = number of levels -1). As requirement in Taguchi, the selection of OA should be equal or greater than the total number of DOF for ideal parameter setting. Hence L_{18} OA is selected as a layout experiment since it has lowest array with 17 DOF that can placed all the seven processing parameters. The all seven parameters are assigned according to specific columns; melt temperature (A), mould temperature (B), packing pressure (C), packing time (D), cooling time (E), injection pressure (F) and injection time (G) as tabulated in Table 2.

2.4 Quality Testing
The experiment was conducted by following the combination of parameter setting that is provided in OA. All 18 trials needed to be run to accomplish the experiment in producing moulded gear by using Battenfeld TM750/210 injection moulding machine. The quality measurements of moulded gear are evaluated according to the following:

2.4.1 Shrinkage Test
The shrinkage test of moulded plastic gear has been measured by following a standard procedure of ASTM D955-08. Before the measurement process, all sample of plastic gear is kept at ambient temperature for 24 hours. In this study, the dimension of plastic gear was measured by projector

profile RAX Vision DC 3000 Mitutoyo. The specimen was magnified by projector profile into 2D axis and the coordinates was pointed on 20 edge of tooth gear along addendum and dedendum cirles. Otherwise, the tooth thickness gear was measured by micrometer. A larger magnification display and micrometer provided a high accuracy measurement. Probability error during measuring the geometry of specimen can be reduced. However, as precaution of human error, five samples for each trial were measured. The relative shrinkage of spur gear was calculated based on following equation:

$$S = \frac{D - D_m}{D_m} \times 100\% \quad (1)$$

Where: S = shrinkage, D = measurement using profile projector, D_m = mould dimension,

2.4.2 Tensile Test

The tensile test for circular object has some limitations in term of hold specimen method in symmetry condition. Due to this limitation, split disc fixture was designed and utilized for plastic gear specimen as shown in Figure 3. Based on ASTM D638-10, tensile test was performed by using Instron 3367 series table-mounted universal testing machine with 3mm/min of cross head speed under 50kN tensile load. To ensure the consistency of result, five specimens were tested for each trial.

Table 3. Taguchi's Orthogonal Array L_{18}

Experiment	Factor							
	A	B	C	D	E	F	G	-
1	1	1	1	1	1	1	1	1
2	1	1	2	2	2	2	2	2
3	1	1	3	3	3	3	3	3
4	1	2	1	1	2	2	3	3
5	1	2	2	2	3	3	1	1
6	1	2	3	3	1	1	2	2
7	1	3	1	2	1	3	2	3
8	1	3	2	3	2	1	3	1
9	2	3	3	1	3	2	1	2
10	2	1	1	3	3	2	2	1
11	2	1	2	1	1	3	3	2
12	2	1	3	2	2	1	1	3
13	2	2	1	2	3	1	3	2
14	2	2	2	3	1	2	1	3
15	2	2	3	1	2	3	3	1
16	2	3	1	3	2	3	1	2
17	2	3	2	1	3	1	2	3
18	2	3	3	2	1	2	3	1

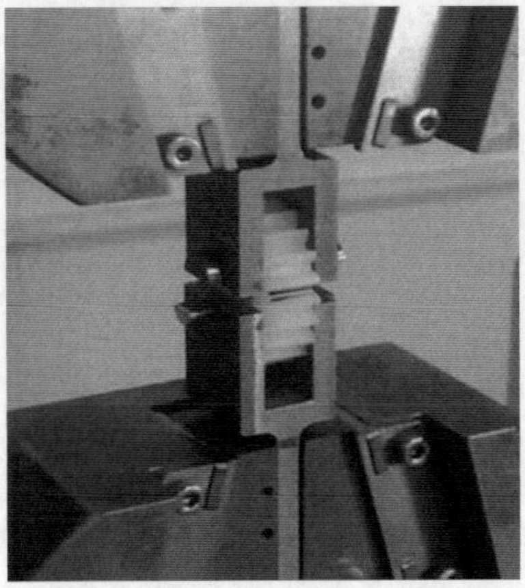

Figure 3. Fixture for Tensile Test

3. Results and Discussion

3.1. S/N Ratio

In examining the results of shrinkage behaviour and tensile properties for the plastic gears produced, the Taguchi method proposed the utilization of the S/N ratio to decide the nature of the attributes employed. As discussed by Ozcelik et al. [8], the signal to noise (S/N) ratio is a measure of achievement targeted at developing products and processes insensitive to noise factors. A high signal to noise (S/N) ratio suggests that the signal is much higher than the arbitrary impact of the commotion factors. The process or part operation persistent with the higher signal to noise (S/N) ratio at all times yields optimal quality attributes with least variance. In the Taguchi method, the quality attributes can be assigned into the smaller the better, the nominal the better and the bigger the better. In this study, the smaller value of the shrinkage behaviour and the larger value for tensile properties are expected. The S/N ratio is computed from the mean square deviation equation as in Equation 2. The results of S/N ratios for both shrinkage behaviour and tensile properties are listed in Table 3.

$$S/N = -10\,log(MSD) \quad (2)$$

MSD = mean-square deviation from the target values of response.

3.2. Analysis of variance (ANOVA)

The aim of ANOVA application for this study is to identify the most influential parameters on dimensional stability and mechanical properties of HDPE plastic moulded gear. The ANOVA was performed individually on S/N ratios for each quality characteristic response in order to obtain the relative contribution of each processing parameter. The results of each quality characteristics of the HDPE gear produced are presented in Table 4. For shrinkage in addendum of the HDPE gear produced, injection pressure and cooling time was found as significance factor with percentage contribution 31.89% and 30.69% respectively. Subsequently packing time contributes 13.41%, followed by mould temperature 6.42% and melting temperature 5.62%. The other factors, packing pressure and injection time have small relative percentage of contribution in the shrinkage defect at addendum area.

IOP Conf. Series: Journal of Physics: Conf. Series **908** (2017) 012049 doi:10.1088/1742-6596/908/1/012049

Table 4. S/N Ratios of Experiment Results

Trial	Shrinkage			Mechanical		
	Addendum (dB)	Dedendum (dB)	Tooth Thickness (dB)	Modulus Young (dB)	Elongation at Max Load (dB)	Ultimate Strength (dB)
1	38.2944	36.7860	29.3726	26.8451	13.0695	11.9234
2	43.7864	40.1624	30.6331	27.9088	9.4182	10.7399
3	44.2118	42.1088	28.9839	27.4025	11.2166	11.5466
4	52.8113	45.5206	29.8294	26.9618	12.9056	11.9979
5	39.1721	37.0065	29.6370	27.5420	10.2981	11.3933
6	47.6995	43.7616	30.3536	27.0902	10.9291	11.4868
7	46.9182	41.9745	29.7179	26.7289	12.9262	11.6653
8	40.9239	44.7115	31.5078	26.6409	13.3553	11.7398
9	39.9659	37.8332	29.5171	26.5851	13.2602	11.6972

Table 5. Percent contribution of shrinkage and tensile properties

Parameters	Percent Contribution			Percent Contribution		
	Addendum	Dedendum	Tooth Thickness	Modulus Young	Elongation at Max Load	Ultimate Strength
Melt Temperature	5.62	1.71	36.84	32.31	37.69	22.95
Mould Temperature	6.42	13.42	9.40	6.02	33.69	37.38
Packing Pressure	2.87	44.57	8.23	9.15	3.27	2.19
Packing Time	13.41	10.20	5.14	5.82	3.84	3.17
Cooling Time	30.69	3.54	13.98	37.44	15.99	13.52
Injection Pressure	31.89	14.21	14.66	6.91	4.22	12.82
Injection Time	3.08	6.52	7.12	0.58	0.68	6.07
Error	6.02	5.82	4.63	1.77	0.62	1.91

For the case of dedendum area, packing pressure is found as the most significant factor with percentage contribution of 44.57%. The remaining factors that insignificant to the dedendum shrinkage are identified as injection pressure, mould temperature, packing time by contributes 14.21%, 13.42% and 10.20% respectively. The result tabulated in Table 4 also indicated that the melting temperature is the most influencing factor for tooth thickness shrinkage with 36.84% of percentage contribution. Additionally, injection pressure and cooling time tend to be significance factors by contributing 14.66% and 13.98% correspondingly. Other factors contribute less than 10% was obtained; i.e. mould temperature with 9.40%, packing pressure with 8.23%, injection time 7.12% and lastly packing time contribute 5.14%.

On the other hand, for the case of tensile properties, the results showed that the mould temperature is the most influence factor by contribute 37.38% for ultimate strength, followed by melting temperature, cooling time and injection pressure with percentage of contribution of 22.95%, 13.52% and 12.82% respectively. The other factors; injection time, packing time and packing pressure are considered insignificant factors and they were less influence on ultimate strength of HDPE moulded gear. For Young's Modulus, the ANOVA analysis revealed that four parameters significantly affected the performance of modulus elasticity HDPE moulded gear. Cooling time is the most influential factor by a contribution of 37.44%, subsequently melting temperature with 32.31% followed by packing pressure with 9.15% and injection pressure 6.91%. The parameters; injection time, packing time and packing pressure are indicated as insignificant factors with percentage contribution 6.07%, 3.17% and 2.19%, correspondingly. For the case of elongation at maximum load, melting temperature shows the highest percentage contribution of 37.69%, followed by mould temperature, cooling time, injection pressure, packing time and lastly packing pressure with percentage contribution of 33.69%, 15.99%, 4.22%, 3.84% and 3.27% respectively. Parameter injection time has displayed less influence on elongation properties by contributing only 0.68%.

The results of significant parameters obtained from the ANOVA is summarized in Table 5. The injection time is found to be an insignificant factor and do not have strong influence toward any of six responses quality characteristics of HDPE moulded gear. Moreover, six parameters were considered as significant parameters which are melting temperature, mould temperature, packing pressure, packing time, cooling time and injection pressure.

Table 6: Result of Significant Parameters

Factor	Response Characteristic					
	Addendum Shrinkage	Dedendum Shrinkage	Tooth Thickness Shrinkage	Ultimate Strength	Modulus Young	Elongation At Max Load
A. Melt Temperature	-	-	Yes	Yes	Yes	Yes
B. Mould Temperature	-	-	-	Yes	-	Yes
C. Packing Pressure	-	Yes	-	-	Yes	Yes
D. Packing Time	-	-	-	-	-	Yes
E. Cooling Time	Yes	-	-	Yes	Yes	Yes
F. Injection Pressure	Yes	-	-	Yes	Yes	Yes
G. Injection Time	-	-	-	-	-	-

4. Conclusion

According to the analysis in the experiment, the Taguchi method has been utilized effectively in identifying the significant parameters for each single response of quality characteristic of HDPE moulded gear. Additionally, Taguchi method has provided a fast and credible method in designing of experiment set up. The credibility of Taguchi method was hindered in solving multi-response issues. Nevertheless, the limitation of Taguchi was overcome by the integration with other methods.

5. References

[1] Shen C, Wang L, Cao W and Qian L 2007 *Polymer-Plastics Technology and Engineering* **46** 219.
[2] López A, Aisa J, Martinez A and Mercado D 2016 *Measurement* **90** 349
[3] Dawoud M, Taha I and Ebeid, S J 2016 *Journal of Manufacturing Processes* **21** 39.
[4] Guo W, Mao H, Li B and Guo X 2014 *Procedia Engineering* **81** 670.
[5] Yang Y, Yang B, Zhu S and Chen X *Journal of Materials Processing Technology* **226** 85.
[6] Mao K, Li W, Hooke C J and Walton D 2010 *Tribology International* **43** 433.
[7] Mathurosemontri S, Uawongsuwan P, Nagai S and Hamada H 2016 *Energy Procedia* **89** 255.
[8] Huszar M, Belblidia F, Davies H M, Arnold C, Bould D and Sienz J 2015 *Sustainable Materials and Technologies* 5 1.

Evaluation of calcination temperature and phase composition ratio for new hyroxyapatite

M N Ahmad Salimi and H S Chin

School of Bioprocess Engineering, Universiti Malaysia Perlis, Jejawi, Perlis.

nabil@unimap.edu.my

Abstract. The demand of production of hydroxyapatite (HA) has been increasing for the purpose of medical and dental application. HA possesses the excellent properties leads to the priority choice for ceramic bone replacement. Synthesis route by wet chemical precipitation is commonly practised in industrial scale. Calcium hydroxide and Orthophosphoric acid are the precursors for production scale. The synthesis of HA is conducted by varying the synthetic condition: stirring rate, calcium-phosphate and calcination temperature. This paper is focused on the properties of HA produced by regulating the synthetic condition so that the qualities of HA can be well performed. Characterization studies were also carried out by Fourier Transform Infrared Spectroscopy (FT-IR) for functional group identification, Scanning Electron Microscope (SEM) for surface morphology analysis and X-Ray Diffraction (XRD) for phase composition and crystallinity respectively. Narrow particle size distribution contributed to better quality of hydroxyapatite for bone replacement. Both calcium-phosphate ratio and calcination temperature would affect the phase composition of calcium phosphate.

Keywords: hydroxyapatite, wet chemical precipitation, stirring rate, calcium-phosphate ratio, calcination temperature

1. Introduction

Ceramic, glass and glass ceramic materials have been introduced and use to substitute fracture bone in many clinical applications since 1970's [1]. Calcium phosphate ceramics have gained the research attention due to their chemical compositions which similar with bone. Calcium phosphate is biocompatible and non-toxic. These excellent properties make them outstanding in medical and dental applications. The most frequently utilized in bone substitutes are hydroxyapatite (HA) and tricalcium phosphate (TCP). Currently, HA is the most commonly applied in bone grafting. . HA has more stable phase to support biological apatite precipitation [2].Hydroxyapatite is non-resorable whereas Tricalcium phosphate is bioresorbable and can be replaced by natural bone. This is the main reason to emphasize on further researches and investigations on Tricalcium Phosphate. Therefore, beta-tricalcium phosphate (Beta-TCP) has been promoted as a biodegradable bone replacement [3].

Several methodologies have been conducted to prepare calcium phosphate ceramics including precipitation technique, sol-gel approach, hydrothermal technique, multiple emulsion technique, biomimetic deposition technique and electro-deposition technique [4]. Wet precipitation method is the most frequently applied in industrial production because of its simplicity, low cost and easy application [5].

For wet precipitation method, processing parameters such as temperature and reaction time, might affect the results of precipitation in terms of purity, surface area and particle size distribution. For instance, the size of HA particles reduce as the reaction time at room temperature increases. In addition, the temperature can also influence the particle size of HA where increasing temperature can produce larger HA particles. On the other hand, lowering temperature causes the particles morphology become more spherical, while raising the temperature can lead to needle-shaped morphology [6].

Among the various processing parameters, this study will focus on two critical parameters which needed to take into conscious consideration during the preparation of calcium phosphates: calcination temperature and phase composition ratio. Calcination temperature could affect the strength and toughness of calcium phosphates while mechanical properties and biodegradation rate strongly depend on the phase composition (Ca/P) ratio [7] . In this research, we are reporting synthesis of HA by wet precipitation method and evaluating the effect of calcination temperature and phase composition ratio on the formation of HA.

2. Methodology

2.1. Wet Precipitation Method

Wet precipitation method was carried out to produce hydroxyapatite. Calcium hydroxide and phosphoric acid were employed as the precursors. Firstly, 100mL of 0.6M H_3PO_4 was added into 100mL 1M $CaOH_2$ solution. Different molar ratio (Ca/P =1.50,1.59,1.67) of calcium phosphate were prepared.NH_4OH was used to maintain the system at pH 11 with continuous stirring (150 rpm, 250 rpm, 300 rpm) for one hour at 90 °C. Later, the mixture was aged for 24 hours at room temperature. After that, precipitate was filtered and eventually the precipitate was dried at 100°C and calcinated with high temperature (600 °C, 700 °C, 800 °C). The equation of the chemical reaction was shown as in equation 1:

$$10CaOH_2 + 6H_3PO_4 \rightarrow Ca_{10}(PO_4)_6(OH)_2 + 18H_2O \quad (1)$$

The flow chart for the synthesis of HA by wet precipitation method was shown in Figure 1. Continuous stirring was required after the reactants are combined as the calcium is slowly incorporate into the apatite structure, helping the material to achieve suitable stoichiometric Calcium/Phosphate (Ca/P) ratio.

2.2. Characterization Studies

The phase composition crystallographic structural analysis was determined by X-ray Diffraction (XRD) with monochromatic Cu Kα = 1.5405Å radiation over the 2 range of 20–80° at a scan rate of 0.02°/min. Fourier Transform Infrared Spectroscopy (FTIR) used to identify the functional group and quality of calcium phosphate compounds. FTIR spectrum was recorded in the range of 4000-400 cm^{-1} region with 2 cm^{-1} resolution.The morphology of synthesized powders were studied by scanning electron microscopy (SEM) operating with accelerating voltage of 20 kV. The magnification ranging from x800 to x2000.

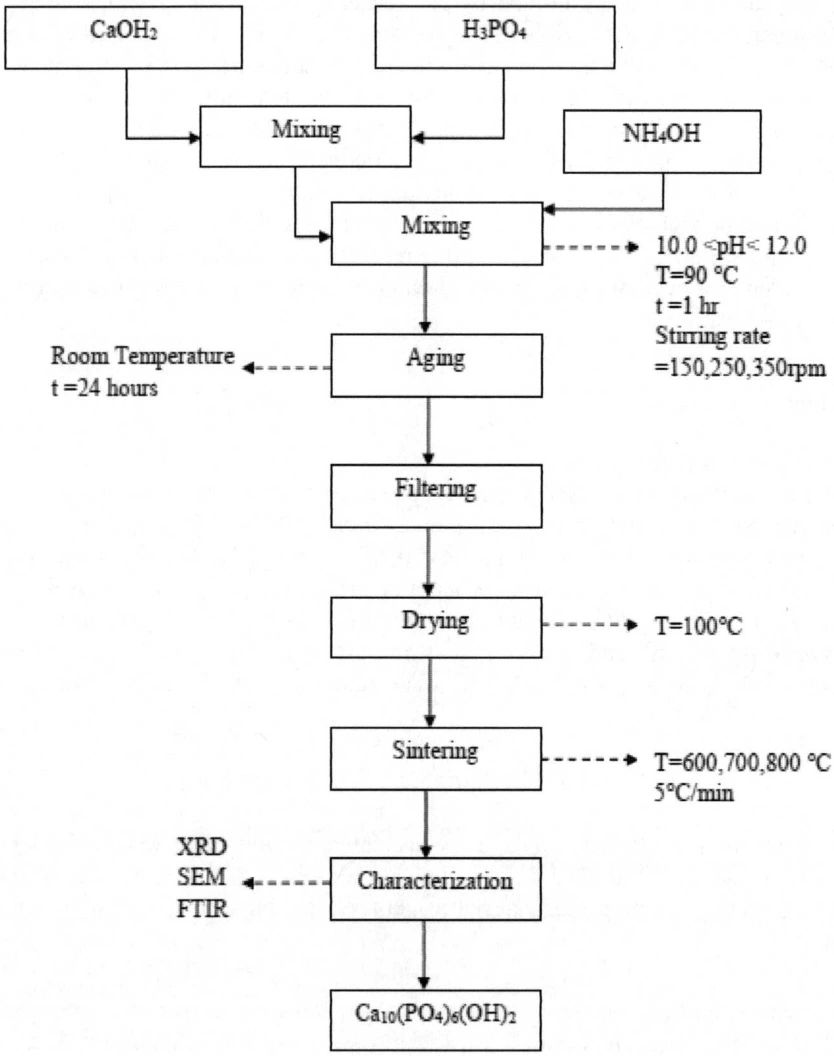

Figure 1. The flow chart of synthesis of hydroxyapatite by using wet precipitation method.

3. Results and Discussion

3.1. Fourier Transform Infrared Spectrometry (FT-IR)

Figure 2 represented the FT-IR spectrum of Hydroxyapatite (HA) powders. FT-IR patterns were presented in Figure 2 confirmed that the formation of HA calcinated at different temperatures of 600 ºC and 800 ºC.

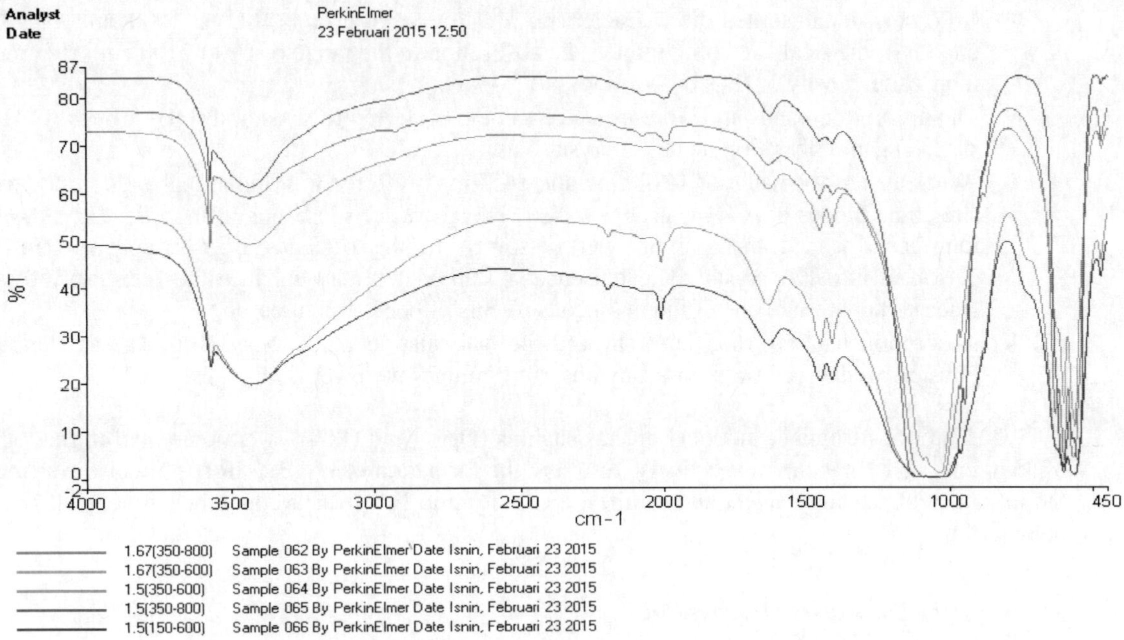

Figure 2: FT-IR spectrum of Hydroxyapatite (HA) powders.

Table 1. Absorption bands of Hydroxyapatite (HA) at different synthetic conditions.

Chemical groups	Absorption band, cm^{-1}
PO_4^{3-}	474;563;602;963; 1000-1100;2000-2100;
OH^-	3570;3435;
CO_3^{2-}	1414;1455;
H_2O	1635;

The absorption bands of chemical functional group characteristic to HA phase can be defined as follows in table 1.

- The presence of $[PO_4]$ groups that possessed tetrahedral apatite structure was proven by absorption bands at 474 cm^{-1}, which characteristic to $[PO_4]v2$ group (v_2 O-P-O) bending variations; double bands at 563 cm^{-1} and 602 cm^{-1} with a high resolution was refer as symmetric and asymmetric deformation modes of $[PO_4]v_4$ group (v_4 O-P-O); absorption band at 963 cm^{-1} corresponded to symmetric stretching mode; absorption band in the range of 1000-1100 cm^{-1} demonstrated the characteristic to $[PO_4]v_3$ groups (v_3 P-O) at 1000 cm^{-1} and 1100 cm^{-1} asymmetrical stretching mode. The absorption in the range of 2000-2100 cm^{-1} referred to the weak intensity of $[PO_4]v_3$, v_1 modes.
- The absorption bands at 3570 cm^{-1} and 3435 cm^{-1} referred to structural [OH] groups (O-H) in stretching and liberation modes at crystallites.
- Weak absorption bands at 1414 cm^{-1} and 1455 cm^{-1} referred to symmetrical and asymmetrical stretching modes of $[CO_3]v_3$ groups (C-O). The characteristic bands from inorganic carbonate ion recommended that carbon was dissolved in the organics from atmosphere instead dissolved into HA crystal. The presence of carbonates may enhance the bioactivity of HA since carbonates are one of the main constituents of bone structures.
- Absorption band at 1635 cm^{-1} showed the deformation mode v_2 of H_2O. The existence of physically adsorbed water in the synthesized samples were verified.

The two essential functional groups such as $(OH)^{-1}$ and $(PO_4)^{3-}$ were presented in the region of 3570 cm^{-1} and 1059 cm^{-1} respectively, resulting the formation of HA. FT-IR spectra also confirmed the influence of starting reagent and synthesise condition in HA structure and their functional groups belong to.

3.2. Scanning Electron Microscope (SEM)

Figure 3 showed the SEM images of HA powder synthesized at different synthetic conditions. HA powders were produced with various morphologies, such as sphere, rod and layer by controlling the synthetic conditions. The HA powders formed were highly agglomerated. Besides, the HA powders were appeared as aggregates, rough, granular to dense.

(a) 1.67,350rpm,600°C **(b)** 1.67,350rpm, 800°C

(c) 1.67, 150rpm, 600°C

(d) 1.67, 350rpm, 800°C

Figure 3. SEM images of HA powder synthesized at different synthetic conditions

From the images shown, the morphology of HA powders had small particles distribution and large agglomerates. This results were similar with Santos et al. [8] and Mobasherpour et al.[9] . The shape of HA powders produced were not obviously spherical. This might cause by the crushing procedure during synthesis of HA powder. The crushing was done by mortar and pestle which the dispersion coefficient was not balance. Thus, the HA powders had large agglomerates.

3.3. X-Ray Diffraction (XRD)

Figure 4(a) and Figure 4(b) showed the XRD pattern produced at different condition. XRD patterns in Figure 4 confirmed that the hydroxyapatite was produced in this sample and no other traces of calcium phosphate impurities was discovered.

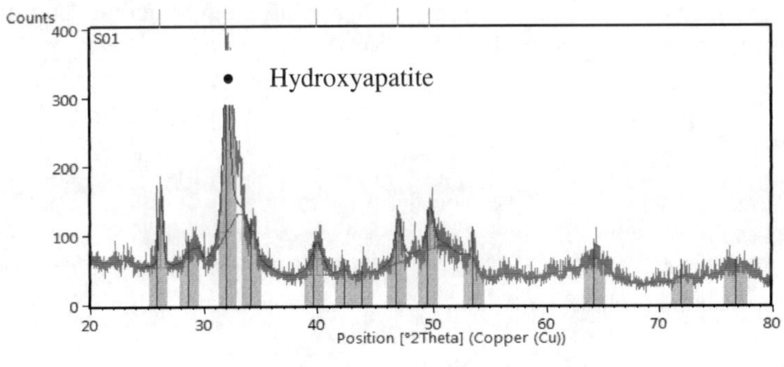

(a) 1.67, 350rpm, 600°C

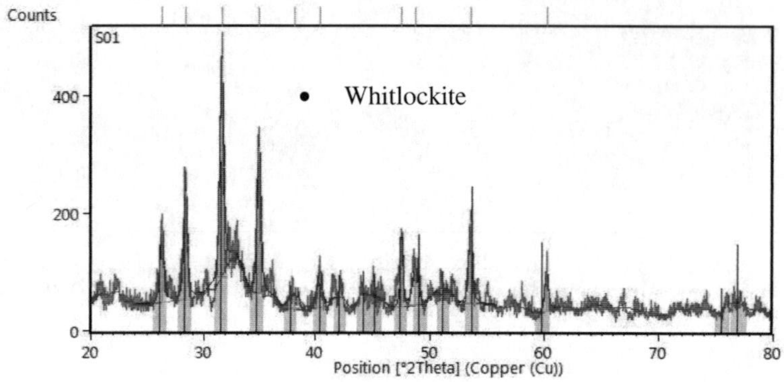

(b) 1.50, 350rpm, 800°C

Figure 4. XRD pattern of HA powder

Based on the result, no foreign elements were detected. Whitlockite is the mineral name of tricalcium phosphate. The chemical formula of whitlockite is $Ca_3(PO_4)_2$. Calcium-phosphate ratio and calcination temperature would affect the phase composition of calcium phosphate. Hydroxyapatite was transformed into secondary bone mineral, whitlockite.

4. Conclusion

Chemical precipitation technique was applied in this present work due to its simplicity, great reproducibility in addition to affordable benefits it provided upon commercial size. In this investigation, calcium hydroxide and orthophosphoric acid were used as starting reagents. The functional group detected were $(OH)^{-1}$ and $(PO_4)^{3-}$, resulting the formation of Hydroxyapatite (HA). The HA powders produced in this present approach were large agglomerates and narrow particle size distribution, resulting better quality of hydroxyapatite for bone replacement. In conclusion, both calcium-phosphate ratio and calcination temperature would greatly influence the phase composition of HA. Besides, secondary phases were only discovered when calcium deficiency HA and the calcination temperature at 800 °C.

5. References

[1] D.Bellucci, A.Sola, M. Gazzarri, F.Chiellini, and V. Cannillo , A new hydroxyapatite-based biocomposite for bone replacement. *Materials Science & Engineering: C*, vol **33**, no 3, pp. 1091–1101, 2013.

[2] Y-K.Jun, S-H., Hong, and Y-M. Kong, Effect of Co-Precipitation on the Low-Temperature Sintering of Biphasic Calcium Phosphate. *Journal of the American Ceramic Society*, vol **87**, no 7, pp. 2295-2297, 2006.

[3] J.S.Al-Sanabani, A.A Madfa, and F.A.Al-Sanabani,Application of calcium phosphate materials in dentistry. *International Journal of Biomaterials*, vol **2013**(2013), pp.1-12,2013.

[4] A.K.Nayak, (2010). Hydroxyapatite Synthesis Methodologies : An Overview, *International Journal of ChemTech Research*, vol **2**,No 2,pp. 903–907,2010.

[5] P.Gentile, C.J.Wilcock, C.A.Miller, R.Moorehead and P.V.Hatton, Process optimisation to control the physico-chemical characteristics of biommimetic nanoscale hydroxyapatites prepared using wet chemical precipitation, *Materials*, vol **8**,pp. 2297-2310, 2015.
[6] Q.Tian,L.R.Castaneda, Z.S.Dunn, & H.Liu,Bioceramics for orthopaedic device applications: Hydroxyapatite. In: Thian, E.S.,Huang,J.,& Aizawa,M.: Nanobioceramics for Healthcare Applications. World Scientific Publishing Europe Ltd,pp.49-78,2017.
[7] S.Ramesh, C.Y.Tan, M.Hamdi, I. Sopyana, and W.D. Teng, W. D,The influence of Ca / P ratio on the properties of hydroxyapatite bioceramics, International Conference on Smart Materials and Nanotechnology in Engineering,vol **6423**,2007.
[8] M.H.Santos, M.De Oliveira, L.Palhares., D.F.Souza, H.S.Mansur, and W.L.Vasconcelos,Synthesis Control and Characterization of Hydroxyapatite Prepared by Wet Precipitation Process, *Material Researach*, vol. **7**,no.4, pp. 625–630. 2004
[9] I.Mobasherpour, M.S. Heshajin, A. Kazemzadeh, M. Zakeri, Synthesis of nanocrystalline hydroxyapatite by using precipitation method,*Journal of Alloys and Compounds*, vol.**430**, Issue 1-2,pp.330–333,2007

IOP Conf. Series: Journal of Physics: Conf. Series **908** (2017) 012051 doi:10.1088/1742-6596/908/1/012051

Design optimization of rear uprights for UniMAP Automotive Racing Team Formula SAE racing car

M Azmeer[1], M H Basha[1*], M F Hamid[2], M T A Rahman[1] and M S M Hashim[1]

[1]Mechanical Engineering Programme, School of Mechatronic Engineering, Universiti Malaysia Perlis, Pauh Putra Campus, 02600 Arau,Perlis,Malaysia.
[2]School of Mechanical Engineering, Universiti Sains Malaysia, Engineering Campus, 14300, Nibong Tebal, Penang, Malaysia

hafifbasha@unimap.edu.my

Abstract. In an automobile, the rear upright are used to provide a physical mounting and links the suspension arms to the hub and wheel assembly. In this work, static structural and shape optimization analysis for rear upright for UniMAP's Formula SAE racing car had been done using ANSYS software with the objective to reduce weight while maintaining the structural strength of the vehicle upright. During the shape optimization process, the component undergoes 25%, 50% and 75 % weight reduction in order to find the best optimal shape of the upright. The final design of the upright is developed considering the weight reduction, structural integrity and the manufacturability. The final design achieved 21 % weight reduction and is able to withstand several loads.

1. Introduction

Formula SAE is a design competition for students that provide opportunity to enhance engineering and project management skills by applying learned classroom theories in a challenging competition. The goal for teams is to develop a small formula-style race car that conforms to the rules and regulations that have been set-up by the organizer. In the competition, a lighter car accelerate quicker and requires shorter stopping distance than heavier car so, reduction in vehicle mass becomes increasingly important to improve acceleration performance. Furthermore, FSAE competition has no minimum weight requirement, so the weight reduction benefits essentially have no limits. A reduction in the weight of the vehicle also improved the vehicle handling performance [4]. Therefore, it would be a great advantage for the team if the weight of the vehicle components is reduced.

Upright has a greater weight compared to the tie rod, control arm and ball joint, it will give a greater impact in reducing the weight of upright compared to other parts [3]. Therefore, a weight reduction on the upright is essential. Upright is an important component in a vehicle which it is used to hold the hub for the wheel and connects the upper and lower control arms. Furthermore, it also carries the power thrust from tie rod to the stub axle and hence it must be very strong, rigid and also as light as possible. The objective of this work is to reduce the weight and cost, while maintaining the structural strength of the upright. A shape optimization had been performed and the strength of the upright had been observed.

IOP Conf. Series: Journal of Physics: Conf. Series **908** (2017) 012051 doi:10.1088/1742-6596/908/1/012051

2. Methodology

2.1. Base Design

The CAD model of the base design upright has been developed using CATIA V5. The upright with themass of 3.027 kg consists of stub hole, tie-rod mounting points, upper arm and lower mounting points as shown in Figure 1. The material used is steel, with its properties as shown in table 1.

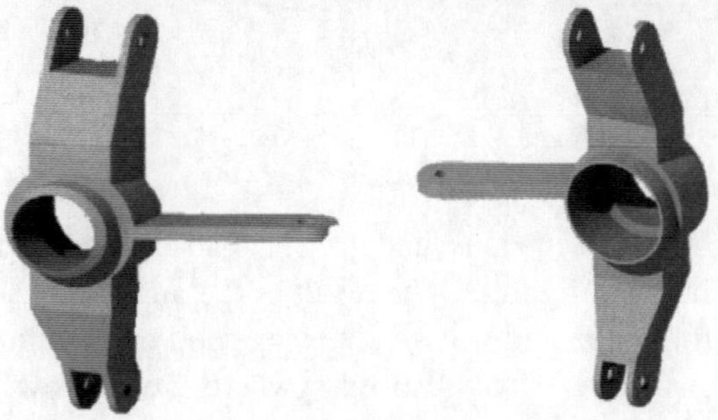

Figure 1. Base design of upright

Table 1. Property of steel

Property	Value
Young's Modulus	$2x10^{11}$ Pa
Poisson Ratio	0.3
Bulk Modulus	1.667×10^{11} Pa
Shear Modulus	7.692×10^{10} Pa

2.2. Static structural analysis

By using converted ".STP" file format, the CAD model of the upright is imported to ANSYS WORKBENCH and the meshing process started. Fine element size and hundred percent relevance were selected to obtain better mesh quality. Analytical calculation was done to determine the magnitude of force and moment under 2g braking, 2g cornering and 2g bumping. Table 2 shows the combination of these forces and moments form a critical load which was used to observe the maximum deformation and stress of the upright.

Table 2. Forces and moments values

Condition	Moment and Force
Axial force (Braking), 2g	559.150 N.m
Lateral force (Cornering), 2g	615.026 N.m
Vertical force (Bumping), 2g	2001.24 N
Critical situation (Combination of Braking, Cornering and Bumping)	Moments 559.150 N.m, 615.026 N.m Force 2001.24N

IOP Conf. Series: Journal of Physics: Conf. Series **908** (2017) 012051 doi:10.1088/1742-6596/908/1/012051

2.3. Shape optimization

Shape optimization is a method to modify structural shape into optimal shape. Topology analysis is used to find the best optimal shape under specified constraints. Shape optimization of the upright was done using 25%, 50% and 75% target reduction in order to determine which exact region and trusses that need to be kept or remove without making the structure fail. The optimized mass for every reduction were also recorded.

3. Result and discussion

3.1. Static structural analysis of the base design

Static analysis of the upright has been done under critical load condition which all the force and moment happen at once as shown in figure 2. The result of the analysis is shown in table 3. Based on this result, the maximum deformation and stress value was good enough to avoid failure in structure. Furthermore, the distribution of stress was also everywhere well below the yield strength of the material.

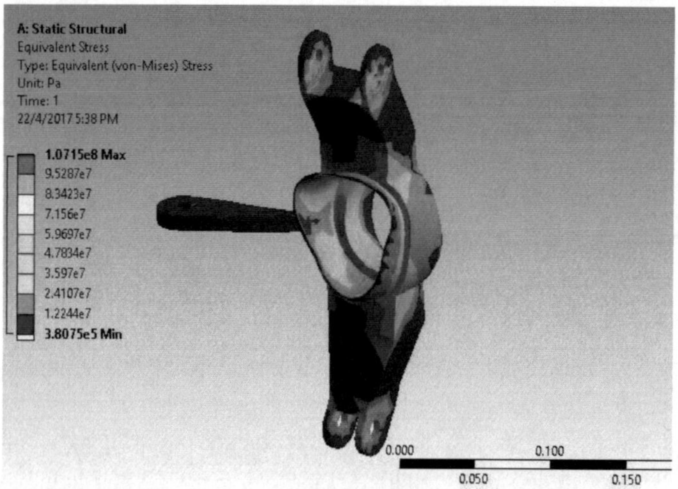

Figure 2. Stress analysis (Von-Misses) of the upright

Table 3. Static structural analysis result of the base design upright

Result	Value
Maximum stress	1.0715×10^{8} Pa
Total Deformation	2.8618×10^{-5}m
Safety Factor	Min 2.3332 Max 15
Equivalent Elastic Strain	0.00057939

3.2. Mass reduction

The base mass of the upright is 3.027 kg. As shown on figure 3, by starting with target reduction of 25%, the mass of upright is reduced to 2.3023kg and no significant difference in the shape of upright formed. While, for target reduction of 50% as shown in figure 4, the mass of upright is reduced to 1.5292 kg and some different in the shape begin to formed. Lastly, for target reduction of 75%, the mass of upright is reduced to 0.77621 kg and significant different in the shape had formed. Based on figure 5, the shape formed after the material been removed for 75% percent reduction can be clearly seen. The shape formed indicates that the design maintain it strength by using "X" shape trusses. Therefore, the "X" shape trusses will be used in designing a new optimized design of upright.

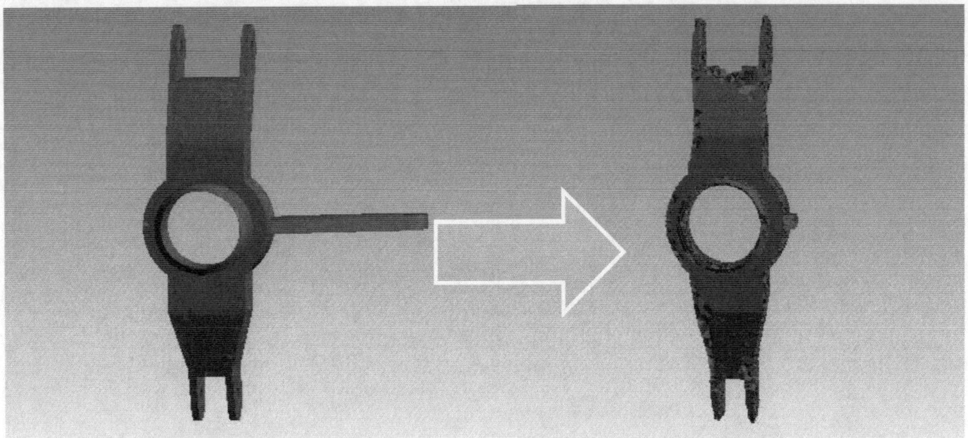

Figure 3. Shape optimization for 25% target reduction

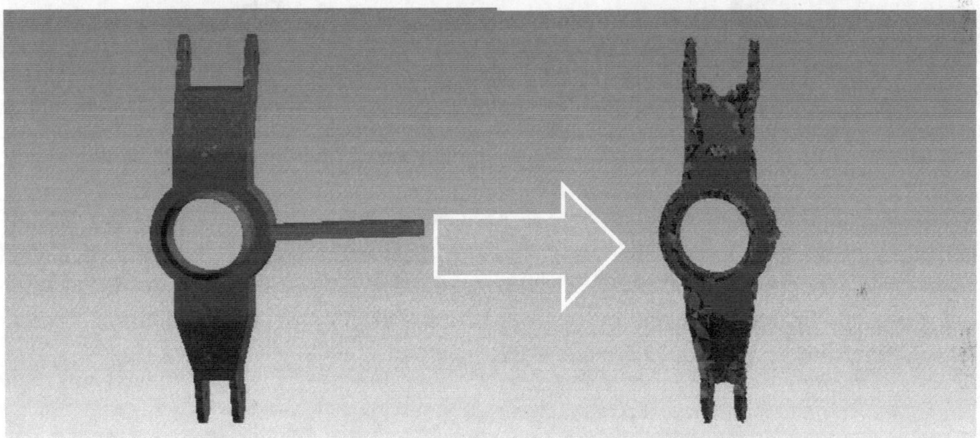

Figure 4. Shape optimization for 50% target reduction

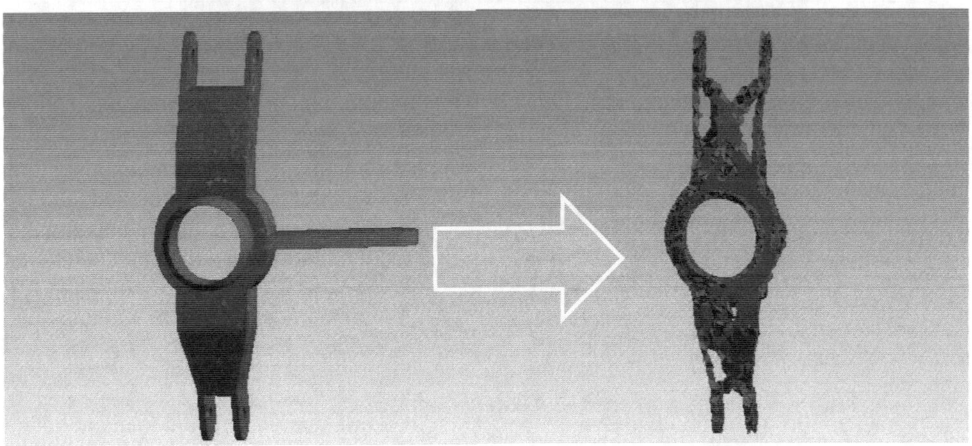

Figure 5. Shape optimization for 75% target reduction

3.3 New Optimized Design

Based on figure 3 ~ 5, the design cannot be used directly because shape optimization is only a tool to indicate the potential geometry changes and allow user to determine where material can be removed without adversely affecting the strength of the overall structure. So, a new design must be created to replace the trusses design in Figure 5. For fabrication purposes, the upright design as shown in figure 6 had been choose because of its advantages in low cost, easy to fabricate and availability of material. Besides that, with the help of "X" shaped trusses, the mass of the new optimized design has become 2.3863kg which is about 21% target reduction. This mass reduction show that shape optimization is very useful tool in making an improvement to a certain design.

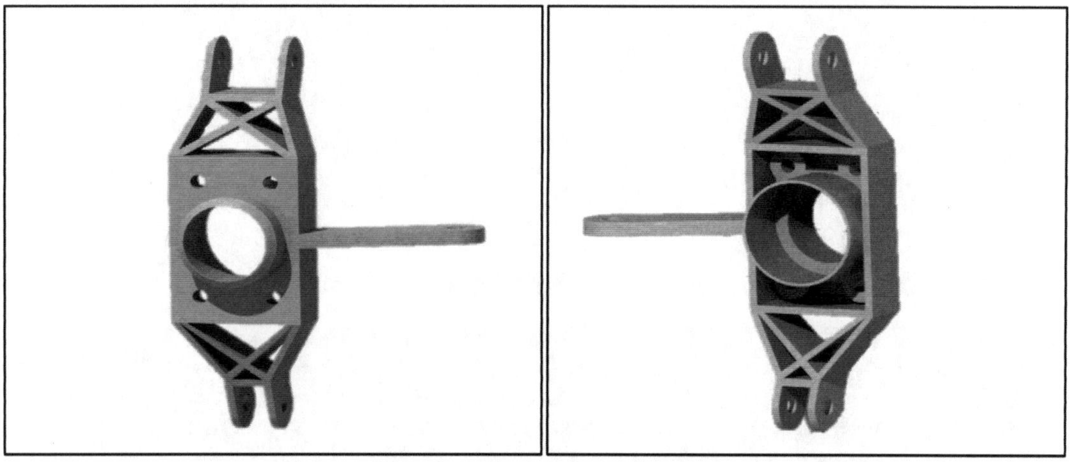

Figure 6. New optimized design

3.4 Static Structural Analysis of New design

Structural analysis of the new optimized design is done as shown in figure 5, with the same condition as stated in table 2. The result of the analysis is shown in table 4. Based on the results, the maximum stress value was good enough to avoid failure in structure. Furthermore, the distribution of stress was also everywhere well below the yield strength of the material.

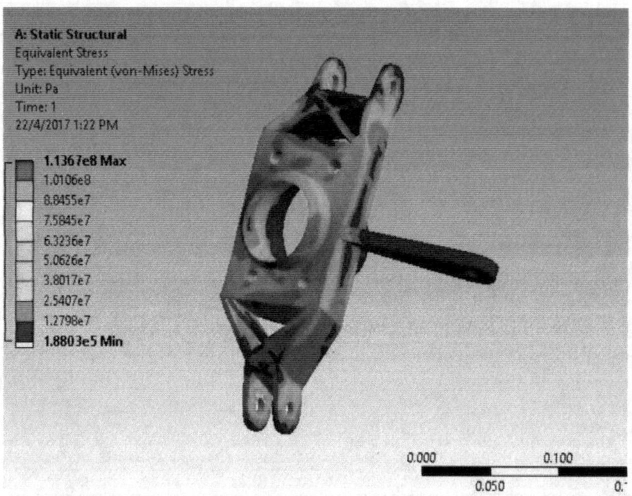

Figure 7. Stress analysis (Von Misses) of the new optimized upright

Table 4. Static structural analysis result of the new optimized upright

Result	Value
Maximum stress	1.1367×10^8 Pa
Total Deformation	6.4769×10^{-5}m
Safety Factor	Min 2.1993 Max 15
Equivalent Elastic Strain	0.00059006
Mass of upright	2.3863

4. Conclusion
Shape optimization had been performed and the mass of the upright was reduced by 21%. Maximum stress and deformation are within control. The shape optimization gives small change in deformation when several loads are applied, which means that change of volume and shapes does not influence significantly to stiffness of the structure. Therefore, the total weight of the vehicle has been reduced, as well as, costs and materials used. This will improve the fuel efficiency, handling and carbon emission.

5. References
[1] Solanki, V. V, & Patel, P. N. S. (2014). Static Analysis and Geometric Optimization of Independent Suspension Link, *2*(2), 47–57.

[2] Hetawal, S., Giridhar, N. S., & Baskar, P. (2014). Suspension Geometry Optimization and Analysis of Steering Upright for Weight Reduction, *3*(5), 403–407.

[3] Kim, G., Han, S., & Lee, K. (2014). Structural Optimization of an Upright with Consideration of Stiffness and Durability Requirements, *2014*(2), 1-7.

[4] Deshmukh, S. J., & Bhadange, P. J. (2016). Optimization of Steering Upright for Off-Road, *4*(Xii), 579–584.

[5] Sharma, M. P., Mevawala, D. S., Joshi, H., & Patel, D. A. (2014). Static Analysis of Steering Upright and Its Shape Optimization, 34–38.

[6] Tagade, P. P. (2013). Optimization and Finite Element Analysis of Steering Upright, 17–21.

[7] Lumpur, U. K., Jernang, J. T., Lumpur, U. K., Jernang, J. T., Lumpur, U. K., Jernang, J. T., & Tua, D. (2014). Model Development And The Shape Optimization Analysis of a Rear Upright for Race Car, *3*(12), 172–177.

[8] Anitha, P., & Shankar, V. H. (2016). Design and Topology of a Steering Upright Joint Using FEA, *6*(11), 311–316.

[9] Teja, G. P., Sreeram, K. Y., Student, P. G., Pradesh, A., Pradesh, A., & Pradesh, A. (2016). Weight Optimization of Steering Upright Joint Using FEA, *3*(12), 276–281.

[10] Guruprasadnayak, R. B. T. (2015). Shape and Material Optimization of the Steering Upright Component, *3*(2), 130–133.

Design and analysis of hydraulic ram water pumping system

N S M Hussin, S A Gamil, N A M Amin, M J A Safar, M S A Majid, M N F M Kazim and N F M Nasir

School of Mechatronic Engineering, Pauh Putra Main Campus, University Malaysia Perlis,02600 Arau, Perlis, Malaysia

nasrulamri.mohdamin@unimap.edu.my

Abstract. The current pumping system (DC water pump) for agriculture is powered by household electricity, therefore, the cost of electricity will be increased due to the higher electricity consumption. In addition, the water needs to be supplied at different height of trees and different places that are far from the water source. The existing DC water pump can pump the water to 1.5 m height but it cost money for electrical source. The hydraulic ram is a mechanical water pump that suitable used for agriculture purpose. It can be a good substitute for DC water pump in agriculture use. The hydraulic ram water pumping system has ability to pump water using gravitational energy or the kinetic energy through flowing source of water. This project aims to analyze and develop the water ram pump in order to meet the desired delivery head up to 3 meter height with less operation cost. The hydraulic ram is designed using CATIA software. Simulation work has been done using ANSYS CFX software to validate the working concept. There are three design were tested in the experiment study. The best design reached target head of 3 m with 15% efficiency and flow rate of $11.82 l/min$. The results from this study show that the less diameter of pressure chamber and higher supply head will create higher pressure.

1. Introduction

The hydraulic ram is used for agriculture purpose and it can be a good substitute for DC water pump. Hydraulic ram water pumping system is operated automatically without using any external energy. The device is called as hydraulic ram pump has the ability to pump the water to a certain height with no external energy. It uses only the potential energy of the water such as spring, stream or river which caused by the gravitational force [1]. Basically, the system of the hydraulic ram pump consists of two basic parts which are waste valve and the delivery valve as shown in Figure 1. The system also consists of an air chamber and an air valve. Sequences processes of hydraulic ram pump are intermittent due to the opening and closing behavior of the waste and delivery valves. The behavior of hydraulic ram pump relies on water hammer phenomenon which represent on the closing and opening of the waste and delivery valves. The process begin when water entering the derive pipe coming from specific elevation height with high pressure. Therefore, the waste valve is closed by water momentum [2]. Thus, high pressure is created that will cause the delivery valve to open allowing pressurized water to rise in vacuum air chamber. Therefore, the air chamber will pressurize the water causing delivery valve to close and air valve will open allowing water to rise through delivery pipe reaching the desired place [3].

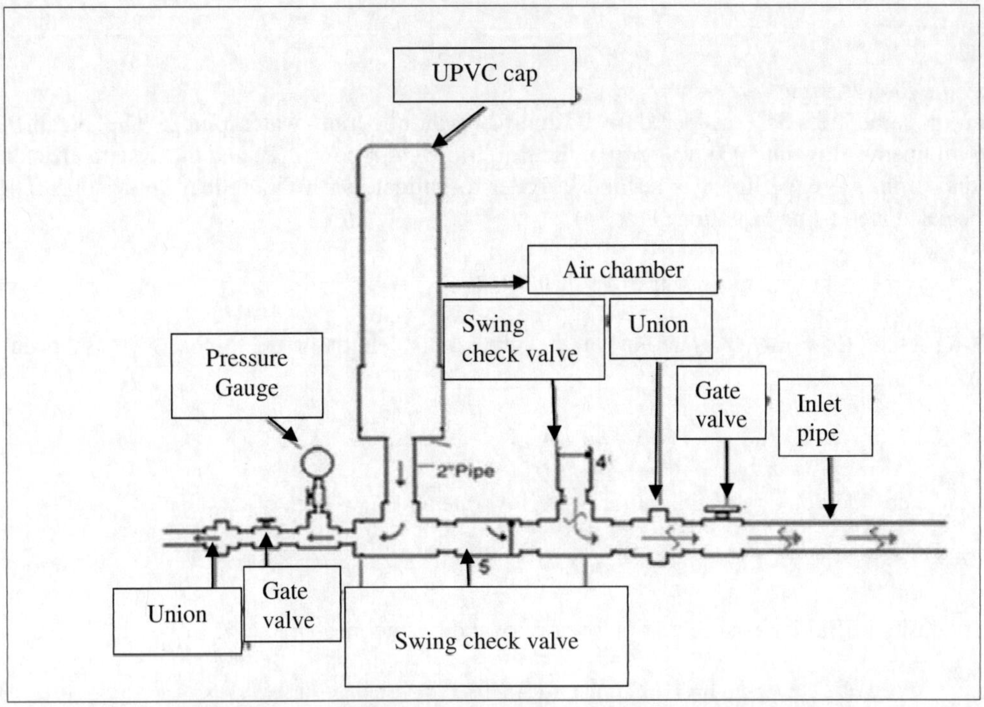

Figure 1. Hydraulic water pumping system configuration [4].

2. Literature Review

In 2007, Shuaibu [5] has designed and fabricated hydraulic ram pump that capable to lift the water to reach 2 m of head pump from depth of 2 m below the surface without any external energy required. The supply head was 1.5m and delivery head was 2.87m. The result shows the volume flow rate of the derived pipe and the power was 4.5238×10-5m^2/s and 1.273 kW respectively. The efficiency of this water ram is about 57.3%. The author found that the total cost for the design and fabrication of hydraulic ram pump is cheaper than the current pumps.

In addition, the other authors [6] have investigated hydraulic rams with self-built valves with two size of pressure chamber for high water supply. The aim of this investigation was to construct a reliable and low-cost ram. An efficiency of 44% was achieved for the performance of the hydraulic ram. However, there was no significant difference in in term of system efficiency for the design with the large (3.6l) and medium size (2.3l) pressure chamber (33.1% and 32.6%, respectively).

Besides, the result from the previous study shows that the pressure of the output increases as the drive pipe length increases. The drive pipe length of 2 meter the discharge at the outlet is maximum compared to the other two cases .this design also produces a power of 0.73 kW and the efficiency of 59.5% [4].

Furthermore, Balgude [7] presented the techniques and guidelines in designing a hydraulic ram pump. There are three essential factors need to considers in order to determine the quantity delivered by the hydraulic pump. These factors could be represented in the distance between the heights of the water source to the ram pump level, the distance between the desired place height and the ram pump and the volume of water source which are the basic factors of hydraulic ram pump.

The current pumping system is powered by household electricity (AC water pump) and the cost will be increased due to the higher electricity consumption. In some cases in the agricultural field, the water needs to be supplied at different height of trees and different places that are far from the water source. Besides, the existing DC water pump only can pump the water to 1.5 m height. Therefore, a water ram pumping system is proposed to deliver the water at height higher than 1.5 m.

3. Methodology

3.1. Hydraulic Ram Design Parameter

A number of parameters is considered in designing hydraulic ram water pump. The parameters including volumetric flow rate (Q), velocity of the fluid flow (V_d), power (P) and the system efficiency (η). The data from the experiment is needed in order to validate with theoretical knowledge. These data can be calculated using Equation (1) to (4).

$$Q = \pi r^2 L n/60 \tag{1}$$

Where Q represent flow rate, r is radius of the pipe, L is length of the pipe and n is speed of revolution.

$$V_d = Q/ A_d \tag{2}$$

Where V_d indicates the velocity of fluid, Q if fluid flow rate and A_d is area of the pipe.

$$P=\rho ghQ \tag{3}$$

Where ρ is density of fluid, g is acceleration of gravity and h is net head

$$\eta= (H\ Q_{out})/(F\ Q_{in}) \times 100 \tag{4}$$

Where H is a delivery head, Q_{out} is the outlet flow rate, F is the supply head and Q_{in} is the inlet flow rate.

3.2. Simulation study

The simulation study divided into design and analysis stage. The hydraulic ram pumping system modeling have been design using CATIA software as shown in Figure 2. The analysis study for pressure, velocity and volume of water and air inside the system have been done using ANSYS CFX software.

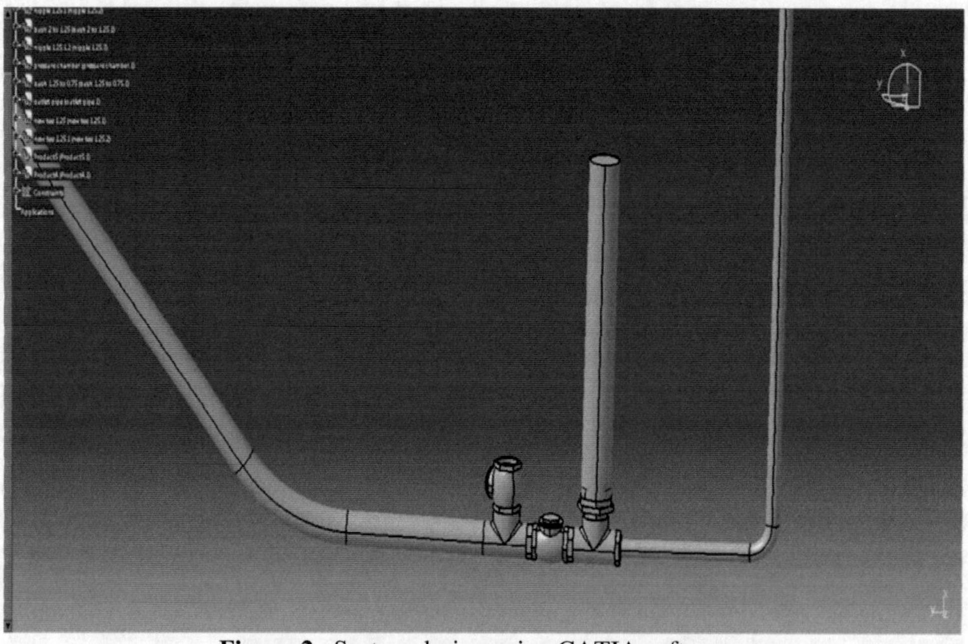

Figure 2. System design using CATIA software.

3.3. Experimental work

The experiment rig as shown in Figure 3 have been set up at Universiti Malaysia Perlis Main Campus, Arau, Perlis. There are 3 designs with different diameter and volume of the pressure chamber have been tested for this experiment. All the parameters of the designs as listed in Table 1 are similar except the diameter and volume of the pressure chamber.

Figure 3. Experimental set up

Table 1. Design Specification

Parameter	Design 1	Design 2	Design 3
Supply head (m)	1.8	1.8	1.8
Delivery head (m)	3	3	3
Minimum delivery head (m)	2.85	2.75	2.6
Number of pushes on waste valve	1	7	9
Drive pipe length (m)	2.25	2.25	2.25
Delivery pipe length (m)	4	4	4
Inlet pipe diameter (mm)	32	32	32
Outlet pipe diameter (mm)	19.05	19.05	19.05
Pressure chamber diameter (inch)	2	3	4
Pressure chamber volume (l)	1.013	4.56	4.05

4. Results and discussion

4.1. Simulation results

Simulation analysis has been done using ANSYS CFX software. The analysis including the pressure, velocity behavior and also the volume of water inside the hydraulic ram pump system. Figure 4 shows the pressure simulation results for the hydraulic ram pump. The inlet side has the highest pressure with red colour which means high pressure area. Then the water pressure dropped relatively after falling down to the critical part of the system causing the waste valve to close and check valve to open to allow water to enter the pressure chamber. Inside the pressure chamber water will be pressurized in order to go out to through outlet pipe to be raised to certain height.

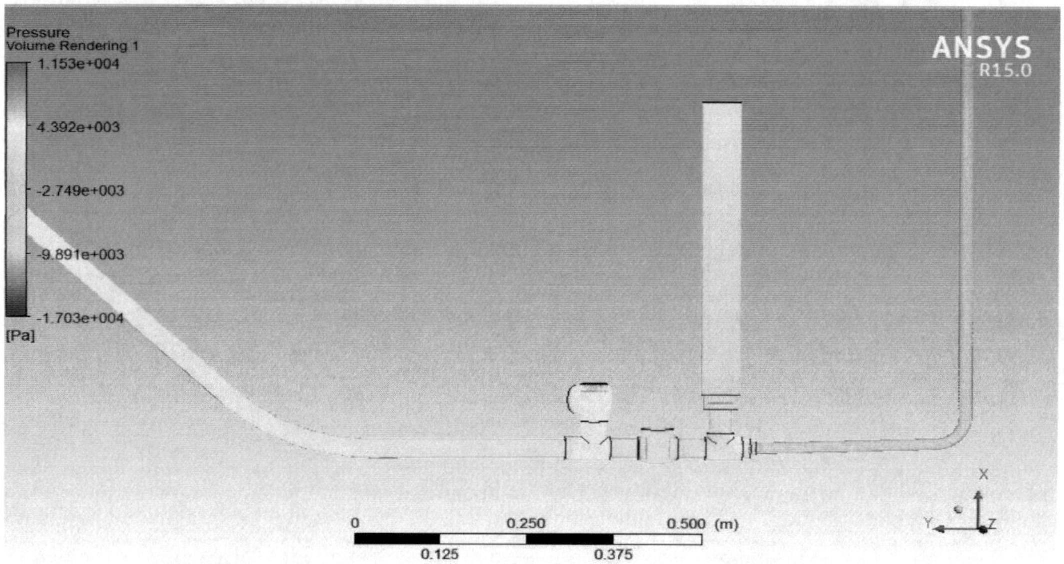

Figure 4. Pressure simulation result using ANSYS CFX software.

Figure 5 illustrates the velocity behavior of the hydraulic ram pump. Water enter the inlet pipe with relatively high velocity which indicated by green color. After that velocity drops until reaching the pressure chamber. Water velocity will be increased by going up to the outlet pipe which indicated by blue color.

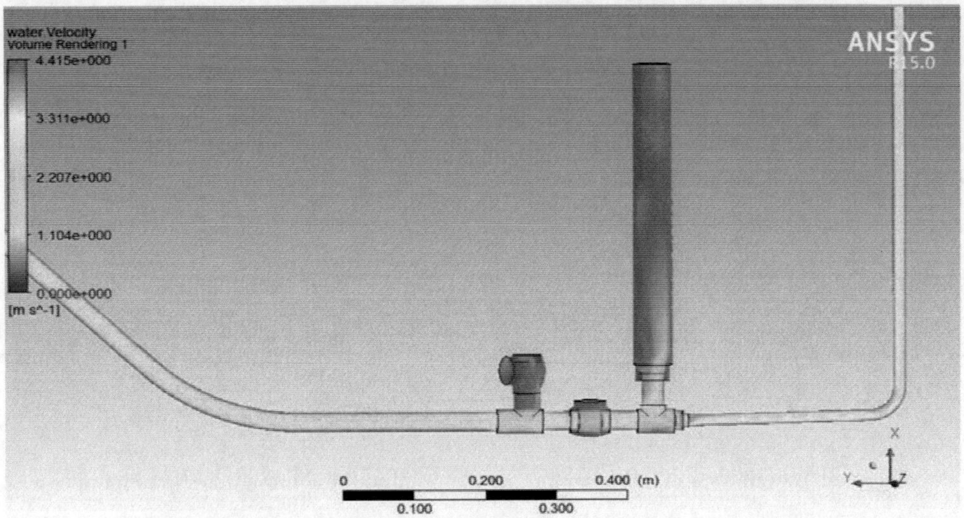

Figure 5. Velocity simulation result using ANSYS CFX software.

Figure 6 shows the volume of water inside the hydraulic ram pump system which indicated by blue colour. Also the figure illustrates the air volume inside the pressure chamber by white colour. Water enters to pressure chamber which will compress the air until a point that the air cannot be compressed anymore so that will pressurize the water causing it to get out to outlet pipe.

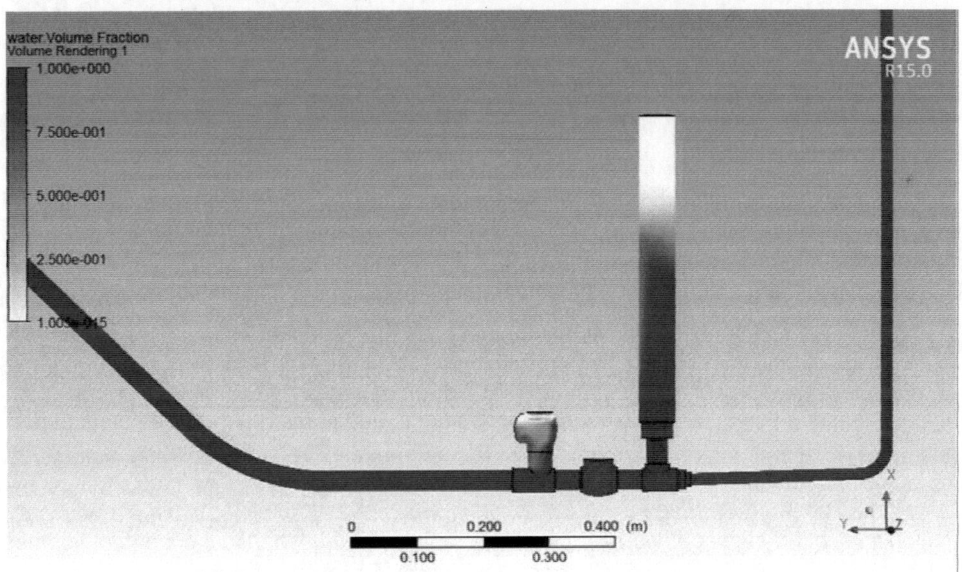

Figure 6. Simulation on volume of water and air inside the system.

4.2. Experimental results

The experiment results are divided into 3 design configuration as shown Table 2. Design 1 has reached target head of 3 m with 15% efficiency and flow rate of 11.82 L/m. Design 2 also has achieved the target head with 14% efficiency and 10.8 L/m flow rate. For design 3, even though it could reach target head but it has less efficiency with 13% and flow rate of 10 L/m. Therefore, Design 1 is the best design.

IOP Conf. Series: Journal of Physics: Conf. Series **908** (2017) 012052 doi:10.1088/1742-6596/908/1/012052

Table 2. Experimental result

Parameter	Design 1	Design 2	Design 3
Flow rate,Q	11.82 l/min	10.82 l/min	10.2 l/min
Velocity, V	0.69 m/s	0.65 m/s	0.59 m/s
Reynolds number (Re)	1.83×10^4	1.72×10^4	1.59×10^4
Friction factor (*f*)	0.027	0.027	0.028
Head loss (H_L)	0.138 m	0.13 m	0.11 m
Pressure drop (ΔP)	1.35 kPa	1.2 kPa	1.06 kPa
Pumping Power (P)	2.3 W	2.2 W	2 W
Potential energy (PE)	32 J	32 J	32 J
Efficiency (η)	15%	14.4%	13.1%

4.3. Discussion

The experiment work has been done inside the area with limited source of water which is different from the running source (river, pond etc.). There are 3 basic designs have been set for this experiment. All the parameters of the designs are similar except the pressure chamber diameter and volume. There are several factors were considered in the experiment which are the height of supply head, pressure chamber diameter and source of the supply source. From the observation of this experiment, the pressure chamber diameter is a critical point to increase the water pressure. The smaller pressure chamber diameter, the greater water pressure. In addition, supply head plays an essential role in hydraulic ram pump system. If the supply head is increased, the water pressure will be increased. Another important factor is the quantity of available source. The more water quantity, the more pressurized water will be. In the pre experiment attempts, water has been failed to get the target delivery head due to small amount of water source. While in the actual experiment, water could reach the target delivery head due to the large amount of water source.

Unfortunately, there were some undesired issues during the experiment which can cause other people not to use the hydraulic ram pump if these issues are not solved. One of these issues is the large amount of the waste water through the waste valve. The tank of supply source could contain 50 gallon but the outlet net volume is less than 20 gallon. So, about 30 gallon is wasted. Which means the hydraulic ram pump is more suitable for running source place rather than stable and limited amount of source. This problem also can be solved by putting a system can collect and return the wasted water to the tank. In addition, another problem occurs for the hydraulic ram pump has been observed during the experiment is the high noise resulted from the waste valve behavior by open and close. This action caused by the waste valve creates high noise which actually is unwanted by the user.

5. Conclusion

The hydraulic ram water pumping system was fabricated and tested. The simulation study also has been done using CATIA software and ANSYS CFX software to validate the working concept. Three design were tested in the experiment study. The best design reached target head of 3 m with 15% efficiency and flow rate of $11.82 l/min$. The results from this study show that the less diameter of pressure chamber and higher supply head will create higher pressure.

6. Acknowledgments

The authors would like to thank the Ministry of Higher Education of Malaysia and Universiti Malaysia Perlis for providing financial support under the research grant no. FRGS (9003-00564).

7. References

[1] Rennie L C and Bunt E A1990 The Automatic Hydraulic Ram Experimental results *Journal of Power and Energy 1990-1996 (vols 204-210)* **204** p 23-31

[2] Seth J 2013 *How to build a Hydraulic Ram Pump* LandToHouse.com

[3] Young B W1996 *Journal of Power and Energy 1990-1996 (vols 204-210)* **210** p 245-248.

[4] Sampath S S, Sawan S, Allan M P, Waleed J and Chithirai P S M 2015 Estimation of Power and Efficiency of Hydraulic Ram Pump with Re-circulation System *International Journal of Computer-aided Mechanical Design and Implementation* **1** p 7-18

[5] Shuaibu M 2007 Design and Construction of a Hydraulic Ram Pump *Leonardo Electronic Journal of Practices and Technologies*, **1583** p 59-70

[6] Matthias I, Suchard S, Johannes F J M, Johannes M and Wolfram S 2015 Hydraulic Ram Pumps for Irrigation in Northern Thailand *Agriculture and Agricultural Science Procedia* **5** p 107-114

[7] Rohan D B, Swapnil P R, Pradeep S B and Manoday R R 2015 Designing of Hydraulic Ram Pump *International Journal of Engineering And Computer Science* **4** p 11966-11971

IOP Conf. Series: Journal of Physics: Conf. Series **908** (2017) 012053 doi:10.1088/1742-6596/908/1/012053

Effects of machining conditions on the specific cutting energy of carbon fibre reinforced polymer composites

A I Azmi [1], A Z Syahmi[1], M Naquib [1], T C Lih [2], A F Mansor [1] and A N M Khalil [2]

[1] Faculty of Engineering Technology, Universiti Malaysia Perlis (UniMAP), UniCITI Alam Campus, Padang Besar, 02000 Perlis, Malaysia
[2] School of Manufacturing Engineering, Universiti Malaysia Perlis (UniMAP), Pauh Putra Campus, Pauh, 026000 Perlis, Malaysia

azwaniskandar@unimap.edu.my

Abstract. This article presents an approach to evaluate the effects of different machining conditions on the specific cutting energy of carbon fibre reinforced polymer composites (CFRP). Although research works in the machinability of CFRP composites have been very substantial, the present literature rarely discussed the topic of energy consumption and the specific cutting energy. A series of turning experiments were carried out on two different CFRP composites in order to determine the power and specific energy constants and eventually evaluate their effects due to the changes in machining conditions. A good agreement between the power and material removal rate using a simple linear relationship. Further analyses revealed that a power law function is best to describe the effect of feed rate on the changes in the specific cutting energy. At lower feed rate, the specific cutting energy increases exponentially due to the nature of finishing operation, whereas at higher feed rate, the changes in specific cutting energy is minimal due to the nature of roughing operation.

1. Introduction

Machining operations in manufacturing industries generate a number of waste resources that contribute toward a substantial environmental impact and manufacturing costs. Among waste resource issue that attracted significant attention in both academia and industries is energy efficient machining through optimization and reduction strategies. These strategies are highly challenging due to the complexity of cutting mechanics; the interrelationships among efficiency, energy, speed, cost, and quality; and the diversity of energy-consuming units of a machine tool and its auxiliaries [1]. In addition to that, it has been well reported that energy efficiency in machining operations is often very low. Zhao et al. [2] and Jun et al. [3] have reported that the energy used during actual machining only accounts approximately 15% - 30% of the total energy requirement. Yet, it should be emphasized that more energy is consumed due to the poor machinability of the material. The increase of machining forces and localized heat generation near the cutting zone have a major influence on the growth of tool wear, accelerates the power and energy demand for the machining operation. Among poor machinability metallic materials include titanium alloys, nickel based alloys, stainless steels, whereas for polymer composites such as fibre reinforced composites (FRPs) are also classified as difficult-to-cut materials. In recent years, the development and application of carbon fibre reinforced composites (CFRPs) are occurring at an increasingly fast pace to replace the conventional metallic materials in in various application parts such as automobile, aerospace, and marine. Superior properties of CFRPs include low density, high strength, high stiffness, good toughness, good fatigue, creep, wear and corrosion resistance, low friction coefficient and good dimensional stability [4]. Despite of these superiorities, the fibre reinforcements of the CFRP composites are highly abrasive that accelerate the wear on cutting tool and hence probably requires high force and energy to shear or machine them. Furthermore, previous understanding of machining CFRP is that, it involves brittle fractures and lack of plastic deformation which make them difficult to obtain the required quality and strict final dimensional accuracy in machining. Thereby, proper selection of machining parameters and tools are inevitable in ensuring efficient cutting process for excellent surface finish and qualities, subsequently help to achieve minimal energy usage. From the literature, there is abundant amount of research works focusing on the characterising, optimising and predicting the machinability of CFRP composites. Among the pioneering work was reported by Sakuma and Seto [5] that studied the performance of different tool materials and geometries during machining of FRP composites. Later around 2002-2010, major research works of machining via turning process for CFRP composites and glass fibre reinforced composites (GFRP) can be found through the work of Davim et al [6-8], Palanikumar et al. [9-11] and others. Various solutions are proposed among which statistical, analytical and experimental were employed to determine effective and efficient machining of those composites. However, the limitation of those works is that the aspects of energy and power consumption were not taken into account.

Only until recently, Marathe and Javali [4] reported the effects of machining parameters on the specific cutting energy and delamination damage of glass fibre-polyester reinforced composites in drilling operation. The authors showed that the specific cutting energy for this type of composite is approximately 0.81-0.83 J/ mm^3. Feed rate was the main factor that contributed to the lowest cutting energy per unit volume or material removal. It is to note here that the value of specific cutting energy for polymer composite is relatively low as compared to that of metal [12]. In another earlier work, Rosario et al. [13] studied the energy required during the dry drilling of PEEK GF30, a thermoplastic material, polyetheretherketone, reinforced with glass fibre. An analysis of variance (ANOVA) showed that the type of drill is a more influential factor and the optimum conditions are with the tool made of tungsten carbide (WC) with diamond point under the higher cutting conditions of speed and feed rate [4]. From the aforementioned review, it can be concluded that rarely efforts have been made towards understanding the effects of cutting parameters on the specific cutting energy of CFRP composites via turning process. Through this study, it enables the identification of process parameters and evaluation of the process efficiency at which the different cutting mechanisms occurred.

2. Experimental setup and methodology

2.1. Materials and cutting inserts

Two types of CFRP composites were used in this experimentation, Figure 1(a-b). The first is a 12 mm diameter of 2 m length pultruded CFRP composites, made from uni-directional carbon fibres with epoxy matrix. Another one is a CFRP hollow pipe of 60 mm outer diameter and 57 mm inner diameter. This CFRP pipe is made of 3K 2/2 twill weave type carbon fibres with epoxy matrix. The length of the tube is also 2 m. It is to note that the fibres used are of standard modulus fibres. For both of the samples, its diameter has a tolerance of ± 1 mm. All of these CFRP composites were supplied by EasyComposites UK Ltd. For machining test purposes, both of the CRFP composites were trimmed to the length of 75 mm. For the cutting inserts in turning process, PVD cemented coated carbide insert from SUMITOMO was used with a tool holder ETJNR-2020K16, Figure 1(c). A cutting length of 50 mm was carried out on the workpiece for each of the cutting conditions. During each condition, a new side of cutting edge insert was used in order to alleviate any effect of worn tool on the measurement of machining force. All cutting process was carried out in a dry condition. This is due to the recommendation from previous reported research that dry condition is much preferable and the use of cutting fluids has insignificant effect on the cutting performance.

IOP Conf. Series: Journal of Physics: Conf. Series **908** (2017) 012053 doi:10.1088/1742-6596/908/1/012053

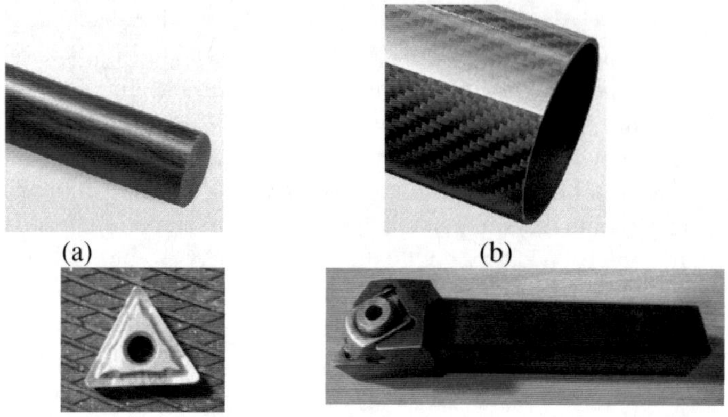

(a) (b)

(c)

Figure 1. (a) Pultruded CFRP composite (b) Woven CFRP composite pipe (c) Insert and its tool holder

2.2. Machining Conditions

The workpiece specimens were machined at the cutting speeds ranging from 20 m/min to 100 m/min for the pultruded CFRP composites, whereas 50 to 250 m/min for the composite pipes. The feed was varied from 0.15 mm/rev to 0.60 mm/rev corresponding to every cutting speed for the pultruded CFRP composites. As for the composite pipes, feed was varied from 0.10 mm/rev to 0.50 mm/rev. Depth of cut was 0.2 mm for the pultruded CFRP and 0.1 mm for the CFRP pipes, respectively. Details of cutting conditions are shown in Table 1.

Table 1. Machining conditions for this experimentation

Depth of cut, a_p (mm)	feed, f (mm/rev)	V_c (m/min)	Condition
0.2	0.15, 0.30, 0.45, 0.60	20, 40, 60, 80, 100	Dry
0.1	0.10, 0.20, 0.40, 0.50	50, 100, 150, 200, 250	Dry

2.3. Machine Tools and Data Acquisitions

All straight turning process was performed on Chevalier FCL-608 CNC turning machine. The machine is capable of handling a maximum spindle speed up to 6000 RPM, maximum cutting diameter of 260 mm and maximum cutting length of 290 mm. Figure 2(a) depicts the CNC turning machine used in this study. Since the main objective of this study is to determine the specific cutting energy, the cutting force profile of the workpiece was measured with a 3-channel dynamometer (Kistler® 9129) as depcited in Figure 2(b). Kistler Dynoware software with Kistler® 5070 charge amplifier and data acquistion module were used to acquire and sample the cutting force data at appropriate sampling frequency of 2kHz.

(a) (b)

Figure 2. (a) Chevalier FCL-608 CNC (b) Kistler® 9129 force dynamometer with charge amplifier

2.4. Determination of specific cutting energy

According to Zhao et al. [2], energy consumption of a machining process can be evaluated at different levels, namely; machine tool, spindle, and process levels. Based on these three levels, it was decided that the energy consumption at the process level is being evaluated. In this level, only the energy consumed by actual material removal is considered, which is machine tool independent. As the energy consumption at the level governs the chip formation process and surface generation, therefore, this energy is considered when selecting process parameters where the objective is to balance energy consumption with surface integrity [2]. It is well known that specific energy is widely used to evaluate the energy consumption for a machining process. It is defined as energy consumption to remove a unit volume of work material. The formula for the specific cutting energy is given as:

$$k = \frac{P_m}{MRR} \tag{1}$$

Where
k – specific cutting energy in J/ mm^3
P_m – Cutting Power (J/s or Watt), given by:
$P_m = F_c \times V_c$, F_c – *cutting force*, V_c – *cutting velocity*

MRR – Material Removal Rate (mm^3), given by:

$$MRR = \pi \cdot \frac{D_i^2 - D_f^2}{4} \cdot \frac{f \cdot V_c}{\pi \cdot D_{ave}} \cdot \frac{1000}{60} \tag{2}$$

Where
D_i – initial diameter, D_f – final diameter
f – feed
V_c – cutting speed
D_{ave} – average diameter

3. Results and discussion

Figure 3-4 show the relationship between cutting power and material removal rate during turning of pultruded CFRP composites and CFRP composite pipes, respectively. From these graphs, it is apparent that there is a positive linear relationship between cutting power and the MRR. In fact, the linear fitting of the experimental data shows a good relationship between cutting power and MRR since the R^2 is in the range of 0.95 – 0.99 (which is very close to 1.0) regardless of the type of CFRP composites used.

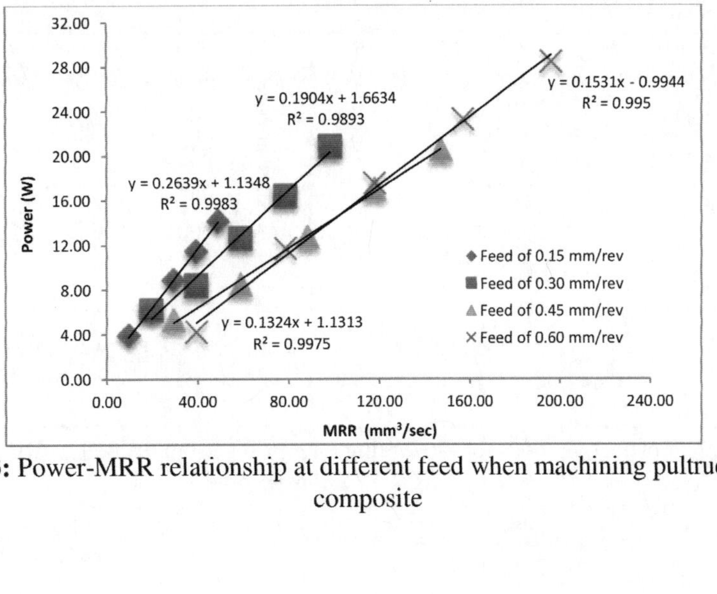

Figure 3: Power-MRR relationship at different feed when machining pultruded CFRP composite

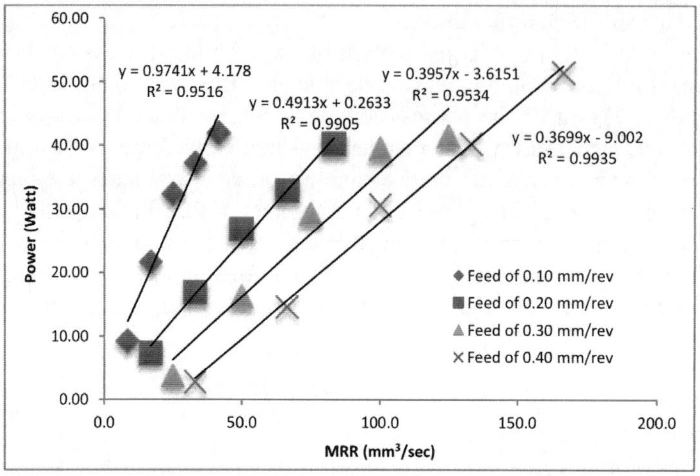

Figure 4: Power-MRR relationship at different feed when machining CFRP composite pipe

It is important to highlight here that the specific cutting energy constant, k, is represented by the slope of cutting power-MRR trend line. For both of the material, the values of k's are tabulated in Table 2. It can be observed that the specific cutting energy constant reduces from 0.264 J/mm^3 to 0.153 J/mm^3 for the pultruded CFRP composite. Meanwhile, the specific energy constant also reduces with respect to the feed rate particularly for the CFRP composite pipe. The changes are from 0.974 J/mm^3 to 0.370 J/mm^3. The relationships of this specific cutting energy with the feed rate are plotted in graph shown in Figure 5. It can be seen that relationship can be curved-fitted well with a power law function. The value of R^2 for the pultruded CFRP composite is 0.851, whereas for the CFRP composite pipe is 0.949. As apparent, the specific cutting energy is initially high at lower feed but reduces exponentially and almost plateau when feed is at higher level. This result is consistent with the findings of Balogun and Mativenga [14] when machining AISI 1045 steel alloy and titanium 6Al-4V alloys. The authors asserted that fairly plateau specific cutting energy at higher feed is due to a higher un-deformed chip thickness, which represent typical of roughing operations (high feed rate). Whereas, exponentially higher at low feed (typical of finishing operations) is due to low un-deformed chip thickness [14]. The finding of our study also confirms that the specific cutting energy is significantly influenced by the feed. It is also in agreement with the theoretical specific cutting force model [14].

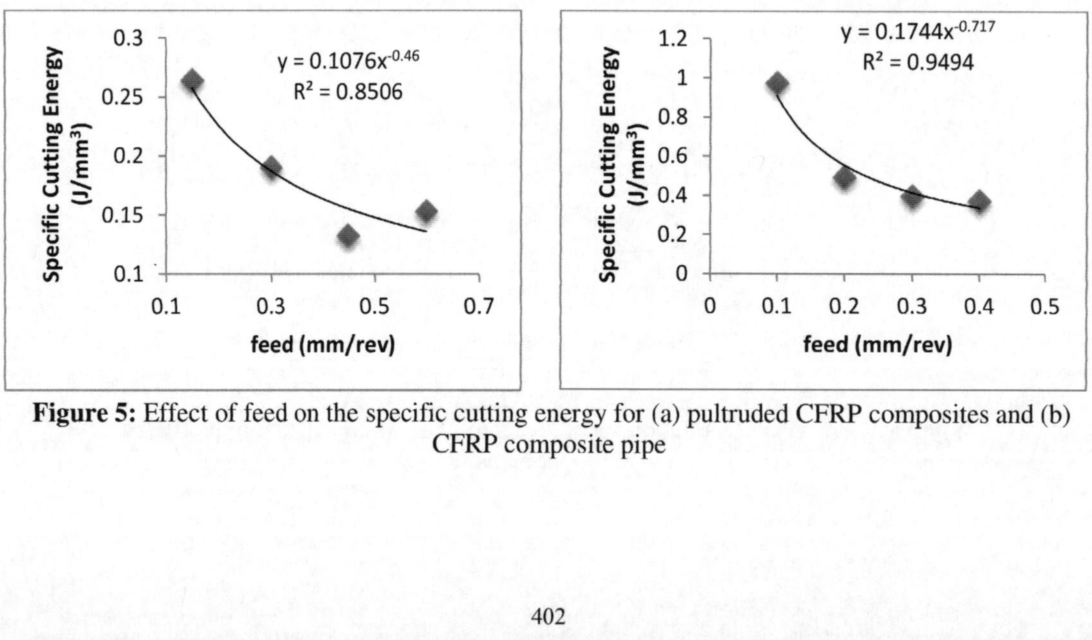

Figure 5: Effect of feed on the specific cutting energy for (a) pultruded CFRP composites and (b) CFRP composite pipe

Table 2. Specific cutting energy of the tested material at different feed

Pultruded CFRP composite		CFRP composite pipe	
feed (mm/rev)	k (J/mm^3)	feed (mm/rev)	k (J/mm^3)
0.15	0.264	0.1	0.9741
0.30	0.190	0.2	0.491
0.45	0.132	0.3	0.396
0.60	0.153	0.4	0.370

4. Conclusion

This paper has discussed the power, MRR and specific cutting energy of pultruded CFRP composites and CFRP composite pipes during turning process at different parameter settings. Based on the experimental results, the effects of turning parameters on the MRR and specific cutting energy of CFRP composites have been determined. The increase in cutting power was due to the MRR since the experimental data shows a positive linear relationship and R^2 was very close to 1.0 regardless of the type of CFRP composites used. The experimental results also revealed that the specific cutting energy of the both CFRP composites were strongly influenced by the feed rate. The specific energy constant can be reduced with respect to the feed rate particularly for the pultruded CFRP composite and CFRP composite pipe. Lastly, the relationships of specific cutting energy with the feed rate in this present study were good agreement with the theoretical specific cutting force model.

5. References

[1] Newman ST, Nassehi A, Imani-Asrai R, Dhokia V 2012 *CIRP J. Manuf. Sci. Technol.* **5** 127

[2] Zhao GY, Liu ZY, He Y, Cao HJ and Guo YB 2017 *Energy*. 133 142

[3] Xie J, Liu F and Qiu H 2016 *Int J Adv Manuf Technol.* **85** 1339

[4] Marathe AB, and Javali AM 2016 *J Thermo Plastics Compos.* **29(9)** 1261

[5] Sakuma K. and Seto T. M. 1985 *Bull. JSME*, **28 (245)** 2781

[6] Davim JP, Mata, F, Gaitonde VN and Karnik SR 2010 *J Thermo Plastics Compos.* **23(1)** 5

[7] Davim JP and Mata, F 2008 *Mater & Design* **29(8)** 1568

[8] Davim JP and Mata, F 2005 *J Mater. Process. Technol.* **170(1)** 436

[9] Palanikumar K, Mata, F and Davim JP 2008 *J Mater. Process. Technol.* **204(1)** 436

[10] Palanikumar K 2007 *Mater & Design* **28(10)** 2611

[11]Palanikumar K 2006 *J Reinf Plast. Compos* **26(16)** 1739

[12] Sheikh Ahmad J 2008 *Mach. Polymer Compos*

[13] Rosario D, Manuel G and Maria R 2013 *Proc. Eng.* **63** 687

[14] Balogun VA and Mativenga PT 2014 *J Clean Prod.* **69** 260

Acknowledgments
Authors wishing to acknowledge assistance or encouragement from colleagues, special work by technical staff or financial support from UniMAP and Malaysia Ministry of Higher Education (MOHE).

IOP Conf. Series: Journal of Physics: Conf. Series **908** (2017) 012054 doi:10.1088/1742-6596/908/1/012054

Design of refrigeration system using refrigerant R134a for macro compartment

M F H Rani[1], Z M Razlan[1], A B Shahriman[1], C K Yong[1], A Harun[2], M S M Hashim[1], M K Faizi[1], I Ibrahim[1], N S Kamarrudin[1], M A M Saad[1], I Zunaidi[1], W K Wan[1] and H Desa[1]

[1] School of Mechatronic Engineering, Universiti Malaysia Perlis, Pauh Putra Campus, 02600 Arau, Perlis, Malaysia.
[2] School of Microelectronic Engineering, Universiti Malaysia Perlis, Pauh Putra Campus, 02600 Arau, Perlis, Malaysia.

zuradzman@unimap.edu.my

Abstract. The main objective of this study is to analyse and design an optimum cooling system for macro compartment. Current product of the refrigerator is not specified for single function and not compact in size. Hence, a refrigeration system using refrigerant R134a is aimed to provide instant cooling in a macro compartment with sizing about 150 × 150 × 250 mm. The macro compartment is purposely designed to fit a bottle or drink can, which is then cooled to a desired drinking temperature of about 8°C within a period of 1 minute. The study is not only concerned with analysing of heat load of the macro compartment containing drink can, but also focused on determining suitable heat exchanger volume for both evaporator and condenser, calculating compressor displacement value and computing suitable resistance value of the expansion valve. Method of optimization is used to obtain the best solution of the problem. Mollier diagram is necessary in the process of developing the refrigeration system. Selection of blower is made properly to allow air circulation and to increase the flow rate for higher heat transfer rate. Property data are taken precisely from thermodynamic property tables. As the main four components, namely condenser, compressor, evaporator and expansion valve are fully developed, the refrigeration system is complete.

1. Introduction

Due to the warm weather, especially in Malaysia, it had become a common practice to go for a cold drink under the hot weather. Thus, cold drinks are preferable by people and the demand for them never ceases since this is a good way to cool the body temperature down [1]. Therefore, there are people starting to commercialize them and looking for a better solution to prepare cold drinks in shortest time either designing new approaches in chilling [2] or developing high efficiency cycle in refrigeration [3]. However, the refrigerator, chiller and freezer available at electrical shops are in standard size, not having much variation, usually in big size and designed to perform combined tasks. Their volume and cooling capacity are the reason why those refrigerators are not suitable to provide an immediate cooling to a size of drink bottle or can for people. For normal cases, once a drink can is placed into the freezer, the cooling process may take up about 15 to 20 minutes, in order to lower the temperature of the drink to the desired temperature, which is about 8°C for good tasting [4]. An alternative that people are still practicing is to go to a nearby shop or vending machine to purchase. However, this is still inconvenient for consumers to go out only to get a bottle or can of cold drink. At present, there are some mini fridge available with storage capacity of approximately 0.3 m^3 to 0.5 m^3 or in other words, it could store up to 12 bottles or drink cans and able to provide cooling in a few minutes to reach the desired temperature of cold drink. In this research, developing a macro compartment refrigeration system of size about of 150 × 150 × 250 mm is concerned, which will be able to cool a single bottle or drink can in only 1 minute as illustrated in Figure 2.

2. Refrigeration cycle analysis

The pressure – enthalpy diagram is analysed by selecting a particular region of the Mollier diagram of refrigerant R134a as shown in Figure 1. Basically, point 1 to 2 represent the compressor work input, point 2 to 3 represent heat dissipation from the condenser to the surroundings, while point 3 to 4 is the expansion process by capillary tube. Evaporator (point 4 to 1) works vice versa to the compressor, which extracts heat from the surrounding into the cycle [4].

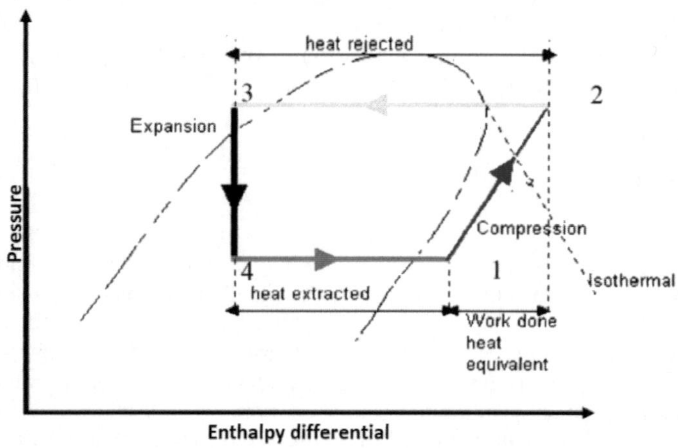

Figure 1. Pressure – enthalpy diagram.

3. Refrigerant landscape for HVAC industry

R134a has been very successfully used in screw chillers where short pipe lengths minimise costs associated with larger tubing. R134a also finds a niche where extra high condensing temperatures are needed in many transport applications [4]. By applying working principle of air conditioning, refrigerant R134a is chosen to be working fluid in removing heat from a bottle or drink can. According to Montreal Protocol [5], this refrigerant has replaced former refrigerant R22 in the refrigeration industry as it is potential for ozone layer depletion. Hydrofluorocarbons, HFCs such as R134a have no ozone depletion potential (ODP) but they do have global warming potential (GWP), which affects the environment if the refrigerant escapes to the atmosphere. A catastrophic chiller failure leading to a refrigerant leak may cause HFC refrigerant to escape into the atmosphere, yet, such occurrences are generally very low.

There is a claim that lower GWP has been always better, but a lower GWP could come with trade-offs. The lower GWP refrigerants, i.e., R32 and R152a have lower efficiency than the refrigerants we are using today, which is not preferable for the HVAC market. For chillers, apart from electricity generation to run the equipment, the refrigerant containment is also critical. There is no direct impact on global warming from refrigerants as long as they are contained and not released into the atmosphere. Selecting a chiller with an A1 refrigerant solely on the basis of GWP could be a losing proposition. If the refrigerant is less efficient than R134a or R410A, the chiller could lose efficiency and contribute more to global warming through higher energy usage leading to increased carbon emissions. Thus, Class A refrigerants, i.e., R134a and R410a are a safer and smarter choice for refrigeration purpose as they offer the lowest ASHRAE toxicity classification ("A") [6].

4. Methodology

4.1. Identification of heat load produced by a drink can

A drink can is assumed to be at room temperature initially before it is placed into the macro cooling compartment for cooling process to take place. Several drinks are selected to determine the heat load produced by a drink can. Based on ASHRAE Handbook: Refrigeration [7], specific heat capacity for selected beverages are considered in calculating the cooling capacity for the macro compartment. As summarized in Table 1, the heat load of the selected beverages is then calculated when being cooled from 30°C to 8°C. Since water possesses higher heat load among them, it is selected as reference for cooling capacity value in the entire process of refrigeration analysis [8]. The cooling capacity required for the evaporator must be higher in order to overcome the heat load. Hence, a safety factor of 10% is added as the cooling capacity of the evaporator.

Table 1. Specific heat capacity and heat load of selected beverages.

Beverage	Specific heat above freezing (kJ/kg. K)	Heat load, Q (kJ)
Cola	1.760	15.84
Cream soda	1.790	16.11
Orange juice	1.730	15.57
Water	4.187	62.81

4.2. Compressor displacement volume evaluation

Compressor displacement value could be obtained by evaluating the mass flow rate, $\dot{m}$ and volumetric flow rate, $\dot{V}_{th}$ of refrigerant R134a [9]. Based on the schematic of Mollier diagram in Figure 3, enthalpy difference compressor and evaporator (point 1 and 4), Δh is crucial in determining both values.

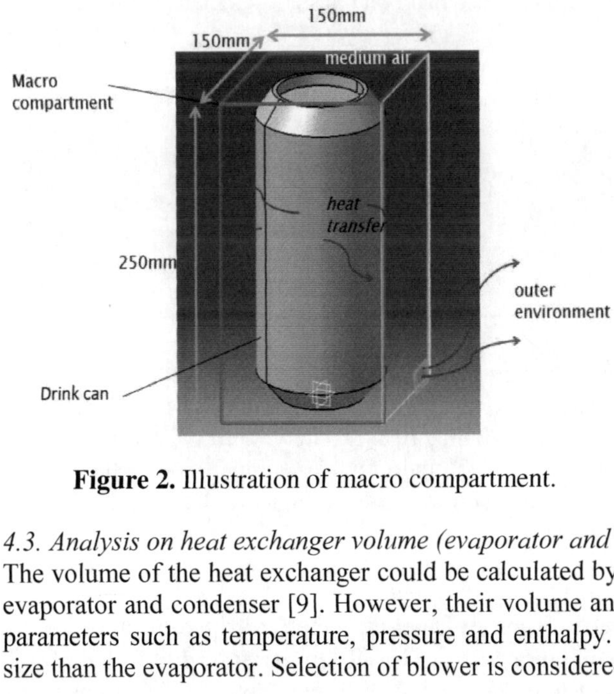

Figure 2. Illustration of macro compartment.

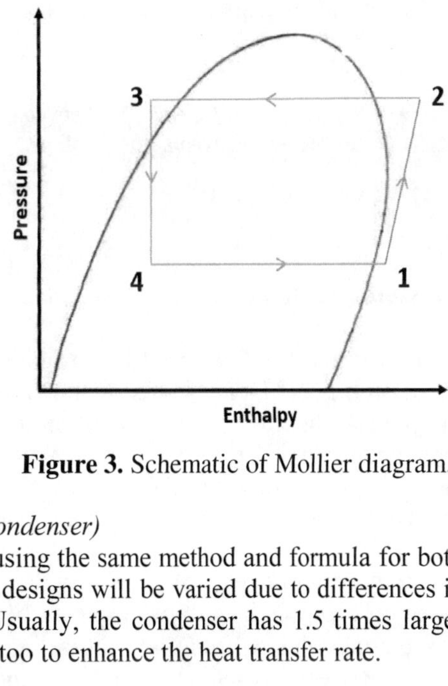

Figure 3. Schematic of Mollier diagram.

4.3. Analysis on heat exchanger volume (evaporator and condenser)

The volume of the heat exchanger could be calculated by using the same method and formula for both evaporator and condenser [9]. However, their volume and designs will be varied due to differences in parameters such as temperature, pressure and enthalpy. Usually, the condenser has 1.5 times larger size than the evaporator. Selection of blower is considered too to enhance the heat transfer rate.

4.4. Analysis on suitable resistance value for expansion valve

For the purpose of optimization, suitable resistance value for expansion valve must be analysed to reduce the pressure in the refrigeration system especially between condenser and evaporator [9]. As shown in Figure 3, the pressure difference between condenser and evaporator (point 3 and 4), ΔP is regulated by temperature sensor and pressure sensor in the expansion valve. The expansion device regulates the mass flow rate of refrigerant and removes pressure from the liquid refrigerant to allow expansion from liquid to vapour in the evaporator. Table 2 summarizes essential equations that were utilized during refrigeration system analysis for macro compartment by using refrigerant R134a.

Table 2. Equations involve in refrigeration system analysis for macro compartment.

Parameter and equation	Nomenclature
• Heat load $Q = mc_p[T_i\text{-}T(t)]$	m = Mass of drink, (kg) c_p = Specific heat capacity of drink, (kJ/kg. K) T_i = Initial temperature of drink, (°C) T(t) = Final temperature of drink at time (t), (°C)
• Required heat transfer rate $\dot{Q}_{load} = \frac{Q}{t}$	Q = Heat load (W) t = Targeted time to transfer heat (s)
• Suggested cooling capacity $\dot{Q}_{cooling} = \dot{Q}_{load}+10\%$	$\dot{Q}_{load}$ = Required heat transfer rate (W)
• Compressor displacement volume $V = \frac{60\dot{V}_{th}}{\eta_m N}$	$\dot{V}_{th}$ = Volumetric flow rate of refrigerant, (m^3/s) η_m = Mechanical efficiency N = Compressor motor revolution, (rpm)
• Overall heat transfer coefficient $\frac{1}{U} = \frac{1}{h_i}+\frac{1}{h_o}$	h_i = Internal heat transfer coefficient, (W/m^2. K) h_o = External heat transfer coefficient, (W/m^2. K)
• Condenser heat transfer surface area $A_{s.\ cond} = 1.5\times A_{s.\ evap}$	$A_{s.\ evap}$ = Evaporator heat transfer surface area, (m^2)
• Pressure difference between condenser and evaporator $\Delta P = P_3\text{-}P_4$	P_3 = Pressure at condenser, (MPa) P_4 = Pressure at evaporator, (MPa)

5. Result and discussion

5.1. Cooling capacity required for refrigeration system in macro compartment
Based on the heat load of water as shown in Table 1, at least 1046.75 W amount of heat transfer rate must be established in the macro compartment. By considering 10% of safety factor, about 1150 W of cooling capacity is suggested to implement a refrigeration system in macro compartment. The cooling capacity will decide the value of compressor displacement volume, design of the heat exchanger volume and the resistance value of the expansion valve.

5.2. Compressor displacement volume
Based on Figure 6, heat transfer occurs between air and refrigerant, where refrigerant increases in temperature when gaining heat from the air while air decreases in temperature due to heat loss. 0°C of refrigerant is set in the inlet of evaporator in order to obtain a desired outlet temperature of air around 8°C. By considering suggested cooling capacity, $\dot{Q}_{cooling}$ as well as enthalpy value at the inlet and outlet of the evaporator (h_1 and h_4), an amount of 295 cc/stroke of pumping power is required to allow 0.0076 kg/s of mass flow rate of refrigerant in the cooling system of macro compartment.

5.3. Design of heat exchanger volume (evaporator and condenser)

Overall heat transfer coefficient in macro compartment is found to be 32.31 W/m^2. K, which covers a total surface area of 0.0588 m^2 of heat transfer at evaporator and 0.0882 m^2 of heat transfer at condenser respectively. For desired heat transfer to take place, at least 2.9475 m length of copper tube must be designed at evaporator while a minimum length of 4.4212 m at the condenser. By designing the mean diameter of a coil spring to be 0.07 m, the evaporator is designed to have 14 number of turns with a height of approximately 0.18 m as shown in Figure 4. On the other hand, as shown in Figure 5, two layers of copper tube are required at condenser since the length of copper tube at condenser is longer compared to evaporator.

By designing the copper tube with 20 number of turns, its length is reduced to 0.11 m. Full designs of heat exchanger for internal flow are shown in Figure 8 (a) and Figure 8 (b) while the illustration of drink bottle in the evaporator is illustrated in Figure 8 (c). For external flow, as shown in Figure 8 (e), a compact fan [10] with maximum 0.0067 m^3/s air flow volume is selected to improve the heat transfer rate that covers 0.0144 m^2 area of the evaporator. Due to severe loss, such as friction, the air flow rate discharged from the fan is assumed to have 80% efficiency, which supply 0.00536 m^3/s of air flow rate only into the evaporator.

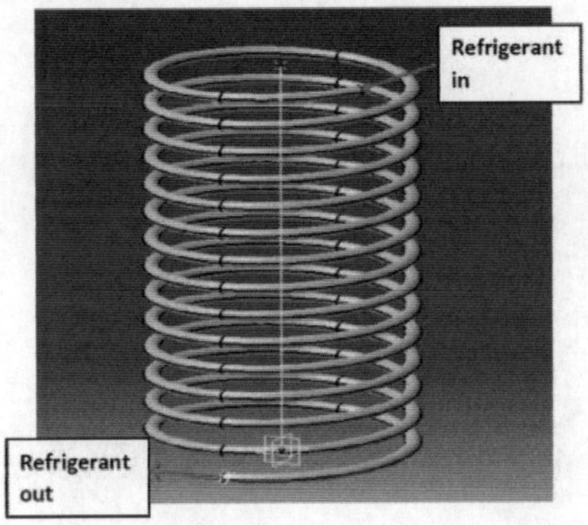

Figure 4. Copper coil design at evaporator.

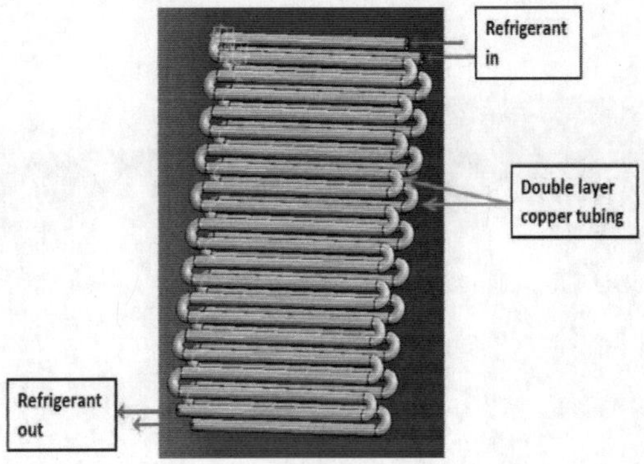

Figure 5. Copper tube design at condenser.

5.4. Resistance value of expansion valve

Since the condenser is placed outdoors, the condenser coil is contacting with the outdoor ambient air, which is not constant in temperature all the time. Thus, any changes to the outdoor temperature will affect the pressure at the condenser, thereby influence the overall refrigeration cycle. As illustrated in Figure 7, the expansion valve must contain a pressure difference, ΔP of 1.0251 MPa between condenser and evaporator in order to optimize the overall refrigeration cycle.

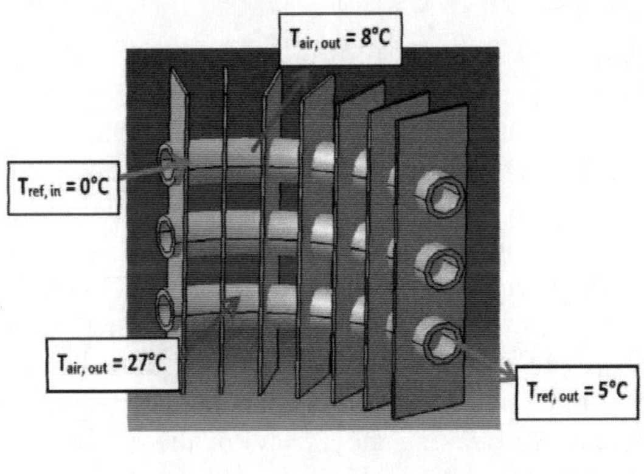

Figure 6. Heat exchange at evaporator.

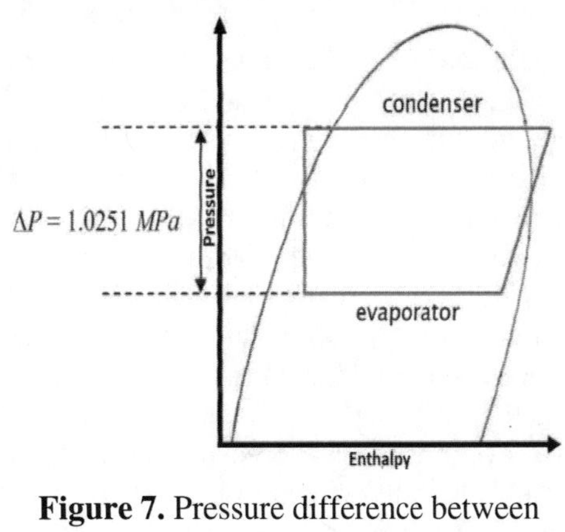

Figure 7. Pressure difference between

IOP Conf. Series: Journal of Physics: Conf. Series **908** (2017) 012054 doi:10.1088/1742-6596/908/1/012054

condenser and evaporator.

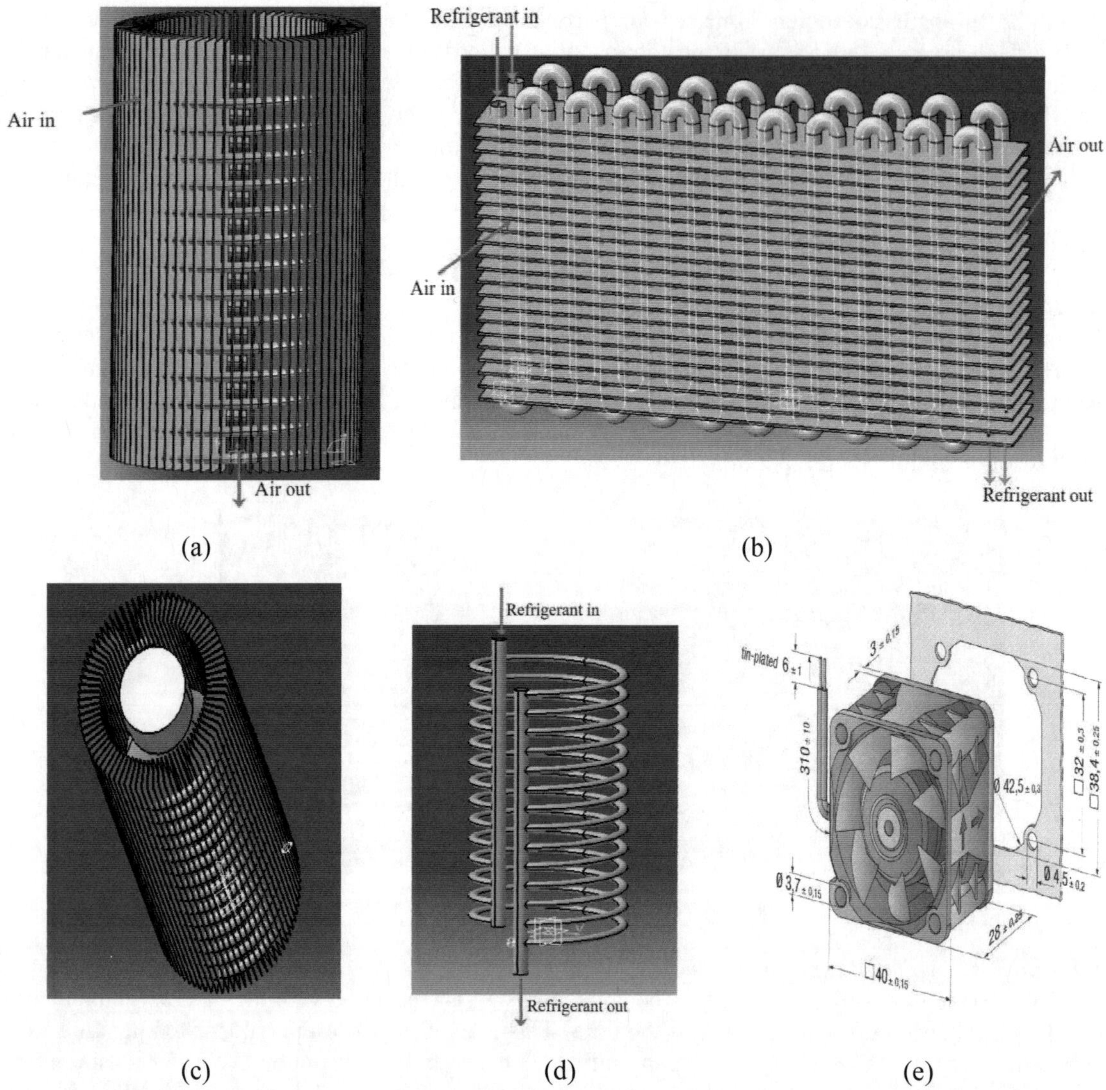

Figure 8. (a) Design of evaporator with fin, (b) Design of condenser with fin, (c) Illustration of drink bottle in evaporator, (d) Alternative design of evaporator, (e) DC axial fan series 420 J.

Besides of redesigning the evaporator by spring helical shape as illustrated in Figure 8 (d), the rate of heat transfer also could be improved by adding shaker or rotator mechanism at the platform where the drink can is placed. An artificial induced convection current circulation is created once the can is shaken or rotated. This situation will help the drink can to achieve equilibrium heat transfer and cool the drink can faster at all corners. If this method is proven to be successful, the macro compartment will take shorter time to cool down the drink bottle or drink can.

The heat load of drink can determines the cooling capacity of the refrigeration system. As higher heat load is accumulated, higher cooling capacity is required for the refrigeration system to be designed. The heat load produced by drink bottle or drink can in macro compartment was calculated to be 62.805 kJ, where the water becomes the reference drink. The mass flow rate of the refrigerant depends on the cooling capacity and enthalpy difference. By obeying the 1st law of thermodynamic (law of conservation energy), the mass flow rate of the refrigerant R134a needs to be calculated precisely since it is kept constant throughout the system. The evaporator coil was designed with spring helical shape to fit the geometry of drink bottle or drink can. The diameter of cooling coil of evaporator was determined to be 0.07 m with 14 turns and height approximately 0.18 m. The wire diameter of copper tube used for both evaporator and condenser is 6.35 mm. Meanwhile, the copper tube used for condenser was designed to be double layers, with 20 turns alternately and a length of 0.11 m for better heat transfer to the outdoor ambient air. Fins were added to allow mix flow of the heat exchangers for effective heat transfer. The pressure difference between the inlet and outlet of expansion valve was determined to be 1.0251 MPa. All design parameters of refrigeration system by using refrigerant R134a for macro compartment are summarized in Table 3.

Table 3. Summarize of design parameters of refrigeration system by refrigerant R134a for macro compartment.

Parameter	Value
Required heat transfer rate, $\dot{Q}_{load}$	1046.75 W
Suggested cooling capacity, $\dot{Q}_{cooling}$	1150 W
Mass flow rate of refrigerant, $\dot{m}_{ref}$	0.0076 kg/s
Volumetric flow rate of refrigerant, $\dot{V}_{th}$	4.187
Compressor displacement volume (Pumping power), V_c	295 cc/stroke
Cross sectional area of flow passing through copper tube, A_c	1.77×10^{-5} m^2
Reynolds number	76256
Nusselt number	311
Heat transfer coefficient inside copper tube, h_i	808.27 W/m^2. K
Heat transfer coefficient outside copper tube, h_o	30.32 W/m^2. K
Overall heat transfer coefficient, U	32.31 W/m^2. K
Mass flow rate of air, $\dot{m}_{air}$	0.0082 kg/s
Minimum heat capacity ratio, C_{min}	8.2574 W/K
Number of transfer units, NTU	0.23
Total surface area for heat transfer at evaporator, $A_{s,\ evap}$	0.0588 m^2
Minimum length of copper tube at evaporator, L_{evap}	2.9475 m
Number of turns of copper tube at evaporator, N_{evap}	14 turns
Suggested height of copper coil at evaporator, H_{coil}	18 cm
Total surface area for heat transfer at condenser, $A_{s,\ cond}$	0.0882 m^2
Minimum length of copper tube at condenser, L_{cond}	4.4212 m
Number of turns of copper tube at condenser, N_{cond}	20 turns
Pressure difference between point 3 and 4, ΔP	1.0251 MPa

6. Conclusion

An optimum refrigeration system for macro compartment that fits a single drink bottle or drink can is possible to be developed in compact approach. The heat load of drink can and cooling capacity of macro compartment determine the design of the four main components in refrigeration cycle, i.e., compressor, condenser, expansion valve and evaporator. The compressor converts the liquid refrigerant into refrigerant vapour by regulating the refrigerant's pressure. As heat exchanger, evaporator and condenser allow heat transfer to occur between refrigerant and air. Thus, a suitable volume of heat exchanger is important to optimize the efficiency of the heat transfer process. In the macro compartment, the evaporator coil is designed with spring helical shape to fit the geometry of drink bottle or drink can. The condenser coil is designed with double layer to maximize the heat transfer rate to the outdoor ambient air to provide effective cooling to the refrigerant. The expansion device decreases the pressure of the liquid refrigerant to allow expansion from liquid to vapour in the evaporator.

7. Acknowledgments

Authors wish to acknowledge School of Mechatronic Engineering, Universiti Malaysia Perlis for the lab facilities. Special thanks to those who contributed to this project directly or indirectly.

8. References

[1] R. Eccles, L. Du-Plessis, Y. Dommels, and J. E. Wilkinson, "Cold pleasure. Why we like ice drinks, ice-lollies and ice cream," *Appetite*, vol. 71, pp. 357–360, 2013.
[2] N. Arena, P. Sinclair, J. Lee, and R. Clift, "Life cycle engineering of production, use and recovery of self-chilling beverage cans," *J. Clean. Prod.*, vol. 142, pp. 1562–1570, 2016.
[3] M. Yang, C. W. Jung, and Y. T. Kang, "Development of high efficiency cycles for domestic refrigerator-freezer application," *Energy*, vol. 93, pp. 2258–2266, 2015.
[4] G. F. Hundy, A. R. Trott, and T. C. Welch, *Refrigeration, Air Conditioning and Heat Pumps*. 2016.
[5] S. Kasera and S. C. Bhaduri, "Performance of R407C as an Alternate to R22: A Review," *Energy Procedia*, vol. 109, pp. 4–10, 2017.
[6] D. Applied, "Myths and facts of HVAC Refrigerants," pp. 1–4, 2016.
[7] American Society of Heating, "2006 ASHRAE Handbook: Refrigeration," 2006.
[8] Y. A. Cengel and A. J. Ghajar, *Heat and Mass Transfer: Fundamentals and Applications*, 5th Edition. 2015.
[9] HITACHI, Basic knowledge of room air conditioner. HITACHI, 1996.
[10] E. PAPST, "EBM PAPST DC axial fans Series 420 J," 2014.

Influence of processing methods on flexural strength in interwoven hemp/ PET reinforced POM hybrid composite.

T Y Chen, S F Khan and W H Tan

School of Mechatronics Engineering, Universiti Malaysia Perlis (UniMAP), 02600 Arau, Perlis, Malaysia.

sfkhan@unimap.edu.my

Abstract. One of the challenges of utilizing natural fibre as reinforcement in polymer composite is poor interfacial bonding with thermoplastic matrix. In this study, hemp and polyethylene terephthalate (PET) fibre were selected to develop the interwoven fabric as reinforcement and polyoxymethylene (POM) was chosen as the matrix. The interwoven Hemp/PET fabric was used to produce hybrid composite using hot-press moulding technique. Different production methods of using POM layers and POM pellets were investigated. Three point bending test was used to determine the flexural modulus and compared. The results show that the different process in producing Hemp/PET/POM hybrid composite did not give significant effect towards the flexural property of composite. The modulus of elasticity for both specimens produced with POM layers and POM pellets is 2.24 GPa and 2.15 GPa, respectively. Modulus of elasticity of composite with POM layers is higher than POM pellets by 4%. This may due to POM thermoplastic characteristic which can be reversibly melted and re-solidified without significant changes to the mechanical properties.

1. Introduction

Composite industries have been trying to minimize the used of petroleum-based reinforcement in composites due to environmental issues. Substitution of petroleum-based or synthetic reinforcement needs to be made with the increment of high performance engineering products. In this research, interwoven hemp/polyethylene terephthalate (PET) fibre reinforced polyoxymethylene (POM) hybrid composite's flexural properties are studied as there is lacking information regarding the mechanical properties of such composite. Natural fibre has many appealing properties such as high strength to weight ratio, light weight, corrosion resistance, non-irritating, low energy consumption during production, carbon dioxide free of emission when burned, and biodegradable [1,2]. Hemp fibre is used in this research because it is a fast grown plant [3] with extremely strong and durable characteristics. However, natural fibre becomes less attractive compared to synthetic fibre due to low resistant to moisture absorption, low degradation temperature and insufficient of good interfacial bonding with hydrophobic matrix. PET is used as interweave reinforcement because it is light weighted and it makes a good barrier for gas, moisture, and alcohol. In addition, PET is strong and impact-resistant. Interwoven of hemp and PET significantly changes to become stronger and more durable. Furthermore, POM is used as matrix because it has advantages such as low fiction coefficient, high abrasion resistance, good electrical and dielectric properties, high heat resistance, and low water absorption.

2. Methodology

2.1. Materials

Hemp and PET fibres used in this research were form into twisted yarn as its enhanced the composite's mechanical properties [4]. Hemp and PET yarns were used to produce interwoven fabric. In this research POM, hemp and PET fiber were supplied by Innovative Pultrusion Sdn. Bhd.

2.2. Equipment

A hot-press machine and a rectangular mild steel mould were used to produce hybrid composites. A universal material testing machine of model Shimadzu AG-IS Ms 250kN was used to conduct flexural tests on the specimens from the composite. Besides that, a wooden frame was used for weaving of hemp/ PET fabric. In addition, a rope making equipment was used to twist PET fibre into yarn.

2.3. Production of yarn and Interwoven Hemp/ PET Fabric

Twisted yarn of PET fibre was produced using the same method as in rope making. The PET yarn was made by twisting 3 strings that consist of 10 filaments. The yarns were used in producing interwoven hemp/ PET fabric where hemp was used as warp yarn and PET as weft yarn. The diameter for both PET and Hemp yarn is 2mm. Hemp yarn was tied on the nails of wooden frame before PET yarn was woven over and under across the hemp yarn as shown in Figure 1.

Figure 1 (a) Completed hemp/PET fabric. (b) Close-up of Interwoven hemp/ 2mm PET fabric.

2.4. Production of POM layers

Mould release agent was sprayed onto the mould to ease removal of POM layers. The mould was preheated for 20 minutes at 180°C as shown in Figure 2. POM in pellets form was weighted and placed into the heated mould. Two layers of POM were produced simultaneously by using a metal sheet as separator. It was then put into the hot-press machine and hot-press for 20 minutes at 180°C with 5000psi pressure. After 20 minutes, the mould was naturally cooled to room temperature before the POM layers were removed. Figure 3 shows a sample of POM layer.

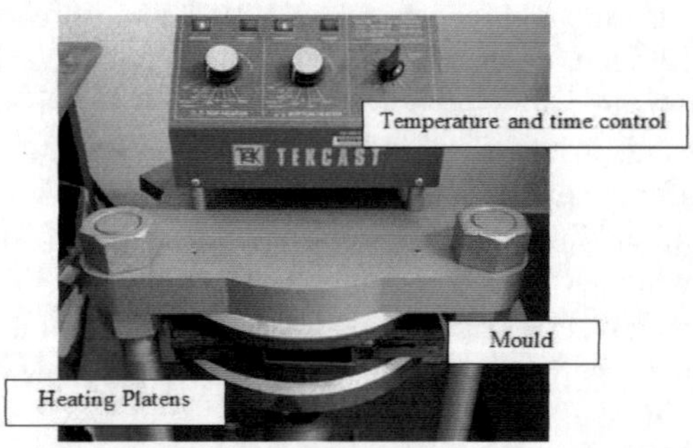

Figure 2 Mould places into hot-press machine.

Figure 3 POM sheet.

2.5. Production of Interwoven Hemp/ PET Fibre Reinforced POM Hybrid Composite using different methods.

The mould was preheated for 20 minutes at 180°C in the hot-press before 2 POM layers were placed into the mould with interwoven hemp/ PET fabric put in between the POM layers. The entire assembly was the placed into the hot-press for 20 minutes at 180°C with 5000psi pressure. The mould was then let to cool naturally to room temperature before removing the hybrid composite. In the second method POM pellets of equal amount as used to produce the POM layer were used directly by placing the pellets into the bottom of the mould cavity before placing the hemp/PET fabric. Then another equal amount of POM pellets were placed on top of the hemp/PET fabric. Thus producing sandwiched layers of pellets/fabric/pellets. The entire assembly moulding was hot-pressed for 20 minutes at 180°C with 5000psi pressure. The composite was removed after cooling naturally to room temperature. **Figure 4** shows a sample of interwoven hemp/ PET fibre reinforced POM hybrid composite**.**

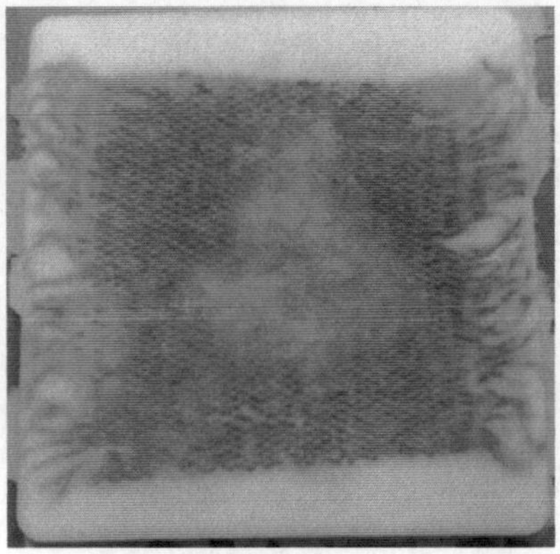

Figure 4 (a) Top side Hybrid composite (b) Bottom side

2.6. Specimen Cutting

According to ASTM D790, 90mm length with 40mm overhand length at supports, 13mm width and 5mm thick were used as the dimension of specimen in this research. X-axis of the specimen was hemp fiber and y axis was PET fiber as shown in Figure 5 . Figure 6 shows samples of specimens.

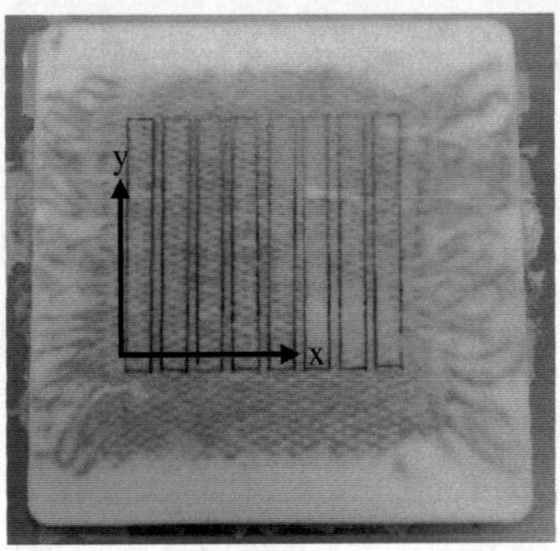

Figure 5 Specimens dimension as drawn on composite.

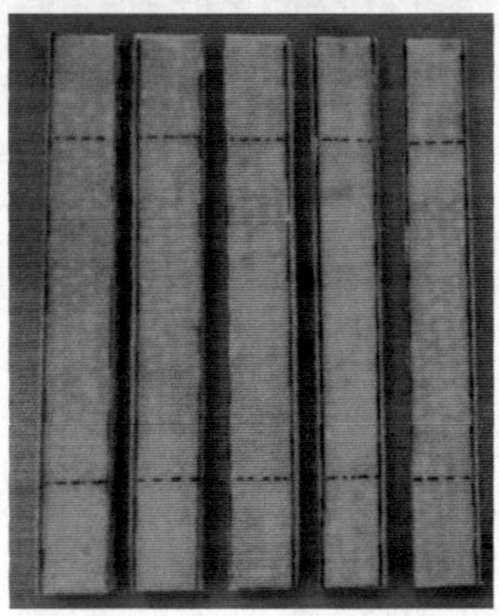

Figure 6 Specimens cut to ASTM D790.

2.7. Flexural Test

Flexural test was conduct to evaluate the force needed to bend a beam under three point loading conditions by using a universal testing machine (UTM) according to the ASTM D790 standard as shown in Figure 7[5]. A speed rate of 1mm/ min and a load cell of 100kN were used. When the specimen breaks, the test was stopped. In order to obtain a satisfactory result, five specimens from each composite were tested.

2.8. Experimental flowchart.

The experimental flowchart is given in Figure 8. The experiment is started with the preparation of interwoven fabrics and production of POM layers. Then the production of Hemp/PET/POM hybrid composite is performed. Finally three point bending test in accordance to ASTM D790 is carried out.

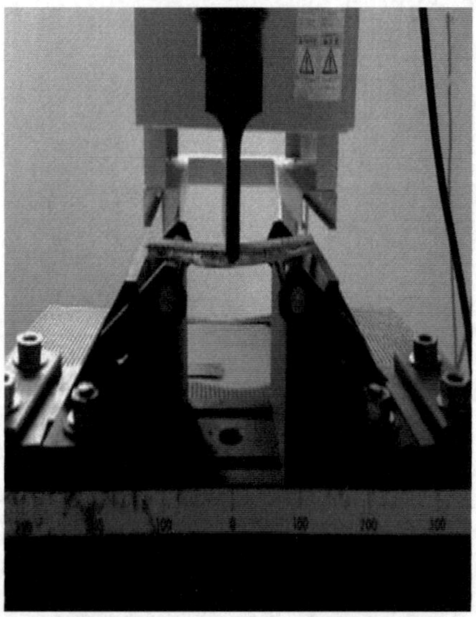

Figure 7 Experimental setup of flexural test.

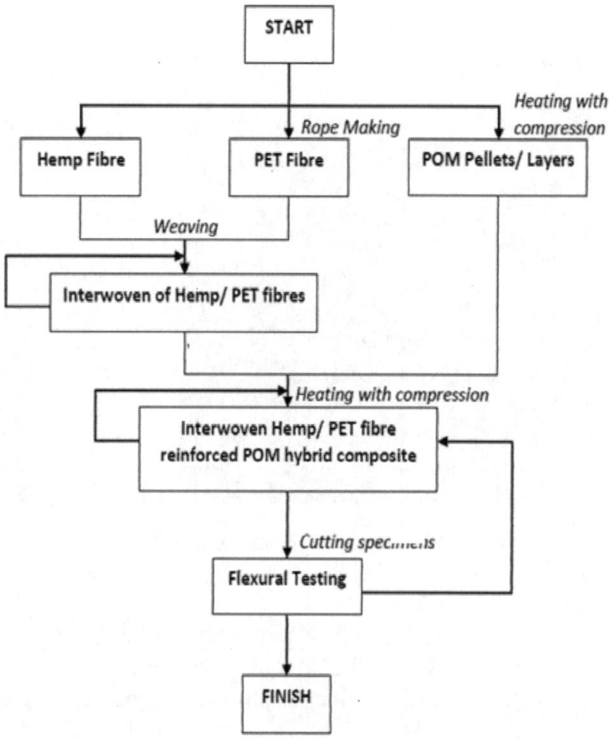

Figure 8 Flow chart of complete experiment

3. Results and Discussion

The statistical analysis of 5 specimens each for the two different production methods is as shown in Table 1. The result in the three point bending test from Table 1 indicates that using POM layer in the production of hybrid composite gave higher values. Figure 9 and Table 2 show that composite produced by using POM layers has higher flexural strength compared to POM pellets which is 37.38±4.24MPa and 34.79±4.05MPa; respectively. The difference between the two types is 6.4%. Similarly in Table 2, the flexural strain for composite using POM layers and POM pellets was determined to be 0.016±0.005mm/mm and 0.013±0.004mm/mm respectively. The difference between the two types was 17.6%. This higher values could have been contribute from the fact that the composite produced using POM pellets have higher void as shown in Figure 11.

Table 1 T-test results of flexural test on composite using POM Layer and POM Pellets.

	POM Layer				POM Pellets			
	Max. Force N	Max. Displace-ment mm	Max. Stress MPa	Max. Strain mm/mm	Max. Force N	Max. Displace-ment mm	Max. Stress MPa	Max. Strain mm/mm
n =5								
Average	90.00	4.436	37.38	0.016	83.75	3.587	34.79	0.013
Standard Deviation	10.22	1.442	4.24	0.005	9.74	1.189	4.05	0.004
Standard Error of the Mean	4.57	0.645	1.90	0.002	4.35	0.532	1.81	0.002
95% Confidence Level	9	1.3	3.7	0.004	8.5	1	3.5	0.004

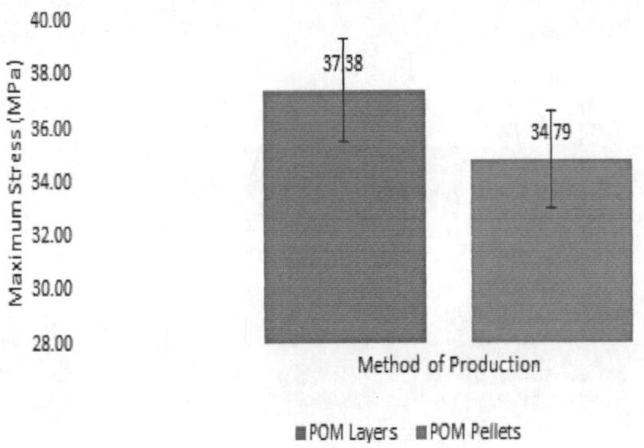

(a)

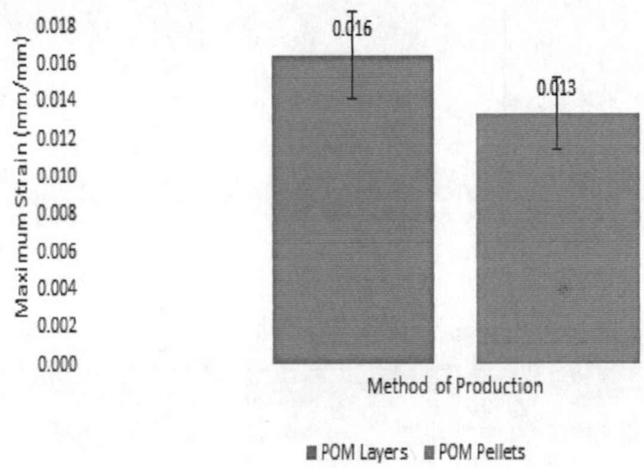

(b)

Figure 9 (a) Maximum Stress and (b) maximum strain with error bar.

As shown in Figure 10 and Figure 11, both composites contained bubbles/ air gaps which were caused by trap air bubbles or insufficient of amount POM. The production method of using POM pellets gave a more severe defect. This high porosity in the composite produced contributed to lower flexural strength.

Figure 10 Defects in composite produced with POM layers.

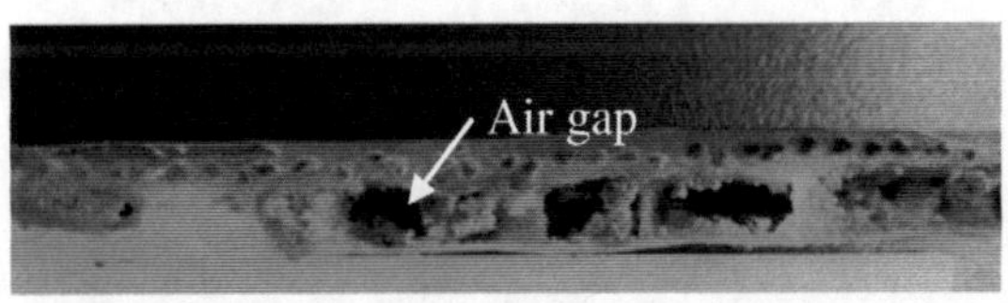

Figure 11 Defects in composite produced with POM pellets.

IOP Conf. Series: Journal of Physics: Conf. Series **908** (2017) 012055 doi:10.1088/1742-6596/908/1/012055

Furthermore, both composites show similar trend of stress-strain graph as shown in Figure 12. Modulus of elasticity for composite using POM layers was obtained at 2.24GPa and the specimen produced with POM pellets was 2.15GPa. A difference of 4.2% between these results was considered relatively small and insignificant. Hence heating of POM pellets to create POM layers and later reheating the POM layers in the production of composite in this study does not alter the Modulus of elasticity of POM as most commercially available POM has a modulus of elasticity of 2.58GPa-2.85GPa [6-8]. As POM is thermoplastic, it can be reversibly melted by heating and re-solidified without significant changing in mechanical and chemical properties [9, 10]. Thus, the production methods by using POM layers or POM pellets have insignificant effect to the production of hybrid composite using POM as matrix as POM can be reheated and re-solidified with minimal changes to the Modulus of Elasticity.

Table 2. Mechanical Properties of Hybrid Composite from different production methods.

Method of Production	Maximum Stress (MPa)	Maximum Strain (mm/mm)	Modulus of Elasticity (GPa)
POM Layers	37.38±4.24	0.016±0.005	2.24
POM Pellets	34.79±4.05	0.013±0.004	2.15

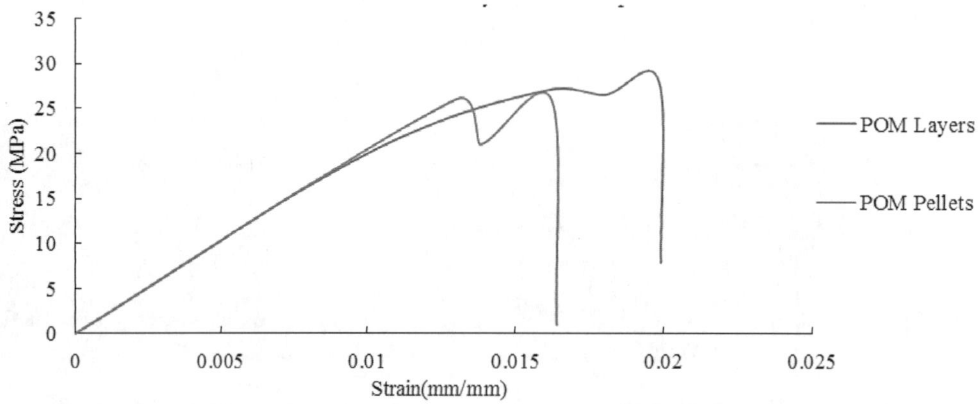

Figure 12 Stress-Strain Curve of Hybrid Composite from different production method.

4. Conclusion

From this study, the production methods of using pre-made POM layers and using POM pellets directly shows that the thermoplastic characteristics of POM is able to give similar result without significant differences. Therefore, by using POM pellets directly can decrease the time and energy consumed during the hybrid composite production.

5. Acknowledgment

The authors would like to express profound gratitude and deep appreciation to the School of Mechatronics Engineering and Universiti Malaysia Perlis in providing researching platform well as providing the facilities and equipment.

IOP Conf. Series: Journal of Physics: Conf. Series **908** (2017) 012055 doi:10.1088/1742-6596/908/1/012055

6. References

[1] Gupta V B and Bashir Z 2002, PET Fibers, Films, and Bottles *Poly(Butylene Terephthalate) - Handbook of Thermoplastic Polyesters Homo-polymers, Copolymers, Blends, and Composites* ed. Fakirov S, (KGaA, Weinheim: Wiley-VCH Verlag GmbH & Co.) Sections 1–4.13

[2] www.newworldencyclopedia.org/entry/Composite_material. (Access January 2017.)

[3] Deitch R 2003, *Hemp: American History Revisited: The Plant with a Divided History*. (New York: Algora Publishing).

[4] Dan-Mallam Y, Abdullah M Z, and Megat Yusoff P S M 2014, The effect of hybridization on mechanical properties of woven kenaf fiber reinforced polyoxymethylene composite *Polym. Compos.* **vol. 35**, no. 10, pp. 1900–1910,.

[5] ASTM D790-15e2 2015, *Standard Test Methods for Flexural Properties of Unreinforced and Reinforced Plastics and Electrical Insulating Materials*, (West Conshohocken, PA: ASTM International)

[6] https://www.alibaba.com/countrysearch/CN/pom-pellets.html (Access Nov 2016.)

[7] https://www.matbase.com/material-categories/natural-and-synthetic-polymers/thermoplastics/engineering-polymers/material-properties-of-polyoxymethylene-homopolymer-pom-h.html#properties (Access Nov 2016.)

[8] https://plastics.ulprospector.com/generics/2/c/t/acetal-properties-processing (Access Nov 2016.)

[9] Klein R 2011, *Laser Welding of Plastics, First Edition* (KGaA, Weinheim: Wiley-VCH Verlag GmbH & Co.), p 9

[10] Henry A 2014, Thermal Transport in Polymers *Annual Review of Heat Transfer* (Redding, CT: Begell House, Inc.) p487

IOP Conf. Series: Journal of Physics: Conf. Series **908** (2017) 012056 doi:10.1088/1742-6596/908/1/012056

Conversion of rice husk into fermentable sugar by two stage hydrolysis

M N Salimi[1,2], S E Lim[1,2], A H M Yusoff[1] and M F Jamlos[1]

[1]School of Bioprocess Engineering, Universiti Malaysia Perlis, Perlis, Malaysia.

nabil@unimap.edu.my; siew_eng917@hotmail.com

Abstract. Rice husks, a complex lignocellulosic biomass which comprised of high cellulose content (38-50%), hemicellulose (23-32%) and lignin (15-25%) possesses the potential to pursue as low cost feedstock for production of ethanol. Dilute sulfuric acid at concentration of 1, 2, 3 (%, v/v) were used for pretreatments at varied hydrolysis time (15-60 min) and enzymatic saccharification at range of 45-60°C and pH 4.5-6.0 were evaluated for conversion of rice husk's cellulose and hemicellulose to fermentable sugars. The maximum yield of fermentable sugars from rice husks by dilute sulfuric acid (2%, 60 minutes) was 0.0751 g/l. Total fermentable sugar was identified using dinitrosalicylic acid (DNS) method and expressed in g/l. Enzymatic hydrolysis for conversion of cellulose to fermentable sugar has been studied by applying response surface methodology (RSM) and Analysis of Variance (ANOVA). Two independent variables namely initial pH and incubation temperature were considered using Central Composite Design (CCD). The determination coefficient, R^2 obtained was 0.9848. This indicates that 98.48% capriciousness in the respond could be clarified by the ANOVA. Based on the data shown by Design Expert software, the optimum condition for total sugar production was at pH 6.0 and temperature 45°C as it produced 0.5086 g/l of total sugar.

1. Introduction

Today, more than 90% of world's energy supply came from fossil fuels [1]. As consequences from increasing fossil fuel consumption, depletion of fossil fuels and environment impacts has raised public awareness in shifting to new energy resources. The theory of supply and demand accounts that prices will rise as hydrocarbon sources diminish. Therefore, higher prices lead to development of alternative energy source which is more available and affordable.

Research reveals that one of the important energy alternatives is utilization of lignocellulosic feedstock [2]. Lignocellulosic resources can be predominantly processed to produce liquid or gaseous fuel which is known as biofuel. Examples of biofuel are such as bioethanol, methanol, biodiesel and gaseous fuels like hydrogen and methane [3]. Unlike fossil fuel, bioethanol is a form of renewable energy that can be made from any lignocellulosic feedstock containing sugar or carbohydrates. Bioethanol is a promising energy alternative due to the potential to replace gasoline, as it increases the fuel octane level of gasoline and burns cleaner than gasoline. In Brazil, most of the ethanol used as motor fuel is produced from sugar cane [4]. As compared to energy of non-renewable sources, energy of renewable sources are sustainable, clean and eco-friendly.

Among the lignocellulosic feedstocks, rice husk is one of the most abundant agriculture by-products worldwide where its production reaches million tons annually [5]. This inedible lignocellulosic material is the outermost layer of paddy grains, commonly removed in milling process

IOP Conf. Series: Journal of Physics: Conf. Series **908** (2017) 012056 doi:10.1088/1742-6596/908/1/012056

as biowaste, which occupies up to 20-22% of total weight of milled paddy [2]. Rice husk is suitable for bioethanol production owing to high cellulose content. Significant improvement in engine performance, including brake power, torque, brake mean effective pressure, volumetric efficiency and fuel consumption by using ethanol produced from rice husk is observed when blends with gasoline [6]. Consequently rice husk is recommended to be exploited for low cost bioethanol production on a commercial scale, which eventually solve for disposal issue and reduce waste treatment cost [7].

Conversion of lignocellulosic materials to ethanol includes two processes: hydrolysis of cellulose into fermentable reducing sugar and fermentation of the sugars to ethanol. Commercial cellulase enzymes could be used in conversion of lignocellulose into fermentable reducing sugars. The conversion is carried out by highly specific cellulase enzyme involving mixture of endoglucanases, exoglucanases, β-glucosidases and cellobiohydrolases [8]. Various research have been completed on degradation of rice husk cellulose to yield monomeric sugars using enzymes [9]-[12].

The efficiency of enzymatic hydrolysis depends on various factors including type of pretreatment and catalytic action of enzymes [8]. Lignin content, availability of enzymatic site of cellulose fibers and crystallinity properties reduce the accessibility of cellulase enzyme to cellulose, thus affect the efficiency of enzymatic hydrolysis. Therefore, lignocellulosic materials require a pretreatment to solubilize the hemicellulose and enhance the cellulose accessibility, which can be carried out by different techniques such as alkalis, acids, heat, pressure, solvents, etc. [13]. In present study, methods were developed to optimize acid pretreatment and enzymatic hydrolysis of rice husk cellulose into fermentable sugars. The hydrolysis efficiency, in terms of total reducing sugar yields obtained from pretreatment and enzymatic hydrolysis were evaluated.

2. Experiment

2.1. Materials

Rice husks were collected from a rice processing mill, Padi Beras Nasional Berhad (BERNAS) located at Simpang Empat, Perlis. Rice husks were washed thoroughly with distilled water to eliminate adhering soil and dust. The husks were dried at 100°C overnight. The dried husks were then grinded into powder form and sieved by passing through a 1 mm screen to standardize the particle size range of 1 mm. The sample was kept in a tightly closed container at room temperature.

2.2. Methodology

2.2.1. Sulfuric acid pretreatment. In optimization of acid hydrolysis of rice husk, grinded rice husk was pretreated with dilute sulfuric acid (1%, 2% or 3%, v/v) and the sample was soaked in water bath at 90°C at different retention time (15-60 min). The liquid fraction was tested for glucose concentration while the solid residue was washed several times with distilled water to remove retaining acid solution. Then, the residue was dried in the oven at 50°C overnight.

2.2.2. Enzymatic hydrolysis. The solid residue resulted from optimal condition in acid pretreatment was proceeded for enzymatic hydrolysis. Commercial enzyme cellulase produced by *Aspergillus niger* was used. Enzyme dose of 0.04 ml/g husk was added into the hydrolysis process. The solid residue was suspended in 100 ml sodium citrate buffer and incubated in a rotating shaker for 24 hour at agitation rate of 150 rpm. The incubation pH and temperature were adjusted according to the parameters selected for optimization.

2.3. Design of experiment

The optimization of sugar production from enzymatic hydrolysis was conducted by Design Expert 7.1.5 software by using Response Surface Methodology (RSM) through Central Composite Design (CCD). The objective of RSM is optimization [14]. In this case, variables tested were pH and temperature, thus response surface designed experiment is used to determine the optimal settings for each variables. A total of 13 experiment runs are required for this study and total sugar concentration was taken as the response of the designed experiments. The effect of pH and temperature for hydrolysis of rice husks were studied in the range between pH 4.5 to pH 7.5 and 30°C to 60°C, respectively, shown in Table 1.

Table 1. Parameters range for enzymatic hydrolysis optimization.

Parameters	Minimum	Maximum
pH	4.5	7.5
Temperature (°C)	30	60

2.4. Model fitting and statistical analysis

The results obtained from CCD were fitted to a second order model equation to explain the reliance of total sugar concentration on the effect of pH and temperature in terms of coded values A and B, respectively. The experimental data was analyzed by using Design Expert software and further discussed in following regression analysis, ANOVA and graphical analysis sections.

2.5. Fermentable sugar analysis

The response variable in this research is concentration of glucose produced from each runs, which was determined by using DNS method [15]. Sample of 1.0 ml was withdraw and added with dinitrosalicylic acid (DNS) reagent (dilution factor will be used if needed), then the sample was place in hot water bath at 90°C for 5 minutes. Absorbance reading was taken by using UV-Vis spectrophotometry once the sample cooled down to room temperature. Blank is prepared from mixture of DNS and distilled water.

3. Result and discussion

3.1. Optimization of acid pretreatment

The optimization of acid pretreatment of rice husks was done by manipulating two variables, including concentration of sulfuric acid and hydrolysis time duration. Purpose of this optimization is to examine the optimum condition to break down rigid lignin structure and matrix conformation of cellulose for accessible of enzymatic hydrolysis in next step. The target is to produce maximum yield of sugar. The sugar concentration produced from acid dose (1.0-3.0 %, v/v) on the pretreatment at 90°C for 15 – 60 minutes were shown in Figure 1. Pretreatment with an acid dose of 2.0% (v/v) for 60 minutes generated 0.0751 g/l sugars, which is the highest yield among the acid dose up to 3.0%. The effects of acid dose and hydrolysis time are illustrated.

From Figure 1, each hydrolysis at specified acid concentration followed a trend where concentration of total sugar increased when the hydrolysis time increasing. This indicated that longer time is required by acid to carry out hydrolytic chemical reaction, which is the cleavage of β-1-4-glycosidic bond, before converting cellulose into glucose. However, it can be observed that the sugar production increased when the acid dose increased from 1.0% to 2.0% but decreased when the acid dose increased from 2.0% to 3.0%. This result is corresponded to the research of Saha et al. (2005) when the total sugar they obtained with pretreatment with 2, 3 and 4 (%, v/v) sulfuric acid followed by enzymatic saccharification were decreased, respectively. Thus, the highest sugar recovery efficiency from the optimization of acid hydrolysis performed is 2.0% sulfuric acid and hydrolysis time of 60 minutes, where 0.0751 g/l sugar is produced. Therefore, this optimum condition is used to proceed enzymatic hydrolysis in the next step. The hydrolyzed material was washed with distilled water to eliminate retained acid and the solid residue is further processed for optimization enzymatic hydrolysis.

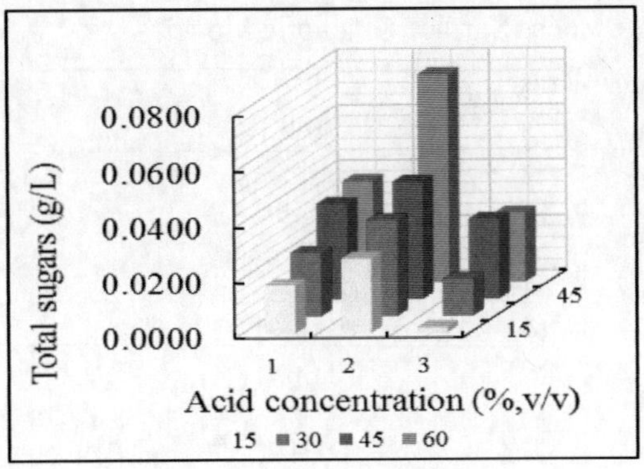

Figure 1. Effect of treatment time (15 – 60 min) on the release of total sugar from rice husks by dilute H2SO4 pretreatment (1.00 – 3.00%) at 90°C.

3.2. Optimization of enzymatic hydrolysis

3.2.1. Regression Analysis.

Quadratic model was selected as suggested by the software and was used to explain the mathematical relationship between the independent variables and the dependent response. The regression model for both coded and actual factors were obtained where positive sign represents synergistic influence while negative sign represents antagonistic influence.

Final equation in terms of coded factors:
Total sugar concentration = 0.52 + 0.15 A + 0.445 B – 0.018 AB – 0.16 A^2 – 0.18 B^2 (1)

Final equation in terms of actual factors:
Total sugar concentration = – 9.29736 + 1.94747 pH + 0.15726 Temperature – 0.07886 pH Temperature – 0.14483 pH^2 – 0.08127 $Temperature^2$ (2)

3.2.2. ANOVA.

Analysis of variance (ANOVA) is a collection of statistical models used to analyze the differences among group means and their variation. The analysis of variance of the selected quadratic model was presented in Table 2 and Table 3.

The magnitude of p-value (prob>F) is used to determine statistical significance of a model. A p-value lower than 0.01 indicating only a 1.0% chance that a 'model F-value' could occurred. The model F value obtained in Table 2 is 90.80 with p-value less than 0.0001, indicates that the model generated was statistically significant with only 0.01% chance of noise occurrence.

The significant of each coefficient can be distinguished from the F-test and p-value. In this study, the variables A, A^2 and B^2 had significant effect on the total sugar yield as their p-values are less than 0.0001. The variable B and AB were insignificant because their p-value is greater than 0.05. The p-value of lack of fit is less than 0.0001, indicates that this model term is significant. However, this can be explained by claiming that two factors interaction model or linear model normally features significant lack of fit [16].

The analysis of the coefficient of determinates is shown in Table 3. The coefficient of determination value (R^2) is very high at 0.9848 indicates that the equation was highly reliable. The "Predicted R-Squared" of 0.9046 is in reasonable agreement with the "Adjusted R-Squared" of 0.9740 because the difference between them is less than 0.2. The "Adequate Precision" measures the signal to noise ratio. A ratio greater than 4 is desirable. The ratio of 22.650 indicates an adequate signal and this model can be used to navigate the design space.

The coefficient of variance (CV) is the degree of accuracy. The value of CV obtained in the study was 11.41%. CV is a function of standard deviation and normally standardizes the scale of data. In general, low CV value means the factors are highly reliable during optimization.

Table 2. Analysis of variance (ANOVA) for Response Surface Quadratic Model.

Source	Sum of Squares	Degree of Freedom	Mean Square	F Value	p-value Prob > F	
Model	0.55	5	0.11	90.80	< 0.0001	significant
A-pH	0.17	1	0.17	143.21	< 0.0001	
B-Temperature	0.117	1	0.117	0.53	0.4892	
AB	0.063	1	0.063	1.06	0.3379	
A^2	0.18	1	0.18	153.80	< 0.0001	
B^2	0.23	1	0.23	195.38	< 0.0001	
Residual	0.418	7	0.060			
Lack of Fit	0.356	3	0.119	7.57	0.0399	significant
Pure Error	0.063	4	0.058			

Table 3. R-squared value for model developed by Design Expert.

Standard deviation	C. V. (%)	R-squared	Adjusted R-Squared	Predicted R-Squared	Adequate Precision
0.035	11.41	0.9848	0.9740	0.9046	22.650

3.2.3. Graphical Analysis. Predicted versus Actual Plot reveals how well the model fit the data. The plot is also used to compare the fit against the null model. The actual value indicates the data obtained from the experimental run while the predicted value is the value evaluated from Design Expert software. Figure 2 shows a great correspondence between the predicted and actual sugar concentration. Most of the points in the Figure 2 are scattered along the straight line although some points shows slightly deviation. Points that vertically distant from the straight lines indicating possible existence of outliers. The result overall considered as good fitted to the quadratic model.

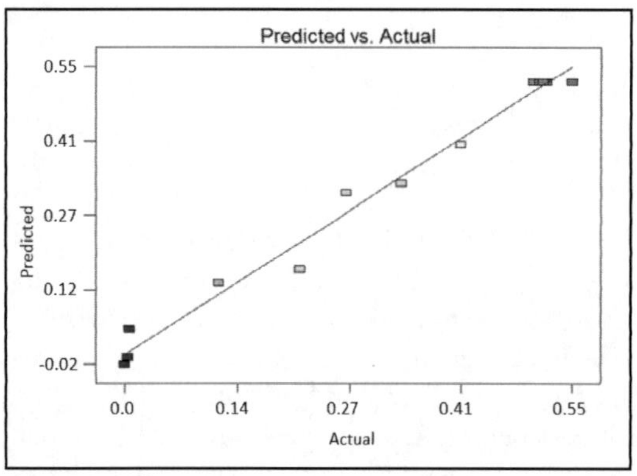

Figure 2. Predicted versus the actual total sugar concentration.

3.2.4. Interaction between parameters. The response surface was used to obtain the maximum response which is the total sugar production at optimum condition at each design variables. Variables pH and temperature were considered in this study for the interaction effect between the variables and effect on the total sugar concentration.

Two dimensional (2D) contour plot and three dimensional (3D) response surfaces are used as graphical tool to display the interaction between variables on fermentable sugar concentration. The shape of the response surface indicate the nature and extent of the interaction between factors. Figure 3 and Figure 4 demonstrate the correlation of the pH and temperature. In Figure 3, it shows egg-shaped contour plot which indicates that the associations between the related variables are vital. Figure 4 illustrating a surface with maximum point located inside the experimental region [17]. The red region in the contour represented the maximum predicted value.

According to the 2D contour plot and 3D response surface, total sugar increased when the pH value increased. However, the sugar concentration started to decline when pH approach neutral state, where pH near 7 is considered as optimum for many enzymes, enzyme is less active at pH lower or higher than this optimum. This variation in enzyme activity with changing pH is due to the influence of pH on acidic and basic side chains in active site of enzyme which can cause substrate cannot bind to the active site or it cannot undergo catalysis [18].

From analysis of variance, effect of temperature is not significant to the sugar production, this has been proven by the plots where sugar is produced in the range between 34.39°C to 55.61°C. This might be due to the adaptability of cellulase complex for the temperature range from 30°C to 60°C which was selected in this study. However, there was slightly decreased in enzymatic hydrolysis efficiency at temperature beyond 50°C, this could partially be explained by the loss of enzyme activity due to thermal inactivation [19]. It is observed that the highest sugar concentration was produced at the range of pH 6 to pH 7 and 40°C to 50°C, where this result is also found in study of Ong et al. (2012) by using rice straw as substrate [20].

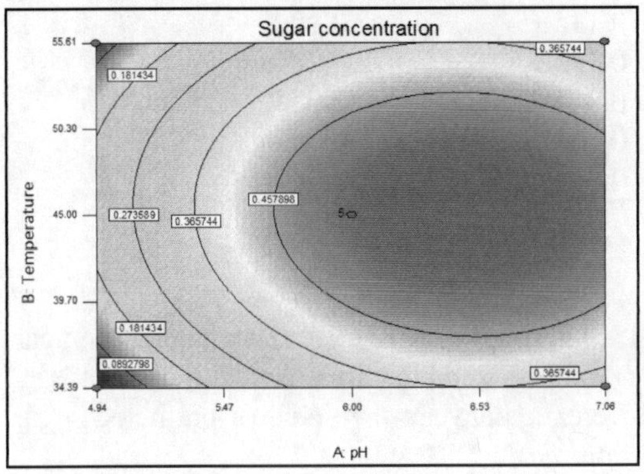

Figure 3. Contour plot of interaction between pH and temperature on sugar concentration.

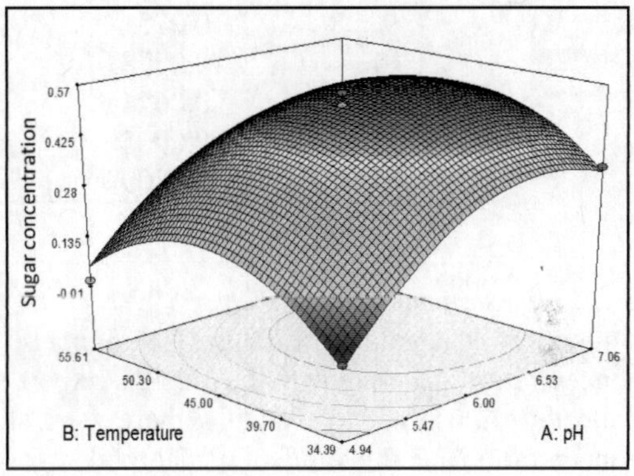

Figure 4. 3D plot interaction between pH and temperature on sugar concentration.

3.3. Total glucose production

Total production of fermentable sugar were retrieved by adding up sugar yield from acid hydrolysis and enzymatic hydrolysis. Table 4 showed the concentration of sugar for both hydrolysis. The values obtained mostly were contributed by enzymatic hydrolysis. Therefore, it implies that enzymatic hydrolysis is more effective in producing fermentable sugar than acid hydrolysis.

There is difference in amount of fermentable sugar from both acid hydrolysis and enzymatic hydrolysis. The sugar concentration in acid hydrolysis is almost the same for each run due to the no changes in variables addressed. The highest sugar produced from the acid hydrolysis was 0.0905 g/l which was obtained from sample 2 and 13. The lowest sugar produced in this hydrolysis is 0.0559 g/l obtained in run 6, which differed 0.0346 g/l from the highest value.

The sugar concentration in enzymatic hydrolysis shows significant difference for each runs where the highest and lowest amount of sugar produced were 0.5471 g/l and 0.0019 g/l, from run 6 and run 12 respectively. However, sugar produced from enzymatic hydrolysis is overall higher than that of acid hydrolysis, except for run 12 and 13. The difference in amount of sugar produced is affected by the variables manipulated in enzymatic hydrolysis, in this case, pH and incubation temperature.

The highest total sugar yield is obtained in run 6, where 0.6030 g/l of sugar is achieved. In this hydrolysis, the condition applied is pH 6.0 and 45°C. Meanwhile, similar condition but using different substrate, Ong et. al (2012) recovered 3.62 g/l sugar from rice straw [20]. The amount of sugar obtained in present study is apparently lesser than the research reported. However, the difference may due to the period of hydrolysis of rice straw is longer and variation of cellulose content in the substrate.

Table 4. Total sugar production from acid and enzymatic hydrolysis.

Run	Fermentable Sugar Concentration (g/l)		Total Sugar Concentration (g/l)
	Acid Hydrolysis	Enzymatic Hydrolysis	
1	0.0616	0.3391	0.4007
2	0.0905	0.5009	0.5914
3	0.0867	0.5086	0.5953
4	0.0578	0.4103	0.4681
5	0.0655	0.5124	0.5779
6	0.0559	0.5471	0.6030
7	0.0732	0.2697	0.3429
8	0.0790	0.5163	0.5953
9	0.0751	0.1156	0.1907
10	0.0578	0.2138	0.2716
11	0.0674	0.0058	0.0732
12	0.0905	0.0019	0.0925
13	0.0809	0.0077	0.0886

3.4. Validation model

Validation is conducted to signify the capability of RSM through the predicted fermentable sugar concentration generated by the Design Expert software. Table 5 demonstrates the validation of data collected. According to the table, there were slightly difference between the value of predicted sugar concentration and the value of actual total sugar concentration by 1.64%.

$$\%\ \text{Error} = \frac{\text{Theoritical value} - \text{Experimental value}}{\text{Theoritical value}} \text{x}100\% \tag{3}$$

Table 5. Result of validation for verification of optimization.

Run	pH	Temperature (°C)	Predicted total sugar concentration (g/l)	Actual total concentration (g/l)	Percentage error (%)
1	6.00	45	0.517073	0.5009	3.1278
2	6.00	45	0.517073	0.5086	1.6386
3	6.00	45	0.517073	0.5163	0.1495
Average				0.5086	1.64

4. Conclusion

Rice husk is one of the most abundant agriculture wastes in the world. From a view point of the biomass energy, rice husks, comprises about 70% mass organics which could be effectively used as environmentally friendly fuels to supply electric power or biomass ethanol. This study has qualitatively and quantitatively discussed the production of total fermentable sugar by two steps hydrolysis. Rice husk was pretreated by using dilute sulfuric acid at different hydrolysis time. The optimized acid hydrolysis condition is obtained by using 2.0% (v/v) sulfuric acid at 60 minutes. The rice husk residue produced from this optimized condition undergo enzymatic hydrolysis with the presence of cellulase enzyme. The combined of both hydrolysis processes gave a high yield of total sugar production. Statistical study was performed by Response Surface Methodology (RSM) through Central Composite Design (CCD) for the enzymatic hydrolysis and result showed that the model is reliable and significant. The optimal condition was identified by the Design Expert software which conducted the enzymatic hydrolysis at pH 6 and at temperature of 45°C. When the rice husk was hydrolysed at this optimum condition, approximately 0.50 g/l of sugar was recovered. The determination coefficient (R^2) obtained in this study was 0.9848 which correspond to 98.48% and the standard deviation is 0.035. The magnitude of R^2 indicated there was a correlation between the experimental and predicted data. The model F value of 90.80 with p-value less than 0.001 imply that this model was statistically significant.

5. Acknowledgments

This research was able to be completed because of supports from many sides. Deepest appreciation to the dedication of supervisor, Dr. Midhat Nabil for his constant guidance throughout the research. Appreciation to Padi Beras Nasional Berhad (BERNAS) located at Simpang Empat, Perlis, for supplying the materials for research. Appreciation also to Universiti Malaysia Perlis for funding and providing the facilities in laboratory. Thank profusely to all the technicians of School of Bioprocess Engineering who assisted in analysist of samples.

6. References

[1] Miller G T and Spoolman S E 2016 *Living in the Environment* (Canada: Cengage Learning)
[2] Bazargan A, Bazargan M and McKay G 2015 *Renew.Ener.* **77** 512-520
[3] Balat M and Balat H 2010 *Appl.Ener.* **87** 1815-1835
[4] Oth Mullis 1993 *Alcohol Fuels: Impacts from Increased Use of Ethanol Blended Fuels* (United States: Diane Publishing)
[5] Yavari S, Malakahmad A and Sapari N 2016 *Environ.Sci.Pollut.Res.* **23** 17928-17940
[6] Nyachaka C J, Yawas D S and Pam G Y 2013 *Int.J.Engir.Res.Tech.* **2** 3516-3533
[7] Kumar S, Sangwan P, Dhankhar R Mor V and Bidra S 2013 *Res.J.Chem.Env.Sci.* **1** 126-129
[8] Sartori T, Tibolla H, Prigol E, Colla L M, Costa J A and Bertolin T E 2015 *Biomed.Res.Int.* **2015**:342716
[9] Saha B C, Iten L B, Cotta M A and Wu Y V 2005 *Biotechnol.Prog.* **21** 816-822
[10] Saha B C, and Cotta M A 2007 *Enz.Mic.Tec.* **41** 528-532
[11] Wei G-Y, Lee Y-J, Kim Y-J, Jin I-H, Lee J-H, Chung C-H. and Lee J-W 2010 *Appl.Biochem.Biotechnol.* **162** 1471-1482
[12] Dutta N, Mukhopadhyay A, Dasgupta A K and Chakrabarti K 2014 *Bior.Tech.* **153** 269-277
[13] Dagnino E P, Chamorro E R, Romano S D, Felissia F E and Area M C 2013 *Ind.Crop.Prod.* **42** 363-368.
[14] Myers R H, Montgomery D C and Anderson C M 2009 *Response Surface Methodology: Process and Product Optimization Using Designed Experiments 3rd ed* (New York: Wiley)
[15] Mizutani S, Iijima S, Morikawa M, Shimizu K, Matsubara M, Ogawa Y, Izumi R, Matsumoto K and Kobayashi T 1987 *J.Ferm.Tech.* **65** 325-331
[16] DeLoach R and Ulbrich N 2007 *A Comparison of Two Balance Calibration Model Building Methods* (Nevada: American Institute of Aeronautics and Astrautics) p 147
[17] Bezerra M A, Santelli R E, Oliveira E P, Villar L S and Escaleira L A 2008 *Talanta* **76** 965-977
[18] Seager S L and Slabaugh M R 2010 *Organic and Biochemistry for Today 7th ed* (Belmont: Brooks/Cole Cengage Learning)
[19] Pandiyan K, Tiwari R, Singh S, Nain P K S, Sarika R, Arora A, Singh S B, Nain L 2014 *Enzy.Res.* **2014:**764898
[20] Ong L G A, Chan C H and Chew A L 2012 *J.Med.Bioeng.* **1** 14-16

IOP Conf. Series: Journal of Physics: Conf. Series **908** (2017) 012057 doi:10.1088/1742-6596/908/1/012057

Finite element analysis of 2-Station hip himulator

M I M Fazli[1], A Yahya[2], A Shahrom[2], S W Nawawi[1], M R Zainudin[1] and M S Nazarudin[1]

[1]Faculty of Electrical Engineering, Universiti Teknologi Malaysia, 81310 Skudai, Johor Bahru, Malaysia
[2]Faculty of Biosciences and Medical Engineering, Universiti Teknologi Malaysia, 81310 Skudai, Johor Bahru, Malaysia

mikram33@live.utm.my

Abstract. This paper presented the analysis of materials and design architecture of 2-station hip simulator. Hip simulator is a machine used to conduct the joint and wear test of hip prosthetic. In earlier work, the hip simulator was modified and some improvement were made by using SolidWorks software. The simulator consists of 3DOF which controlled by separate stepper motor and a static load that set up by manual method in each station. In this work, finite element analysis (FEA) of hip simulator was implemented to analyse the structure of the design and selected materials used for simulator component. The analysis is completed based on two categories which are safety factor and stress tests. Both design drawing and FEA was done using SolidWorks software. The study of the two categories is performed by applying the peak load up to 4000N on the main frame that is embedded with metal-on-metal hip prosthesis. From FEA, the value of safety factor and degree of stress formation are successfully obtained. All the components exceed the value of 2 for safety factor analysis while the degree of stress formation shows higher value compare to the yield strength of the material. With this results, it provides information regarding part of simulator which are susceptible to destruct. Besides, the results could be used for design improvement and certify the stability of the hip simulator in real application.

1. Introduction

Human body consist of joints which enable every part of the body to move freely up to certain limits. Hip joint or hip articulation is one of important joint in human motion. The hip joint which also known as ball-and-socket joint have large degree of freedom and provide wide range in human mobility. Therefore, dysfunction of this joint will certainly limit the motions of people who suffer from this problem. Hip prosthesis is an orthopaedic procedure which the hip joint is replaced by prosthetic or artificial implant to replace arthritic or dysfunctional joint surface [1]. However, there are lots of postoperative complication occur regarding the hip prosthetic such as dislocation [1], and wear debris [2] which affect the safety and efficacy of patient [3] and eventually reduce the lifespan of hip prosthetic.

Therefore, hip simulator has been developed so that the lifespan of metal-on-metal hips prosthesis can be tested before being implant to the patient by surgery. The hip simulator will generate similar joint posture and joint contact force as in daily life motion. Besides, the simulator will provide useful information regarding the biomechanics of hip joint and evaluated the limits of performance of the biomaterials used in hip implant.

Hip joint simulator is not a recent finding in the field of orthopaedic. Various design of hip simulator has been developed decades ago to regenerate with high accuracy in vivo condition. The simulators differ from each other from a few parameters which are number of station, loading type, sequence of motion, degree of freedom (DOF), type of lubricant, and configuration of hip implant whether it is anatomical or not [4, 5].

The most crucial process in designing mechanical system is to ensure the endurance of the parts and components especially for application that use high force or load. There is very high risk taken by establish a product or system without verify the performance of the structure and materials. Finite element analysis (FEA) is one of the method used by engineer and scientist to model and numerically solve complex structural or system. Development of system or structure in virtual environment will help to predict and enhance the reliability of the product. Besides, it will help in reducing the cost for prototyping and testing which eventually reduce the material usage.

In this paper, structural and materials analysis in term of safety factor and stress formation are implemented on the 2-station hip simulator design using SolidWorks software.

2. Hip simulator model and finite element modelling

Figure 1 shows the design of 2-station hip simulator from different perspective; isometric, front and side. Basically, the simulator is designed with some modification based on specification that are obtained from literature. The system of simulator is expected to replicate human joint motion [6] and provide testing on hip prosthesis in term of wear and dislocation.

Figure 2 represents the principle diagram of main frame in the proposed design hip simulator. The analysis of the simulator in this work will focus only on the main frame which are the upper part and the lower part as shown in figure 3. This is because both parts experience more load than other parts in the simulator. Besides, the hip implant part is embedded in innermost of the frame and become the fixed geometry during the simulation of the whole main frame.

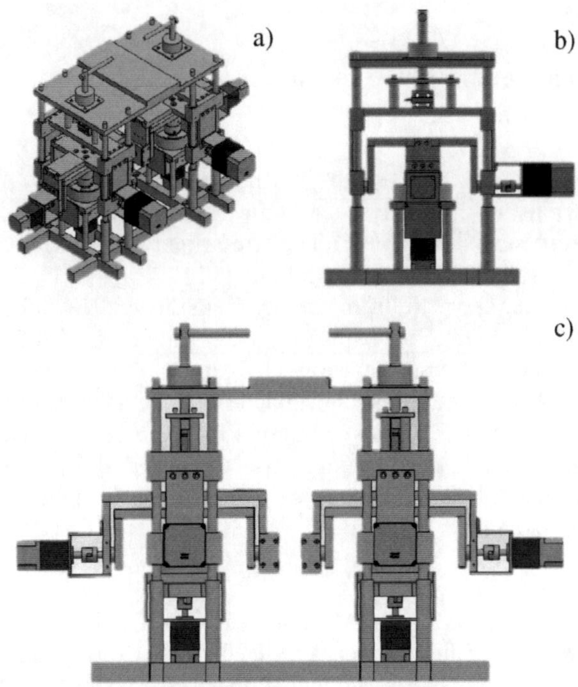

Figure 1. Design of 2-station hip simulator: a) isometric view b) side view c) front view.

IOP Conf. Series: Journal of Physics: Conf. Series **908** (2017) 012057 doi:10.1088/1742-6596/908/1/012057

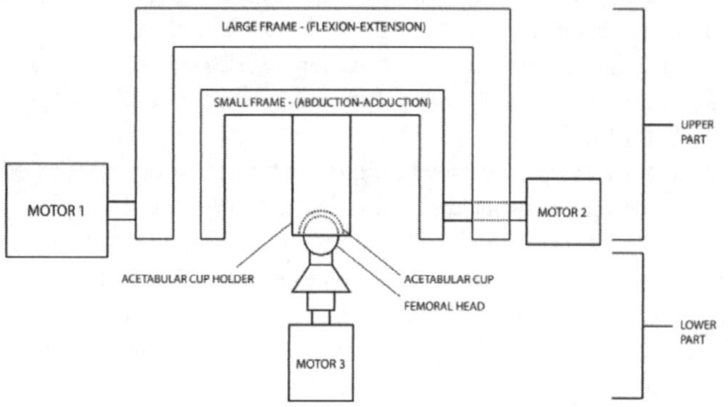

Figure 2. Principle diagram of main frame.

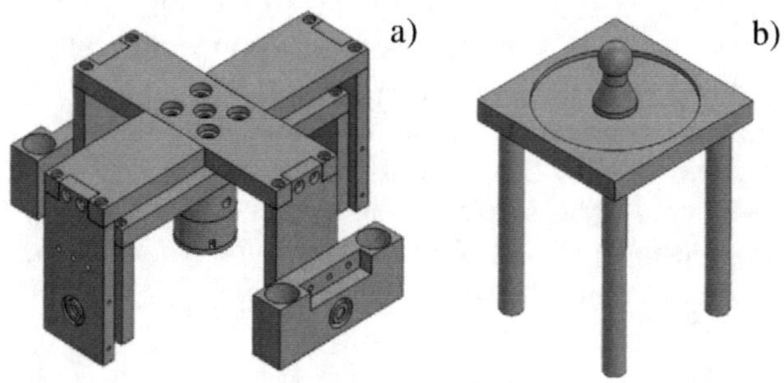

Figure 3. Isometric view a) upper part b) lower part

The SolidWorks software is used to execute the finite element analysis. Aluminium alloys 7050-T73510 is the chosen materials for this analysis and future manufacturing. The details specification of the selected materials is represented in table 1. The stress limit of material can resist without being deformed is indicated by the value of yield strength in table 1.

The analysis start by isolating the main frame model design from the whole model of hip simulator. Next, the materials chosen is applied to all part in the main frame. The fixed geometry is chosen and the load applied in the area shown by purple arrow in figure 4. The value of load is set based on the previous report in [7]. 2000 N load is applied at each side of the holder pushing downward in figure 4 a) while 4000 N load applied to femoral head in figure 4 b). The stress formation and factor of safety (FOS) value is obtained from the analysis. The main frame is tested with variety of state which imitate human hip joint motion such as static state, flexion, extension, abduction, and adduction. The parameter for the motion is taken from graph in [4].

3. Result analysis

The stress formation is defined by looking at the colour formed from the analysis. Red represent the highest deformation followed by yellow, green and the lowest deformation which is blue. Besides, factory of safety result is determined by the number shown in the upper left of the result. If the value of safety factor exceeds 2, the part is found to inherit low deformation scale.

From the simulation, the result for stress formation and FOS were successfully obtained. Based on the stress formation analysis, all the parts in the main frame shows that the frame can withstand high pressure load. Furthermore, the factor of safety value shows value more than 2 for all states.

Table 1. Specification of aluminium alloy 7050-T73510.

	Value	Units
Elastic Modulus	7.2e+010	N/m^2
Poisson's Ratio	0.33	N/A
Shear Modulus	2.69e+010	N/m^2
Density	2830	Kg/m^3
Tensile Strength	495000000	N/m^2
Yield Strength	435000000	N/m^2

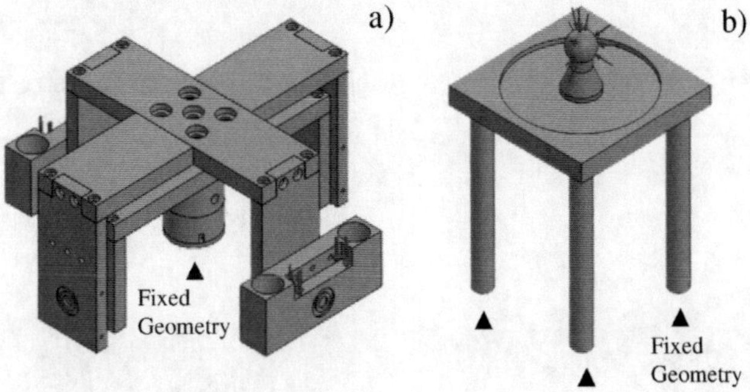

Figure 4. Load applied area a) upper part b) lower part

3.1. Stress formation

Figure 5 shows the stress formation during static state which means large and small frame are inclined 90 degrees from the base. The middle area of the large frame shows only small area at the edge of the frame with a higher stress. The value of von Mises at the red area is 116,590,080 N/m2 which is lower than the yield strength of the aluminium alloy used in the analysis. Therefore, it can be concluded that the frame can perform well during static state without being deformed.

In figure 6, the frame is executing flexion state which is the largest motion in this study. The inclination angle from the base is 65 degree resulting the formation of angle of flexion which is 25 degree. The result shows two edges in the centre of the large frame have highest stress formation. The von Mises value at the red area is 131,382,512 N/m2 which is higher than the value during static state. However, the value still lower than the yield strength of materials used in the simulation.

In figure 7, the result obtained in extension state was almost similar as in flexion state. The difference is the other two edges has higher stress but the value of von Mises not as high as in flexion state which is 127,775,424 N/m2. From both flexion and extension state, it can be seen that only two edges experienced higher force at a time. Although the von Mises value is lower than the yield strength of the material, but it need to be considerate for selection of other materials in the future.

Due to small angle executed in the main frame during abduction and adduction state, there is not much difference can be seen from these states and static state. These three states which are static state, abduction state, and adduction state show same result but only small difference in von Mises value. The abduction state performs an angle of 7 degree while adduction make a 4 degree in other direction. In abduction state, the max von Mises value is 110,270,600 N/m2 and for adduction is 110,082,592 N/m2. Although the value shown is slightly different, but it shows that the larger angle performs by the frame will result in higher stress in the centre of the large frame.

As shown in figure 10, the lower part has the least stress formation in the whole area. There is no red area which mean none of the component will destruct when high load is applied. The highest von Mises value in the analysis is 5,743,119 N/m2 which is the lowest among all main frame result.

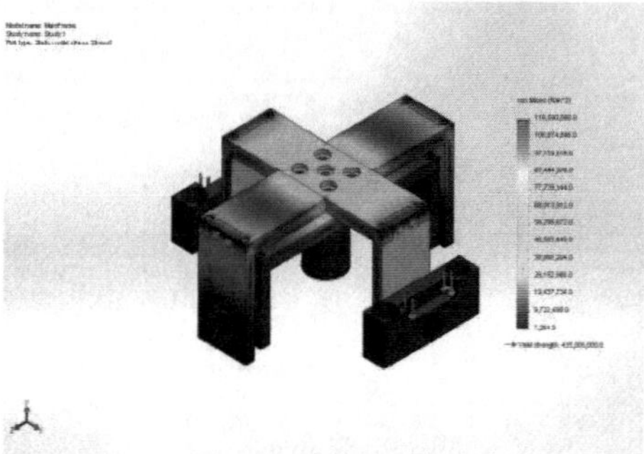

Figure 5. Static state stress formation

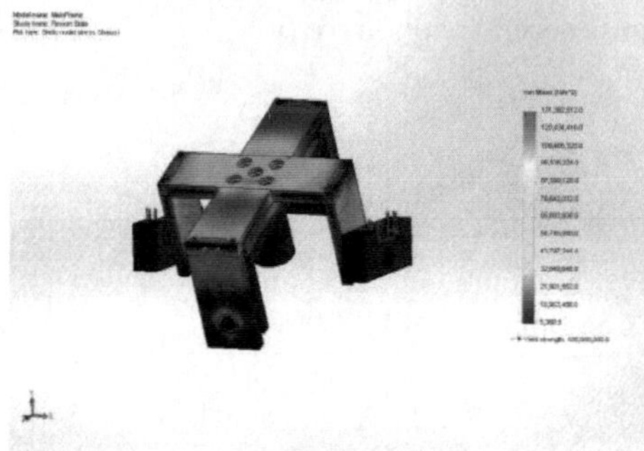

Figure 6. Flexion state stress formation

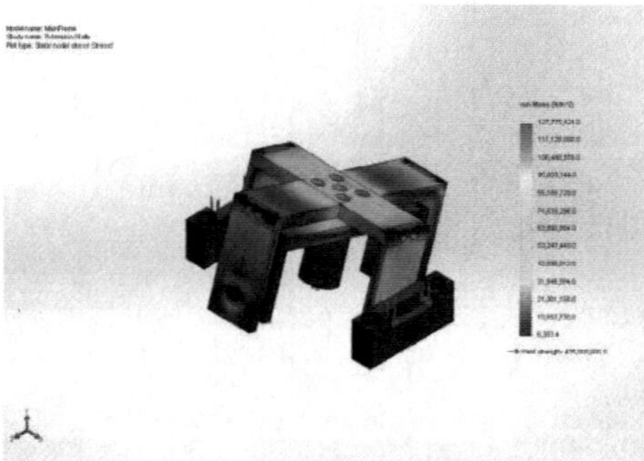

Figure 7. Extension state stress formation

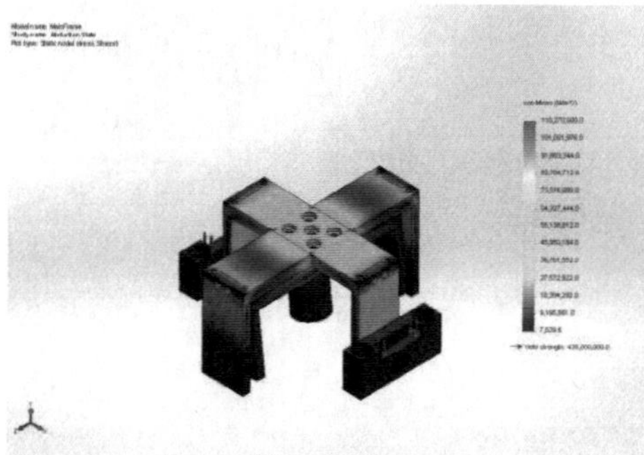

Figure 8. Abduction state stress formation

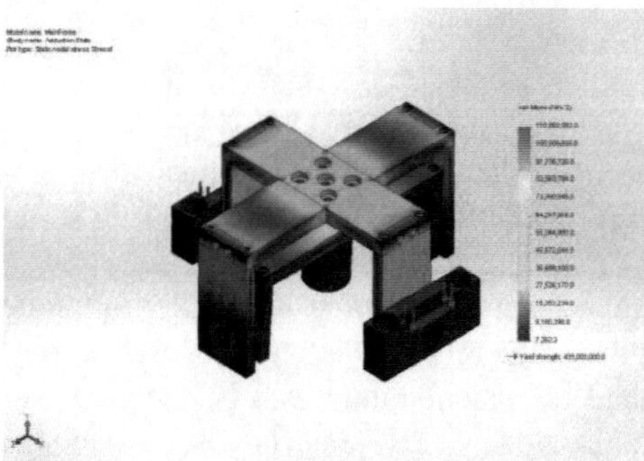

Figure 9. Adduction state stress formation

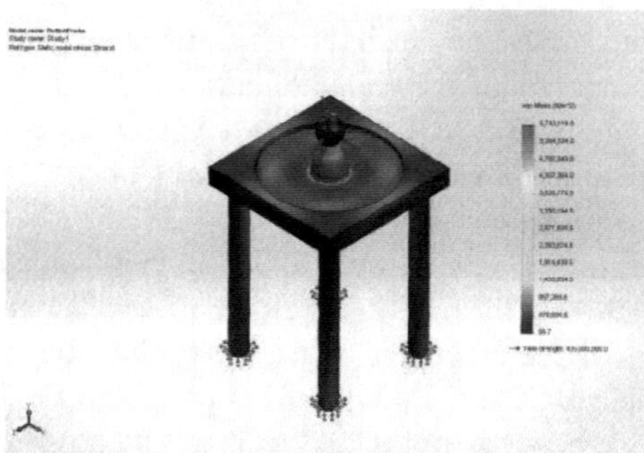

Figure 10. Lower part static state stress formation

3.2. Factor of Safety

From the analysis, safety factor value for all the state are above 2. As shown in figure 11, the FOS value is 2.9 during static state when the load applied. The colour appears in the frame is not important in this simulation because the limit is defined manually from 1 to 100. Reducing the limit will also change the colour appear on the frame analysis.

During the large inclination angle in flexion state, the FOS value is 2.7 which is lower than in static state as illustrated in Figure 12. Similarly, in the stress formation, larger inclination angle will result in high deformation scale thus lowering the FOS value. However, for the extension state, the FOS value was higher than flexion state. The value obtained from extension state in figure 13 is 2.5 which is the lowest FOS value in this analysis. Although the value is the lowest but it still exceeds requirement that needed in this analysis.

For the abduction and adduction state, the FOS value has the same value which is 3.3 as shown in figure 14 and figure 15. Due to slightly different in inclination angel, the result may vary depends on the safety factor distribution.

Furthermore, Figure 16 represents the FOS value for the lower part of the frame. This part is the most stable part based on the simulation analysis. The FOS value give a result of 69 which indicate the part is strong enough to resist any deformation. It is safe to say that lower part is more than capable of receiving load more than 4000 N.

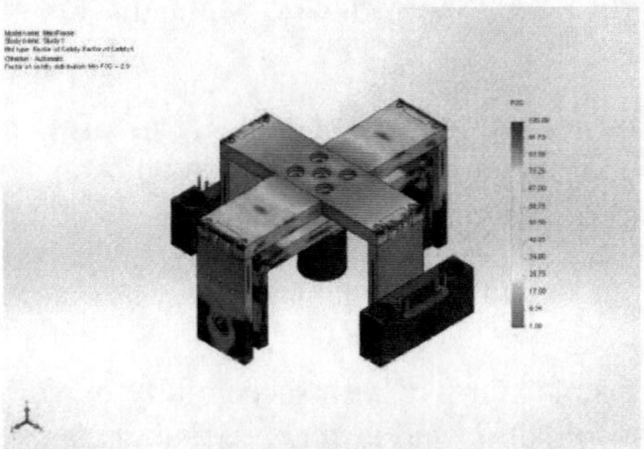

Figure 11. Safety factor value for static state

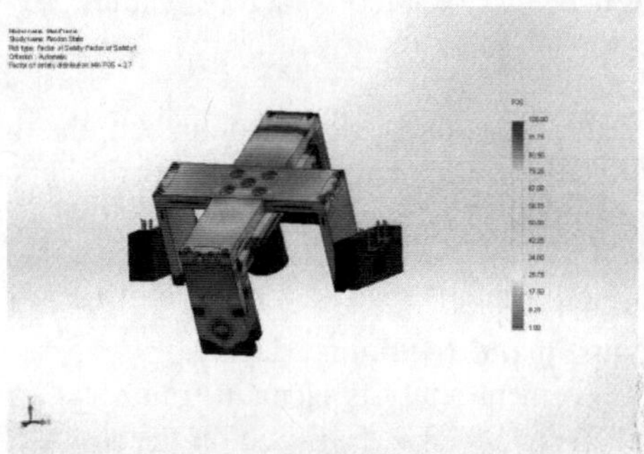

Figure 12. Safety factor value for flexion state

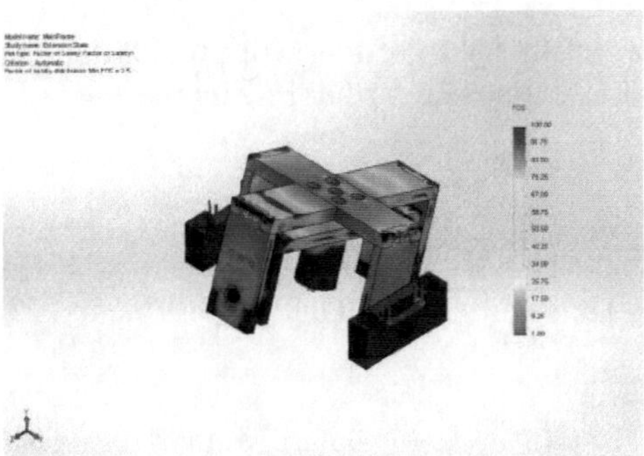

Figure 13. Safety factor value for extension state

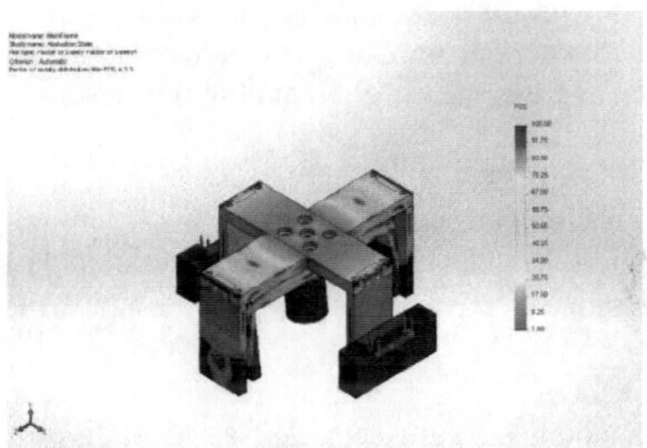

Figure 14. Safety factor value for abduction state

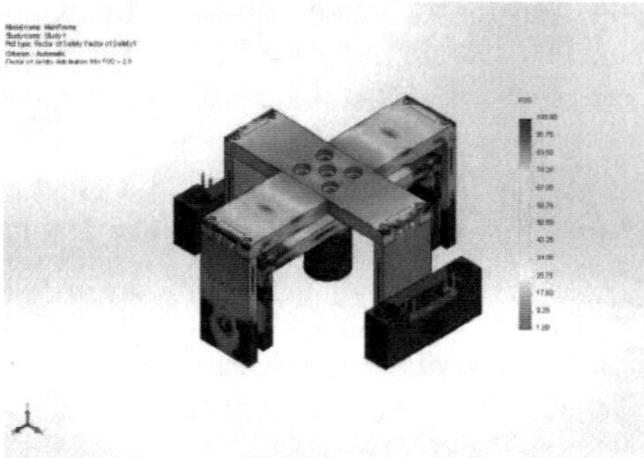

Figure 15. Safety factor value for adduction state

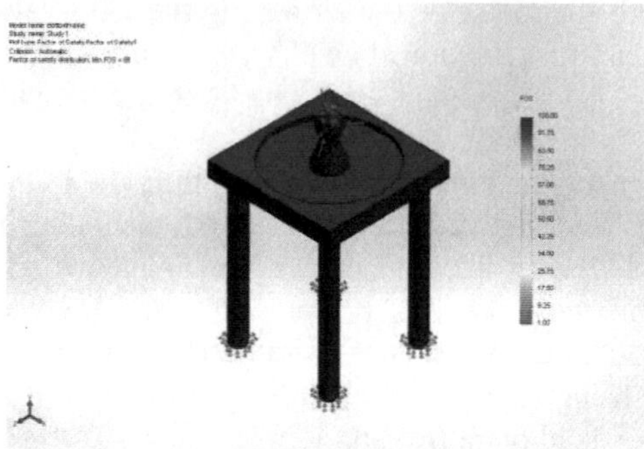

Figure 16. Safety factor value for lower part

The hip simulator is designed to test the hip prosthesis or hip implant in term of wear and dislocation. From the analysis result, it can be shown that the simulator capable to withstand higher load up to 4000 N which is greater than the peak load of previous simulator. Further study can be executed for testing heavier load and enhance the stability of the structure.

4. Conclusion and recommendation

The finite element analysis of main frame 2-station hip simulator has been successfully implemented using SolidWorks software. Based on the result of stress formation and factor of safety value, it can be concluded that selected material which is aluminium alloys 7050-T73510 is a suitable material for the design. Besides, no further major modification need to be done. The result shows with the load applied up to 4000 N, deformation scale of the parts is low and the von Mises value does not exceed yield strength of the selected material. Factor of safety showing value results more than 2 for all the parts which indicate that the parts will not permanently deformed. These results will provide information and used as reference for determining the structure of hip simulator in further research.

5. Acknowledgments

The authors would like to thank the Universiti Teknologi Malaysia (UTM) and Ministry of Education (MoE) Malaysia for financial support through Research University Grant (GUP) with Vote No.13H23.

6. References

[1] Kiguchi K, Yamashita A, Sasaki M, Ueno M, Kobayashi T, Mawatari M and Hotokebuchi T 2008 *Proc. of Int. Conf. on Control, Automation and Systems (Seoul)* (South Korea: IEEE) p 1942–1945

[2] Bragdon C R, Jasty M, Muratoglu O K, O'Connor D O and Harris W H 2003 *Journal of Arthroplasty* **18** 553–561

[3] Trommer R M and Maru M M 2016 *Revista Brasileira de Ortopedia (English Edition)* **52** 251-259

[4] Galanis N I and Manolakos D E 2011 *Proc. of the World Congress on Engineering (London)* vol 3, ed S. I. Ao and Len Gelman *et al* (U. K: Newswood Limited) p 2–7

[5] Affatato S, Leardini W and Zavalloni M 2006 *11th BIOLOX Symposium Proc.* (Rome) vol 11 (Darmstadt: Steinkopff) p 171–180

[6] Saikko V and Calonius O 2002 *Journal of Biomechanics* **35** 455–464

[7] Calonius O and Saikko V 2002 *Journal of Biomechanics* **35** 1439–1450

Analysis of a front suspension system for UniART FSAE car using FEA

M N A Zaidie[1], M S M Hashim[1,2], M Tasyrif[1], M H Basha[1], I Ibrahim[1,2], N S Kamaruddin[1,2], Shahriman A B[1,2]

[1]School of Mechatronic Engineering, Universiti Malaysia Perlis, 02600 Arau, Perlis, Malaysia.
[2]Motorsport Technology Research Unit (MoTech), Universiti Malaysia Perlis, 02600 Arau, Perlis, Malaysia.

sanihashim@unimap.edu.my

Abstract. In recent years, many research works from institutions that participated in Formula SAE had highlighted on suspension systems. The aim is to improve the system in term of performance and robustness. However, every suspension system for a racing car is tailored to the car itself. Thus, this paper proposes a new design for front suspension system for UniART FSAE car. The new design was than being compared to the previous suspension system for enhancement. The analysis covered in this paper based on several conditions such as braking, cornering and bumping condition and was carried out using finite element analysis. Each main component for the suspension system such as lower arm, upper arm and knuckle has been analysed in term of strength and performance. From the results, the proposed new design of the suspension system has improved in term of strength and performance compared to the previous suspension system.

1. Introduction

Suspension system typically plays a crucial and importance role in vehicle comfort and stability [1]. Generally, a suspension system arrangement of a vehicle connection the unsprung mass includes wheels, brakes, knuckle and large portion of the heaviness of control arm and dampers, with the sprung mass including the total body [2]. The suspension has primarily two function which is give the transfer limitation of road irregularities to the sprung mass and also maintain the tire always contact with the road at all times [2]. In other words, it give protection for passenger from shock and vibration producing by tires and road interaction as well as give smooth handling and stability [3].

Looking into the automotive industries, generally suspension system needs to comply with safety demand as important priority to experience under or over steer, bounce and gripping when the vehicle faces under certain condition [4]. Furthermore, it must resist roll of chassis and also to keep the wheels take after any uneven road condition by separating the chassis from the unpleasantness of the road [5].

In racing competition, it is crucial for the suspension system to be able to withstand severe conditions. For Universiti Malaysia Perlis Automotive Racing Team (UniART) racing car, the suspension system has undergone failure during the racing competition. Thus, a new front suspension system will be designed based on several conditions and all possible engineering parameters that have been indicated in FSAE regulations for UniART racing car. The aim for a new design is to enhance the strength and performance through the entire system by referring to the previous design as reference. The analysis is conducted based on the related conditions such as cornering, braking and bumping conditions. Besides that, the worst condition will be taken into account for the analysis. Each individual component of a suspension system will be analysed for strength and performance analysis for the fully system.

2. Methodology

2.1. Suspension geometry and rules

The suspension geometry section concentrates on some of the basic guideline which have been highlighted by the SAE organization. Therefore, FSAE suspension system should be focus on the constraints of the competition. For example, the wheelbase parameters must have dimension at least 1525mm or 60 inches whereas for wheel track must be no less than 75% of the larger track [6]. These two parameters basically influence the performance impact of the racing car itself. Figure 1 shows the previous design and the new design for the suspension system. The new suspension system has redesigned to also reduce the weight of the suspension system.

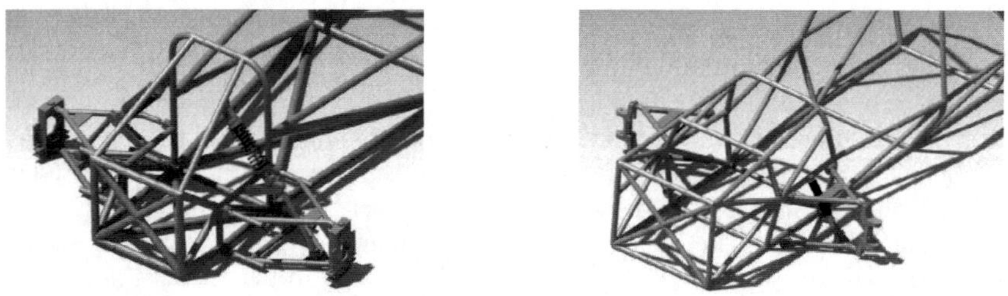

(a) (b)

Figure 1: Suspension system for UniART FSAE car. (a) Previous design and (b) new design

2.2. Analysis in ANSYS Workbench

All the suspension components were designed using CATIA. ANSYS Workbench was utilized for further analysis of strength and performance of the suspension system. Material assigned for this for the suspension components is mild steel. Three main conditions are set for the analysis which are braking, cornering and bumping condition. Table 1 tabulates the parameters that are important for the analysis purposes. First condition consideration is due to braking condition which the g-force applied is 1.5g at the horizontal position (x-direction). Secondly, the cornering condition which g-force applied at 2g and lastly for the bumping condition that the g-force applied at 2.5g.

Table 1: List of parameter for analysis

Parameters	Values
Weight (kg) *	340kg
Length car (mm)	1600mm
g-cornering	2.0
g-braking	1.5
g-bumping	2.5

3. Result and Discussion

Static stress analysis has been carried out by using ANSYS for the previous and new design of the suspension system. The analysis focuses on major components of suspension system, which are A-arm

IOP Conf. Series: Journal of Physics: Conf. Series **908** (2017) 012058 doi:10.1088/1742-6596/908/1/012058

and knuckle, and also the full suspension system for a better evaluation between the previous and new design.

3.1 Lower arm analysis

The results of equivalent stress for lower arm are shown in figure 2. For each of the condition, the maximum equivalent stresses are compared as shown in figure 3. From the results, the maximum equivalent stress for the new design is lower than the previous design and the value is not exceeding the yield stress of the material, which is better compared to the previous design. The percentage of reduction for braking, cornering and bumping condition are 28.1%, 38.1% and 66.3% respectively.

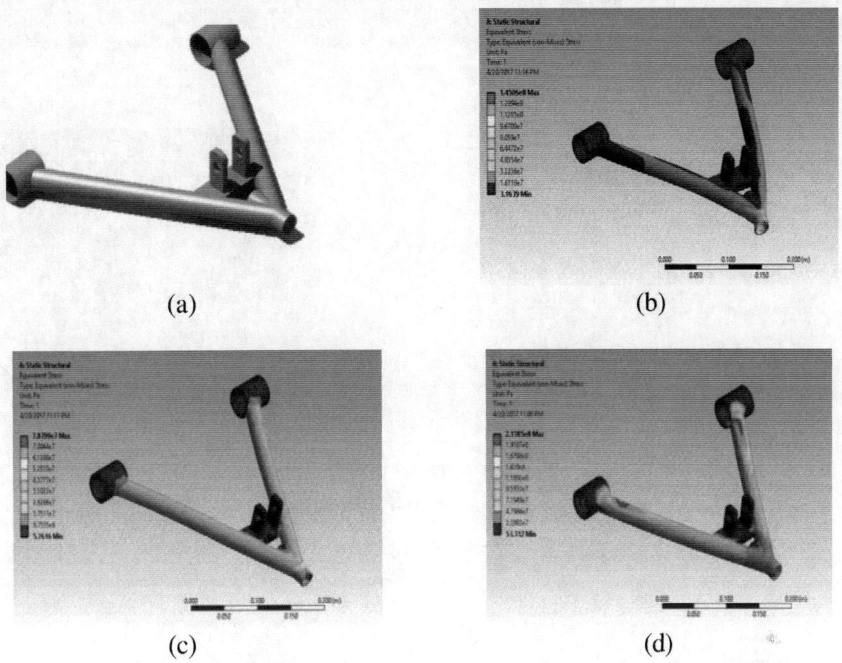

Figure 2: Result for lower arm under various conditions (a) original, (b) braking, (c) cornering and (d) bumping.

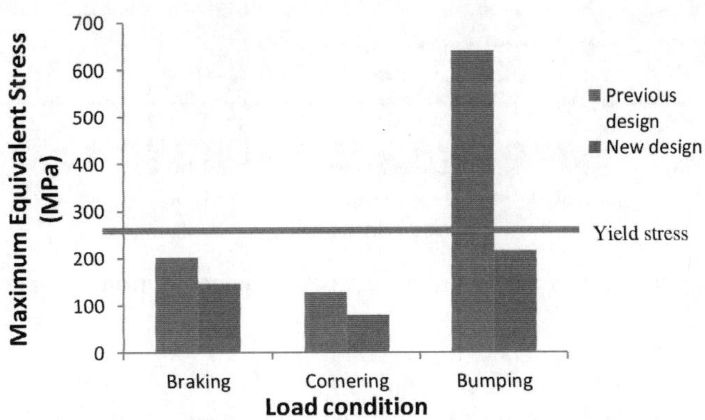

Figure 3: Comparison of equivalent stress for each condition for lower arm

3.2 Upper arm analysis

The results of equivalent stress for lower arm are shown in figure 4. For each of the condition, the maximum equivalent stresses are compared as shown in figure 5. From the results, the maximum equivalent stress for the new design is lower than the previous design and the value is not exceeding the yield stress of the material, which is better compared to the previous design. The percentage of reduction for braking, cornering and bumping condition are 43.4%, 68.2% and 71.9% respectively.

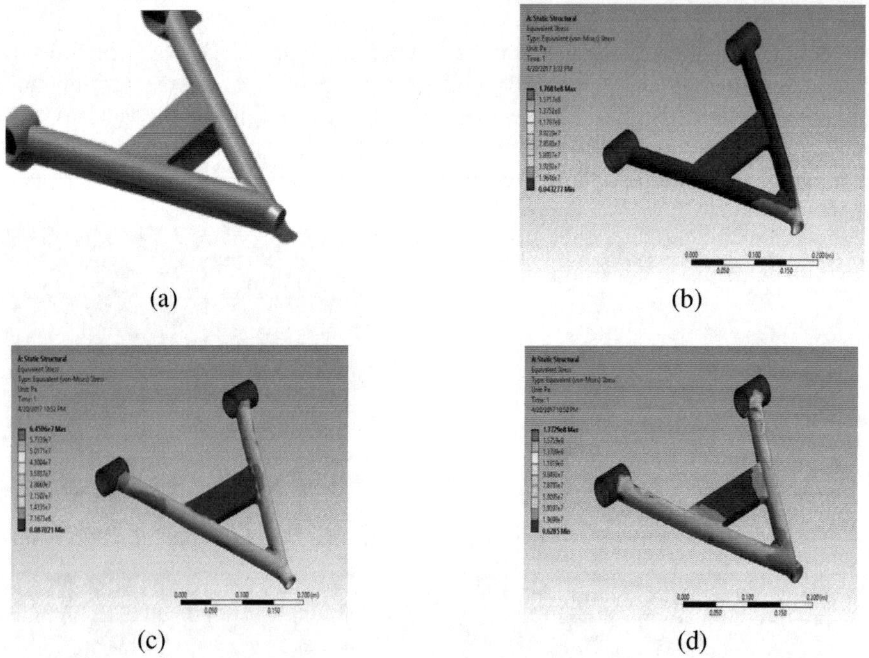

Figure 4: Result for upper arm under various conditions (a) original, (b) braking, (c) cornering and (d) bumping.

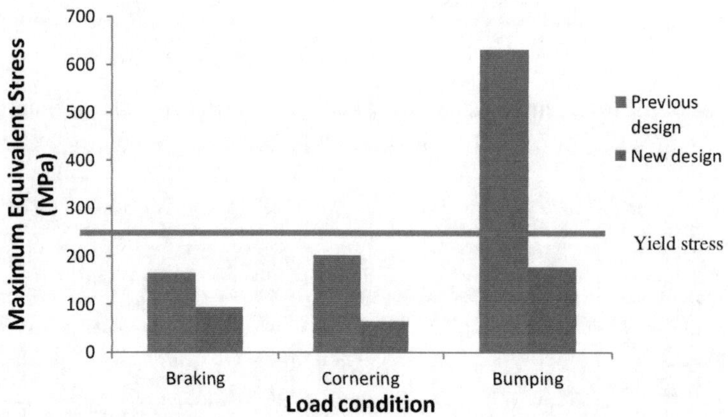

Figure 5: Comparison of equivalent stress for each condition for upper arm

3.3 Knuckle analysis

A good design for the knuckle is when it can withstand under critical condition, which is a combined loading. Figure 6 shows the design comparison between the previous and new design of the knuckle. From the results in figure 7, the maximum equivalent stress for the previous knuckle exceeding the yield stress of material. While the value for the new design is below the yield stress, which is good. The percentage of reduction between the previous and new design is about 30%.

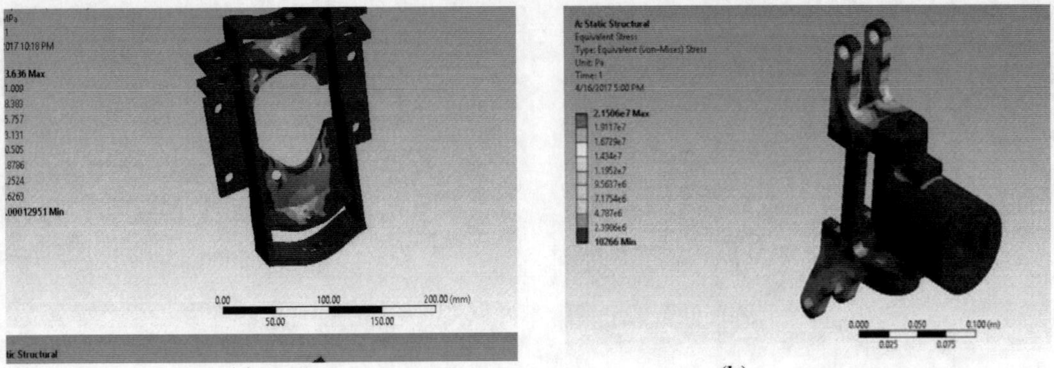

(a) (b)

Figure 6: Result for knuckle under critical condition (a) previous design and (b) new design

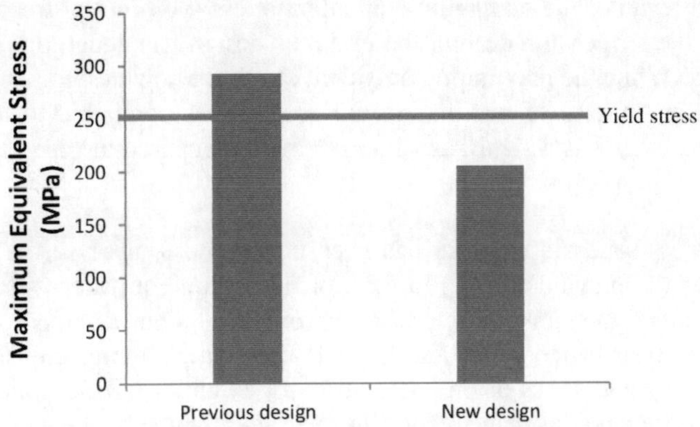

Figure 7: Comparison of equivalent stress for the previous and new design under critical condition

3.4 Full suspension system

In order to obtain an overall performance and strength of the suspension system, each component was assembled and the analysis on the full suspension system was carried out. A force of 1667.7 N was applied to the system, which equivalent to the bumping condition. Figure 8 shows the previous and new full suspension system. The pushrod in the previous design was removed and the absorber is now directly attached between lower arm and chassis.

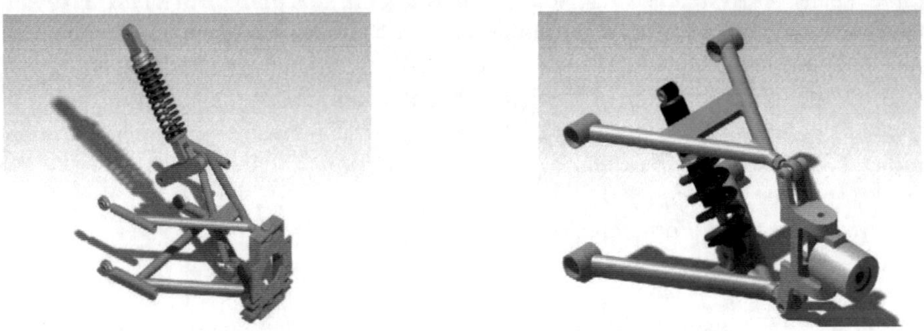

(a) (b)

Figure 8: Front suspension system. (a) Previous design and (b) new design

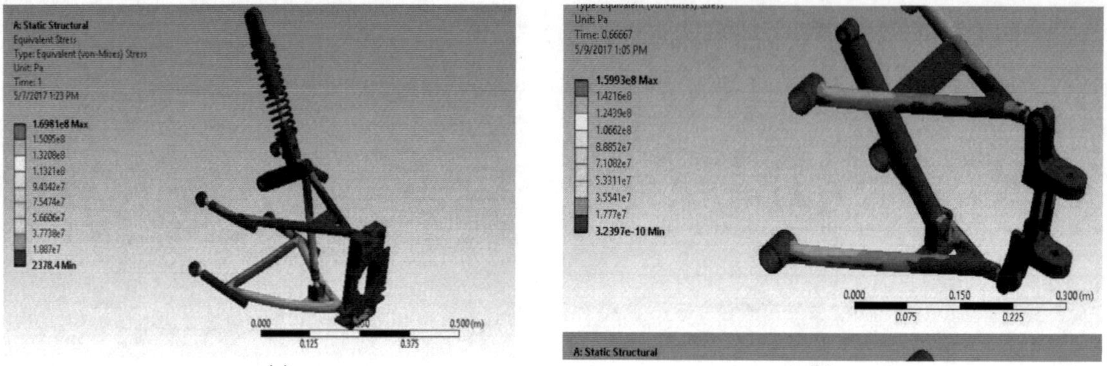

(a) (b)

Figure 9: Result for suspension system (a) previous design and (b) new design

Figure 9 shows the result of the maximum equivalent stress (von-mises stress) for both designs. When a force applied on the suspension design, the previous design has undergone the maximum equivalent stress of 169.8MPa. While the maximum equivalent stress for new design is about 160MPa. This gives about 6.2% reduction in term of maximum equivalent stress compared to the previous design. This shows that the new design has a significant improvement compared the previous.

4. Conclusion

The analysis of front suspension system has been conducted to analyse the performance of the new design compared to the previous design in term of maximum equivalent stress. From the results, the overall performance of the new design has improvement significantly compared to the previous design. From the analysis of the a-arm and knuckle component, the results show that the components are not exceeding the yield stress of the material in all conditions. This indicates the components can withstand any condition and the chances for failure are minimal.

5. References

[1] Eskandary P K 2014 Interconnected Air Suspensions with Independent Height and Stiffness Tuning

[2] Johansson A 2011 Development Of A Hi-Per Strut Front Suspension Chalmers University Of Technology.

[3] Pathmasharma S, Suresh J K, Viswanathan P and Subramanian R 2013 Analysis of Passenger Car Suspension System Using Adams *International Journal of Science, Engineering and Technology Research (IJSETR)*, **Volume 2,** Issue 5, pp. 1186–93

[4] Hassan M Z, Kamarul M, Aziz H A, Delbressine F, and Rauterberg M 2016 Numerical Analysis of Spring Stiffness in Vehicle Design Development Stage *International Journal of Applied Engineering Research* **Vol. 11**, No. 7, pp. 973–4562

[5] Gillespie T D 1992 *Fundamentals of Vehicle Dynamics Analysis Vol. 400*. Society of Automotive Engineer pp. 519

[6] SAE International *2015 Formula SAE ® Rules* pp. 178

Assembly line balancing with resource constraints using new rank-based crossovers

N H Kamarudin, M F F Ab. Rashid

Faculty of Mechanical Engineering, Universiti Malaysia Pahang, 26600 Pekan, Malaysia.

ffaisae@ump.edu.my

Abstract. Assembly line balancing (ALB) is about distributing the assembly tasks into workstations with the almost equal workload. Recently, researchers started to consider the resource constraints in ALB such as machine and worker, to make the assembly layout more efficient. This paper presents an ALB with resource constraints (ALB-RC) to minimize the workstation, machine and worker. For the optimization purpose, genetic algorithm (GA) with two new crossovers is introduced. The crossovers are developed using ranking approach and known as rank-based crossover type I and type II (RBC-I and RBC-II). These crossovers are tested against popular combinatorial crossovers using 17 benchmark problems. The computational experiment results indicated that the RBC-II has better overall performance because of the balance between divergence and guidance in the reproduction process. In future, the RBC-I and RBC-II will be tested for different variant of ALB problems.

1. Introduction

Assembly line balancing (ALB) plays a vital function in a production system. The installation of an assembly line is a long-term decision and requires large capital investments. It is important that such a system is designed and balanced so that it works as efficiently as possible [1]. The simplest version of ALB problem is known as simple assembly line balancing problem (SALBP) which also known as One-sided ALB [2]. SALBP deals with a serial assembly line which processes a unique model of a single product. In previous research, a lot of attention has been given to this type of problem.

However, in the majority of the previous works, researchers make assumptions where any of assembly tasks can be processed or assembled in any workstations. This is certainly true for the product which only requires a common or simple tool to be assembled. However, when the complexity of a product increased, it requires a special tool, machine or highly skilled labor to assemble that particular component. Therefore, the limitation of resources will be another constraint for the industry. In fact, the issue of line balancing with the minimum number of resources has always been a serious problem in the industry [3]. This problem is known as assembly line balancing with resource constraints (ALB-RC)

Previously, researchers had studied the line balancing with resource constraints. [4] started the ALB-RC by considering two resources and solve the problem using integer programming. Next, [5] proposed a model to support generalized constraints problem. [6] later on model and optimize the ALB with worker skill constraint. The purpose is to match the assembly task with the level of the worker skill. Besides that, [7] optimize the multi-objective ALB with general resources using domination concept.

IOP Conf. Series: Journal of Physics: Conf. Series **908** (2017) 012059 doi:10.1088/1742-6596/908/1/012059

Researchers also implemented different algorithms to optimize the ALB-RC problem. [8] combined priority rule-based method (PRBM) and genetic algorithm (GA) to optimize this problem. The PRBM is used to generate initial chromosomes for GA. Meanwhile, [9] implement the hybrid multi-objective genetic algorithm (MOGA) to optimize the problem to obtain Pareto front. In addition, [10] implemented elitist non-dominated sorting GA (NSGA-II) to optimize this problem.
This paper extends the existing ALB-RC by considering the worker selection, besides the workstation number and tool resource. In this problem, the engineer has a different option for workers with different ability to conduct assembly task. The problem is later optimized by GA with new crossover operators.

2. ALB-RC Problem Modelling

The ALB problem is represented using a precedence graph. The number inside the node represents the assembly task. The directed edge means the precedence between task *i* and *j*. In ALB, the assembly tasks need to be assigned into workstations, so that the workstation time is almost equal. To presents the ALB-RC, the following example is used. Figure 1 shows a precedence graph that represents an assembly process. Each of nodes represents the assembly task while the arrows represent the precedence.

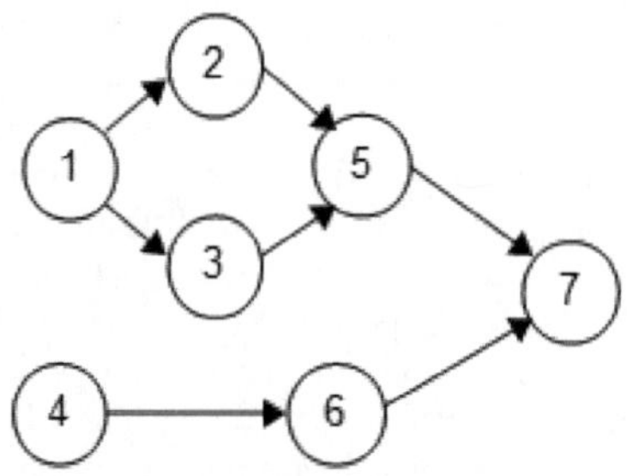

Figure 1. Example of precedence graph

Table 1 meanwhile shows the assembly information which includes the task time, tool and also worker. The worker columns with tick mark meaning that the worker is able to conduct a specific assembly task. To assemble an assembly task, only one worker is required.

Table 1. Assembly information for Figure 1

Task	Time	Tool	Workers						
			1	2	3	4	5	6	7
1	18	A	/		/	/		/	/
2	22	B		/		/		/	
3	9	B	/	/	/		/		/
4	7	A			/		/	/	/
5	12	A	/			/		/	
6	6	B		/	/		/		
7	20	A	/	/		/			/

For clarity of the ALB-RC evaluation, let consider a feasible assembly sequence f_1 = [1 4 3 2 6 5 7]. For this example, the maximum cycle time, ct_{max} is 34 time unit. It means that the workstation time cannot exceed the ct_{max} or otherwise, the demand for the product cannot be fulfilled. In table 2, the Worker row shows the entire workers that capable to conduct a specific assembly task.

IOP Conf. Series: Journal of Physics: Conf. Series **908** (2017) 012059 doi:10.1088/1742-6596/908/1/012059

Table 2. Example of a feasible assembly sequence

Sequence	1	4	3	2	6	5	7
Time	18	7	9	22	6	12	20
Tool	A	A	B	B	B	A	A
Worker	1,3,4,6,7	3,5,6,7	1,2,3,5,7	2,4,6	2,3,5	1,4,6	1,2,4,7

The assembly tasks were assigned into the workstation as shown in table 3. The station time row shows cumulative time to conduct assembly process for all tasks in a specific station. The Tool row shows the required tool to conduct assembly process in a specific station. Meanwhile, the worker selection is made based on the number of workers frequency in a workstation. For example, in workstation 1, workers 3 and 7 have the highest frequency. In this case, the worker is select randomly.

Table 3. Assembly task and workstation assignment

Workstation	1	2	3
Task	1, 4, 3	2, 6	5, 7
Station time	34	28	32
Tool	A,B	B	A
Worker	3	2	1

Based on the presented approach, the objective function can be measured as follow:

Number of workstation = 3
Number of tool = 4
Number of worker = 3

3. Genetic Algorithm

Genetic algorithm is an optimization technique that mimics the survival for the fitness concept. Solutions with better fitness have larger possibilities to remain in the population, while the solution with bad fitness will be eliminated from the population [11]. In general, GA consists of five main steps; Initialization, Evaluation, Selection, Crossover and Mutation. The algorithm is coded using permutation number to represent the assembly tasks. However, due to the randomness of permutation, the generated number may violate the precedence relation in assembly. Therefore, topological sort based on the earliest task appearance is implemented to decode the solution.

The purpose of selection step is to choose the chromosome to be placed in the mating pool. The selected chromosome will be the parent of the children in a new generation. The selection process is conducted using Roulette wheel selection (RWS) mechanism. Meanwhile, for the crossover, we introduced two crossover operators, named Rank based crossover type I and II (RBC-I and RBC-II). The proposed crossovers are compared with popular crossover operator for permutation problem, i.e. ordered crossover (OX), partially matched crossover (PMX) and Moon crossover [12].

3.1. Proposed Rank Based Crossover

Both of the proposed crossovers taken into account the best chromosome from the population in the reproduction process. In both of the proposed crossover, each of the assembly tasks will be given a rank according to their position in the chromosome. Then the rank for parent and best chromosome will be summed up to form a new rank. The child solution will be generated based on the new rank. By using this approach, the new child will inherit the gene from their parent and also the best solution.

3.1.1. Rank based crossover type-I (RBC-I)

In RBC-I, the parent rank (R_1) and the best solution rank (R_{best}) will be added (Figure 2). Next, the rank is sorted according to the parent (P_1) and the best solution (X_{best}). The sorted parent rank (R'_1) and the best rank (R'_{best}) is added to form sorted offspring rank (R'_{O1}). Finally, the P'_1 is sorted according to the R'$_{O1}$ to generate offspring solution, O_1. In the case where the rank is tied, the selection is made randomly. The numerical procedure for RBC-I is presented in Figure 2.

1. Assign rank to the P_1 and X_{best}

P_1	4	2	5	3	1
R_1	1	2	3	4	5

X_{best}	2	3	1	4	5
R_{best}	1	2	3	4	5

2. Sort rank according to P_1 and X_{best}

P'_1	1	2	3	4	5
R'_1	5	2	4	1	3

X'_{best}	1	2	3	4	5
R'_{best}	3	1	2	4	5

3. Sum up R'_1 and R'_{best} as R'_{O1}

P_1	1	2	3	4	5
R'_1	5	2	4	1	3
R'_{best}	3	1	2	4	5
R'_{O1}	8	3	6	5	8

4. Sort P_1 according to the sum rank to generate O_1

O_1	2	4	3	1	5
R_{O1}	3	5	6	8	8

Figure 2. Numerical procedure for RBC-I

3.1.2. Rank based crossover type-II (RBC-II)

The RBC-II applied the same rank concept as in RBC-I, but this crossover considers two parents. The early steps where the rank is assigned and sorted is the same with RBC-I as shown in figure 3. To calculate the rank for offspring solutions (R'$_O$), the following formula is used in RBC-II.

$$R'_O = C_{best}\,(R'_{best}) + C_1(R'_1) + C_2(R'_2) \tag{1}$$

C_{best}, C_1 and C_2 are the coefficients for the X_{best}, P_1 and P_2 respectively. The C_{best} is fixed at 0.2. Meanwhile, the C_1 and C_2 coefficient is depend on the offspring. To generate offspring 1 (O_1), the C_1 and C_2 are as follow.

$$C_1 = 0.7(1 - C_{best}) \tag{2}$$

$$C_2 = 0.3(1 - C_{best}) \tag{3}$$

On the other hand, to generate offspring 2 (O_2), the following coefficients are used.

$$C_1 = 0.3(1 - C_{best}) \tag{4}$$

$$C_2 = 0.7(1 - C_{best}) \tag{5}$$

The offspring solutions (O_1 and O_2) are generated by sorting the R'$_O$ in the ascending orders. As in RBC-I, in the event of tie rank, the selection is made randomly. The numerical example for RBC-II is shown in figure 3.

1. Assign rank to the P_1, P_2 and X_{best}

P_1	4	2	5	3	1
R_1	1	2	3	4	5

P_2	1	5	2	4	3
R_2	1	2	3	4	5

X_{best}	2	3	1	4	5
R_{best}	1	2	3	4	5

2. Sort rank according to the P_1, P_2 and X_{best}

P'_1	1	2	3	4	5
R'_1	5	2	4	1	3

P'_2	1	2	3	4	5
R'_2	1	3	5	4	2

X'_{best}	1	2	3	4	5
R'_{best}	3	1	2	4	5

3. Calculate rank for offspring
$R'_O = C_{best}(R'_{best}) + C_1(R'_1) + C_2(R'_2)$

O'_1	1	2	3	4	5
R'_{O1}	3.64	2.04	3.84	2.32	3.16

O'_2	1	2	3	4	5
R'_{O2}	2.36	2.36	4.16	3.28	2.48

4. Sort O' according to R_O to generate offspring

O_1	2	4	5	1	3
R_{O1}	2.04	2.32	3.16	3.64	3.84

O_2	1	2	5	4	3
R_{O2}	2.36	2.36	2.48	3.28	4.16

Figure 3. Numerical example of RBC-II

4. Computational Experiment

A computational experiment has been conducted to measure the performance of RBC-I and RBC-II. For this purpose, a set of ALB benchmark problem by Scholl is used [13]. The benchmark set consist of 17 problems that varies in term of the size. The benchmark test problem is divided into three categories; small ($n \leq 20$ task), medium ($20 < n \leq 70$) and large ($n > 70$). For comparison purpose, the RBC-I and RBC-II are compared with popular crossover operators for the combinatorial problem. The comparison crossovers are the ordered crossover (OX), partially matched crossover (PMX) and Moon crossover. The OX and PMX are among popular crossover operator for the combinatorial problem. Meanwhile, the Moon crossover is used since it was claimed to be the best crossover for the combinatorial problem [12]. For the computational experiment, the population size is set to 30, maximum generation is 300, probability of crossover is 0.7 and probability of mutation is 0.2. The optimization is run for ten times to reduce the pseudorandom effect. Table 4 presents the best fitness obtained by GA using different crossover strategies.

Based on the results in table 4, all the crossovers were able to search for an optimum solution for the small size problems. However, when the problem size is increased, the RBC-I and RBC-II has better performance compared with other crossovers, except in Hahn problem. In medium size problem, the RBC-II has better fitness in 83% of the problem. Meanwhile, in large size problem, the RBC-I and II individually has better fitness in 50% of the problem.

To have better view from the computational experiment result, a standard competition ranking approach is used. The crossover with the best fitness will be given rank 1, followed by the next as rank 2 etc. If the crossover performance is equivalent, the following rank is ignored. The summary of the standard competition ranking is presented in table 5.

Table 4. Optimization results

No	Problem	No. of Task	Given Cycle Time	Crossover type				
				OX	PMX	Moon	RBC-I	RBC-II
1	Mertens	7	8	**8.7500**	**8.7500**	**8.7500**	**8.7500**	**8.7500**
2	Bowman	8	20	**8.5000**	**8.5000**	**8.5000**	**8.5000**	**8.5000**
3	Jaeschke	9	18	**3.5000**	**3.5000**	**3.5000**	**3.5000**	**3.5000**
4	Mansoor	11	48	**2.9286**	**2.9286**	**2.9286**	**2.9286**	**2.9286**
5	Jackson	11	13	**2.2857**	**2.2857**	**2.2857**	**2.2857**	**2.2857**
6	Buxey	29	54	4.0457	4.0576	4.0517	3.7789	**3.5181**
7	Sawyer	30	75	1.6800	1.6800	1.6800	1.8000	**1.4400**
8	Gunther	35	69	2.7952	2.8952	2.7810	2.6738	**2.5881**
9	Kilbridge	45	69	3.8831	3.9642	3.9599	3.8789	**3.7999**
10	Hahn	53	2004	5.6467	**5.5190**	5.5815	5.6495	5.6440
11	Warnecke	58	111	3.4024	3.5080	3.8168	3.5628	**3.1858**
12	Wee Mag	75	56	4.1857	4.1303	4.0446	4.0053	**3.8964**
13	Arc83	83	6540	2.4743	2.5198	2.5879	**2.4320**	2.5132
14	Lutz 2	89	19	3.0760	2.9237	3.0062	**2.5626**	2.9492
15	Mukherje	94	263	3.2866	3.3370	3.2747	3.4274	**3.0903**
16	Arc111	111	6540	6.7874	6.6003	6.5334	6.4993	**5.5394**
17	Barthol2	148	170	3.2254	3.1278	3.1912	**2.9669**	3.1034

Table 5. Summary of standard competition ranking

Crossover	Rank 1	Rank 2	Rank 3	Rank 4	Rank 5	Average rank
OX	5	3	3	2	4	2.8235
PMX	6	2	2	4	3	2.7647
Moon	5	3	3	4	2	2.7058
RBC-I	8	5	0	1	3	2.1764
RBC-II	13	1	3	0	0	1.4117

Based on table 5, the RBC-II was most frequently ranked as 1, followed by RBC-I and PMX. Meanwhile, the OX has the most frequently ranked as 5. The average rank for each of crossover is then calculated. Based on the average rank the best crossover is RBC-II. The RBC-II is only ranked from 1 until 3. In the meantime, RBC-I is in the second best according to the average rank. For RBC-I, besides ranked as 1 and 2, this crossover was also ranked as 5 in three cases. On the other hand, the OX is the worst crossover based on the average rank.

The RBC-I and II have shown better performance because of the involvement of the best chromosome in the reproduction process. This makes the search direction is more guided compared with other crossovers. In the OX, PMX and Moon crossovers, the reproduction process solely depend on the parents. Even though the parents were selected among the best, the variation in the chromosomes makes the search direction become too diverse.

Meanwhile, in the comparison between RBC-I and II, the RBC-I is too dependent on the best solution because a single parent is mated with the best solution for the regeneration. This makes the chance for the chromosome to trap in local optima is slightly higher. In RBC-II, the regeneration process involved a pair of parents and the best solution. Two chromosomes from parents make the regeneration is not too relied on the best solution. Furthermore, the generated offspring only inherit

20% of the gene from the best solution (since $C_{best} = 0.2$). This makes the RBC-II able to generate more varied offspring, but in the guided mode.

5. Conclusions

This paper presents an assembly line balancing with resource constraints. In particular, besides balancing the assembly workload in the station, this work also consider to minimize the number of machines and workers in an assembly line. For optimization purpose, two crossover operators for genetic algorithm were introduced. The proposed crossovers were based on the assembly sequence rank, known as Rank-based crossover type I and II (RBC-I and RBC-II). In different with other crossover operators, the RBC-I and II consider the best chromosome in the regeneration process.

The computational experiments using 17 benchmark problems indicated that the RBC-II has better overall performance compared with comparison crossovers in the genetic algorithm. The RBC-II performance is because of the balance between divergence and guidance during the reproduction process in the crossover. In future, an industrial case study will be conducted to validate the problem modeling and the RBC-II performance.

6. Acknowledgments

This research is supported by Ministry of Higher Education Malaysia and Universiti Malaysia Pahang under grant number RDU140103.

7. References

[1] Becker C and Scholl A Feb 2006 A survey on problems and methods in generalized assembly line balancing *Eur. J. Oper. Res.*, **vol. 168** no. 3 pp. 694–715

[2] Ağpak K and Zolfaghari S Sep. 2014 Mathematical models for parallel two-sided assembly line balancing problems and extensions *Int. J. Prod. Res.* pp. 1–13

[3] Hamta N, Fatemi-Ghomi S M T, Jolai F, and Shirazi M A Jan.2013 A hybrid PSO algorithm for a multi-objective assembly line balancing problem with flexible operation times, sequence-dependent setup times and learning effect *Int. J. Prod. Econ.*, **vol. 141** no. 1 pp.99–111

[4] Ağpak K, Gökçen H, Ağpak K and Gökçen H Apr. 2005 Assembly line balancing: Two resource constrained cases *Int. J. Prod. Econ.*, **vol. 96** no. 1, pp. 129–140

[5] Corominas A, Ferrer L, and Pastor R Jun. 2011Assembly line balancing: General resource-constrained case *Int. J. Prod. Res.*, **vol. 49** no. 12, pp. 3527–42

[6] Koltai T and Tatay V 2013 Formulation of workforce skill constraints in assembly line balancing models *Optim. Eng.*, **vol. 14** no. 4, pp. 529–45

[7] Jusop M and Rashid M F F A 2016 Optimisation of assembly line balancing type-E with resource constraints using NSGA-II *Key Eng. Mater.* **vol. 701** pp. 195–99

[8] Quyen N T P, Chen J C, and Yang C L 2016 Hybrid genetic algorithm to solve resource constrained assembly line balancing problem in footwear manufacturing *Soft Comput.* pp. 1–17

[9] Triki H, Mellouli A, and Masmoudi F 2017 A multi-objective genetic algorithm for assembly line resource assignment and balancing problem of type 2 (ALRABP-2) *J. Intell. Manuf.*, **vol. 28** no. 2 pp. 371–385

[10] Jusop M and Rashid M 2017 Optimization of Assembly Line Balancing with Resource Constraint using NSGA-II : A Case Study *Int. J. Appl. Eng. Res.*, **vol. 12** no. 7 pp. 1421–26

[11] Delice Y, Aydoğan E K, and Özcan U 2016 Stochastic two-sided U-type assembly line balancing: a genetic algorithm approach *Int. J. Prod. Res.*, **vol. 54** no. 11 pp. 3429–51

[12] Moon C, Kim J, Choi G, and Seo Y Aug. 2002 An efficient genetic algorithm for the traveling salesman problem with precedence constraints *Eur. J. Oper. Res.*, **vol. 140** no. 3 pp. 606–617

[13] Scholl A 1993, Benchmark Data Sets by Scholl *Assembly Line Balancing Data Dets & Research Topics*,[Online]Available: http://assembly-line-balancing.mansci.de/salbp/benchmark-data-sets-1993/.

IOP Conf. Series: Journal of Physics: Conf. Series **908** (2017) 012060 doi:10.1088/1742-6596/908/1/012060

Design of biped hip simulator using SolidWorks

M R Zainudin[1], A Yahya[2], M I M Fazli[1], A Syahrom[2], F K C Harun[2], M S Nazarudin[1]

[1] Faculty of Electrical Engineering, Universiti Teknologi Malaysia, 81310 Skudai, Johor Bahru, Malaysia
[2] Faculty of Biosciences and Medical Engineering, Universiti Teknologi Malaysia, 81310 Skudai, Johor Bahru, Malaysia

mrezza4@live.utm.my

Abstract. The increasing number of people who underwent both hip implant surgery based on World Health Organization (WHO) has received massive attention from researchers lately to develop various types of hip simulators in order to test the hip implant. Various number of hip simulator have been developed with different functions and capabilities. This paper presents the design development of biped hip simulator using SolidWorks software by taking into consideration some improvement and modifications. The finite element method is used to test the design whether it is safe to be used or not. The biped hip simulator has been successfully designed and ready to be fabricated as the endurance testing shown a positive results. The von Mises stress induced in the material is an alloy steel which is 2,975,862.3 N/m^2 lower than the yield strength. Thus, the design is safe to be used as it obey the safety criterion.

1. Introduction

A few last decades, hip joint prosthetics have been widely used in world's orthopedic surgery where 400 000 operations were performed every year and more than 100 designs of orthopedic construction have been proposed. Based on that, the development of hip simulator has received massive attention from researchers. A simulator generally described as machine that used to test a joint replacement under condition approximating the motion that occur in human body. Simulator test can be used to conduct protocol that replicates particularly extreme conditions and establishing the performance limit of the material [1].

There are few simulators that have been developed in order to achieve similarity between simulation and in-vivo conditions [2]. The simulators are differ from each other from a lot of parameters such as movements, number of channels, degree of freedom, configuration of the femoral head and acetabular component, and loading condition. Wear and dislocation of the artificial hip joint are still serious problems to be solved in hip joint replacement [3].

Basically, there are a lot of hip simulators that have been developed and differ in their sophistication level and plays a crucial role in the pre-clinical validation of biomaterials used for hip implants. Most of them were focused on testing wear and load on hips implant in which unfortunately, the metal-on-metal hips articulation suffers from friction due to the moving metal surface thus, shortening the surviving of the joint [4-5].

What is a biped hip simulator? Generally, the biped hip simulator is a simulator for testing both right and left hip implant as it is built in pair (consist of both right and left hip joint). This is because, both right and left performances is assumed to be symmetrical, but some research reported there are differences in the action of lower extremities although in able-bodied. Unfortunately, it actually affect the performance of both right and left hip joints since the wear mechanism were difference between them. Therefore, biped hip simulator is designed that able to be used to test for wear rate and load effect on hip implant especially for both hip replacement surgery. This paper introduces biped hip simulator design and simulation by using SolidWorks software. Finite element method for design endurance testing will also be discussed further in this paper.

2. Design Methodology

Biped hip simulator design was drawn using SolidWorks sofware and the design is based on the current available hip simulator with some improvement in term of size and performance.

2.1. Design selection process

The selection criteria are based on manufacturability, complexity, stability, cost, maintenance, compact size, light weight, and lubrication space. Design selection is important as it will results to a competitive design, better product-process coordination, and minimizing product fabrication.

In selection concept, there are two-stage design selection methodology in which the first stage is called concept screening and the second stage is called concept scoring. Each is supported by a decision matrix, which is used to rate, rank, and selects the concepts. Concept selection is often performed in two stages as a way to manage the complexity of evaluating dozens of design concepts. Furthermore, the design of biped hip simulator is an improvement and modification made from various available hip simulator that have been developed. Therefore, a correct procedure will ensure smooth design process.

2.2. Design Concept Screening

Concept screening is based on a method developed by Struart Pugh in the 1980 and is often called Pugh design concept selection [6-7]. The purpose of this stage is to narrow the number of design concept quickly and to improve the concepts. Six design concepts were chosen from others work in literature which are wobble plate, actuator (reference), hydraulic, compressor, pumping and linear concept.

Table 1. The concept-screening matrix for the biped hip simulator design concept.

	Design					
Selection Criteria	A (wobble plate)	B (actuator) Reference	C (linear)	D (hydraulic)	E (compressor)	F (pumping)
Manufacturability	+		+	+	+	+
Complexity	0	D	+	-	-	-
Stability	+	A	+	0	-	0
Cost	0	T	0	0	-	0
Lubrication space	+	U	0	+	+	-
maintenance	0	M	0	0	0	0
Small size	-		0	-	-	+
Light weight	+		-	-	-	-
Sum +'s	4	0	3	2	2	2
Sum 0's	3	8	4	3	1	3
Sum -'s	1	0	1	3	5	3
Net score	3	0	2	-1	-3	-1
Rank	1	3	2	4	5	4
Continue?	YES	YES	YES	NO	NO	NO

The design A (wobble plate), C (linear) , D (hydraulic), E (compressor) and F (pumping) are compared to reference design B (actuator) in order to narrow down the number of design concept as to make it easy to select the best concept. Symbol (+) means better than, (0) means same as and (-) means worse than reference design concept as shown in table 1. Based on table 1, it can be seen clearly that, the chosen design concept are design A, B and C due to the highest net score results from the subtraction of Sum of (+) and Sum of (-). The net score result of the design A, B and C are 3, 0 and 2 respectively.

2.3. Design Concept Scoring

Concept scoring is used when increased resolution shows a better differentiation among the competing concept. In this stage, it goes to weight the relative importance of the selection criteria and focuses on more refined comparisons with respect to each criterion. The design concept scores are determined by the weighted sum ratings. In this part, there are three concepts design (A, B and C) which will be rated to choose the biped hip simulator design concept for developing a new biped hip simulator design.

Based on table 2, it can be seen clearly that, the best selection of the design concept of biped hip simulator is design C (linear concept). This is because, design C has the highest weighted score compared to design A and B with 3.95 score. It is based on the rating given as depicted in table 3 to each selection criteria after considering some aspect or factors. Design C is the chosen design concept and will be realized into a real design by using SolidWorks software.

Table 2. The concept-scoring matrix of the biped hip simulator design.

		Design					
		A (wobble plate)		B (actuator) Reference		C (linear)	
Selection Criteria	Weight (%)	Rating	Weighted score	Rating	Weighted score	Rating	Weighted score
Manufacturability	10	3	0.3	1	0.1	3	0.3
Complexity	10	3	0.3	3	0.3	3	0.3
Stability	10	3	0.3	3	0.3	3	0.3
Cost	15	2	0.3	5	0.75	5	0.75
Lubrication space	5	2	0.1	3	0.2	3	0.2
maintenance	15	2	0.3	4	0.6	5	0.75
Small size	20	3	0.6	1	0.2	3	0.6
Light weight	15	2	0.3	3	0.45	5	0.75
	Total score	2.9		2.9		3.95	
	Rank	3		2		1	
	Decision	NO		NO		YES	

Table 3. Performance rating of for the design concept

Relative Performance	Rating
Much worse than reference	1
Worse than reference	2
Same as reference	3
Better than reference	4
Much better than reference	5

IOP Conf. Series: Journal of Physics: Conf. Series **908** (2017) 012060 doi:10.1088/1742-6596/908/1/012060

3. Results

3.1. Biped hip simulator design

The biped hip simulator is designed with some improvement and modification made based on specification that are obtained from literature on others work. The system of simulator is expected to imitate the motions and loads seen by the human hip joint and able to test the contact load and wear on the artificial hip for pitted surface filled with lubricant in order to prolong the lifespan. The simulator can be used to test the lifespan of hip implant especially for both hip joint replacement surgery, to investigate the different of contact load and wear on human dominant and non-dominant limb apart from reducing the manufacturing cost as it is developed in compact size. Figure 1 shows the design of biped hip simulator that was drawn using SolidWorks from the chosen design C after undergo the design selection process. SolidWorks software is a solid modeller and utilizes a parametric feature based approach to create models and assemblies that easy to use, affordable, and available on the Windows desktop. Based on that features, SolidWorks software is used to draw and simulate the biped hip simulator design.

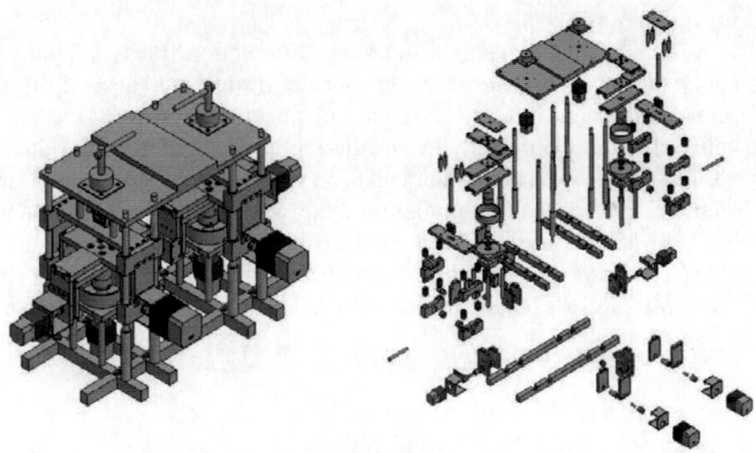

Figure 1. Design of biped hip simulator before and after assemble

Designing of the biped hip simulator design is tedious as it need to be designed part by part as each part plays a crucial role in making it completely functioning as desired. From each part then, it need to be assembled to make a complete design as shown in figure 1. Figure 2 shows the illustration of the main frame in the biped hip simulator design. The hip implant part is embedded in innermost of the frame and become the fixed geometry of the whole main frame. The design consists of 6 stepper motor in which it will be controlled by design control system to make it run as illustrated in figure 2. This is because, the simulators provides the motions of flexion-extension, abduction-adduction, and internal-external rotation.

Besides, the load will be applied manually by rotating the handle at the top of the design as shown in figure 1 and the load cell sensor will read the applied load value. The maximum load applied for the simulator is up to 3 kN. In addition, the simulator can be reconfigured in term of the load exerted and angle of movement by controlling the speed of the motor to imitate human daily life motion. In addition, the biped hip simulator design is based on the ISO standard 14242-1 and applicable to the human normal walking cycles during the gait cycles [8].

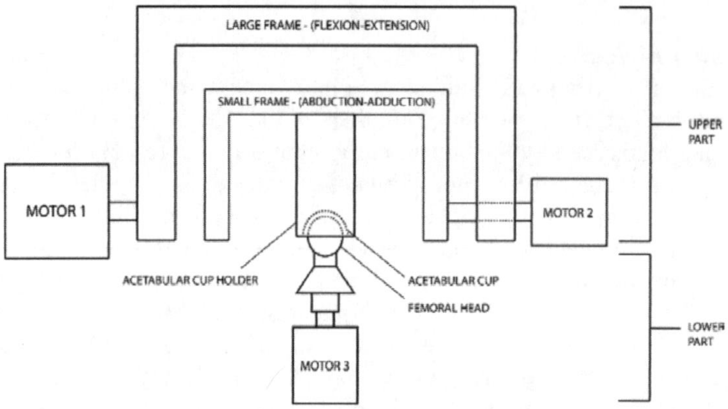

Figure 2. Illustration of the main frame.

3.2. Design endurance simulation

Endurance simulation is performed by using SolidWorks SimulationXpress for finite element method (FE) [9]. This is really important to determine whether the design are successfully done and can be fabricated. All aspects of a design have some degree of uncertainty and for the reasons, a factor of safety (FOS) is employed. The FOS of unity means that failure is eminent and it does not mean that a part or assembly is safe to be used or being fabricated. In current practice, it is advisable to justify 1 < FOS < 8 and FOS must be greater than 1 in order to make sure that the design is safe and functioning properly as desired. Von Mises stress is used in the simulation to check whether the design can withstand a given load condition (3kN for biped hip simulator). The design is considered fail, if the maximum value of von Mises stress induced in the material is more than the strength of the material.

Figure 3. Simulation result of a part of biped hip simulator design using SolidWorks

The design was simulated part by part in order to obtain a safe and good final product. Every single part of the design need to be simulated as it can help to predict how the part will perform under load condition and help to detect potential problems early in the design stage before it is being fabricated. Materials used for this simulation design is alloy steel and force exerted is 3 kN as it is the maximum load that the human hip joint can withstand. Figure 3 shows a simulation result of a part of biped hip simulator design using SolidWorks. The FOS must be more than 1 in order to have a safe design and if it is less than 1, then some readjustment must be made in terms of material selections, design size and others.

IOP Conf. Series: Journal of Physics: Conf. Series **908** (2017) 012060 doi:10.1088/1742-6596/908/1/012060

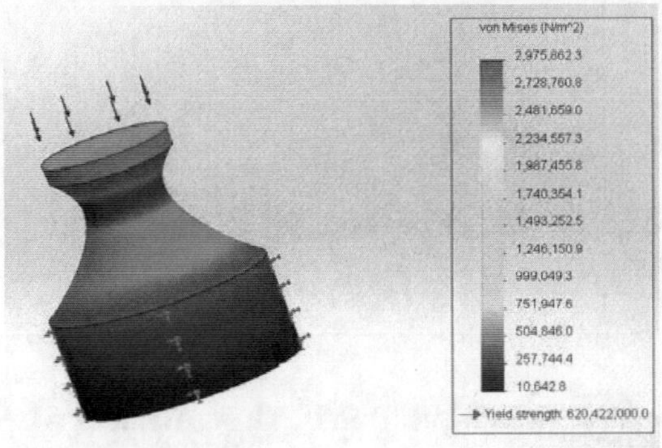

Figure 4. Von Mises stress induced in the biped hip simulator hip cup and the yield strength.

In figure 4, it can be seen clearly that von Mises stress induced in the material lower than the yield strength which means that the design is safe to be used as it is obeyed the safety criterion. The value of von Mises at red area is 2,975,862.3 N/m^2 in which red represents the highest deformation followed by yellow, green and blue However, if the von Mises stress induced in the material exceeds the yield strength, some adjustments need to be done in term of the material selections, size of the defect part of the design and other related factors. The steps are repeated for other part of the design so that the overall design is safe to be fabricated.

4. Conclusion
Biped hip simulator has been successfully designed and simulate using SolidWorks software. Designing biped hip simulator has been done carefully as it deals with a very small components and parts. Based on the results, the design capable to provide the motions of flexion-extension, abduction-adduction, and internal-external rotation. It is found that, the biped hip simulator design is ready to be fabricated as the endurance testing shown a positive results.

5. Acknowledgments
Authors are grateful to the Universiti Teknologi Malaysia for facilitating this research and for their financial support through GUP grant Q.J130000.2545.13H23.

6. References
[1] Trommer R M, Maru M M, Oliveira F, Nyakanen V P S, Gouvea C P, Archanjo B S, Martins F, Rui S, and Achete C A 2015 *J.Biotribology* **4** 1-11
[2] Duncan D, Cath H, Magnus F and Graham H I 2004 *J. Arthroplasty* **19** 118–23
[3] Gang C, Yang L, Lanlan F, Xianlei S, and Jianhua Y 2015 *J. Meachanism and Machine Theory* **86** 62-72
[4] Loving L, Lee R, Herrera L, Paul A, and Edward J 2013 *J. Arthroplasty* **28** 1041-46
[5] Patrick A V, Ulrich G, and Peter A 2015 *J. Biomechanics* **48** 3227-33
[6] Stuart B 2009 The Systems Engineering Tool Box (Winston Churchill) pp 1–15.
[7] Bill H and Stuart P 1990 *Successful Product Design* (Butterworth-Heinemann Ltd, Guildford) p 118
[8] Oliveira A, Lima R G, Cueva E G, and Queiroz R D 2011 *J. Wear* **271** pp 2340-45
[9] Iveta G, Tomas L, and Jozef B 2014 *Procedia Engineering* **96** pp131-35

IOP Conf. Series: Journal of Physics: Conf. Series **908** (2017) 012061 doi:10.1088/1742-6596/908/1/012061

Mechanical properties of friction stir welded butt joint of steel/aluminium alloys: effect of tool geometry

W. M. Syafiq[1], M. Afendi[1], R. Daud[1], M. N. Mazlee[2], M. S. Abdul Majid[1], Y. S. Lee[1]

[1]School of Mechatronic Engineering, Pauh Putra Campus, Universiti Malaysia Perlis, 02600 Arau, Perlis, Malaysia
[2]Centre of Excellence for Frontier Materials Research, School of Materials Engineering, Universiti Malaysia Perlis, 01000 Kangar, Perlis, Malaysia

wmsyafiq10@gmail.com

Abstract. This paper described the mechanical properties from hardness testing and tensile testing of Friction Stir Welded (FSW) materials. In this project, two materials of aluminium and steel are welded using conventional milling machine and tool designed with different profile and shoulder size. During welding the temperature along the weld line is collected using thermocouples. Threaded pins was found to produce stronger joints than cylindrical pins. 20 mm diameter shoulder tool welded a slightly stronger joint than 18 mm diameter one, as well as softer nugget zone due to higher heat input. Threaded pins also contributed to higher weld temperature than cylindrical pins due to increase in pin contact surface. Generally, higher temperatures were recorded in aluminium side due to pin offset away from steel.

1. Introduction

In present days, many large and complex structures are formed due to the needs of the public. Most of the structures are using metals such as steel and aluminium as raw materials because of their strength and toughness. There are times where there is a need to join two materials together. Therefore, a joining technique is needed to hold the structure together strongly. Among all the joining techniques there is a process called welding where a small portion of the parts-to-be-joint's volume will be melted, later it will stick together as it cools. There are several types of welding processes such as tungsten inert gas (TIG) welding, friction stir welding (FSW), electron beam (EB) welding and fiber laser welding [1].

However, there is a serious problem when two dissimilar alloys such as aluminium and steel are joined because there will be a formation of intermetallic compounds (IMCs) in the weld zone. Such occasion occurs by the differences between the physical and chemical properties of the two alloys such as melting point, thermal conductivity, coefficient of linear expansion and heat capacity. Even though there are few different techniques have been applied, obtaining a solid joint between aluminium and steel looks very tough [2].In 1991, a welding process was invented by The Welding Institute (TWI) in United Kingdom to join the aluminium alloys. It is called as friction stir welding (FSW) which is a solid state joining process where the frictional heat is produced between the rotational tool and the material to fuse the two working metals [3]. In comparison with other conventional joining methods, FSW holds the advantages with reduced porosity, cracking, distortion, and reinforcement dissolution. Therefore, FSW can produce low-cost and high performance joints. Besides that, due to its low energy consumption, no gas emission and no consumable materials needed, FSW is known as an environmental-friendly technology in joining process[4].

In this experiment, two specimens of aluminium and steel are prepared to weld using friction stir welding process on a conventional milling machine. Mechanical properties of the friction stir welded butt joint such as tensile strength and hardness are evaluated. The influence of tool geometry and welding speed on aforementioned mechanical properties as well as the resulting temperature are investigated.

2. Experimental Procedures

To conduct this experiment, 5 mm thick plates of aluminium alloy 6061-T6 and mild steel were prepared. FSW was performed on a conventional belting milling machine (MASTIKA). FSW tools and a clamping jig were designed and fitted onto the milling machine in order to enable FSW on it. FSW tools with different geometries were designed for the purpose of this study. Tools with different shoulder diameter (18 mm and 20 mm) and different tool pin profile (cylindrical and threaded) were chosen to investigate their effects on the joint. Joints with different parameters were welded as presented in table 1.

Table 1. Specimen with different parameters.

Specimen	Rotational Speed (rpm)	Travel Speed (mm/min)	Pin Profile	Shoulder Diameter (mm)
A	650	50	Threaded	18
B	650	50	Cylindrical	18
C	650	50	Cylindrical	20
D	650	70	Cylindrical	18
E	650	70	Threaded	18

Four thermocouples (CH001, CH002, CH003, CH004) were inserted into the jig to measure the temperature distribution along the weld line. Holes were drilled 3 mm below the welded samples in the jig to fit four thermocouples. The positions of thermocouples are shown in figure 1.Thermocouple CH002 is located 3 mm into the steel side, while thermocouple CH004 is directly on the joint line border between steel and aluminium alloy. Thermocouples CH001 and CH003 are located 6 mm and 10 mm into the aluminium alloy side respectively. After the FSW process, the specimen was cut using EDM wire cut to obtain a dumbbell shape according to ASTM E8-E8M-13a[5]that will be used for tensile testing. Tensile testing was done on a universal testing machine (SHIMADZU TCE-N300) with a crosshead speed of 1 mm/min. Vickers Micro hardness test was done on a cross section of the joint, with readings taken at mid thickness of the joint. The indenter was a square-based diamond pyramid and a load of 200 N was applied with dwell time of 15 seconds.

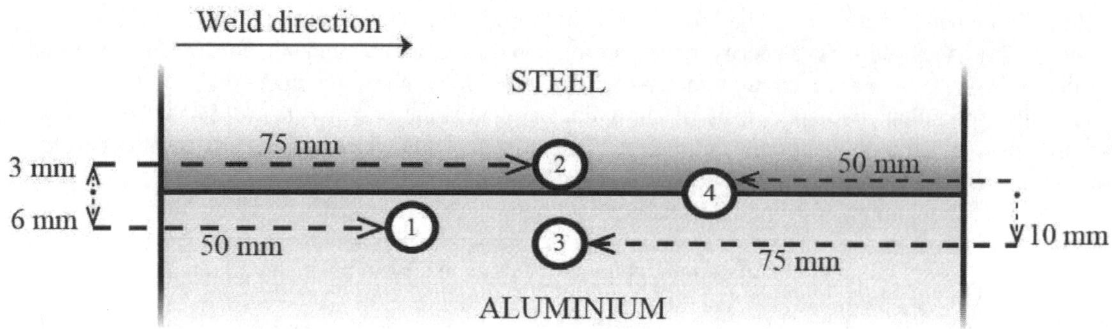

Figure 1. Position of thermocouples with respect to plates to be welded.

3. Results and Discussion

3.1. Tensile Strength

The tensile strength results of all welded joints are tabulated in table 2, and the influence of different tool geometries on the tensile strength is illustrated in figure 2. From these results, it was seen that the highest ultimate tensile strength (UTS) was obtained by Specimen A at 117.5 MPa. The lowest UTS was recorded at 35.7 MPa by Specimen D. Generally, joints that were welded by a threaded pin (Specimens A and E) outperform joints welded with a cylindrical pin (Specimens B and D) with other parameters kept constant. Joint that was welded with a bigger shoulder diameter of 20 mm (Specimen C) recorded a small improvement over the joint welded with 18 mm shoulder diameter (Specimen B) whilst keeping other parameters constant. It is also noted that when the tool travel speed was increased from 50 mm/min to 70 mm/min, welded joints recorded lower UTS values, as can be seen comparing the UTS values of Specimen B with Specimen D, as well as Specimen A with Specimen E.

Table 2. Ultimate tensile strength (MPa) for tensile testing specimens.

Specimen	Ultimate Tensile Strength (MPa)
A	117.5
B	86.5
C	91.1
D	35.7
E	77.7

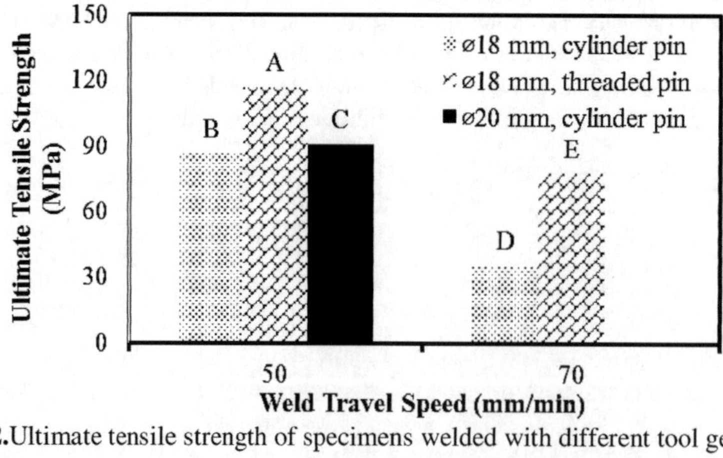

Figure 2. Ultimate tensile strength of specimens welded with different tool geometries.

3.2. Hardness

The hardness profiles of the welded specimens are presented in figure 3. Note that0 is located at the joint line border between the two dissimilar plates. Steel and aluminium alloy sidesare assigned negative and positive direction respectively. At 5 mm into steel side, it is seen that Specimen C recorded the highest hardness value at 202.8 HV, while the other specimens recorded values close to each other, with Specimen D recorded the lowest value at 165.1 HV. In the nugget zone, i.e. 5 mm into the aluminium alloy side, Specimens D and C recorded the maximum and minimum hardness values, with measurements of 90.1 HV and 76.9 HV respectively.

For all hardness profiles, similar patterns were obtained that hardness increased from distance of -25 mm to -5 mm. Values decreased from -5 mm to 5 mm when crossing the joint line from steel to aluminium alloy, before increasing again at 25 mm. Maximum and minimum hardness values were obtained at -5 mm and 5 mm respectively for all hardness profiles.

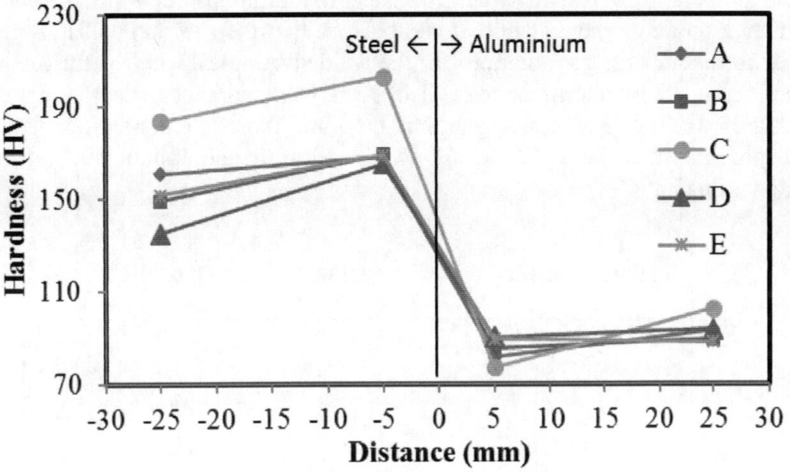

Figure 3. The hardness profiles of all the specimens taken at mid thickness.

3.3. Temperature

3.3.1. Maximum Temperature of Welds

Four thermocouple channels (CH001, CH002, CH003, CH004) were used to record temperature variation during welding. Only the maximum temperature recorded by each thermocouple during the entire welding run is considered, and the term "highest maximum temperature" refers to the highest temperature recorded during FSW of the specimen by any thermocouple. The highest maximum temperatures recorded during welding for Specimens A to E are 218.2 °C, 198.5 °C, 247.1 °C, 168.6 °C and 182.1 °C respectively.

For Specimens B to E, the highest maximum temperatures were recorded on thermocouple CH001, whereas thermocouple CH002 recorded the highest temperature during the FSW of Specimen A. From Table 3, it is noted that Specimen C was welded with the highest temperature at 247.1 °C, while Specimen D was the least hot at 168.6 °C. It is noted that the holes to fit the thermocouples were drilled 3 mm below the top surface of the jig, therefore the temperatures recorded by the thermocouples are not the exact temperatures on the workpieces, however the difference in temperatures recorded when performing FSW with different parameters was clearly exhibited.

Table 3. Maximum temperature and its respective thermocouple recorded during FSW.

Specimen	Maximum Temperature(∘C)	Thermocouple
A	218.2	CH002
B	198.5	CH001
C	247.1	CH001
D	168.6	CH001
E	182.1	CH001

3.3.2. Temperature across the Weld Joint from Steel to Aluminium

Figure 4 shows the temperature across the weld from steel to aluminium for all the specimens. Most of the plots show fluctuation along the distance. It can be seen that thermocouple CH001 recorded most of the highest maximum temperatures achieved for the welded samples, whereas the lowest maximum temperatures were recorded by thermocouple CH003. It is noted that even though thermocouple CH004 is located under the tool pin's passing path at the joint line, the recorded temperatures are less than the values recorded by thermocouple CH001 positioned in the aluminium side, a few millimetres away from the pin passing path.

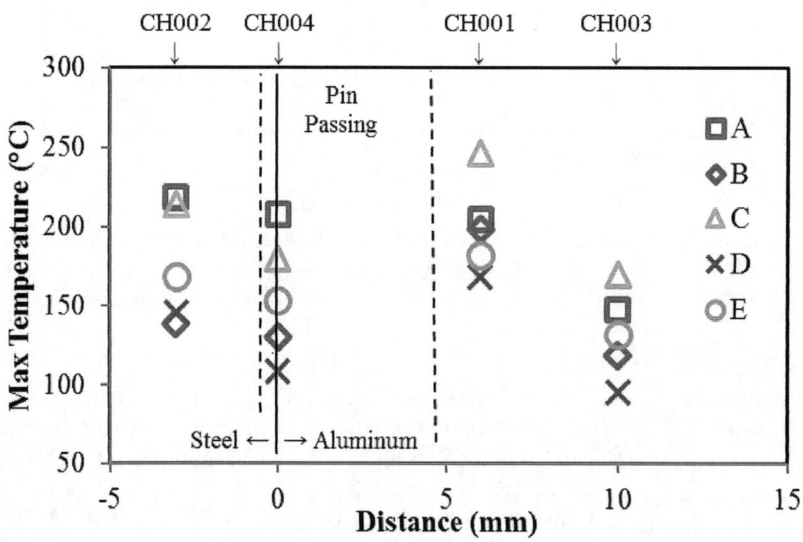

Figure 4. Maximum temperatures recorded by thermocouples during FSW of specimens.

3.4. Discussion

From the tensile strength result, it is seen that joints welded with threaded pins have higher ultimate tensile strength than joints welded with normal cylindrical pins. Specimen A and Specimen E were welded as a comparison to Specimens B and D, as both A and E are welded with threaded pins as opposed to B and D that were welded with cylindrical pins. From figure 2 it is seen that joints welded with a threaded pin were stronger. The design of the thread was such that during welding, plasticized material will be pushed from the top towards the bottom of the pin. The presence of threads is known to improve material consolidation and mixing, as the plasticized material's movement around the pin is better facilitated with the presence of threads, compared to an unthreaded cylindrical pin [6]. Increasing tool shoulder diameter from 18 mm to 20 mm also had a positive influence on the UTS,

IOP Conf. Series: Journal of Physics: Conf. Series **908** (2017) 012061 doi:10.1088/1742-6596/908/1/012061

albeit only by a small margin. A larger shoulder diameter will be able to push more plasticized material into the void left behind by the traveling pin, therefore improving material consolidation[7].

Increasing the weld travel speed from 50 mm/min to 70 mm/min deteriorated the joint's strength, as demonstrated by Specimens B and D. The increase in weld travel speed effectively reduced the time spent by the tool to interact with the plasticized material at joint line. The decrease in heat input also affected material flow, which resulted in less than satisfactory contact between the plasticized material and the steel faying surface [8], [9]. However, under the same welding speed, adding a thread to improve heat generation as well as material flow improved the UTS of the joint as shown by Specimens D and E. The heat generated by a threaded pin is larger than a cylindrical pin, due to the increased pin surface area in contact with the workpiece [6], [10]. As a result, the material flow was improved, better contact between aluminium and steel was achieved, thus the UTS was also improved.

The hardness profiles of all specimens peaked in the steel side close to the joint line. This indicates work hardening effect of the tool shoulder onto the steel surface, as well as the stirring of the tool pin [11]. The increased contact area between the workpiece surface and the 20 mm diameter tool shoulder while welding Specimen C resulted in a higher degree of work hardening onto the steel side. All specimens welded with 18 mm diameter tool shoulders recorded similar hardness values with each other. The hardness values went down to a minimum when crossing the joint line border from steel to aluminium alloy at 5 mm, with Specimen D recording the highest hardness value here.

It is expected for the hardness values to decrease as the profile crossed the joint line, as the two dissimilar materials did not mix in the nugget zone. At 5 mm into the aluminium alloy side, the nugget zone which is the zone where the tool pin passed is found. The hardness of the nugget zone is lower than the base metal aluminium alloy due to the dissolution of hardening second phase particles during the stirring and heating of the FSW process [11], [12].In FSW, severe plastic deformation from stirring and heat from friction causes dynamic recrystallization in the nugget zone, which resulted in fine equiaxed grains [13]. Specimen D was welded with a higher welding speed that reduced the heat input into the joint. This contributed to less grain growth and subsequently higher hardness values, which is in accordance to the Hall-Petch relation [14], [15].

The heat input per unit length in a friction stir welding process is given by the expression below [16], [17]:

$$Q/v = \frac{4}{3}\pi^2 \mu P \frac{\omega}{v} {R_s}^3 \quad (1)$$

In Equation 1, $\mu$is the friction coefficient, P is the applied pressure, ω is the tool rotational speed, v is the tool travel speed and R_s is the tool shoulder radius It is seen from this expression that tool shoulder geometry has a large influence on the heat generation, which was why the maximum temperature recorded during the welding of Specimen C is the highest of all welded specimens. The increase in shoulder diameter contributed to an increase in contact area between the rotating shoulder and the workpiece. Specimen D was welded with a smaller 18 mm shoulder diameter tool, as well as a higher travel speed of 70 mm/min. These factors contributed to Specimen D's lowest maximum temperature. It is noteworthy that specimens welded with a threaded pin recorded higher maximum temperatures that their cylinder pin counterparts. This is due to threaded pins generating more heat, as discussed previously.

In terms of temperature across the joint during welding, it was seen from figure 9 that thermocouple CH001 which was located in the aluminium side 3 mm away from the pin passing path recorded higher temperatures than CH004, even when CH004 is located under the pin passing path at the joint border line. This phenomenon is due to the tool pin mostly interacting with the aluminium alloy side. In FSW of aluminium alloy and steel, the tool pin is mostly plunged into the aluminium alloy side in order to protect the tool from excessive heat resulting from friction between the tool and steel [18]. Majority of the severe plastic deformation occurred at the aluminium alloy side, which is why the thermocouple CH001 recorded higher temperatures than CH004. Thermocouple CH003 recorded the lowest maximum temperatures due to it being located furthest away from the pin passing path.

4. Conclusion

This paper investigated the influence of different tool geometries such as shoulder diameter and pin profile on mechanical properties and temperature of the weld. From the results obtained, the following conclusions are made.

1. In terms of pin profile, threaded pins were found to produce joints with better tensile strength than joints welded by tools with cylindrical pins. Specimens A and E both outperform Specimens B and D that were welded with similar welding parameters except for the tool pin profile. The improved material flow offered by the threads improved material consolidation.
2. Increasing the tool shoulder diameter from 18 mm to 20 mm had a positive influence on joint strength, if only slightly. Specimen C benefited from the larger shoulder diameter thus better material consolidation.
3. Increasing tool travel speed was found to negatively impact on the joint strength, as demonstrated by Specimens B and D due to less time for the tool to interact with the steel faying surface and plasticized material.
4. FSW with a larger shoulder diameter tool was found to increase the maximum temperature during the welding, which is in agreement with the heat generation equation by Frigaard [17]. This is due to the increased area of contact between the rotating tool shoulder and the workpiece.
5. Joints welded by tools with threaded pins were found to be hotter than the ones welded with the straight cylindrical pins due to increased contact area between rotating tool pin and the workpiece.
6. Increasing the tool travel speed reduced the maximum temperature recorded at a particular location due to less time spent by the tool to interact with the workpiece per unit length.

5. References

[1] Kano S, Oba A, Yang H L, Matsukawa Y, Satoh Y, Serizawa H, Sakasegawa H, Tanigawa H, and Abe H 2016 *Nucl. Mater. Energy* **9** 300–5
[2] Dehghani M, Amadeh A and Akbari Mousavi S A A 2013 *Mater. Des.* **49** 433–41
[3] Takahara H, Tsujikawa M, Chung S W, Okawa Y, Oki S and Higashi K 2008 *Mater. Trans.* **49** 1911–4
[4] Salih O S, Ou H, Sun W and McCartney D G 2015 *Mater. Des.* **86** 61–71
[5] ASTM Int., "Standard test methods for tension testing of metallic materials 1," *Astm*, no. C, pp. 1–27, 2009.
[6] Ilangovan M, Rajendra Boopathy S and Balasubramanian V 2015 *Def. Technol.* **11** 174–84
[7] Elangovan K and Balasubramanian V 2008 *Mater. Des.* **29** 362–73
[8] Liu X, Lan S and Ni J 2014 *Mater. Des.* **59** 50–62
[9] Ghosh M, Gupta R K and Husain M M 2014 *Metall. Mater. Trans. A* **45** 854–63
[10] He X, Gu F and Ball A 2014 *Prog. Mater. Sci.* **65** 1–66
[11] Kakiuchi T, Uematsu Y and Suzuki K 2016 *Procedia Struct. Integr.* **2** 1007–14
[12] Yazdipour A and Heidarzadeh A 2016 *J. Alloys Compd.* **680** 595–603
[13] Krasnowski K, Hamilton C and Dymek S *Arch. Civ. Mech. Eng.* **15** 133–41
[14] Kadlec M, Růžek R and Nováková L 2015 *Int. J. Fatigue* **74** 7–19
[15] Sato Y S, Urata M, Kokawa H and Ikeda K 2003 *Mater. Sci. Eng. A* **354** 298–305
[16] Frigaard Ø, Grong Ø and Midling O T 2001 *Metallurgical and Materials Transactions* A **32** 1189-200
[17] Dehghani M, Akbari Mousavi S A A and Amadeh A 2013 *Trans. Nonferrous Met. Soc. China (English Ed)* **23** 1957–65
[18] Kimapong K and Watanabe T 2004 **83** *Weld. J.* 277–82

Evaluation of RBD palm stearin as alternative lubricant for cold forward extrusion process

A N Farhanah, S Syahrullail, M A Ajruddin

Faculty of Mechanical Engineering, Universiti Teknologi Malaysia, 81310 UTM Skudai, Johor

nurulfarhanahazman@gmail.com

Abstract. A rapid growth in awareness level over environmental problems have brought researchers to find a substitute lubricant to reduce the use of mineral based lubricants. In this paper, the lubrication performance of RBD palm stearin was evaluated in order to be utilized as alternative lubricant for metal forming lubricant. Commercial paraffinic mineral oil VG460 and VG95 were used for comparison purposes. The experiment was conducted using cold work extrusion apparatus that consists of taper die (made of tool steel SKD11) and a pair of billet (made of pure aluminium A1100). The results obtained from these experiments showed that RBD palm stearin can reduce extrusion load, produce a lower surface roughness value and have smooth surface profile compared to commercial paraffinic mineral oils. It is demonstrated that RBD palm stearin have promising lubrication performance and can be considered as alternative lubricant in metal forming process.

1. Introduction

In the metal forming process, lubrication is greatly important to reduce friction and wear at tool/workpiece interface, prevent adhesion, scratching, galling and material transfer [1] in order to ease metal flow, increase tool life, energy consumption, heat evolution and improve product quality of the finished product [2]–[4]. The lubricant's effectiveness in metal forming process relies on parameters such as die-punch system, tool/workpiece materials, contact pressure, sliding velocity, surface temperature and surface finish [5], [6].

Traditionally, commercial mineral based lubricant has been used as metal forming lubricant. However, the world are facing environmental crisis and create the awareness among researchers to find an alternative lubricant to replace the mineral based lubricant. Vegetable based lubricant was the most suitable candidate to replace mineral based lubricant due to their biodegradability, non-toxicity and environmental friendly [7]–[9]. Moreover, vegetable oils have excellent boundary lubrication, high viscosity index, high flash point and low volatility due to their polar ester structure [10], [11].

In response to this issues, this research was conducted to evaluate the palm oil based lubricant as an alternative lubricant for metal forming lubricant. Refined, bleached and deodorized (RBD) palm stearin was selected as tested lubricant and commercial paraffinic mineral oil VG460 and VG95 were used for comparison purposes. The lubrication performance of the tested lubricants were evaluated using a cold work forward extrusion process. The evaluations of the lubrication performance will be in term of extrusion load, surface roughness, surface profile and flow angle.

IOP Conf. Series: Journal of Physics: Conf. Series **908** (2017) 012062 doi:10.1088/1742-6596/908/1/012062

2. Methodology

2.1Experimental apparatus and procedure

A plain strain extrusion apparatus was designed and the experimental setup was shown in Figure 1. It consisted mainly of three components: container wall, taper die and workpiece (billet). Tapper die used in this research is fabricated from alloy tool steel SKD11 which had a chemical composition of C (1.55wt%), Si (0.30wt%), Mn (0.35wt%), Cr (11.75wt%), Mo (0.75wt%), V (0.95wt%), S (0.005wt%) and P (0.020wt%) [12]. The tapper dies were subjected to necessary heat treatment process before each experiment. This tool steel are high carbon and high chromium alloy tool steel. The taper die has to be sufficiently hard to sustain the force implied to the billet that is in contact with taper die during the extrusion process. The taper die has an edge 45 degree at die half angle (figure 2(a)). Before each experiment, the surface of the tapper dies which is in contact with billet (see figure 2(b)) should be polished to prevent the debris or any impurities that adheres on the surface of the tapper dies. The optimum value of arithmetic surface roughness, Ra should be 0.15 μm.

The billet used in this research was made of pure aluminium A1100 which had a chemical composition of Si (0.08wt%), Fe (0.33wt%), Mg (0.0016wt%), Cu (0.054wt%), Ti (0.013wt%), Zn (0.013wt%) and Al (99.00wt% min) [12]. The billets, with 4.5 mm thickness, 15 mm width and 80 mm length, were formed by using NC wire cut electric discharge machining device. The billets were subjected to an annealing treatment at elevated temperature of 320°C to give the best condition for cold forming [4] and to relieve the residual stresses present in the billets prior to cutting and surface milling process [13]. A pair of billet were stacked together at which one side of the billet were marked with square gridline (see figure 2(c)) using marking machine as the observation plane for plastic flow angle analysis.

The components (container wall, taper die and billet) were assembled together and the stacked paired billets are put in the rig to be placed on the load cell afterward for load extrusion (y-axis) during each run. The displacement for ram stroke (x-axis) is recorded using the displacement sensor that is attached to the holder of hydraulic extrusion machine. Once the displacement reached at 40 mm piston stroke, the extrusion stopped. This range of ram stroke for displacement is expected to reach the steady state condition. Finally, the billets are taken out of the plane strain extrusion apparatus and then separated for surface roughness and flow angle measurement.

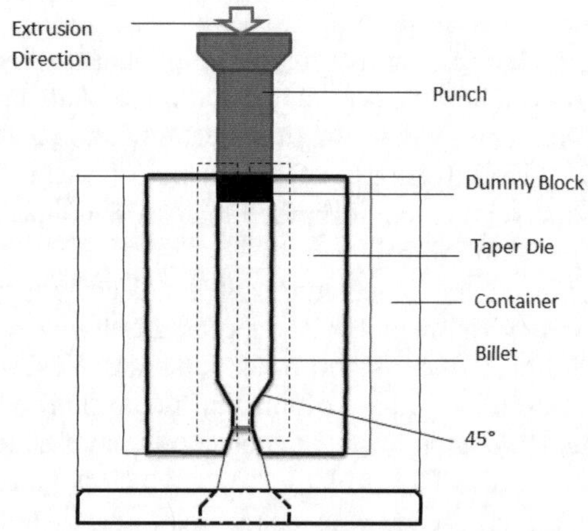

Figure 1. Schematic plane of plain strain extrusion apparatus.

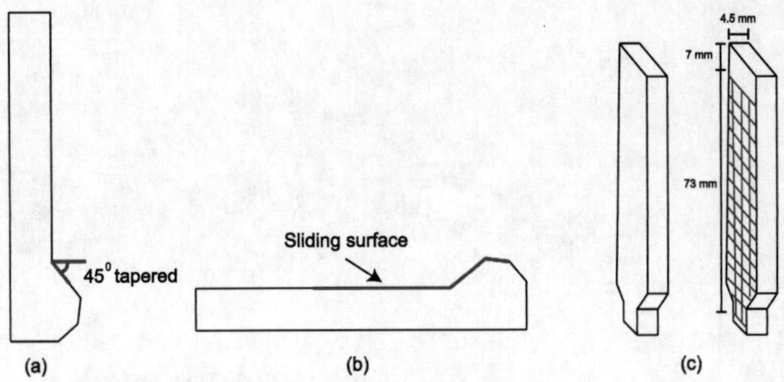

Figure 2. (a) Tapper die angle; (b) Surface area of tapper die to be polished; (c) Schematic drawing of billet.

2.2 Lubricants

In this research, the alternative extrusion lubricant assessed was refined, bleached and deodorized (RBD) palm stearin. While the conventional extrusion lubricants, paraffinic mineral oil VG95 and VG460, were used for comparison purposes. The physical properties of the tested lubricants were given in table 1. Before each experiment, 15 mg of the tested lubricant (equivalent to one drop of lubricant) was applied on the experimental surface of the taper die.

Table 1. Physical properties of the tested lubricants.

Lubricants	**Density (g/cm^3)**	**Viscosity index (VI)**	**Kinematic viscosity (mm^2/s)**		
			27°C	**40°C**	**100°C**
RBD PS	870	171	48.29	38.01	8.56
PMO VG95	850	192	249.95	71.75	13.4
PMO VG460	850	95	1374.6	411.25	28.1

3. Results and discussion

3.1 Extrusion load

Figure 3 shows the curved of extrusion load that is required to extrude billet lubricated by RBD PS, VG95 and VG 460. This figure shows that the extrusion load has reached the steady state load at point of 18 mm of piston stroke until 40 mm. It is noticed that the steady state only reached by billet extruded by lubricating oil of RBD PS and VG460. The extruded billet of VG95 on the other hands gradually increases at piston stroke from 18 mm to 30 mm, then decreases gradually to 36 mm of piston stroke and starts to maintain until 40 mm. The maximum extrusion load were plotted in a bar graph (figure 4) where the values for RBD PS, VG460 and VG95 are 60.78 kN, 57.20 kN and 71.37 kN respectively. The results showed that the value of the maximum extrusion load of RBD PS only has a slight different with VG460, and much lower than VG95.

The comparative results of VG460 shows a lesser extrusion load than VG95 due to its viscosity. Since VG460 has higher viscosity than that of VG95, a thick lubricant film will be formed to minimize metal-to-metal contact between billet and taper die which contributed to the lesser friction and reduction of the extrusion load. However, although RBD PS has lower viscosity than that of VG95 it produce lesser extrusion load. This phenomenon is attributed to the fatty acid in RBD PS that plays a significant role in maintaining the lubricant film to protect the contact surfaces [14]. From this findings, it is recommended that RBD PS is possible to be used as alternative lubricant for extrusion process which in line with statement made by Wang et al. [15], any lubricant that able to minimize the load to a small range as mineral lubricant and works to the same lubricating performance is good to be considered as substitute mineral lubricant.

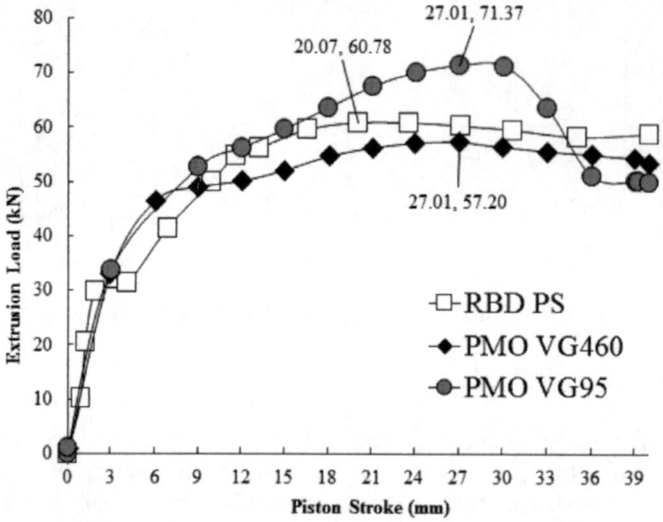

Figure 3. Extrusion load versus piston stroke curve.

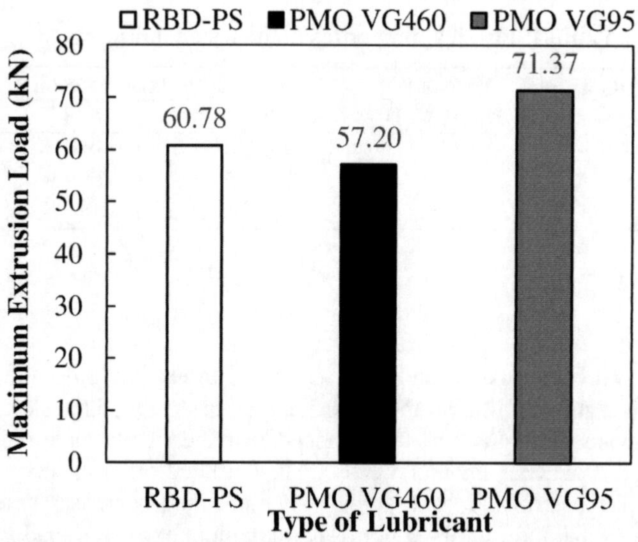

Figure 4. Maximum extrusion load.

3.2 Surface roughness

Figure 5 showed the arithmetic surface roughness, Ra of the billet at product area taken at 0 mm, -2 mm and -4 mm point coordinate of surface area. The surface roughness of the billet was measured perpendicular to the direction of extrusion. From this figure, it's clearly shows that surface roughness of the billet lubricated with RBD PS was smoother than the other conventional mineral lubricants, VG460 and VG95. The presence of polar fatty acid structure in RBD PS was able to create effective boundary lubrication that allows strong interaction with metallic surfaces [16], thus provide better performance than the conventional mineral lubricants.

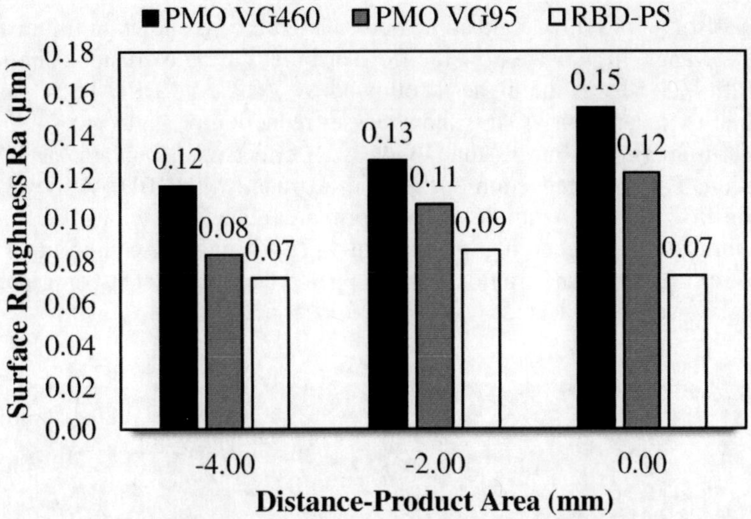

Figure 5. Surface roughness of the billet at product area.

3.3 Surface profile

Figure 6 represents the surface profile of the billet at three point area: product area (0 mm), deform area (8 mm) and undeform area (14 mm). From the observation, the surface profile of billet lubricated with RBD PS have smooth surface compared to surface profile lubricated with VG460 and VG95. While, it was observed that there is parallel groove along the other surface profile of the billet which more parallel groove were found on the billet's surface lubricated with VG460. It was also found that there is no severe wear occur on the surface product area. Deform area is the transition of normal surface to extruded surface and has mixed topography between undeform area and product area. There is an obvious change in topography at deform area lubricated by VG460 and VG95, while when lubricated with RBD PS the change of topography is insignificant. This findings was in line with the surface roughness values along the span of product area to deform area are constant (0.07 µm to 0.08 µm). At undeform area, the surface area lubricated with different tested lubricants was the same since no extrusion effect took place.

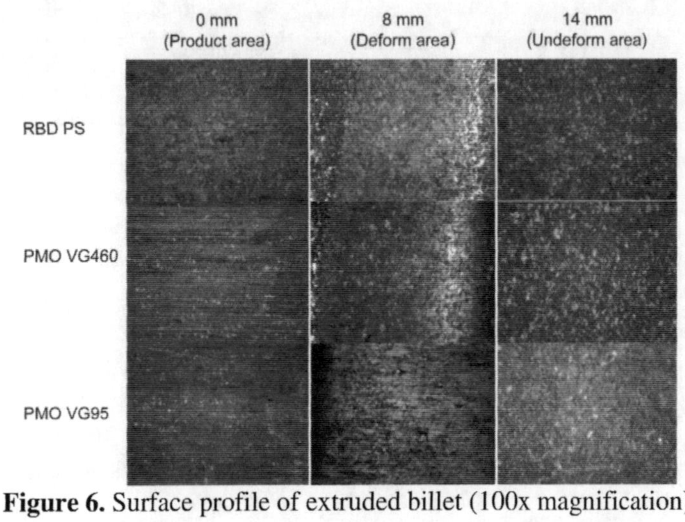

Figure 6. Surface profile of extruded billet (100x magnification).

3.4 Flow angle

The mutual comparison of the flow angle at the deformation area is shown in figure 7. The flow angle is a measurement of angle to indicate how the flow of billet during extrusion. The flow angle of the billet extruded with VG460 was the highest, followed by VG95 and RBD PS. Comparing the angle reduction, the billet extruded with VG460 shows lesser reduction of angle (43.74° of flow angle at 4 mm distance deform area). This means that VG460 can effectively maintain the lubricant film thus reducing the friction. The angle reduction of the billet extruded with RBD PS shows higher reduction of angle (27.38° of flow angle at 4 mm distance deform area) due to its semi solid condition as well as the breaking of lubricant film. According to Ong et al. [17], a semi solid condition of the lubricant will results in high contact pressure and friction. This happen when the billet is being pressed extensively during loading that caused the sudden drop of angle at local point.

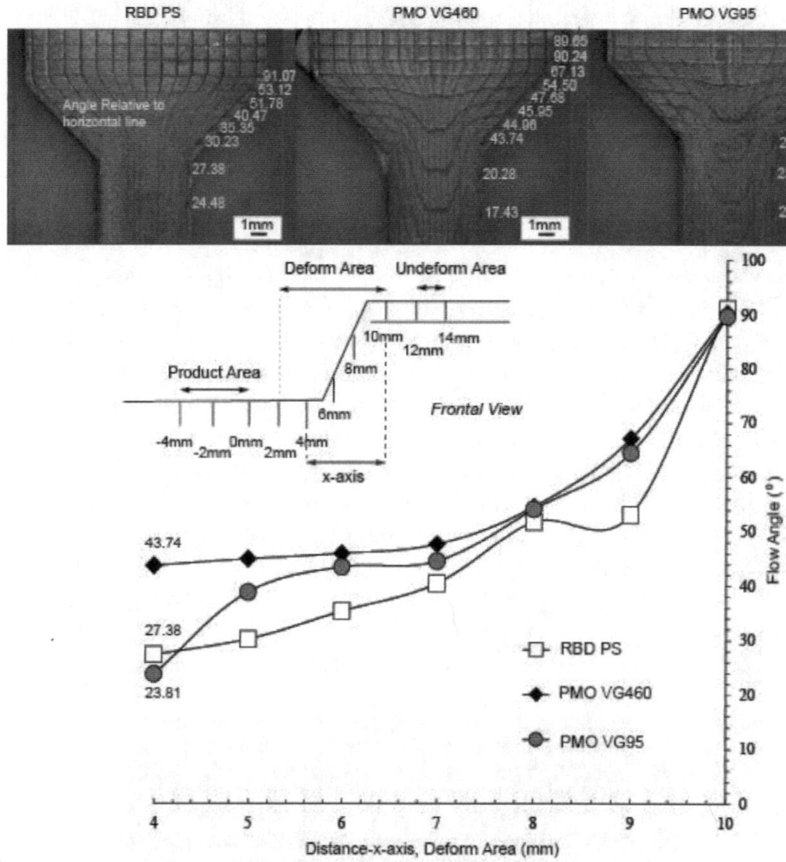

Figure 7. Flow angle at deform area.

4. Conclusion

This research was conducted to evaluate RBD palm stearin as alternative lubricant for metal forming process. Commercial paraffinic mineral oil VG460 and VG95 were used for comparison purposes. The results showed that RBD palm stearin was able to minimize the extrusion load and produce a smooth surface product with a low surface roughness value compared to the commercial mineral lubricants. From the observation of surface profile, the surface profile of billet lubricated with RBD palm stearin was smoother compared to the commercial mineral lubricants and there is no severe wear occur on the surface product area. However, the angle reduction of billet extruded with RBD palm stearin shows a higher reduction. From these findings, it is recommended that RBD palm stearin is possible to be used as an alternative lubricant for metal forming process.

IOP Conf. Series: Journal of Physics: Conf. Series **908** (2017) 012062 doi:10.1088/1742-6596/908/1/012062

5. Acknowledgments
The authors would like to express their thanks to the Research Management Centre (RMC) of Universiti Teknologi Malaysia for the Research University Grant, GUP (17H96) and Ministry of Education of Malaysia and Ministry of Higher Education for their support.

6. References
[1] Syahrullail, S, Azwadi, C S N and Ing, TC 2011 *Int. Rev. Mech. Eng.* 5(1) 21–27.
[2] Hu Z M and Dean T A 2000 *Int. J. Mach. Tools Manuf.* **40** 1637–1649..
[3] Abdulquadir B L and Bolaji Adeyemi M 2008 *Ind. Lubr. Tribol.* **60** 242–48.
[4] Nurul M A, Syahrullail S and Musa M N 2015 *J. Teknol.* **74** 35–43.
[5] Lazzarotto L, Dubar L, Dubois A, Ravassard P and Oudin J 1998 *J. Mater. Process. Technol.* **80–81** 245–50.
[6] Jayaseelan V and Kalaichelvan K 2014 *Procedia Eng.* **97** 166–71.
[7] Golshokouh I, Ani F N, Masjuki H H and Syahrullail S 2013 *Int. Rev. Mech. Eng.* **7(1)** 72–80.
[8] Golshokouh I, Ani F N, Kianpour E, and Syahrullail S 2013 *Life Sci. J.* **10** 110–119.
[9] Farhanah A N and Syahrullail S 2016 *J. Tribol.* **10** 1–15.
[10] Singh A K and Gupta A K 2006 *J. Synth. Lubr.* **23** 167–76.
[11] Aiman Y and Syahrullail S 2017 *J. Tribol.* **13** 1–20.
[12] Syahrullail S, Zubil B M, Azwadi C S and Ridzuan MJ 2011 *Int. J. Mech. Sci.* **53** 549–55.
[13] Tiernan P, Hillery MT, Draganescu B and Gheorghe M 2005 *J. Mater. Process. Technol.* **168** 360–66.
[14] Fox N J, Tyrer B and Stachowiak G W 2004 *Tribol. Lett.* **16** 275–81.
[15] Wang L, Zhou J, Duszczyk J and Katgerman L 2012 *Tribol. Int.* **56** 89–98.
[16] Maleque M A, Masjuki H H and Sapuan S M 2003 *Ind. Lubr. Tribol.* **55** 137–143.
[17] Ong H C, Mahlia T M, Masjuki H H and Norhasyima R S 2011 *Renew. Sustain. Energy Rev.* **15** 3501–15.

IOP Conf. Series: Journal of Physics: Conf. Series **908** (2017) 012063 doi:10.1088/1742-6596/908/1/012063

Fabrication of mandible fracture plate by indirect additive manufacturing

M Aizat and S F Khan

Mechanical Engineering Programme, School of Mechatronics Engineering, Universiti Malaysia Perlis (UniMAP), Pauh Putra, Perlis, Malaysia.

sfkhan@unimap.edu.my

Abstract. Bone fracture is a serious skeletal injury due to accidents and fragility of the bones at a certain age. In order to accelerate fracture healing process, fracture bone plate is use to hold the fracture segment for more stability. The purpose of this study is to fabricate mandibular fracture plate by using indirect additive manufacturing methods in order to reduce time taken during bending and shaping the fracture fixation plate that conform to the anatomy of the fractured bone site. The design and analysis of the plates are performed using CATIA and ANSYS software. The 3D-CAD data were sent to an additive manufacturing machine (fused filament fabricated) to generate master pattern using PLA and the mould were fabricated using Plaster of Paris. A melt ZAMAK 3 was poured directly into the moulds, and left it until completely harden. 3point bending test was performed on the prototype plate using universal testing machine. Stress-strain curve shows the graph exhibited a linear relationship of stress-strain up to a strain value of 0.001. Specimens give a maximum yielding stress and then break before the conventional deflection. Since the maximum flexural stress and the breaking stress are far apart with a plateau stating at strain value of 0.003mm/mm in most specimens, the specimen's failure types are considered plastic failure mode. The average thickness and width are 1.65mm and 2.18mm respectively. The flexural modulus and flexural strength are 189.5GPa and 518.1MPa, respectively.

1. Introduction

In order for a fracture to heal, the bones must gently to be fixed in suitable position and strong enough to withstand body weight and protected [1]. The fracture plate or bone plate is designed for numerous bone fractures, osteoporosis, and other musculoskeletal problems caused by accidents or natural at specific ages. This problem need to be solved by using temporary, permanent and biodegrable fracture plate. The role of fixation plate is to hold fractured bone segments in the correct position without allowing the tensile stress at fracture face but have some critical compression stress caused in it to speed the healing process [1]. However, most of the available fixation plates are standardised that do not fit conformably to the fracture site. Intraoperative bowing of plates can be tedious. Twisting reproduction plates relies on upon the multifaceted nature of resection and specialist's expertise. However, by fabricating the desired plates according to the structure of the fractured site using additive manufacturing technologies as oppose to bending the plates on the 3D models before the surgery will reduces operating times [2,3]. Lethaus et al used the technique of AM on 20 case studies that required mandibular resection and reconstruction because of a tumour or osteoradionecrosis. The average time to bend a plate on a non-sterile model was 0.42 hour (range 0.25-0.68 hour) and can reduce operating surgery time up to 1.4 hours [4]. Additive Manufacturing (AM) is a suitable name to represent the innovations that produce 3D artefacts by stacking successive layers of materials including plastic, metal, and concrete. Common to AM systems is the utilization of a computer, 3D modelling software (Computer Aided Design or CAD), machine hardware and layering material. Once a CAD sketch is produced, the AM equipment reads in data from the CAD file and lays downs or adds successive layers of liquid, powder, sheet material or other, in a layer-upon-layer fashion to fabricate a 3D object. The approach in producing artefacts using AM techniques is the opposite of subtractive manufacturing processes. Many design guides have been developed for these subtractive processes. However, when designing artefacts that are to be manufactured by utilizing AM as the manufacturing process, designers need a different approach. Most AM systems have the benefits of freedom in geometrical shape definition as well as tooling independent. Conventional manufacturing's cost is highly dependent to the geometry complexity, therefore with lower cost combined with design freedom; both manufacturer and customer will benefit. With the help from AM, designers have unrestrained freedom to design for customization rather than to design for manufacturing and assembly. Therefore, products can be designed to fit the user rather than the other way around [5].

2. Methodology

2.1. Materials
PLA was used in this research in a form of master pattern because it easy to melt during to eliminate from mould casting made by Plaster of Paris (POP). In order to fabricate prototype plate, ZAMAK 3 is used in this research as alternative material to obtain the material strength and mechanical properties.

2.2. Equipment
A fused filament fabrication (FFF) 3D printer was used in this research to generate master pattern. A gas furnace was used to heat ZAMAK 3 into molten form before pouring into the created mould. This will generate metal prototype plates. A universal material testing machine was used to conduct flexural tests on the prototype plate specimens.

2.3. Design of Experiment
Figure 2.1(a) shows the intended design of the fixation plate with sprue and air vent required during molding process. The master pattern of mandibular fracture fixation plate was design by using CATIA V5R16. When the design is completed, the plate was analysed using ANSYS Workbench 14.5 to determine which plate are the best. To print out the plate, industrial polylactic acid (PLA) polymer was utilised as the built material in fused filament fabrication (FFF) printer. The melting point of PLA is between 180°C – 220°C [5]. Figure 2.1(b) show the orientation of the printed part during 3D printing. The time taken to print out the master pattern is 44 minutes. In order to get the precise dimension of the plate, the percentage of plate thickness is increase to 40% due to shrinkage that occurs during printing.

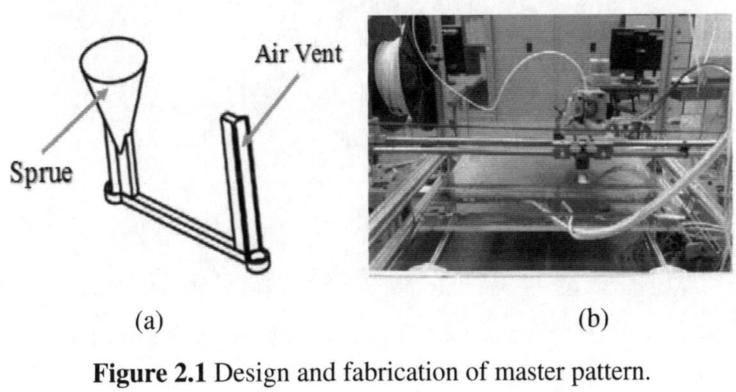

(a) (b)

Figure 2.1 Design and fabrication of master pattern.

2.4. Casting Process
Plaster of Paris (POP) powder was used in this process in order to the casting mould that will produced the prototype of mandibular fixation plate. POP can be executed at room temperature and the maximum working temperature of POP is 1200°C [6]. Therefore, the setting process of POP can be made by mixed the powder of POP with water. The plaster is then poured over the master pattern (PLA) into the container provided. The container can be removed after 20 minutes. The process is repeated to produce the second piece of mould base. The casting can be seen in figure 2.2 (a).In order to remove the PLA master pattern, gas furnace was utilized to melt the PLA at 250°C for 7 hours to completely removed the PLA. Figure 2.2 (b) and (c) shows the process of melting the master pattern and the expected mould that was created to pour ZAMAK 3 in the next stage.

(a) (b) (c)

Figure 2.2 Casting result (a), Eliminate of master pattern (b), and Expected mould (c).

2.5. Fabricating Prototype Plates
ZAMAK 3 was heated by using gas furnaces. The furnaces temperature was set to 384°C to melt the ZAMAK 3 within 30 minutes [7]. Once the ZAMAK 3 is melted, the pouring process can be executed. The two pieces of the mould were place in a container that is filled with soil as shown in figure 2.3(a) before the pouring process is performed. The pouring process of ZAMAK 3 into the mould as shown in figure 2.3(b) must be perform within 10 seconds to maintain ZAMAK 3 flow ability and avoid the melt to thicken. It takes 30 minutes to completely solidify and cool down before the mould can be broken to get the desired physical plate as shown in figure 2.3(c). The physical plate need to undergo finishing process by using grinding and drilling machine as shown in figure 2.3(d) to eliminate unwanted sprue and vents as well as to get a smooth surface.

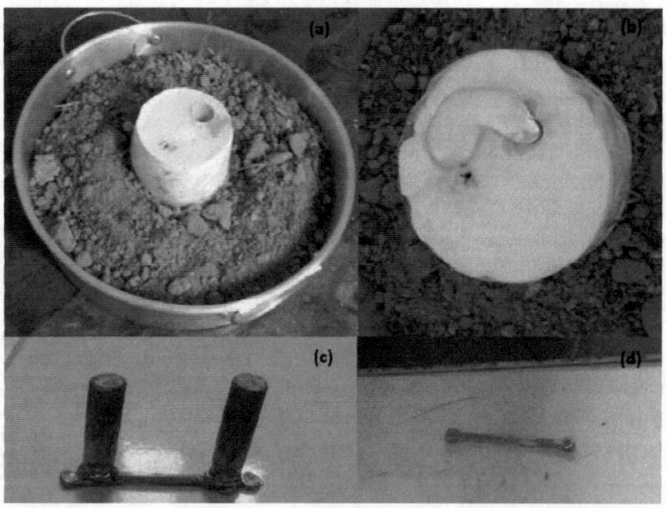

Figure 2.3 Expected mould (a) Poured process (b) Physical plate before finishing process (c) Physical plate after finishing process (d)

2.6. Flexural Test

Flexural test was conducted to evaluate the force needed to bend a beam under three point loading conditions by using a universal testing machine (UTM) according to the ASTM D790 standard as shown in figure 2.4 [8]. The test was conducted using a load cell of 150N with loading speed of 1mm/min. When the specimen breaks, the test was stopped. In order to obtain a satisfactory result, five specimen plates were tested by taking their average of the collected data.

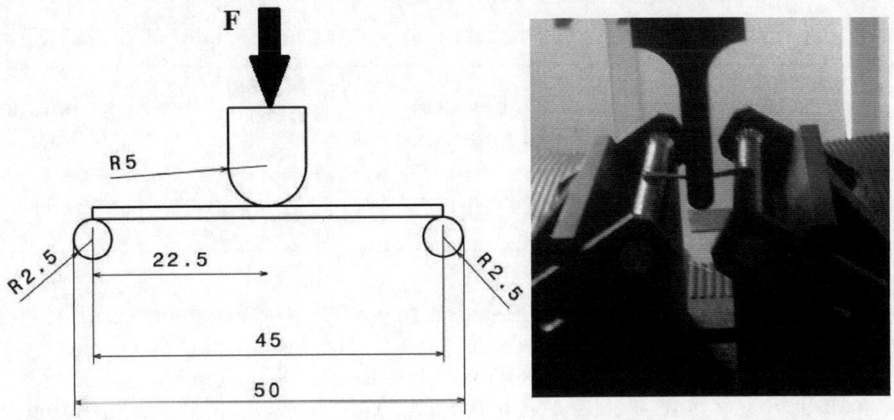

Figure 2.4 Bending test diagram and the actual test on a universal testing machine rig

3. Result and Discussion

In total, five tests of specimen plates are fabricated from ZAMAK 3 and casting mold that has cavities measuring 1.5mm x 2mm x 44.7mm each. All measurements for ZAMAK 3 specimen plates were taken and the range of thickness and width for the specimens are from 1.5mm to 1.8mm and 2.0mm to 2.3mm respectively.

The overall length is 45mm for all the specimens and the span length in the three points bending test is 50mm. The mechanical test is based on 3-point bending testing procedure of ASTM D790 and was conducted for all 5 specimens. The results are as shown in figure 3.1 and table 3.1.

From figure 3.1(a) and (b), the graph exhibited a linear relationship of stress-strain up to a strain value of 0.001. Specimens give a maximum yielding stress and then break before the conventional deflection. Since the maximum flexural stress and the breaking stress are far apart with a plateau stating at strain value of 0.003mm/mm in most specimens, the specimens' failure types are considered to be plastic failure mode. The fracture images are as shown in figure 3.3. The reason why specimen at T3 has such a large deflection and strain as compare to the others 4 specimens is because the thickness of that specimen is higher.

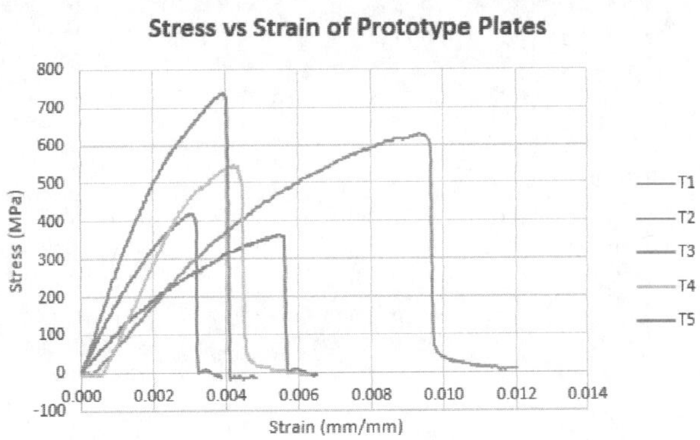

Figure 3.1 Stress-Strain curves of ZAMAK 3 plates

As shown in table 3.1, there is no significant variation in terms of dimensions and mechanical properties of ZAMAK 3 test specimens using the fabrication process. The confidence interval in repeating the manufacturing of the specimen is as calculated using t-test with p-value <0.05.

The average thickness and width are 1.65mm and 2.18mm, respectively. The flexural modulus and flexural strength are 189.478GPa and 518.138 MPa, respectively.

Table 3.1 Mean value for 95% confidence interval for casting test samples of ZAMAK 3 plates.

	Dimension		**Max. Flexural Load Fmax (N)**	**Flex Strength (MPa)**	**Flex Modulus (GPa)**
	Thickness (mm)	**Width (mm)**			
ZAMAK 3	1.65±0.14	2.18±0.13	32.34±9.11	518.14±154.12	189.48±77.79

The result of ANSYS analysis was indicated in figure 3.2. The maximum result is label by red region while the minimum result is label by blue region. The mesh sizing is defined as 1mm this simulation. The maximum deflection is 0.27807mm when load 150N is applied on the midspan of the specimen as indicated in figure 3.2(a). The maximum stress is 364.96MPa as indicated in figure 3.2(b). The percentage error of maximum flexural strength between experimental and theoretical of ZAMAK 3 plates is 41.97%.

(a)

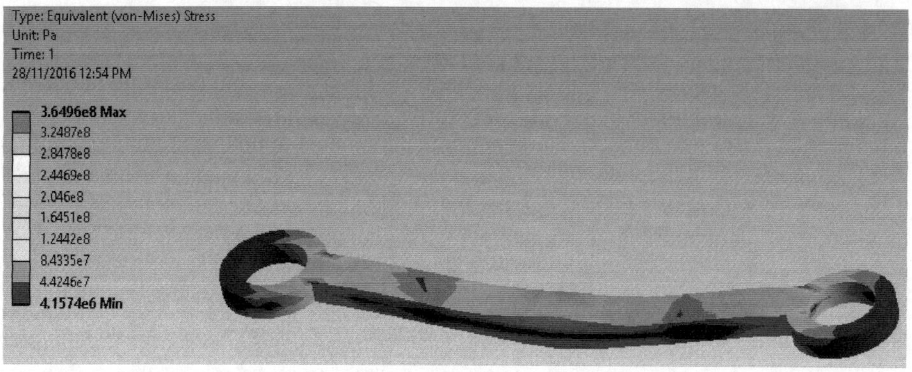

(b)

Figure 3.2 Deflection result after load applied (a) Stress result after load applied (b)

In order to understand further the behavior of ZAMAK 3 during three points bending test, images are captured at 100X magnification by using scanning electron microscope (SEM) at bending fracture failure surface. Images also showed the porosity region A and material consolidation region B experiencing ductile failure (crystal-like surface) as indicated in figure 3.3. There are two caused of porosity that occur during casting process such as gas formation and solidification shrinkage.

Large gas related voids caused by trapped mold or core gases in the molten metal are called blows or blowholes. They generally are large enough to look like bubbles with smooth interior surfaces and are always buoyant and will float near the top of the casting, as it was oriented in the mold. Shrinkage related casting voids are caused by sections of the casting that solidify later than the surrounding sections and do not have enough metal flow into the section for a complete fill. This is generally because the area is too hot for too long due to a thick section or a tight corner.

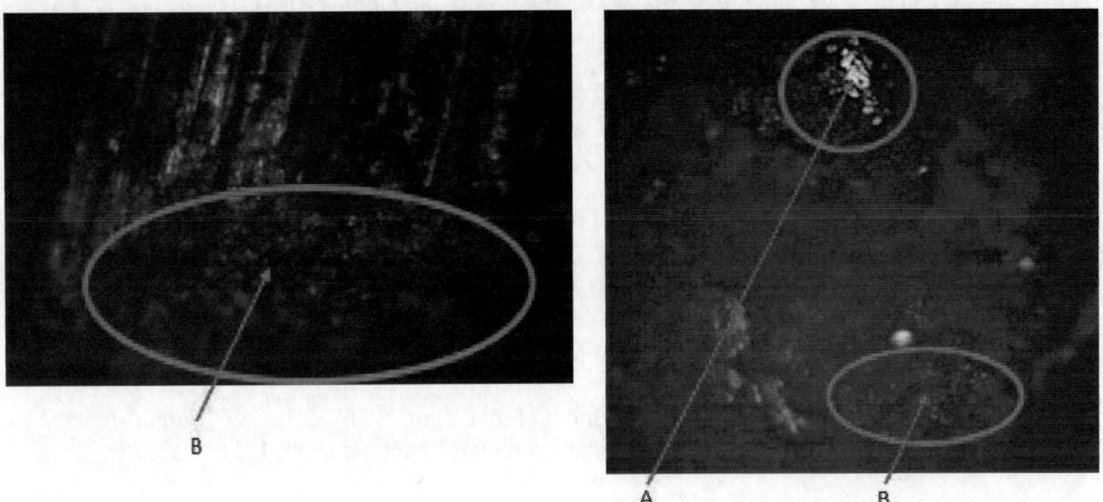

(a)

(b)

Figure 3.3 Fracture of ZAMAK 3 plates obtained from bending test showing the region of tension.

4. Conclusion

In this research, the prototype plate was performed by using indirect additive manufacturing method in this experimental. By using mould created from FDM or FFF, the cost of mould can be reduced as AM fabrication cost is not exponentially related to part complexity. Through this method, it can provide low cost of producing the plates if compared to rapid manufacturing methods.

The application of ZAMAK 3 in this experimental is to evaluate the dimensional accuracy and repeatability of using indirect additive manufacturing method to fabricate fracture plate and to understand the effect on the material strength and mechanical properties of the material. To apply ZAMAK 3 material in implant surgery are not relevant because the corrosion and weak of the strength if compared to Titanium. It is costly to perform this experimental by using Titanium material.

In term of customizing the fracture fixation plate, the plate can be design based on acquired data from CT scan. Thus, reduce the time taken by high skill of surgeon to customize the plate during surgery rather by conventional methods.

Definitely to produce mandibular plate in fixation treatment can be achieved by using indirect additive manufacturing method without spend a lot of money in producing by using EBM machine just to fabricate a small part. Besides that, this indirect additive manufacturing method also can be applied in other engineering field such as automotive field.

IOP Conf. Series: Journal of Physics: Conf. Series **908** (2017) 012063 doi:10.1088/1742-6596/908/1/012063

5. Acknowledgment
The authors gratefully acknowledge the financial support for this research from Short Term Research Grant (STG 9001-00481) under Universiti Malaysia Perlis (UniMAP) and the Ministry of Education Malaysia as well as profound gratitude and deep appreciation to School of Mechatronics Engineering, Universiti Malaysia Perlis.

6. References
[1] Rüedi T P and Murphy W M (eds.). 2000, AO principles of fracture management (Stuttgart, Germany: Thieme Medical Publishing.

[2] Parthasarathy J 2014, 3D modeling, custom implants and its future perspectives in craniofacial surgery, *Annals of Maxillofacial Surgery*, **4** **1** 9

[3] Tack P, Victor J, Gemmel P and Annemans L 2016 3D-printing techniques in medical setting: a systematic literature review *BioMedical Engineering Online*(2016)

[4] Lethaus B, Poort L, Böckmann R, Smeets R, Tolba R, and Kessler P 2012, Additive manufacturing for microvascular reconstruction of the mandible in 20 patients, *Journal of Cranio-Maxillofacial Surgery*, **40** **1** 43–6

[5] Khan S F, Dalgarno K W, Siregar R A 2015, Indirect Additive Manufacturing (AM) of Apatite-Wollastonite (A-W) Glass-Ceramic for Medical Implants, *Applied Mechanics and Materials*, **786**, 354-60

[6] PLA - RepRapWiki, Reprap.org, 2017. [Online]. Available: http://reprap.org/wiki/PLA. [Accessed: 19- May- 2017].

[7] Plaster mold casting, En.wikipedia.org, 2017. [Online]. Available: https://en.wikipedia.org/wiki/Plaster_mold_casting. [Accessed: 19- May- 2017].

[8] Zamak 3, Dynacast, 2017. [Online]. Available: https://www.dynacast.com/zamak-3. [Accessed: 19- May- 2017].

[9] E. I. Materi 2015 Standard Test Methods for Flexural Properties of Unreinforced and Reinforced Plastics and Electrical Insulating Materials 1 pp. 1–11

IOP Conf. Series: Journal of Physics: Conf. Series **908** (2017) 012064 doi:10.1088/1742-6596/908/1/012064

In-House Fabrication and Electrical Characterization of Planner si-Nanogap

Th. S. Dhahi[1], Tijjani Adam [2], Uda Hashim[3]

[1]Department of physics, University of Basra, Basra, Iraq.
[2]Faculty of Technology, Universiti Malaysia Perlis (UniMAP), Kampus Uniciti Alam Sg. Chuchuh, 02100 Padang Besar (U), Perlis, Malaysia.
[3]Institute of Nano Electronic Engineering, University Malaysia Perlis,Perlis, 01000 Kangar, Malaysia.

Abstract. Nanogap is increasingly known to be beneficial, dependable and higher sensing technology. Another possible purpose is to examine a bioactivity and study the reaction of single molecule. It is important to carefully recognize the differences between the sensor surface and electrode in order to incorporate the biological system with nanogap. Also, it crucial to examine the dielectric properties between the planar nanogap with and without a sample. Electrical concentration between the electrodes could be increased due to integrating of microfluidic channel when the sample is being used. This paper is a report on an electrical point of view of planar nanogap capacitor device with comparison of different excitation frequency with and without microfluidic channel. By using 40 nm Si nanogap devices, the sensitivity of nanogap was compared by dropping deionized water and pH 7 onto the target. Experiments were carried out in wide range of frequencies from 1 Hz to 1 MHz at room temperature with 30 mV input signal (0 V, DC, Offset). Both effects of excitation frequency on capacitance sampling with 10 μm microfluidic integration were analysed.

1.0 Introduction

There had been the greater opportunity in the field of healthcare in terms of the ability of manipulating the matters at atomic and molecular level in order to create material with altered and new properties. There had been a rapid area of researches to manipulate the matters at atomic level. Fortunately, today the technology had showed us the way to create matters with varied, new properties as well as the function of it. In the medical field, it is a great achievement which permit us to use it in therapy diagnostics, and many areas of healthcare research and development as well as clinical applications. Again, in the field of medicine, the way we would treat the diseases in the future will be surprisingly change due to the use of nanotechnology. The functionality of the device being use at present time could be improved drastically due to availability of nanotechnologies. As the results, at the present time detecting the diseases before it occurs is possible therefore doctors could be able to control not making the diseasing worst. In past many techniques, had been identified in order to apply it but today it had been in realities[1-5].

IOP Conf. Series: Journal of Physics: Conf. Series **908** (2017) 012064 doi:10.1088/1742-6596/908/1/012064

The components used in order to produce sensors are nanogap, nanoparticles, nano thin film etc. with various geometries and assembling it direct us into a powerful general platform for the ultrasensitive direct electrical detection of biological and chemical species, a device for detecting disease such as cancer, dengue etc, cells at earliest stage. The sensor renders the electric signal when it had found the particular target[6-10]. It had been design to bind with complementary target in the cell. Furthermore, various devices had been tested to be accurate in detecting the target molecules such as nano fibers, magnetic nanoparticles, gold nanoparticles and carbon nanotubes. It has sensors on it which has ability to detect the proteins and DNA of certain diseases with very less time needed, for instance less than half an hour. Between the devices capacitive devices had been known for most sensitive due to its integral or inbuilt capability to react to minute changes and capacitor is a passive component that depot energy in an electrostatic field but not in form of chemical[11].

Typically, capacitive which is grounded with transducer comprises two parallel metallic or sometime non-metallic plates that unusually divided by the permeable material for instance air, liquid which are normally describe as dielectric polarizer. In a typical capacitor, the space between the two plates non-transferable but transducing it, In the normal capacitor, the distance between the two plates non-movable, but transducing it variable this capability of varying the dielectric with allow to measure noticeable changes [1-3]. The operational mechanism is as follows. When a target molecule sample comes in close contact with the receptor/DNA, negligible but opposing partial charges appear in both plates (electrode) and can be observed by impedance analyser [4].

The Electrical bio-detection is solely relying on the changing electrical profile such currents fluctuation and voltages to identify the binding. With this complementary, mismatch and non-complementally bimolecular are studied. Due to the inherent capability of electrical detection methods in allowing Interaction with the targeted ssDNA to create a field across the transducer and increase the capacitive effect, the nanogap capacitor is special kind of capacitor that based on electrical double layer (EDL) theory that has been recognised by chemist since the nineteenth century when Von Helmholtz first developed and modelled his colloidal suspension work [5]. Previous designs of planar nanogap are still lack in geometry and structure [6-7]. Yexian Wu and co-worker using ebeam to fabricate heterogeneous metal nanogaps but still the contact surface area is not well defined [8]

The geometry of the nanogap electrode seems to be distorted and far different form AUTOCAD design due to many process and chemical involved during fabrication. This project managed to get a very sharp trench nanogap having wider surface area as designed. Simulation and electrical characterization can be conducted to get better understanding of nanogap at 40 nm gap size. According to the Boltz, the nanogap can be fabricated with gap spacing in the range of 15 to 300 nm, allowing the electrical double layer at each electrode to overlap each other at sufficiently low ionic strength [9]. By closing the air between the nanogap with microfluidic is perhaps to reduce the noise and increase the capability to store electric charge in a highly reversible way. Even though nanogap allows to readily interact with biological or ionic entities, which have smaller dimensions, other component might be integrated to complete the tuning signal process. The advantages of this device are the small time of analysis, the low reagent costs, and the reduced amount of chemical waste. The application of portable, easy to use, and highly sensitive for real time detection could offer significant advantages over current analytical methods[12-17].

Nanogap in lab on chip is very useful in many fields such as chemical analysis, environmental monitoring, medical diagnosis and cellomics, synthetic chemistry (rapid screening and microreactors for pharmaceutics) DNA labyrinths, single cell detection in bioanalytical applications [10].

IOP Conf. Series: Journal of Physics: Conf. Series **908** (2017) 012064 doi:10.1088/1742-6596/908/1/012064

2.0 Methods

In this project, we developed a simple device containing planar nanogap attached with PDMS microfluidic having micromixer in the middle for fast processing of sample. Two different sample events can be injected at the same time in a second. With well-defined structure and contact surface area of electrode, this device has a very good selectivity and sensitivity as biosensor [18-22].

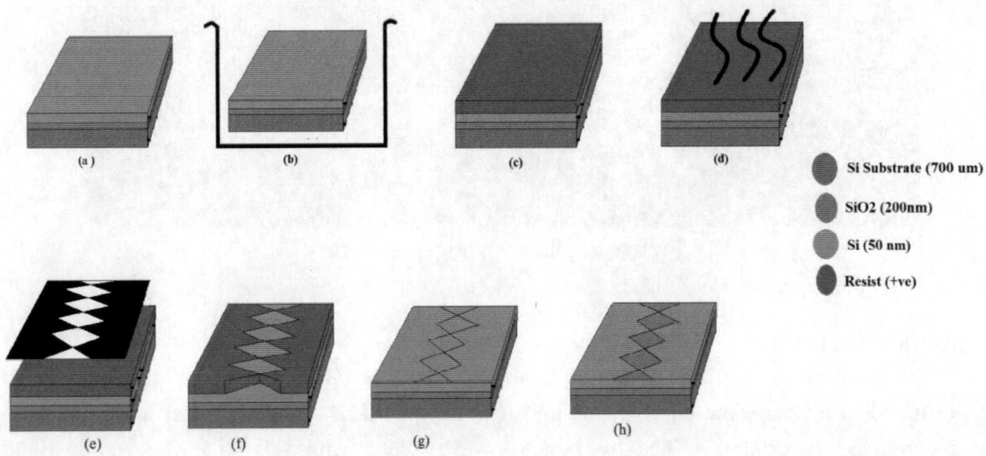

Figure 1. Fabrication flow of nanogap

First, the design of device was accomplished by using AUTOCAD as shown in Figure 1(a). Alignment marks were placed to ease the Ti/Au pad fabrication while red line indicates an oxide trench where Si nanogap electrode stands to tune the signal. Figure 1 (b) highlights the nanogap fabrication steps. Fabrication process of nanogap was realized by using Boron doped SOI substrate. After cleaning the substrate using ultrasonic machine, EZP520-87 resist is diluted in anisole is spin coated at Speed (rpm):4000 (10s), 500 (5s), 0 (5s) to get approximately 100 nm thickness of resist. The resist was baked at 180 °C for 2 minutes as shown in Figure 1 (d). The trench pad and gap line are then patterned using EBL with the same dose of 600 $\mu C/cm^2$ at 100 kV. But there were fabricated at different area step size; dose area performed for trench pad is 6.4µs-320s and 4.2µs-210s for gap. Then, the device was developed by using MIBK: IPA (+) ve developer for 30 s. The trench pad was etched separately by using RIE followed by gap line was etched by using DRIE. As a result, a sharp and higher aspect ratio of nanogap was achieved. In this paper, 40 nm nanogap having thickness of 50 nm were characterized. Finally, Ti/Au pad were simply fabricated on top of electrode by using conventional lithographic technique. To complete the function of device, PDMS microfluidic were fabricated on the Si mould. By applying negative resist, microfluidic mould was fabricated on the silicon. PDMS microfluidic could be matched with nanogap by using plasma bombardment at 27°C. To realize the geometrical aspect of device, nanogap was observed using SEM. Figure 2 has shown a very point that 40 nm nanogap was lie down on the oxide trench.

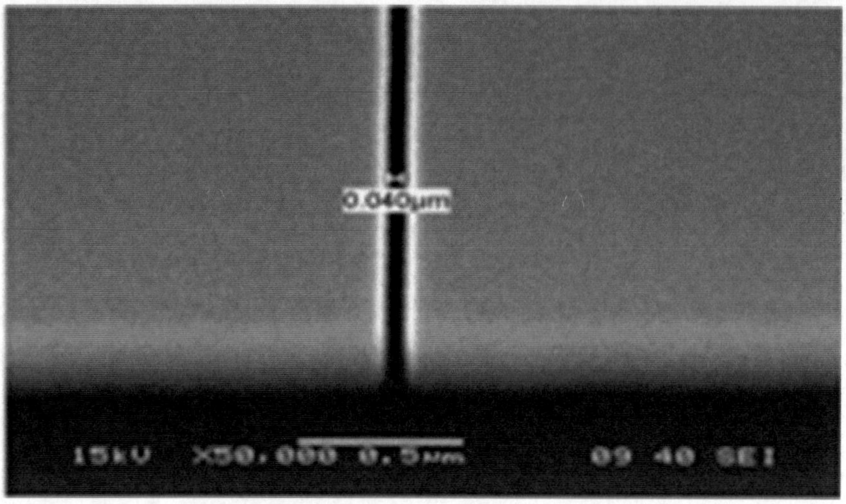

Figure 2. planar nanogap ~40nm

3.0 Results and Discussions

The structure is designed as gap and since the nanogap is designed as capacitor, the dielectric Analyzer was programmed to measure the capacitance and loss tangent. Loss tangent is defined as the ratio of capacitive reactance to resistance. The frequencies were swept from 1Hz to 1 MHz using dielectric analyzer with AC Volt [Vrms]=0.300 or 30 mV input signal (0 V D.C Offset). Reference plane is full calibrated to reference plane and probing station was fully vacuum. Below is typical nanogap measurement at 40 nm showing a near to 33 Pf ceramic capacitor. Air actual is deviates from actual model due to lack of supply vacuum. Primary capacitance model value shown in the graph is 37 nm as shown in Figure 3 (a). Loss tangent for air, deionized water and ph 7 is $<10^{-1}$ and below 10 kHz as shown in Figure 3 (b), (c) and (d). The factors that ultimately determine the best capacitor is the loss lie between these values. Note that the presence of dispersion peaks and the inverse power dependence of the capacitance with the frequency at the whole range [11-12]. First peak indicates the double layer and the second peak surely the probe contact. After dry measured capacitance, deionized water and followed by ph 7 were injected separately to the inlet of sample to see any changes. Moreover, the graph in Figure 3 (c) shows the value of capacitance tunes form the event. After triplicating the data, the final results show very a stable and accurate reading. However, air and deionized water reading is almost similar. 32 nF and 34 nF. Deionized water really has no ions. Of course, it still has about 10^{-7} molar each of H+ and OH- ions similarly to pH 7, at room temperature. In the AC field. Each H_2O dipole changes orientation with the field. If the frequency of the electric field is increased, eventually its becomes so high that the water dipoles cannot turn fast enough to keep up. Then all that happens is that nuclei and the electron cloud of each water molecule move a little response to the field. Measured at a very high frequency for water molecules are not free to rotate. So, this device only valid at low frequency up to 1Mhz.

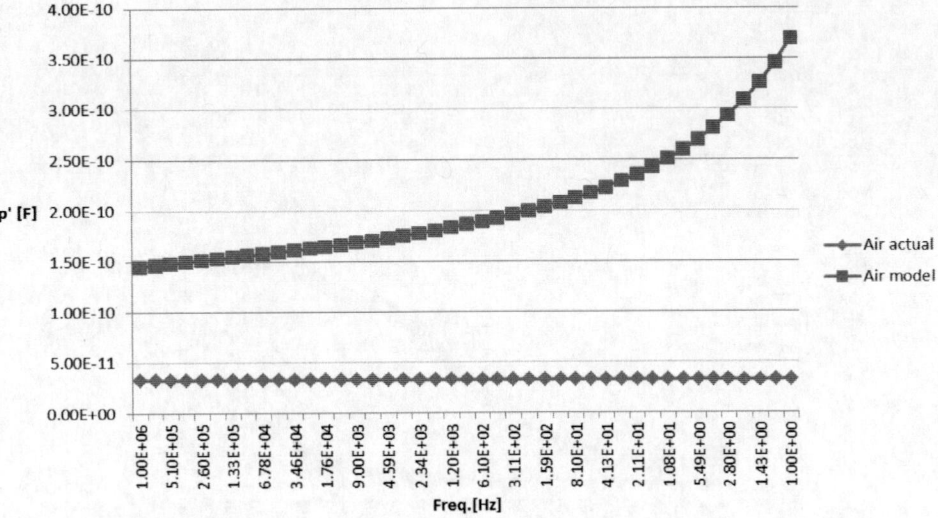

(a)

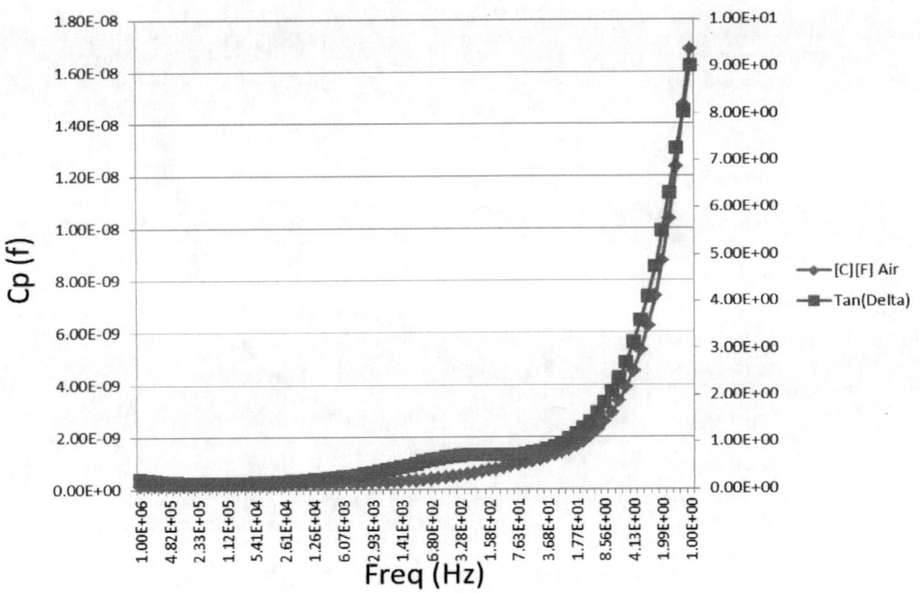

(b)

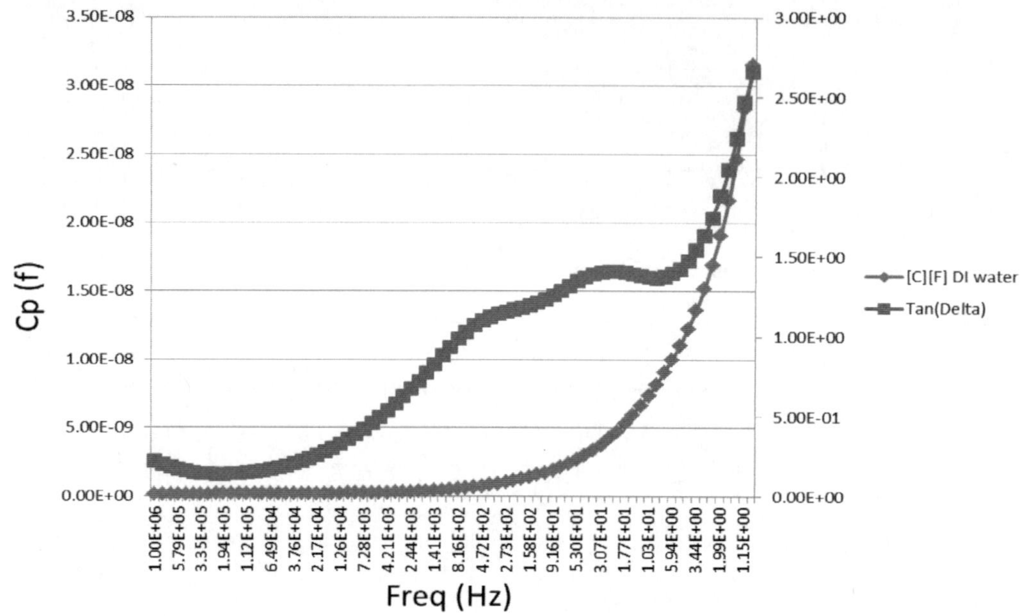

(c)

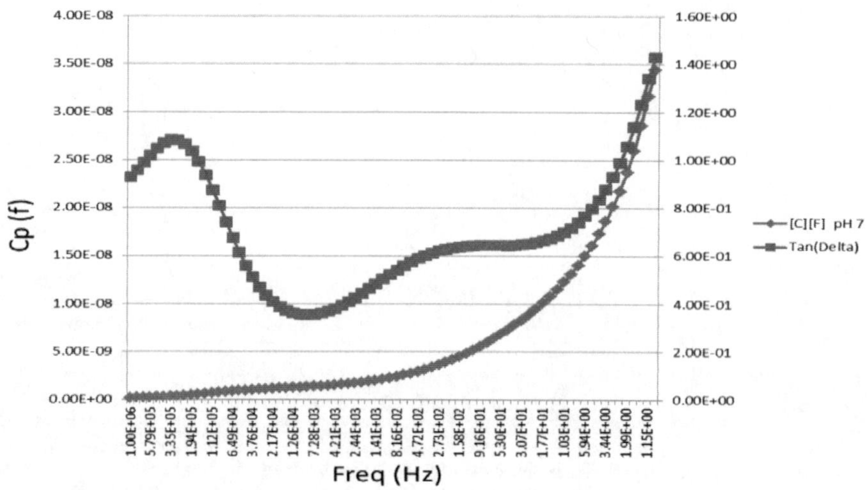

(d)

Figure 3: (a) Dry measured capacitance for 40 nm nanogap model and 33 Pf ceramic capacitor model,

From Figure 4, the response characteristic indicted by the three curves indicates the double layer and the second peak surely the probe contact. After dry measured capacitance, deionized water and followed by ph 7 were injected separately to the inlet of microfluidic to see any changes. Graph in Figure 4 shows the value of capacitance tunes form the event. After triplicating the data, the final results show very a stable and accurate reading. However, air and deionized water reading is almost similar. 32 nF and 34 nF. Deionized water really has no ions. Of course, it still has about 10^{-7} molar each of H+ and OH- ions similarly to pH 7, at room temperature. In the AC field. Each H_2O dipole changes orientation with the

field. If the frequency of the electric field is increased, eventually its becomes so high that the water dipoles cannot turn fast enough to keep up. Then all that happens is that nuclei and the electron cloud of each water molecule move a little response to the field. Measured at a very high frequency for water molecules are not free to rotate. So, this device only valid at low frequency up to 1Mhz.
(b,c,d) capacitance and loss tangent of 40 nm device.

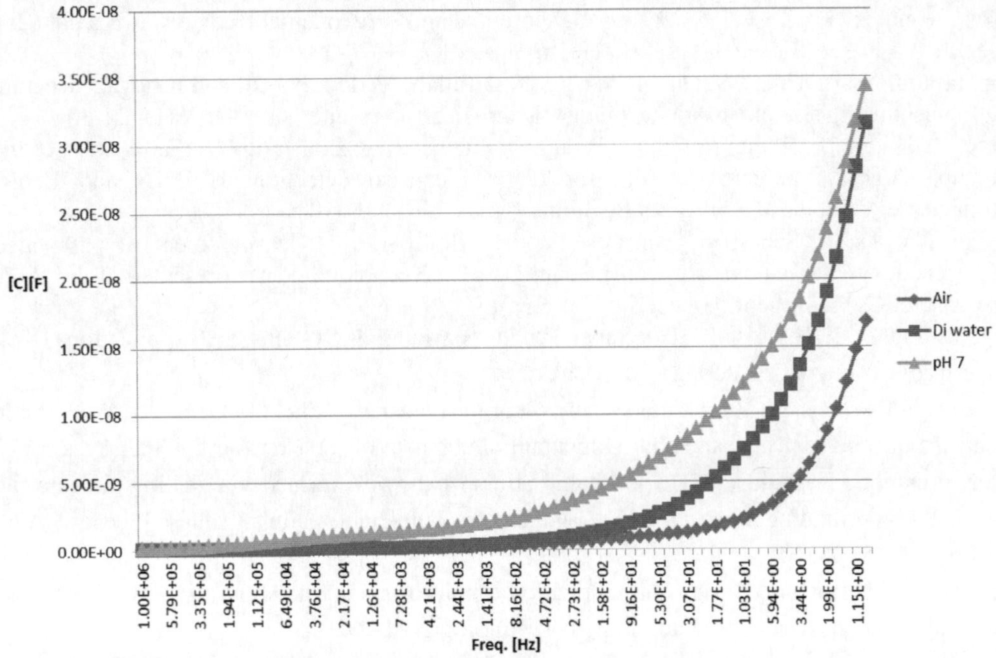

Figure 4: Capacitance reading for air= 17 n F , deionized water= 32 nF, and ph 7 =34 nF

4.0 Conclusions

The paper demonstrated the fabrication of Nano gap which is increasingly known to be beneficial, dependable and higher sensing technology. Another possible purpose is to examine a bioactivity and study the reaction of single molecule. It is important to carefully recognize the differences between the sensor surface and electrode in order to incorporate the biological system with nanogap. Also, it crucial to examine the dielectric properties between the planar nanogap with and without a sample. Electrical concentration between the electrodes could be increased due to integrating of microfluidic channel when the sample is being used. This paper is a report on an electrical point of view of planar nanogap capacitor device with comparison of different excitation frequency with and without microfluidic channel. By using 40 nm Si nanogap devices, the sensitivity of nanogap was compared by dropping deionized water and pH 7 onto the target. Experiments were carried out in wide range of frequencies from 1 Hz to 1 MHz at room temperature with 30 mV input signal (0 V, DC, Offset). Both effects of excitation frequency on capacitance sampling with 10 μm microfluidic integration were analyzed.

References

[1.] Tijjani.Adam, U.Hashim, Highly sensitive silicon nanowire biosensor with novel liquid gate Control for detection of specific single-stranded DNA molecules Biosensors and Bioelectronics67 (2015) 656–661.

[2.] Md. Eaqub Ali, Th.S. Dhahi, Rasel Das, U. Hashim, DNA hybridization detection using less than 10-nm gap silicon nanogap structure, Sensors and Actuators A 199 (2013) 304–309.
[3.] Chen, X., Guo, Z., Yang, G.-M., Li, J., Li, M.-Q., Liu, J.-H., & Huang, X.-J. (2010). Electrical nanogap devices for biosensing. Materials Today, 13(11), 28-41. doi
[4.] M Wesam Al-Mufti, U Hashim, M Rahman, Tijjani Adam, M Arshad, Studying Effect Dimensions of Design and Simulation Silicon Nanowire Filed Effect Biosensor, Applied Mechanics and Materials 754, 854-858, 2015.
[5.] Yi, M., Jeong, K.-H., & Lee, L. P. (2005). Theoretical and experimental study towards a nanogap dielectric biosensor. Biosensors and Bioelectronics, 20(7), 1320-1326
[6.] Uda Hashim, N. T., Thikra S. Dhahi, Azizullah Saifullah. (2011). Polysilicon nanogap structure development using size expansion technique. Microelectronics International, 28(3), 24-30
[7.] Wang, C., Huang, J., Wang, J., Gu, C., Wang, J., Zhang, B., & Liu, J. (2009). Fabrication of the nanogapped gold nanoparticles film for direct electrical detection of DNA and EcoRI endonuclease. Colloids and Surfaces B: Biointerfaces, 69(1), 99-104
[8.] Yexian Wu, T. A., Sebastian Gautsch, Nico De Rooij. (2011). Development of passivated Heteregeneous metal nanogaps using E-Beam Overlay Technique. Paper presented at the Proc. Eurosensors XXV, Anthens, Greece.
[9.] Francois Beguin, E. F. (Ed.). (2013). Supercapacitors, Materials,Systems, and Applications (Vol. 4). Germany: Wiley-VCH Verlag GmbH & Co.
[10.] Dimitrios P.Nikolelis, T. V., Arzum Erdem, Georgia-Paraskevi Nikoleli (Ed.). (2014). Portable biosensing of food toxicants and environmental pollutants (Vol. 1). New York: CRC Press.
[11.] Tijjani Adam U. Hashim and Th S. Dhahi, Silicon nanowire formed via shallow anisotropic etching, Si-ash-trimming for specific DNA and electrochemical detection, Chinese Physics B Vol. 24, No. 6 (2015) 06810
[12.] Tijjani Adam, U Hashim, Design and fabrication of micro-mixer with short turns angles for self-generated turbulent structures, Microsystem Technologies, 1-8, 2015
[13.] T Adam, U Hashim, TS Dhahi, KN Khor, PS Chee, PL Leow, Electrochemical Etching: An Ultrasonic Enhance Method of Silicon Nano Porous Fabrication, Wulfenia Journal 20 (1), 45-55, 2013
[14.] Tijjani Adam, H Uda, M Eaqub, PL Leow, The electroosmosis mechanism for fluid delivery in PDMS multi-layer microchannels, American Scientific Publishers All rights reserved.2013
[15.] U Hashim, T Adam, J Lung, P Ling, Fabrication of Microchannel and Micro Chamber for Microfluidic Lab-on-Chip.Australian Journal of Basic & Applied Sciences 7 (1)2013
[16.] T Adam and U. Hashim Light Observation in Polymer: A Study of Silicon-Based Organic Polymer Using Visible Spectroscopy, Australian Journal of Basic and Applied Sciences 7 (1), 76-80, 2013.
[17.] U Hashim, MW Al-Mufti, T Adam, Silicon Nanowire Geometry: Investigation of Interaction Site Potential in Semiconductor-DNA Interaction, Australian Journal of Basic and Applied Sciences 7 (5), 242-245, 2013
[18.] MW Al-Mufti, U Hashim, T Adam, Simulation of Nano lab on chip devices by using COMSOL MultiphysicsJournal of Applied Sciences Research 9 (2), 1056-1061, 2013
[19.] T Adam, U Hashim, KL Foo, TS Dhahi, T Nazwa, Technology development for nano structure formation: Fabrication and characterization, Advanced Science Letters 19 (1), 132-137, 2013
[20.] T Adam, U Hashim, PL Leow, KL Foo, PS Chee, Selection of optimal parameters in fabrication of poly (dimethylsiloxane) microfluidics using taguchi method, Advanced Science Letters 19 (1), 32-36,2013
[21.] T Adam, U Hashim, KL Foo, Microfluidics design and fabrication for life sciences application, Advanced Science Letters 19 (1), 48-53, 2013
[22.] T Adam, U Hashim, ME Ali, PL Leow, The electroosmosis mechanism for fluid delivery in PDMS multi-layer microchannel, Advanced Science Letters 19 (1), 12-15, 2013

Dependence of lattice strain of magnetite nanoparticles on precipitation temperature and pH of solution

A H M Yusoff[1,2], M N Salimi[1] and M F Jamlos[2]

[1]School of Bioprocess Engineering, Universiti Malaysia Perlis, 02600 Arau, Malaysia.
[2]School of Computer and Communication Engineering, Universiti Malaysia Perlis, 02600 Arau, Perlis.

huzaifah.usm@gmail.com[2]

Abstract. Magnetite nanoparticles were prepared by co-precipitation method at different precipitation temperature and pH of solution. The produced inverse spinel cubic structure can be denoted to magnetite structure as deduced from X–ray diffraction (XRD) result. X-ray peak broadening analysis was utilized to determine the crystallite sizes and lattice strain present by the Williamson-Hall method. The results showed that the average crystallite size and tensile strain of the nanoparticles was first decrease before rising again at a certain temperature. The same phenomenon of size was experienced on the pH variation of the solution with compressive and tensile strain was displayed. Is was observed that the almost perfect structure (very low strain) and relatively small nanoparticles was synthesized at temperature of 70 °C and pH of 11.94 with 8.89 nm in size.

1. Introduction

Among the various magnetic nanoparticles, magnetite is being widely studied due to its ability to magnetically guidance their movement into the targeted site. This attractive property has made magnetite a potential material to use in bioseparations[1], magnetic resonance imaging [2], hyperthermia [3], drug delivery agent [4] and magnetofection[5]. Thus, understanding to control nanoparticles motion is a beneficial knowledge to cure various disease at maximum performance. Magnetism property of magnetite exhibit size-dependent properties. Research to have magnetism property appropriate to the ideal size that suited for biomedical application has been greatly discussed. Besides size, magnetism also strongly influenced on their crystal imperfection [6]. Lattice distortion, grain boundary, dislocations and faulting are some example that lead to distribution of imperfect crystallinity of nanoparticles [7].

Having a perfect structure (unstrained particle) and additionally small size of particle are two remarkable parameter to yield well magnetism properties. It is noticed that the defects in structural properties can cause peak broadening and shift the Bragg peak angle of the X-ray diffraction (XRD) data. Thus, the crystallinity defect and the strain distribution in the sample can be resolved by analysing the peak width. Williamson-Hall (WH) method is one of the existing way to extract the size and strain from the diffraction peak. Hence, in the present study, the well-known Scherrer method and WH analysis was employed for estimating crystallite size and lattice strain.

This paper accounts for the synthesis of magnetite nanoparticles using a simple co-precipitation route. The role of precipitation temperature and pH condition was studied by evaluating the strain and stress distribution as less work is done in respect of this.

2. Methodology

2.1. Magnetite nanoparticles preparation

Iron (III) chloride ($FeCl_3$, 98%pure, anhydrous), iron (II) chloride ($FeCl_2.4H_2O$, 99%), sodium hydroxide (NaOH) were obtained from Acros Organics. All chemical reagents applied in this work were of analytical grade and the chemicals used without further purification.

The magnetite nanoparticles were prepared through co-precipitation method from $FeCl_3$ and $FeCl_2.4H_2O$. Both iron precursors were dissolved in treated deionized water (DI) at a mole ratio of 2:1. The resulting yellowish solution was refluxed at different temperature (30 °C, 50 °C, 70 °C and 90 °C) for 20 minutes prior NaOH was added. The stirring rate was maintained at 700 rpm throughout of the experiment. pH of the mixture was manipulated by addition of pH (pH = 11.33, 11.94, 12.15 and 12.25). Synthesis processed are performed under oxygen free condition to prevent the formation of maghemite. The black suspension formed were collected by centrifuge at 3000 rpm for 15 minutes and washed several times with DI water. Lastly, the precipitated was redispersed and dried in a vacuum oven for overnight at 60 °C for further characterization.

2.2. Characterization

The sample crystal structure were determine by PanalyticalX'pert Pro X-ray diffractometerusing Cu Kα radiation (λ = 0.15406 nm) at room temperature. The prepared Fe_3O_4 powders were pressing over the sample holder and the analysis was recorded over 2θ range of 20° to 80° at a scan rate of 1°/min.

3. Results and Discussion

All the X-ray diffraction (XRD) pattern under study were compared with the standard reference data of magnetite (JCPDS Card No. 89-0691). The significant diffraction peaks were appeared at Bragg angles that corresponding to Miller indices of (220), (311), (400), (511) and (440) which consistent with the standard pattern. Thus, all the samples possessed the spinel structure of Fd3m group. No additional peaks were observed indicated that no impurities phases such as maghemite or goethite except the only magnetite phase was found in the final product. During the nanoparticles preparation, growing crystallites of magnetite may take a specific crystallographic direction that normally noticeable through the highest intensity of the diffraction data. The preferred direction of crystallite growth was influence by the ions compositions and the synthesis condition. Earlier studies [8] show that the growth orientation is found to be preferentially in the direction of (110) prior change to (104) crystallographic direction. Thermal treatment during nanoparticle preparation were observed to have that serious change of growth orientation where hematite phase was evolved after calcination at 650 °C and above. Another study reveal that the calcination time [9] and high pH of solution [10, 11] also change the final phase into maghemite or goethite, respectively. Besides that, Wang et al. [12] suggest that the change of growth orientation may relate to the strain developed in nanoparticle grains. Nonetheless, in present case, the preferred growth orientation shows along the (311) crystallographic direction which verified the only magnetite phase was formed.

The average crystallite size and lattice strain distribution in the material can be find out through the peak broadening. But, the breadth of the Bragg peak must be refined due to the convolution of the profiles with instrumental error and sample dependent effects. Thus, line broadening of standard material such as silicon was employed to correct and eliminate both contributions in β_D calculation through the following equation:

$$\beta_D = \sqrt{\beta^2_{measured} - \beta^2_{instrumental}} \qquad (1)$$

Mostly, the trend of change of crystallite size is determined by a well-known Scherrer formula. Values of full width at half maximum (FWHM) of diffraction peaks at 2θ = 35.4° was chosen for calculation of the crystallite size using Scherrer's formula:

$$D_S = \frac{K\lambda}{\beta_D \cos\theta} \qquad (2)$$

IOP Conf. Series: Journal of Physics: Conf. Series **908** (2017) 012065 doi:10.1088/1742-6596/908/1/012065

where D_s is the crystallite size (nm), K is the shape parameter which is 0.94 for spherical shape, λ is X-ray wavelength (λ = 0.15406 nm for Cu Kα radiation), β_D is the full width at half maximum for the diffraction peak under consideration (rad) and θ is the corresponding Bragg angle (°). The average crystallite size (D_S) from the Scherrer method is tabulated in Table 1.

In addition, Williamson-Hall (WH) plot was also employed to determine the crystallite size and effect of strain in the as-synthesized magnetite nanoparticles. Each XRD result were fitted with Lorentziancurve shape and the total broadening (β_{hkl}) is obtained as the following equations:

$$\beta_{hkl} = \beta_S + \beta_D \quad (3)$$

where β_s is crystallite size broadening and β_D is lattice strain broadening. A Williamson-Hall's mathematical relation is employed that relate between strain-induced broadening in the sample and average crystallite size as follows:

$$\beta_{hkl} = \frac{K\lambda}{D\cos\theta} + 4\varepsilon\tan\theta \quad (4)$$

By rearrangingthe Equation(4):

$$\beta_{hkl}\cos\theta = \frac{K\lambda}{D_{WH}} + 4\varepsilon\sin\theta \quad (5)$$

where ε is lattice strain and D_{WH} is crystallite size computed from WH relation. This strain was assumed to have a uniform strain in all crystallographic directions. A plot of 4sinθ versus $\beta_{hkl}\cos\theta$ were plotted in figure 1 and figure 2 for different reaction temperature and different pH of solution, respectively. The slope from the straight line represent the value of strain in the sample while y-intercept was utilised for D_{WH} measurement corresponding to zero strain. The D_{WH} and lattice strain computed through the WH method were shown in table 1. The small difference of crystallite size between D_S and D_{WH} is owing to the crystal strain involvement[7]. Scherrer method ignores the microstrain in the lattice that originating from the imperfection of structure whereas Williamson-Hall method isolates the instrumental and sample broadening effects in calculation.

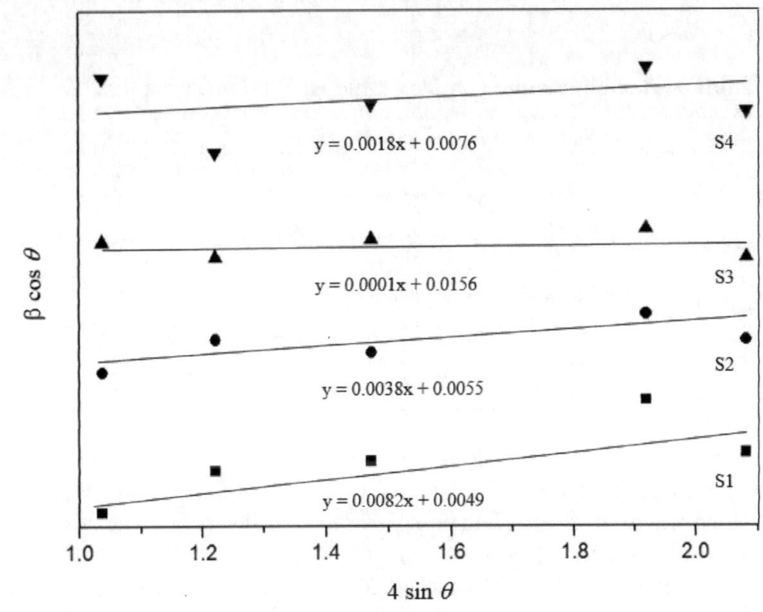

Figure 1. The WH analysis of magnetite nanoparticles prepared at 30 °C (S1), 50 °C (S2), 70 °C (S3) and 90 °C (S4).

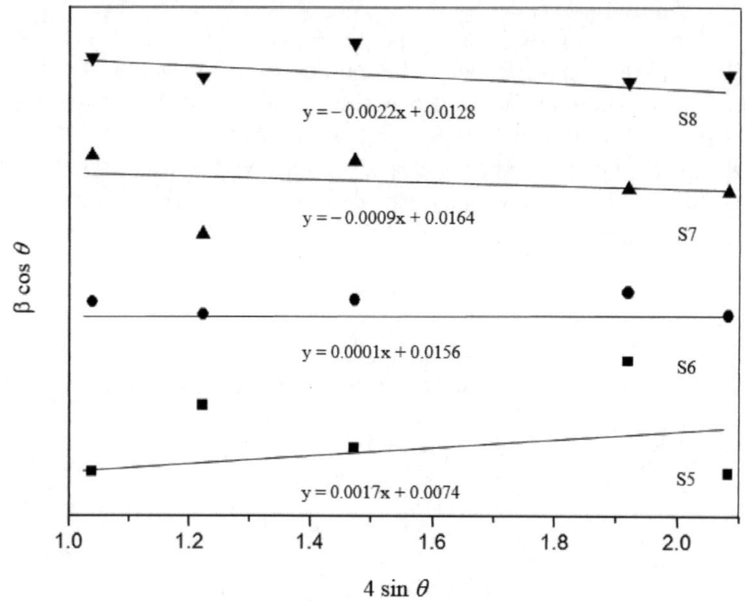

Figure 2. The WH analysis of magnetite nanoparticles prepared at pH = 11.33 (S5), pH = 11.94 (S6), pH = 12.15 (S7) and pH = 12.25 (S8).

The strain observed in all samples were relatively small as compared to another literature. Synthesis technique is one of the factor that contribute to the microstrain broadening of diffraction peak. Vives et al. [13] studied the line profile of iron ball milled powders and obtained high internal microstrain in their sample. It is obvious that the samples were allowed to receive high rotation speed (150, 250 and 350 rpm) that can induce high residual stress in the compacted part. Contrary to the chemical synthesis technique that not required heavy mechanical process and thermal stress such as quenching during the sample preparation. Our result were consistent with Jafari et al. [8] and Khorsand et al. [14] which both prepare the nanoparticles using co-precipitation method and sol-gel method, respectively.

Table 1. A slightly more complex table with a narrow caption.

Sample code	Processing condition	D_S (nm)	D_{WH} (nm)	Strain($\times 10^{-4}$)
S1	30 °C	21.18	28.30	82
S2	50 °C	11.78	25.21	38
S3	70 °C	10.60	8.89	1
S4	90 °C	30.26	18.24	18
S5	pH 11.33	17.66	18.74	17
S6	pH 11.94	13.15	8.89	1
S7	pH 12.15	10.10	8.45	–9
S8	pH 12.25	17.67	10.83	–22

It is noted that inhomogeneous strain either compressive or tensile strain give rise to peak widths broadening and shift the 2θ peak position. As can be observed in Figure 1 and Figure 2, the lattice strain has a positive slope when the crystal lattice experience tensile forces whereas lattice shrinkage due to the compressive strain was indicated by the negative slope. Samples at different precipitation temperature showed a continuous lattice expansion where the strain decreases when temperature was keep increasing. The increase in temperature is likely to drive proper modification of atom arrangement in the lattice since the atoms receive more energy at high temperature. However, tensile stress was increases again that arising from the Ostwald ripening-type phenomenon. The phenomenon

IOP Conf. Series: Journal of Physics: Conf. Series **908** (2017) 012065 doi:10.1088/1742-6596/908/1/012065

is the predominant growth mechanism at higher precipitation temperature that results in greater final diameter of particles. Increase in strain may be associated with the change in lattice strain number where the number was increasing during the growth phase correspondingly to their size.

On the other hand, samples at variation of pH showed the contribution of both type of strain. pH at 11.33 and 11.94 showed a reduction of tensile strain whereas sample at pH 12.15 and 12.25 showed an increment in compressive strain. Lian et al. [15] reported that goethite was formed first before magnetite formation at very high pH. It is noted that goethite has an orthorhombic crystal structure and this is expected to induce intense atomic competition during magnetite formation. Structural clashes generate lattice shrinkage in the structure and keep increasing at higher pH.

4. Conclusion

Simple way to obtain magnetite nanoparticles by co-precipitation method which enhance their structural properties at small particle size was presented. The result demonstrate that the parameter under study have significant influence on crystallite size and strain distribution of synthesized nanoparticles. For particles with less strain distribution resulted in less surface imperfection. Generally, surface defects have decreased to a certain temperature because of energy gain in the system. While, same pattern was observed at change of pH but the compressive strain evolved can be explained by the overlap structure formed at high pH. Therefore, respect to this study, it could be concluded that such well structure of magnetite can be obtained at 70 °C and pH = 11.94 with 8.89 nm in size.

5. References

[1] Nam J, Huang H, Lim H, Lim C and Shin S 2013 *Anal. Chem.* **85** 7316
[2] Zhang Y, Zhang L, Song X, Gu X, Sun H, Fu C and Meng F 2015 *Journal of Nanomaterials* **2015** 6
[3] Nguyen N T, Tran D L, Nguyen D C, Nguyen T L, Ba T C, Nguyen B H, Ba T D, Pham N H, Nguyen D T, Tran T H and Pham G D 2015 *Current Applied Physics* **15** 1482
[4] Niu C, Wang Z, Lu G, Krupka T M, Sun Y, You Y, Song W, Ran H, Li P and Zheng Y 2013 *Biomaterials* **34** 2307
[5] Das M, Wang C, Bedi R, Mohapatra S S and Mohapatra S 2014 *Nanomedicine* **10** 1539
[6] Upadhyay S, Parekh K and Pandey B 2016 *J. Alloys Compd.* **678** 478
[7] Cullity B D 1957 *American Journal of Physics* 25 394
[8] Jafari A, Shayesteh S F, Salouti M and Boustani K 2015 *Indian Journal of Physics* **89** 551
[9] Duhan S and Devi S 2010 *International Journal of Electronics Engineering* **2** 89
[10] Gnanaprakash G, Mahadevan S, Jayakumar T, Kalyanasundaram P, Philip J and Raj B 2007 *Mater. Chem. Phys.* **103** 168
[11] Andrade Â L, Souza D M, Pereira M C, Fabris J D and Domingues R Z 2010 *Quim. Nova* **33** 524
[12] Wang Z, Zhou B-x, Chen B, Zhu W, Wen B, Wu L, Tang H-k, Fang Z-q, Li Q and Yao M-y S 2017 *Corros. Sci.* **122** 26
[13] Vives S, Gaffet E and Meunier C 2004 *Materials Science and Engineering: A* **366** 229
[14] Khorsand Zak A, Abd. Majid W H, Abrishami M E and Yousefi R 2011 *Solid State Sciences* **13** 251
[15] Lian S, Wang E, Kang Z, Bai Y, Gao L, Jiang M, Hu C and Xu L 2004 *Solid State Commun.* **129** 485

IOP Conf. Series: Journal of Physics: Conf. Series **908** (2017) 012066 doi:10.1088/1742-6596/908/1/012066

Development of Matlab Simulink model for dynamics analysis of passive suspension system for lightweight vehicle

M S Jamali[1], K A Ismail[1], Z Taha[2] and M F Aiman[1]

[1] School of Manufacturing Engineering, UniMAP, 02600 Arau, Perlis, Malaysia
[2] Faculty of Manufacturing Engineering, UMP, 26600 Pekan, Pahang, Malaysia

jamali@unimap.edu.my

Abstract. In designing suitable isolators to reduce unwanted vibration in vehicles, the response from a mathematical model which characterizes the transmissibility ratio of the input and output of the vehicle is required. In this study, a Matlab Simulink model is developed to study the dynamic behaviour performance of passive suspension system for a lightweight electric vehicle. The Simulink model is based on the two degrees of freedom system quarter car model. The model is compared to the theoretical plots of the transmissibility ratios between the amplitudes of the displacements and accelerations of the sprung and unsprung masses to the amplitudes of the ground, against the frequencies at different damping values. It was found that the frequency responses obtained from the theoretical calculations and from the Simulink simulation is comparable to each other. Hence, the model may be extended to a full vehicle model.

1. Introduction

The suspension system of a vehicle is primarily designed to provide good handling ability and to prevent occupants from discomfort, fatigue and health problems due to various road profiles and irregularities. However, it is difficult to achieve an optimal condition of ride comfort and handling simultaneously when using a passive suspension system [1]-[3]. The passive type suspension consists of spring and damper as the main components where no energy is added into the system compared to an active suspension system. An active suspension is capable of improving the ride and handling condition but it requires external power to function which results to added complexity in the design, cost and weight of the system. As a result, a passive type suspension system is preferred for a lightweight electric vehicle due to its simplicity and less cost involved [4]-[5].

Noise, vibration and harshness (NVH) engineering has been practiced in the automotive industry since 1980s due to customers demand for improved vehicle refinement, ride and quality. This discipline has become a crossover point for engineering practice, ergonomics and psychology where prediction of comfort levels in production vehicle is one of the objectives. This has demanded for the development of predictive models to evaluate the ride comfort level without using extensive prototyping [6]. Most of the numerical analysis studies related to vehicle vibration were based on mass-spring-damper systems. Generally, the studies were done by adopting parameters of ready-made systems and necessary modification is done to existing systems to overcome the problems. This approach is limited because of the complexity of the system. However, the use of simulation software to analyse vibration characteristics for structures has enabled designers to analyse and determine the components and suitable parameters that are able to overcome the vibration problems.

The purpose of this study is to investigate the vehicle passive suspension performance and its ability to provide good driving condition by comparing the transmissibility ratios for different damping values. A mathematical model for the vehicle suspension system is developed in order to simulate its dynamic performance. This model will help to find the best suspension system parameters in providing an optimum driving comfort level.

2. Mathematical modeling

A vehicle development most likely commences with a mathematical model in order to optimise its dynamic and seating responses before a prototype is constructed. Furthermore, the model is used as an evaluation method that considers the sensitivity of human body to different frequencies, directions, and durations of vibration, and to obtain subjective evaluations that can contribute to design improvement [7].

A complex vehicle body supported by several elastic suspensions can be modelled as a simplified lumped parameter system. In this chapter, the mathematical equations governing the behaviour of the lumped parameters system for a lightweight solar vehicle are derived and discussed. The rigid body of the vehicle or the sprung mass is connected to other masses such as the wheels through massless springs and dampers that represent the suspension system. The wheels or the unsprung masses are connected to the ground through massless springs and dampers simulating the tyres. This simplified model is an approximation since the suspensions and the tyres actually have their own mass and natural frequencies. Apart from that, the body and the linkages can be considered to be non-rigid bodies [8].

The equations of motion of a vibrating system are usually presented in the form of a set of ordinary differential equations for a discrete system and partial differential equations for a continuous system. The fundamental law from which most vehicle dynamics analyses begin is the Newton's second law that can be applied to translational and rotational systems [9]-[10]. The law states that, for a translational systems, the sum of the external forces acting on a body in a given direction is equal to the product of its mass and acceleration in a particular direction, as represented by the following equation,

$$\sum F_i = M \cdot a_i \quad (1)$$

where F_i is the forces in the *i*-direction, M is the mass of the body and a_i is the acceleration in the *i*-direction.

2.1. Development of model for electric vehicle

Various components inside a solar vehicle are assembled together with the body to become a unit vehicle and moving together which makes it possible to represent them as one lumped mass at the centre of gravity (CG) of a system with appropriate total mass and inertia properties. Fundamentally, all road vehicles have common ride isolation properties that use primary suspension system at each wheel to support a sprung mass. This system dynamic behaviour is the first level of isolation from the unevenness of the road. An isolation system can be passive or active depending on whether there is external power to function as an isolator. Normally, a passive isolator consists of a resilient member (stiffness) such as metal spring, elastomer spring or pneumatic spring, and energy dissipator (damping). An active control isolator applies actuator such as hydraulic piston, piezoelectric device or electric motor to provide a force to the system whose vibration properties are to be changed [11]-[12]. Even though an active control system performs better than a passive isolator, it is costly and less reliable than the latter.

2.1.1. Quarter vehicle model

The essential dynamics can be represented by two DOF quarter vehicle models as shown in figure 1. The sprung mass is supported by a primary suspension that is placed on the unsprung mass of the axle. The equations of motion for this model represent the motions of unsprung mass, M_u and sprung mass, M_s according to their free body diagram, and solving the forces acting on the bodies at equilibrium.

IOP Conf. Series: Journal of Physics: Conf. Series **908** (2017) 012066 doi:10.1088/1742-6596/908/1/012066

The motion of each mass is represented by the displacement, velocity and acceleration, in the vertical direction, represented by Z , Z and Z respectively from the static equilibrium position. The input displacement due to road profile is given by Z_r while Z_u and Z_s are displacement of M_u and M_s respectively.

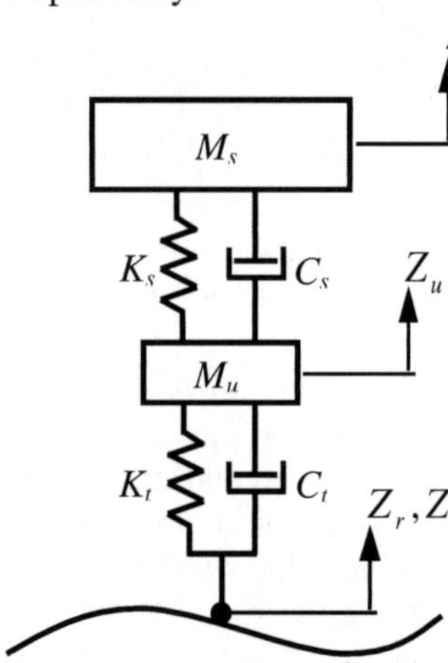

Figure 1: Two DOF quarter-car model.

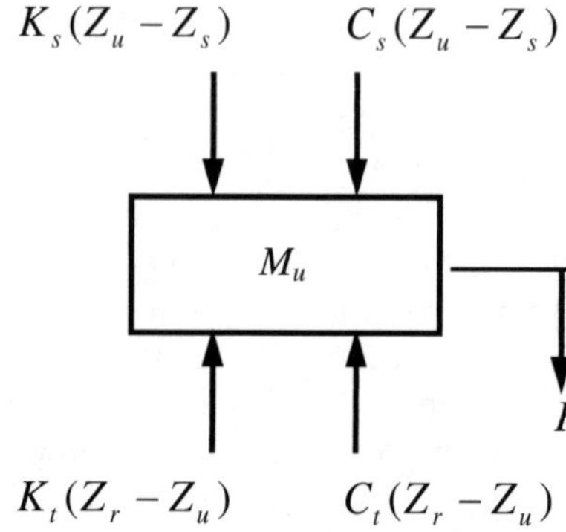

Figure 2: Free body diagram for unsprung mass, M_u.

Firstly, the equation of motion for each mass is developed when the system is at equilibrium. The free body diagram for mass, M_u as in figure 2 is used to develop the equation by assuming that the system is linear and Z_r is greater than Z_u, and the latter is greater than Z_s. The stiffness and damping constants of the shock absorber are K_s and C_s respectively while K_t and C_t represents the stiffness and damping constants of the tyre respectively. The equation of motion for the M_u is given by

$$M_u Z_u = K_t(Z_r - Z_u) - K_s(Z_u - Z_s) + C_t(Z_r - Z_u) - C_s(Z_u - Z_s) - F_u \quad (2)$$

where F_u is the force acting on M_u. Similarly, the equation of motion for M_s is developed based on the free body diagram as shown in figure 3.

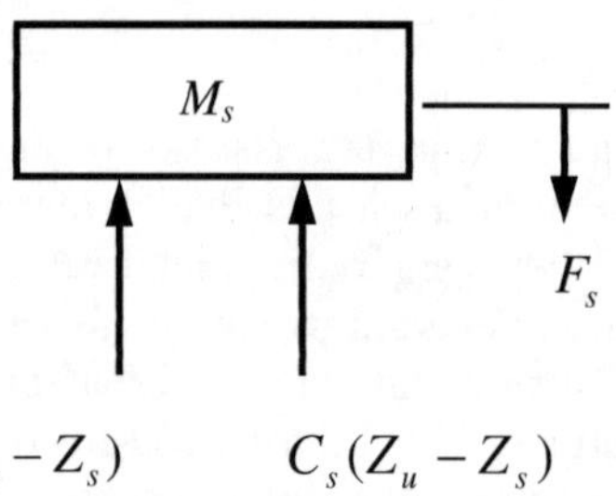

Figure 3: Free body diagram for sprung mass, M_s.

The equation of motion for M_s is given by

$$M_s Z_s = K_s(Z_u - Z_s) + C_s(Z_u - Z_s) - F_s \quad (3)$$

where F_s is the force acting on M_s. Therefore, the equations of motion for the two DOF model as in figure 1 can be written in the form

$$\begin{bmatrix} M_s & 0 \\ 0 & M_u \end{bmatrix}\begin{Bmatrix} \ddot{Z}_s \\ \ddot{Z}_u \end{Bmatrix}+\begin{bmatrix} C_s & -C_s \\ -C_s & C_s+C_t \end{bmatrix}\begin{Bmatrix} \dot{Z}_s \\ \dot{Z}_u \end{Bmatrix}+\begin{bmatrix} K_s & -K_s \\ -K_s & K_s+K_t \end{bmatrix}\begin{Bmatrix} Z_s \\ Z_u \end{Bmatrix}=\begin{Bmatrix} -F_s \\ C_t\dot{Z}_r+K_tZ_r-F_u \end{Bmatrix} \tag{4}$$

which is of the form

$$[M]\{\ddot{Z}\}+[C]\{\dot{Z}\}+[K]\{Z\}=\{F(t)\} \tag{5}$$

where $[M]$ is the mass matrix, $[C]$ is the viscous damping matrix, $[K]$ is the stiffness matrix, and $\{\ddot{Z}\}$, $\{\dot{Z}\}$, $\{Z\}$ and $\{F(t)\}$ are vectors representing accelerations, velocities, displacements and external forces respectively. The matrix equation can be simplified by neglecting some parameters such as the weight of the masses due to static equilibrium position and also the damping constant of the tyre which is usually very small [10]. The simplified equation is given by

$$\begin{bmatrix} M_s & 0 \\ 0 & M_u \end{bmatrix}\begin{Bmatrix} \ddot{Z}_s \\ \ddot{Z}_u \end{Bmatrix}+\begin{bmatrix} C_s & -C_s \\ -C_s & C_s \end{bmatrix}\begin{Bmatrix} \dot{Z}_s \\ \dot{Z}_u \end{Bmatrix}+\begin{bmatrix} K_s & -K_s \\ -K_s & K_s+K_t \end{bmatrix}\begin{Bmatrix} Z_s \\ Z_u \end{Bmatrix}=\begin{Bmatrix} 0 \\ K_tZ_r \end{Bmatrix} \tag{6}$$

if F_s and F_u as in Eq. (4) are equal to M_sg and M_ug respectively where g is the gravitational force. The frequency response of the system to a harmonic motion can be obtained by stating a harmonic input of the type $Z_r = Z_{r0}e^{i\lambda t}$. The outputs are harmonic and can be expressed as $Z_u = Z_{u0}e^{i\lambda t}$ and $Z_s = Z_{s0}e^{i\lambda t}$ where all amplitudes Z_{r0}, Z_{u0} and Z_{s0} are complex numbers to account for the different phasing of the response and excitation at frequency λ [8]. The amplitude ratio of the output and input is given by

$$\frac{|Z_{s0}|}{|Z_{r0}|} = K_t\sqrt{\frac{K_s^2+C_s^2\lambda^2}{f^2(\lambda)+C_s^2\lambda^2 g^2(\lambda)}} \tag{7}$$

and

$$\frac{|Z_{u0}|}{|Z_{r0}|} = K_t\sqrt{\frac{(K_s-M_s\lambda^2)^2+C_s^2\lambda^2}{f^2(\lambda)+C_s^2\lambda^2 g^2(\lambda)}} \tag{8}$$

where

$$f(\lambda) = M_sM_u\lambda^4-[K_tM_s+K_s(M_s+M_u)]\lambda^2+K_sK_t$$
$$g(\lambda) = (M_s+M_u)\lambda^2-K_t$$

2.1.2. The MATLAB Simulink model

Simulink is a block diagram environment for multi-domain simulation and model-based design [13]. Each block implements some function on its inputs and produces the results. The blocks are chosen and connected based on the mathematical equations derived for the model of the vehicle. A Simulink block diagram for the quarter vehicle model has been constructed based on equation (2) and equation (3) as shown in figure 4 and its subsystems as in figure 5 and figure 6 [14]-[16].

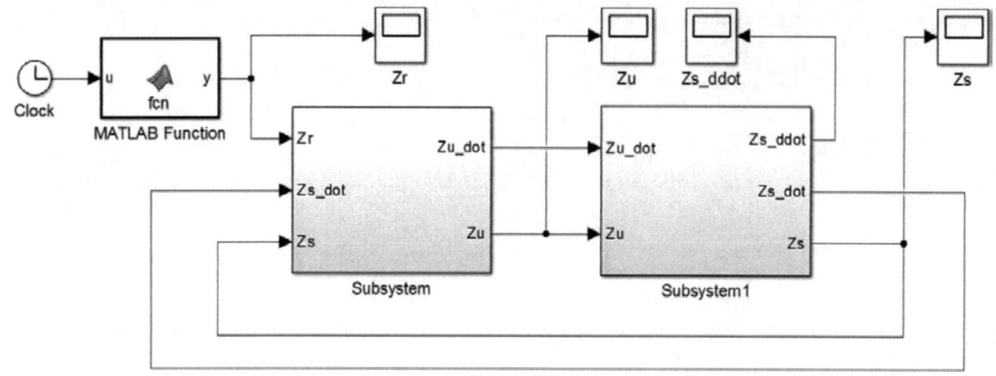

Figure 4: Simulink quarter vehicle model.

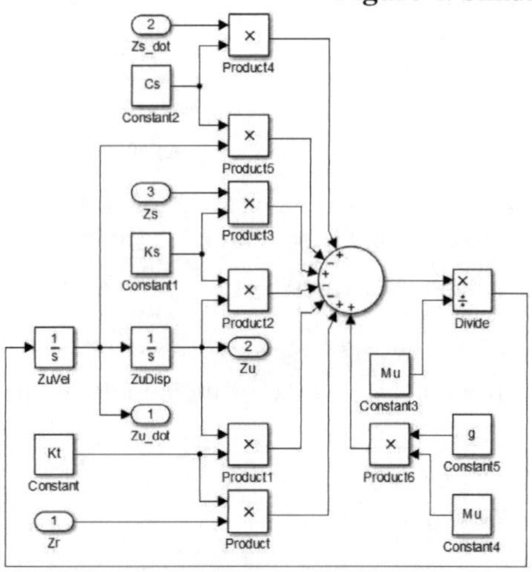

Figure 5: Simulink quarter vehicle model subsystem from figure 4.

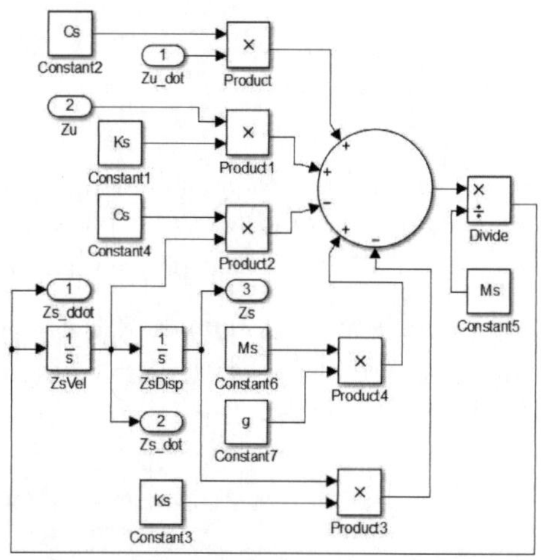

Figure 6: Simulink quarter vehicle model subsystem from figure 4.

There are different types of road excitation that can be applied to the system such as step, sine wave and speed bump signals. The MATLAB Function as in figure 4 allows a sinusoidal excitation to be developed and used as input to the model. The mathematical expression of the excitation is given by:

$$\left.\begin{array}{lll} y=0 & \text{for} & 0 \le u < 10 \\ y=0.01\sin(\lambda^{*}u) & \text{for} & 10 \le u \le 30 \\ y=0 & \text{for} & u > 30 \end{array}\right\} \qquad (9)$$

3. Results and discussion

The frequency responses related to both M_s and M_u are plotted as shown in figure 7 and figure 8 for a system with K_t = 10 kN/m, K_s = 2.5 kN/m ($K_t = 4K_s$) and M_s = 150 kg, M_u = 10 kg ($M_s = 15M_u$). The values of the masses, M and stiffness constants, K are based on an actual three-wheeled solar vehicle previously developed. The plots show the amplitude ratio of the output and input against the non-dimensional frequency $\lambda^* = \lambda(M_s/K_s)^{0.5}$ obtained using different values of C_s. In figure 7, two natural frequencies can be observed for $C_s = 0$ withhigher peak being at approximately 4 Hz. It is also noticed thatthe peak is highest for the case of $C_s = \infty$ where the whole system is considered to be rigid. All of the peaks occurred at frequencies below 10 Hz except for the second natural frequency of Cs = 0 which has its highest peak at 20 Hz. The lowest peak of the ratio happens when the optimum damping value of $C_{s\text{-}opt}$ was used.

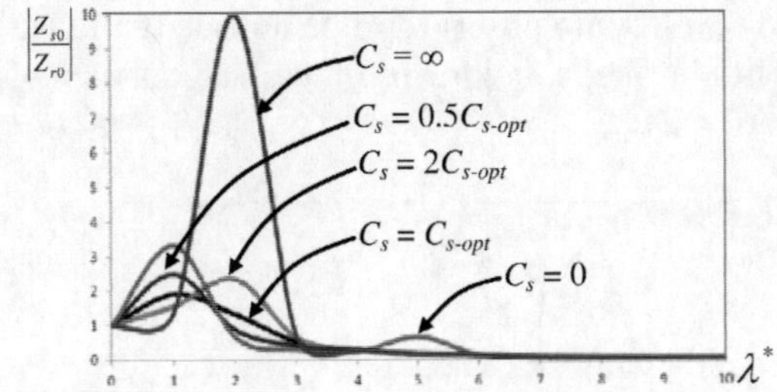

Figure 7: Ratios between amplitudes of the displacement M_s, Z_{s0} to the amplitude of the displacement of the ground, Z_{r0} for different values of damping of the shock absorber.

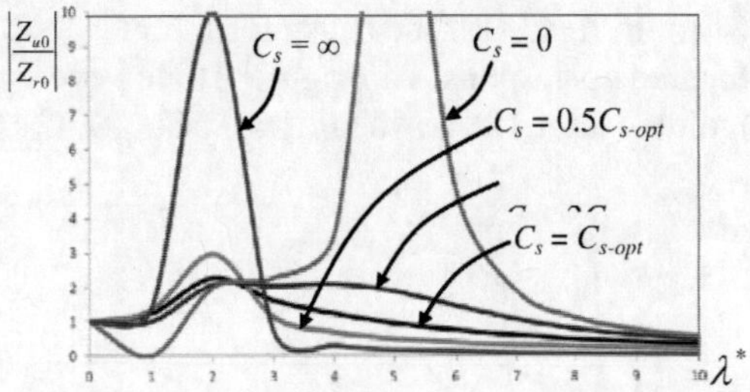

Figure 8: Ratios between amplitudes of the displacement M_u, Z_{u0} to the amplitude of the displacement of the ground, Z_{r0} for different values of damping of the shock absorber.

In figure 8, it is observed that the highest peaks occurred when $C_s = 0$ and $C_s = \infty$. All of the peaks are approximately at 8 Hz except for the case of $C_s = \infty$ in which the recorded peak is at 20 Hz. The ratio of amplitudes Z_{u0} to Z_{r0} (for M_s) is higher than the ratio of amplitudes Z_{s0} to Z_{r0} (for M_u) especially at frequencies above 10 Hz for $C_s = C_{s\text{-}opt}$, $2C_{s\text{-}opt}$ and $0.5C_{s\text{-}opt}$. The frequency responses of figure 7 and figure 8 are multiplied by λ^{*2} which give the non-dimensional ratio between the accelerations of the two masses and the displacement of the input as shown in figure 9 and figure 10.

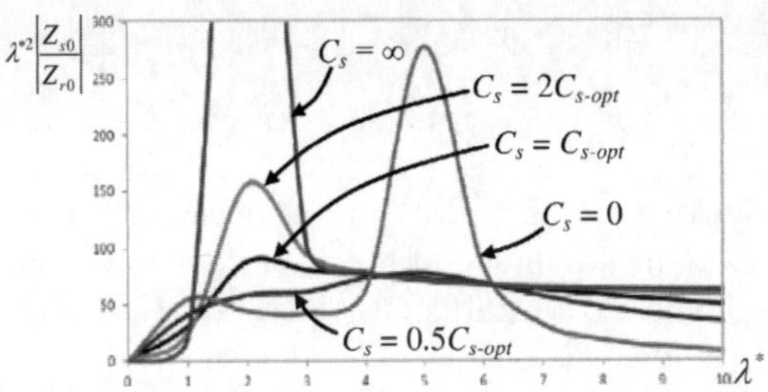

Figure 9: Ratios between the acceleration of M_s to the amplitude of the displacement of the ground, Z_{r0} for different values of damping of the shock absorber.

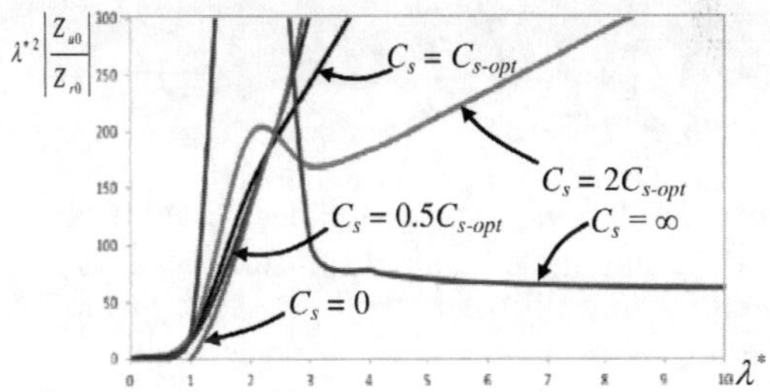

Figure 10: Ratios between the acceleration of M_u to the amplitude of the displacement of the ground, Z_{r0} for different values of damping of the shock absorber.

In figure 9, the ratio of the acceleration $\ddot{Z}_s$ to Z_{r0} increases with decreasing C_s for frequencies below 6 Hz and between 16 Hz and 25 Hz while the ratio increases with increasing C_s for frequencies between 6 Hz and 16 Hz and from 25 Hz upwards. In figure 10, the ratio of the acceleration $\ddot{Z}_u$ to Z_{r0} increases from 4 Hz up until 12 Hz with increasing C_s but the ratio decreases with increasing C_s beyond that. The optimum value of C_s can be found by trying to keep the acceleration as low as possible in a large field of the frequencies. In the above cases, the value of $C_{s\text{-}opt}$ is obtained from

$$C_{s-opt} = \sqrt{\frac{K_s M_s}{2}} \sqrt{\frac{K_t + 2K_s}{K_t}} \tag{10}$$

Figure 9 also shows that this value of damping is effective in maintaining low acceleration in a wide field of frequencies. Figure 11 and figure 12 show the frequency responses of the ratios between amplitudes of displacement M_s, Z_s and displacement M_u, Z_u to the amplitude of the displacement of the input signal or the ground, Z_r for different values of C_s using MATLAB Simulink. Both set of graphs exhibit similar output patterns with figure 7 and figure 8. In figure 11, the ratio between Z_s and Z_r increases with decreasing C_s at 4 Hz with one significant peak at 8 Hz when $C_s = \infty$. Similarly as in the previous cases, two peaks are obtained if $C_s = 0$. However, the magnitude of the ratios are different

IOP Conf. Series: Journal of Physics: Conf. Series **908** (2017) 012066 doi:10.1088/1742-6596/908/1/012066

and the lowest peak occurred at a different frequency. Furthermore, no peak is observed if $C_s = 2C_{s\text{-}opt}$ at 8 Hz. In figure 12, the peaks for $C_s = C_{s\text{-}opt}$, $2C_{s\text{-}opt}$ and $0.5C_{s\text{-}opt}$ are observed to be not so significant compared to responses in figure 8. However, the amplitude of the ratios are of the same order of magnitude after 12 Hz when $C_s = 0.5C_{s\text{-}opt} > C_s = C_{s\text{-}opt} > C_s = 2C_{s\text{-}opt}$.

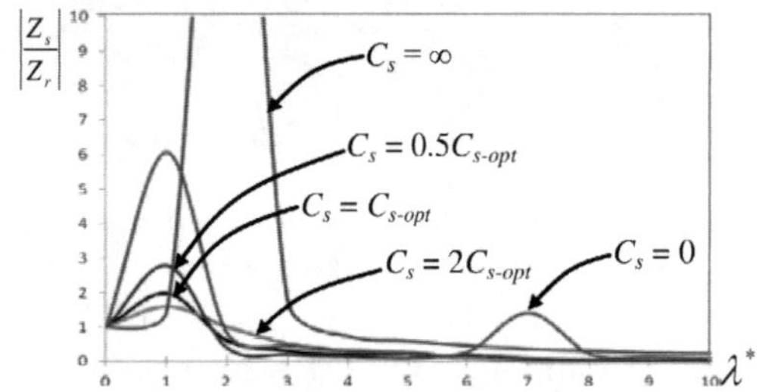

Figure 11: Ratios between amplitudes of the displacement M_s to the amplitude of the displacement, Z_r for different values of C_s using Simulink.

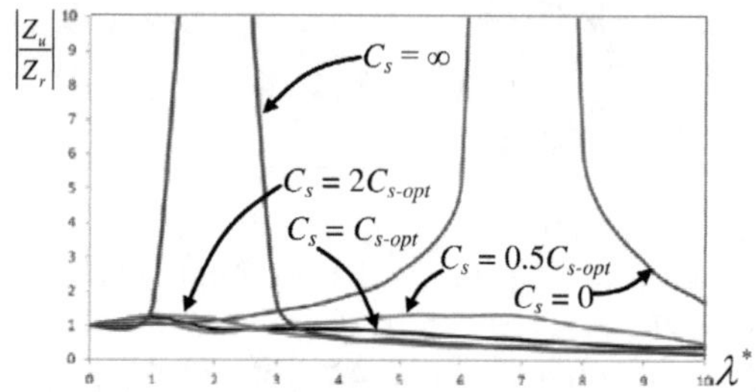

Figure 12: Ratios between amplitudes of the displacement M_u to the amplitude of the displacement, Z_r for different values of C_s using Simulink.

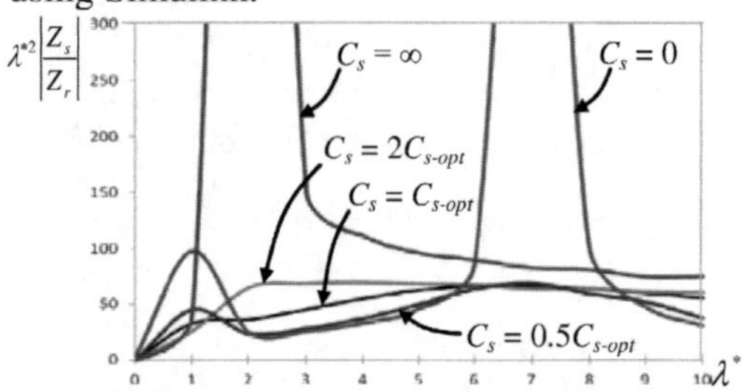

Figure 13: Ratios between the acceleration of M_s to the amplitude of the displacement of the ground, Z_r for different values of C_s using Simulink.

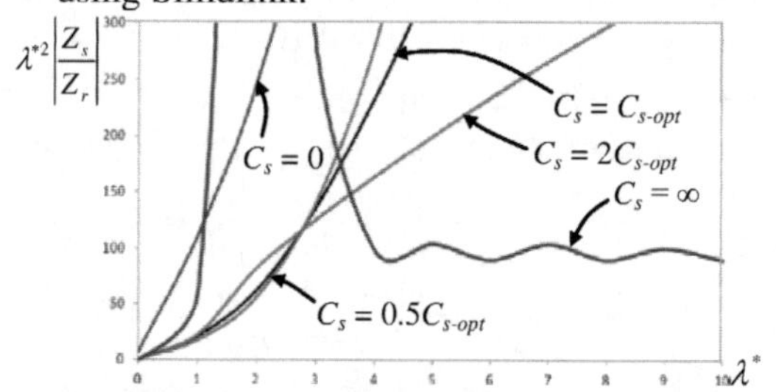

Figure 14: Ratios between the acceleration of M_u to the amplitude of the displacement of the ground, Z_r for different values of C_s using Simulink.

Figure 13 and figure 14 show the frequency responses of the ratios between acceleration $\ddot{Z}_s$ to Z_r and acceleration $\ddot{Z}_u$ to Z_r for different values of C_s using MATLAB Simulink. The order of amplitude for the ratios of $\ddot{Z}_s$ to Z_r, obtained below 4 Hz, between 4 Hz and 24 Hz, and above the 24 Hz are almost similar, as shown in figure 13 with the responses in figure 9. Similarly in figure 14, the ratios of $\ddot{Z}_u$ to Z_r increase with increase in frequency except for $C_s = \infty$.

4. Conclusion

A quarter-vehicle model for an electric lightweight vehicle was developed in order to analyse the performance of the vehicle's passive suspension system in isolating disturbances from the roughness of the road. Differential equations were used to plot the frequency responses of the system in terms of the ratios of the displacements M_s and M_u to the displacement of input Z_r representing the road disturbance. The ratios of the accelerations for M_s and M_u to the input Z_r are also plotted against the non-dimensional frequency. Simulink model developed from the differential equations were also used to analyse the frequency responses of the system with different values of the damping constant, C_s. It can be seen that the most suitable value of the damping is $C_s = C_{s\text{-}opt}$ since in all analyses the responses are minimum at most level of frequency range as shown. It is also important to show that the frequency responses obtained from the calculation and simulation using Simulink are comparable and agree with the frequency responses given by Genta, 1997 [8]. However, the differences that can be observed from the graphs might be due to the inaccuracies in the characteristics of the input signal applied to the simulation.

5. Future scope

A full-vehicle model for a lightweight electric vehicle will be developed using Simulink to analyse the vibration responses using suspension system parameters similar to the off-the-shelf components. Furthermore, the input signal can be easily modified based on actual road surface roughness and the vibration responses from the involvement of more masses can be obtained easily.

6. References

[1] Shen Y, Chen L, Yang X, Shi D and Yang J 2016 Improved design of dynamic vibration absorber by using the inerter and its application in vehicle suspension *J. Sound and Vibration* **361** pp 148-158

[2] Banerjee S, Balamurugan V and Krishnakumar R 2014 Ride dynamics mathematical model for a single station representation of tracked vehicle *J. Terramechanics* **53** pp 47-58

[3] Els P S, Theron N J, Uys P E and Thoresson M J 2007 The ride comfort vs handling compromise for off-road vehicles *J. Terramechanics* **44** pp 303-317

[4] Mehrdad N K and Vahid A 2007 Comparison between optimized passive vehicle suspension system and semi active fuzzy logic controlled suspension system regarding ride and handling *Proc. World Academy of Science, Engineering and Technology* **21** pp 57-61 ISSN 1307-6884

[5] Kruczek A and Stribrsky A 2004 A full-car model for active suspension – some practical aspects *ICM Proc. of the IEEE Int. Conf. on Mechatronics* pp 41-45

[6] Griffin M J 1990 *Handbook of Human Vibration* Elsevier Academic Press London UK

[7] Griffin M J 2007 Discomfort from feeling vehicle vibration *Vehicle System Dynamics* **45** 7-8 pp 679-698

[8] Genta G 1997 *Motor vehicle dynamics : modeling and simulation* World Scientific Publishing Co. Pte. Ltd. Singapore

[9] Den Hartog J P 1948 *Mechanics* Dover Publications Inc. UK

[10] Gillespie T D 1992 *Fundamentals of vehicle dynamics* Society of Automotive Engineers, Inc. Warrendale PA

[11] Inman D J 2001 *Engineering Vibration* 2nd ed. Prentice Hall International Inc. New Jersey

[12] Rao S S 2004 *Mechanical Vibrations* 4th ed. Pearson/Prentice Hall Upper Saddle River New Jersey

[13] Markel T, Brooker A, Hendricks T, Johnson V, Kelly K, Kramer B, O'Keefe M, Sprik S and Wipke K 2002 ADVISOR : a systems analysis tool for advanced vehicle modeling *J. of Power Sources* pp 255-266

[14] Griffiths K R 2012 *An Improved Method for Simulation of Vehicle Vibration Using a Journey Database and Wavelet Analysis for the Pre-Distribution Testing of Packaging* (Doctoral dissertation) retrieved from opus.bath.ac.uk/38300/1/UnivBath_PhD_2012_K_Griffiths.pdf

[15] Phalke T P and Mitra A C 2017 Analysis of ride comfort and road holding of quarter car model by Simulink *Materials Today : Proc.* **4** pp 2425-2430

[16] Talukdar R P and Talukdar S 2016 Dynamic analysis of high speed maglev vehicle-guideway system using Simulink *Procedia Engineering* **144** pp 1094-1101

Design and development of injection moulding machine for manufacturing maboratory

R A Siregar [1], S F Khan[2] and K Umurani[1]

[1]Mechanical Engineering Department, Universitas Muhammadiyah Sumatera Utara, Indonesia
[2]School of Mechatronic Engineering, University Malaysia Perlis, Malaysia

rakhmadarief@umsu.ac.id

Abstract. This paper presented the design process and manufacturing of a benchtop and inexpensive injection moulding machine for use as learning and teaching equipment in a manufacturing laboratory. The design use a vertical plunger type of injection equipped with clamping system. The maximum volume of barrel is 290 cc combined with injection plunger 60 mm provides ideal capacity for lab. The design concept process and preliminary test result are discussed. The flow rate increases with increase of motor speed and the packing time decrease with increase of motor speed. At 2500rpm the flow rate is 0.42m/s and pack time is 15 second.

1. Introduction

Commonly used materials in teaching the process of producing near net shape components in manufacturing lab are low cost materials such as plastic and polymer-composite although sometimes low-melt temperature metals or alloys and ceramics are used as well. All these materials can be produced, machined, fabricated and assembled by equipment such as, CNC machining, 3D printing, casting, and injection moulding machines. The injection moulding process is primarily a sequential operation that results in the transformation of plastic pellets into a moulded part. Identical parts are produced through a cyclic process involving the melting of a pellet or powder resin followed by the injection of the polymer melt into the hollow mould cavity under high pressure Common process to produce polymeric components is by using injection moulding machine. In injection moulding process, polymer melts must have high viscosity and cannot simply be poured into a mould like metal die casting. Instead a large force must be used to inject the polymer into the hollow mould cavity. The process of clamping, injection, cooling and ejection are essential knowledge for practical activities in manufacturing lab. The commercial injection moulding machine can provide that knowledge at ease however not all manufacturing labs have sufficient budget or space for it.

Many research have been carried out to improve the process of injection moulding based on theoretical, computer based simulation models and experimental trials [1]. Specific optimisations have been done to the machine for specific manufacturing product, such as a brake booster valve body, bird cage and automotive part [2-4]. The work on optimization that employed Taguchi method for minimizing warpage and sink index in injection moulded thermoplastic was carried out by Erzurumlu et al. In their study they considered mold temperature, melt temperature, packing pressure, rib cross section and rib layout angle and material PC/ABS, POM, and PA66. Other optimization approach that is artificial neural network and fuzzy logic have been used for initial process parameter setting that provides proper setting for good quality molded parts [5-6].

This work aims to provide the needs of proper knowledge of injection moulding process for undergraduate students learning in a manufacturing lab by developing a machine that is relatively low cost. This machine is expected to provide similar experience as commercial injection moulding machines in addition to research needs.

2. Design Concept

The injection moulding machine is designed for lab work purposes with main considerations for inexpensive, small size, capable of producing small number of products for prototype testing, able to accommodate up to medium size specimens, and have similar operational procedure as in commercial large injection moulding machine. In order to accommodate those constrains, the injection moulding machine is designed as a benchtop size, vertical type and having four main components that can be modified for experimental works.

Three concepts of vertical type are considered and sketched roughly in figure 1. Concept 1 utilize screw component to inject the heated plastic to die, while concept 2 and concept 3 uses plunger. The different between concept 2 and 3 is on movement of the plunger. Concept 2 utilize hydraulic instead of motors as in concept 3. In concept 3, the injection pressure is controlled by the speed of motor using motor speed drive. A simple decision making tool is employed and odd numbers from 1 to 9 are used to denote the magnitude value presented in table 1. Based on the weighted decision in table 1, concept 3 has the highest rating for material, manufacturing process and product result. These criteria will be most likely to be able to simulate the condition of injection moulding as in an commercial environment. Thus, concept 3 was selected for prototyping development.

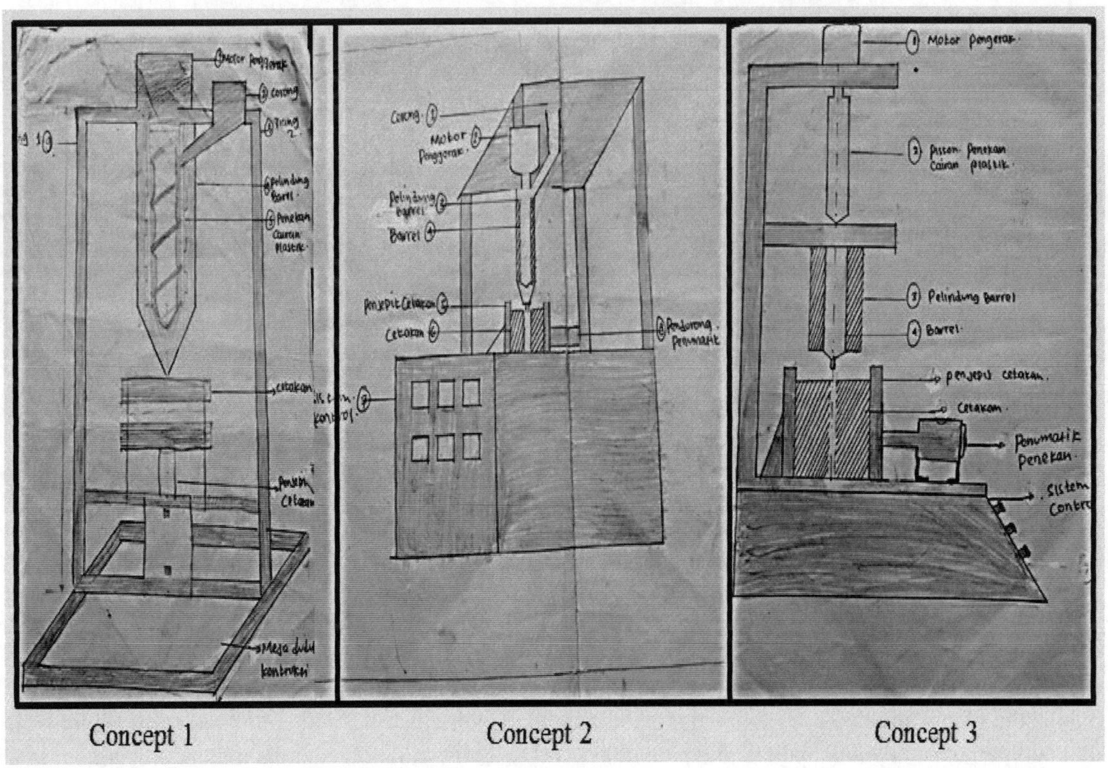

Figure 1. Alternative concepts of vertical injection moulding machine.

Table 1. Weighted decision matrix for conceptual design of injection moulding machine.

	Concept 1			Concept 2			Concept 3		
	Magnitude	Score	Rating	Magnitude	Score	Rating	Magnitude	Score	Rating
Material	7	0.15	1.05	3	0.15	0.45	9	0.15	1.35
Construction	3	0.2	0.6	9	0.2	1.8	7	0.2	1.4
Manufacturing process	9	0.1	0.9	5	0.1	0.5	9	0.1	0.9
Time to produce	5	0.2	1	7	0.2	1.4	7	0.2	1.4
Product result	7	0.2	1.4	7	0.2	1.4	9	0.2	1.8
Cost	9	0.15	1.35	5	0.15	0.7	7	0.15	1.05
			6.3			6.05			7.9

The four main sub-assemblies for concept 3 are drawn with CATIA as in figure 2. It was designed such that these sub-assemblies can be modified later to accommodate experimental parameters. Components such as motors, barrels, heater, clamper and plungers are usually needs some adjustments during experimental work. Similar experimental works by Dimla et. al.[8] is able to be performed using this machine by adding a cooling channels.

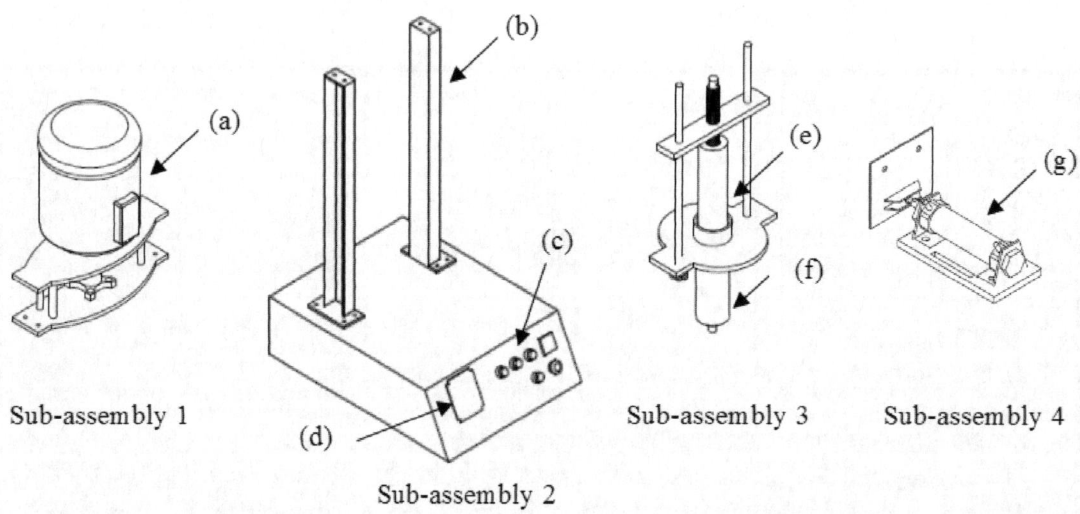

Figure 2. Four main sub-assembly of injection moulding machine (a) motor (b) main pillar (c) switch panel (d) LCD panel (e) heater (f) barrel (g) clamper.

The machine weighs 73 kg and has 35 x 60 cm footprint with height of main pillar at 95 cm. The total production cost of the machine is about Rp. 10 million (US$750). The highest cost contributors are motors, controllers and heater while frame and machining is relatively inexpensive due to the precision works is not so high. The results of construction work can be seen in figure 3(b) that compared side by side with CAD rendered drawing figure 3(a).

(a) (b)

Figure 3 Injection moulding machine (a) CAD drawing using CATIA (b) photograph of injection moulding machine.

3. Preliminary Test Results and Discussion

In order to test the performance of injection pressure and its flow rate, a simple mould was developed. The mould is designed with two outputs to evaluate the time to pack. The simple mould together with its test result is shown in figure 4. Figure 4(c-d) shows the relation between motor speed that control injection pressure and time to pack the mould with liquid plastic. While average flow rate is calculated based on it volume and time. It is considered in good moulding area of injection capacity.

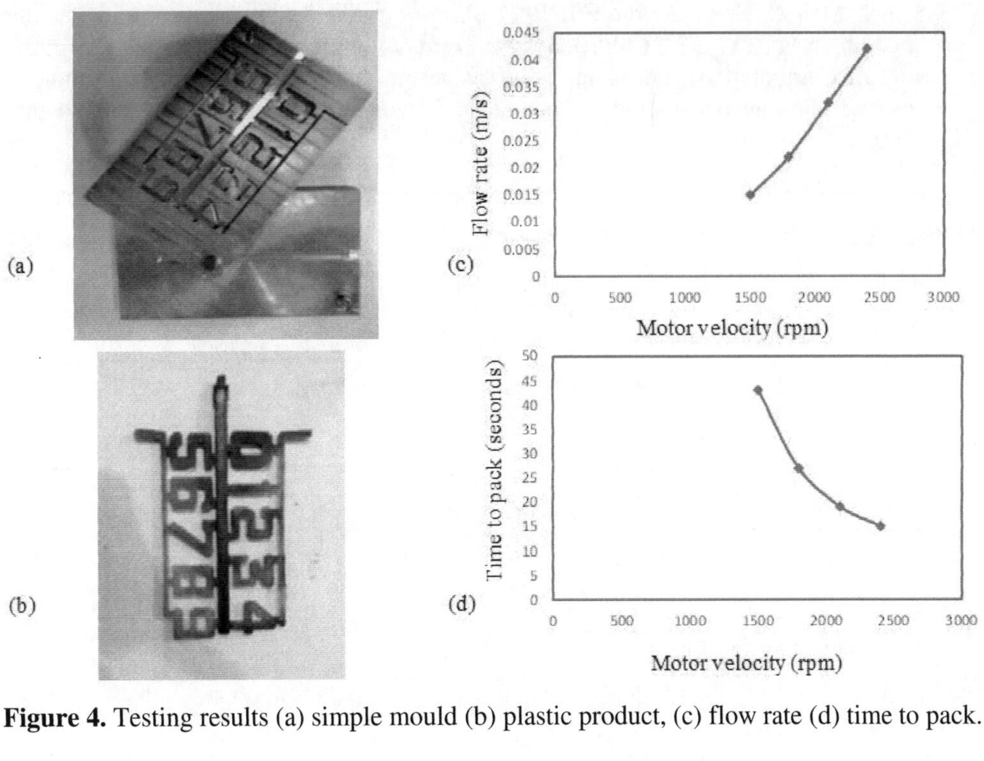

Figure 4. Testing results (a) simple mould (b) plastic product, (c) flow rate (d) time to pack.

IOP Conf. Series: Journal of Physics: Conf. Series **908** (2017) 012067 doi:10.1088/1742-6596/908/1/012067

4. Conclusion

This paper present the design and development of an injection moulding machine for manufacturing lab that have features of low cost, bench top size, and have similar proses as in commercial injection moulding machine. The design concept and preliminary test of the prototype were discussed and validated in terms of its performance. The machine is well suited for use in undergraduate teaching labs as well as research needs.

5. Acknowledgement

The authors wish to express gratitude to Mechanical Engineering Department, Universitas Muhammadiyah Sumatera Utara, Indonesia and School of Mechatronic Engineering, University Malaysia Perlis, Malaysia for the providing assistant and facilities in carrying out this research project.

References

[1] Singha G and Vermaa A 2016, A brief review on injection moulding manufacturing process *Proc. Int Conf. of Materials Processing and Characterization* (Materials Today: Proceedings) pp 1423-33.

[2] Siswanto, Adi Yusuf Setiawan and Sugiarto 2014, Perancangan Mesin Molding Sebagai Alternatif Pemanfaatan Limbah Plastik Untuk Bahan Dasar Sangkar Burung *Proc. Pekan Ilmiah Mahasiswa* (Jakarta: Electronic Diltilabmas Ditjen DIKTI in Bahasa Indonesia).

[3] Yi-qi W, Jae-gyu K, and Jung-il S 2013, Optimization of plastic injection molding process parameters for manufacturing a brake booster valve body *J. Materials and Design* **56**, pp. 313-7.

[4] Saman, Alias Mohd, Abdul Halim Abdullah, and Mohd Asri Mohd Nor 2009 Computer simulation opportunity in plastic injection mold development for automotive part *Proc. Int Conf. In Computer Technology and Development* (Kota Kinabalu: Proceedings) pp 495-8.

[5] Erzurumlu Tuncay, and Babur Ozcelik 2006 Minimization of warpage and sink index in injection-molded thermoplastic parts using Taguchi optimization method *Materials & design* **27,** pp. 853-61

[6] Mok S L, and Che Kit Kwong 2002, Application of artificial neural network and fuzzy logic in a case-based system for initial process parameter setting of injection molding. *J Intelligent Manufacturing* **13**, **3** pp 165-76.

[7] Chris M S, Alan A D and George F L 1994 Multi objective optimization of a plastic injection molding process *IEEE T Control System Technology* 2 pp 157–68.

[8] Dimla D E, Camilotto M, and Miani F 2005, Design and optimisation of conformal cooling channels in injection moulding tools *J. Materials Processing Technology* **164** pp 1294-300.

Analysis of airbox performance improvement for Modenas model by geometrical analysis

A B Shahriman[1,2,3], A K M Syafiq[1,2,3], M S M Hashim[1,2], D Hazry[3], Z M Razlan[2,3], K Wan[2,3], R Daud[2], E M Cheng[2], S K Zaaba[2], K A Hamid[4], W S W Teh[4], N A Shuaib[2] and A A Ishak[4]

[1]Unit Penyelidikan Teknologi Sukan Permotoran, MoTECH, Universiti Malaysia Perlis (UniMAP), Pauh Putra Campus, 02600 Arau, Perlis, Malaysia.
[2]School of Mechatronic Engineering, Universiti Malaysia Perlis (UniMAP), Pauh Putra Campus, 02600 Arau, Perlis, Malaysia.
[3]Centre of Excellence for Unmanned System (UAS), Universiti Malaysia Perlis (UniMAP), Perlis, Malaysia.
[4]Modenas Research and Development, Gurun, Kedah.

shahriman@unimap.edu.my, syafiq.akbar89@gmail.com

Abstract. Airbox system consist of filter element which greatly influenced the quality air into the engine system. The main objective of this study is to improve the performance of airbox system that attached in MODENAS CT115S engine system. Airbox, an empty air chamber, accumulates the outside air and feed it to the cylinder of the engine for combustion process and contains filter element. The function is to provide cleaner air for greater combustion rate and cleaner emission. The performance of airbox can be influenced by changing the location of the filter element and the geometry of airbox. By changing these parameters, the performance of the airbox system can be enhanced. In this study, the geometry of the airbox inlet has been changed in terms of shape and size. The inlet shape for Design 1 in round shape with a diameter of 43.60 mm. Meanwhile for Design 2, the geometry of airbox inlet was changed to a rectangular shape. Between these two designs, Design 1 provides a better result in terms of mass flow rate (10.41 kg/s), velocity (9.84 m/s) and pressure difference (83.44 kPa). By increasing the performance of airbox, the performance of engine can be increased.

1. Introduction

The iteration of engine performance is the major concern in the automotive industry. The aim of this study is to improve the performance of airbox for MODENAS CT115S engine system in terms of mass flow rate, the velocity of the air and the pressure difference of airbox. Airbox system is one of the components in the engine system. The airbox is a void chamber at the inlet of most ignition engines. The airbox gathers the air from surrounding and feed it to the admission hoses in every intake part. The Airbox needs to supply clean air to the carburetor before the mixing process. The supply air and the fuel are mixed before being introduced into the cylinder for the combustion process. The specific amount of air is provided according to the type and application of the engine. Different types of engine application requiredifferent amount of air [1]. For the normal engine, a stable and continuous air flow is needed to provide comfortable driving. Meanwhile, for the racing application, a greater amount of air is entailed in order to produce greater combustion rate and generates the better amount of torque and power of the engine [2].

Pressure difference also known as pressure drop is the vital parameter in constructing the new design of airbox. In intake system, the capacities of pressure drop inside the airbox system can be accomplished by redesign geometry of the airbox design. The air from airbox will be used for overall intake system, which includes the cylinder head and intake port. From this study, the geometry of the airbox is redesigned and simultaneously, the mass flowrate, air velocity and pressure difference are investigated [3].

2. Methodology

This study involves three stages namely sketching, designing and analysis. For first two stages, Catia V5 software was used. For analysis, Ansys Fluent 14.5 was used as the simulation tool. The airbox model consists of inlet part, body part and outlet part. All these parts were measured carefully before the sketching process. The dimensions were measured as accurate as possible in producing benchmarking result for comparison purpose. The first drawing was the actual design, which acts as benchmarking.

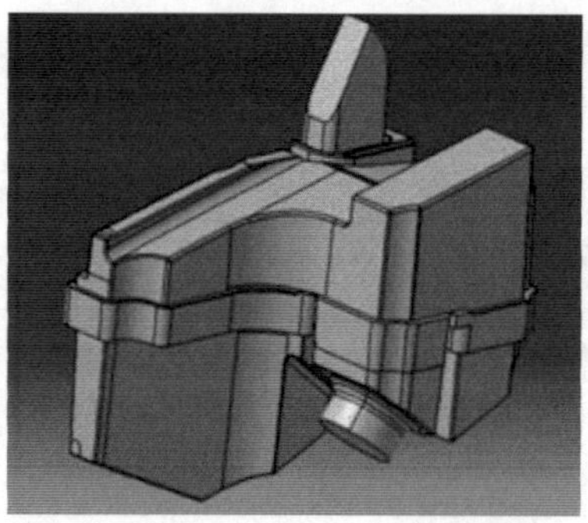

Figure 1. Airbox CATIA model.

For improvement of design, two designs were proposed. Both designs have different geometry of inlet part, with different shape and size of inlet part. The volume of airbox remained constant and the diameter of outlet part was verified but the round shape of outlet part remained. Table 1 describes detail design for all the designs.

TABLE 1. The design detail of airbox.

Parameter	Actual Design	Design 1	Design 2
Volume (l)	1.53	1.53	1.53
Inlet Dimension (mm)	43.56 × 18.53	Ø 43.60	33 × 35
Outlet Dimension (mm)	Ø36.60	Ø43.60	Ø36.6

For Design 1, the shape proposed was in round shape, with a diameter of the design was 43.60 mm. The volume of Design 1 is kept constant as the actual one, (1.53 l). The diameter of the outlet part was increased from actual dimension of 36.60 mm to 43.60 mm. For Design 2, the volume of the airbox and the diameter of outlet part were kept constant. The geometry of the inlet part was changed to a rectangular shape, with a dimension of 33 mm × 35 mm. This dimension was selected in order to investigate the effect of different shape of the inlet part and dimension changes to both intake and outlet part.

After the sketching and designing stage, the analysis was carried out by using Computational Fluid Dynamic (CFD) module in Ansys Fluent 14.5 in order to scrutinize the flow characteristics for all the designs including the actual one. Before the simulation, the best description of meshing was selected to avoid error during the analysis. The selected meshing is shown in Figure 2 and Figure 3.

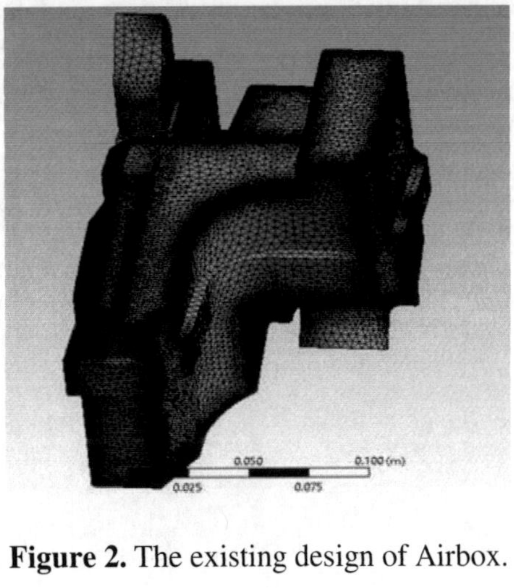

Figure 2. The existing design of Airbox.

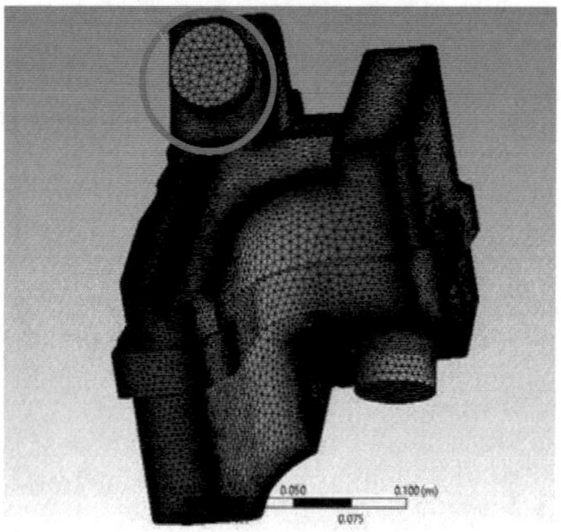

(a)

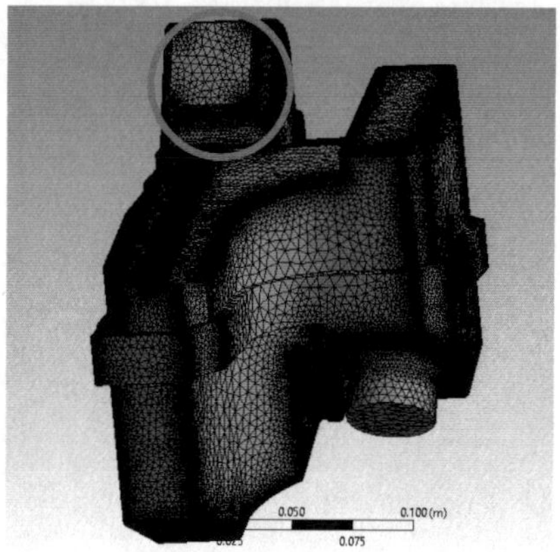

(b)

Figure 3 (a). Design 1; **Figure 3(b).** Design 2.

The streamlines of airflow inside the airbox were observed for each airbox design. Besides, the velocity profile was also examined together with mass flow rate, velocity magnitude and the pressure difference during post processing.

3. Results And Discussions

3.1 General Flow Field

Figure 4 shows the performance of flow field inside the airbox system. The flow described the motion of dirty air from the atmosphere that been drawn into the airbox. The inlet acts as the starting point and the outlet part set as the output point. The air flow circulated inside the airbox produces streamline profiles and path line as the result of the simulation. The result of streamlines of air flow can be seen in Figure 4. From the results, Design 2 produces better air flow streamlines compared to the other designs, with the number of streamlines greater than others.

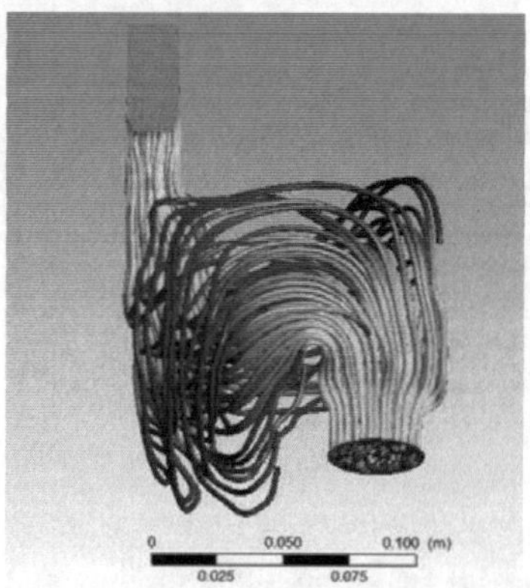

a)

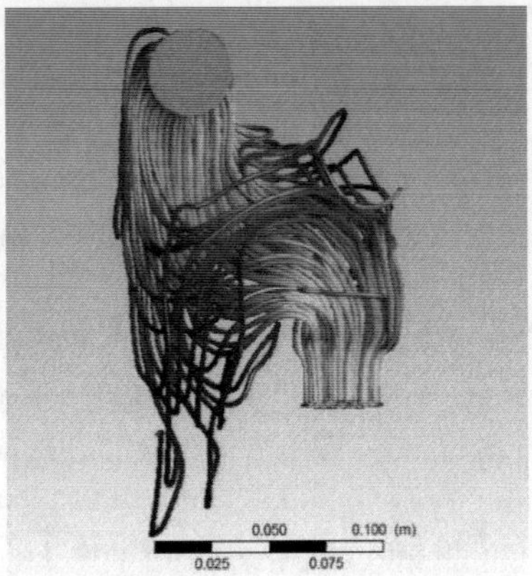

b)

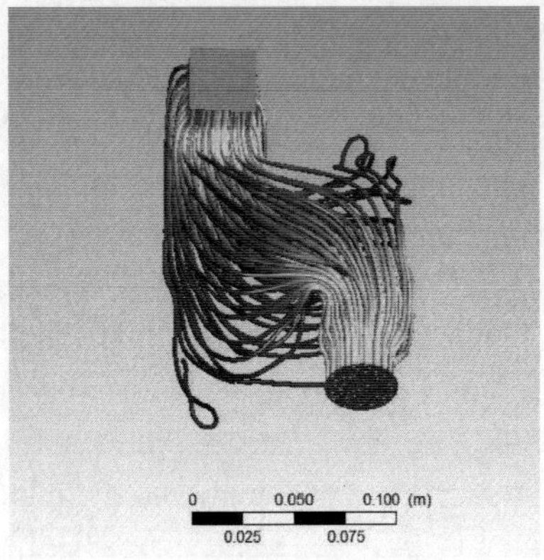

c)

Figure 4. The airflow streamlines of airbox; **a)** Existing Design; **b)** Design 1; **c)** Design 2

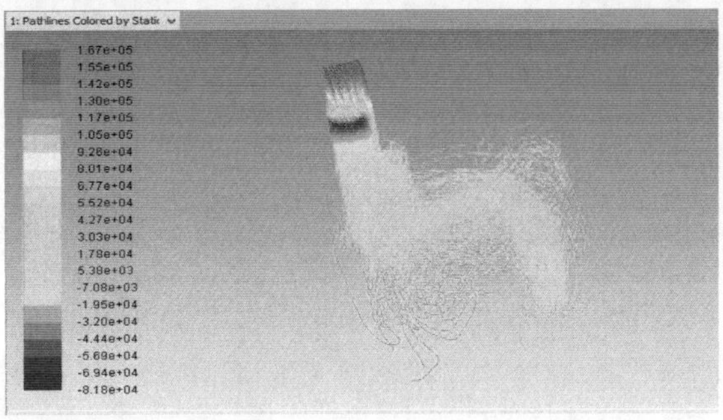

(a)

(b)

(c)

Figure 5. Streamlines colored by velocity magnitude for; (a) Existing design; (b) Design 1; (c) Design 2

4. Result Analysis

The analysis of mass flow rate, velocity and the pressure difference determines the improvement percentage and the best design. The value of mass flow rate for all designs is different to each other influenced by the design of airbox. For the original design, the maximum of mass flow rate occurred during medium valve lift. For the improvement designs, both have maximum flow rate during low valve lift. Figure 6 represents the result of mass flow rate for all designs.

The value of mass flow rate for the original design was 8.46 kg/s, which was the lowest one. For Design 1 and Design 2, the mass flow rate was 10.41 kg/s and 9.73 kg/s respectively. The variation of geometrical design influenced the result [4]. Design 1 has round shape with the diameter of both inlet and outlet is similar, (43.06 mm) and the improvement rate was 23.05%. Design 2 improved the value of mass flow rate by 15.01%. Design 1 was the best design in enhancing the flow rate value.

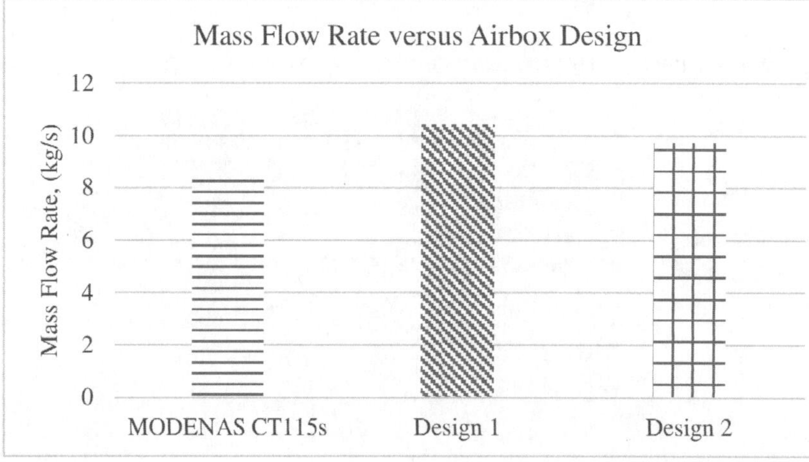

Figure 6. Mass flow rate comparison between the existing air box and new designs.

Velocity magnitude were constant at the inlet for all the designs. All designs experienced the same trend, with the outlet velocity magnitude decreased compared to the inlet. The differences between inlet and outlet were calculated and the smallest difference percentage was chosen as the best design.

Figure 7 represent the results for velocity magnitude for all designs and shows the constant velocity magnitude of 10m/s for all analysis. Pressure differences and the geometric design produce the friction in the air and cause the changes of velocity for the outlet of every air box. The result gives the value of velocity, which is the most approximate to 10m/s is the best result of the analysis.

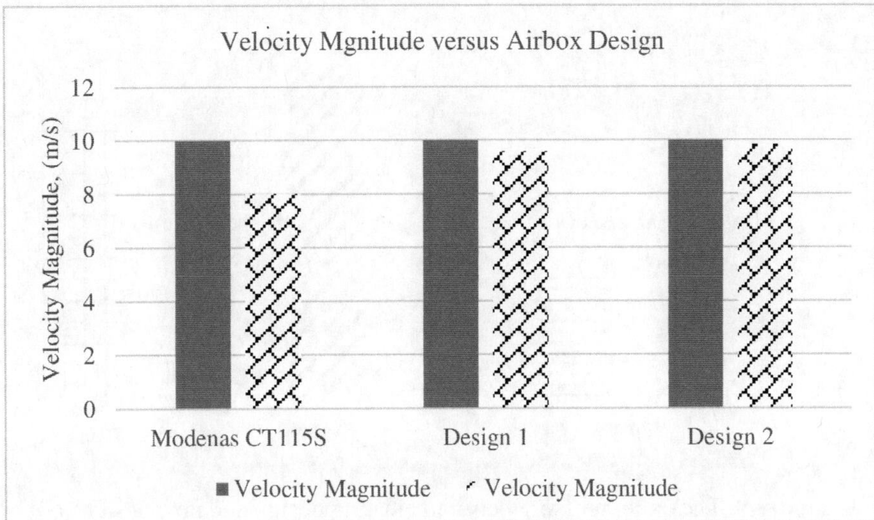

Figure 7. The comparison of velocity magnitude among the airbox.

The outlet velocity magnitude for the original one was the lowest one, with velocity dropped from 10 m/s to 8.022 m/s (19.78%). For Design 1, the velocity at the outlet was the highest with the value of 9.84 m/s. The difference percentage was only 1.6%. Meanwhile, the outlet velocity for Design 2 was 9.58 m/s, 4.2% drop from inlet velocity. The velocity drop was due to the existing of friction acting on air flow inside airbox and when the air flowing out from the outlet [5]. The value of friction was different according to the design of airbox. The round shape improves the flow of airflow and reducing the friction that acts on the movement air [6].

Performance of the airbox was also analyzed in term of pressure difference. The initial pressure at the inlet was set at the atmospheric pressure, 100 kPa. After the simulation completed, the value of pressure at the outlet was obtained for the pressure drop calculation. For existing design, the outlet pressure was 73.97 kPa, with the pressure difference was 26.03 kPa. For Design 1, the pressure at the outlet was 83.44 kPa and the outlet pressure was drop by 16.56 kPa. The outlet pressure for last design, Design 2 was 81.25 kPa, with the pressure difference was 18.75 kPa. From all these values of the pressure difference, Design 1 recorded the lowest pressure difference among all the designs.

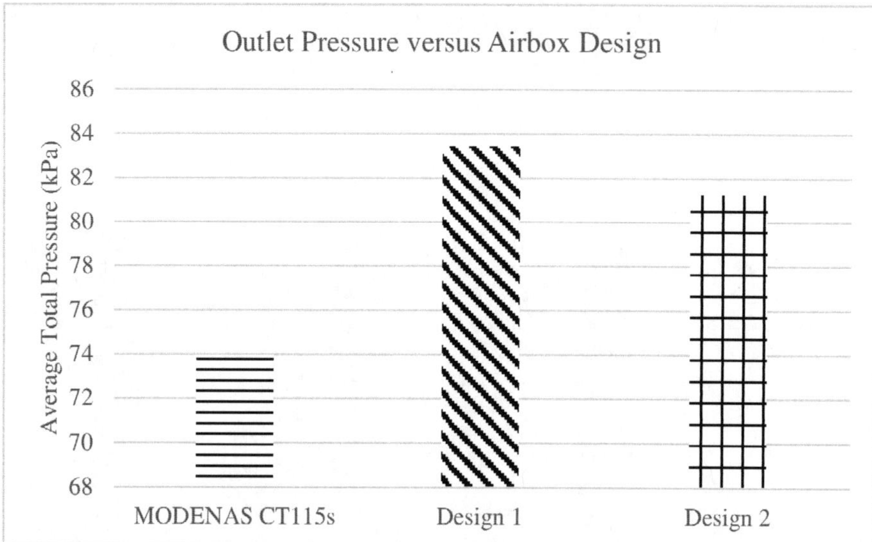

Figure 8. The outlet pressure between existing design and new designs of the airbox.

5. Discussion

The changes of inlet geometry influence the value of mass flow rate, velocity magnitude and the pressure difference that produced by airbox system. Basically, the inlet with round shape can enhance the air the flowing into the system. This is because the air flow is smoother compared to other shapes. It is really important to improve the air movement to enter the airbox system through inlet part by reducing the obstacles, sharp edge and friction. This is because all these blockages can reduce the amount of air that will be supplied to the overall engine system for the combustion process. When the amount of air reduced, the combustion rate will be dropped [7]. Therefore, it is important to generate the sufficient of air mass flow rate to be supplied for the combustion process.

By adjusting the inlet geometry to round shape and has similar diameter with outlet part, the pressure difference can be reduced [8]. This is because the pressure is needed to push the air from airbox to the cylinder effectively. If the pressure difference is too big, the required pressure to drive the air from airbox to cylinder is not adequate. Therefore, the small pressure difference is entailed in order to ensure the sufficient pressure for the engine system. To get 100% efficiency of the airbox is difficult since the losses occurred when the air passed through the element system inside airbox system [9].

6. Conclusion

By redesign inlet geometry to the round shape and the dimension was 43.6 mm for both inlet and outlet, the performance of airbox was enhanced. The value of mass flow rate, velocity magnitude and pressure difference for air that passing through the airbox were increased compared than the other design. This is because the round shape for inlet part reducing the friction, smoother the air flow entering the airbox, minimizing the air losses and increase the overall performance. The greater amount of air that supply to the cylinder from airbox will generate greater combustion rate and boost the overall engine performance. Design 1 was the best design among all airbox designs.

7. Acknowledgments
The authors acknowledge the support from UniMAP staffs and MODENAS staff for all the knowledge and experience sharing. Besides that, the deepest gratitude to ICADME team and UniMAP for this opportunity.

8. References

[1] M.A. Ceviz, M. Akın., "Design of a new SI engine intake manifold with variable length plenum," Energy Conversion and Management, vol. 2, no. 2, Mar. 2010.

[2] Katlego Webster Makgata., "Computational Analysis And Optimization Of The Inlet System Of A High-Performance Rally Engine ," Degree Of Master Of Mechanical Engineering, pp. 1–110, May 2005.

[3] A.Singhal, M.Parveen, IAENG "Air Flow Optimization via a Venturi Type Air Restrictor," Proceedings of the World Congress on Engineering 2013, July 2013.

[4] Sayin, C., Gumus, M., & Canakci, M. Effect of fuel injection timing on the emissions of a direct-injection (di) diesel engine fueled with canola oil methyl ester–diesel fuel blends. Energy and Fuels, 24(4), pp 2675-2682.2010.

[5] Yusaf, T., Hamawand, I., Baker, P., & Najafi, G. 'The effect of methanol-diesel blended ratio on ci engine performance'. International Journal of Automotive and Mechanical Engineering, 8, pp 1385-1395. 2013

[6] D. Ramasamy, S. Mahendran, Zamri.M and S.Vijaya, " Design Optimization of Air Intake System (AIS) of 1.6L Engine by Adding Guide Vane". Faculty of Mechanical Engineering, Universiti Malaysia Pahang, Pekan, Pahang, 2009.

[7] Saad, I., & Bari, S. "Cfd Investigation Of In-Cylinder Air Flow To Optimize Number Of Guide Vanes To Improve Ci Engine Performance Using Higher Viscous Fuel," International Journal Of Automotive And Mechanical Engineering, vol 8, , 2013.

[8] Yusaf, T., Baker, P., Hamawand, I., & Noor, M. M.. Effect of compressed natural gas mixing on the engine performance and emissions. International Journal of Automotive and Mechanical Engineering, 8,pp 1416-1429. 2013.

[9] Shannak, B., Damseh, R., & Alhusein, M. Influence of air intake pipe on engine exhaust emission. Forschung im Ingenieurwesen, 70(2), pp 128-132. 2005.

Study of intake manifold for Universiti Malaysia Perlis automotive racing team formula student race car

A Norizan, M T A Rahman, N A M Amin, M H Basha, M H N Ismail and A F A Hamid

School of Mechatronic Engineering, Universiti Malaysia Perlis, Pauh Putra Campus, 02600 Arau, Perlis, Malaysia.

tasyrif@unimap.edu.my

Abstract. This paper describes the design differences between the intake manifold and restrictor used in racing cars that participate in the Formula Student (FSAE) competition. To fulfil the criteria of rules and regulation of the race, each race car must have a restriction device that has a maximum diameter of 20 mm installed between the throttle body and intake manifold. To overcome these problems, a restrictor has been designed and analysed using the steady state analysis, to reduce the loss of pressure in the restrictor. Design of the restrictor has a fixed parameter of the maximum diameter of 20mm. There are some differences that have been taken to make the comparison between the design of the restrictor, the diameter of the inlet and outlet, the curvature of the surface, convergence and divergence angle and length of the restrictor. Intake manifold was designed based on the design of the chassis, which shall not exceed the envelope defined by the FSAE competition. A good intake manifold design will affect the performance of the engine. Each design have made an analysis designed to ensure that each cylinder engine gets its air evenly. To verify the design, steady state analysis was made for a total mass flow rate and the velocity of air leaving a runner in each engine. Data such as the engine MAP reading was recorded by using Haltech ECU Management Software as reference purposes.

1. Introduction

UniART is Universiti Malaysia Perlis Automotive Racing Team has established in 2009 under the School of Mechatronics Engineering, Universiti Malaysia Perlis. The main objective is to nurture student's automotive and technical skills, to prepare the student for new challenges in the industry. UniART has participating many event including FSAE-Asean. The formula car has developed by using Suzuki GSX-R 600 2005 engine model. FSAE regulation mentioned that all internal combustion formula car must use a single throttle body to operate the engine and must have a restrictor between throttle body and intake manifold. The maximum throat of restrictor is 20mm diameter. During FSAE-Asean competition, the UniART race car have a problem where the engine cannot idle below the 2000 RPM. The car need 4000 RPM to maintain idle speed. This problem will cause high fuel consumption because of the high RPM when idle. The team has investigated this issue and found that this problem is from the 20mm restrictor. To overcome this problem, the team has decided to redesign the 20mm restrictor and the intake manifold system. The 20mm restrictor needs to redesign before proceeding with the design of the intake manifold. To understand the design consideration, the research has been made to study the behaviour of the air flow pattern when entering the restrictor and the intake

IOP Conf. Series: Journal of Physics: Conf. Series **908** (2017) 012069 doi:10.1088/1742-6596/908/1/012069

manifold. The design has been made by drawing in cad software, CATIA V5R21. The computational fluid dynamic (CFD) simulation was conducted by using ANSYS Workbench 15, CFX. CFD has a very important role in all analyses of fluid prior to producing the actual product in addition to reducing the cost of manufacture [1].

2. Design of Restrictor

In order to comply the FSAE rules and regulations, a single 20mm diameter of restrictor must be placed between the throttle body and the intake manifold inlet. The air flow must be passed through this restrictor before entering the intake manifold. Restrictor main purpose is to limit the power produced by the engine of a vehicle [2].

2.1. Venturi Restrictor and Calculation

The venturi tube or simply a venturi is a tubular setup of varying pipe diameter through which the fluid flows. Venturi effect is when the fluid through the shrinking surface area, it will increase the speed of the fluid and the fluid pressure is reduced. [3]. Based on governing fluid dynamics law, a fluid's velocity must increase as it passes through a constriction to fulfil the principle of conservation of mechanical energy. An equation for the drop in pressure due to the venturi effect may be derived from a combination of Bernoulli's principle and the continuity equation. It generally has a coefficient of drag around 0.85 [4]. The equation 1 show below is related with the mass flow rate to the flow area, A, total pressure, P_t, and temperature, T of the flow, M, Mach number, γ, ratio of specific heats of the gas, and R, the gas constant [5]. By taking a Mach number is 1, so the maximum flow rate can be achieved. [6]

$$m = \frac{APt}{\sqrt{Tt}}\sqrt{\frac{\gamma}{R}}M\left(1+\frac{\gamma-1}{2}M^2\right)^{-\frac{\gamma+1}{2(\gamma-1)}} \quad (1)$$

2.2. Calculations

The values taken for substitution in Equation 1 are:

P_t = 101325 Pa
T = 300K
γ = 1.4
R (air) = 0.286 kJ/Kg.K
A = 0.001256 m^2
M = 1 (Choking Conditions)

With the value shown above in Equation 1, the result of the calculation at choking is = 0.0703 kg/s. The results will be used in the output and the pressure at the inlet is ambient. The goal of this analysis is to maximize pressure recovery in output.

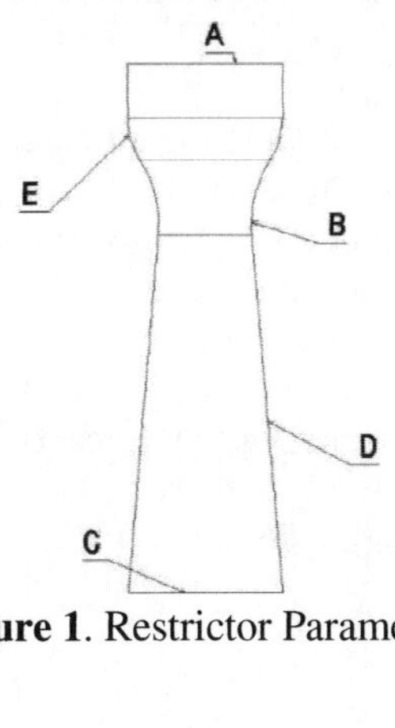

Figure 1. Restrictor Parameter.

Figure 1 shows the parameters that will be used to determine the differences between the designs made for the pressure loss at the restrictor outlet. (A) is the inlet restrictor, (B) is a choke diameter of 20mm maximum. (C) is the output and (D) is the divergence angle of the restrictor, while the (E) is type of shape, either from a convergence angle or using a curve of the restrictor. Table 1 below shows the values used to design the restrictor.

Table 1. Restrictor Design Parameter.

	A(mm)	B(mm)	C(mm)	D(°)	E(°)	Length(mm)
Design 1	38	18	50	10	14	192
Design 2	38	20	38	7	-	178.47
Design 3	38	20	38	7	-	180.8
Design 4	40	20	40	10	-	167

Figure 2 shows the design of a different restrictor. (A) choke with a diameter of 18mm and a convergence angle of 14 degrees and an angle of divergence is 10 degrees. (B) is the choke area 20mm diameter and divergence angle is 7 degrees. At the inlet, oval-shaped curve is connecting to the choke and continue to the point of divergence area. (C) choke area 20mm diameter and divergence angle is 7 degrees. At the inlet, in the form of semi-circular curve and straight to the point of divergence. (D) the choke area 20mm diameter and divergence angle is 10 degrees. At the inlet, the curve is semi-circular shape and has a smooth curve that is connected to the choke and continue to the point of divergence area.

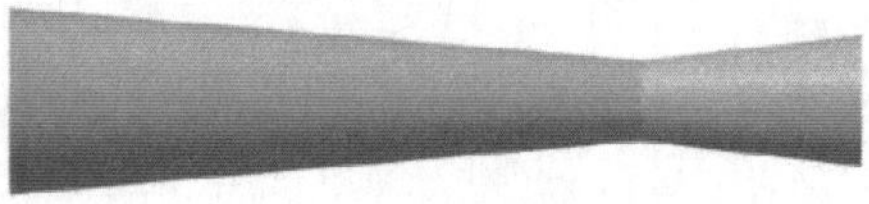

(a) Design 1

(c) Design 2

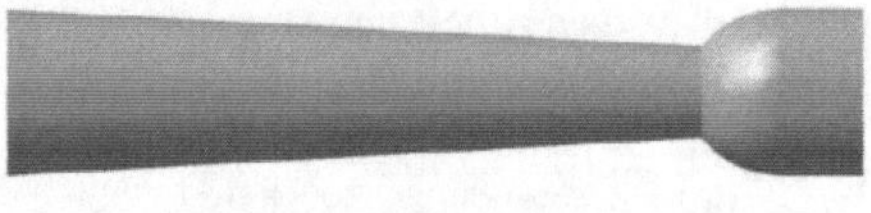

(b) Design 3

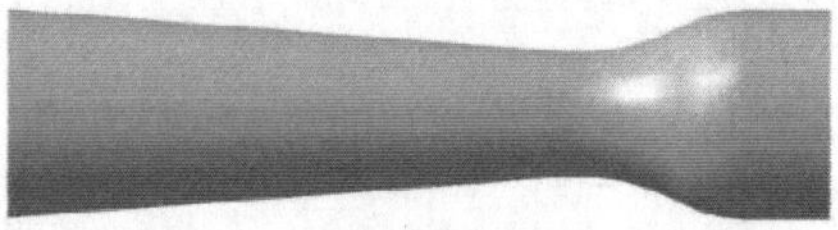

(d) Design 4

Figure 2. Restrictor Design with Difference Parameter.

3. Design of Intake Manifold

The function of the intake manifold is to distribute the air enters the engine evenly. Evenly distribution is very important for better engine performance and optimum efficiency [7]. The design of the intake manifold is very important for better engine performance. The length and diameter of the runner, the total volume of the intake manifold plenum, and shape runner is among the factors to be considered in the design of the intake manifold. some improvements that can be made in the intake manifold is to direct the airflow straight towards the runner. [8]. Intake manifold geometric design greatly affects the volumetric efficiency of an engine, therefore directly impacts on the performance of the vehicle. [9]. Better understanding of the flow, inlet velocity and pressure loss through the intake manifold system is very important to optimize the flow in the intake manifold, [10].

IOP Conf. Series: Journal of Physics: Conf. Series **908** (2017) 012069 doi:10.1088/1742-6596/908/1/012069

3.1. Intake Manifold Data Value.

The engine reading has been recorded using Haltech Engine Management ECU Software. The map reading has been recorded to make a reference when analyze the intake manifold design. Table 2 shows the data that has been recorded by using Haltech Engine Management ECU software, when the data is retrieved, the engine speed to idle speed, 3300RPM where the main problems that need to be repaired. Vacuum value during idling is -42,300 Pascal, total pressure at the inlet is fixed which is at ambient pressure, and the amount of 101325 Pascal outlet pressure is 59,025 Pascal.

Table 2. Engine Data Value.

Engine Speed	3300 RPM
Vacuum Level	-42300 Pa
Pressure Inlet	101325Pa
Pressure Outlet	59025 Pa

(a) Intake Manifold Design 1

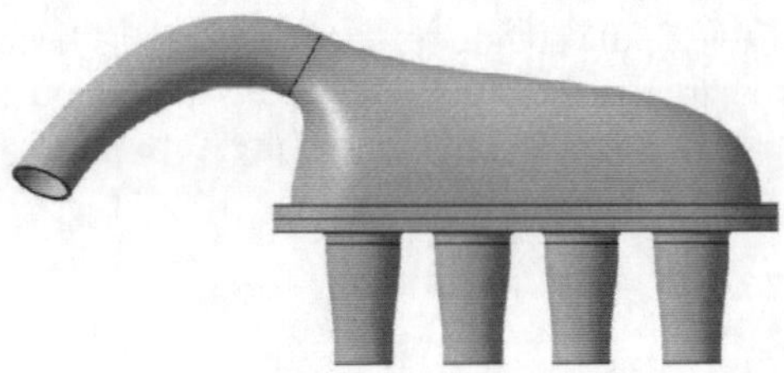

(b) Intake Manifold Design 2

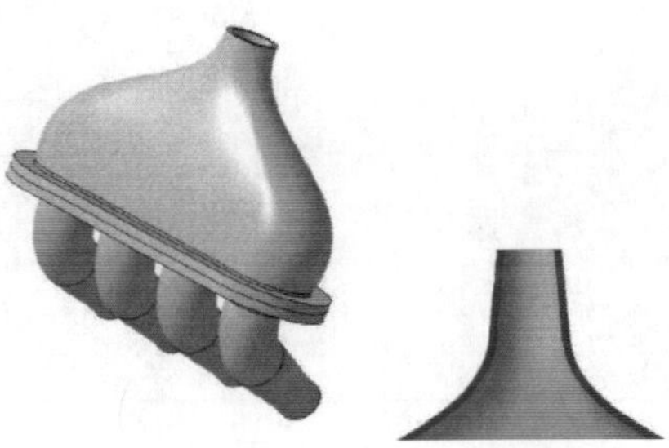

(c) Intake Manifold Design 3

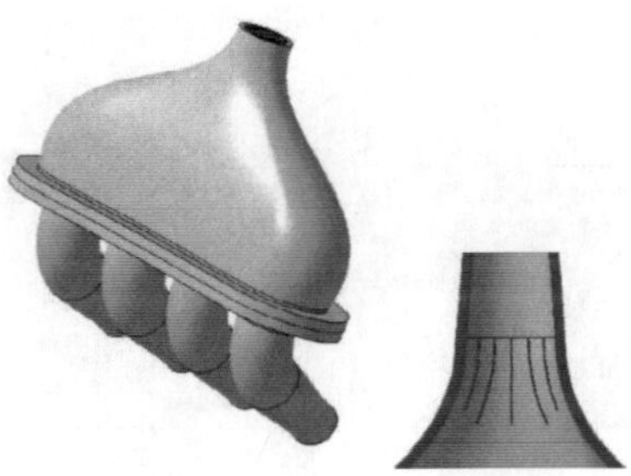

(d) Intake Manifold Design 4

Figure 3. Variation of Intake Manifold Design.

Figure 3 shows the various types of the intake manifold without restrictor. (a) The intake manifold has two branch paths after the inlet. Shaped plenum box and quarter round. (b) The inlet of the intake manifold edges, shape longitudinal and tapered at the ends. (c) The intake manifold inlet of the central and symmetrical. (d) The intake manifold inlet of the central and symmetrical but having additional fins at the inlet of the intake manifold.

IOP Conf. Series: Journal of Physics: Conf. Series **908** (2017) 012069 doi:10.1088/1742-6596/908/1/012069

4. Data Collection and Analysis

4.1. Restrictor Design Result.

By simulating the design of restrictor and intake manifold, the restrictor design result comparison has been made with comparing the pressure at the outlet. Table 3 shows the results of the analysis carried out using Ansys. Analysis performed on each design using same parameters as the inlet pressure is equal to ambient pressure, 101325 Pascal and the outlet is as it has been calculated that 0.0703kg / s. The Design 1 has the more pressure loss while the Design 4 pressure loss at least when compared with all four of the design.

Table 3. Restrictor Pressure Drop Result.

Restrictor	Inlet(Pa)	Outlet(Pa)	Pressure Drop(Pa)
Design 1	101325	95062	6263
Design 2	101325	91739	9586
Design 3	101325	93721	7604
Design 4	101325	98286	3039

From Table 3, the contour of the pressure has been shown in Figure 4 below. Figure 4 shows the change in pressure contour analysis results based on the mass flow rate that has been calculated. Therefore, the restrictor that have a low pressure loss have been selected to proceed with the analysis of the intake manifold as well.

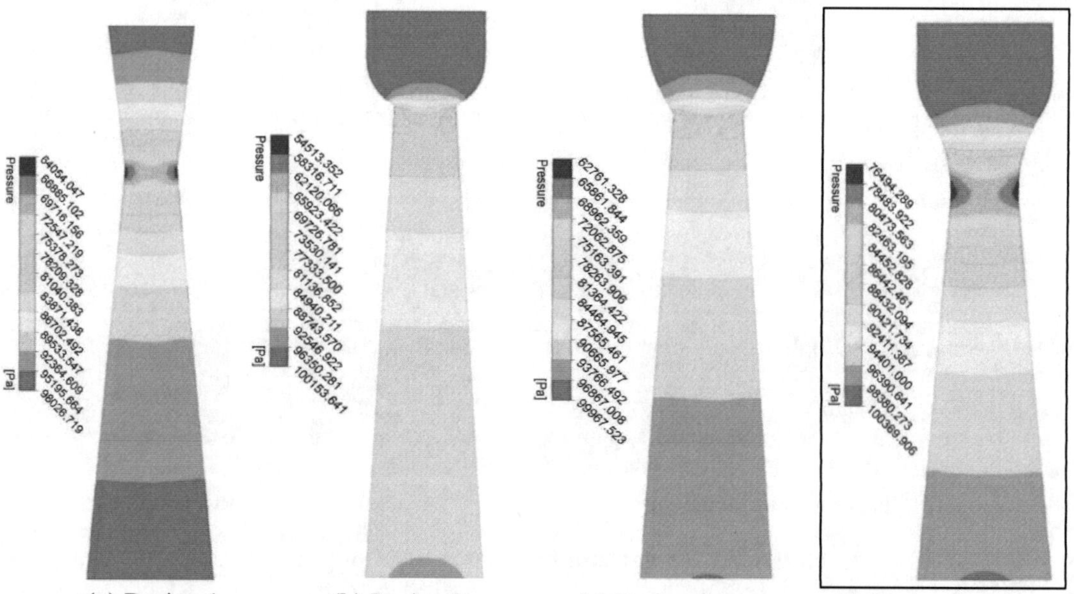

(a) Design 1 **(b)** Design 2 **(c)** Design 3 **(d)** Design 4

Figure 4. Variations of Pressure Contour in Each Restrictor Design.

4.2. Intake Manifold Design Result

Based on the result in Table 3, the Design 4 has been selected. The results of the analysis of the intake manifold and restrictor were recorded as in the table below.

IOP Conf. Series: Journal of Physics: Conf. Series **908** (2017) 012069 doi:10.1088/1742-6596/908/1/012069

Table 4. Intake Manifold Mass Flow Rate Result.

	Mass Flow Rate (kg/s)			
Cylinder	**1**	**2**	**3**	**4**
Design 1	0.07883	0.06659	0.06862	0.07827
Design 2	0.04241	0.04186	0.07784	0.14447
Design 3	0.01330	0.09130	0.10160	0.01490
Design 4	0.13085	0.15248	0.13999	0.14159

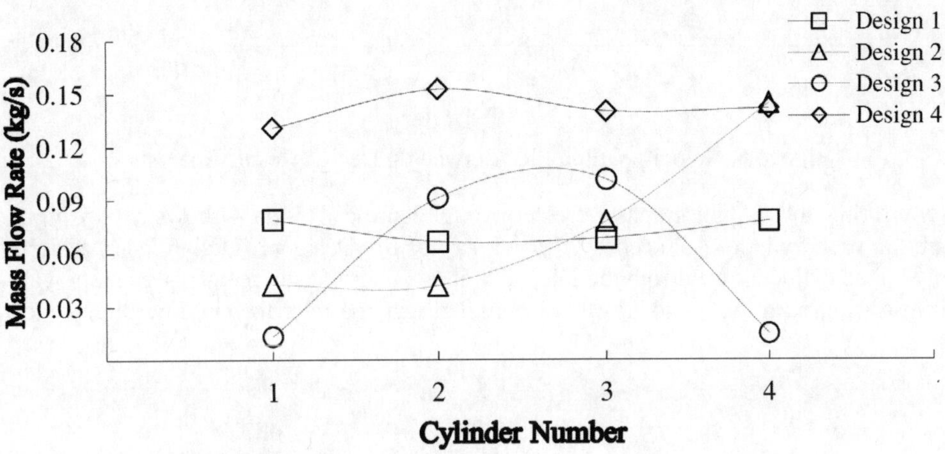

Figure 5. Mass Flow Rate Graph Plot Between 4 Design Result Analysis.

Figure 5 shows the results of the data that has been recorded in table 4. Based on observation, Design 1 has almost uniform mass flow rate value in each cylinder. From Design 2, Cylinder 3 and 4 only have high mass flow rate than Cylinder 1 and 2. Design 3, the cylinder closest to the inlet only will get the high mass flow rate high, the Cylinder 2 and 3. Design 4, the mass flow rate is almost evenly on all cylinders and is higher than the design. Table 5 shows the outlet velocity data sets obtained during the analysis is conducted.

Table 5. Intake Manifold Velocity Analysis Result.

	Velocity (m/s)			
Intake Manifold	**1**	**2**	**3**	**4**
Design 1	97.63	99.03	98.93	97.51
Design 2	33.00	34.93	61.76	109.56
Design 3	10.76	76.89	83.73	12.12
Design 4	103.73	106.03	105.26	104.41

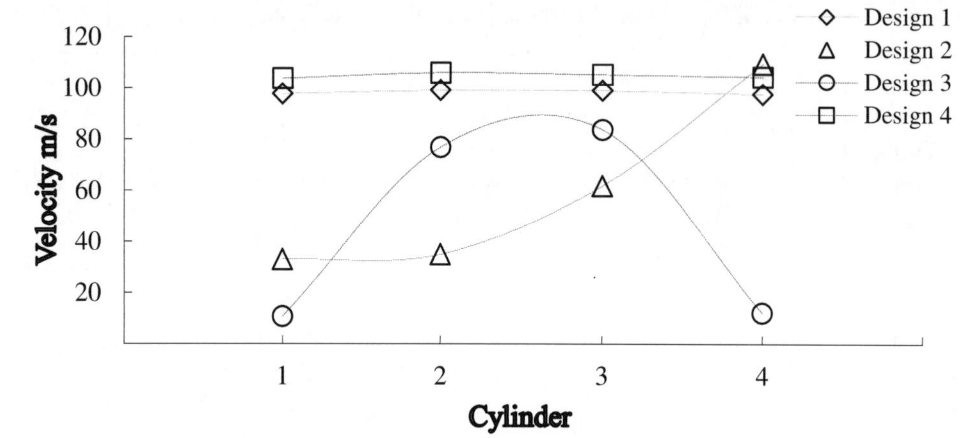

Figure 6. Velocity Graph Plot Between 4 Design Result Analysis.

By referring able 5, plot graphs were shown as Figure 6. Design 4 shows a very high velocity and evenly on each cylinder followed by the velocity of the Design 1. Design 2 get a high velocity value only on the cylinder away from the inlet as it approaches the surface of the plenum wall. Design 3 is also just acquired a high velocity in the cylinder near the inlet for air flow directly toward the cylinder.

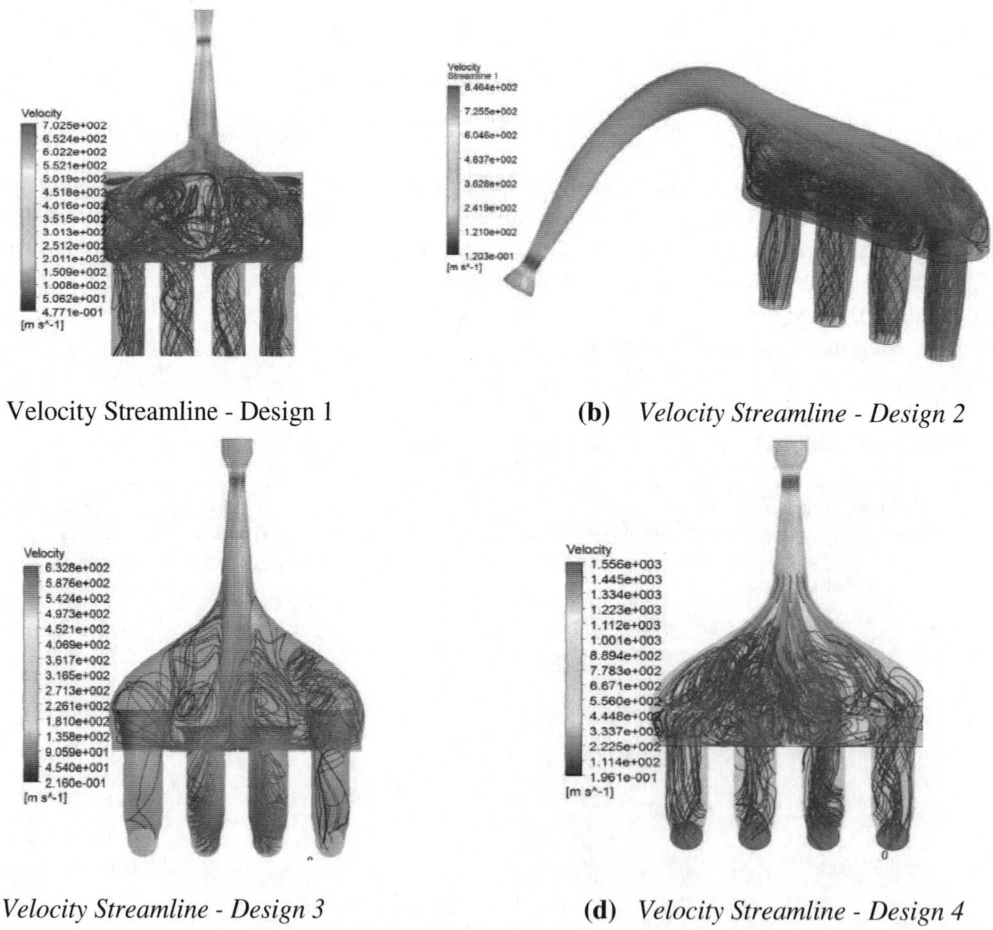

(a) Velocity Streamline - Design 1

(b) *Velocity Streamline - Design 2*

(c) *Velocity Streamline - Design 3*

(d) *Velocity Streamline - Design 4*

Figure 7. Variations of Velocity Streamline Pattern in Each Intake Manifold Design.

Figure 7 shows a variety streamline velocity in each intake manifold design. (a) Velocity streamline seems equally due to the inlet manifold has been split into two paths after the restrictor. (b) Velocity streamline at the end of the intake manifold higher than the cylinder near the inlet. (c) Velocity streamline is more focused on the area near the inlet compared to far away from the inlet. (d) Velocity streamline more evenly when added fins at the inlet of the intake manifold.

5. Conclusion

From the result of the study, the restrictor converging boundary must have smooth edge corner to make sure the air flow maintains with their flow path. The pressure recovery of restrictor must be maximum as much as possible to minimize the pressure drop. Based on simulation that has been made, to get a little pressure loss on the outlet restrictor, the restrictor design geometry must be focused to achieve the objective restrictor design. The air flow in the intake manifold must distributed evenly to improve the volumetric efficiency while improve the engine performance. It is necessary to determine the velocity of throttle body inlet and Manifold Air Pressure for another engine and intake system with sensitive experimental studies. For future validations, high calibrated sensors should be used to improve the accuracy of the experimental data obtained.

6. Acknowledgments

This work has been supported by the UniMAP Automotive Racing Team (UniART) and School of Mechatronic Engineering, Universiti Malaysia Perlis.

7. References

[1] P. A. Shinde, "Research and optimization of intake restrictor for Formula SAE car engine," *International Journal of Scientific and Research Publications,* vol. 4, 2014.

[2] N. Persaud, "DESIGN AND OPTIMZATION OF A FORMULA SAE COOLING SYSTEM," Department of Mechanical and Industrial Engineering, University Of Toronto, 2007.

[3] ASME MFC-7M, Measurement of Gas Flow by Means of Critical Flow Venturi Nozzles, New York: American Society if Mechanical Engineers, 1987.

[4] "Venturi Description - Flow through a Venturi," [Online]. Available: en.wikipedia.org/wiki/Venturi_effect.

[5] National Aeronautics and Space Administration, [Online]. Available: https://www.grc.nasa.gov/www/k-12/airplane/mflchk.html. [Accessed 17 March 2017].

[6] A. G. F. Y. a. A. B. Harry C. Watson, "Optimizing the Design of the Air Flow Orifice or Restrictor for Race Car Applications," *SAE TECHNICAL PAPER SERIES,* 2007.

[7] S. Singla, "Study of Design Improvement of Intake Manifold of Internal Combustion Engine," *International Journal of Engineering Technology, Management and Applied Sciences www.ijetmas.com,* 2015.

[8] M. &. K. P. Kumar, "CFD Simulation of Air Flow in Intake Manifold of SI Engine," 2013.

[9] C. B. W. Ryan Ilardo, "DESIGN AND MANUFACTURE OF A FORMULA SAE INTAKE SYSTEM USING," *Design, Research, and Education for Additive Manufacturing Systems Laboratory,* 2009.

IOP Conf. Series: Journal of Physics: Conf. Series **908** (2017) 012070 doi:10.1088/1742-6596/908/1/012070

Buoyancy induced Couette-Poiseuille flow in a vertical microchannel

M Narahari

Fundamental and Applied Sciences Department, Universiti Teknologi PETRONAS, 32610 Seri Iskandar, Malaysia

marneni@utp.edu.my

Abstract. The fully developed buoyancy-induced (natural convective) Couette-Poiseuille flow in a vertical microchannel is investigated with the velocity slip and temperature jump boundary conditions. Closed form analytical solutions for the velocity and temperature fields are obtained. The effects of the fluid-wall interaction parameter, wall-ambient temperature difference ratio, Knudsen number, mixed convection parameter, and the dimensionless pressure gradient on the velocity, temperature, volume flow rate, heat flux between the plates and the Nusselt number have been discussed in detail through graphs. The outcomes of the investigation indicate that the volume flow rate increases with increasing values of mixed convection parameter, wall-ambient temperature difference ratio, and Knudsen number.

1. Introduction

In recent years, demands from all around the world calling for more functions and increased portability devices has driven the technology advancement towards devices miniaturization. This process has led to the generation of micro-electro-mechanical systems (MEMS) such as micro-reactors, micro-engines, micro-opto-electromechanical systems, micro-fluidic devices and biomedical applications that involve DNA sequencing and bio-MEMS. All microfluidic devices consist of various microchannel geometries whose characteristic dimension range from 1 micrometre to 1 millimetre. Generally, a parameter called Knudsen number is used to measure the effect of small length scale of microgeometries on the flow condition. Knudsen number represents the ratio of mean free path to the characteristic dimension of the system which was defined in four flow regimes by Beskok and Karniadakis [1] as $Kn < 0.001$: continuum regime, $0.001 < Kn < 0.1$: slip regime, $0.1 < Kn < 10$: transition regime, and $Kn > 10$: free molecular regime. As the Kn increases, rarefaction effects become much more significant and it is inappropriate to address the microscale rarefied gas flow problems by applying the macroscale solution due to the great behavioural difference. The no-slip velocity and no-temperature-jump boundary conditions are no longer valid for microscale devices. In the present study, gas no longer reaches the temperature of the two plates and the velocity of the surface due to the moving hot plate and a stationary cold plate. Hence, velocity slip and temperature jump boundary conditions are considered. Chen and Weng [2] investigated the natural convection flow in a vertical microchannel with asymmetric wall temperatures. They discovered that the volume flow rate was higher and the heat transfer rate was lower in microscale than in macroscale. Dongari et al. [3] used the second-order slip boundary condition to investigate the gas flow in long microchannels. It was proven that employing the second order accurate boundary condition in the study of gas flow through microchannels provide a better insight of the behavior of pressure and volume flux compared to the first order. Avci and Aydin [4] examined the mixed convective heat transfer in a vertical microchannel. In this study, they discovered that an increase of mixed convection parameter increased the velocity near the hot wall while it decreased the velocity near the cold wall. Avci and Aydin [5] presented the exact analysis of the mixed convection flow in a microchannel with asymmetric wall heat fluxes. Their results indicated that the velocity slip increased and the maximum velocity decreased with increased Knudsen number. Aydin and Avci [6] also studied the laminar forced convection in a microchannel for two types of thermal boundary conditions namely constant heat flux (CHF) and constant wall temperature (CWT). They found that the Nusselt number was decreased with the increase of Knudsen number in both cases. The similarity solutions were derived by Wang et al. [7] for flow and heat transfer in microchannels. They employed the similarity transformation to transform the governing equations and Homotopy analysis method (HAM) to solve the resulting ordinary differential equations. Kushwaha and Sahu [8] analyzed the gaseous flow in a micropipe by using the second order velocity slip and temperature jump boundary conditions to solve the governing equations. They reported that Nusselt number has lower value in the case of second-order slip model as compared to the first order slip model. Buonuomo and Manca [9] conducted the study of transient natural convection in microchannel by using a finite-difference method. They identified that mass flow rate increased with increased Knudsen number while the Nusselt number decreased with increased Knudsen number. Moslehi and Saghafian [10] numerically examined the developing laminar mixed convection flow in a microchannel under the influence of a uniform magnetic field by employing the finite volume method. They found that an increase in the Hartmann number decreased the velocity for both Knudsen number equals to 0 and 0.1, and for all values of mixed convection parameter.

Even though there are plenty of studies that focused on the microchannel flows, studies that focus on shear-driven microchannel flows remains limited. Kabov et al. [11] experimentally investigated the flow of a locally heated shear-driven liquid film moving under the friction of gas in a channel. Kabov et al. [12] also extended the experimental study to the evaporation and flow dynamics of thin, shear-driven liquid films in microgap channels. Through the two experiments, it was found that shear-driven fluid serves as a more suitable option for cooling applications compared to gravity-driven liquid films as the critical heat flux of the shear-driven fluid is significantly higher than gravity-driven fluid. Recently, Narahari and Rajashekhar [13] investigated analytically the natural convective Couette flow in a vertical microchannel. Their results showed that the fluid velocity as well as the volume flow rate increased with increased Grashof number. However, the shear-driven mixed convection flow in a microchannel has not been addressed in the literature even though such kind of flow is encountered in micromotors, comb mechanisms, and microbearings. The objective of the present paper is to investigate the fully developed natural convective Couette-Poiseuille flow of a rarefied gas in a vertical parallel plate microchannel with a moving hot plate and a stationary cold plate. The effects of the fluid-wall interaction parameter (*In*), wall-ambient temperature difference ratio (ξ), Knudsen number (Kn), mixed convection parameter (Gr/Re), and the dimensionless pressure gradient (P_1) on the velocity (*U*), temperature (θ), volume flow rate (*M*), heat flux between the plates (*Q*) and the Nusselt number (Nu) are determined and discussed graphically.

2. Mathematical analysis

Consider the laminar rarefied gaseous flow in a vertical microchannel with cooler and hotter plates temperatures maintained at T_1 and T_2, respectively. The x – axis is taken along the plates in the upward direction and the y – axis is normal to the plates. The cooler plate located at $y=0$ is fixed while the hotter plate at $y=b$ is moving with uniform velocity u_p in the upward direction in its own plane. Both ends of the channel are open to ambient of temperature T_0 and density ρ_0 as shown in figure 1. The fluid properties are assumed to be constant, internal heat generation and viscous dissipation are neglected. Then under the Boussinesq approximation, the hydrodynamically fully developed steady state flow can be represented by the following equations (Avci and Aydin [4]):

$$0 = -\frac{dp}{dx} + \rho g \beta (T - T_0) + \mu \frac{\partial^2 u}{\partial y^2} \tag{1}$$

$$u\frac{\partial T}{\partial x} = \frac{k}{\rho c_p}\frac{\partial^2 T}{\partial y^2} \tag{2}$$

The corresponding velocity slip and temperature jump boundary conditions (Chen and Weng [2]) are

$$u = \frac{2 - F_v}{F_v}\lambda\frac{\partial u}{\partial y} \quad \text{at } y = 0 \tag{3}$$

$$u = u_p - \frac{2 - F_v}{F_v}\lambda\frac{\partial u}{\partial y} \quad \text{at } y = b \tag{4}$$

$$T = T_1 + \frac{2 - F_t}{F_t}\frac{2\gamma_s}{\gamma_s + 1}\frac{\lambda}{\Pr}\frac{\partial T}{\partial y} \quad \text{at } y = 0 \tag{5}$$

$$T = T_2 - \frac{2 - F_t}{F_t}\frac{2\gamma_s}{\gamma_s + 1}\frac{\lambda}{\Pr}\frac{\partial T}{\partial y} \quad \text{at } y = b \tag{6}$$

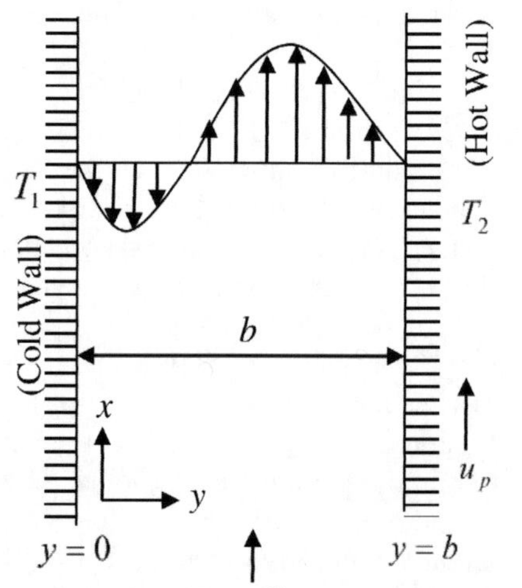

Figure 1. Schematic diagram of the physical model.

Introducing the following dimensionless quantities:

$$\left.\begin{array}{l} X = \dfrac{x}{\operatorname{Re} b}, Y = \dfrac{y}{b}, U = \dfrac{u}{u_p}, \theta = \dfrac{T - T_0}{T_2 - T_0}, Gr = \dfrac{g\beta(T_2 - T_0)b^3}{\upsilon^2}, \Pr = \dfrac{\mu c_p}{k}, P = \dfrac{p}{\rho u_p^2}, \\ \operatorname{Re} = \dfrac{u_p b}{\upsilon}, \beta_\upsilon = \dfrac{2 - F_\upsilon}{F_\upsilon}, \beta_t = \dfrac{2 - F_t}{F_t}\dfrac{2\gamma_s}{\gamma_s + 1}\dfrac{1}{\Pr}, \operatorname{Kn} = \dfrac{\lambda}{b}, \xi = \dfrac{T_1 - T_0}{T_2 - T_0}. \end{array}\right\} \tag{7}$$

The dimensionless form of equations (1) to (6) are, respectively, as follows

$$\frac{dP}{dX} = \frac{Gr}{\operatorname{Re}}\theta + \frac{d^2 U}{dY^2} \tag{8}$$

IOP Conf. Series: Journal of Physics: Conf. Series **908** (2017) 012070 doi:10.1088/1742-6596/908/1/012070

$$\mathrm{Pr}U\frac{\partial\theta}{\partial X}=\frac{\partial^2\theta}{\partial Y^2} \tag{9}$$

$$U=\beta_v\mathrm{Kn}\frac{\partial U}{\partial Y}\quad \text{at } Y=0 \tag{10}$$

$$U=1-\beta_v\mathrm{Kn}\frac{\partial U}{\partial Y}\quad \text{at } Y=1 \tag{11}$$

$$\theta=\xi+\beta_t\mathrm{Kn}\frac{\partial\theta}{\partial Y}\quad \text{at } Y=0 \tag{12}$$

$$\theta=1-\beta_t\mathrm{Kn}\frac{\partial\theta}{\partial Y}\quad \text{at } Y=1 \tag{13}$$

where $p=p'-p''$ is the pressure difference, p' is the static pressure, p'' is the hydrostatic pressure, x and y are rectangular coordinates, ρ is the density, g is the gravitational acceleration, β is the thermal expansion coefficient, T is the fluid temperature, T_0 is the inlet fluid temperature, μ is the dynamic viscosity, u is the velocity component in the x-direction, k is the thermal conductivity, c_p is the specific heat at constant pressure, F_t is the thermal accommodation coefficient, F_v is the tangential momentum accommodation coefficient, λ is the molecular mean free path, b is the channel width, u_p is the plate velocity, T_1 and T_2 are the cold and hot walls temperatures, respectively, $\gamma_s=c_p/c_v$ is the ratio of specific heats, c_v is the specific heat at constant volume, Pr is the Prandtl number, X and Y are dimensionless rectangular coordinates, U is the dimensionless velocity component in X-direction, P is the dimensionless pressure, Gr is the Grashof number, Re is the Reynolds number, ν is the kinematic viscosity, θ is the dimensionless temperature, β_t and β_v are the dimensionless variables, ξ is the wall-ambient temperature different ratio, $\mathrm{Kn}=\lambda/b$ is the Knudsen number.

For fully developed thermal flow, equation (9) becomes

$$0=\frac{\partial^2\theta}{\partial Y^2} \tag{14}$$

Integrating equation (14) with respect to the boundary conditions (12) and (13), we have,

$$\theta(Y)=A_1Y+A_2 \tag{15}$$

where

$$A_1=\frac{1-\xi}{1+2\beta_v\mathrm{Kn}\,In} \tag{16}$$

$$A_2=\frac{\xi+(\xi+1)\beta_v\mathrm{Kn}\,In}{(1+2\beta_v\mathrm{Kn}\,In)} \tag{17}$$

and $In=\dfrac{\beta_t}{\beta_v}$ is the fluid-wall interaction parameter.

Solving the momentum equation (8) subject to the boundary conditions (10) and (12), gives

$$U=\frac{P_1Y^2}{2}-\frac{Gr}{\mathrm{Re}}\left(A_1\frac{Y^3}{6}+A_2\frac{Y^2}{2}\right)+B_1Y+B_2 \tag{18}$$

where

$$P_1 = \frac{dP}{dX} \tag{19}$$

$$B_1 = \frac{1}{(1+2\beta_v \text{Kn})} - \frac{P_1}{2} + \frac{Gr}{6\text{Re}}\left[A_1 \frac{(1+3\beta_v \text{Kn})}{(1+2\beta_v \text{Kn})} + 3A_2\right] \tag{20}$$

$$B_2 = \frac{\beta_v \text{Kn}}{(1+2\beta_v \text{Kn})} - \frac{P_1 \beta_v \text{Kn}}{2} + \frac{\beta_v \text{Kn}}{6}\frac{Gr}{\text{Re}}\left[A_1 \frac{(1+3\beta_v \text{Kn})}{(1+2\beta_v \text{Kn})} + 3A_2\right] \tag{21}$$

The non-dimensional volume flow rate 'M' is given by

$$M = \frac{M'}{\text{Re}\, u_p b^2} = \int_0^1 U\, dY = \frac{P_1}{6} - \frac{Gr}{\text{Re}}\left(\frac{A_1}{24} + \frac{A_2}{6}\right) + \frac{B_1}{2} + B_2 \tag{22}$$

where M' is the volume flow rate. The total heat absorbed by the fluid while traversing the channel (or) the vertical heat flux is given by

$$Q = \frac{Q'}{\rho c_p u_p (T_2 - T_0) b} = \int_0^1 U\theta\, dY = \frac{P_1}{24}(3A_1 + 4A_2) + \frac{A_1 B_1}{3} + A_2 B_2 + \frac{1}{2}(A_1 B_2 + A_2 B_1) - \frac{Gr}{6\text{Re}}\left(\frac{A_1^2}{5} + A_2^2 + A_1 A_2\right) \tag{23}$$

where Q' is the vertical heat flux and Q is the dimensionless vertical heat flux. The non-dimensional mean/bulk temperature 'θ_m' is given by

$$\theta_m = \frac{T_m - T_0}{T_2 - T_0} = \frac{\int_0^1 U\theta\, dY}{\int_0^1 U\, dY} = \frac{Q}{M} \tag{24}$$

The convective heat transfer coefficient (h) is given by

$$h = \frac{-k \left.\frac{\partial T}{\partial y}\right|_{y=b}}{T_m - T_2} = \frac{\frac{k}{b}\left.\frac{\partial \theta}{\partial Y}\right|_{Y=1}}{(1-\theta_m)} \tag{25}$$

which can be obtained from the following Nusselt number (Nu),

$$\text{Nu} = \frac{hb}{k} = \frac{1}{(1-\theta_m)}\left.\frac{\partial \theta}{\partial Y}\right|_{Y=1} = \frac{MA_1}{(M-Q)} \tag{26}$$

3. Results and Discussion

Using the analytical approach, closed-form solutions are obtained for the velocity and temperature fields. In the present investigation, the range of Knudsen number Kn is $0 \le \text{Kn} \le 0.1$, where $\text{Kn} = 0$ represent the macroscale case while $\text{Kn} > 0$ represent the microscale case. The effects of the fluid-wall interaction parameter (*In*), wall-ambient temperature difference ratio (ξ), Knudsen number (Kn), mixed convection parameter (Gr/Re), and the dimensionless pressure gradient (P_1) on the velocity (*U*), volume flow rate (*M*), heat flux between the plates (*Q*) and the Nusselt number (Nu) are projected through graphs. The effect of Knudsen number (Kn) on the velocity distribution along the y-axis is shown in figure 2. The velocity is distributed the most at the center of two walls. Kn appeared to have a higher impact on the velocity distribution and increases the flow velocity as well as slip velocity at the plates with increasing Kn. Thus the rarefaction effect is to increase the slip velocity at both plates in the natural convective Couette-Poiseuille flow. Figure 3 illustrates the effects of dimensionless pressure gradient on the velocity distribution along the y-axis. A similar trend can be observed where velocity is the highest in the middle of the channel. The velocity increases with increasing negative value of P_1 (pressure assisted flow) and an opposite trend can be observed with increasing positive value of P_1 (pressure opposed flow). From figure 4 it is clear that the velocity distribution under the effect of wall-ambient temperature difference ratio is increasing and the maximum velocity is shifting towards the middle of the channel with the increase of ξ. When $\xi = 0$, the peak of the velocity is distributed nearer to the moving hot wall. Also, the slip velocity is increasing with increasing value of ξ and it is slightly low at the moving hot plate as compared to the fixed cold plate. Figure 5 shows the effect of mixed convection parameter (Gr/Re) on the velocity distribution. It can be seen that the velocity is slightly higher at the moving hot wall. However, as the Gr/Re rises from 0 to 500 the velocity rises sharply and the highest velocity is distributed at the center of the microchannel. Further, the velocity slip is increasing considerably with increasing value of Gr/Re.

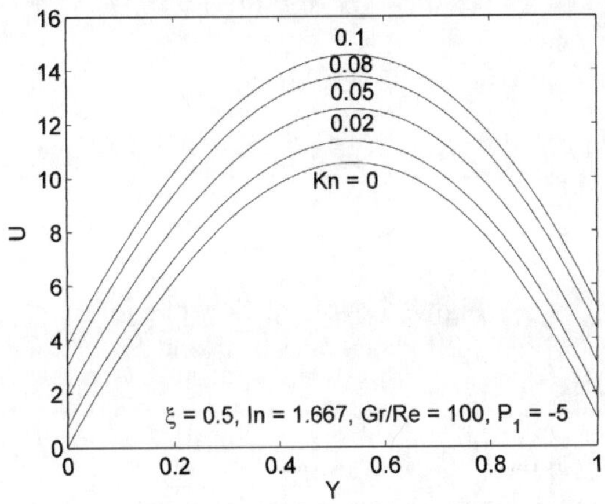

Figure 2. Velocity distribution at various values of Kn.

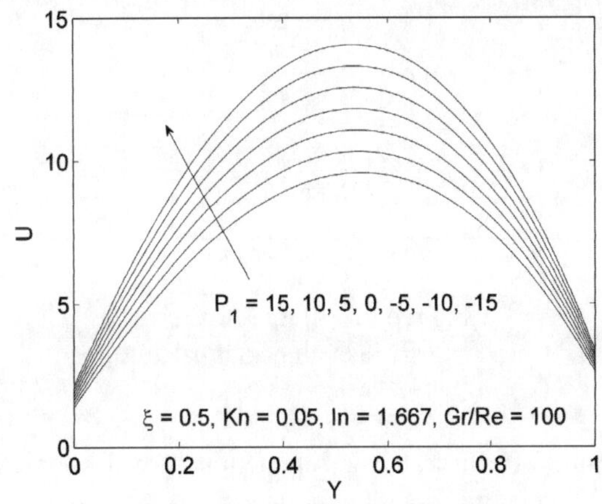

Figure 3. Velocity distribution at various values of P_1.

The effects of Gr/Re and Kn on the volume flow rate are depicted in figure 6. The volume flow rate is constant with Kn when Gr/Re = 0 and for other values of Gr/Re the volume flow rate is increasing with increasing Gr/Re and Kn. However, the rises of volume flow rate with Kn is more significant at higher values of Gr/Re. Figure 7 shows the effects of *In*, ξ and Kn on the volume flow rate (*M*). The figure clearly reveals that *M* increases monotonically as Kn and ξ increases. Also, the volume flow rate is constant at any value of *In*. Figure 8 depicts the variation of Gr/Re and Kn on the vertical heat flux. It is observed that when Gr/Re is equal to zero, the vertical heat flux remain constant even with the increase of Kn. The vertical heat flux starts to increase linearly with Kn when Gr/Re increases and the vertical heat flux pronounced more at higher values of Gr/Re. The impact of *In* on *Q* is shown in figure 9. The vertical heat flux (*Q*) slightly decreases with increasing values of fluid-wall interaction parameter (*In*), however, this decrease is negligible at higher values of *In*.

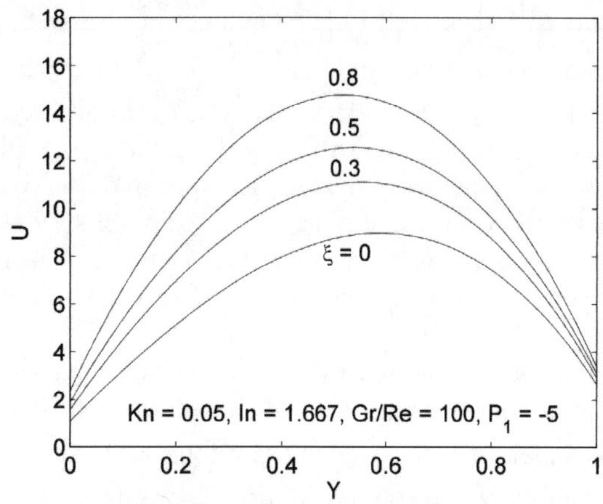

Figure 4. Velocity distribution at various values of ξ.

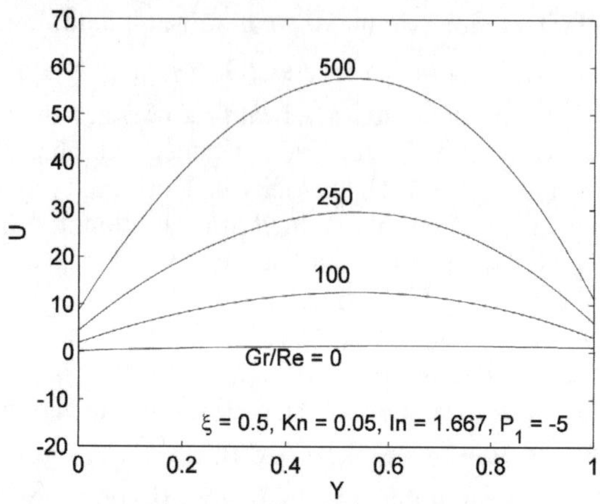

Figure 5. Velocity distribution at various values of Gr/Re.

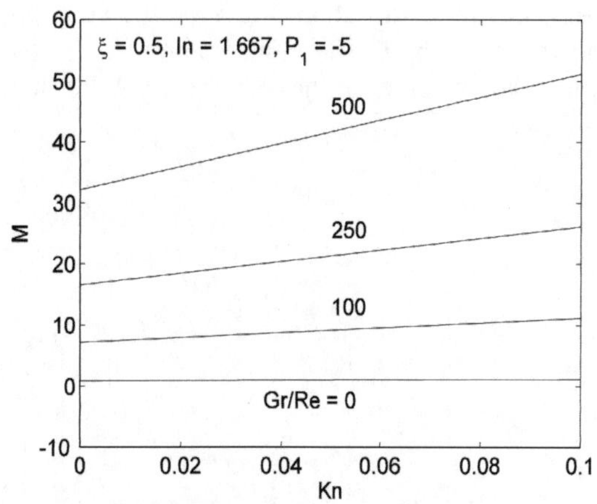

Figure 6. Volume flow rate at various values of Gr/Re.

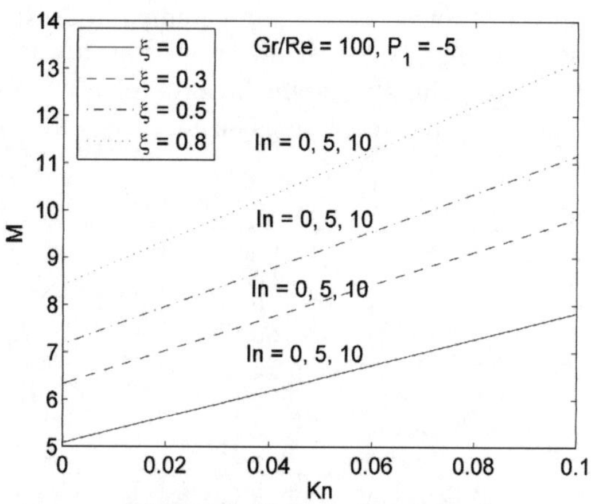

Figure 7. Volume flow rate for different values of ξ and *In*.

Figure 10 illustrate the interaction of wall-ambient temperature difference ratio (ξ) on vertical heat flux (Q) and it can be seen that an increase in ξ substantially increases the vertical heat flux. Also, the rise of vertical heat flux with Kn pronounced more at higher values of ξ. Figure 11 depicts the Nusselt number under the impact of Gr/Re and Kn. The figure clearly reveals a declining curve of Nu with increasing values of both Kn and Gr/Re. Nu is significantly higher when Gr/Re = 0 as compared to Gr/Re = 100, 250 and 500. The decrease of Nu becomes almost negligible at higher values of Gr/Re. Figure 12 depicts the influence of dimensionless pressure gradient on the Nusselt number. It is clear that Nu decreases with increasing negative value of P_1 (pressure assisted flow) and Nu increases with increasing positive value of P_1 (pressure opposed flow). Thus the pressure assisted flow causes a decrease in the Nusselt number while pressure opposed flow increases the Nusselt number. The effect of ξ on the Nusselt number is presented in figure 13. It is observed that the Nusselt number is maximum at $\xi = 0$ and it decreases with increasing values of ξ as well as Kn. Thus the rarefaction effect is to decrease the Nusselt number in the buoyancy induced Couette-Poiseuille microchannel flow.

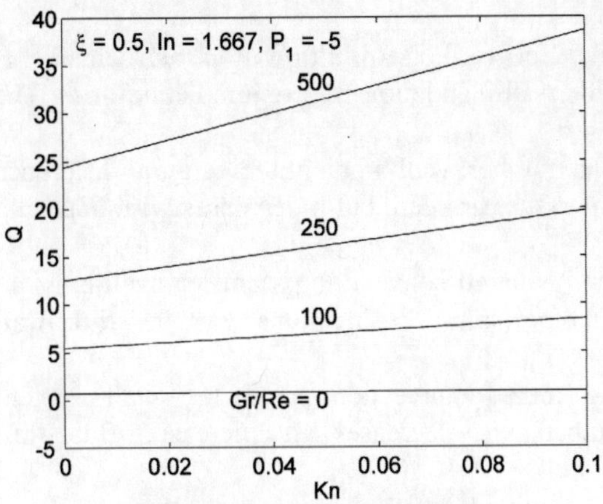

Figure 8. Vertical heat flux for different values of Gr/Re.

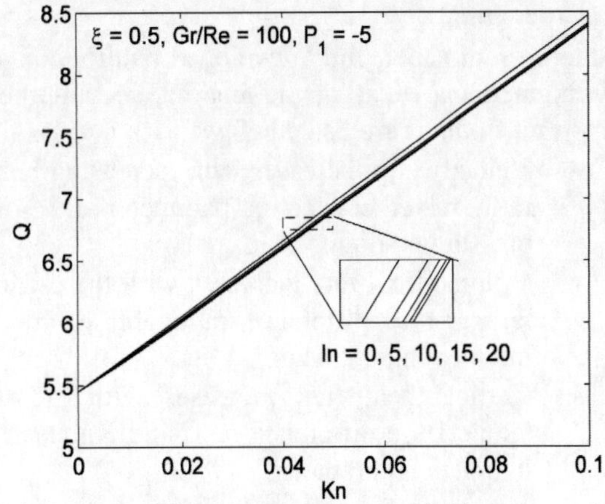

Figure 9. Vertical heat flux for different values of *In*.

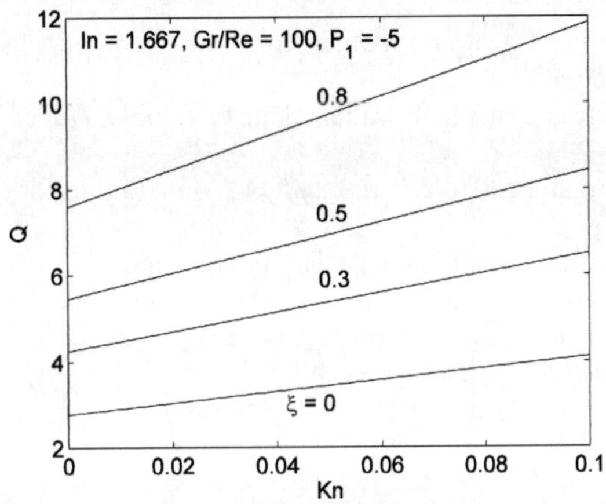

Figure 10. Vertical heat flux for different values of ξ.

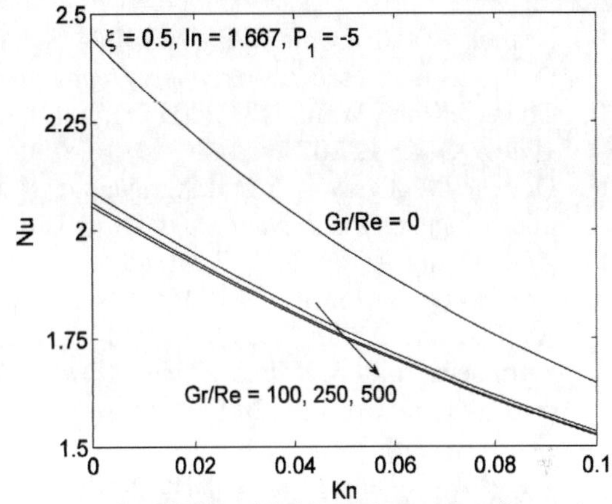

Figure 11. Nusselt number for different values of Gr/Re.

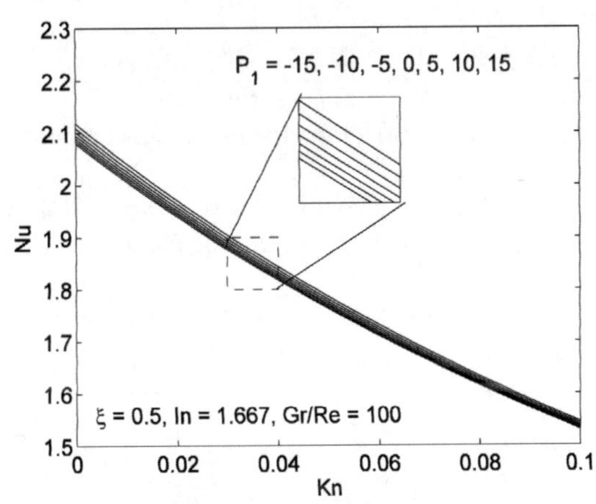

Figure 12. Nusselt number for different values of P_1.

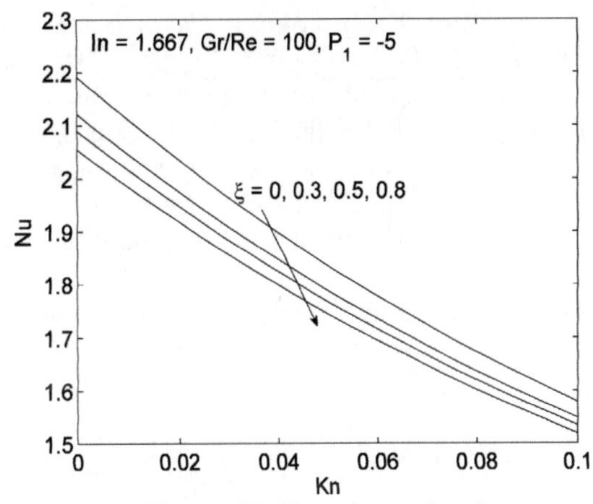

Figure 13. Nusselt number for different values of ξ.

4. Conclusions

In the present paper, fully developed natural convective Couette-Poiseuille flow in a microchannel is investigated analytically by taking into account the velocity slip and temperature jump conditions. The important findings are listed below:

- Velocity slip increases with increasing Knudsen number, wall-ambient temperature difference ratio, mixed convection parameter and aiding pressure gradient but it decreases with opposing pressure gradient.
- Volume flow rate increases with the escalation of mixed convection parameter, wall-ambient temperature difference ratio and Knudsen number but it is invariant with the fluid-wall interaction parameter.
- Vertical heat flux increases with increasing mixed convection parameter, wall-ambient temperature difference ratio and Knudsen number but it decreases with increasing fluid-wall interaction parameter.
- The Nusselt number decreases with the rise of mixed convection parameter, wall-ambient temperature difference ratio, aiding pressure gradient and Knudsen number but the Nusselt number increases with the opposing pressure gradient.

References

[1] Beskok A and Karniadakis G E 1994 Simulation of heat and momentum transfer in complex micro-geometries *J. Thermophys. Heat Transf.* **8** 647
[2] Chen C K and Weng H C 2005 Natural convection in a vertical microchannel *J. Heat Transf.-Trans. ASME* **127** 053
[3] Dongari N, Agrawal A and Agrawal A 2007 Analytical solution of gaseous slip flow in long microchannels *Int. J. Heat Mass Transf.* **50** 3411
[4] Avci M and Aydin O 2007 Mixed convection in a vertical parallel plate microchannel *J. Heat Transf.-Trans. ASME* **129** 162
[5] Avci M and Aydin O 2007 Mixed convection in a vertical parallel plate microchannel with ssymmetric wall heat fluxes *J. Heat Transf.-Trans. ASME* **129** 1091
[6] Aydin O and Avci M 2007 Analysis of laminar heat transfer in micro-Poiseuille flow *Int. J. Therm. Sci.* **46** 30
[7] Wang Z C, Tang D W and Hu X G 2011 Similarity solutions for flows and heat transfer in microchannels between two parallel plates *Int. J. Heat Mass Transf.* **54** 2349
[8] Kushwaha H M and Sahu S K 2014 Analysis of gaseous flow in a micropipe with second order velocity slip and temperature jump boundary conditions *Heat Mass Transf.* **50** 1649
[9] Buonomo B and Manca O 2012 Transient natural convection in a vertical microchannel heated at uniform heat flux *Int. J. Therm. Sci.* **56** 35
[10] Mosleh M and Saghafian M 2015 MHD mixed convection slip flow in a vertical parallel plate microchannel heated at asymmetric and uniform heat flux *J. Mech. Sci. Technol.* **29** 5317
[11] Kabov O A, Lyulin Yu V, Marchuk I V and Zaitsev D V 2007 Locally heated shear-driven liquid films in microchannels and minichannels *Int. J. Heat Fluid Flow* **28** 103
[12] Kabov O A, Zaitsev D V, Cheverda V V and Bar-Cohen A 2011 Evaporation and flow dynamics of thin, shear-driven liquid films in microgap channels *Exp. Therm. Fluid. Sci.* **35** 825
[13] Narahari M and Pendyala R 2015 Natural convective Couette flow in a vertical parallel plate microchannel *Applied Mechanics and Materials* **705** 182

IOP Conf. Series: Journal of Physics: Conf. Series **908** (2017) 012071 doi:10.1088/1742-6596/908/1/012071

Curing characteristics tensile and physical properties of rice straw filled standard Malaysian rubber

A Alnaid[1], N Z Noriman[1], O S Dahham[1], R Hamzah[1], T Adam[1], M N Al-Samarrai[1], M Mohammed[2] and U A A Azlan[3]

[1] Center of Excellence Geopolymer and Green Technology (CEGeoGTech), Faculty of Engineering Technology (FETech), Universiti Malaysia Perlis (UniMAP), Level 1 Block S2, UniCITI Alam Campus, Sungai Chucuh, Padang Besar, 02100, Perlis, Malaysia.
[2] Center of Excellence Geopolymer and Green Technology (CEGeoGTech), School of Materials Engineering, Universiti Malaysia Perlis (UniMAP), Kompleks Pengajian Jejawi 2, 02600 Arau, Malaysia
[3] Faculty of Engineering Technology, Universiti Teknikal Malaysia Melaka (UTeM), Hang Tuah Jaya, 76100 Durian Tunggal, Melaka, Malaysia

E-mail: niknoriman@unimap.edu.my

Abstract. In this research, the effects of Rice Straw (RS) reinforced Standard Malaysian Rubber (SMRL) on curing characteristics, tensile properties and physical properties were investigated. All compounds were prepared using two roll mill at five different RS loading (10, 20, 30, 40, 50 phr). In addition, two different size of RS, fine size (FS) at 300 μm and coarse size (CS) at 10 mm were used. The properties such as cure characteristics, tensile properties and physical properties were determined. Results indicated that the fine size of RS filled SMRL contributed to the better properties such as tensile, hardness and crosslink density compare to coarser size of RS filled SMRL at same loading.

1. Introduction

Recently, the compounding of several materials with the crude rubber is commonly used in industries particularly in tires manufacturing and many studies on their processing, morphology and properties were conducted [1]. One of the most important materials that could added to the rubber compound are fillers, such as carbon black, which had been prove to be worthy filler as it gives higher tensile strength [2]. Despite rubber, particularly natural rubber (NR) have notable properties, such as high tensile strength, high flexibility and elasticity, poor conductivity, high abrasion resistance reinforcement, fillers are needed in rubber compounding to improve the properties of the elastomer itself. Paddy field has become one of Malaysia's accommodations. Thus, the wastage of the rice straw from paddy may potentially become the next natural reinforcement filler. Researchers and scientists still attracted in making rice straw their main attention to study. Liu et al has studied on the flexural properties of rice straw and starch [3]. They stated that composite from straw had better interface and higher flexural properties. During harvest, the moisture content of straw is usually exceeded 60%. However, in dry weather straw can quickly dry to its equilibrium moisture content, which is around 10-12%. Thus, it is low in density and lightweight. In addition, rice straw has low protein but high ash content. The main sources of carbohydrate in rice straw are lignin, hemicellulose and cellulose. Silica content that is high in rice straw does give effect on the economic [4]. In other research, rice straw has been reinforced with high-density polyethylene composite to find the effect of coupled compatibilizing and toughtening treatment. This is due to the elastomer that alone fails to effectively improve the strength and modulus of final composites [5]. The addition of natural filler such as rice straw may reduce the cost of rubber manufacturing. In term of properties, rice straw may give good properties towards rubber compounding as good as synthetic filler used.

Our earliest work has studied the influences of different size and content of rice husk on the cure characteristics, tensile and physical properties of Epoxidized Natural Rubber/ rice husk Compounds [6]. Results showed that the addition of rice husk into rubber has enhanced the cure characteristics of the compounds by reducing the scorch and cure time. Furthermore, the physical properties (crosslinking density and hardness) were enhanced with rice husk addition. Nevertheless, the tensile strength and elongation at break showed deterioration with rice husk increasing.

In our current research, we have studied the effects of RS as a natural fiber on the cure characteristics, tensile and physical properties of SMR L/RS compounds at different RS size and content

2. Materials and methods

2.1. Materials

Standard Malaysian Rubber (SMR L) was used in this research as a raw material, which supplied by Rubber Research Institute of Malaysia (RRIM). While other chemicals, such as sulfur, zinc oxide, stearic acid and N-cyclohexyl-2-benzothiazole sulfonamide (CBS) was supplied by Anchor Chemical Co. (M) Ltd. Rice Straw was taken at Paddy field at Wang Ulu - Kangrar - Perlis. Two different size of rice straw (300 μm and 10 mm) were prepared by using Crusher model RT34 (Chyun Industrial Co. Ltd.).

2.2. Preparation and testing of SMRL/RS compounds

The formulation of SMRL/RS compounds is shown in Table 1. Mixing and pre-blending of rubber with chemicals and additives was done using two-roll mill X (S) K - 160 X 320 based on ASTM D 3184-89 at room temperature. Cure characteristics were studied using Monsanto Moving Die Rheometer (MDR 2000) according to ASTM D 2240-93 at 160 C with about 4 g samples of the respective compounds. Electrically hot press machine was utilized to compression accordingly to respective cure time at 160 C and under a pressure of 30 tonne.

Hardness test was carried out for vulcanized radial shape samples that have 6 mm thickness using Shore-A Durometer (Asker, Kobunshi Keiki Co. Ltd). The tensile test was done using dumbbell-shaped mold cutting from the vulcanized rubber sheets of 2 mm thickness. 5 samples were tested for each rubber compound loading. Universal testing machine (Instron 5582) was used for tensile test according to ASTM D412 at room temperature. ASTM D3616 was used for swelling density. Samples were weighted before immerse in a glass containing 30 ml of toluene and kept in dark place, then weight again after 48 hours. The calculations that used for swelling data were Flory-Rehner equations to calculate the crosslink density of rubber compounds [7].

Table 1. The formulation of SMRL/RS compounds.

Ingredients	Phr					
	R0	R10	R20	R30	R40	R50
SMRL	100	100	100	100	100	100
ZnO	5	5	5	5	5	5
Stearic Acid	2	2	2	2	2	2
CBS	1	1	1	1	1	1
Sulfur	2	2	2	2	2	2
*RS (FC &CS)	0	10	20	30	40	50

IOP Conf. Series: Journal of Physics: Conf. Series **908** (2017) 012071 doi:10.1088/1742-6596/908/1/012071

3. Characterization

Figure1 shows the distribution of raw rice straw morphology in SEM. Low magnification displays that the RS surface is rough enough to entrapped rubber molecule during compounding. Moreover, it can be observed that there are some spaces between rice straw which indicates less compact distribution in RS molecules. This might result in uneven distribution of rice straw in rubber compounding. Thus, it might reduce some result such as tensile strength. From 2000x magnification, it is clearly observe that lignin was surrounded the rice straw particle as it shows glow at the end of the structure.

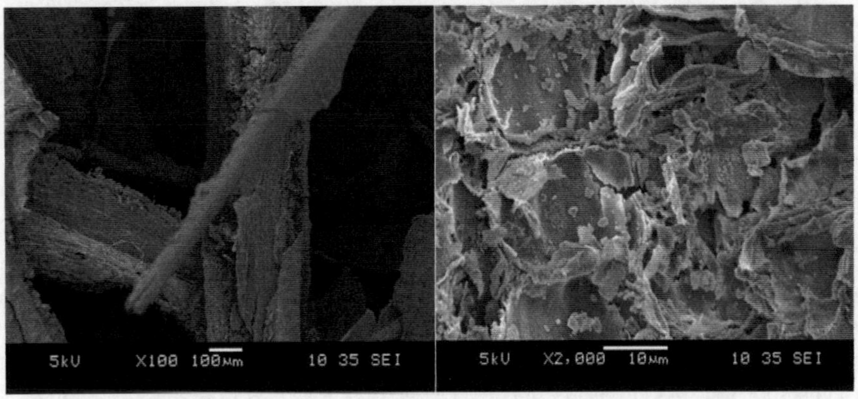

Figure 1. SEM image of raw RS at low (X100) and high (X2000) magnification.

4. Results and discussion

4.1. Cure characteristics

The effect of rice Rice Straw (RS) loading and sizes on scorch time (t_2), cure time (t_{90}), minimum torque (M_L) and maximum torque (M_H) are shown in Table 1. Results indicate that t_2 continuously decrease with increase in RS phr loading on SMRL vulcanizates. The decrease on t2 was due to the restriction of the mobility and deformability of the matrix with mechanical restrain indicating that SMRL vulcanize took less time to reach the onset time of vulcanization as rice straw loading increase. However, the RS coarse size was higher t_{90} than fine size at same loading, which means the tendency of rice straw filler with coarser size requires longer time to coalesce and form agglomerates between each other to distribute in the SMRL rubber. t_{90} was faster when RS loading increase in the compounding.

At same loading, the coarse size of RS need longer time to cure compare to finer size of rice straw loading. This is due to the poor interaction between the rubber matrix and bigger particle size of RS that relate poor dispersion during the compounding. ML, increased with increasing of RS loading due to increase in viscosity. This indicates that the processability of the blends becomes more difficult [8]. M_L of RS coarser size was higher than fine size. This might due to the coarser size gives viscosity and rigidity to the rubber vulcanized. M_H shows same trend of M_L. The increment of M_L relates to the increasing of stiffness of the rubber compound. This might due to the restriction of the chain mobility when RS added. However, the fine size of RS shows lower M_H than coarser size at same loading. This was attributed to better dispersion of RS finer size, which made better compounding process [9].

Table 2. The cure characteristic of SMR L/RS compounds.

	RS	t_2 (min)	t_{90} (min)	M_L (dNm)	M_H (dNm)
R0	Control	3.49	6.34	3.20	12.34
R10	FS	2.28	6.25	5.56	22.28
	CS	2. 41	6.75	6.00	36.16
R20	FS	2.15	6.10	7.99	23.00
	CS	2. 31	6.35	11.29	36.94
R30	FS	0.59	5.24	11.04	25.62
	CS	0. 67	5.79	12.41	37.41
R40	FS	0.39	5.15	12.82	26.00
	CS	0. 43	5.65	13.50	38.44
R50	FS	0.28	5.06	13.44	40.21
	CS	0. 32	5.45	13.99	28.8

4.2. Tensile and physical properties

The values of tensile strength (Ts), elongation at break (Eb), modulus at 100% elongation (M100), crosslink density and hardness of the compounds are listed in Table 3 respectively. It can be seen obviously that Ts decreased gradually with increasing RS loading, which might due to incompatibility between fiber and matrix [10,11]. Nevertheless, Ts for fine size was higher than coarser size at same loading. It is believe that the finer size is well distributed in the rubber compounding.

Eb decreased with increasing RS loading as well causing decrease the resilience and toughness, which in turn led to reduce the mobility of the chain inside rubber compound. Therefore, the compound became stiffer and harder. At same RS loading, fine size shows lower Eb than coarser size due to lower viscosity of rubber vulcanized and easily to be fractured. M100 values increased gradually as RS content increased. According to Sharifah, the increment of M100 values is directly proportional to the crosslink density [12].

The increases in M100 values were due to improvement in crosslinking density. These results indicate the fact that the incorporation of RS into the rubber matrix enhanced the stiffness of the composites [13,14]. Moreover, RS fine size was higher M100 than coarser size. This contributed to the high RS distribution that gives larger surface area to come in contact with rubber matrix. The values of hardness values also increased constantly. Theoretically, both M100 and hardness values increases as filler loading increase. These results indicate the fact that the incorporation of RS into the rubber matrix improved the stiffness and brittleness of the rubber compounds.

Figure 2 shows the swelling percentage of the rubber compounds at different RS loading and size. It can be clearly seen the coarse size of RS, showed higher swelling percentage as compared with fine size of RS. This was might be due to the tendency of RS coarse size to agglomerate, which in turn led to increase the voids of the rubber compounds causing toluene penetration. Therefore, swelling percentage increased. Though, the high RS loading, particularly (50 phr) reduced the voids of the rubber compounds, which led to reduce the toluene penetration through rubber compounds [13].

Table 3. The tensile and physical properties of SMR L/RS compounds.

	RS	Ts (MPa)	Eb (%)	M100 (MPa)	Crosslink density X 10^{-4} (Mole/cm^3)	Hardness (Shore A)
R0	Control	18.4	1450	0.45	0.70	33.4
R10	FS	15.1	1220	1.25	0.81	43.2
	CS	11.0	1350	1.20	0.77	41.5
R20	FS	14.0	1175	1.45	0.98	47.1
	CS	8.8	1290	1.35	0.87	45.8
R30	FS	12.2	1015	1.90	1.14	53.2
	CS	7.6	1160	1.85	1.05	52.0
R40	FS	9.3	900	2.00	1.23	57.1
	CS	4.5	985	1.95	1.19	56.4
R50	FS	5.9	800	2.10	1.34	61.0
	CS	3.0	825	2.00	1.27	58.9

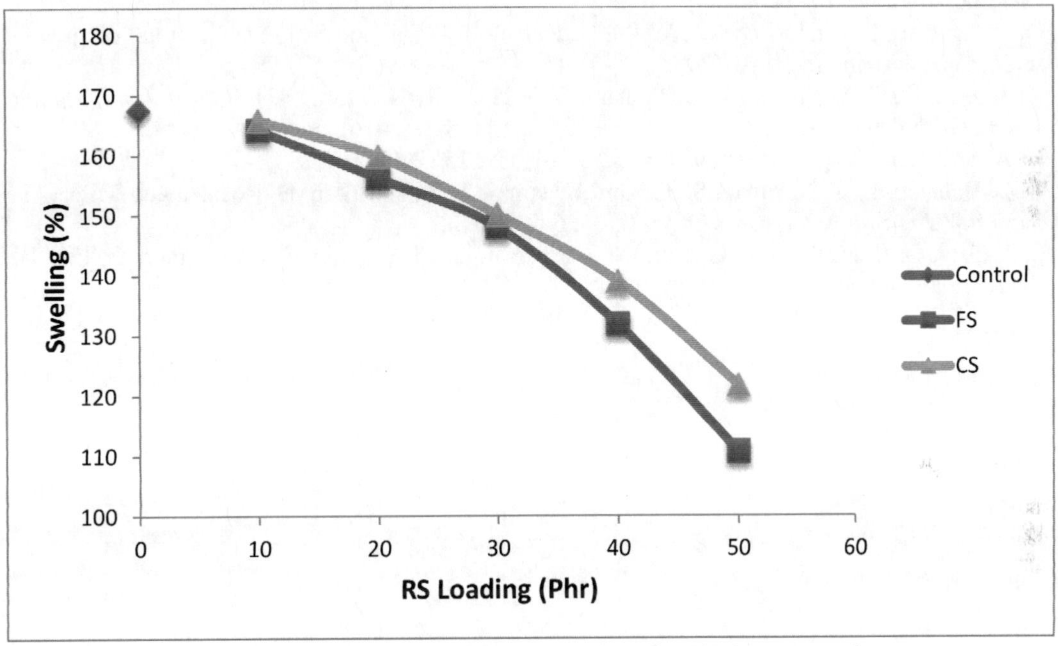

Figure 2. Swelling percentage of SMR L / RS compounds

5. Summary

The addition of RS in SMRL has effected on the properties of the rubber compounds. From curing characteristics, it is shows that higher RS loading gave shorter scorch and cure time in SMRL compounds and increased the torque. On the other hand, it is shows that increment in RS content improved modulus, hardness and crosslink density but reduced the tensile strength and elongation at break. However, the fine size of RS gave better properties than coarser size at same loading.

Acknowledgments

The authors would like to acknowledge the Fundamental Research Grant Scheme (FRGS) Phase 1/2015 (9003-00523) under Ministry of Higher Education, Malaysia. We also would appreciate Universiti Malaysia Perlis (UniMAP) for providing us all the facilities to make this research.

References

[1] P. Schidrowitz and T. R. Dawson, Cambridge: Institution of the Rubber Industry **xxiv** 406 (1952).
[2] J. Clarke, B. Clarke and P. K. Freakley, Rubber chemistry and technology **74**, 1 (2001).
[3] J. Liu, C. Jia and C. He, AASRI Procedia **3**, 89 (2012).
[4] F. A. A. Mohdy, E. S. A. Halim, Y. M. A. Ayana and S. M. Sawy, Carbohydrate Polymers **75,** 44 (2008).
[5] F. Yao, Q. Wu, H. Liu, Y. Lei and D. Zhou, Journal of Applied Polymer Science **119**, 2214 (2011).
[6] O. S. Dahham, N. Z. Noriman, S. T. Sam, N. M. Al-Samarrai, Z. Shayfull, A. M. Alakrach and S. A. Elnaid, *MATEC Web of Conferences* **78**, 01075 (2016).
[7] P. J. Flory and J. Rehner, Journal Chemistry Physics **11,** 512 (1943).
[8] N. Z. Noriman, H. Ismail and A. R. Azura, Polymer-Plastics Technology and Engineering **47**, 1016 (2008).
[9] O. S. Dahham, N. Z. Noriman, S. T. Sam, M. F. Omar and A. Alakrach, Applied Mechanics and Materials **754**, 693(2015).
[10] O. S. Dahham, N. Z. Noriman, A. W. Mohd Kahar, H. Ismail and S. T. Sam. Journal of applied science and agriculture **10**, 33 (2015).
[11] K. Balaed, N. Z. Noriman, O. S. Dahham, S. T. Sam, R. Hamzah, M. F. Omar, *International Journal of Polymer Analysis and Characterization* **21**, 590 (2016).
[12] H. A. Sharifah, *Composite. Science. Technology* **65**, 525 (2005).
[13] O. S. Dahham, N. Z. Noriman, S. T. Sam, H. Ismail, S. Ragunathan, H. Rosniza and M. N. Al-Samarrai, *Journal of Polymer Materials*, **33**, 647 (2016).
[14] S. T. Sam, O. S. Dahham, P. G. Gan, N. Z. Noimam, J. Y. Kuan, and A. M. Alakrach, (2017). *Solid State Phenomena* **264**, 120 (2017).

Dual direction blower system powered by solar energy to reduce car cabin temperature in open parking condition

N S Hamdan, M F M Radzi, A A M Damanhuri and S N Mokhtar

Faculty of Engineering Technology, Universiti Teknikal Malaysia Melaka
Hang Tuah Jaya, 76100 Durian Tunggal, Melaka, Malaysia

saffreena@utem.edu.my

Abstract. El-nino phenomenon that strikes Malaysia with temperature recorded more than 35°C can lead to extreme temperature rise in car cabin up to 80°C. Various problems will arise due to this extreme rising of temperature such as the occupant are vulnerable to heat stroke, emission of benzene gas that can cause cancer due to reaction of high temperature with interior compartments, and damage of compartments in the car. The current solution available to reduce car cabin temperature including tinted of window and portable heat rejection device that are available in the market. As an alternative to reduce car cabin temperature, this project modifies the car's air conditioning blower motor into dual direction powered by solar energy and identifies its influence to temperature inside the car, parked under scorching sun. By reducing the car cabin temperature up to 10°C which equal to 14% of reduction in the car cabin temperature, this simple proposed system aims to provide comfort to users due to its capability in improving the quality of air and moisture in the car cabin.

1. Introduction

Today car is very important as it provides a common means of transportation whether it is a longer commute to work or shorter trip to run errands around town. As the number of vehicles on the road are increasing day by day, drivers are having difficulties to get an indoor or roof parking space especially during peak hours [1]. Malaysia's standard ofice hour is from 8.00am until 5.00pm. During this period, the temperature inside car cabin that parked under scorching sun will tremendously increased, approaching 80 °C [2]. This will make the driver and passenger become uncomfortable while entering the car. Moreover, the car can also encounter car aging problem and bring damage to the goods and compartments in the car [1]. There is no air ventilation inside the car when left unused, thus, the heat will trapped in the car cabin. As a result, the car user is forced to wait for a period of time around 2 to 5 minutes before getting into the car to let the interior condition to cool down either by rolling down the window or running the air conditioner system at high speed that affect the fuel consumption [3]. The increase of fuel consumption by the air conditioning system subsequently increase CO_2 emisson [4] into the environment.

Solar radiation that enters car cabin through windows or windshield is the biggest influencing factor of car cabin temperature [5]. Installation of windows tint and application of aluminum sunshade are among the conventional methods to reduce the car cabin temperature. Windows tint reduces car cabin temperature by reducing the amount of visible light entering car cabin. A study by Jasni and

Nasir comparing three passive methods to reduce car cabin temperature, which include sunshade, ventilator and windows tint, revealed that windows tint was the most effective method in reducing interior car cabin temperature [4]. However lower visible light transmittance (VLT) levels is needed to achieve lower solar radiation entering the car cabin in order to reduce the car cabin temperature even further [6]. With regulation by authorities with certain specification for windows tint VLT percentage, limitation for temperature reduction in car during parking condition is among the unsolved issues.

Therefore, as an alternative to have higher reduction of car cabin temperature despite windows tint and sunshade, recent researches have focused on reducing car cabin temperature by inventing portable heat rejection devices such as Kulcar system and Portable Car Cooling system which require users to purchase an additional device to provide comfort to reduce car cabin temperature [1] . Comparison study on effect of ventilation in the car cabin showed that ventilation is able to reduce maximum temperature at front compartment of the car as much as 2.3°C and 8.3°C for windshield and dashboard respectively. In the same study, the additional ventilator able to reduce the car cabin temperature up to 8.3°C [7] [8].

The objective of the study is to ventilate car cabin when the car is parked in an open area condition. Hence, in this project, modification is made on automotive air conditioning system by making the existing blower motor to blow in dual direction in order to exhaust out hot air in car cabin during parking condition. The system is powered by solar energy to assure the system is not effecting the automotive battery life span. Four cases were experimentally studied to compare the effectiveness of this sytem in reducing temperature in car cabin.

2. Methodology

2.1 Blower modification

Normal automotive air conditioning blower motor rotates in forward direction and the function is to blow the cooled air that produce by the air conditioning system into the car cabin. Using the polarity concept, the poles of wiring system for the blower is changed to allow the blower motor rotates in reversed direction when car is in parking mode. In this condition, the blower with reversed direction act as vacuum to exhaust the hot air that trapped inside the car cabin, to the environment.

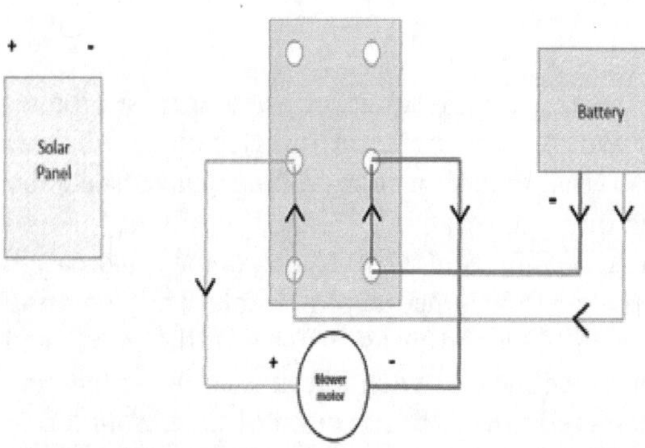

Figure 1. Forward direction circuit with DPDT switch.

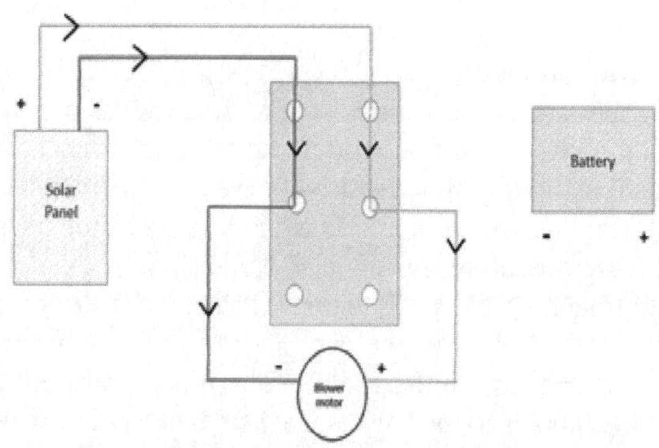

Figure 2. Backward direction circuit with DPDT switch.

The polarity change is applied on double pole double throw (DPDT) switch to allow the blower motor to rotate in dual direction which is forward and backward direction. The connection using DPDT switch are as shown in figure 1 and figure 2. The solar generator system is used to ensure the blower motor is running continuously to remove the trapped hot air without using any power from the car's main battery. The solar panel is a photovoltaic panel which operates at 40 watt and 2.5ampere. The dimension of the solar panel is 2 feet length times 2 feet width. The mechanism of the system is shown in figure 6.

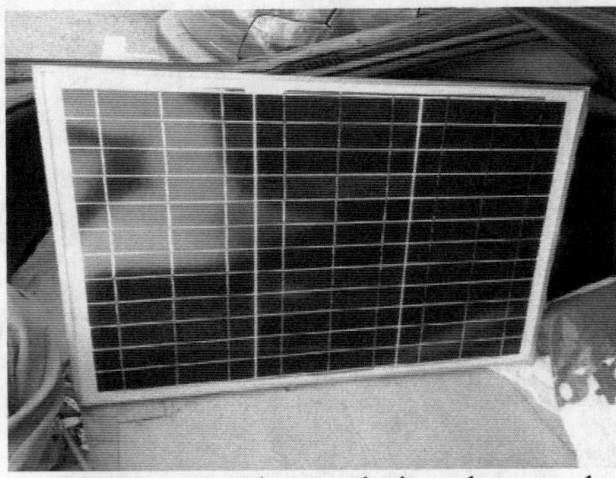

Figure 3. Photovoltaic solar panel.

Figure 4. DPDT Switch.

Figure 5. Connection modified blower with solar panel.

Figure 6. Reversed blower vacuum out the hot air that trapped inside car cabin.

2.2 Experiment

Experiment works is done on a sedan car with tinted window which comply with Road Transport Department Malaysia (JPJ) specification as shown in figure 7 [9]. The car is parked at an open parking area at residential field in Klang, Selangor. Data is collected from 8am to 5pm daily for 2 consecutive weeks using a thermocouple. Solar panel for the dual blower system is located on the roof of the car and temperature sensor is placed hanging on top of the front passenger sit . Four different cases is analysed to compare the temperature variation for different cases as shown in table 1 below. The sunshade was made from aluminum foil and applied at the front windshield with the dimension of 1180mm x 490mm. Rear side windows and back windshiled are left bare with only tint films.

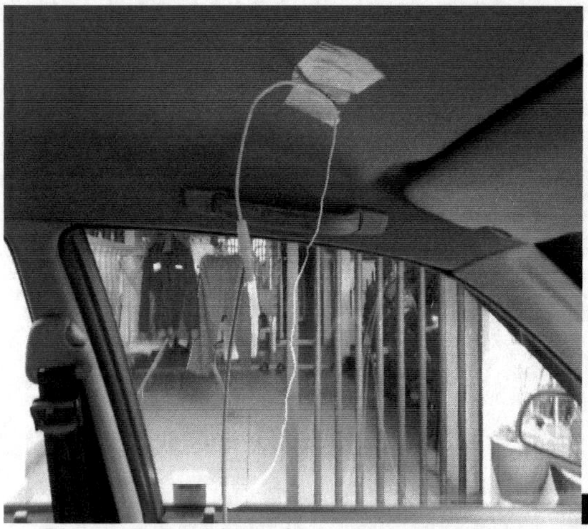

Figure 7. Position of temperature sensor

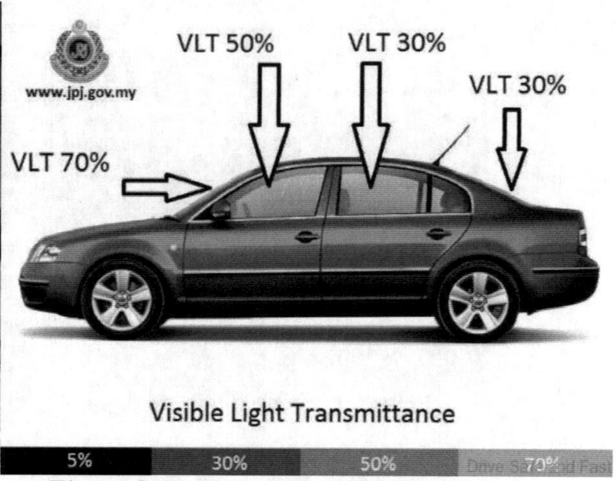

Figure 8. Road Department Malaysia Tinted Window Specification. [9]

Table 1. Four different cases for temperature variation.

Cases	Condition
Case 1	Tinted window
Case 2	Tinted with front sun shade
Case 3	Tinted with modified blower
Case 4	Tinted with front sun shade and modified blower

3. Results and Discussion

The experiment was conducted between 8am until 5pm which is according to office hour in Malaysia for three days for each case. Ambient air temperature was recorded for all the experiment conducted days to show temperature variation of the day.

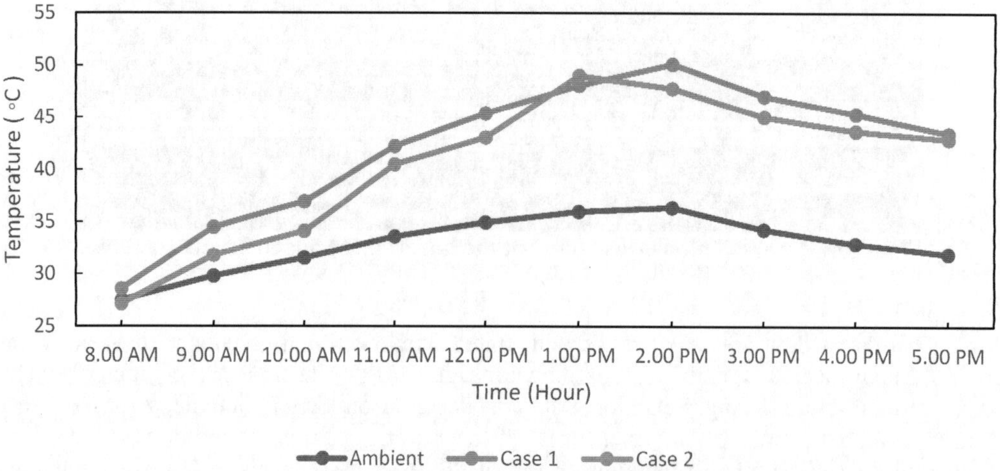

Figure 9. Comparison for case 1 and case 2.

The temperature for both cases started to increase from 8am to 1pm. Temperature for case 1 reached the highest temperature of 48°C at 1pm and continue to decrease after 1pm. Meanwhile, the temperature for case 2 reached the highest point of 50°C at 2pm and continue to decrease after 2pm. From the ambient temperature trend, it can be observed that the highest temperature recorded is in between 1 pm to 2pm. From figure 9, it is clear that with addition of sunshade in case 2, temperature reduction in the car cabin up to 4% compared to case 1 can be achieved.

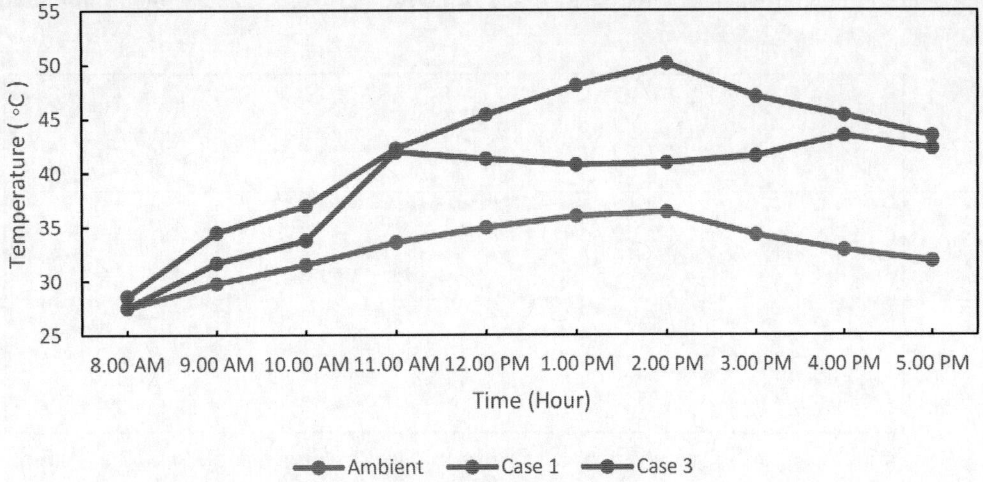

Figure 10. Comparison for case 1 and case 3.

Figure 10 shows comparison between case 1 and case 3. Identical temperature changes pattern is observed for case 1 and ambient temperature. The car cabin temperature for both cases increased until 11am. After 11am, although the ambient temperature increased, the car cabin temperature for case 3 decreased gradually until 2pm. The modified blower speed depends on amount of sun radiation to the solar panel. From 11 am to 2 pm, the sun provides direct radiation to the solar panel and blower run at its maximum speed allowing more reduction in the car cabin temperature. However, the car cabin temperature started to increase again from 2pm until 4pm as the modified blower loss its speed and capability to exhaust hot air from car cabin. This is as a result of the solar panel located flat on top of the car roof. It received less sun radiation when the sun radiation incident decreased starting from 2pm onwards. Case 3 reached its highest temperature at 4pm with temperature recorded is 43°C. Based on figure 10, with addition of modified blower system to the car, temperature reduction up to 14% can be achieved.

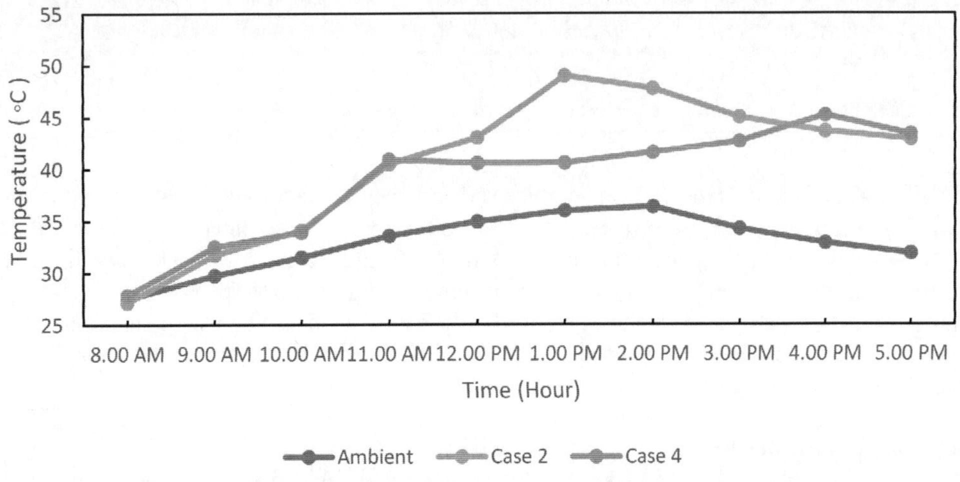

Figure 11. Comparison of case 2 and case 4.

Similar trend with other cases is observed where the temperature increased from 8am to 11am. After 11am, the car cabin temperature for case 4 decreased gradually and increased again after 1pm. It reached its highest temperature at 4pm with temperature recorded is around 45°C. From figure 11, it is observed that addition of modified blower system to tinted window and front shade contribute to 9% of temperature reduction when compared to tinted window and front shade without any additional ventilation.

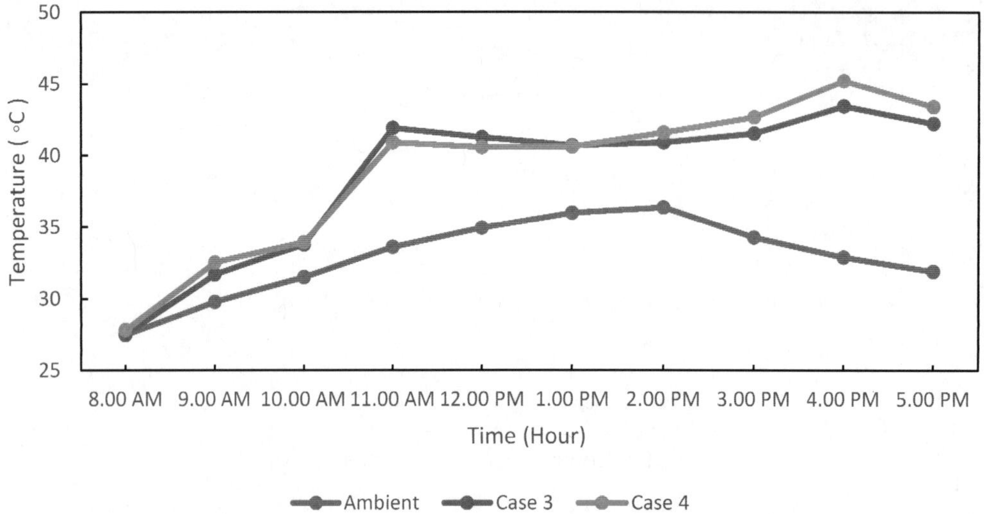

Figure 12. Comparison of case 3 and case 4

Based from graph in figure 12, much the same pattern of car cabin temperature changes could be seen for both cases. Yet, from 2pm to 5pm, case 4 showed higher temperature compared to case 3 at around 4% difference in temperature. As the speed of blower system decreased starting from 2pm, case 4 which used additional front shade, prevent more heat loss through front windshield than case 3 which contribute to higher temperature in car cabin after 2pm.

Table 2. Temperature reduction percentage between cases

	Case 1	Case 2	Case 3	Case 4
Case 1	-	4%	14%	-
Case 2	4%	-	-	9%
Case 3		-	-	4%
Case 4	-	9%	4%	-

Table 2 above summarize the temperature reduction percentage when the maximum car cabin temperature is compared between different cases. From the above table, tinted windows with modified blower showed the highest temperature reduction when compared with case 1, which is tinted window without any additional method to reduce car cabin temperature. Case 4 recorded the second highest temperature reduction percentage when compared with case 2 as much as 9%. This revealed that the application of sunshades prevents the heat trapped inside the car to be transferred through windows when the modified blower starts to loss its ability to exhaust hot air.

4. Conclusion and Future works

As conclusion, this study presents that the implementation of modified blower to reduce car cabin temperature was successful with temperature reduction up to 14% compared to car cabin without any ventilation. As recommendation, a charge controller to regulate the current before supply it to the solar panel is proposed to maintained the modified blower speed at all time. Automatic adjustable solar panel holder to allow maximum solar radiation incident throughout the office hour period will also be the future recommendation.

5. Acknowledgments

The author would like to express appreciation to the team of this project from Faculty of Engineering Technology of Universiti Teknikal Malaysia Melaka (UTeM) and to the Ministry of Higher Education Malaysia (MOHE) for supports including the funding of this research which are under the research grant RAGS/1/2015/TK0/FTK/03/B00123.

6. References

[1] Basar. M. F, Musa M, Faizal M. Y, Razik N. H. A, 2013 "Alternative Way In Reducing Car Cabin Temperature Using Protable Car Cooling System (Car-Cool)," *International Journal of Innovative Technology and Exploring Engineering* **3 (3)**, pp. 140-143.

[2] Sunawar A, Garniwa I, 2017 "Maximum Potential of the Car Cabin Temperature in the Outdoor Parking Condition as a Source of energy in Thermoelectric Generator," *IOP Conference Series: Materials Science and Engineering,* **180 (1)**.

[3] Hussain H. A, Sidik M. F. B. M, Yuganthira R. A. L. M, 2010 "Study on The Thermal Accumulation and Distribution Inside a Parked Car Cabin," *American Journal of Applied Sciences,* **7 (6)** pp. 784-789.

[4] Nasir F. M, Jasni M. A, 2012 "*Experimental Comparison Study of the Passive Methods in Reducing Car Cabin Interior Temperature*," Int. Conference On Mechanical, Automobile and Robotics Engineering, Penang p 229-233.

[5] Saidur R, Masjuki H. H, Hasanuzzaman M, 2009 "Performance of an Improved Solar Car Ventilator," *International Journal of Mechanical and Materials Engineering.* **4(1)**, pp. 24-34.

[6] Razak A. A, M. Fohimi N. A, S. Salim S. A. Z, Noordin N. H, 2011 "Temperature Influence on Total Volatile Compounds (TVOCs) Inside The Car Cabin of Visible Light Transmittance," *ARPN. J. Eng. and App. Sci.,* **11 (14)**, pp. 8629-34.

[7] Rugh J, Chaney L, Lustbader J, Meyer J, 2007 "*Reduction in Vehicle Temperatures and Fuel Use from Cabin Ventilation, Solar-Reflective Paint, and a New Solar-Reflective Glazing*," SAE World Congress, Detroit, Michigan pp. 1-8.

[8] Sudhir C.V, Jalal M. A. D, 2015 "Effect of Solar Ventilation on Air Conditoning System Performance of The Car Parked Under Sun Light" *ARPN Journal of Engineering and Applied Sciences,* **10 (22)**, pp. 10618-26.

[9] JPJ, "The Malaysian Road Transport Department," [Online]. Available: www.jpj.my.

Co-pyrolysis of rice straw and Polyethylene Terephthalate (PET) using a fixed bed drop type pyrolyzer

N I Izzatie[1], M H Basha[1], Y Uemura[2], M S M Hashim[1], N A M Amin[1], M F Hamid[3]

[1] Mechanical Engineering Programme, School of Mechatronic Engineering, Universiti Malaysia Perlis, Pauh Putra Campus, 02600 Arau, Perlis, Malaysia.

[2]Centre for Biofuel and Biochemical Research, Universiti Teknologi PETRONAS, Malaysia.

[3]School of Mechanical Engineering, Universiti Sains Malaysia.

hafifbasha@unimap.edu.my

Abstract. In this work, co-pyrolysis of rice straw and polyethylene terephthalate (PET) was carried out at different temperatures (450,500,550, and 600°C) at ratio 1:1 by using fixed bed drop-type pyrolyzer. The purpose of this work is to determine the effect of pyrolysis temperature on the product yield. As the temperature increased, the pyrolysis oil increased until it reaches certain high temperature (600°C), the pyrolysis oil decreased as of more NCG were produced. The temperature 550°C is considered as the optimum pyrolysis temperature since it produced the highest amount of pyrolysis oil with 36 wt.%. In pyrolysis oil, the calorific value (13.98kJ/g) was low because of the presence of high water content (52.46 wt.%). Main chemicals group from pyrolysis oil were an aldehyde, ketones, acids, aromatics, and phenol and all compound have abundant of hydrogen and carbon were identified. Co-pyrolysis of rice straw and PET produced a higher amount of carbon oxides and recycling back the NCG could increase liquid and char yields.

1. Introduction

The depletion of fossil fuels such as petroleum, coal and natural gas inspire researchers to find solutions by utilizing alternative energy. Biomass can be a sustainable and renewable source of fuel not affiliated with fossil fuels. Thermochemical conversion of biomass to liquid products has the advantage to displace petroleum fuels since it has similar characteristics to petroleum-derived. Biomass is very abundant worldwide and can be easily found in varies forms such as dedicated energy corps, municipal solid waste, wood residues and agriculture residues.

After wheat, rice is the second most important crop in the world with Asia being the largest producer and consumer. The production of the rice straw in Malaysia field at 2011 was 1,933,889.3 tons [1]. Regrettably, the burning of the rice wastes such as rice straw and rice husk remains the cultural practice in Malaysia [2] and it reduced the air quality and will worsen the greenhouse effect.

Thermal decomposition of biomass also known as pyrolysis is the constitutional chemical reaction that is the forerunner of both combustion and gasification processes naturally occurs in short time with the absence of oxygen. The products of biomass pyrolysis include pyrolysis oil, biochar and non-condensable gases (NCG) which consist of carbon dioxide, hydrogen, carbon monoxide, and methane. Pyrolysis is an endothermic process that required sufficient heat transfer surface to complete the process. The pyrolysis products can be applied in the production of heat, chemical, and power generation.

However, instead of using biomass as a single feedstock, the researcher has also experimented adding plastics in the pyrolysis process to enhance the quality and quantity of products. Polyethylene terephthalate (PET), polypropylene (PP), and polystyrene (PS) are non-biodegradable plastics. The disposal of these plastics became a problem as landfills are filled with these non-biodegradable wastes. One of the promising ways to manage these wastes, as well as taking a valuable profit of the energy value of these wastes is by co-pyrolysis. Co-pyrolysis of plastics polymers with biomass could balance the carbon, oxygen, and hydrogen in the feedstock with the strong effect of the dissipation products since plastic has higher thermal stability compared to plastics. Pyrolysis oil could yield up to 50 wt.% at a higher heating rate with shorter residence time [3][4].

Pyrolysis of rice straw has been studied by Tsai and Lee [5]. They found that at a higher heating rate, the larger solid yield can be obtained. Besides, the pyrolysis liquid products still contained a large amount of water (70-86 wt.%). However, the data for the study of co-pyrolysis between rice straw and plastics is not widely available. The objective of work is to study the co-pyrolysis of rice straw and polyethylene terephthalate (PET). The effect of temperature was analyzed at 450,500,550 and 600°C and constant ratio 1:1. Co-pyrolysis process was performed using a drop-type pyrolyzer and products yields of co-pyrolysis were pyrolysis oil, biochar, and non-condensable gases will be characterized.

2. Feedstocks and materials.

The rice straw that is used in work is collected from a rice field in Kangar, Perlis Malaysia. The rice straw was ground to size 0.15-0.50mm and dried in an oven at 105°C for 24 h to reduce the moisture content so it is less than 10wt% [6]. PET plastic was collected and cut into smaller size 0.30 mm x 0.30 mm and both of the feedstocks is stored in the air-tight container. Characteristics of the feedstock as shown in Table 1.

Table 1. Feedstocks properties.

	Rice straw	Polyethylene terephthalate (PET)
Proximate analysis, wt.%		
Moisture	5.93	0.23
Ash	9.56	0.10
Volatile	76.85	99.20
Elemental analysis, wt.%		
C	36.02	58.93
H	6.63	3.98
N	1.12	0.13
O	56.25	36.96
Calorific value (kJ/g)	15.63	36.72

3. Experimental setup

Figure 1 shows the fixed bed drop type pyrolyzer schematic diagram used to carry out the co-pyrolysis experiment. To create an inert condition, the pyrolyzer was linked with nitrogen gas lines. The reactor is the first purged three times to ensure inert conditions. The biomass was put on the holder zone and two ball valves in closing mode. The reactor is heated with an electric furnace and insulated to avoid heat loss during an experiment. A K-type thermocouple was attached to the reactor to record the actual temperature. After the desired reaction temperature is achieved, the biomass will be dropped down by gravity and the pyrolysis process will begin. Two condensers were connected to the reactor to increase the condensation efficiency and immersed in an ice cooled water bath to condense the pyrolysis vapors into pyrolysis oil. A gas bag was connected to the outlet of vapor trap to collocate non-condensable gases (NCG). Meanwhile, the biochar was loaded in the reactor and collected after the experiment finished.

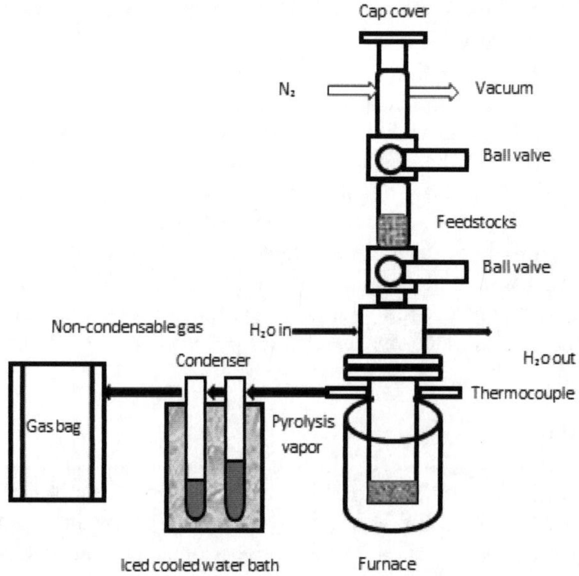

Figure 1. Fixed-bed drop type pyrolyzer schematic diagram.

4. Pyrolysis procedure

15 g of rice straw and PET feedstocks with mass ratio 1:1 was loaded into the biomass holder zone for all experiments. The feedstocks are dropped down to the reactor when the desired temperature achieved. The duration of the experiment was held at the desired temperature for 20 minutes until no significant brownish gas vapor was observed. In this study, the experiment was done and temperatures were considered at 450,500,550 and 600°C and constant ratio 1:1.

Three products yields produced from the co-pyrolysis process were pyrolysis oil, biochar, and non-condensable gases. The pyrolysis oil and biochar we determined by calculated the weight changes at before and after of the reactor and condensers respectively. The yield of non-condensable gases was calculated using an equation [7]:

$$\text{Pyrolysis oil (wt.\%)} + \text{biochar (wt.\%)} + \text{non-condensable gases (wt.\%)} = 100 \quad (1)$$

5. Products analysis

To determines the quality of co-pyrolysis products, the product analysis is done on the pyrolysis oil in which temperature yield is on the maximum liquid product. pyrolysis oil yields collected. The elemental analysis of elements of C, H, N, O, and S was carried out on pyrolysis oil and biochar using Perkin Elmer 2400 CHN Analyzer.

The water content of the pyrolysis oil was measured by a volumetric Karl Fisher titrator (Mettler Toledo, V30) with combrititant 5 Keto (Merck) as a titrant and Combisolvent Keto (Merck) as a solvent. The calorific values of the pyrolysis oil and biochar were also verified by using a bomb calorimeter IKA-WERKE (2000). Thermogravimetric analysis (TGA) was performed using a TA Instruments Model QA50 at heating rate 10°C/ min with 60 mL/min of N_2.

The organic compound of pyrolysis oil was identified by using a Gas Chromatography /Mass (GC-MS). The GC-MS analysis was carried out with an Agilent Technology model 7890A series. A

BPX-5 capillary column was selected with 30mm of diameter, 0.25 µm of its film thickness. The oven temperature was held at 35°C for 2 min before and raised to 250°C with a heating rate 20°C/min and held for 20 min. Helium with 99.99% purity used as a carrier gas with 1.5 mL/min of flowrate. The injector and detector temperatures were set constantly at 280°C.

The composition of non-condensable gas formed from pyrolysis process was collected in the gas bag and tested using a Shimadzu GC-8A Gas Chromatography /Thermal Conductivity Detector (GC-TCD), with a Davidson Grade 12 Silica gel packed column for CO_2 detection and a Molecular Sieve 5A packed column to detect the presence of CH_4, O_2, CO, and H gases.

6. Results and discussions

6.1 Products yields

Figure 2 shows the co-pyrolysis effect of temperatures (450,500,550 and 600°C) on products yields. With the increase of temperature from 450°C to 550°C, the yield of pyrolysis oil and NCG increased while the yield of biochar decreased. However, the yield of pyrolysis oil dropped at temperature 600°C with 26.67 wt.%). The secondary cracking reaction was taking place by the conversion of pyrolysis vapors and char into NCG at elevated temperatures. Maximum pyrolysis oil is produced at temperature 550°C with 36wt.%, 19.33 wt.% and 44.67 wt.% of pyrolysis oil, biochar, and NCG respectively. This temperature considered as the maximum temperature for co-pyrolysis of rice straw and PET.

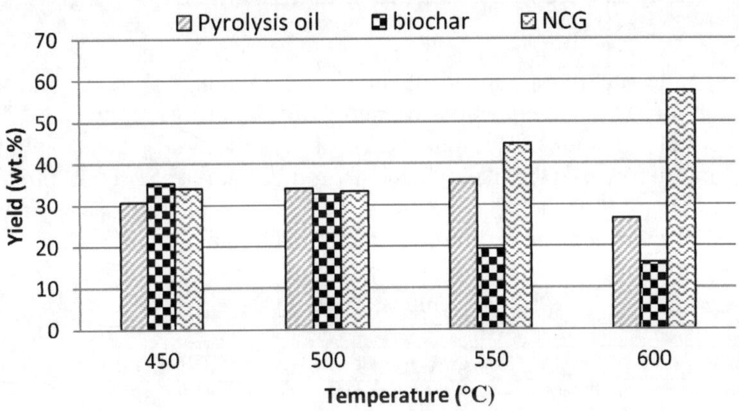

Figure 2. Co-pyrolysis products yield by different temperatures at ratio 1:1.

6.2 Proximate and elemental analysis

Table 2 shows the proximate and ultimate analysis of co-pyrolysis products for pyrolysis oil and biochar. There was some reduction of moisture content in the biochar yield compared to feedstocks (Table 1) due to dehydration during the heating process. The amount of volatile matter decomposed into vapors and reduced the amount of volatile matter in biochar. It is also noted that by mixing with PET resulted in an increase in calorific value for products yield. Carbon and hydrogen content in co-pyrolysis oil and biochar from rice straw and PET is higher because plastics contained high C and H [8] compared from feedstocks only. Higher carbon content and calorific values in biochar benefits to agriculture session as soil improver [9]

Table 2. Proximate and elemental analysis of co-pyrolysis products.

Proximate analysis, wt.%	RS: PET		RS only	
	Biochar	Pyrolysis oil	Biochar	Pyrolysis oil
Moisture	2.8	-	-	-
Ash	12.9	-	-	-
Volatile	48.70	-	-	-
Elemental analysis, wt.%				
C	73.69	46.19	57.50	9.47
H	1.71	13.71	2.61	11.43
N	0.74	0.14	1.09	0.45
O	23.86	39.96	38.80	78.65
Calorific value (kJ/g)	27.64	13.98	10.84	8.75
Water content (wt.%)	-	52.46	-	66.87

6.3 GC-MS analysis

Pyrolysis oil is a complex mixture of abundant chemical compounds [10]. Table 3 shows the GC-MS chromatograms of pyrolysis oil composition from co-pyrolysis of rice straw and PET at 550°C. More than 100 peaks were detected and only high degree of probability and peaks area greater that 0.1% were listed. Referring to Table 3, the main chemical group in the pyrolysis oil produced from the co-pyrolysis of rice straw and PET included aldehyde, ketones, acids, aromatic and phenol. Compounds consisting abundant of hydrogen and carbon were identified in the pyrolysis oil.

Table 3. Composition of pyrolysis oil

Compounds	Group	Rt (min)	% Area
Acetaldehyde (C_2H_4O)	Aldehyde	1.487	5.20
Ethanol (C_2H_6O)	Alcohol	1.535	3.36
Acetone (C_3H_6O)	Ketone	1.579	4.98
Formic acid (CH_2O_2)	Acid	1.670	1.83
2-propanone (C_3H_6O)	Ketone	2.086	13.37
Propanoic acid ($C_3H_6O_2$)	Acid	2.184	3.15
Acetoin ($C_4H_8O_2$)	Ketone/hydroxyl	2.329	1.43
1-Hydroxy-2-butanone ($C_4H_8O_2$)	Ketone	2.751	2.70
Cyclopentanone (C_5H_8O)	Ketone	2.967	0.50
Pyridine (C_5H_5N)	Aromatic	3.253	0.23
Butanone (C_4H_8O)	Ketone	3.642	0.40
2-cyclopentanone (C_5H_6O)	Ketone	3.718	0.44
Butyrolactone ($C_4H_6O_2$)	Ketone	4.334	0.62
Benzaldehyde (C_7H_6O)	Aldehyde	5.090	0.20
Phenol (C_6H_6O)	Phenol	5.252	0.11
Acetophenone (C_8H_8O)	Ketone	6.608	0.39
Benzoic acid (C_7H_6O)	Acid	8.174	16.14
Ethanone ($C_8H_{14}O$)	Ketone	12.021	0.28

6.4 Non-condensable analysis

Figure 3 shows the composition analysis of non-condensable gases which comprises of nitrogen (N_2), hydrogen (H), methane (CH_4), carbon dioxide (CO_2), and carbon monoxide (CO). Decarboxylation formation from CO_2, and CO mainly from the reaction between CO_2 and steam which porous char surface of the biomass [11]. Co-pyrolysis process produces a high amount of carbon oxides with 35 vol.% of carbon monoxide and 27 vol.% of carbon dioxide. Some of the researchers suggested the utilization of NCG could be recycled back as heat source into pyrolysis reaction [8][9]. Advantages of recycling the NCG was it has potential to increase the liquid and char yields. Besides, it can also increase the high heating value and pH of pyrolysis oil as well as decreased the density and viscosity [14]

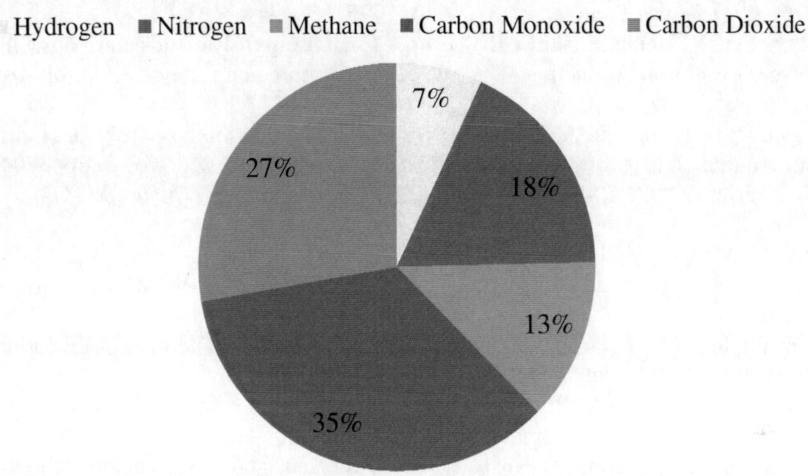

Figure 3. Composition analysis of non-condensable gases.

7. Conclusions

In this work, co-pyrolysis of rice straw and PET experiments were performed using fixed bed drop type pyrolyzer in order to assess the effect of different temperatures (450,500,550 and 600°C) on products yields. As the reaction temperature rises, the pyrolysis oil yield increases, until in reaches 550°C, where it obtained the highest yield of pyrolysis oil with 36 wt.%. The relatively low calorific value was recorded for pyrolysis oil with 13.98 kJ/ g due to high water content with 52.46 wt%. However, comparing the calorific value of the co-pyrolysis with the pyrolysis of rice straw alone, a calorific value significantly increased at 60 wt.%. The GC-MS analysis detected aldehyde, ketone, acid, aromatic and phenol groups in pyrolysis oil. NCG from co-pyrolysis of rice straw and PET which comprises (7 vol.%)H, (18 vol.%)N_2, (13 vol.%)CH_4, (35 vol.%) CO and (27 vol.%) CO_2.

Acknowledgement

The authors would like to thank the Ministry of Higher Education Malaysia for the financial support under the Research Acculturation Grant Scheme (RAGS) under the grant number of RAGS/1/2015/TK03/UNIMAP/03/2

References.

[1] S. M. Shafie, H. H. Masjuki, and T. M. I. Mahlia, "Life cycle assessment of rice straw-based power generation in Malaysia," *Energy*, vol. 70, pp. 401–410, 2014.
[2] H. Nori, R. A. Halim, and M. F. Ramlan, "The effects of nitrogen fertilization levels on the straw nutritive quality of Malaysian rice varieties," *J. Agron.*, vol. 5, no. 3, pp. 482–491, 2006.
[3] M. A. Sukiran, C. Chow Mee, and N. K. A. Bakar, "Bio-oils from pyrolysis of oil palm empty fruit bunches," *Am. J. Appl. Sci.*, vol. 6, pp. 869–875, 2009.
[4] L. Zhou, Y. Wang, Q. Huang, and J. Cai, "Thermogravimetric characteristics and kinetic of plastic and biomass blends co-pyrolysis," *Fuel Process. Technol.*, vol. 87, no. 11, pp. 963–969, 2006.
[5] W. T. Tsai, M. K. Lee, and Y. M. Chang, "Fast pyrolysis of rice straw, sugarcane bagasse and coconut shell in an induction-heating reactor," *J. Anal. Appl. Pyrolysis*, vol. 76, no. 1–2, pp. 230–237, 2006.
[6] F. A. Agblevor, S. Besler, and A. E. Wiselogel, "Fast Pyrolysis of Stored Biomass Feedstocks," *Energy & Fuels*, vol. 9, no. 4, pp. 635–640, 1995.
[7] S. R. Naqvi, Y. Uemura, and S. B. Yusup, "Catalytic pyrolysis of paddy husk in a drop type pyrolyzer for bio-oil production: The role of temperature and catalyst," *J. Anal. Appl. Pyrolysis*, vol. 106, pp. 57–62, 2014.
[8] E. Önal, B. B. Uzun, and A. E. Pütün, "Bio-oil production via co-pyrolysis of almond shell as biomass and high density polyethylene," *Energy Convers. Manag.*, vol. 78, pp. 704–710, 2014.
[9] C. E. Brewer and R. C. Brown, "Biochar," in *Comprehensive Renewable Energy*, 2012, pp. 357–384.
[10] M. A. F. Mazlan, Y. Uemura, N. B. Osman, and S. Yusup, "Fast pyrolysis of hardwood residues using a fixed bed drop-type pyrolyzer," *Energy Convers. Manag.*, vol. 98, pp. 208–214, 2015.
[11] R. H. White and M. A. Dietenberger, "Wood Products : Thermal Degradation and Fire," *Encycl. Mater. Sci. Technol.*, vol. 48, pp. 9712–9716, 2001.
[12] M. N. Uddin, W. M. A. W. Daud, and H. F. Abbas, "Effects of pyrolysis parameters on hydrogen formations from biomass: a review," *RSC Adv.*, vol. 4, no. 21, p. 10467, 2014.
[13] Ingemarsson, U. Nilsson, M. Nilsson, J. R. Pedersen, and J. O. Olsson, "Slow pyrolysis of spruce and pine samples studied with GC/MS and GC/FTIR/FID," *Chemosphere*, vol. 36, no. 14, pp. 2879–2889, 1998.
[14] O. D. Mante, F. A. Agblevor, S. T. Oyama, and R. McClung, "The influence of recycling non-condensable gases in the fractional catalytic pyrolysis of biomass," *Bioresour. Technol.*, vol. 111, pp. 482–490, 2012.

Simulation study of air and water cooled photovoltaic panel using ANSYS

Z Syafiqah[1], N A M Amin[2], Y M Irwan[1,3], M S A Majid [2] and N A Aziz[2]

[1]Centre of Excellence for Renewable Energy, School of Electrical System Engineering, Universiti Malaysia Perlis (UniMAP), Malaysia
[2]School of Mechatronic Engineering, Universiti Malaysia Perlis (UniMAP), Malaysia
[3]Centre for Diploma Studies, Universiti Malaysia Perlis (UniMAP), Malaysia

nasrulamri.mohdamin@unimap.edu.my

Abstract. Demand for alternative energy is growing due to decrease of fossil fuels sources. One of the promising and popular renewable energy technology is a photovoltaic (PV) technology. During the actual operation of PV cells, only around 15% of solar irradiance is converted to electricity, while the rest is converted into heat. The electrical efficiency decreases with the increment in PV panel's temperature. This electrical energy is referring to the open-circuit voltage (V_{oc}), short-circuit current (I_{sc}) and output power generate. This paper examines and discusses the PV panel with water and air cooling system. The air cooling system was installed at the back of PV panel while water cooling system at front surface. The analyses of both cooling systems were done by using ANSYS CFX and PSPICE software. The highest temperature of PV panel without cooling system is 66.3 °C. There is a decrement of 19.2% and 53.2% in temperature with the air and water cooling system applied to PV panel.

1. Introduction

Solar PV system is one of the growing renewable energy technologies, besides wind energy and it is expected to become a major renewable energy in the future [1]. PV cells are semiconductor that convert the solar irradiance into electricity. The terms solar irradiance is referring the measure of power density of sunlight receives at any location on earth, and it is measured in watt per metre square (W/m^2) [2]. When the solar irradiance rises, the open-circuit voltage (V_{oc}) and short-circuit current (I_{sc}) will also rise. The short-circuit current (I_{sc}) increases with the increment of solar irradiance, hence improving the power generated by PV panel. Besides solar irradiance, temperature also plays an important role in determining the performance of PV panel. When the PV cell's temperature increase, the rate of photon generation increases. The reverse saturation current will increase rapidly, which lead to reduce in the band gap thus affecting the PV cell's electrical parameters. The open-circuit voltage (V_{oc}) is the parameter that will most affect [3].

As solar irradiance acts as a positive factor in affecting the performance of PV panel, meanwhile temperature will acts as a negative factor. When the solar irradiance increase, the output power generated by PV panel will increas too. However, with the increase of PV panel's temperature, the output power generated will decrease. With the PV panel temperature gain, its efficiency (in terms of power output) decreases because of the negative temperature coefficient (-0.5%/°C rise in temperature) [4]. It can be said that PV panels are more efficient at a lower temperature, thus a cooling system is needed due to lower the PV panel's temperature. Numerous researchs have been conducted to enhance the performance of PV panel [5-7]. It has been proven by Smith [8] that with a cooling system attach to PV panel, the power generated increase by 5%. Sayran [9] and Saurabh [10] had immersed the PV panel into water with different depth due to increase the efficiency of PV panel. It is found out that the efficiency increases by 11% at water depth of 6 cm and 17.8% at depth of 1 cm respectively. While, Teo [11] maintained the PV panel's temperature by 38 °C with a blower that was attached at the backside of PV panel. Besides, the electrical efficiency also increases from 8.6 % to 12.5% with the utilisation of cooling system. Prastish [12] evaluate the effectiveness of PV/T hybrid water collector that generates thermal energy and electricity, hence, producing a higher energy conversion rate of absorbed solar radiation. Authors also emphasize the effect of PV panel temperature towards the electrical, thermal and exergy efficiency. It is concluded that the electrical, thermal and exergy efficiency can be raise if the heat was removed from the PV panel completely.

Khelifa [13] studied solar hybrid photovoltaic thermal (PV/T) using ANSYS Fluent software. The author had simulated the heat transfer phenomenon and fluid flow between the PV cells and the coolant. The results show that with a lower PV cell temperature, the thermal and electrical performance increased. Hiren et al. [4] presented an experimental study and CFD analysis for PV panel with a heat exchanger of water, flowing on top of it. The authors also demonstrated that the refraction of incident solar radiation while striking to PV panel through the water was favorable. The overall electrical efficiency had increased from 6% to 40% by the water cooling system. Another direct water cooled PV panel experimental analysis has been studied by Stefan [14]. It is found that with the help of cooling, monthly energy saving efficiency had increased to approximately 10.3%. Salih [7] improved the net power saving of PV panel by experimented a water spraying technique cooling system with a constant water mass flow rate. The water cooling system has been installed at front of PV panel due to spray the water thus removing the heat. The performance of the water cooling system was studied by observing its electrical performance throughout the day.

This paper presents a study on the performance differences of PV panel with DC fan and water cooling system. The performance study will based on thermal performance, which will be evaluated by ANSYS CFX, while the electricity study will be evaluate using PSPICE software.

2. DC fan and water cooling system analysis

Figure 1 shows the water and air cooling system investigated in this study. These cooling systems were designed to tackle the problem of the negative effect of the temperature towards PV panel. The utilisation of these cooling systems are presumed to reduce the temperature of PV panel, hence increasing its performance (output power generated).

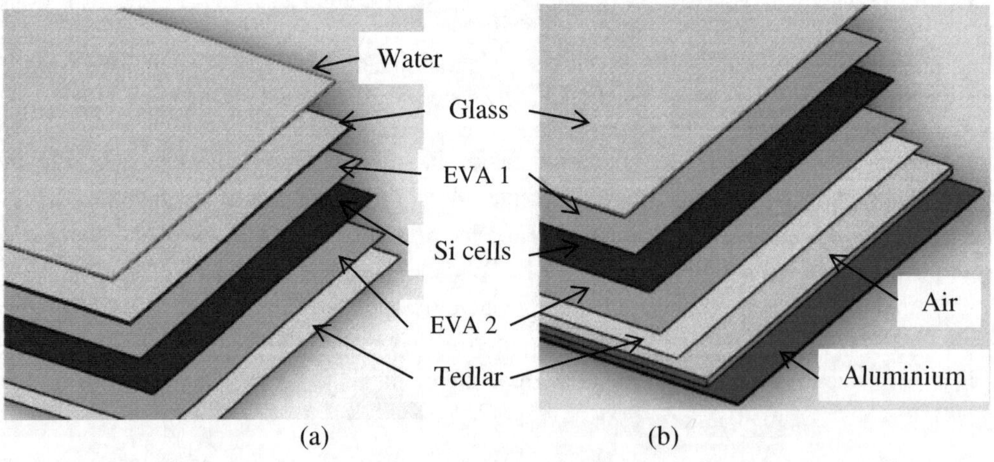

Figure 1. PV panel with (a) DC Fan cooling system (b) water cooling system

A Sharp Solar panel with peak power of 100 W, I_{sc} of 5.78 A and V_{oc} of 22.2 V has been employed for both simulation. The PV panel dimension was based on actual dimension, which is 1200 mm x 540 mm. A thin layer of air and water were added at the back and top of PV panel respectively. The geometry model has been drawn by using Solidwork before imported into ANSYS CFX for further analysis. Normally, PV panel consists of 5 different layers, which are glass surface, Ethylene-Vinyl Acateta 1 (EVA1), silicon, EVA 2 and tedlar. Each layer is differed in thickness, density, thermal conductivity and specific heat capacity.
The heat transfer from the air and water flowing at the back and top of PV panel respectively were modelled by using a computational fluid dynamic (CFD) software. Several assumptions have been made, such as, both working fluids were considered as incompressible fluid. The governing differential equations used to describe the heat transfer and fluid flow of the both PV panel cooling system are given as follow:

Conservation of mass (continuity)

$$\nabla(p\vec{V}) = 0 \tag{1}$$

Conservation of momentum

$$\vec{V}.\nabla(p\vec{V}) = -\nabla + \nabla.(\mu\nabla\vec{V}) \tag{2}$$

Conservation energy for fluid

$$\vec{V}.\nabla(pc_pT_f) = \nabla.(k_f\nabla T_f) \tag{3}$$

Conservation of energy for solid

$$\nabla.(k_w\nabla T_w) = 0 \tag{4}$$

The computations were performed by using the commercial package ANSYS CFX software. The PV panel with air cooling system was acknowledged standard k-Omega model because the flow is turbulent (Re > 2200). Whereas, the PV panel with water cooling was considered as laminar model since the flow is laminar (Re < 2000). The energy equations and flow momentum were solved with a first-order upwind scheme. Both simulations were performed using a convergence criterion of 10^{-4}. A fine hexa mesh grid scheme was used to mesh both cooling system. The PV panel with air cooling system consists of 29208 nodes and 30109 element, while, the nodes and element for PV panel with water cooling are 22272 and 10584 respectively. The meshed model does satisfy the 0.85 of maximum skewness and have a low aspect ratio.

Boundary condition needs to be applied at all PV panel's surface. Under the sun, not all surfaces of PV panel will directly expose to it. Hence, only the outer layer of PV panel will be considered receiving the solar irradiance and wind. The ambient temperature and solar irradiance are varying with time, whereby it started at 8.00 a.m. until 7.00 p.m. The temperature of PV panel is assumed dependent on the ambient temperature. Meanwhile, the initial temperature for air and water were 25 °C. Since both of the cooling system are an active cooling system, a similar power input is chosen which is 2.16 W. For air cooling system, 2 units of DC fan were installed at the back of PV panel with wind speed of 3.07 m/s. As for water cooling system, a water mass flow rate of 0.0556 kg/s was selected. Both PV panels cooling systems will be analysed with the same weather conditions. The comparative performance of it will be analysed in terms of thermal performance using ANSYS CFX as well as electricity performance by using PSPICE software.

Besides PSPICE software, the electricity performance of PV panel with cooling panel will be evaluated by studying its parameter. The parameters involved are I_{sc}, V_{oc}, I-V curve, fill-factor (FF) and efficiency. The maximum power output generate by PV panel can be calculated as follow:

$$P_{mp} = V_{mp}.I_{mp} \tag{5}$$

I_{mp} is the maximum current generated by PV panel, while, V_{mp} is the maximum voltage generated by PV panel. FF represent the ratio of P_{max} from the PV panel to the product of I_{sc} and V_{oc}, by which determines the maximum power from PV panel.

$$FF = \frac{P_{mp}}{I_{sc}.V_{oc}} \tag{6}$$

The performance of PV panel with cooling system also can be compared by using the efficiency parameters. Efficiency is the ratio of power output generated by PV panel (P_{out}), to the input power (P_{in}) from the sun. It defines how much solar irradiance is converted into electrical energy, indicating how efficient the PV cells functions. P_{in} is the product of input solar irradiance (G, in W/m^2) with surface area of PV cell (A in m^2). The conversion efficiency (%) is determined by:

$$Efficiency(\eta) = \frac{I_{mp} \times V_{mp}}{G \times A} \tag{7}$$

3. Results

This section presents and discussed both PV panels with cooling systems impact towards thermal performance and output power generated. Figure 2 displays the temperature of PV panel without and with cooling system. The highest ambient temperature recorded was 36.5 °C with solar irradiance of 1175 W/m^2 at 1.30 p.m. The highest temperature of PV panel with air cooling system is 53.6 °C while for PV panel with water cooling system is 31.3 °C. In the meantime, the highest temperature for PV panel without cooling system is 66.3 °C. There are almost 12.7 °C differences in temperature for PV panel without cooling system and PV panel with air cooling system. It can be said that with the aid of cooling system, the PV panel's temperature can be lowered. PV panel with water cooling system shows the largest decrement in temperature compared to air cooling system, which is almost 35.0 °C differences in temperature when compared to PV panel without cooling system.

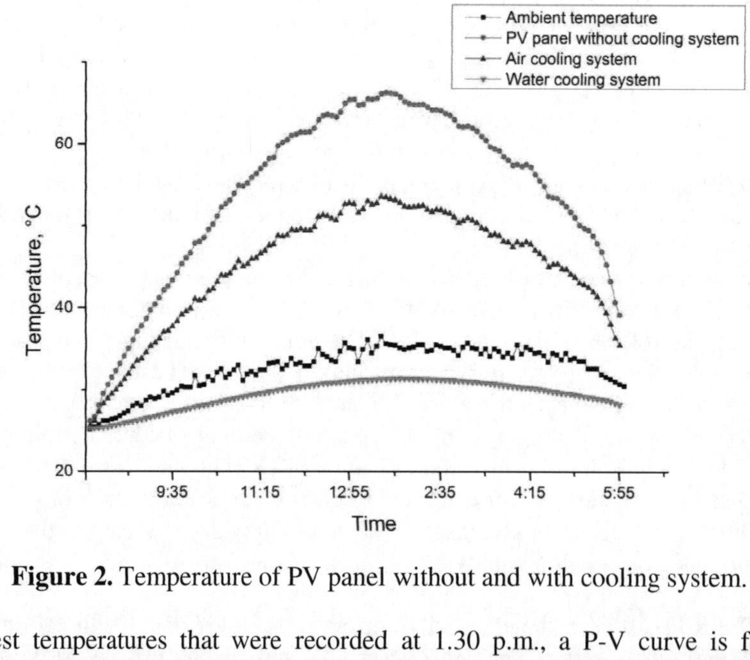

Figure 2. Temperature of PV panel without and with cooling system.

With the highest temperatures that were recorded at 1.30 p.m., a P-V curve is formed by using PSPICE software as illustrated in Figure 3. This is due to evaluate the power generated for each PV panel cooling. Pradhan [2] stated that, when PV panel's temperature increases, it will cause only marginal changes in current, but major changes in voltage. This is proven with the graph shown in Figure 3. As the PV panel without cooling system has the highest temperature, hence its V_{oc} is only

IOP Conf. Series: Journal of Physics: Conf. Series **908** (2017) 012074 doi:10.1088/1742-6596/908/1/012074

19.0 V. But, with water cooling system attach to it, the V_{oc} is enhanced up to 21.5 V. The V_{oc} for PV panel with air cooling system recorded is 20.0 V. Besides, the output power generated by PV panel is the highest with water cooling system, which is 78.5 W, compared with air cooling system, only 71.1 W. Overall, it has an increment in output power generated by PV panel compared to without a cooling system.

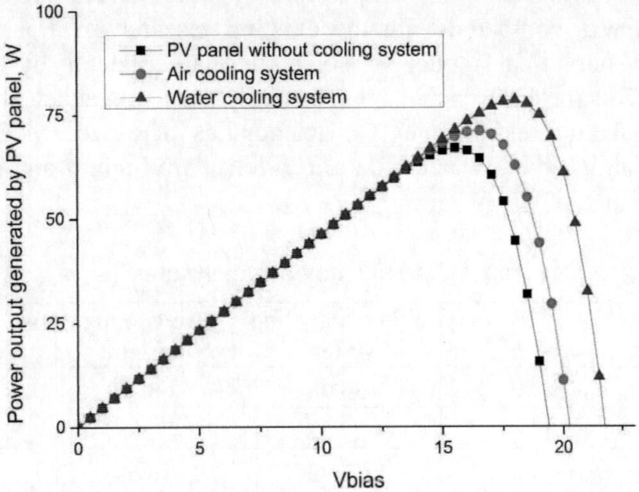

Figure 3. The P-V curve for PV panel without and with cooling system.

Table 1 presents the net output power saving for PV panel without and with the cooling system. The power generated is by using the highest temperature at 1.30 p.m. Both PV panel cooling systems are active cooling system, by which it consumes external energy to operate it. In the other words, it used the power generated by PV panel to operate the cooling system. Therefore, it is important to select a cooling system that does not consume much power in operating it. Even though both cooling systems have the same input power, but water cooling system eventually has the highest output power saving compared to air cooling system, which is 9.38 W. This is because water has the higher specific heat capacity compared to air. The greater the heat capacity, more heat per unit mass is needed in raising the temperature of PV panel [15].

Table 1. Net output power saving for PV panel without and with cooling system.

	Without cooling system	Air cooling system	Water cooling system
PV panel output power generated (W)	66.96	71.13	78.5
Power consumption (W)	0.0	2.16	2.16
Net output power (W)	66.96	68.97	76.34
Net output power saving (W)	-	2.01	9.38
Percentage of net output power saving (%)	-	3.00	14.00

The efficiency of the system is illustrated as in Table 2. The net output power is referred to the output power generated from PV panel after considering the power consumption of each cooling system. There is only slightly difference of net output power of PV panel without cooling system and with air cooling system. The output power from PV panel was obtained from PSPICE simulation. From that, the efficiency of each cooling system can be calculated. It is found out that, the PV panel with water cooling system actually has the highest efficiency compared to others. Since there is only slightly differences in net output power for PV panel without cooling system and with air cooling system, hence its efficiency only differs for about 0.66 %. The efficiency of PV panel with water cooling system increase 3.11 %. These efficiency were calculated based on net output power generated at 1.30 p.m., with a solar irradiance of 1175 W/m^2 and ambient temperature of 36.5 °C. The efficiency of the water cooling system then was compared with the existing research by Bahaidarah et. al [16]. Author developed a numerical model of solar thermal collector by using Engineering Equation Solver (EES). The heat exchanger (cooling panel) was attached at the backside of PV panel, and it is found out that the energy efficiency has increased 9 %. The simulation was based on Dhahran, Saudi Arabia weather data location, with highest ambient temperature of 21 °C and solar irradiance of 979 W/m^2.

Table 2. Efficiency of the system.

	Without cooling system	Air cooling system	Water cooling system
Net output power (W)	66.96	68.97	76.34
Fill factor, FF	0.522	0.538	0.595
Efficiency, η (%)	21.99	22.65	25.07

4. Conclusion

This paper presents the comparison between PV panel with air cooling system and PV panel with water cooling system. Both cooling system is compared with the temperature criteria, as well as output power generated by PV panel. The input power for both cooling is the same which is 2.16 W with DC fan speed of 3.07 m/s and water mass flow rate of 0.0556 kg/s. As a conclusion, PV panel with cooling systems show a decrement in temperature compared to PV panel without cooling system. By using the highest temperature, the P-V curve is formed due to evaluate the output power generated by PV panel at that particular time. The water cooling system is chosen as the best cooling compared to air cooling system, since it has the highest output power saving, compared to it.

5. Acknowledgement

The authors would like to acknowledge the Ministry of Higher Education of Malaysia and the Universiti Malaysia Perlis, Perlis, Malaysia for the financial support under FRGS (9003-00564).

6. References

[1] International Renewable Energy Agency (IRENA) 2012 *Renewable Energy Technologies:Cost Analysis Series, Solar Photovoltaics* pp 4-5

[2] Pradhan A, Ali SM, Jena C 2013 Analysis of Solar PV Cell performance With Changing Irradiance and Temperature (*International Journal of Engineering and Computer Science* vol 2) pp 214-220

[3] Davud MT, Rahim M, Daryush N 2011 The Effect of Temperature on Electrical Parameters of Solar Cells (*International Journal of Advanced Research in Electrical, Electronics and Instrumentation Engineering* vol 2) pp 6404-6407

[4] Hiren DR, Subama M, Ashish M 2014 Computational fluid dynamics analysis and experimental validation of improvement in overall energy efficiency of a solar photovoltaic panel by thermal energy recovery (*Journal of Renewable and Sustainable Energy* vol 6) pp 1-12

[5] Akbarzadeh A, Wadowski T Heat-pipe-based cooling systems for photovoltaic cells under concentrated solar radiation (*Applied Thermal Engineering* vol 16) pp 81-87

[6] Dubey S, Tiwari GN 2008 Thermal modelling of a combined system of photovoltaic thermal (PV/T) solar water heater (*Solar Energy* vol 82) pp 602-612

[7] Salih MS, Osama IA, Kaleid WA 2014 Performance Enhancement of PV Array Based on Water Spraying Technique (*International Journal of Sustainable and Green Energy* vol 4) pp 8-13

[8] Smith K Water 2014 Cooling Method to Improve the Performance of Field Mounted, Insulated and Concentrating Photovoltaic Modules (*Journal of Solar Energy Engineering* vol 136) pp 1-4

[9] Sayran AA, Omar SO, Kamil MY 2014 Improving the Efficiency of Polycrystalline Solar Panel Via Water Immersion Method (*International Journal of Innovative Research In Science, Engineering and Technology* vol 3) pp 8127-8132

[10] Saurabh M, Pratish R, Mary D, Sudhakar K 2014 Performance of a Solar Panel With Water Immersion Cooling Technique (*International Journal of Science,* vol 3) pp 1161-1172

[11] Teo HG, Lee PS, M NA 2012 Hawlader An Active Cooling System for Photovoltaic Modules (*Applied Energy* vol 90) pp 309-315

[12] Pratish R, Mary D, Saurabh M, Sudhakar K, Prakash KS 2014 Performance Evaluation of Solar Photovoltaic/Thermal Hybrid Water Collector *(Impending Power Demand and Innovative Energy Paths)* pp 268-275

[13] Khelifa A, Touafek K, Ben MH, Tabet I, Ben Cheikh EHH, Holoui H 2015 Analysis of a Hybrid Solar Collector Photovoltaic Thermal (PVT) (*International Conference on Technologies and Materials for Renewable Energy, Environment and Sustainability* vol 74) pp 835-843.

[14] Stefan K 2004 Increased Electrical Yield Via Water Flow Over the Front of Photovoltaic Panels (*Solar Energy Materials & Solar Cells* vol 82) pp 131-137

[15] Irwan YM, Leow WZ, Irwanto M, Hassan SIS, Safwati I, Amelia AR 2015 Comparison of Solar Panel Cooling System by Using DC Brushless Fan and DC Water (*Journal of Physics: Conference Series 622*) pp 1-10

[16] Bahaidarah H, Subhan A, Gandhidasan P, Rehman S Performance evaluation of a PV (photovoltaic) module by back surface water cooling for hot climate 2013 *(Energy)* pp 445-453

IOP Conf. Series: Journal of Physics: Conf. Series **908** (2017) 012075 doi:10.1088/1742-6596/908/1/012075

Experimental investigation on AC unit integrated with sensible heat storage (SHS)

N A Aziz[1], N A M Amin[1], M S A Majid[1], A Hussin[2], S Zhubir[1]

[1] School of Mechatronic Engineering, Universiti Malaysia Perlis, 02600 Arau, Perlis, Malaysia
[2] ILP Kepala Batas, Jln Tun Hamdan Sheikh Tahir, 13200 Penaga, Penang, Malaysia

nasrulamri.mohdamin@unimap.edu.my

Abstract. The growth in population and economy has increases the energy demand and raises the concerns over the sustainable energy source. Towards the sustainable development, energy efficiency in buildings has become a prime objective. In this paper, the integration of thermal energy storage was studied. This paper presents an experimental investigation on the performance of an air conditioning unit integrated with sensible heat storage (SHS) system. The results were compared to the conventional AC systems in the terms of average electricity usage, indoor temperature and the relative humidity inside the experimented room (cabin container). Results show that the integration of water tank as an SHS reduces the electricity usage by 5%, while the integration of well-insulated water tank saves up to 8% of the electricity consumption.

1. Introduction

Buildings account for approximately 40% of the worldwide energy demand [1]. The usage of air conditioning system in buildings primarily, has increases load on power used, thus increasing energy consumptions and therefore cause of high electricity tariff. Energy efficiency improvements in buildings is therefore vital to reduce the global energy demand towards sustainable development. These issues which was also amplified by a strong demand for uninterrupted supply has encouraged significant development in renewable energy technologies. Thermal Energy Storage (TES) system is one of the technologies which have attracted profound interest from researchers for their use in eliminating environmental problems and to increase the efficiency of energy consumption in general [2].

Thermal energy storage is defined as the temporary holding of thermal energy in the form of hot or cold substances for later utilisation. TES system has an enormous potential to make the function of thermal energy equipment more effective and to facilitate large-scale energy substitutions. They are highly valuable from an economic perspective. It is an advanced energy technology that has recently attracted increasing interest for thermal applications such as space and water heating, cooling and air-conditioning. It appears to be the most appropriate method for reducing the mismatch between the supply and demand of energy [3]

Over the last decades, numerous research were conducted to improve the energy efficiency in buildings by the integration of TES system [4–5]. PCM thermal storage system was acknowledged as most favourable storage system prior this objective. The integration of PCM thermal storage system into air distribution system for peak shaving purpose was studied [4]. PCMs mixtures of paraffin waxes with the heat of fusion around 90 KJ/kg was chosen as thermal storage medium. From this study, the authors conclude that the integration of TES systems could maintain a constant room temperature without any cold source operation at a certain time.

On the other hand, the possibility of peak shaving control with the incorporation of PCM thermal storage to the ceiling boards was examined [6]. The thermal capacity of PCM ceiling board was measured by using a small chamber, approximately 16m^2 surface area. The PCM ceiling boards was developed by adding the microencapsulated PCM to the commonly used rock wool ceiling board. The mixture of n-paraffins with a melting point of 25 °C was used in this study. During overnight, the cool air from air handling unit (AHU) will flow into the ceiling chamber space and cools the PCM ceiling board and storing cool thermal energy. Whereas, during the peak shaving period, the air from the room returns to the AHU via the ceiling chamber space.

During the peak shaving period, when the thermal load peaks, the air from the room returns to the AHU via the ceiling chamber space. As a result of passing through the cooled-down PCM ceiling board, the warm air returning from the room cools down before returning to the AHU. With such an arrangement air is precooled and the result was that the maximum thermal load using the PCM ceiling board was 85.2% of that using the rock wool ceiling board. The result shows that, the transition rate of the thermal load to the night was 25.1% thus lowering running cost for 91.6% compared to the rock wool ceiling board [6].

In this paper, the effect of TES system integration into the split unit air-conditioning (AC) unit was investigated. In contradict to the previous research that favour the utilisation of latent heat storage (LHS) medium (eg: PCMs), this paper proposes a sensible heat storage (SHS) medium to be retrofitted to the AC suction line, aiming to reduce the energy consumptions simultaneously improving the thermal comfort inside the buildings.

2. Experimental setup

Figure 1 shows the schematic diagram of the constructed thermal storage air-conditioning system. 1 HP split unit air conditioning system was installed in a 20 × 10 × 8.4 feet room (cabin container). Three cooling systems were experimented and studied. The first system is the conventional air conditioning (AC) as shown in Figure 1 (a). While the second system was tested on the same AC unit but it was integrated with the water tank as illustrated in Figure 1 (b). The helical coil in tank thermal storage was constructed from the polyethene water tank (as shown in Figure 2 (a) with the dimension of 2 feet high and 1 feet diameter. 5 feet length of 3/8 inch diameter copper tube was assembled 2 feet from the outdoor unit by the used of 3-way valve. The helical coil was then submerged inside the polyethene water tank which was filled with 98% of water at the room temperature. The third system was upgraded from the second system which the polyethene tank was well insulated to minimise the heat loss to the surroundings. All tests were experimented under the same condition that conducted in a closed room with AC was set at 16 ℃ and full speed blower. Heat load for the

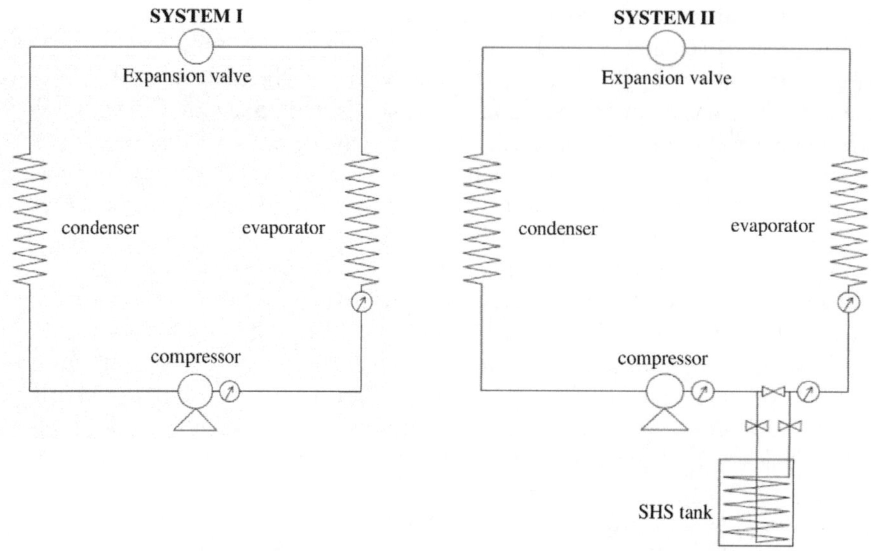

Figure 1. Schematic diagram of the tested system (a) conventional AC system (b) AC – TES integrated system

Figure 2. (a) SHS tank (b) Retrofitted SHS tank to the suction line

The pilot test was first conducted to observe the trends of the outdoor temperature. The average temperature of 10 days pilot test was recorded as 27.6 °C, with the maximum and minimum daily temperature were 39 °C and 23 °C respectively. For all set of tested experiments, it is very difficult to be performed under exactly same outdoor temperature. Therefore, the outdoor temperature was monitored to be in a small margin with the average outdoor temperature, assuming they are in the same condition. PEL: power and energy logger was employed to record the voltage, current and power consumption for all three systems. Meanwhile, the HOBO data logger was utilised to monitor the indoor temperature and relative humidity of the tested room. For each test, the data were logged in 5 minutes interval for 10 days long. The data were then compiled and analysed.

3. Results and discussion

Figure 3 shows the comparisons of average electricity usage per day for all three tested systems with the average outdoor temperature of 27.5 °C. The voltage usage for 1 HP AC system was recorded in the range of 230 – 240 V. With the constraints of similar outdoor temperature and voltage, the conventional AC system was recorded to consume the highest electricity amongst three systems. Meanwhile, comparing the integrated AC – SHS system shows higher consumption than the integrated AC – insulated SHS system. The average power consumption for all three systems was calculated as 420.052 W, 395.828 W and 386.532 W for conventional, AC – SHS, and AC – insulated SHS respectively. These results indicated that the integration of SHS tank would reduce the total energy consumption for at least 5% for uninsulated water tank while it saves up to 8% for insulated water tank integration (at the measured point).

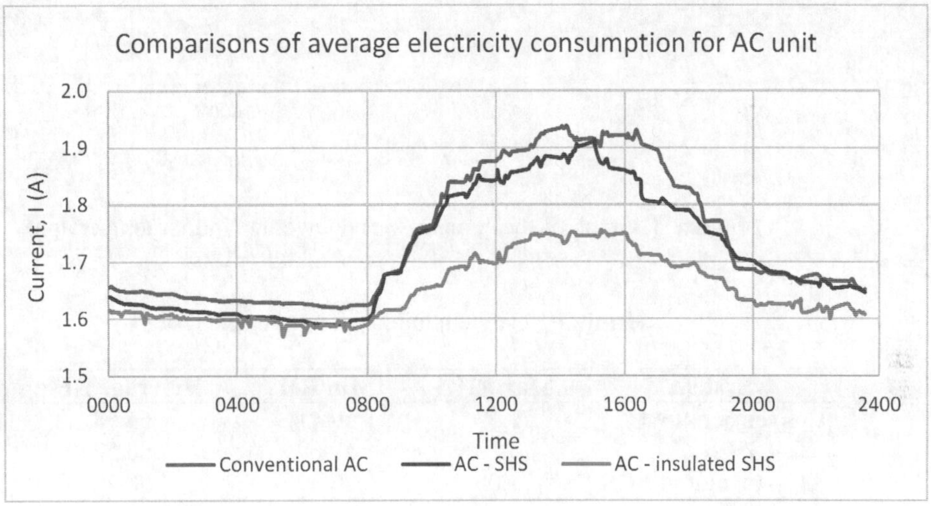

Figure 3. Graph of the comparisons on the electricity consumption

Nevertheless, in order to evaluate the overall performance of all tested systems, the indoor condition of the room must be taken into consideration. Indoor thermal comfort evaluation is an important measure in analysing the overall performance of the system. Providing a better thermal comfort in the building while simultaneously decreasing the power consumption is an important key in designing more efficient cooling system. The term thermal comfort is basically referring to the state of the indoor temperature, which the resident considers as comfortable to stay in. There are many factors affecting the thermal comfort inside the buildings such as air temperature, relative humidity and air speed [6]. In this study, indoor temperature and relative humidity (RH) were monitored and analysed towards this objective. For a country with a tropical climate, suggested indoor temperature and RH are 21 °C – 27 °C and 28% - 80% humidity respectively [7]. It is vital to maintain the RH at a comfortable range, which is high enough to avoid problems associated with dry air but low enough to be comfortable [8].

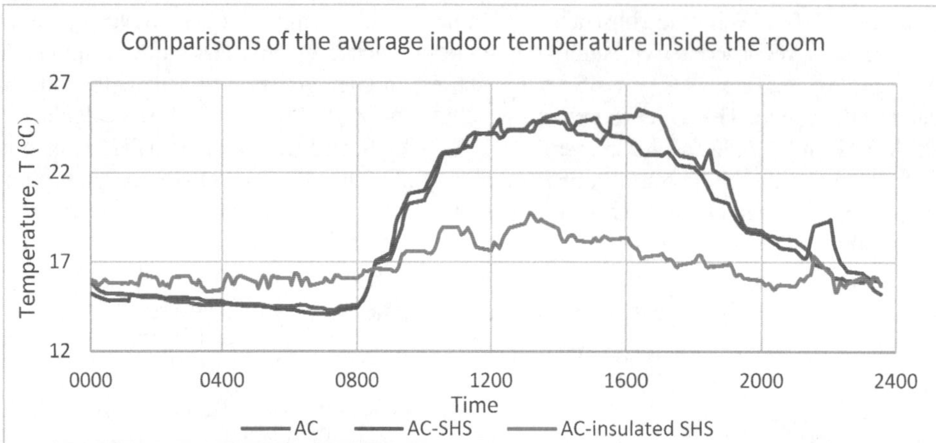

Figure 4. Graph of the comparisons on average indoor temperature

Table 1 Relative humidity for all tested system

System	Max RH	Min RH	Average RH
Conventional AC	71.37	59.41	65.58
AC – SHS	73.06	59.69	66.52
AC – insulated SHS	72.05	56.39	66.48

Figure 4 shows the comparisons on the average indoor temperature, while the minimum, maximum and average relative humidity for all systems were tabulated in Table 1. Referring to Figure 4, it can be seen that the indoor temperature for conventional AC system and the AC – SHS integrated system recorded about the same values. Referring to the plotted graph, it shows that the temperature inside the room for AC – integrated system somehow goes higher than the temperature with the use of conventional AC. These data indicated that the integrated of non-insulated water tank does not show a significant impact towards the thermal comfort of the buildings.

Other than that, the graph shows that the average indoor temperature of the tested room is seen to be less fluctuated with the integration of insulated water tank when compared to the other two systems. The mean temperature reduces about 20% in comparisons with the conventional AC. The average indoor temperature for all three systems are 28.07 °C, 27.46 °C and 22.15 °C respectively. Whereas, the values of RH tabulated in Table 1 shows that an average daily RH were 65.58, 66.52 and 66.48 for all three systems respectively. These data conceded the integration of SHS tank does not show too much significant impact on the thermal comfort of the buildings as all these data were in the range of suggested thermal comfort condition [7]. However, it is clearly proven that indoor temperature and RH were improved by the integration of the insulated SHS tank compared to the use of conventional AC and AC-SHS (water tank) integration.

4. Conclusion

The present work mainly constructed and tested the air conditioning unit performance with and without the integration of sensible heat storage (SHS). The power input usage was recorded and compared for all three tested systems. Results show that the integration of water tank to the AC system would reduce the power consumption by 5% while the integration of well-insulated water tank reduces it by almost 8%. From this study, it is clearly proven that the power consumption, indoor temperature and RH were improved by the integration of the insulated SHS tank compared to the use of conventional AC and AC-SHS (water tank) integration.

5. Acknowledgements
The authors would like to acknowledge the Ministry of Higher Education of Malaysia and the Universiti Malaysia Perlis, Perlis, Malaysia for the financial support under FRGS (9003-00564).

6. References

[1] B. R. Hughes, H. N. Chaudhry, S. A. Ghani, B. Richard, H. Nasarullah, S. Abdul et al." A review of sustainable cooling technologies in buildings" Renew Sustain Energy, 15, 3112–20 (2011)

[2] Mehling, Harald, and L.F. Cabeza. 2008. Heat and Cold Storage with PCM. Springers.

[3] Mahlia, T M I et al. 2014. "A Review of Available Methods and Development on Energy Storage ; Technology Update". Renewable and Sustainable Energy Reviews 33: 532–45 (2014)

[4] M. Yamaha, S. Misaki, The Evaluation of Peak Shaving by a Thermal Storage System Using Phase-Change Materials in Air Distribution Systems HVAC&Research, 12, 861–869 (2006)

[5] T. Kondo, T. Ibamoto, "Research on Thermal Storage using Rock Wool PCM Ceiling Board" ASHRAE Trans, 526–531 (2006)

[6] T.H. Karyono "Report on thermal comfort and building energy studies in Jakarta–Indonesia", Building and Environment 35, pp. 77–90 (1998).

[7] R.J.De Dear, K.G. Leow and A. Ameen."Thermal Comfort in the Humid tropics. Part 1. Climate Chamber Experiments on Temperature Preferences in Singapore", ASHRAE Transctions 97 (1991).

[8] O. K. Akande and M. A. Adebamowo, "Indoor Thermal Comfort for Residential Buildings in Hot-Dry Climate of Nigeria" *Adapting to Change: New Thinking on Comfort,* Cumberland Lodge, Windsor, UK, London: Network for Comfort and Energy Use in Buildings (2010)

Natural convection in an inclined parallelogrammic porous enclosure under the effect of magnetic field

R D Jagadeesha[1], B M R Prasanna[2], D Younghae[3] and M Sankar[4]

[1] Department of Mathematics, Government Science College, Hassan, India
[2] Department of Mathematics, Siddaganga Institute of Technology, Tumakuru, India
[3] Department of Mathematics, KNU-Center for Nonlinear Dynamics, Kyungpook National University, Daegu, 41566, Korea
[4] Department of Mathematics, School of Engineering, Presidency University, Bengaluru, India

msankar@presidencyuniversity.in

Abstract. A numerical study is conducted to understand the effect of magnetic field on the natural convection in a tilted parallelogrammic porous enclosure. The two opposing side walls are differentially heated with a temperature difference specified, while the upper and lower walls are adiabatic. Using the Darcy model to formulate the problem, a finite difference scheme consisting of the Alternating Direction Implicit and Successive Line over Relaxation methods are used to solve the coupled non-linear governing equations. Computations are carried out for a wide range of Rayleigh number ranging from 100 to 5×10^3, inclination of the walls of the parallelogram ϕ from -60^0 to 60^0, angle of inclination of the enclosure α from -60^0 to 60^0, aspect ratio from 0.5 to 2 and Hartmann number from 0 to 20. The flow structure and the corresponding heat transfer characteristics inside the enclosure are presented in detail. The results revealed that both the magnetic force and the inclination angle have significant effect on the flow field and heat transfer in porous medium. The reported results are in good agreement with the available published work in the literature.

1. Introduction

Natural convection of electrically conducting fluid in finite sized enclosures under the influence of a magnetic field applied in one or more directions have been studied by several investigators. In many industrial applications, such as materials processing and crystal growth techniques, the electrically conducting fluid the magnetic field is applied in one or more directions. In such situations, the fluid movement is influenced by an additional force known as Lorentz force which suppresses the fluid flow velocities which in turn affect the heat transport rates.

Due to enormous applications, natural convective heat transfer in the non-rectangular geometries has become an essential phenomenon. Among the non-rectangular enclosure, a parallelogrammic enclosure is an important type of enclosure and aptly describes the physical configuration of many applications. A detailed investigation of free convection in a parallelogrammic enclosure for engineering applications with special attention to the case of parallelogrammic diode cavity can be found in Bairi et al. [1]. Baytas and Pop [2] studied convective flow and heat transport in oblique enclosures filled with porous medium. They transformed the computational domain to a regular shaped domain using a standard transformation. Buoyancy-driven convection in a parallelogrammic enclosure is reported by Hyun and Choi [3] to analyse the transient effects. They analysed the influence of transient term on the flow pattern and the average Nusselt number for various tilt angles and aspect ratios. Later, Costa et al. [4] studied natural convection in parallelogrammic enclosures with variable geometry.

Revnic et al [5] examined the influence of inclination of a square porous enclosure on the convective flow and heat transport characteristics for different combinations of parameters. It has been observed that the flow and thermal patterns, the total heat transport rates are significantly modified by the cavity inclination and Rayleigh number. The influence of magnetic field on convection heat transfer is also investigated in inclined enclosures. Han [6] studied free convection of an electrically conducting fluid in an inclined enclosure under the influence of magnetic field. Ece and Buyuk [7] examined the effects of magnetic field on convective flow and heat transfer in an inclined rectangular enclosure, which is heated from one side and cooled from the adjacent side. This study reveals that the flow circulation became stronger as the Rayleigh number increases and the magnetic field suppresses the flow and the heat transport rates. For a tilted square enclosure, Pirmohammadi and Ghassemi [8] investigated the convection heat transfer by considering the effect of magnetic field. It is shown that for a given inclination angle, as the value of Hartmann number increases, the convection heat transfer reduces. Saleh et al. [9] numerically studied free convection in a porous parallelogrammic enclosure with an applied magnetic field. They concluded that, the variation of magnetic field strength and the cavity inclination angle, will affect the heat transfer performance.

A detailed analysis of the literature reveals that natural convection flow under the influence of magnetic field has not been investigated in an inclined parallelogrammic enclosure. Hence, in this paper, the natural convective flows and the associated heat transport characteristics are analysed in an inclined parallelogrammic enclosure with a magnetic field applied along the x-direction. The primary objective of this study is to analyse the effects of various controlling parameters, such as, angle of inclination of the enclosure, magnetic field strength, aspect ratio, Rayleigh number and the parallelogrammic angle, on the fluid flow and heat transfer.

2. Physical model and basic equations

Consider a parallelogrammic porous enclosure, as shown in Fig.1 along with the coordinate system, containing an electrically conducting fluid. Let *H* and *L* be the height and width of the enclosure respectively, and the enclosure is tilted at an angle α with respect to the x-axis. The inclined left wall is maintained at a higher temperature (T_h), while the right inclined wall is at lower temperature (T_c). However, the top and bottom walls are assumed to be perfectly insulated. The fluid is assumed to obey the Boussinesq approximation and is incompressible. Along the horizontal direction, a magnetic field of constant strength B_0 is applied. The inertial effects are assumed to be neglected and the heating causes by viscous, radiation and Joule heating are not taken into account. Further, the induced magnetic field is neglected on comparison with the applied magnetic field and hence the magnetic Reynolds number is considered to be small. Assuming that the Darcy's law to be hold, the equations of conservation of energy and momentum can be written as,

$$\frac{\partial T}{\partial t^*}+\frac{\partial \psi^*}{\partial y}\frac{\partial T}{\partial x}-\frac{\partial \psi^*}{\partial x}\frac{\partial T}{\partial y}=\alpha m\left(\frac{\partial^2 T}{\partial x^2}+\frac{\partial^2 T}{\partial y^2}\right) \tag{1}$$

$$\frac{\partial^2 \psi^*}{\partial x^2}+\frac{\partial^2 \psi^*}{\partial y^2}=\frac{gK\beta}{\nu}\left(\frac{\partial T}{\partial y}\sin\alpha-\frac{\partial T}{\partial x}\cos\alpha\right)-\frac{\sigma B_0^2 K}{\mu}\frac{\partial^2 \psi^*}{\partial x^2} \tag{2}$$

For the computational convenience, the parallelogrammic shaped computational domain is mapped on to a square domain as shown in Fig. 1 (bottom) using the transformation X = x - ytanϕ, Y = y proposed by Baytas and Pop [2]. Using this transformation and by introducing the following dimensionless variables

$$\xi=\frac{X}{L},\eta=\frac{Y}{H\cos\phi},t=\frac{t^*\alpha_m}{LH\cos\phi},\psi=\frac{\psi^*}{\alpha_m},\theta=\frac{(T-T_r)}{(T_h-T_c)},\text{ where } T_r=(T_h+T_c)/2,$$

the equations (1) and (2) in dimensionless form can be written as

$$\frac{\partial\theta}{\partial t}+\frac{\partial\psi}{\partial\eta}\frac{\partial\theta}{\partial\xi}-\frac{\partial\psi}{\partial\xi}\frac{\partial\theta}{\partial\eta}=\frac{A}{\cos\phi}\left(\frac{\partial^2\theta}{\partial\xi^2}-2\frac{\sin\phi}{A}\frac{\partial^2\theta}{\partial\xi\partial\eta}+\frac{1}{A^2}\frac{\partial^2\theta}{\partial\eta^2}\right), \tag{3}$$

$$\frac{\partial^2\psi}{\partial\xi^2}\left[1+Ha^2\cos^2\phi\right]-2\frac{\sin\phi}{A}\frac{\partial^2\psi}{\partial\xi\partial\eta}+\frac{1}{A^2}\frac{\partial^2\psi}{\partial\eta^2}=Ra\cos\phi\left[\frac{\sin\alpha}{A}\frac{\partial\theta}{\partial\eta}-\cos(\phi-\alpha)\frac{\partial\theta}{\partial\xi}\right] \tag{4}$$

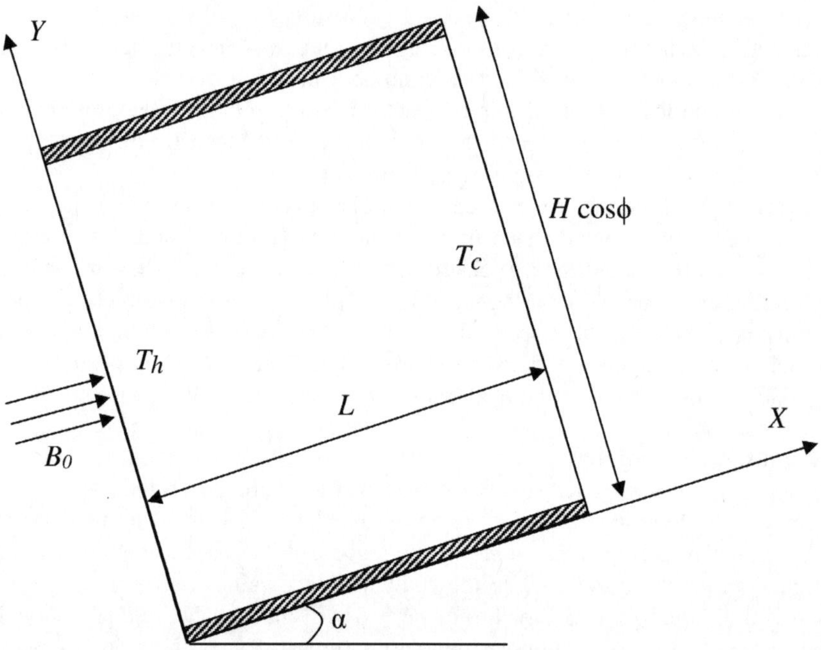

Figure 1. Physical configuration and coordinate system.

The initial and boundary conditions in dimensional form are:

$$\begin{aligned}
t=0:&\quad \psi=0,\theta=0;\quad 0\le\xi\le1, 0\le\eta\le1\\
t>0:&\quad \psi=0,\theta=+0.5;\qquad \xi=0, 0\le\eta\le1\\
&\quad \psi=0,\theta=-0.5;\qquad \xi=1, 0\le\eta\le1\\
&\quad \begin{cases}\psi=0,\\ \dfrac{\partial\theta}{\partial\eta}-A\sin\phi\dfrac{\partial\theta}{\partial\xi}=0\end{cases}\quad \begin{cases}\eta=0 \text{ and } 1,\\ 0\le\xi\le1.\end{cases}
\end{aligned}$$

In the above equations, $A=\dfrac{H}{L}$ is the aspect ratio, $Ra=\dfrac{gK\beta\Delta TL}{\nu\alpha_m}$ is the Darcy-Rayleigh number and $Ha=B_0\sqrt{\dfrac{\sigma K}{\mu}}$ is the Hartmann number.

The measure of heat transfer rate is calculated through the local (Nu) and global $(\overline{Nu})$ Nusselt numbers given as

$$Nu = -\frac{1}{\cos\phi}\left(\frac{\sin\phi}{A}\frac{\partial\theta}{\partial\eta} - \frac{\partial\theta}{\partial\xi}\right), \text{ where } \overline{Nu} = \int_0^1 Nu\, d\eta$$

3. Solution methodology

The non-dimensional governing equations (5) and (6) are solved numerically using finite difference based ADI and SLOR methods subjected to the chosen initial and boundary conditions. The discretized finite difference equations are in Tri-diagonal matrix form and are inverted using the Thomas algorithm. After a careful grid independence analysis, all the simulations are executed by choosing a different grid size with respect to the aspect ratio. A 201×101 grids for $A=0.5$, 101×101 grids for $A=1$ and 101×201 grids for $A=2$ has been chosen. The FORTRAN code has been developed to perform the simulations, and successfully validated against the various benchmark solutions. mm.

4. Results and discussion

An in-depth analysis is made on natural convective flow and associated heat transport phenomena inside an inclined parallelogrammic porous enclosure under the influence of magnetic field applied in the direction perpendicular to gravity. The numerical results are discussed through the streamlines, isotherms and global heat transfer rates for vast range of the parameters of the investigation, namely the Darcy-Rayleigh number (Ra), the inclination angle of the walls of the enclosure (ϕ), the tilt angle of the enclosure (α), the aspect ratio (A) and the magnetic field strength (Ha). The range of dimensionless parameters considered is $0\le Ha\le 20$, $-60^0 \le \alpha \le 60^0$, $-60^0 \le \phi \le 60^0$, $10^1 \le Ra \le 5\times10^3$ and $0.5 \le A \le 2.0$. A systematic study is made on the flow pattern and temperature distributions inside the enclosure under the influence of all the parameters considered. Further, the rate of overall heat transport is measured in terms of the average Nusselt number.

4.1. Effect of Darcy-Rayleigh number

The flow characteristics and thermal pattern in a tilted parallelogrammic enclosure is analysed for the Darcy-Rayleigh number through streamlines and isotherms. Figure 2 illustrates the flow and thermal pattern for different strengths of magnetic field, by fixing the other governing parameters. In the zero-magnetic field case (Ha=0), it is observed that the intensity of the flow is high and the streamline movement is observed mainly in the centre region of the cavity forming a single shallow cell. Further, the flow circulation rate is found to be higher for non-magnetic case, as can be noticed through the maximum stream function value. The isotherms show strong variations for non-magnetic case, but become less variant and stagnant in the core region as the value of Ha increases. At higher value of the magnetic field strength (Ha=20), the intensity of the flow circulation is reduced, due to the retarding force exerted by the magnetic parameter Ha and are accumulated at the bottom and top portion of the boundaries. Also, the direction of streamlines is changed from horizontal to vertical cell, and appears to be a parallel flow. As far the temperature lines are concerned with respect to the magnetic field, the isotherms reveal a conduction-like structure and are parallel to the inclined vertical walls. The effect of Darcy-Rayleigh number on the overall Nusselt number for four different magnetic field strengths (Ha = 0, 5, 10 and 20), but fixing the other governing parameters at ϕ = 300, α =300 and A=1 is illustrated in Fig 3. When Ha = 0, the average Nu curve is increasing monotonically with the increase in the Darcy-Rayleigh number. It can be observed that the average *Nu* decreases with increasing the Hartmann number and the average *Nu* reaches unity for a higher magnetic field (Ha=20).

4.2. Effect of the angle of inclination of the walls

In order to examine the combined effects of the angle of inclination of the walls and magnetic field strength, computations are carried out for three values of *Ha* and fixing the other governing parameters at Ra = 3000, $\alpha = 30^0$ and $A = 1$. Figure 4 illustrates the effects of different values of magnetic field strength (Ha = 0, Ha = 5 and Ha = 20) at the inclination angle $\phi = 30^0$. In the zero-magnetic field case, it can be seen that the stream function attains its maximum value when compared to the presence

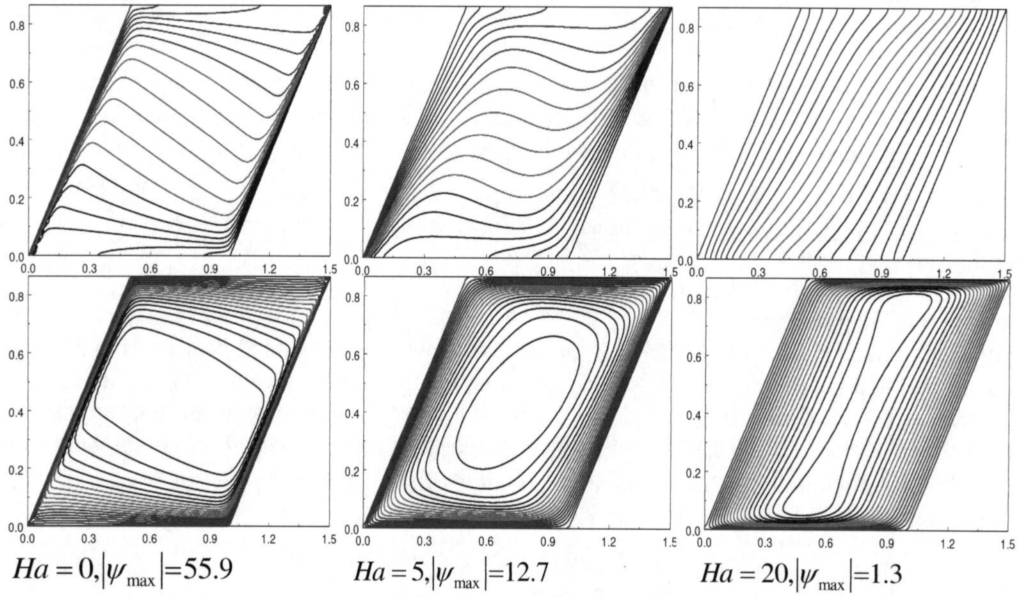

Figure 2. Isotherms and streamlines for different values of Ha and Ra at $A = 1, \phi = 30^0, \alpha = 30^0$.

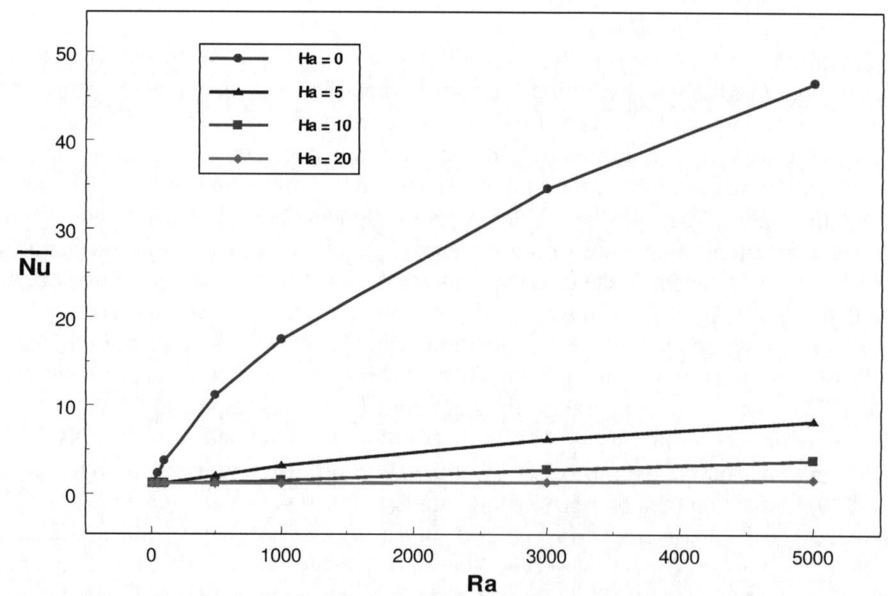

Figure 3. Effect of Ra and Ha on the average Nusselt number for $\phi = 30^0, \alpha = 30^0, A = 1$

IOP Conf. Series: Journal of Physics: Conf. Series **908** (2017) 012076 doi:10.1088/1742-6596/908/1/012076

of magnetic field. Also, the direction of streamlines greatly depends on the inclination of enclosure. In general, as the magnetic field is applied, the flow is remarkably suppressed due to the retarding force generated from the magnetic field force. As the magnetic field strength is increased, the main vortex at the core gets elongated along the entire enclosure. Also, for a positive inclination of side walls (ϕ = 30^0), flow circulation rate is higher compared to the negative inclination. This can be observed through the $|\psi_{max}|$ values of these two cases. A meticulous examination of the change of streamlines and isotherms pattern exhibits an important fact that the magnetic field strength and the inclination angle of sidewalls plays a key role in altering the flow and thermal pattern.

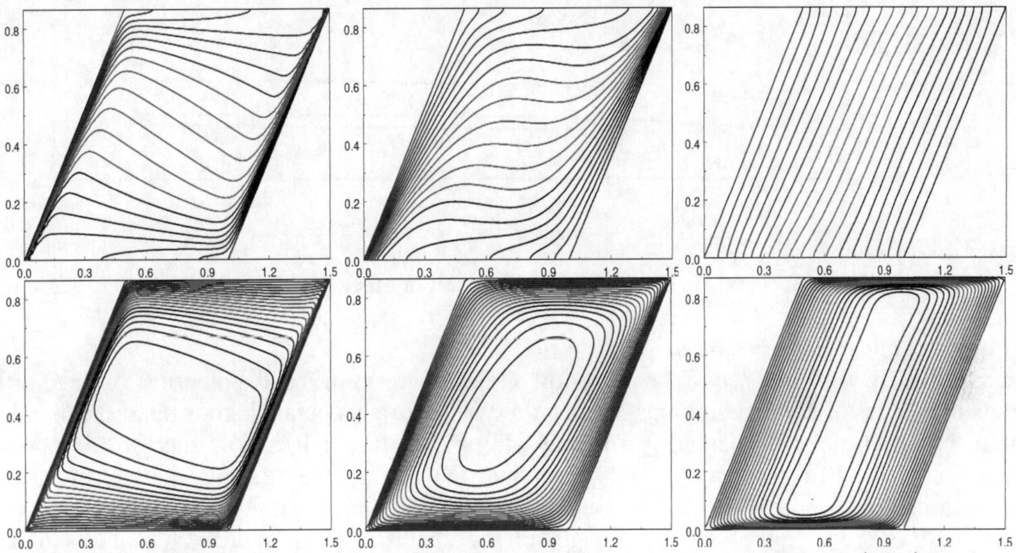

$Ha = 0, |\psi_{max}| = 43.6$ $Ha = 5, |\psi_{max}| = 9.5$ $Ha = 20, |\psi_{max}| = 0.8$

Figure 4. Effect of Hartmann number on isotherms and streamlines for $A = 1, Ra = 3\times10^3, \alpha = \phi = 30^0$.

The effect of inclination angle of the sidewalls on the average Nusselt number for different *Ha* is shown in Fig. 5. When ϕ = 0^0, the maximum heat transfer occurs for all the values of magnetic field, i.e., *Ha* = 0, 5, 10 and 20. The overall heat transport rates are comparatively higher for *Ha* = 0 compared to the other values of *Ha*. With the introduction of magnetic field, the overall heat transfer is suppressed drastically and can be observed from the figure. Further, the variation in average *Nu* is very minimal as the magnetic field is introduced (*Ha* =5, 10 and 20). This indicates that the suppression of heat transport is due to both magnetic field and inclination angle of the sidewalls.

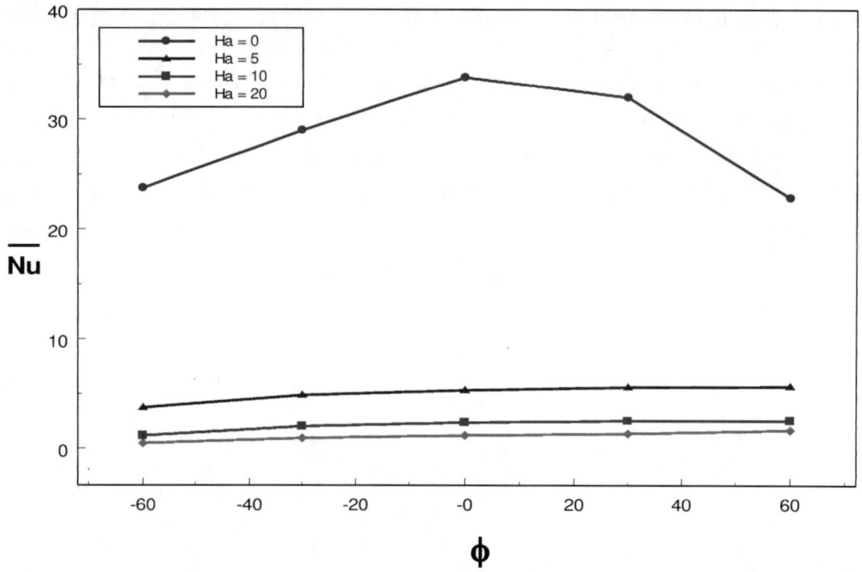

Figure 5. Effect of ϕ and Ha on the average Nusselt number for $Ra = 3\times10^3, \alpha = 15^0, A = 1$.

4.3. Effect of the angle of inclination of the walls

Figure 6 exhibits the influence of the cavity tilt angle on streamlines and isotherms for three different parametric values of *Ha*. For zero- magnetic field case, a strong a parallelogram shaped cell is formed within the enclosure with diagonally rotating eddies. The flow intensity is high in the absence of magnetic field. When a magnetic field is applied, the axis of the streamlines is changed and an elongated cell is observed with a compressed flow in the core region. This is due to the impeding effect of the Lorentz force. As the strength of the magnetic field is further increased, it can be observed that the fluid movement becomes slow and forms a weak parallel flow in the enclosure. The flow is compressed in the core region of the enclosure. The isotherms are initially stratified and became straight lines with the increase in the magnetic field.

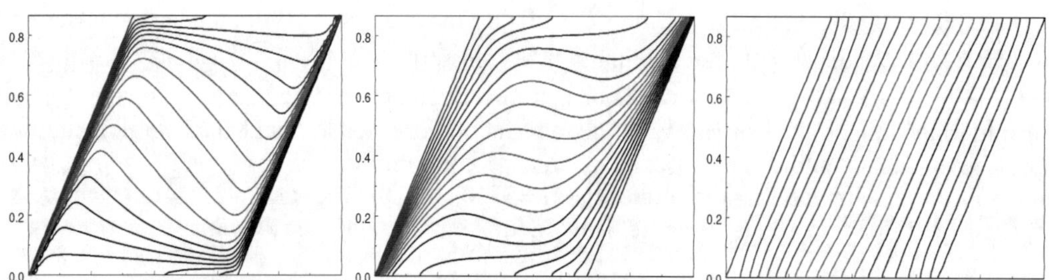

Figure 6. Isotherms and streamlines for different values of Ha at $A = 1, Ra = 3\times10^3, \alpha = 45^0, \phi = 30^0$.

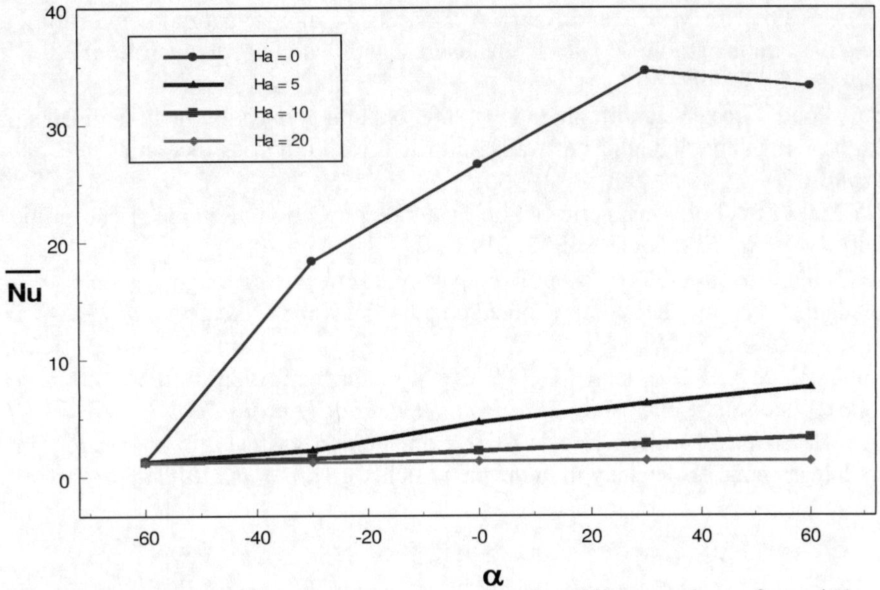

Figure 7. Variation of average Nusselt number for different values of α and Ha.

The variation of average Nusselt number for different values of Ha and α is shown in Fig.7. In the absence of magnetic field the average Nusselt number increases with an increase in α from -60^0 to 60^0. The maximum heat transfer occurs in this case is at α = 30^0. With the introduction of magnetic field, the rate of heat transfer reduces. This shows that the convection in the enclosure is suppressed due to the introduction of magnetic field. As *Ha* is increased, the heat transfer rate dropped down abruptly due to the suppression of convective flow caused by the Lorentz force. It has been noted that the effect of magnetic field is to reduce the overall heat transport considerably for all parameters considered in this analysis.

5. Conclusions

Numerical simulations are carried out to understand the influences of tilt angles of cavity and side walls, magnetic field on buoyancy-driven convection in a porous parallelogrammic enclosure. The flow circulation and heat transfer rates are suppressed by the Hartmann number, but are augmented by the Darcy-Rayleigh number, enclosure inclination. Through the systematic analysis, it has been observed that the heat transfer rate can be effectively controlled by the careful choices of tilt angles of the enclosure and sidewalls.

6. Acknowledgments

The authors acknowledge the UGC for giving the financial support and encouragement to take up the project notification number 1505-MRP/14-15/KAMY028/UGC-SWRO. Further the authors RDJ, BMRP and MS are respectively grateful to the Government Science College, Hassan, SIT, Tumkur and Presidency University, Bangalore and to VTU, Belgaum, Karnataka, India for the support and encouragement.

7. References

[1] Baïri A, Zarco-Pernia E and García de María J M 2014 A review on natural convection in enclosures for engineering applications. The particular case of the parallelogrammic diode cavity *Applied Thermal Engineering* **63** 304 – 22.

[2] Baytas A C and Pop I 1999 Free convection in oblique enclosures filled with a porous medium *Int. J. Heat Mass Transfer* **42** 1047 – 57.

[3] Hyun J M and Choi B S 1990 Transient natural convection in a parallelogram-shaped enclosure *Int. J. Heat Fluid Flow* **11** 129 – 34.

[4] Costa V A F, Oliveira M S A and Sousa A C M 2005 Laminar natural convection in a vertical stack of parallelogrammic partial enclosures with variable geometry *Int. J. Heat Mass Transfer* **48** 779 – 92.

[5] Revnic C, Pop I, Grosan T and Ingham D B 2009 Natural convection in an inclined square cavity with heated and cooled adjacent walls and filled with a porous medium *The Open Tran. Ph. Journal* **1** 20 – 9.

[6] Han C Y 2009 Effect of a magnetic field on natural convection of an electrically fluid in a tilted cavity *J. Korean Phy. Society* **55 (5)** 2193 – 99.

[7] Ece M C and Buyuk E 2006 Natural convection flow under magnetic field in an inclined rectangular enclosure heated and cooled on adjacent walls *Fluid Dynamics Research* **38** 564 – 90.

[8] Pirmohammadi M and Ghassemi M 2009 Effect of magnetic field on convection heat transfer inside a tilted square enclosure *Int. Comm. Heat Mass Transfer* **36** 776 – 80.

[9] Saleh H, Basriati S and Hashim I 2010 Computation of natural convection in a porous parallelogrammic enclosure with a magnetic field *Makara Sains*, **14(2)** 188 – 91.

A study of energy consumption in turning process using lubrication of nanoparticles enhanced coconut oil (NECO)

A F Mansor[1,a], M S Zakaria[1,b], A I Azmi[1,c], A N M Khalil[2,d] and N A Musa[1,e]
[1]Faculty of Engineering Technology, Universiti Malaysia Perlis, Kampus Uniciti Alam, Sg. Chuchuh, 02100 Padang Besar, Perlis.
[2]School of Manufacturing Engineering, Universiti Malaysia Perlis, Kampus Pauh Putra, 02600 Arau, Perlis.

[a] fairuzmansor@unimap.edu.my, [b] syamil@unimap.edu.my, [c] azwaniskandar@unimap.edu.my, [d] nabilkhalil@unimap.edu.my, [e]nuratikahmusa46@gmail.com

Abstract. Cutting fluids play very important role in machining application in order to increase tool life, surface finish and reduce energy consumption. Instead of using petrochemical and synthetic based cutting fluids, vegetable oil based lubricants is safety for operators, environmental friendly and become more popular in the industrial applications. This research paper aims to find the advantage of using vegetable oils (coconut oil) with additional of nano particles (CuO) as lubricant to the energy consumption during machining process. The energy was measured for each run from 2 level factorial experimental layout. Obtained results illustrate that lubricant with enhancement of nanoparticles has capability to improve the energy consumption during the machining process.

1. Introduction

Nowadays, sustainability becomes a major consideration aspect during fulfilling the requirements of humans. One of the most important requirement especially during industrial revolutionary is energy. Over the last decades, energy consumption is drastically increased due to the excessive used and improper management of the resource. In manufacturing sectors, energy is crucial especially involving heavy machineries for production activities. The initiative to establish advanced methods for higher quality of products and productivity are common target for continuous improvement activities in the factories. However, it might affect the energy consumption if no proper monitoring for excessive used. Sarhan et. Al claimed that correct application of lubricants has been proven to greatly reduce friction and hence, reduced power consumption [1]. Related with the issue, this study was implemented that focusing on the efficiency of energy consumption during turning process by using Nanoparticles Enhanced Coconut Oil (NECO) as the lubricant. In general, lubricant or cutting fluid is well-known plays an important role in lubricating and removing heat, gives cooling effect instead of overheating the machine, tool and workpiece during cutting process and removed chips from the cutting zone of flushing effect [2]. NECO lubricant is seen can enhance the efficiency of energy due to the significant impact on the machinability. The previous experimental researches also indicated that lubricant with nano particles can increase the machinability rather than using general lubrication. Moreover, as a way of addressing this issue towards the environmentally friendly manufacturing, minimal quantity lubrication (MQL) is introduced due to its contribution of benefits for health and environment. MQL is based on the residue-free cutting oil usage principle and low flow of cutting fluid as it is mixed with compressed air and it is been used in high pressure. Usually, a small quantity of lubricant is used for less than 100mL/h to reduce the friction at the cutting area [3].

This development of sustainability technology can overcome negative impact of abusing the lubrications or cutting fluids especially when applying mineral oils that can cause environmental pollution, water pollution, soil contamination, and to machining operators. While machine operators are exposed to cutting fluids, they are in high-risk to the diseases of lung cancer, respiratory problem, dermatological and genetic diseases. The initiative to replace the mineral oil based with vegetable oils is important and good practice during the production. Thus, in this research, vegetable oil-based was employed and enhanced with the mixing of nanoparticles in order to achieve higher machinability and better efficiency of energy consumption.

2. Methodology

The method of the experiment was carried out by turning operation on carbon steel S45C, diameter 50mm using CNMG120408 HSS tool insert. Two levels of full factorial experiment with three replications of centre points runs was selected to design the experimental layout with total number of 11 runs, see Table 1 for the range of the parameters. Three numeric parameters were used in the experiment; cutting speed (Vc), feed rate (f), and proportion of nanoparticles to analyse the specific cutting energy. The nanoparticles was copper oxide (CuO) with average size of 100nm.The results were analyzed statistically using Design Expert 7.0.0. The data of the energy consumption was obtained using Picolog current data logger, see Figure 1. CNC turning machine was employed to carry on the experimental using MQL system, brand UNiST.

Table 1: Range of the parameters

Parameters	Low level (-)	Centre level (0)	High level (+)
Cutting speed (m/min)	100	130	160
Feed rate (mm/rev)	0.20	0.25	0.30
Proportion of CuO (wt%)	0.00	0.25	0.50

Figure 1: Picolog current data logger

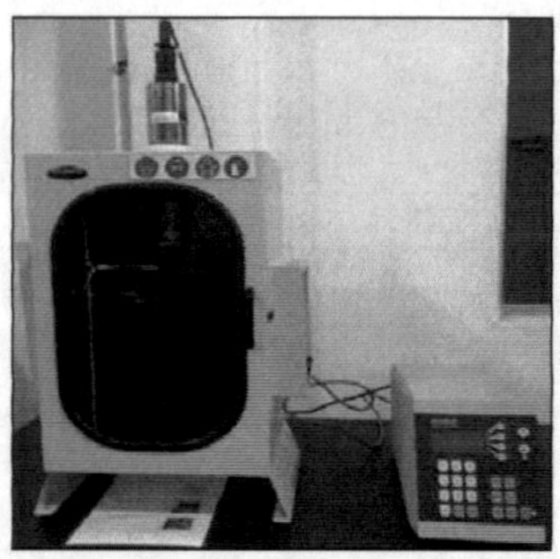

Figure 2: Ultrasonic mixer

The lubrications were prepared based on 3 different of weight proportions of nanoparticles in 150ml of pure coconut oils, which are 0 wt%, 0.25 wt% and 0.5 wt%. The particles were mixed in coconut oils using ultrasonic mixer for 1 hour, see Figure 2. Then, the nanolubricant was poured into MQL system's beaker, attached on machine and located the flexible nozzle on the cutting tips.

During the experimental runs, average currents consumption data were determined from the picolog graphs. Then, the average data were converted into cutting energy. The following equations were used to obtain the required cutting energy for each run, see equation 1 and 2.

$$P = \sqrt{3}\, x\, I\, x\, V \quad (1)$$

$$E = P\, x\, t \quad (2)$$

3. Experimental results

The energy data that collected from the experimental work was shown in Table 2. Results show that the highest energy consumption that recorded was 156 kJ. It can be obtained from the lowest setting of cutting speed and feed rate, and without nanoparticles in the lubricant. Meanwhile, the lowest energy consumption value was 56 kJ when setting the lowest cutting speed, highest feed rate and 0.50 wt % of nanoparticles in the lubricant. The present of nanoparticles in this experimental work can impact the energy consumption of the process. Using Analysis of Variance (ANOVA), the characteristic of main effects and interactions were determined as shown in Table 3. All main effects with interaction between cutting speed and feed rate, and interaction between cutting speed and percentage of nanoparticles significantly influenced the energy consumption. The highest contribution factor was feed rate and proportion of nano particles only contributed 1.98%. However, the nanoparticles impact 12.69% while interacted with cutting speed. The reason was the nanoparticles in the lubricant can reduce wear and friction between cutting tool and the workpiece surface [4]. Thus, the cutting energy was also reduced because the less loading capacity on the spindle during the cutting process.

Figure 3 indicated the behaviour of the energy consumption against the different ranges of parameters setting. The parameters show negative effects for the energy when increasing the setting values. It shows that the cutting energy was getting better by increasing the setting values of cutting speed and feed rate, see Figure 3 (a) and (b). It happened because the instability of cutting occurred when too slow rotation of workpiece and movement of cutting tool. The instability cutting can increase the spindle's load that consumes larger energy in order to keep the performance of the cutting. Meanwhile, proportion of nanoparticles also exhibited the same trend, see Figure 3 (c). The energy can be reduced by adding greater amount of nanoparticles. The reason is the nanoparticles serve as spacer between cutting tool and workpiece [5]. They tend to roll while having movement between the surfaces. In general, lubricant itself has capability to reduce the friction between cutting tool and workpiece. However, nanoparticles can reduce the friction two times greater than pure oil that can significantly improve the lubrication performance [6].

Cutting speed exhibited different pattern when interacted with both of feed fate and %wt of nanoparticles in lubricant as shown in Figure 4 (a). The energy consumption graph tends to go down although at low range of feed rate while increasing the cutting speed. This condition means that the lowest setting of feed rate which giving the highest reading by itself can be compromised with the cutting speed in order to reduce the energy consumption. Meanwhile, %wt of nanoparticles also indicated the same situation when interacted with cutting speed, see Figure 4 (b). However, there was a stagnant effect when applying larger amount of nanoparticles in the lubricant. Too much amount of nanoparticles at an area of contact surface can disturb the movement of the grains and avoiding the rolling effect, hence no more advantage to have the particles in the lubricant. Only the optimized proportion of nanoparticles can be used to significantly decreasing the energy consumption when increasing the speed of the cutting.

Table 2: Experimental layout with data collection

Standard order	Cutting speed	Feed rate	wt% of CuO	Energy (J)
1	-	-	-	156,036
2	+	-	-	80,704
3	-	+	-	81,566
4	+	+	-	57,974
5	-	-	+	117,225
6	+	-	+	91,145
7	-	+	+	56,471
8	+	+	+	75,237
9	0	0	0	76,347
10	0	0	0	70,376
11	0	0	0	71,450

Table 3: Result of ANOVA (Reduced model)

Source	Sum of Square	df	Mean Square	F-Value	P-value	Percentage contribution
Model	7.568E+009	5	1.514E+009	76.71	0.0005	
A-CS	*1.411E+009*	1	*1.411E+009*	*71.50*	*0.0011*	*17.07*
B-FR	*3.778E+009*	1	*3.778E+009*	*191.49*	*0.0002*	*45.72*
C-CuO	*1.638E+008*	1	*1.638E+008*	*8.30*	*0.0450*	*1.98*
AB	*1.166E+009*	1	*1.166E+009*	*59.10*	*0.0015*	*14.11*
AC	*1.049E+009*	1	*1.049E+009*	*53.16*	*0.0019*	*12.69*
Curvature	6.173E+008	1	6.173E+008	31.28	0.0050	
Residual	7.893E+007	4	1.973E+007			
Lack of Fit	*5.867E+007*	2	*2.933E+007*	2.90	0.2567	
Pure Error	*2.026E+007*	2	*1.013E+007*			
Cor Total	8.265E+009	10				

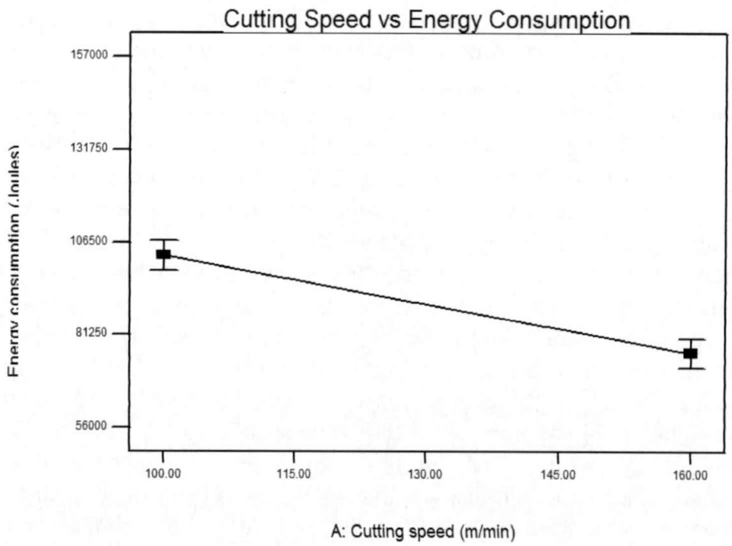

Figure 3 (a): Cutting Speed vs. Energy Consumption

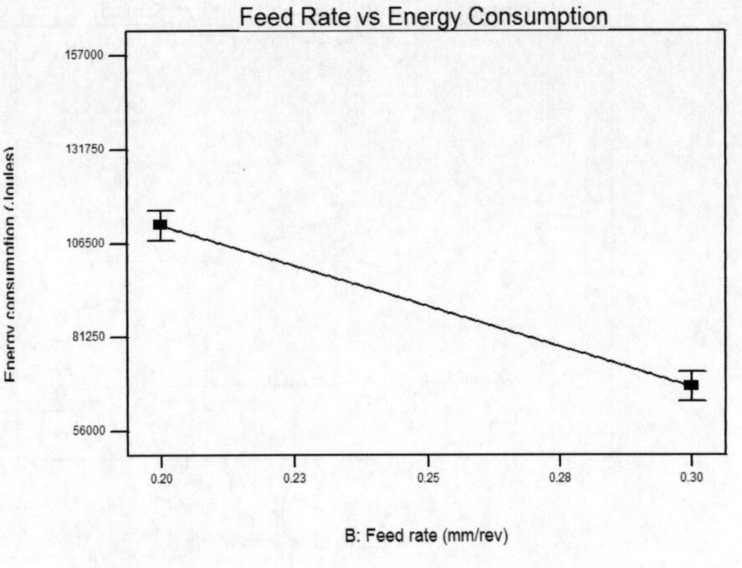

Figure 3 (b): Feed Rate vs. Energy Consumption

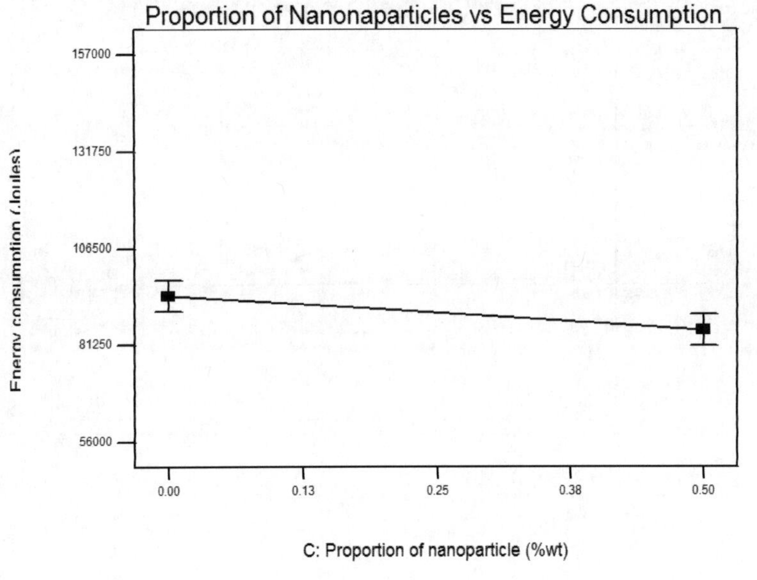

Figure 3 (c): Cutting Speed vs. Energy Consumption

Figure 3: Main effects of the model, (a) A: Cutting speed, (b) B: Feed rate and (c) C: Proportion of Nanoparticles.

IOP Conf. Series: Journal of Physics: Conf. Series **908** (2017) 012077 doi:10.1088/1742-6596/908/1/012077

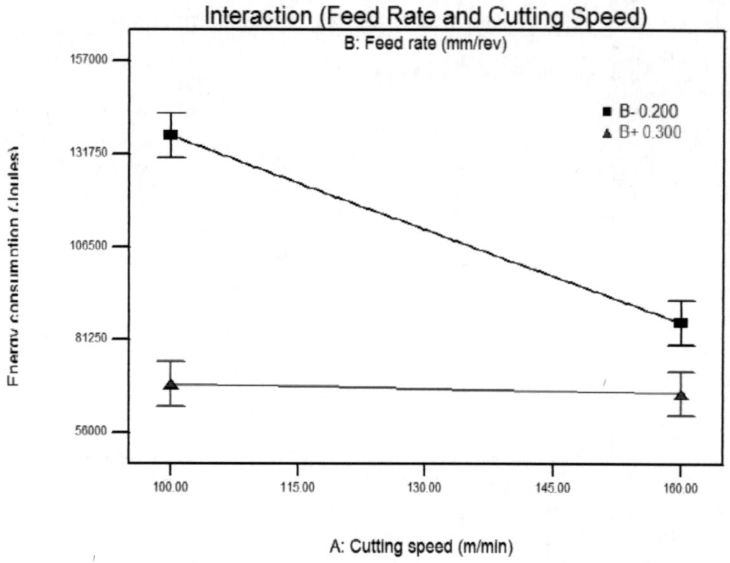

Figure 4(a): Interaction of Feed Rate and Cutting Speed

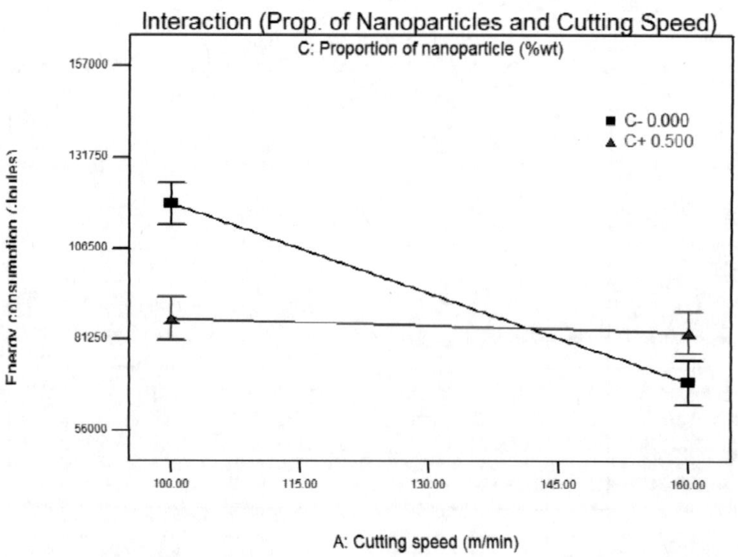

Figure 4(b): Interaction of Proportion of Nanoparticles and Cutting Speed

Figure 4: Significant interactions of the model (a) Interaction of Feed Rate and Cutting Speed, (b) Interaction of Proportion of Nanoparticles and Cutting Speed

IOP Conf. Series: Journal of Physics: Conf. Series **908** (2017) 012077 doi:10.1088/1742-6596/908/1/012077

4. Conclusion

In conclusion, it was found that cutting speed, feed rate and proportion of nanoparticles (CuO) significantly affect the energy consumption. The interaction between cutting speed and proportion of nanoparticles also indicated the same impact. The nanoparticles are able to enhance the lubrication of cutting fluid between the cutting tool and workpiece surfaces by reducing the friction. This condition can decrease the cutting load and able to make an efficient process in term of energy consumption. Overall, NECO lubricant can be used significantly in machining process for sustainability and better environmentally.

5. References

[1] Sarhan, A.A.D., .Sayuti, M., Hamdi, M., 2012. Reduction of power and lubricant oil consumption in milling process using a new SiO2 nanolubrication system. Int. J. Adv. Manuf. Technol, vol. 63, pp. 505-512.

[2] G.E Totten, 2006, Handbook of Lubrication and Tribology volume 1: Application and Maintenance, Second edi. Boca Raton, FL: Taylor & Francis Group.

[3] L. R. Silva, E. C. S. Corrêa, J. R. Brandão, and R. F. De Ávila, 2013, Environmentally friendly manufacturing: Behavior analysis of minimum quantity of lubricant - MQL in grinding process, *J. Clean. Prod. https://doi.org/10.1016/j.jclepro.2013.01.033*.

[4] J. Padgurskas, R. Rukuiza, I. Prosyčevas, and R. Kreivaitis, 2013, Tribological properties of lubricant additives of Fe, Cu and Co nanoparticles, *Tribol. Int.*, vol. 60, pp. 224–232.

[5] Mohd Sayuti, Ooi Ming Erh, Ahmed E.D. Sarhan, Mohd Hamdi, 2014, Investigation on the morphology of the machined surface in end milling of aerospace AL6061-T6 for novel uses of SiO2 nanolubrication system, *J. Clean. Prod.*, vol. 66, pp. 655-663.

[6] C. Nagelreiter, H. Kotisch, T. Heuser, and C. Valenta, 2015, Size analysis of nanoparticles extracted from W / O emulsions, vol. 488, pp. 29–32.

Effect of various bending angles on a passive light pipe for eco-daylighting systems

Y F Ng[1], M S M Hashim[1,2], M A Z Din[1], N A M Amin[1], N S Kamarrudin[1,2], M H Basha[1] and M Tasyrif[1]

[1] School of Mechatronic Engineering, Universiti Malaysia Perlis, 02600 Arau, Perlis, Malaysia.
[2] Motorsport Technology Research Unit (MoTech), Universiti Malaysia Perlis, 02600 Arau, Perlis, Malaysia.

sanihashim@unimap.edu.my

Abstract. Daylighting systems is one of alternative to reduce a high energy consumption that caused by artificial lighting. However the use of passive light pipes in daylighting systems with various bending angles may affect the efficiency of light transfer from the sunlight to the room. Thus this paper is proposed to study the effect of various bending angles on a light pipe. Three bending angles of light pipe, which are 0°, 30° and 45°, were analyzed through the experimental works. A test bed room was constructed in order to simulate daylighting in a room. The results were then visualized in graphs based on the efficiency by considering the maximum average internal illuminance achieved by each light pipe. From the results, it shows that when the bending angle increases, the average internal illuminance decreases. And the highest average internal illuminance was achieved by 0° light pipe.

Keywords: daylighting, light pipe, bending angles, external illuminance, internal illuminance.

1. Introduction

Energy consumption is increasing rapidly in Malaysia from year to year[1]. Artificial lighting is one of the main factor of this high energy consumption which leading to high electricity as well. Building constitutes approximately 30-45% of the global energy demand with statistic result shows that 25-35% is constituted by lighting in office building[2]. Daylighting, the use of natural light for illumination of basic daily activities is used as an sustainable approach in alternative lighting system to reduce the use of artificial lighting effectively and efficiently[3]. Daylight is known as the primary natural light source which is applicable during daytime for indoor illuminance purpose as well as a visual connection between indoor and outdoor[4, 5].

Light pipe which is also known as light tube has becoming well-known result in an increase amount usage of light pipe system[3]. It is commonly used in daylighting technology which is made up by collector, light pipe and diffuser. A collector is used to collect daylight from outdoor which will be reflected in the light pipe consisting of aluminum with reflectance of about 95-99% to be diffused through the diffuser as indoor illuminance for basic daily indoor usage[5].

The use of light pipe has several advantages over the use of windows although both are using daylight which is renewable for illuminance purpose. For typical office workspace, the value of illuminance is in the range of 250 – 500lux [6]. However, the illuminance level will decrease when work plane is far away from window [5]. Light pipe will be able to solve this problem as well as applied to deep plan building or room without window. Therefore the purpose of this study is to observe and analyze the effect of bending angles of light pipe to the value of internal illuminance. The performance of daylight system is measured through the maximum value of internal illuminance.

IOP Conf. Series: Journal of Physics: Conf. Series **908** (2017) 012078 doi:10.1088/1742-6596/908/1/012078

2. Daylight Observation for Malaysia's Climate

Daylight observation on climate was conducted at Campus Pauh Putra, Unversiti Malaysia Perlis (UniMAP), Perlis, Malaysia for four months from September 2016 until December 2016. The data were collected at hourly interval from 8.00 am to 5.00 pm for external illuminance, altitude and azimuth to simulate the working office hours. External illuminance was collected by using light meter. A compass bearing board was constructed for the measurement of azimuth and altitude as shown in **エラー! 参照元が見つかりません。**. Azimuth is measured from the north direction along the horizon while altitude is calculated by using Pythagorean Theorem equation as shown in Equation 1 which is the inverse tan of length of shadow formed by object over the original length of object.

$$tan\,\alpha = \frac{l_o}{l_x} \qquad (1)$$

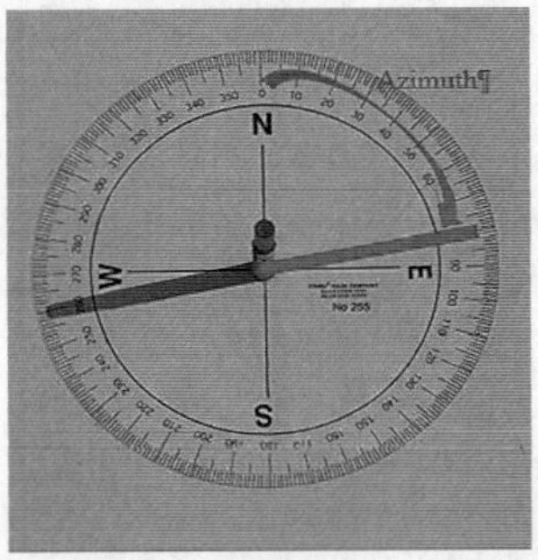

Figure 1: A compass bearing board

Figure and Figure 3 show the average external illuminance obtained from 8.00 am to 5.00 pm and altitude versus azimuth for four months, respectively. The highest average external illuminance of 94,100lux was achieved at 12.00 pm in October and lowest average external illuminance is 14,606lux 5.00 pm in November. This high average external illuminance has exceeded the required indoor illuminance which represent that Malaysia has a potential for using light pipe in daylighting system.

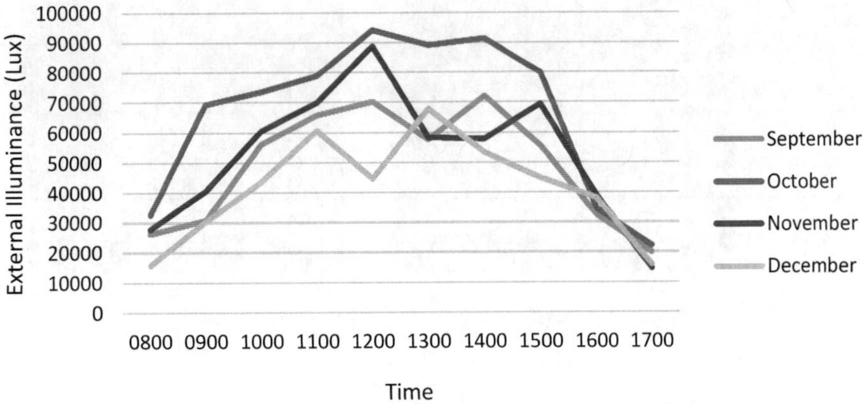

Figure 2: Average external illuminance versus time

IOP Conf. Series: Journal of Physics: Conf. Series **908** (2017) 012078 doi:10.1088/1742-6596/908/1/012078

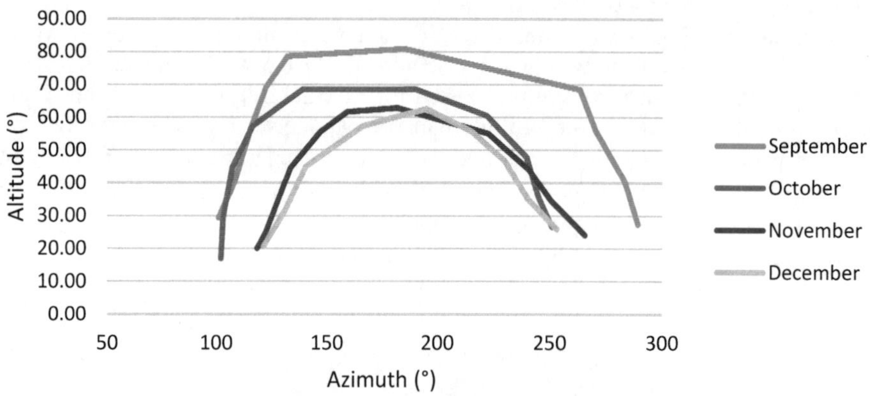

Figure 3: Altitude versus azimuth

3. Experimental setup

A test bed room with dimension of 1.8m (L) × 0.9m (W) × 2.4m (H) was constructed as in Figure 4. It was made of plywood. A door hinge was attached at one side to enable opening of the house for placement of light meter with data logger as well as for the installation of the light pipe during experimental works.

Figure 4: Test bed room for experimental works

Monodraught Sunpipe with bending angles of 0°, 30° and 45° were installed to the test bed room for each of the experimental work as shown in エラー! 参照元が見つかりません。. The experiments were conducted for five days from 8.00 am to 5.00 pm. Both the external and internal illuminance was recorded at hourly interval.

(a) (b) (c)

Figure 5: Actual light pipe with bending angles (a) 0°, (b) 30° and (c) 45°

The efficiency of the light pipe is calculated in percentage by using the maximum average internal illuminance over external illuminance as shown in Equation 2.

$$efficiency = \frac{Maximum\ average\ internal\ illuminance}{External\ Illuminance} \times 100\% \quad (2)$$

4. Results and Discussions

The experimental data was collected for five days from 8.00 am to 5.00 pm for each of the bending angle configuration. The result was plotted by using the average external illuminance and average internal illuminance. Figure 6 - 8 show the average external illuminance and average internal illuminance for each light pipe with bending angles of 0°, 30° and 45°, respectively. The internal illuminance have the similar trend for all light pipes which is low in early morning and late afternoon while higher in the midday. This signifies that the maximum internal illuminance can be achieved when the sun angles is aligned with the angle of light pipe.

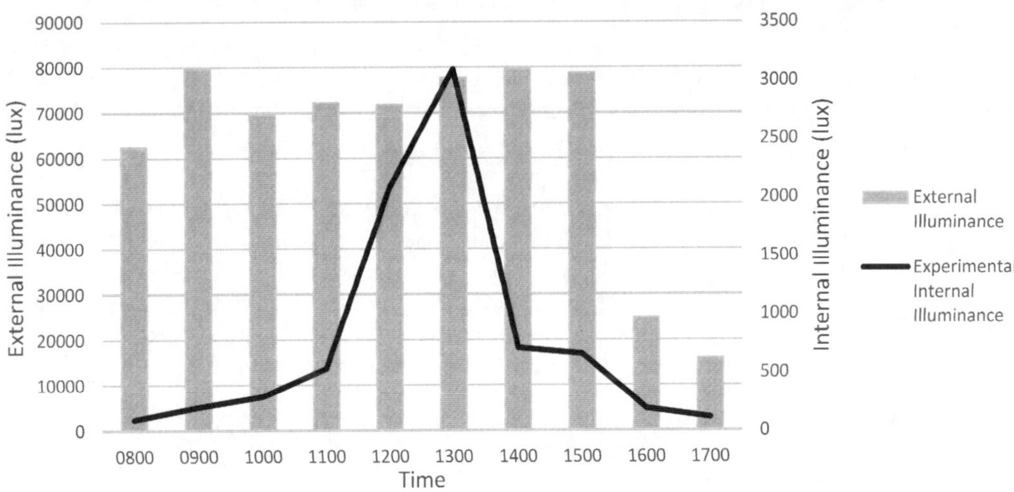

Figure 6: Average external illuminance and average internal illuminance over time for 0° light pipe

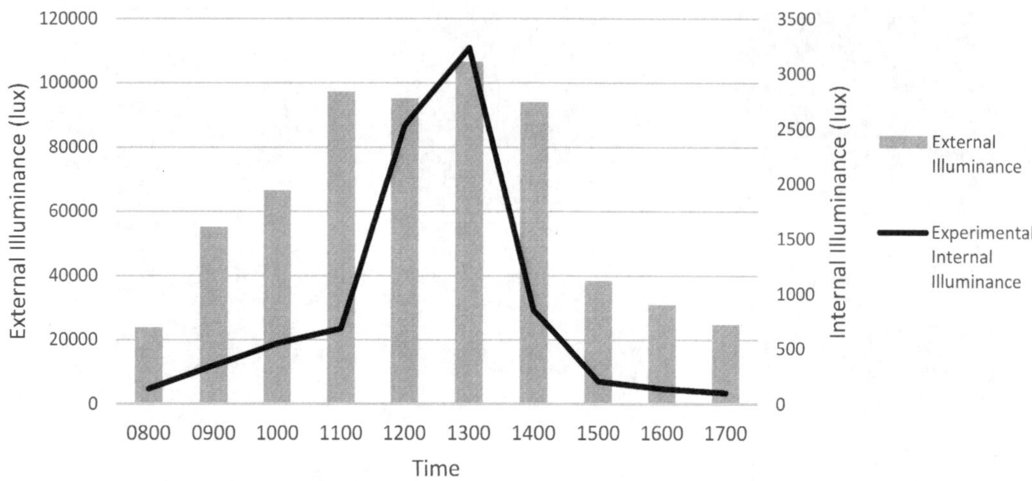

Figure 7: Average external illuminance and average internal illuminance over time for 30° light pipe

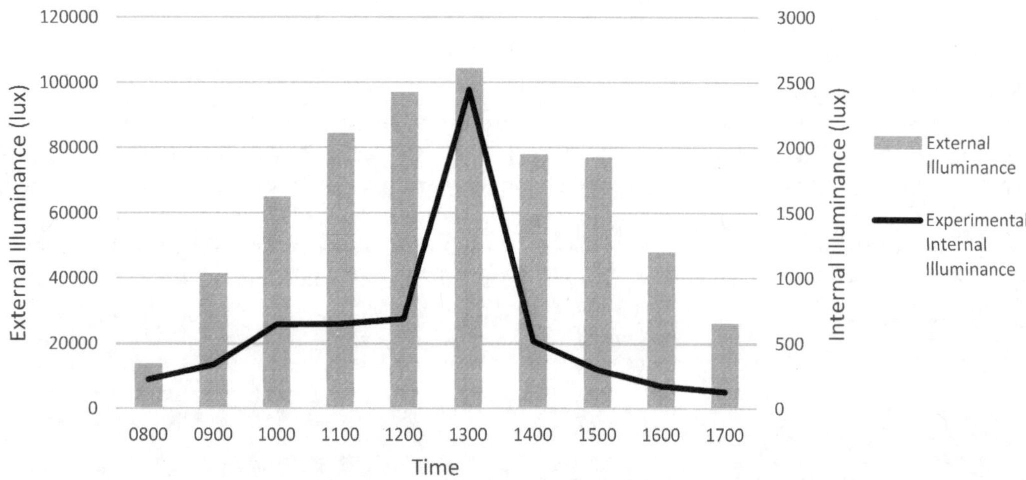

Figure 8: Average external illuminance and average internal illuminance over time for 45° light pipe

From the results, the average optimum lux value for all bending angles of light pipe was achieved at 1.00 pm. This average optimum internal illuminance was used to calculate the efficiency of each light pipe between both experimental and simulation works. The light pipe with 0° bending angle which is also known as the straight light pipe has the highest efficiency at 4.0%. It is followed by bending angle of 30° and 45° with efficiency of 3.0% and 2.4%, respectively. This shows that 0° bending angle light pipe have the least loss of daylight during the transmission from outdoor to indoor.

IOP Conf. Series: Journal of Physics: Conf. Series **908** (2017) 012078 doi:10.1088/1742-6596/908/1/012078

5. Conclusion

This study shows that the external illuminance obtained in Perlis, Malaysia is feasible for the application of an eco-daylighting system. Most of internal illuminance obtained for each hour exceeds the 250lux which is the minimum value for a typical office workspace. It is also observed that the bending angle will affect the internal illuminance. The value of internal illuminance will decrease with the increment of bending angle. The best performance in term of efficiency is achieved by light pipe with 0° bending angle which is known as straight light pipe with maximum average internal illuminance around 3000 lux and efficiency of 4.0%. This is due to the shape itself does not have any bending which result in more successful equivalent reflection.

6. Acknowledgement

This research was funded by Ministry of Higher Education Malaysia (MOHE) through Research Acculturation Grant Scheme (RAGS) no. 9018-00075. In addition, the authors would like to thank Universiti Malaysia Perlis for providing platform to carry out this research.

References

[1] Malaysia Energy Statistics Handbook 2016. Suruhanjaya Tenaga, 2016.

[2] N. Ekren, "Energy saving in lighting system with fuzzy logic controller which uses light-pipe and dimmable ballast," vol. 61, pp. 172–176, 2013.

[3] D. M. Kennedy and F. O. Rourke, "Experimental analysis of a scaled , multi-aperture , light-pipe , daylighting system," *Sol. Energy*, vol. 122, pp. 181–190, 2015.

[4] İ. K and G. Gökmen, "Increase Energy Savings with Lighting Automation Using Light Pipes and Power LEDs," vol. 7, no. 7, pp. 896–903, 2013.

[5] A. A. Kadir, L. H. Ismail, N. Kasim, and M. Kaamin, "Potential Of Light Pipes System In Malaysian Climate," *IOP Conf. Ser. Mater. Sci. Eng.*, vol. 160, p. 12071, 2016.

[6] Recommended Light Levels (https://www.noao.edu/education/QLTkit/ACTIVITY_Documents/Safety/LightLevels_outdoor+indoor.pdf)

IOP Conf. Series: Journal of Physics: Conf. Series **908** (2017) 012079 doi:10.1088/1742-6596/908/1/012079

Connected car: Engines diagnostic via Internet of Things (IoT)

A F A Hamid, M T A Rahman, S F Khan, A H Adom, M A Rahim, N A Rahim, M H N Ismail and A Norizan

School of Mechatronic Engineering, Universiti Malaysia Perlis, Pauh Putra Campus, 02600 Arau, Perlis, Malaysia.

E-mail: azarul.fahmin@gmail.com

Abstract. This paper is about an experiment for performing engines diagnostic using wireless sensing Internet of Thing (IoT). The study is to overcome problem of current standard On Board Diagnosis (OBD-II) data acquisition method that only can be perform in offline or wired method. From this paper it show a method to determined how the data from engines can be collected, make the data can be easily understand by human and sending data over the wireless internet connection via platform of IOT. This study is separate into three stages that is CAN-bus data collection, CAN data conversion and send data to cloud storage. Every stage is experimented with a two different method and consist five data parameter that is Revolution per Minute (RPM), Manifold Air Pressure (MAP), load-fuel, barometric pressure and engine temperature. The experiment use Arduino Uno as microcontroller, CAN-bus converter and ESP8266 wifi board as transfer medium for data to internet.

1. Introduction

Engine diagnostic with Internet of Things (IoT) is a system to overcome the shortcomings of the conventional engine diagnostic systems. Specifically, it is a system that sends data over the On Board Diagnosis (OBD-II) connector using a wireless system that connects to the Internet. The device used for accessing and transmitting the data is simple, low-cost, and easy-to-install. Stakeholders worldwide are struggling to finalize standards to allow specialized wireless on- board units (OBUs) to directly interact with the control area network (CAN-bus) and to communicate with other OBUs and with roadside units, under any propagation condition, intermittent connectivity, and traffic density [1]. There are many wireless network platform that can be used such as Big Data collector through mobile ad hoc network (MANET), wireless sensor networks (WSN) or cyber physical network (CPS) but the better network type is CPS that is the IoT type of network [2]. As for this project, the use of a wireless input and output terminal capable of accessing the Internet to transmit and receive any data from the Electric Control Unit (ECU) and any data from sensors that are attached to the car. Data transmitted from the racing car also can be analysed for purposes unrelated to engine or car component problem. For example, the data can be collected and analysed in real-time to analyst driving patterns, racing car part reliability, and emission characteristics. By using this IOT system, it able to overcome the problem of condition car monitoring. The idea in using IOT system is to perform a wireless dyno test. By using this application, it can perform dyno test without being on the chassis dyno machine. These augmented data can be processed by remote control centres for multiple purposes, such as to inform fleets of vehicles cooperatively driving [3], to detect possible in-vehicle malfunctions, to collect traffic statistics, to get car maintenance tips and service in- formation. This paper presents a proof-of-concept for probe vehicle monitoring and for the exploitation of the collected data. Three system platform are considered: CAN-bus data collection, CAN-bus data conversion and sending raw data from converted CAN-bus data to cloud for storage

2. Methodology

As for this project, there are three system platform is considered. CAN-bus data collection for the purpose of collecting data from ECU to Arduino with using CAN-bus converter. CAN-bus data conversion for converting byte data that receive by Arduino into decimal raw data that can be processed, calculate and easy to understand by human. Also the part of sending converted CAN-bus data to cloud storage via IoT. Diagram 1 shows the flow chart of experiment to achieve the target.

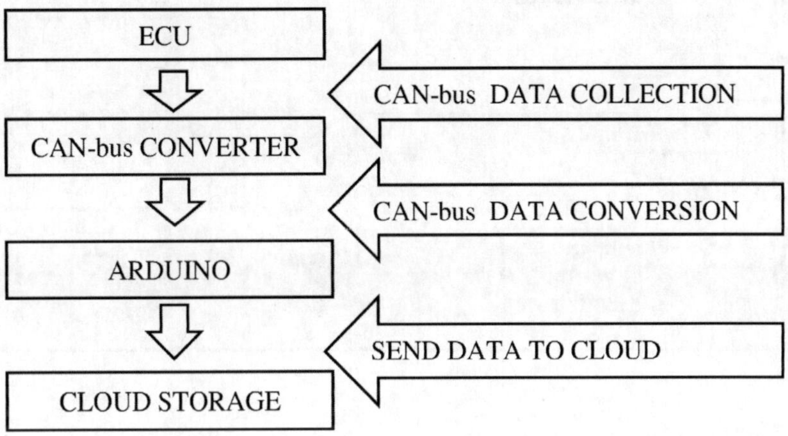

Diagram 1: Block diagram for engines diagnostic via internet of things(IOT).

2.1. CAN-bus data collection.

CAN-bus was first engendered for automotive use, so its most prevalent application is in-conveyance electronic networking. However, as different industries have completed the responsibleness and benefits of will over the past twenty years, they need adopted the bus for a good form of applications. These physical layers relegate bound aspects of the will network, like electrical levels, signaling schemes, cable electric resistance, most baud rate rates, and more. As for automotive, all types of information area unit constantly collected by sensors on board of mundane cars and trucks [4]. These information area unit communicated over a CAN-bus. Depending on the kind of conveyance the data can be read through the OBD plug or by an immediate connection with the CAN-bus itself utilizing a CAN-bus clamp. Several implements were tested for the perpetual monitoring of data on the CAN-bus through the OBD plug. Several communication techniques and set-ups were tested for retrieving the data. A relevant selection CAN-bus data was made and an algorithm was devised in order to collect engine data parameter. During the experiment there two method has been tried. The first method is CAN-bus transceiver TJA1050 as in Figure 1 and Figure 2 that collect data from CAN-bus then send it to microcontroller via serial communication. The figure show block diagram that consist several part that such as part of the board communicate with CAN-bus from ECU via connection of CANH and CANL.

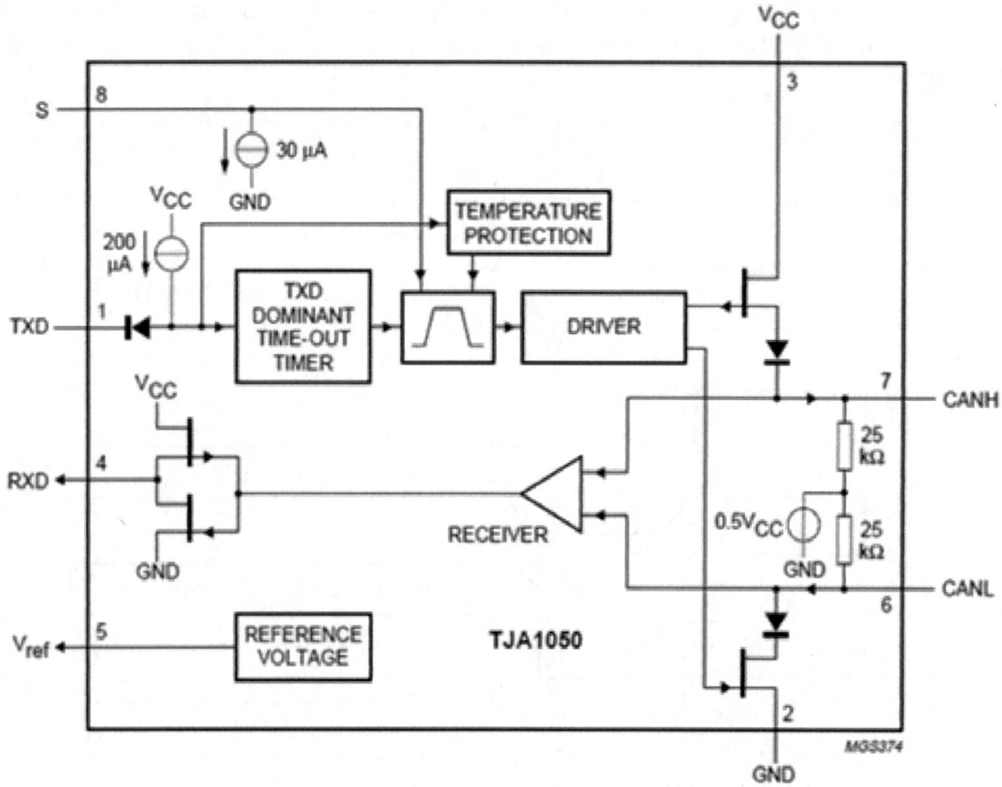

Figure 1: TJA1050 integrated circuit(IC) diagram

Figure 2: TJA1050 board use to collect data from CAN-bus port on ECU.

But the system have a data clocking problem due to the TJA1050 just send data without segregate it by every data packet. Another method that use TJA1050 and MCP2515 IC as in Figure 3 and 4. MCP2515 IC is a serial communication to serial peripheral interface (SPI) bus is tested.

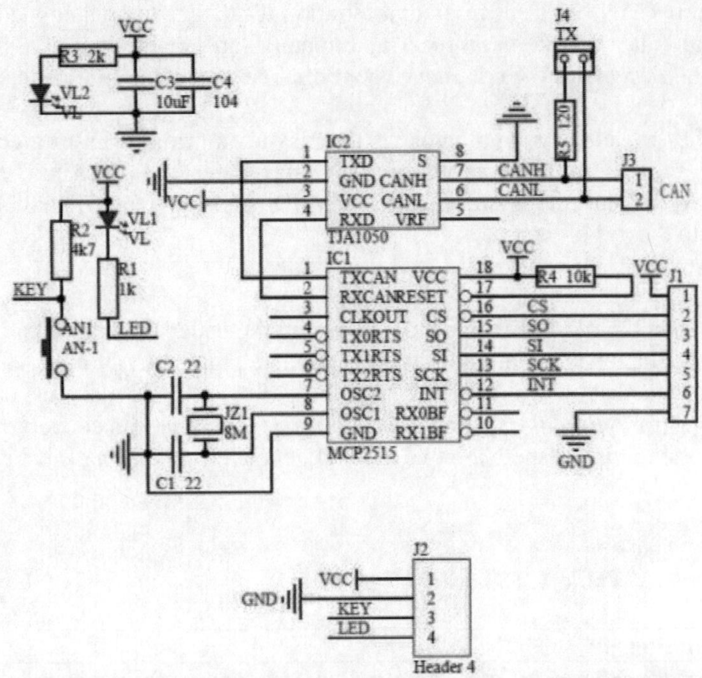

Figure 3: MCP2515 and TJA1050 circuit diagram

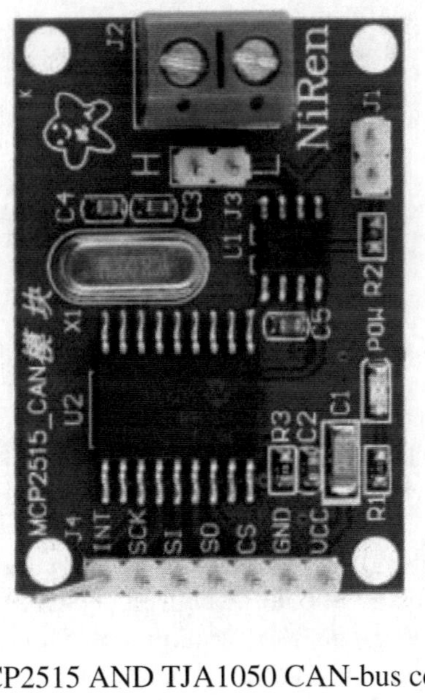

Figure 4: MCP2515 AND TJA1050 CAN-bus converter board.

With this board the system is able to clocking the data packet. Every byte will not messed up among them and it will help the data packet selection and conversion part. CAN-bus contrivances send data across the CAN-bus network in packets called frames [5].

A CAN-bus frame consists of the following sections:-

(a) CAN-bus Frame - an entire CAN-bus transmission: arbitration ID, data bytes, acknowledge bit, and so on or it can be call data packet. Frames additionally are referred to as messages.
(b) SOF (start-of-frame) bit – designates the commencement of a message with an ascendant (logic 0) bit.
(c) Arbitration ID – identifies the message and betokens the message's priority. Frames come in two formats -- standard, which utilizes an 11-bit arbitration ID, and elongated, which utilizes a 29-bit arbitration ID. For this experiment the system get 11-bit ID such as 0x360, 0x3E1 and etc.
(d) Data Field – contains 0 to 8 bytes of data.

2.2. CAN-bus data conversion.

Data acquisition is one amongst the most effective tools to extend the understanding of car behaviour. The network most commonly employed in automotive applications is that the CAN-bus. As mention in CAN-bus data collection the data that being communicate in form of byte [6]. To make the other device understand the data that being transferred the system need a data converter that can convert data in form of byte to decimal [7]. In this experiment there are two method of converting data and five type of data from ECU as in Table 1.

Table 1: Data converting method.

Method	Explanation

Conversion method 1	Test condition based on every byte and program choose the decimal value for every decision.
	If(byte[i] == F){
	Dec[i]=16;
	}
	If (byte[i] == E){
	Dec[i]=15;
	}
	Total=Dec[2]^3+Dec[1]^2

Conversion method 2	String to long integer converter. long int strtol(const char *str, char **endptr, int base);

2.3. Sending raw data to cloud.

There are many sensoring network that has been introduced such as mobile ad-hoc network (MANET) Global System for Mobile Communications (GSM) and wireless detector networks (WSN). A lot of recently, the cyber physical system (CPS) has emerged as a promising direction to counterpoint human-to-human, human- to-object, and object-to-object interactions within the physical world in addition as within the virtual world [2]. CPS with Representational State Transfer (REST) architecture is used as a medium for sending data to cloud. Basically REST is associate art form, associated an approach to communications that's usually utilized in the event of internet services. the utilization of REST is commonly most well-liked over the additional heavyweight SOAP (Simple Object Access Protocol) vogue as a result of REST doesn't leverage the maximum amount information measure, that makes it a far better acceptable use over the net [8]. The SOAP approach needs writing or employing a provided server program (to serve data) and a shopper program (to request data).REST is commonly utilized in mobile applications, social networking websites, mash up tools, and automatic business processes. Flexibility is provided by distribution resources their own distinctive Universal Resource Identifiers (URIs). As a result of every verb encompasses a specific that means (GET, POST, place and DELETE), REST avoids ambiguity. As in Table 2, this experiment done with two method that is HTTPresquest with cloud data.sparkfun,com and Fusion Tables with google cloud storage.

Table 2: Sending data to cloud method.

Method	Explaination
Sending data to cloud method 1	HTTPrequest method that need a publickey and privateKey for channeling the path of network communication and fill it in one url to be access by wifi module ESP8266. http://data.sparkfun.com/input/[publicKey]?private_key=[privateKey]&barometric=[value]&fuel=[value]&map=[value]&rpm=[value]&temperature=[value]
Sending data to cloud method 2	Fusion table in google cloud system api: apiKey=vtyff75656fvgc insertRow=? insertColumn=?

2.4. Hardware setup.

After platform system experiment is done .The final hardware setup as in Figure 5 is made. The ECU will transfer data in CAN-bus communication to the Arduino UNO microcontroller via MCP2515 CAN-bus converter. MCP2515 CAN-bus converter will connect to the Arduino through Serial Peripheral Interface (SPI) and CAN Hi/Low to the ECU Haltech Sprint 500. Data from ECU will be in form data packet that consist CAN-id and another data in Hex value. The Arduino will be need to separate that data packet to obtain the engine raw data. Data that obtain from ECU will be sending to cloud by using ESP8266 wifi module in REST architecture method. ESP8266 is communicating with Arduino by software serial that can make algorithm sending the data by HTTP request script that will fill data.sparkfun.com cloud.

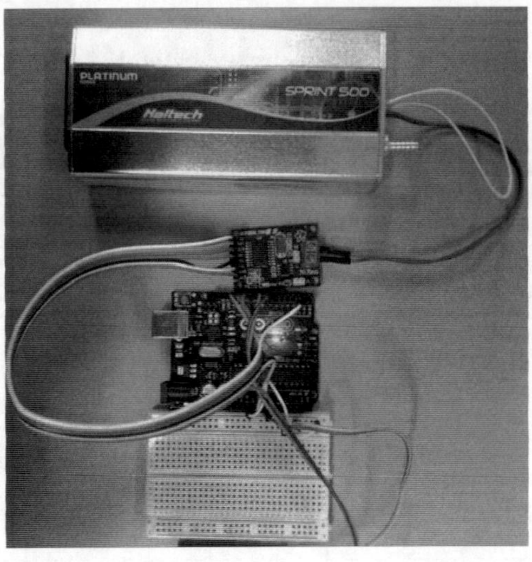

Figure 5: Hardware connection that consist arduino and can converter.

3. Result

There are five type of data that are considered in this experiment that is Revolution per Minute (RPM), Manifold Air Pressure (MAP), Barometric pressure, Load-fuel and engine temperature. The result come with two different method of CAN-bus data collection, CAN-bus data conversion and sending raw data to cloud. From this experiment also there are two user interface(UI) for end user uses that is UI that give by the cloud storage provider and the web interface that consist graph and meter created with using google open chart in the H.T.M.L script. For CAN-bus data collection result is obtain from offline data acquisition when the vehicle is in static condition After the experiment it should get an output like in Figure 6. In the figure it show result of data collected from ECU via CAN-bus.

```
ID: 0x360 DATA: 0x25 0x4E 0x0A 0xEC
ID: 0x361 DATA: 0x12 0x72 0x68 0x41
ID: 0x362 DATA: 0x00 0x01 0xDD 0x45
ID: 0x363 DATA: 0x00 0x00 0x00 0x00
ID: 0x368 DATA: 0x00 0x00 0x00 0x00
ID: 0x369 DATA: 0x24 0x37 0x03 0x10
ID: 0x36A DATA: 0x00 0x00 0x00 0x00
ID: 0x36B DATA: 0x00 0x00 0x00 0x00
ID: 0x36C DATA: 0x00 0x00 0x00 0x00
ID: 0x36D DATA: 0x00 0x00 0x00 0x00
ID: 0x370 DATA: 0x00 0x00 0x00 0x00
ID: 0x371 DATA: 0x00 0x00 0x00 0x00
ID: 0x372 DATA: 0x09 0x59 0x01 0x10
ID: 0x373 DATA: 0x00 0x00 0x00 0x00
ID: 0x374 DATA: 0x00 0x00 0x00 0x00
ID: 0x3E0 DATA: 0x78 0x61 0x28 0x71
```

Figure 6: Data collection from Arduino and CAN converter board.

For CAN-bus data conversion is obtain from offline data acquisition with combining data collection and data conversion algorithm. The experiment got two type of output from different method. In Table 3 the system is tested with method 1 and the data seems unstable due to conversion algorithm is wrong. For Table 4 the data is stable and has same value when compared with the software that provide the ECU manufacturer.

Table 3: Data conversion method 1 result.

Time	RPM	Manifold Air Pressure	Load-Fuel	Barometric Pressure	Temperature
24/01/2017 15:53:10	0	6912	65280	6912	7424
24/01/2017 15:53:12	6912	0	0	0	0
24/01/2017 15:53:14	6912	0	0	0	0
24/01/2017 15:53:19	0	0	0	0	0
24/01/2017 15:53:24	0	0	0	0	0
24/01/2017 15:53:29	0	0	0	0	0
24/01/2017 15:53:35	0	0	0	0	0
24/01/2017 15:53:41	0	0	0	0	0
24/01/2017 15:53:43	0	818176	784128	818176	0
24/01/2017 15:53:48	818176	0	0	0	0
24/01/2017 15:53:53	0	0	257536	0	0
24/01/2017 15:53:58	0	0	0	0	0

Table 4: Data conversion method 2 result.

	RPM	Manifold Air Pressure	Load-Fuel	Barometric Pressure	Temperature
25/01/2017 12:42:57	4096	-50.32	25	13	65
25/01/2017 12:42:57	4128	-54.33	29	13	65
25/01/2017 12:42:58	4223	-56.67	34	14	65
25/01/2017 12:42:58	4096	-50.32	25	13	65
25/01/2017 12:42:58	4096	-50.32	25	13	65
25/01/2017 12:42:58	4096	-50.32	25	13	65
25/01/2017 12:42:59	4096	-50.32	25	13	65

After this platform experiment is done there are too many error in method 1 data converting due to detection of SOF as 0 value at the beginning of data packet and this experiment run smoothly with method 2. In sending raw data to cloud that obtain from quarter mile test the result is analyze to show data transmitting timing difference between two method that tested. Figure 7 and 8 show the transmit data timing diagram for two different method. In figure 7 it use method 1 to transmit the data and the data collecting from ECU is every 2ms then it's transmit data to the cloud in every 12ms. Figure 8 for the method 2 shows that the system capable to collect data form ECU every 2ms and transmit it to the cloud for every 20ms.

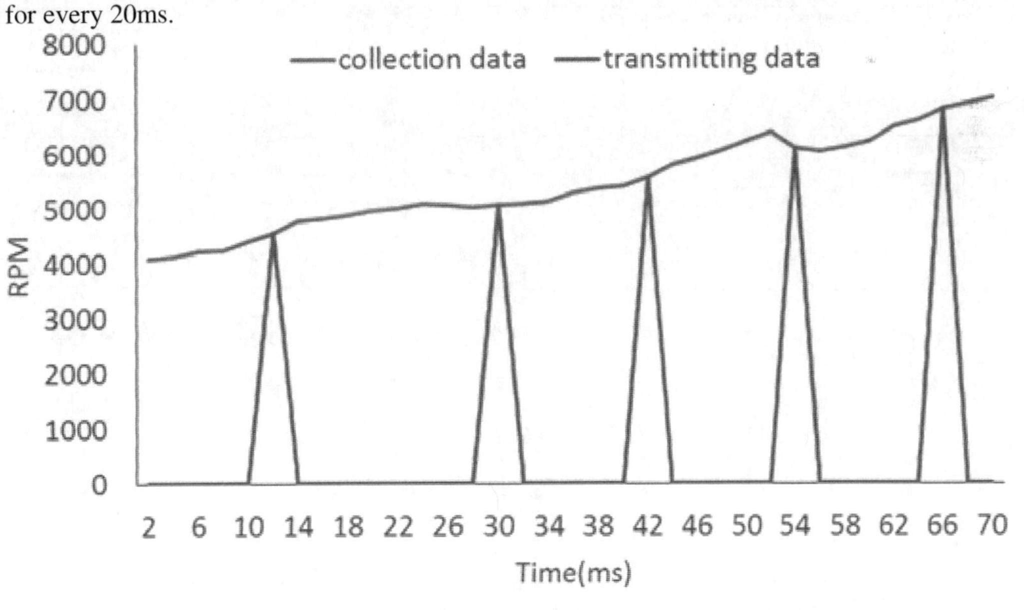

Figure 7: Data transmitting timing diagram method 1

IOP Conf. Series: Journal of Physics: Conf. Series **908** (2017) 012079 doi:10.1088/1742-6596/908/1/012079

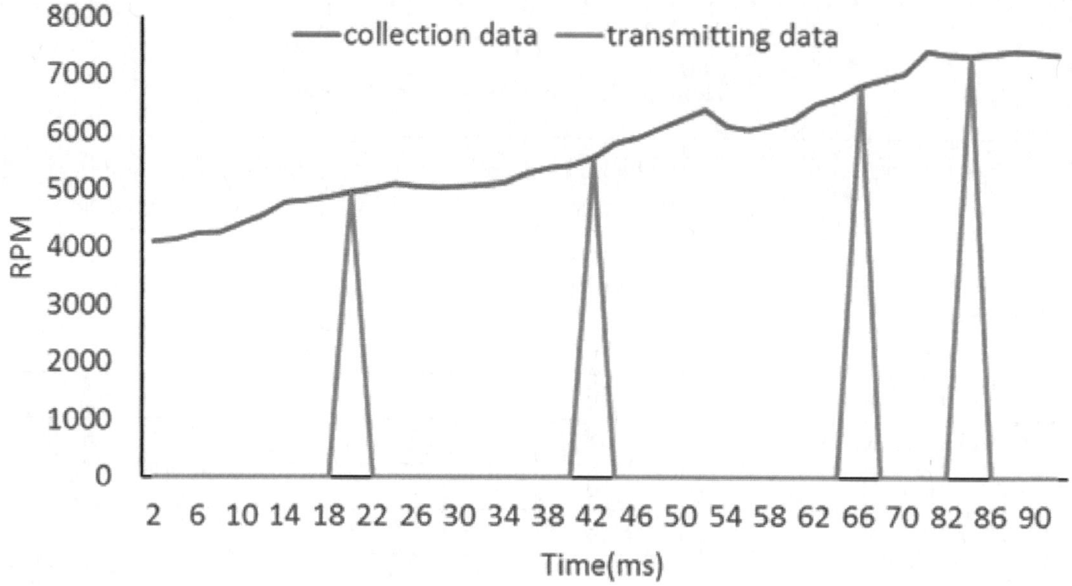

Figure 8: Data transmitting timing dagram method 2

When the experiment is done it show sending data to cloud method 1 is better in case of size of data packet, data packet transferring time and the setup was easier than Fusion Tables in method 2. Faster data packet transferring time will increase processing decision based on condition was set in algorithm [9]. Figure 9 and 10 is a user interface for the cloud and website.

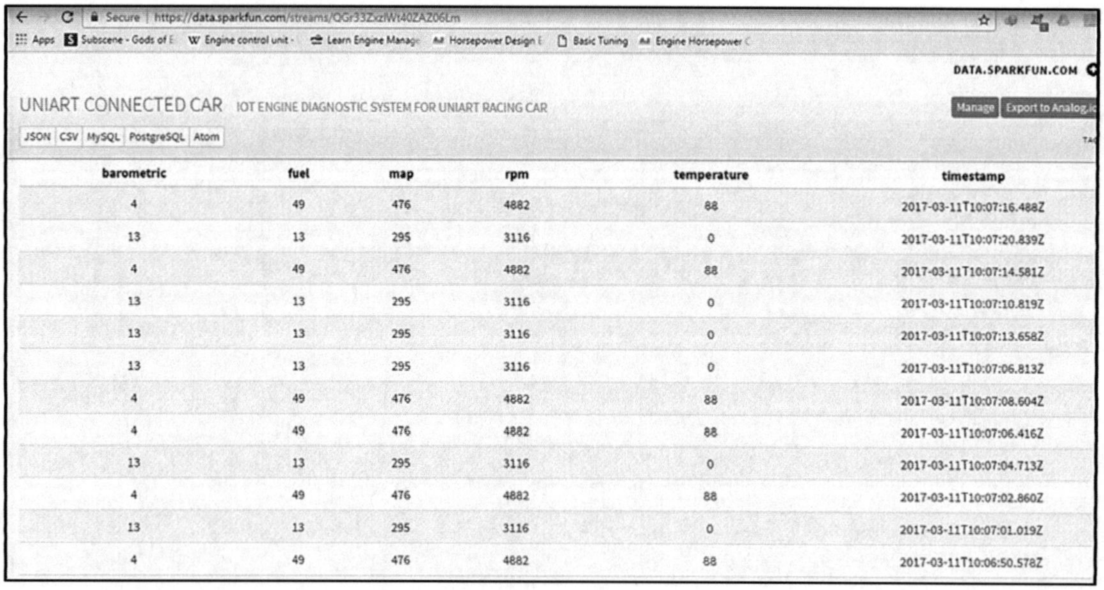

barometric	fuel	map	rpm	temperature	timestamp
4	49	476	4882	88	2017-03-11T10:07:16.488Z
13	13	295	3116	0	2017-03-11T10:07:20.839Z
4	49	476	4882	88	2017-03-11T10:07:14.581Z
13	13	295	3116	0	2017-03-11T10:07:10.819Z
13	13	295	3116	0	2017-03-11T10:07:13.658Z
13	13	295	3116	0	2017-03-11T10:07:06.813Z
4	49	476	4882	88	2017-03-11T10:07:08.604Z
4	49	476	4882	88	2017-03-11T10:07:06.416Z
13	13	295	3116	0	2017-03-11T10:07:04.713Z
4	49	476	4882	88	2017-03-11T10:07:02.860Z
13	13	295	3116	0	2017-03-11T10:07:01.019Z
4	49	476	4882	88	2017-03-11T10:06:50.578Z

Figure 9: UI from data.sparkfun.com.

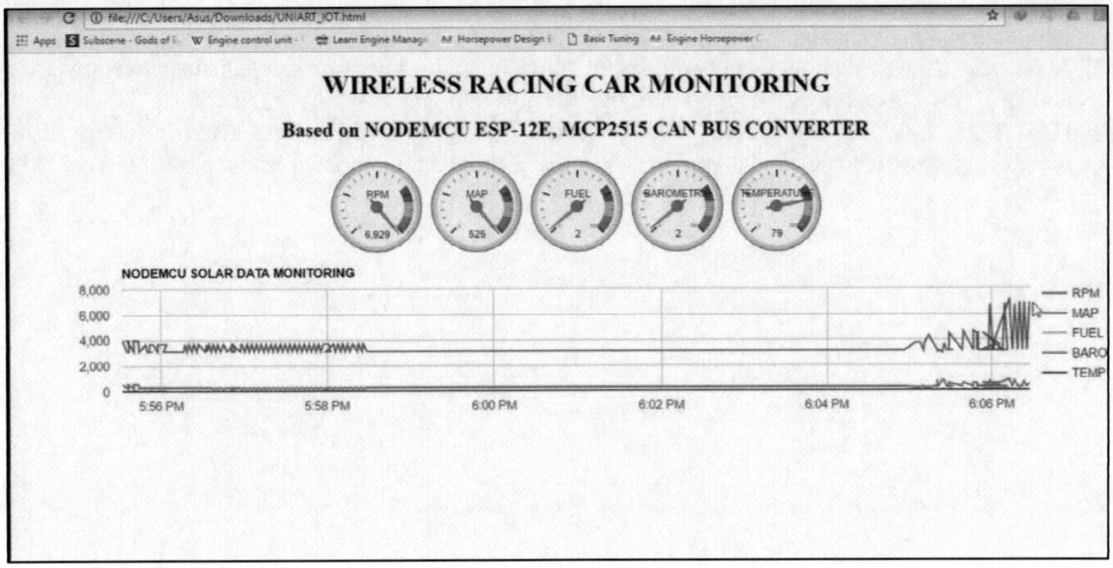

Figure 10: UI from H.T.M.L script.

4. Conclusion

From the experiment result it can be conclude that the best way to collecting data from ECU is by using the MCP2515 TJA1050 CAN-bus converter board due to data clocking for every frame of data packet. Then the wireless layer network is CPS by using REST architecture with HTTP request method and for the improvement the study need to be include the Big Data cloud storage study for better data management and volume of data. Furthermore, security system for CPS will be useful for protection of network layer in crowd sensing.

Acknowledgments

Thanks to UniMAP Automotive Racing Team (UniART) for trusting, help and continue give a funding for develop this system.

References

[1] Van Geem, C. Bellen, M. Bogaerts, B. Beusen, B. Berlmont, B. Denys, T. Hellinckx,. "Sensors on Vehicles (SENSOVO) - Proof-of-concept for Road Surface Distress Detection with Wheel Accelerations and ToF Camera Data Collected by a Fleet of Ordinary Vehicles". Transportation Research Procedia, 14, 2966–2975, 2016.

[2] F.J Wu, Y.F. Kao, Y.C. Tseng. "From wireless sensor networks towards cyber physical systems". Pervasive and Mobile Computing, 7(4), 397–413, 2011.

[3] O. Briante, C. Campolo, A. Iera, A. Molinaro, S.Y. Paratore, & G. Ruggeri. "Supporting augmented floating car data through smartphone-based crowd-sensing". Vehicular Communications, 1(4), 181–196, 2014.

[4] K. Ismail, A. Muharam, M. Pratama. "Design of CAN bus for research applications purpose hybrid electric vehicle using ARM microcontroller". Energy Procedia, 68, 288–296, 2015.

[5] J. Xia, C. Zhang, R. Bai, L. Xue. "Real-time and reliability analysis of time-triggered CAN-bus". Chinese Journal of Aeronautics, 26(1), 171–178, 2013.

[6] X. Hong, W. Jianhua. "An extendable data engine based on OPC specification". Computer Standards and Interfaces, 26(6), 515–525, 2004.

[7] G. Sánchez-Martínez. E. Koutsopoulos, N.H.M. Wilson. "Optimal allocation of vehicles to bus routes using automatically collected data and simulation modelling". Research in Transportation Economics, 59, 1–9, 2016.
[8] A. Navarro, A. Da Silva. "A metamodel-based definition of a conversion mechanism between SOAP and RESTful web services". Computer Standards and Interfaces, 48, 49–70, 2016.
[9] H.L. Yuan, J. Wang, J. Liu, S.L. Li. "Research of Zigbee and big data analysis based pulse monitoring system for efficient physical training". Procedia Computer Science, 80, 2357–2361, 2016.

IOP Conf. Series: Journal of Physics: Conf. Series **908** (2017) 012080 doi:10.1088/1742-6596/908/1/012080

Preliminary study of TEC application in cooling system

A C Sulaiman[1], N A M Amin[1], M S Saidon[2], M S A Majid[1], M T A Rahman[1] and M N F M Kazim[1]

[1]School of Mechatronic Engineering, Universiti Malaysia Perlis, 02600 Arau, Perlis, Malaysia.
[2]Department of Electrical Engineering Technology, Faculty of Engineering Technology, Universiti Malaysia Perlis, Kampus UniCITI Alam, Sungai Chuchuh 02100 Padang Besar, Perlis, Malaysia.

nasrulamri.mohdamin@unimap.edu.my

Abstract. Integration of thermoelectric cooling (TEC) within a space cooling system in the lecturer room is studied. The studied area (air conditioned surrounding) is encapsulated with wall, floor, roof, and glass window. TEC module is placed on the glass window. The prototype of the studied compartment is designed using cabin container. The type and number of TEC module are studied and the effects on the cooling performance are analyzed as it is assumed to be tested within an air conditioned lecturer room. The experimental and mathematical modeling of the cooling system developed. It is expected that the mathematical modeling derived from this study will be used to estimate the use of the number of TEC module to be integrated with air conditioner unit where possible.

1. Introduction

In the last two decades, the world has witnessed severe energy crises in developing countries, especially during summer primarily due to cooling load requirements of buildings. About 40% of the energy has been consumed and this is expected to further increase due to the standards of living improvement and the increase of the world population. The use of air conditioner has increasingly penetrated the market during the last few years and greatly contributed in the upsurge of absolute energy consumption [1-3]. There are two types of system that are used for cooling and heating system in the building namely as passive and active types. Passive cooling involves the use of natural processes for heating or cooling to achieve balanced interior conditions. The flow of energy in passive design is by natural means: radiation, conduction, or convection without using any electrical devices, for example, ambient air is transferring the heat by convection, sky (upper atmosphere) transfers heat by long-wave radiation, water transfers heat by evaporation and groud transfers heat by conduction. An example of the active refrigeration systems classification is shown in figure 1.

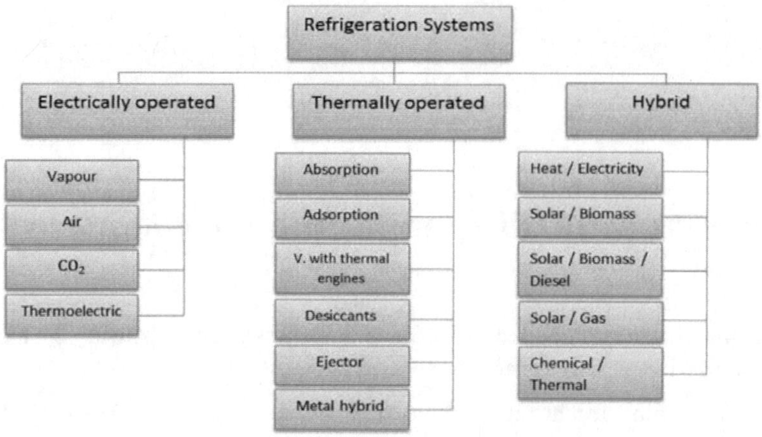

Figure 1. Classification of active refrigeration systems [4].

As for the research project in hand, the thermoelectric module (TEM) as shown in figure 2 is expected to be used for refrigeration, air conditioning system or power generation application. TEM is an alternative device used for cooling and heating via Peltier effect. J.C.A Peltier discovered this in 1834. TEM is a phenomenon of cooling and heating when two junctions of thermoelectric material which are bismuth and Telluride are passed through by current.

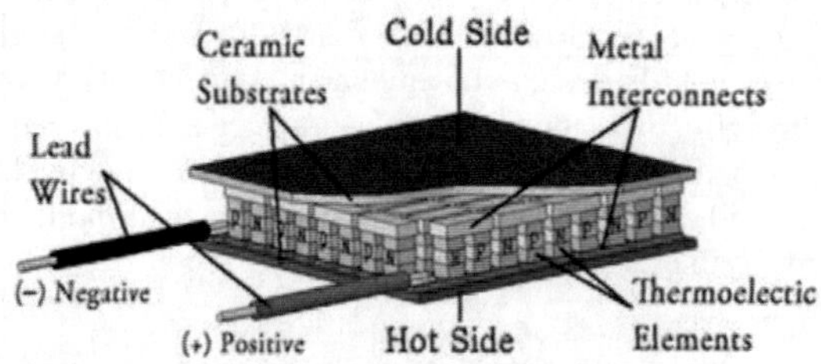

Figure 2. Thermoelectric module (TEM) [5].

During operation, the current flow will causes heat to be transferred from one side to another to create hot and cold sides. If the direction of the current is reversed, the cold and hot sides are changed. Besides, the cooling power for TEM can be varied by adjusting its operating current. On the other hand, thermoelectric cooling (TEC) system can absorbs and dissipates heat depending on the current polarity so that TEM is not required in any moving parts or production compared to conventional cooling system and generator. TEM cooling system also has numerous advantages such as no noise produce due to any vibrational impact, use no coolant, high reliability and environmental-friendly. The single stage cooler of TEC consists of two ceramic plates with p and n-type semiconductor materials (bismuth, telluride) between the ceramic plates. The elements of semiconductor material are connected electrically in series and thermally in parallel. Usually, the thermoelectric couples are electrical in series and thermal in parallel. The module can contain such several hundreds of thermoelectric couples.

Recently, the demand for thermal comfort of the cooling system makes the consumption of energy becomes higher. TEC is an alternative method to reduce electrical consumption for air conditioned room. In addition, the performance of current air conditioning is not very efficient in the long term as it may causes higher energy consumption. Research on TEC module, as in figure 3, has been carried out extensively for some application such as for electronic cooling, medical technology, commercial product and military instrument. In last decade, many researchers have expressed concerns about the potential growth of the thermoelectric system and it may give bigger impact to solve our energy challenge.

Figure 3. Thermoelectric cooling [6].

2. Thermoelectric Cooling (TEC) Application in Building

The research project reported elsewhere has studied the TEC system as shown in figure 4, for 2.8m x 2.7m x 2.7m building space with a variation pulse current ratio, cooling load, pulse width and dissimilar pulse shapes. From that variation, the TEC system provides the average cooling power of 600W with the COP of 1.01 by changing the pulse operation of the thermoelectric cooling system from normal mode operating condition [7].

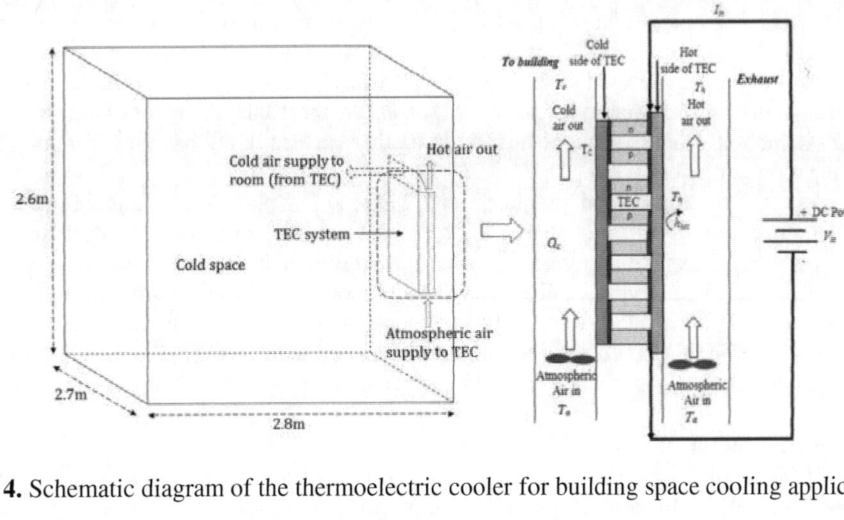

Figure 4. Schematic diagram of the thermoelectric cooler for building space cooling application [7].

In this configuration of thermoelectric radiant air conditioning (TE-RAC) system, the researcher has studied by consuming critical temperature, the design data, and thermal characteristic of thermoelectric radiant panel. They presented that the complex relations between different parameter by considering the temperature decreasing in difference surface from the TEM. Figure 5 shows the novelty of the thermoelectric module adopted in radiant air conditioning to remove the heat load of the room by thermal convection and radiation. The optimal value used of TEM set as 16 per square meter with 4mm thickness of thermoelectric radiant panel that fit for cooling this building. The COP of TE-RAC is larger than the

COP of conventional TE air-conditioning system, which illustrates that the TE-RAC system is a promising approach to enlarge the market of TE technology in the large cooling field [8].

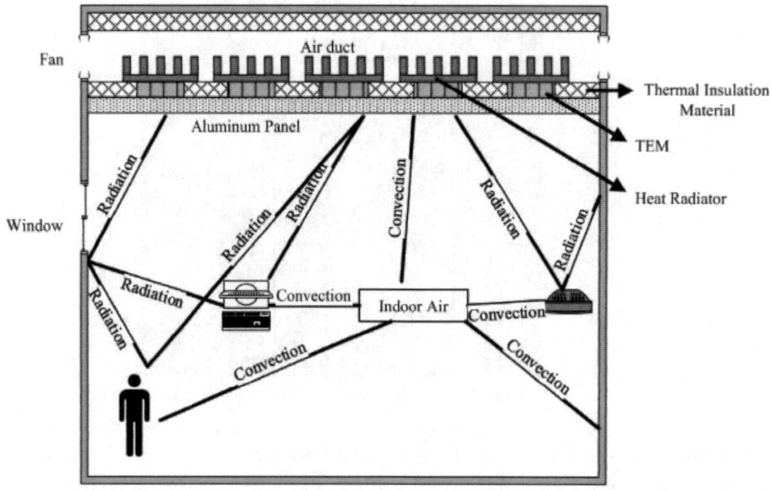

Figure 5. The scheme of thermoelectric radiant air-conditioning system [8].

From the literature review conducted, the research on the performance of the TEC module is particularly involved in the cooling or heating system of various applications. Previous researchers have investigated the TEC system placed on one side of the room wall which is the atmospheric air that is supplied to TEC system and the cold air supplied to the room while the hot air will directly out from the room [7]. The thermoelectric module is placed electrically and thermally in parallel and is sandwiched between heat radiator and aluminum panel to substitute the radiant air conditioning system and the thermoelectric radiant panel exchanges heat with another surface in the room by radiation and exchange heat with the room air by convection [8].

In this paper, the cooling system integrated with TEC module placed at the area of the glass window of an air conditioned room will be studied and the effects on air conditioner compression work will be analyzed. This study looks at the possibility of integrating the TEC within a cooling system to minimize the air conditioner compression work at air conditioned room. The principal aim of the study is to derive a mathematical model that is able to correlate the use of TEC and the effects on the air conditioner compression work. The proposed temperature of the studied area is at 25.5°C inside the air conditioned room and the cooling system at the area of the glass window or its surrounding.

3. Experimental set up

A 60Watt TEC module in an air conditioned room is utilized and cooled by a 1 HP AC unit (split unit type). The air conditioned room volumetric space is limited to 3m x 4m x 2m (length x width x height) and the area of glass window attached to the wall is 1.12m x 0.85m (length x width). The cooling system construction and configuration was studied. For future work, the derivation of the mathematical model system will be implemented based on the first and second order. The step response experiment will be implemented using MATLAB software. The step response experiment and the mathematical model will be verified with the real system. Data collection from the experiment can be described using general structure known as ARMAX and ARX model. These models will be used to obtain the experiment data and to evaluate whether the selected model will serve its purpose by determining the best model.

The general structure for ARMAX model:

$$A(q^{-1})y(t) = B(q^{-1})u(t) + C(q^{-1})e(t) \tag{1}$$

$$\begin{aligned} &where \\ &A(q^{-1}) = 1 + a_1 q^{-1} + \ldots a_{na} q^{-na} \\ &B(q^{-1}) = b_1 q^{-1} + \ldots b_{nb} q^{-nb} \\ &C(q^{-1}) = 1 + c_1 q^{-1} + c_{nc} q^{-nc} \end{aligned}$$

Equation (1) can be written as the difference equation:

$$\begin{aligned} &y(t) + a_1 y(t-1) + \ldots + a_{na} y(t-na) = b_1 u(t-1) + \ldots + b_{nb} u(t-nb) \\ &+ e(t) + c_1 e(t-1) + \ldots + c_{nc} e(t-nc) \end{aligned} \tag{2}$$

The general structure for ARX model:

$$y(t) + a_1 y(t-1) + \ldots + a_{na} y(t-na) = b_1 u(t-1) + \ldots + b_{nb} u(t-nb) + e(t) \tag{3}$$

Meanwhile, the steady state simulation is done using Ansys CFX software to study the effects of TEC to cooling space area. To evaluate the optimal number of TEC module used in the cooling system, a simulation has been constructed in Ansys CFX (Figure 6). The tests were operated under transient and steady state analysis. Transient analysis is involving time function load while steady state is for determine the effect of steady thermal loads on a system. The steady state analysis also can determine the temperature, heat flow rates, thermal gradient and heat fluxes.

4. Simulation model run in the Ansys-CFX

The steady state simulation is conducted to investigate the effects of heat exchange from the area of the TEC region by assuming adiabatic walls. In this preliminary study, a TEC module has been set with 20°C in an air conditioned room. By considering the uniformity of surface temperature, the simulation is conducted. The changes of air flow from a TEC module in the air conditioned room wall with a certain period of time was depicted in Figure 6. Figure 6 (a) shows the air flow from a TEC module after 30 minutes is not affect the air surrounding of the air conditioned room. But after a certain timestep show in figure 6 (b) and figure 6 (c), the cold air from TEC module slightly cooling the surface of the room. When the steady state condition occurred, the room is covered of cold air that shows in figure 6 (d). From the results, the TEC module successfully affects the cooling of the room with steady state mode. Later, the simulations will be conducted for real mode of air conditioning room including heat losses from wall, floor, and roof. Therefore, more literature review will be done to set data of these losses.

IOP Conf. Series: Journal of Physics: Conf. Series **908** (2017) 012080 doi:10.1088/1742-6596/908/1/012080

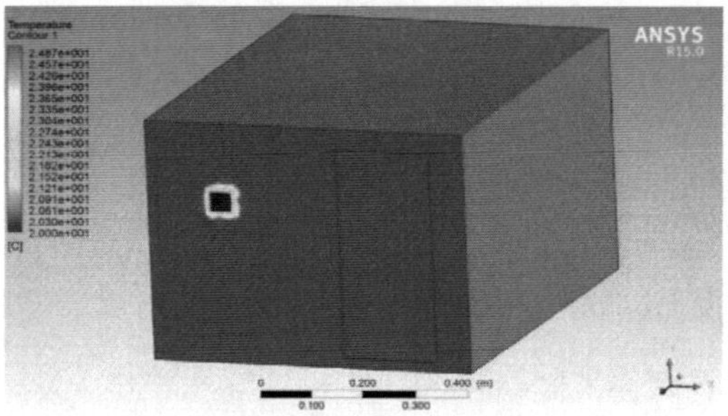

(a) 10 timesteps

(b) 500 timesteps

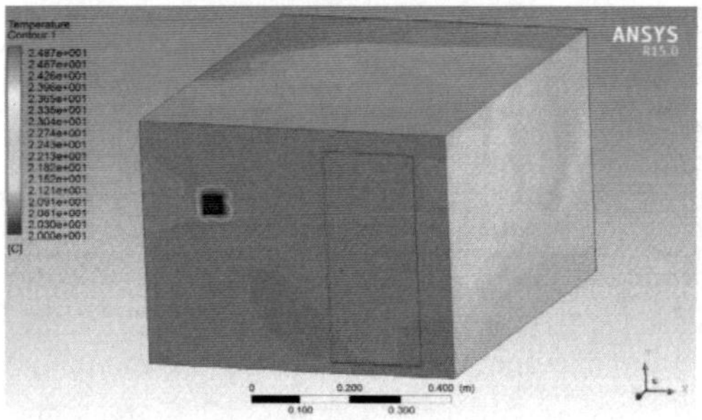

(c) 800 timesteps

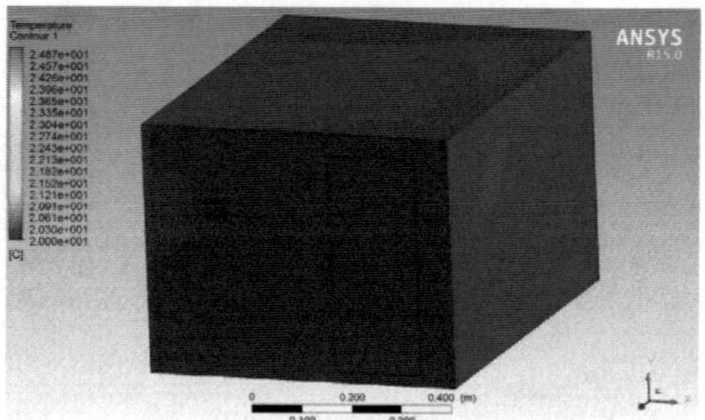

(d) 1000 timesteps

Figure 6. The thermal changes from certain period of time inside the prototype modeling.

5. Conclusion

A preliminary testing on a TEC module effects in an air conditioned room has been studied. The optimum numbers of TEC module might be put into consideration, when it is expected to help the cooling system performance in air conditioned room. The temperature versus ampere data using the number of TEC and different types of TEC will be done on the future. The correlation between the number of TEC and the different types of TEC to the compression of air conditioner unit will be presented in the numerical form/ mathematical model. Lastly, the mathematical model of this research work may be used for a better design of an air conditioned room consists of an air conditioner unit and glass windows.

6. Acknowledgements

The authors would like to acknowledge the Ministry of Higher Education of Malaysia and the Universiti Malaysia Perlis, Perlis, Malaysia for financial support under FRGS (9003-00564).

7. References

[1] Mattia D R, Vincenzo B, Federico S and Luca A. T 2014 *Applied Energy* **128** p 217-29
[2] Xiangfei K, Shilei L, Yiran L, Jingyu H and Shangbao L 2014 *Energy and Buildings* **81** p 404-15
[3] Adeel W and Zia U D 2013 *Renewable and Sustainable Energy Reviews* **18** p 607-25
[4] Clito F A A 2006 *Applied Thermal Engineering* **26** p 1961-71
[5] *Einebinsenweisheit.com* 2016 Available: http://www.einebinsenweisheit.com/p-peltier-element.html
[6] *Artofcircuits.com* 2016 Available: http://artofcircuits.com/product/tec1-12706-12v-6a-60-watt-thermo-electric-cooler-assembled-kit

[7] Manikandan S, Kaushik S C and Ronggui Y *Energy Conversion and Management* 2017 **140** p 145-56
[8] Limei S, Zhilong T, Qiang H, Cheng T and Huanxin C 2017 *Applied Thermal Engineering* **112** p 688-97

Numerical study of double-diffusive convection in a vertical annular enclosure with a baffle

Pushpa B V[1], Prasanna B M R[2], Younghae Do[3] and M.Sankar[4]

[1]Department of Mathematics, Sapthagiri College of Engineering, Bangalore, India.
[2]Department of Mathematics, Siddaganga Institute of Technology, Tumakuru, India.
[3] Department of Mathematics, KNU-Center for Nonlinear Dynamics, Kyungpook National University, Daegu, 41566, Korea.
[4]Department of Mathematics, School of Engineering, Presidency University, Bengaluru, India.

msankar@presidencyuniversity.in

Abstract. This paper numerically examines the influence of a circular thin baffle on thermosolutal convection in a vertical annular enclosure. The inner and outer cylindrical walls, and the baffle are retained with different temperatures and concentrations, while the upper and lower boundaries are kept at adiabatic and impermeable. The model equations are solved using an implicit finite difference scheme consisting of ADI and SLOR methods. Numerical simulations are performed to understand the size and position effects of the baffle on the thermosolutal convection and are successfully captured through our results. It has been observed that the baffle size and location has very important role in controlling the thermosolutal convective flow and the corresponding heat and mass transport characteristics. Further, our results are in good agreement with the available benchmark results for limiting cases.

1. Introduction

Based on the past investigations on double-diffusive convection or thermosolutal convection, it is realized that the thermosolutal convective flows are basically examined in bounded and unbounded geometries. Each of the two geometries has many important applications in their own domain. For thermosolutal convection in an enclosure, the direction of two buoyant forces can either enhances or suppresses the thermal and solutal transport rates. In addition, the ratio of thermal and solutal buoyancies known as the buoyancy ratio, an important parameter for double-diffusive convection, has an important role in altering the flow fields and the corresponding thermal and solutal transport rates. Double-diffusive convection is of great interest among the researchers due to its wide applications in natural process and engineering such as dispersion of chemical contaminants through water saturated soil, grain storage installation, metallurgy and oceanography.

Amongst the bounded geometries, the cylindrical annulus designed by two vertical, concentric tubes attracted many researchers due to the sheer existence of this geometry in wide range of engineering and industrial applications. Beji et al. [1] used the control volume method to analyse the thermosolutal convection in a vertical porous annulus. Bennacer et al. [2] studied the influence Darcy number on thermosolutal convection in a porous annulus using Darcy-Brinkmann formulation. Benzeghiba et al. [3] reported the numerical results on heat and mass transport processes in a vertical annular enclosure formed by two concentric cylinders filled with a fluid saturated porous medium. In a vertical annulus, Chen et al. [4] reported the numerical results on double-diffusive convection and found that for the lower buoyancy ratios, the thermal buoyancy effect dominates the flow and for higher buoyancy ratios the flow is dominated by solutal buoyancy effect. Sankar et al. [5, 6] reported the influence of a discrete thermal and solutal source on thermosolutal convection in a vertical porous annulus. The location of discrete heat and solute source found to control the heat and mass transport inside the porous enclosure in a significant manner.

In a differentially heated enclosure, it is found that the heat and mass transport processes could be controlled effectively by manipulating the flow regimes as the transport processes are highly sensitive to the different flow regimes. Among the available techniques, a simple way to alter the flow regimes in a cavity is by fixing baffles or fins to the heated or cooled walls. Shi and Khodadadi [7] discussed the influence of a fin on buoyancy-driven convection in a rectangular cavity. They found that the flow patterns are modified by the fin, and presence of the fin also acts as an extra heating to the fluid. Convection in a square enclosure with a baffle fixed at the hot wall has been numerically investigated by Tasnim and Collins [8] and found that the presence of baffle increases the heat transfer rate. Bilgen [9] investigated the influence of a thin fin on convective heat transfer in differentially heated square cavity and shown that the heat transfer rate enhances with Rayleigh number, but suppressed with fin length and conductivity ratio. Recently, Khanafer et al. [10] reported natural convection to analyse the effect of a porous baffle fixed to the hot wall of a square enclosure and reported that the presence of porous fin alters the thermal transport and optimum fin location and angle is identified to obtain higher heat transport rate. Based on the detailed literature survey, it has been found that the impact of a circular baffle on double-diffusive convective flow, the associated heat and mass transport characteristics have not been investigated in a vertical annulus. Hence, the influence of baffle length and baffle position on thermosolutal convection in an annular enclosure has been thoroughly investigated.

2. Physical model and basic equations

The physical configuration chosen for the investigation is the vertical annular enclosure designed by an inner cylinder and an outer cylinder with radii r_i and r_o as given in Fig.1. Here the inner and outer cylinders have the same center forming an annular region with D as the annulus width and H as the height of the annulus. We considered the flow to be two-dimensional and the aspect ratio of the annulus has been taken as unity ($H = D$). A highly conductive horizontal baffle of length l is attached to different locations h on the inner wall. The horizontal (top and bottom) walls of the annulus are assumed to be thermally insulated, whereas the inner (left) cylinder is preserved at a high temperature and concentration and the outer (right) cylinder is preserved at a lower temperature and concentration. In addition, the temperature and concentration of the baffle is maintained at the same temperature and concentration of the inner cylinder. Further, the laminar flow and constant thermophysical properties assumptions are invoked in this analysis. The thermal and solutal buoyancy forces in the momentum equation are incorporated through the Boussinesq approximation. In addition, the Newtonian fluid is chosen by neglecting the viscous dissipation with the gravity acting in the negative z-direction, and on all solid boundaries no-slip conditions are assumed to be valid. For these assumptions, the equations governing the mass, momentum, energy and species in the dimensionless form are (see Sankar et al. [5, 6]):

$$\frac{\partial T}{\partial t}+U\frac{\partial T}{\partial R}+W\frac{\partial T}{\partial Z}=\nabla^2 T \qquad (1)$$

$$\frac{\partial C}{\partial t}+U\frac{\partial C}{\partial R}+W\frac{\partial C}{\partial Z}=\frac{1}{Le}\nabla^2 C \qquad (2)$$

$$\frac{\partial \zeta}{\partial t}+U\frac{\partial \zeta}{\partial R}+W\frac{\partial \zeta}{\partial Z}-\frac{U\zeta}{R}=Pr\left[\nabla^2\zeta-\frac{\zeta}{R^2}\right]-Pr\,Ra_T\left[\frac{\partial T}{\partial R}+N\frac{\partial C}{\partial R}\right] \quad (3)$$

$$\zeta=\frac{1}{R}\left[\frac{\partial^2\psi}{\partial R^2}-\frac{1}{R}\frac{\partial \psi}{\partial R}+\frac{\partial^2\psi}{\partial Z^2}\right] \quad (4)$$

$$U=\frac{1}{R}\frac{\partial \psi}{\partial Z},\quad W=-\frac{1}{R}\frac{\partial \psi}{\partial R},\quad \text{where}\quad \nabla^2=\frac{\partial^2}{\partial R^2}+\frac{1}{R}\frac{\partial}{\partial R}+\frac{\partial^2}{\partial Z^2}. \quad (5)$$

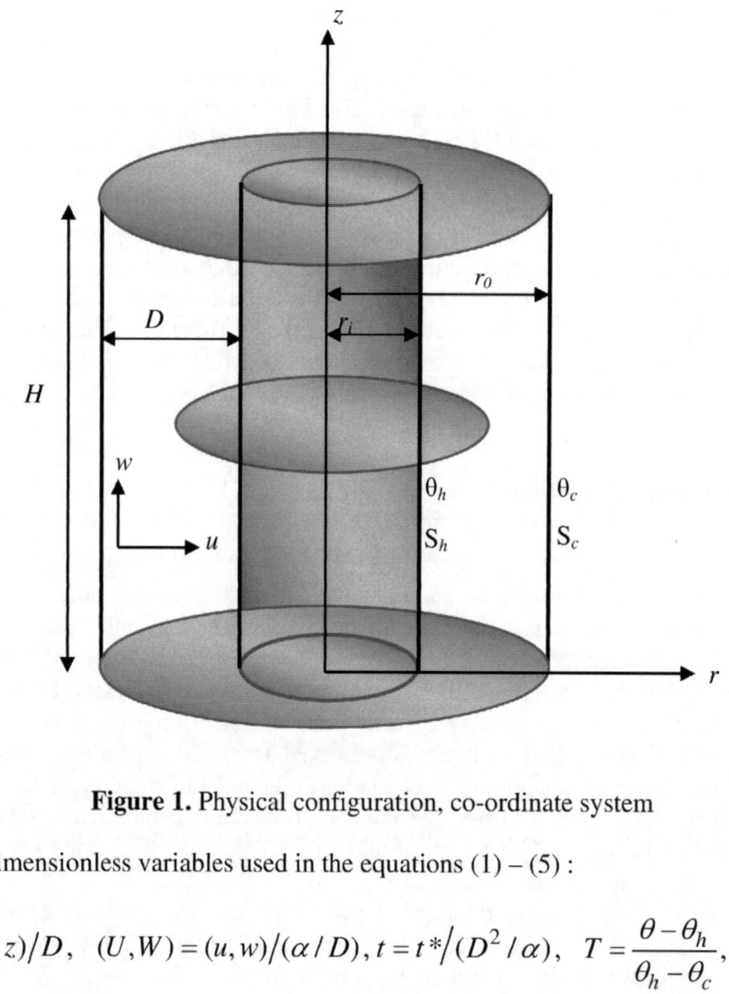

Figure 1. Physical configuration, co-ordinate system

The following dimensionless variables used in the equations (1) – (5) :

$$(R,Z)=(r,z)/D,\quad (U,W)=(u,w)/(\alpha/D),\; t=t^*/(D^2/\alpha),\quad T=\frac{\theta-\theta_h}{\theta_h-\theta_c},\; C=\frac{S-S_h}{S_h-S_c}$$

$$P=p/(\rho_0\alpha^2/D^2),\quad \zeta=\zeta^*/(\alpha/D^2),\psi=\psi^*/(D\alpha),\text{where } D=r_o-r_i.$$

The non-dimensional parameters occur in the present study are: $\lambda=\frac{r_0}{r_i}$ Radius ratio, $A=\frac{H}{D}$ Aspect ratio, $\varepsilon=\frac{l}{D}$ Baffle length, $L=\frac{h}{H}$ Baffle location, $Le=\frac{\alpha}{\alpha_D}$ Lewis number, $\Pr=\frac{\nu}{\alpha}$

Prandtl number, $N = \frac{Ra_S}{Ra_T}$ Buoyancy ratio. The heat and mass transport rates are estimated in terms of local and average Nusselt and Sherwood numbers. The local Nusselt (*Nu*) and Sherwood (*Sh*) numbers are respectively given by $Nu = -\frac{\partial T}{\partial R}$ and $Sh = -\frac{\partial C}{\partial R}$. The average Nusselt and Sherwood numbers are then calculated from the relation $\overline{Nu} = \frac{1}{A}\int_0^A Nu dZ$ and $Sh = \frac{1}{A}\int_0^A Sh dZ$. The initial and boundary conditions in dimensionless form are:

$$t = 0: \quad U = W = T = C = 0, \psi = \zeta = 0$$

$$t > 0: \quad \psi = \frac{\partial \psi}{\partial R} = 0, T = C = 1; \qquad \text{at outer wall and baffle}$$

$$\psi = \frac{\partial \psi}{\partial R} = 0, T = C = 0; \qquad \text{at inner wall}$$

$$\psi = \frac{\partial \psi}{\partial Z} = 0, \frac{\partial T}{\partial Z} = \frac{\partial C}{\partial Z} = 0; \qquad Z = 0 \text{ and } Z = A.$$

3. Solution methodology

The model equations and the appropriate boundary conditions are numerically solved by developing an in-house code, which is also magnificently validated with the available benchmark solutions. In particular, the two-step Alternating Direction Implicit method is used for the vorticity, energy and species equations. The second-order accurate central differencing approximation has been used for second order derivatives and first order linear derivatives, the nonlinear advection terms are discretized with second upwind difference technique and the forward differencing scheme is used for time derivative. But, the numerical solution of stream function equation is obtained through the SLOR method after carefully choosing the relaxation factor. After discretization, the partial differential equations are reduced to tridiagonal matrix form and are solved using the well-known Thomas algorithm. By using the central difference approximation, velocity components are evaluated. The overall Nusselt and Sherwood numbers are determined from the Simpson's rule.

4. Results and discussion

Numerical simulations are carried out to develop the basic information to understand the effects of buoyancy ratio, Lewis number on the thermosolutal convective flow in the annulus geometry with a thin baffle of different lengths placed at different locations. Since the detailed investigation of this problem by considering all parameter ranges is not the main interest, only five main governing parameters, namely the buoyancy ratio (N), Lewis number (Le), Rayleigh number (Ra), baffle length (ε) and baffle location (L) are varied and the remaining parameters are kept at constant. The simulations are discussed through the streamlines, isotherms and isoconcentrations. The rates of thermal and solutal transport in the enclosure are calculated in terms of the average Nusselt and Sherwood numbers.

The influence of buoyancy ratio on streamlines, isotherms and isoconcentrations for three values of *N* (N = -5, 0 and 5) is presented in the figure 2 for fixed values of $Ra_T = 10^6$, Le = 1, $\varepsilon = 0.3$, and L = 0.5. These values of N constitutes three important regimes in the thermosolutal convection flow, namely concentration-driven opposing flow (N = -5), thermal-driven flow (N = 0) and temperature dominated aiding flow (N = 5). For the opposing flow (N = -5), the direction of flow, thermal and solutal fields are counter clockwise with a small rotating vortex observed below the baffle. The streamlines are almost parallel above the baffle and the counter rotating cells cover the entire region of the cavity. For the case of pure thermal-driven flow (N=0), the flow and thermal fields resembles that of buoyancy-driven flow and the direction of flow is changed with clockwise rotating vortex appearing near top wall. The isotherms and isoconcentrations appear similar as the influence of solutal buoyancy is absent. Further, as the value of N is raised to 5 units ($N=5$), the flow intensity is very stronger from the collective thermal and solutal buoyancy acting in the same direction as indicated by the magnitude of maximum stream function. The isotherms and isoconcentrations reveal the strong stratified structure forming a thick thermal and solutal boundary layer along the hot and cold walls, and along the baffle.

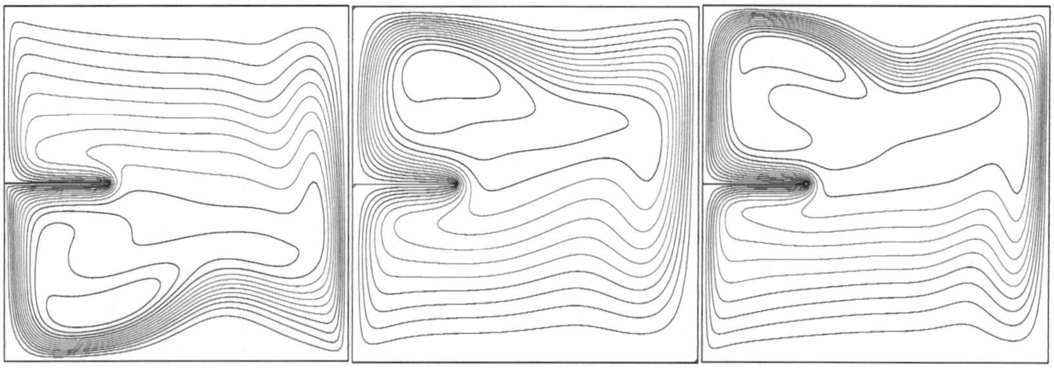

(a)

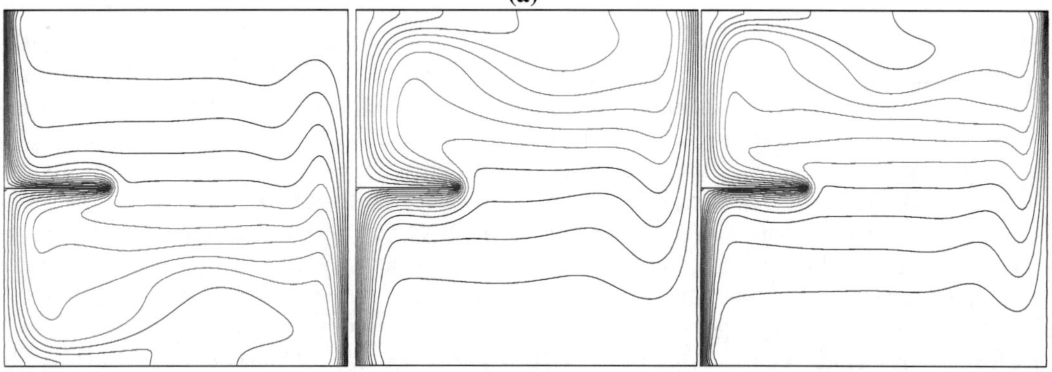

(b)

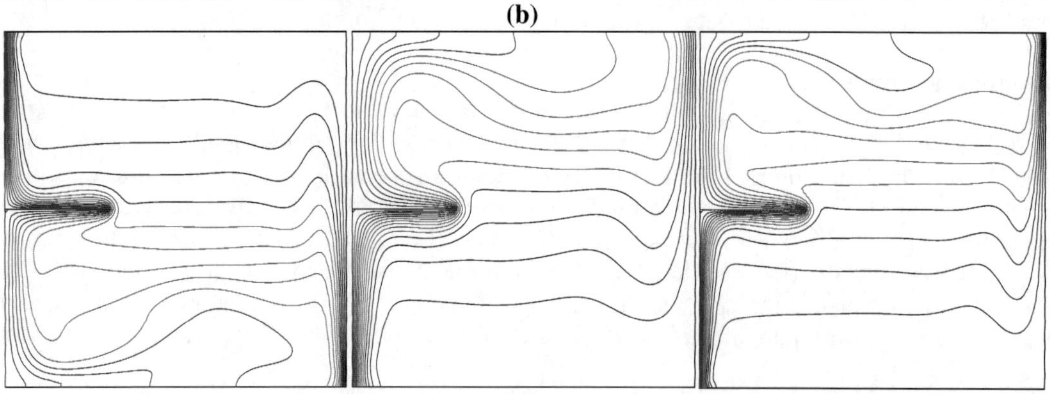

(c)

Figure 2. Effect of buoyancy ratio on streamlines (top), isotherms (middle) and isoconcentrations (bottom). (a) N= -5, $|\psi_{max}| = 46.7$, (b) N= 0, $|\psi_{max}| = 31.4$, (c) N= 5, $|\psi_{max}| = 48.5$.

The effect of buoyancy ratio on the heat and mass transfer rates is an important quantity in double-diffusive convection. Hence, figures 3-5 represents the effect of buoyancy ratio on the average Nusselt number and average Sherwood number for different values of Ra_T, baffle size and baffle location by fixing the Lewis number as *Le* = *1*. In particular, the values of thermal Rayleigh number, baffle length, and baffle location are varied respectively as $10^4 \le Ra_T \le 10^6$, $0.25 \le \varepsilon \le 0.75$ and $0.25 \le L \le 0.75$. For Lewis number equals to unity, the thermal and solutal diffusivities are in same proportion in the mixture. As a result the average Nusselt and Sherwood number profiles appear identical for all values of thermal Rayleigh number. However, the thermal and solutal transport rates are significantly varied with the thermal Rayleigh number. A general observation of the results reveals that the higher thermal Rayleigh number produces maximum thermal and solutal transport rates. Also, the total heat and mass transfer rates are higher in aiding flow regimes compared to opposing flow regimes.

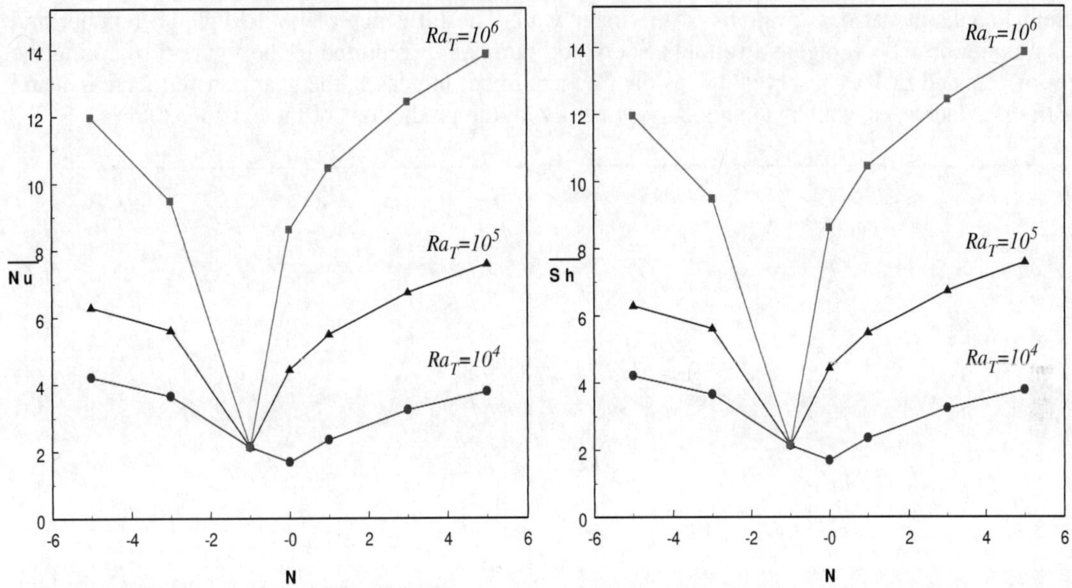

Figure 3. Influence of Rayleigh number on the average Nusselt and Sherwood numbers for different buoyancy ratio with $Le = 1$, $\varepsilon = 0.5$ and $L = 0.5$.

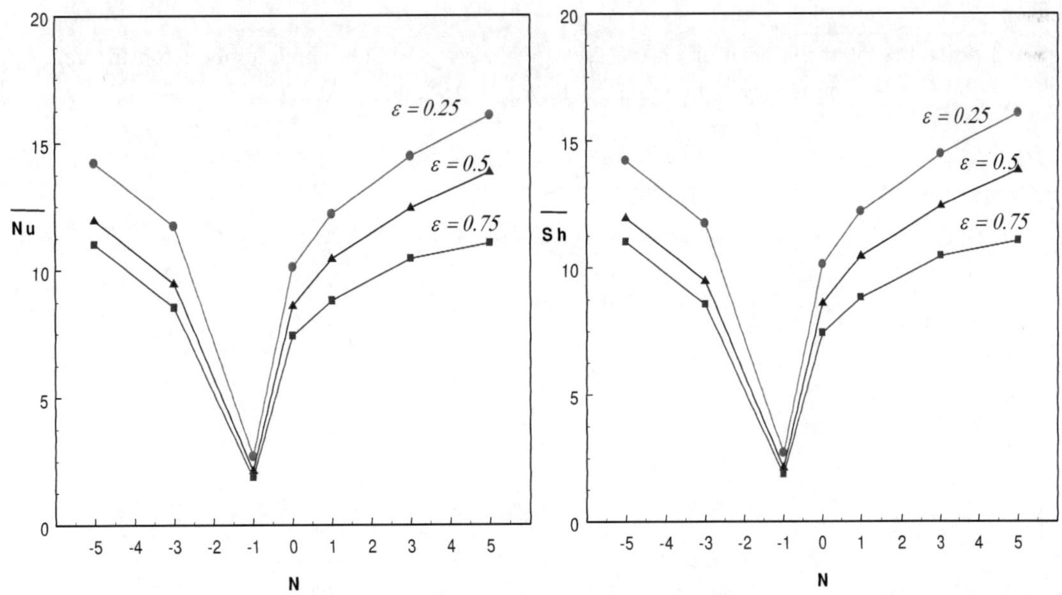

Figure 4. Influence of baffle length on the average Nusselt and Sherwood numbers for different buoyancy ratio with $Le = 1$, $Ra_T = 10^6$ and $L = 0.5$.

The influence of size and location of baffle on the thermosolutal convection is the main objective of our investigation. Hence, the effect of different combinations of baffle length and baffle position on the heat and mass transfer rates are presented in figures 4 and 5. A thorough analysis of the results reveals that, for all the values of the Ra_T, baffle length and baffle location, the heat and mass transfer rates decreases with the buoyancy ratio for N < - 1, while for N > - 1, the heat and mass transfer rates are increasing with buoyancy ratio. Also it is observed that the heat and mass transfer increases with thermal Rayleigh number and baffle location, however, decreases with baffle length. From the simulations of the present analysis, it is observed that the double-diffusive convective flow and the associated heat and mass transfer characteristics can be effectively controlled by placing baffle of different lengths at various locations of the inner wall. The influence of both Rayleigh number and Lewis number on the average Nusselt and Sherwood numbers is reported in the figure 6 by fixing the values of N, ε and L. It is observed that as the Lewis number increases, the heat transfer decreases and mass transfer increases, which is in good agreement with the predictions of the existing studies.

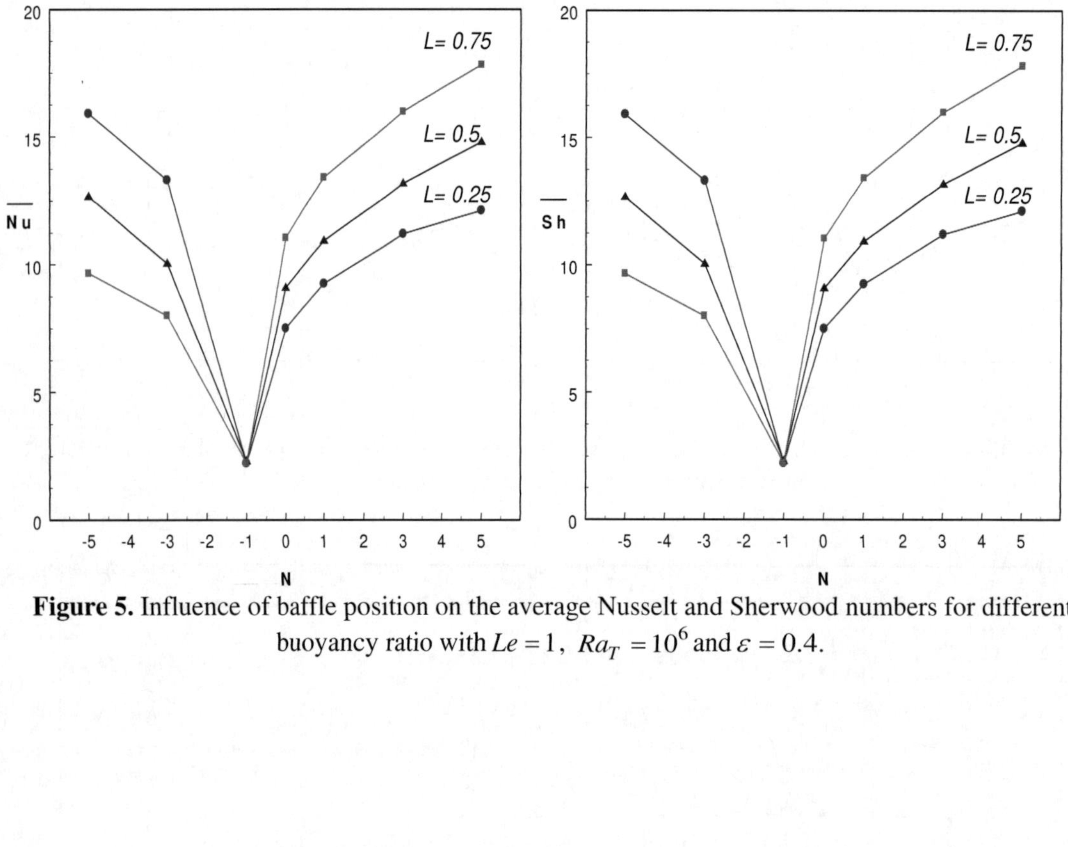

Figure 5. Influence of baffle position on the average Nusselt and Sherwood numbers for different buoyancy ratio with $Le = 1$, $Ra_T = 10^6$ and $\varepsilon = 0.4$.

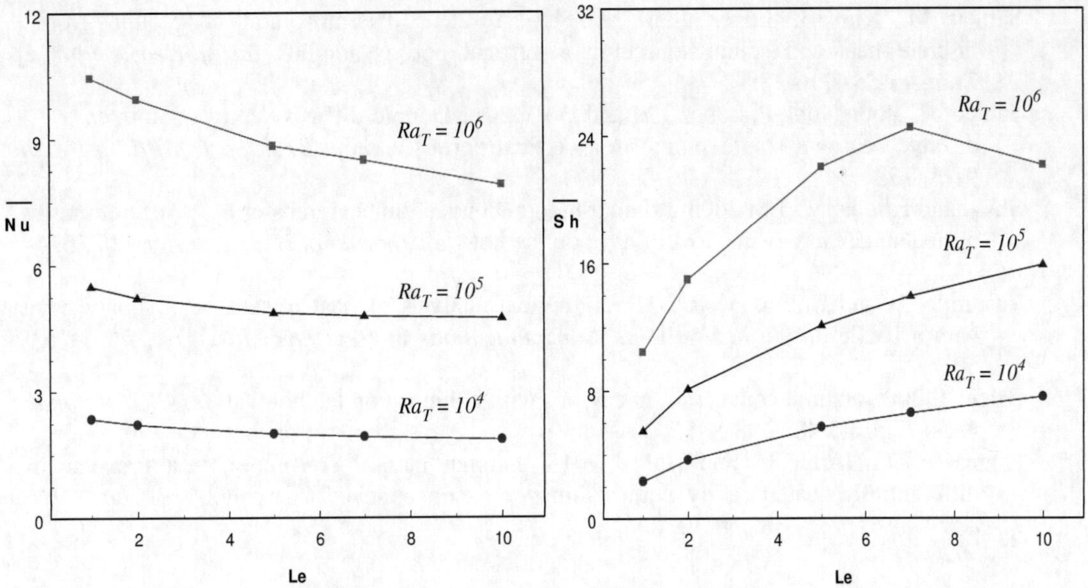

Figure 6. Effect of Rayleigh number on the average Nusselt and Sherwood numbers for different Lewis numbers with $N = 1$, $\varepsilon = 0.5$ and $L = 0.5$.

5. Conclusions
Double-diffusive convection in a vertical annular cavity is numerically investigated by considering a thin baffle having the same temperature and concentration of the inner cylinder. Flow circulation, thermal and solutal distributions are considerably altered by the baffle size and location. The numerical results reveal that the heat transfer rate increases with the thermal Rayleigh number, but decreases with the Lewis number. However, the mass transfer increases with both the Rayleigh and Lewis numbers. The heat and solute transport increases with an increase in baffle location, and higher heat and mass transport is observed for the baffle location near top of the cavity. The heat and mass transport decreases with the increasing baffle length due to blockage caused by the baffle to the fluid movement. For all thermal Rayleigh numbers, baffle sizes and baffle locations, the heat and mass transfer rates are found to be higher in aiding flow regimes due to combined buoyancy effect. Further, it has been observed that the heat and mass transfer rates can be effectively controlled by varying the size and location of the baffle.

6. Acknowledgments
The authors PBV, BMRP and MS are respectively grateful to the Sapthagiri College of Engineering, Bangalore, SIT, Tumkur and Presidency University, Bangalore and to VTU, Belgaum, Karnataka, India for the support and encouragement. Y. D. was supported by the National Research Foundation of Korea (NRF) grant funded by the Korea government (MSIP) (No. NRF-2016R1A2B4011009).

7. References
[1] Beji H, Bennacer R, Duval R and Vasseur P 1999 Double-Diffusive natural convection in a vertical porous annulus *Numerical Heat Transfer, Part A* **36** 153-70.

[2] Bennacer R, Beji H, Duval R and Vasseur P 2000 The Brinkman model for thermosolutal convection in a vertical annular porous layer *Int. Communications in Heat and Mass Transfer* **27 (1)** 69-80.

[3] Benzeghiba M, Chikh S and Campo A 2003 Thermosolutal convection in a partly porous vertical annular cavity *Journal of Heat Transfer* **125** 703-15.

[4] Chen S, Tölke J and Krafczyk M 2010 Numerical investigation of double-diffusive (natural) convection in vertical annuluses with opposing temperature and concentration gradients *Int. Journal of Heat and Fluid Flow* **31 (2)** 217-26.

[5] Sankar M, Kim B, Lopez J M and Do Y 2012 Thermosolutal convection from a discrete heat and solute source in a vertical porous annulus *Int. J Heat and Mass Transfer* **55** 4116-28.

[6] Sankar M, Youngyong P, Lopez J M and Do Y 2012 Double-diffusive convection from a discrete heat and solute source in a vertical porous annulus *Transport in Porous Media* **91(3)** 753–75.

[7] Shi X and Khodadadi J M 2003 Laminar natural convection heat transfer in differentially heated square cavity due to a thin fin on the hot wall *Journal of Heat Transfer* **125** 624-34.

[8] Tasnim S H and Collins M R 2004 Numerical analysis of heat transfer in a square cavity with a baffle on the hot wall *Int. Communications in Heat and Mass Transfer* **31** 639-50.

[9] Bilgen E 2005 Natural convection in cavities with a thin fin on the hot wall *Int. J Heat and Mass Transfer* **48** 3493-505.

[10] Khanafer K, AlAmiri A and Bull J 2015 Laminar natural convection heat transfer in a differentially heated cavity with a thin porous fin attached to the hot wall, *Int. J Heat and Mass Transfer* **87** 59-70.

IOP Conf. Series: Journal of Physics: Conf. Series **908** (2017) 012082 doi:10.1088/1742-6596/908/1/012082

Efficient methodology of route selection for driving cycle development

A R Mahayadin[1,2], Shahriman A B[1,2], M S M Hashim[1,2], Z M Razlan[1,2], M K Faizi[1,2], A Harun[1,3], N S Kamarrudin[1,2], I Ibrahim[1,2], M A M Saad[1,2], M F H Rani[1,2], I Zunaidi[4], M Sahari[5], M S Sarip[5] and M Q H A Razali[5]

[1] Motorsports Technology Research Unit (MoTECH), Universiti Malaysia Perlis, Pauh Putra Campus, 02600 Arau, Perlis, Malaysia
[2] School of Mechatronic Engineering, Universiti Malaysia Perlis, Pauh Putra Campus, 02600 Arau, Perlis, Malaysia
[3] School of Microelectronic Engineering, Universiti Malaysia Perlis, Pauh Putra Campus, 02600 Arau, Perlis, Malaysia
[4] Technopreneur at UniMAP Sdn Bhd, No 26, Blok D, Bangunan PPPIT, Pengkalan Jaya, Jalan Kangar - Alor Setar, Taman Mutiara, 01000 Kangar, Perlis, Malaysia
[5] Malaysia Automotive Institute, Block 2280, Jalan Usahawan 2, Cyber 6, 63000 Cyberjaya, Selangor, Malaysia

shahriman@unimap.edu.my

Abstract. Driving cycle is a series of data points representing the speed of vehicle versus time and used to determine the performance of vehicle in general. One of the critical portions of driving cycle development is route selection methodology. This paper describes the efficient methodology of route selection for driving cycle development. Previous data from JKR Road Traffic Volume Malaysia (RTVM) in 2015 is studied and analysed to propose the methodology in route selection. The selected routes are then analysed by using Google Maps. For each region, four (4) routes are selected for each urban and rural. For this paper, the selection of route is focused on northern region of Malaysia specifically in Penang. Penang is chosen for this study because it is one of the developed state in Malaysia that has many urban and rural routes. The methods of route selection constructed in this study could be used by other region to develop their own driving cycles.

1. Introduction

Driving cycle is a series of data points representing the speed of vehicle versus time that are produced by various countries and organizations to determine the performance of vehicles in terms of fuel consumption and carbon emissions [1]. Every driving cycle features the specific route conditions of a particular place and differs widely. Therefore, the driving cycles developed in a certain country or region may not applicable for other regions unless the driving characteristics are evidently similar. Thus, much research work is aimed towards developing driving cycles for a specific country or region [2, 3, 4, 5].

A vehicle driving cycle for a certain region is developed in order to represent the driving pattern that the vehicles would experience repetitively during the journeys when driving in the considered region. Hence, a typical driving cycle for a city must be obtained from traffic data along the travelled routes of those vehicles. The number of such possible route can be huge and it is impractical to

IOP Conf. Series: Journal of Physics: Conf. Series **908** (2017) 012082 doi:10.1088/1742-6596/908/1/012082

conduct actual assessment of the vehicle speed characteristics on all the routes. An effective way to resolve this problem is to select a number of routes that represent the dominant traffic conditions throughout the region [6]. To do so, an approach is needed to perform a smart selection of routes for the data collection. This paper proposes a comprehensive methodology of route selection in order to determine the most preferable routes for the collection of vehicle speed-time data.

2. Route Selection Methodology

In order to select the routes that can best represent certain traffic condition, the actual situations that happen along each route must be identified and ascertained. Previous data from Malaysian Public Works Department (JKR) Road Traffic Volume Malaysia (RTVM) in 2015 is studied and analysed to propose the methodology in route selection for the northern region of Malaysia specifically in Penang. There are 26 census stations in Penang that have been included in the data collection of traffic flow from JKR in 2015 [7]. The approach is based on the determination of the traffic condition at peak hours and the analysis of the traffic volume. The steps needed to complete this approach are outlined in the following sections.

2.1. Determine ratio of peak hour to 16 hours traffic volume

The first step in the route selection method is to determine the ratio of peak hour to 16 hours traffic volume. As shown in

Table **1**, the data from JKR RTVM in 2015 for Penang region is extracted and the ratio is calculated by dividing the peak hour traffic flow with 16 hours traffic flow for each census station number. The ratio represents the volume of traffic specifically at peak hours compared to the numbers in 16 hours traffic. This method is preferable as the ratio represents real usage of the routes during peak hours. This in essence shows the specific volume rather than throughout the whole 16 hours.

Table 1. Raw data from JKR RTVM in 2015 for Penang region.

Census Station Number	Level of Service (LOS)	Traffic Flow (Peak Hour)	Traffic Flow (16H)	Road Length (KM)	Type of Carriageway	Ratio
PR101	A	992	6,688	12.9	T 1-1	0.148
PR102	B	1,663	18,230	19.0	T 1-1	0.091
PR103	F	3,549	34,613	8.1	T 1-1	0.103
PR104	F	3,121	27,980	12.1	T 1-1	0.112
PR105	A	1,247	10,709	25.4	T 1-1	0.116
PR106	C	1,085	10,955	21.7	T 1-1	0.099
PR107	F	2,426	22,437	8.1	T 1-1	0.108
PR108	F	4,753	44,252	20.9	T 1-1	0.107
PR109	F	5,611	49,173	15.8	T 1-1	0.114
PR110	C	2,948	37,662	9.7	T 2-2	0.078
PR111	A	461	3,924	20.9	T 1-1	0.117
PR112	F	3,712	41,603	3.2	T 2-2	0.089
PR116	F	5,818	65,716	16.9	T 1-1	0.089
PR117	F	11,690	128,962	8.1	K 2-2	0.091
PR113	F	1,872	20,751	24.2	T 1-1	0.090
PR114	F	1,232	13,014	30.6	T 1-1	0.095
PR115	F	3,351	42,147	32.2	T 1-1	0.080
PR201	A	3,007	24,774	11.1	K 2-2	0.121
PR202	F	6,646	67,612	65.5	K 2-2	0.098
PR207	F	3,053	32,022	11.0	T 1-1	0.095
PR208	E	8,868	97,112	0.5	K 3-3	0.091
PR203	B	6,461	62,464	57.2	K 3-3	0.103
PR204	F	2,054	18,614	45.1	T 1-1	0.110
PR205	A	1,046	10,843	38.6	T 1-1	0.096

PR206	C	1,482	12,493	16.9	T 1-1	0.119
PR209	C	6,109	54,699	7.0	K 3-3	0.112

2.2. Ranking up the route based on Level of Service (LOS) followed by ratio of peak hour to 16 hours traffic volume on the specific LOS

Level of Service (LOS) of a traffic facility is an approach introduced to describe the quality of traffic service to a given flow rate. It is introduced by Highway Capacity Manual (HCM) to indicate the level of quality that can be obtain under different operation characteristics and traffic volume. By referring to figure 1, there are six LOS letters defined by HCM included A, B, C, D, E and F. LOS-A denotes the best quality of service whereas LOS-F denote the worst [7].

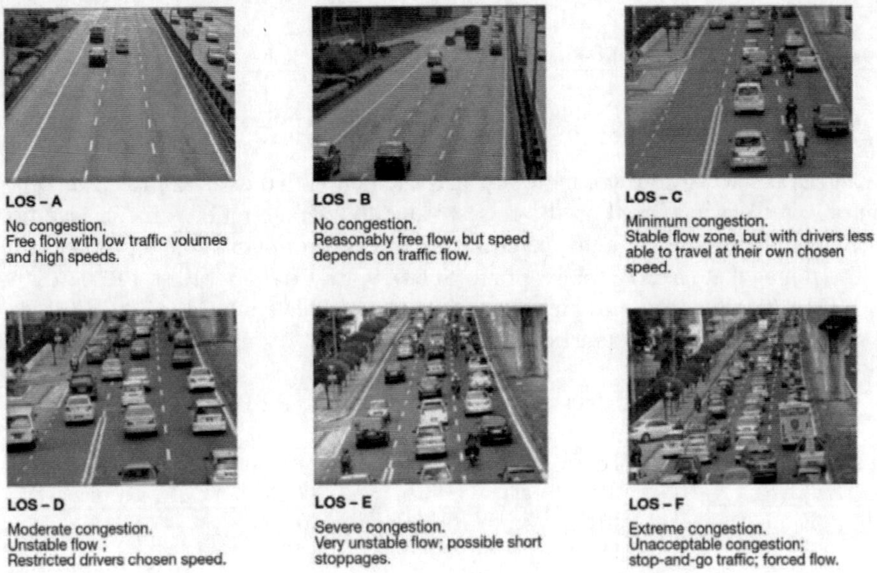

Figure 1. Description of Level of Services (LOS) [7].

Based on the table 2, the census station number is then ranked according to LOS; starting by LOS-F followed by LOS-E until LOS-A. After that, it is ranked by the highest to the lowest ratio of peak hour to 16 hours traffic volume on the specific LOS. Table 2 shows ranked data based on LOS and the ratio for Penang region.

Table 2. Ranked data based on LOS and ratio for Penang region.

Census Station Number	Level of Service (LOS)	Traffic Flow (Peak Hour)	Traffic Flow (16H)	Road Length (KM)	Type of Carriageway	Ratio
PR109	F	5,611	49,173	15.8	T 1-1	0.114
PR104	F	3,121	27,980	12.1	T 1-1	0.112
PR204	F	2,054	18,614	45.1	T 1-1	0.110
PR107	F	2,426	22,437	8.1	T 1-1	0.108
PR108	F	4,753	44,252	20.9	T 1-1	0.107
PR103	F	3,549	34,613	8.1	T 1-1	0.103
PR202	F	6,646	67,612	65.5	K 2-2	0.098
PR207	F	3,053	32,022	11.0	T 1-1	0.095
PR114	F	1,232	13,014	30.6	T 1-1	0.095
PR117	F	11,690	128,962	8.1	K 2-2	0.091
PR113	F	1,872	20,751	24.2	T 1-1	0.090
PR112	F	3,712	41,603	3.2	T 2-2	0.089

PR116	F	5,818	65,716	16.9	T 1-1	0.089
PR115	F	3,351	42,147	32.2	T 1-1	0.080
PR208	E	8,868	97,112	0.5	K 3-3	0.091
PR206	C	1,482	12,493	16.9	T 1-1	0.119
PR209	C	6,109	54,699	7.0	K 3-3	0.112
PR106	C	1,085	10,955	21.7	T 1-1	0.099
PR110	C	2,948	37,662	9.7	T 2-2	0.078
PR203	B	6,461	62,464	57.2	K 3-3	0.103
PR102	B	1,663	18,230	19.0	T 1-1	0.091
PR101	A	992	6,688	12.9	T 1-1	0.148
PR201	A	3,007	24,774	11.1	K 2-2	0.121
PR111	A	461	3,924	20.9	T 1-1	0.117
PR105	A	1,247	10,709	25.4	T 1-1	0.116
PR205	A	1,046	10,843	38.6	T 1-1	0.096

2.3. Select the urban route based on the ratio of peak hour to 16 hours traffic volume closest to average value

As LOS-F denotes the worst traffic conditions, it is then classified as urban route in Penang region. Four (4) routes were selected based on the closest value to average ratio with consideration of 30-32 samples per route to be examined in the experimental works. From table 3, the average ratio for LOS-F is 0.099. The routes that have the closest ratio to 0.099 are PR103 (0.103), PR202 (0.098), PR207 (0.095) and PR114 (0.095). By examining these four (4) selected routes, the 120 – 128 data samples will be exist which are enough for statistical analysis.

Table 3. Ranked data of LOS-F based on ratio for Penang region.

Census Station Number	Level of Service (LOS)	Traffic Flow (Peak Hour)	Traffic Flow (16H)	Road Length (KM)	Type of Carriageway	Ratio
PR109	F	5611	49173	15.8	T 1-1	0.114
PR104	F	3121	27980	12.1	T 1-1	0.112
PR204	F	2054	18614	45.1	T 1-1	0.110
PR107	F	2426	22437	8.1	T 1-1	0.108
PR108	F	4753	44252	20.9	T 1-1	0.107
PR103	F	3549	34613	8.1	T 1-1	0.103
PR202	F	6646	67612	65.5	K 2-2	0.098
PR207	F	3053	32022	11	T 1-1	0.095
PR114	F	1232	13014	30.6	T 1-1	0.095
PR117	F	11690	128962	8.1	K 2-2	0.091
PR113	F	1872	20751	24.2	T 1-1	0.090
PR112	F	3712	41603	3.2	T 2-2	0.089
PR116	F	5818	65716	16.9	T 1-1	0.089
PR115	F	3351	42147	32.2	T 1-1	0.080
					Average	0.099

2.4. Select the rural route based on the ratio of peak hour to 16 hours traffic volume closest to the average value

Based on JKR justification, LOS-A represent no congestion on a particular traffic condition. Hence, LOS-A could be considered as rural route in Penang region. Four (4) routes were selected with consideration of 30-32 samples per route to be examined in the experimental works. From Table 4, the average ratio for LOS-A is 0.120. The routes that have the closest ratio to 0.120 are PR201 (0.121), PR111 (0.117), PR105 (0.116) and PR205 (0.096). By examining these four (4) selected routes, the 120 – 128 data samples will be exist which are enough for statistical analysis.

Table 4. Ranked data of LOS-A based on ratio for Penang region.

Census Station Number	Level of Service (LOS)	Traffic Flow (Peak Hour)	Traffic Flow (16H)	Road Length (KM)	Type of Carriageway	Ratio
PR101	A	992	6688	12.9	T 1-1	0.148
PR201	A	3007	24774	11.1	K 2-2	0.121
PR111	A	461	3924	20.9	T 1-1	0.117
PR105	A	1247	10709	25.4	T 1-1	0.116
PR205	A	1046	10843	38.6	T 1-1	0.096
					Average	0.120

3. Result and Discussion

A comprehensive route selection methodology for Penang region has been deliberate in the previous section. There are four (4) routes each for urban and rural road have been selected. For this section, each route will be discuss thoroughly including description of location and also self-assessment of the route geography and exact location using Google Maps.

3.1. Selected urban and rural route

Based on Table 3, the average ratio of peak hour to 16 hours traffic volume for LOS-F is 0.099. Four (4) census station number which has the ratio closest to the average value are PR103, PR202, PR207 and PR114. Table 5 shows the description of location for selected urban route.

Table 5. Selected urban route.

Census Station Number	Description of Location
PR103	Butterworth – Bagan Ajam – Telok Ayer Tawar
PR202	George Town – Bayan Lepas – Gelugor (Jalan Keliling Pulau)
PR207	Paya Terubong – Ayer Hitam
PR114	Butterworth – Tasek – Val D'or

Based on table 4, the average value of ratio of peak hour to 16 hours traffic volume for LOS-A is 0.120. Four (4) census station number which has the ratio closest to the average value are PR201, PR111, PR105 and PR205. Table 6 shows the description of location for selected rural route.

Table 6. Selected rural route.

Census Station Number	Description of Location
PR201	George Town – Telok Bahang (Jalan Keliling Pulau)
PR111	Butterworth – Bukit Mertajam – Junjong (Through Machang Bubok – Tasek)
PR105	100m from Ara Kuda – Lunas Junction
PR205	George Town – Telok Bahang – Balik Pulau (Jalan Keliling Pulau)

3.2. Self-assessment of the route geography and exact location using Google Maps

Based on the previous section, the urban route that selected are PR103, PR202, PR207 and PR114. On the other hand, the rural routes that were selected are PR201, PR111, PR105 and PR205. Figure 2 shows the location of selected urban and rural route in Penang. Blue circles denote selected urban route whereas green circles denote selected rural route. From the figure, we can see that these routes were not in same networks or chains. Hence, these routes are acceptable to be selected to represent urban and rural routes in Penang region.

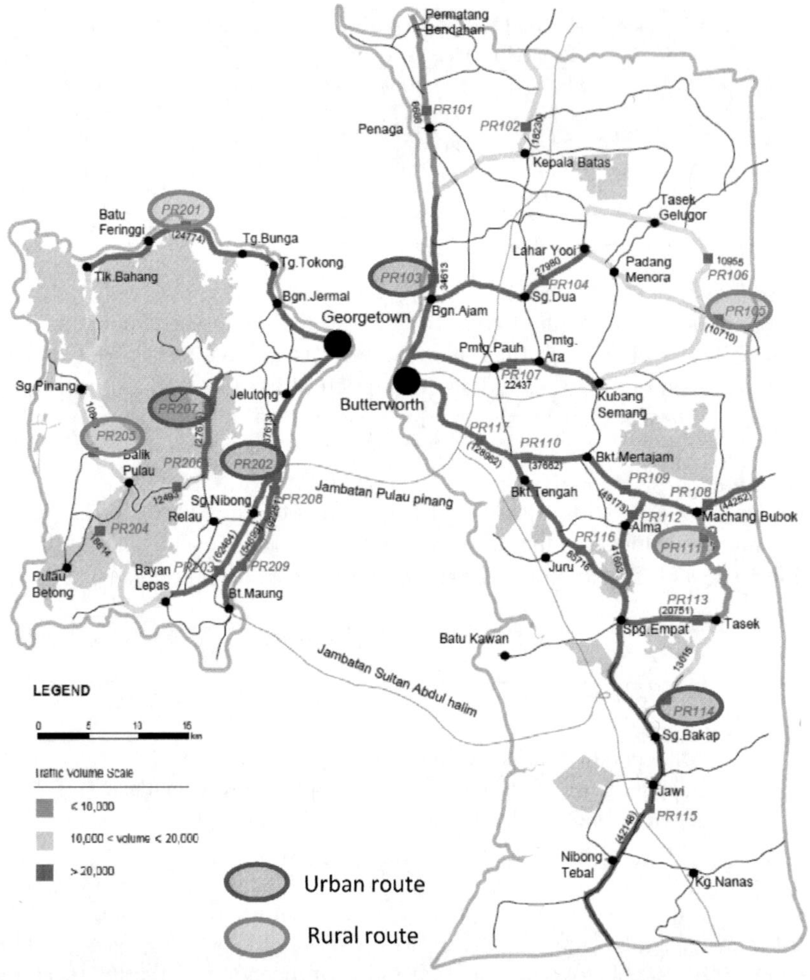

Figure 2. Location of selected urban and rural route in Penang.

Figure 3 to figure 10 show the exact location of each selected routes using Google Maps. The screenshot is taken at peak hour for each location. There are 4 colour intensities that has been introduced by Google Maps to show the congestion level of the certain routes. Green colour denotes fast traffic flow, yellow colour denotes slower traffic movement followed by red.

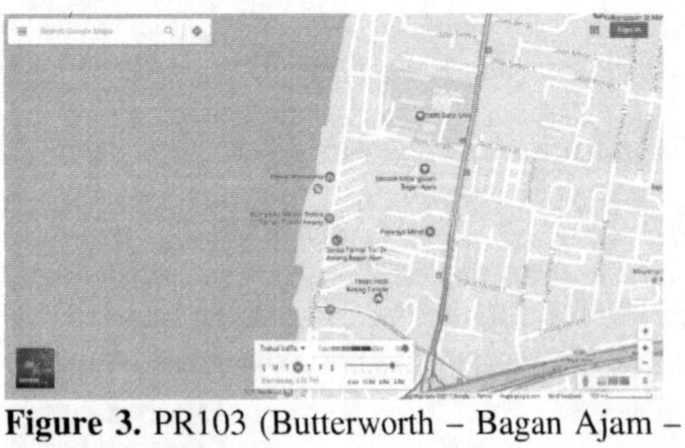

Figure 3. PR103 (Butterworth – Bagan Ajam – Telok Ayer Tawar)

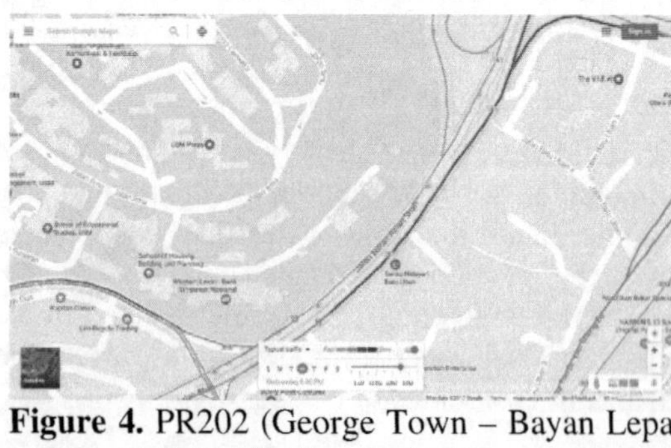

Figure 4. PR202 (George Town – Bayan Lepas – Gelugor (Jalan Keliling Pulau)

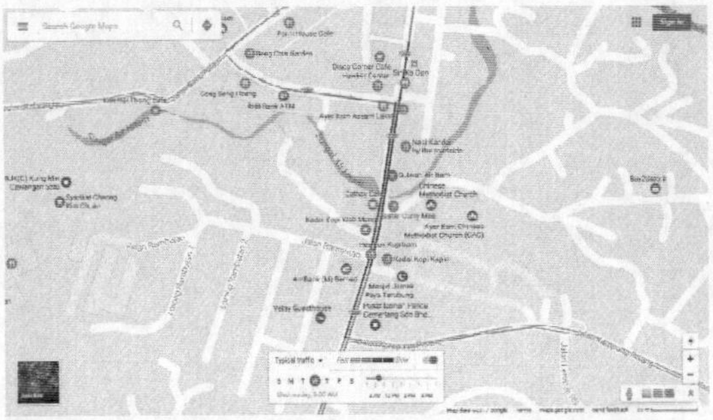

Figure 5. PR207 (Paya Terubong – Ayer Hitam)

Figure 6. PR114 (Butterworth – Tasek – Val D'or)

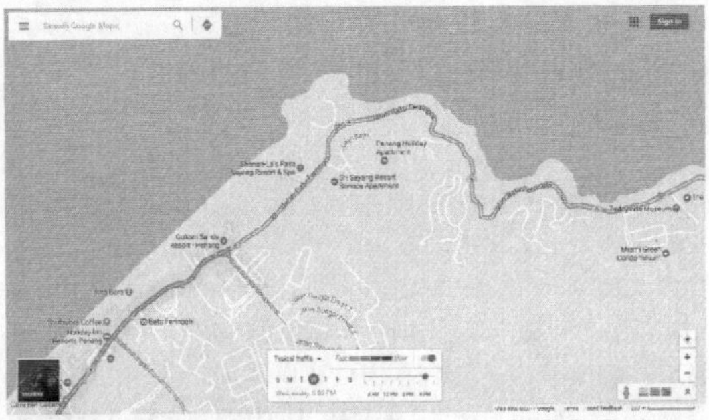

Figure 7. PR201 (George Town – Telok Bahang (Jalan Keliling Pulau)

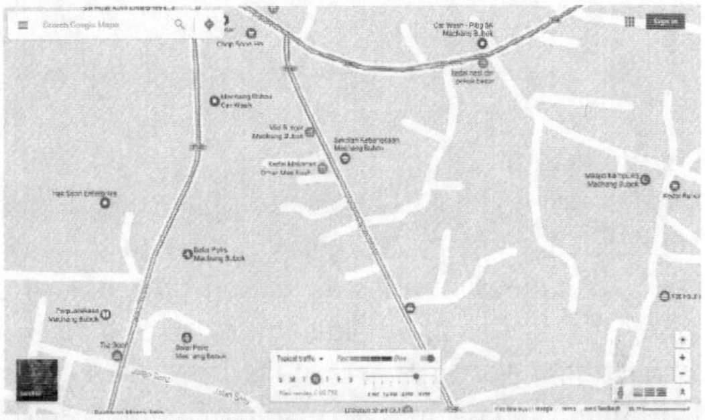

Figure 8. PR111 (Butterworth – Bukit Mertajam – Junjong (Through Machang Bubok – Tasek)

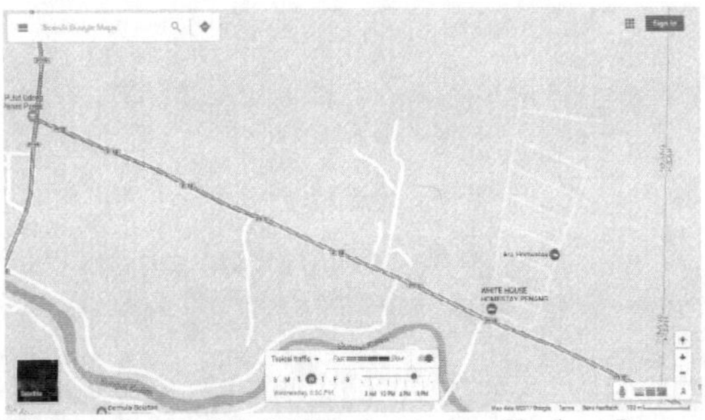

Figure 9. PR105 (100m from Ara Kuda – Lunas Junction)

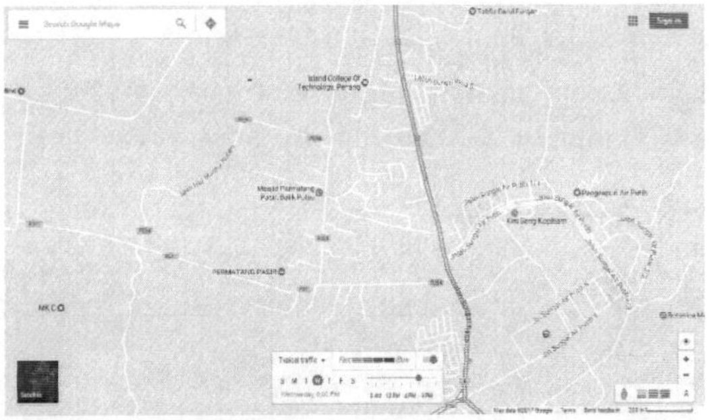

Figure 10. PR205 (George Town – Telok Bahang – Balik Pulau (Jalan Keliling Pulau)

3.3. Routes in the same network and chain

The route that has the same networks or chains must be avoided. If any, the other route that closest to the average value of ratio should be selected. Figure 11 shows the example of routes that located in the same network and chain. PR207 and PR206 are in the same network and chain, although different congestion level (differentiate by colour). Moreover, PR208 and PR209 are also in the same network and chain and in the same congestion level. These kind of routes could not be selected both. It just can be select only one of them.

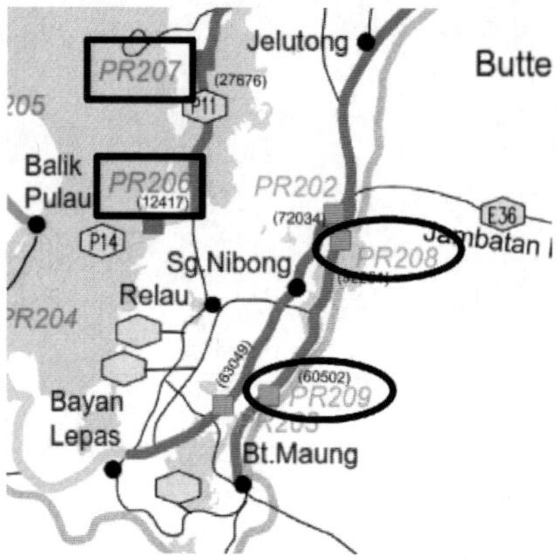

Figure 11. Example of routes in the same network and chain.

4. Conclusion

A methodology of route selection has been thoroughly described in this study. By referring to the data from JKR RTVM in 2015 specifically in Penang, the method by using ratio of peak hour to 16 hours traffic volume is developed. Based on the approach, this method is a preferable method as the ratio represents real usage of the routes during peak hours and ranking the ratio would rank the road in congestion level order. Besides, the average ratio is obtained to represents the data in normally distributed approach. The method of route selection that has been constructed in this study is very effective and could be used to determine routes to be chosen at other region for driving cycle's development.

5. Acknowledgments

The author acknowledge the financial support from Malaysia Automotive Institute (MAI) under research grant of Malaysia Driving Cycle.

6. References

[1] Brundell-Freij K and Ericsson E 2005 Influence of street characteristics, driver category and car performance on urban driving patterns *Transp. Res. Part D Transp. Environ.*, vol. 10, no. 3, pp. 213–229

[2] Lin J and Niemeier D A 2003Regional driving characteristics, regional driving cycles," *Transp. Res. Part D Transp. Environ.* **vol. 8** no. 5, pp. 361–81

[3] Hung W. T., Tong H Y, Lee C P, Ha K, and Pao L Y 2007 Development of a practical driving cycle construction methodology: A case study in Hong Kong," *Transp. Res. Part D Transp. Environ.*, **vol. 12** no. 2, pp. 115–28

[4] Kamble S H, Mathew T V, and Sharma G K 2009 Development of real-world driving cycle: Case study of Pune, India," *Transp. Res. Part D Transp. Environ.*, **vol. 14** no. 2, pp. 132–40

[5] T. Nutramon and C. Supachart 2009 Influence of driving cycles on exhaust emissions and fuel consumption of gasoline passenger car in Bangkok *J. Environ. Sci.*, **vol. 21** no. 5, pp. 604–11

[6] Tamsanya S, Chungpaibulpatana S, and Limmeechokchai B 2009 Development of a driving cycle for the measurement of fuel consumption and exhaust emissions of automobiles in Bangkok during peak periods *Int. J. Automot. Technol.*, **vol. 10** no. 2, pp. 251–64

[7] Highway Planning Unit 2016 *Road Traffic Volume Malaysia 2015* Ministry of Works Malaysia, Highway Planning Division p. 262.

IOP Conf. Series: Journal of Physics: Conf. Series **908** (2017) 012083 doi:10.1088/1742-6596/908/1/012083

Analysis of photovoltaic with water pump cooling by using ANSYS

Z Syafiqah[1], N A M Amin[2], Y M Irwan[3], M Z Shobry [2] and M S A Majid[2]

[1]Centre of Excellence for Renewable Energy, School of Electrical System Engineering, Universiti Malaysia Perlis (UniMAP), Malaysia
[2]School of Mechatronic Engineering, Universiti Malaysia Perlis (UniMAP), Malaysia
[3]Centre for Diploma Studies, Universiti Malaysia Perlis (UniMAP), Malaysia

nasrulamri.mohdamin@unimap.edu.my

Abstract: Almost all regions in the world are facing with problem of increasing electricity cost from time to time. Besides, with the mankind's anxiety about global warming, it has infused an ideology to rapidly move towards renewable energy sources since it is believed to be more reliable and safer. One example of the best alternatives to replace the fossil fuels sourced is solar energy. Photovoltaic (PV) panel is used to convert the sunlight into electricity. Unfortunately, the performance of PV panel can be affected by its operating temperature. With the increment of ambient temperature, the PV panel operating temperature also increase and will affect the performance of PV panel (in terms of power generated). With this concern, a water cooling system was installed on top of PV panel to help reduce the PV panel's temperature. Five different water mass flow rate is tested due to investigate their impact towards the thermal performance and heat transfer rate.

1. Introduction

The overall performance of PV panel is dependent on the PV panel's temperature. When the ambient temperature increases, the temperature of PV panel will rise, thus affecting the performance of PV panel. With the help of water cooling method, it can ensure that PV panel can operate at high efficiency. The cooling agent, which is water, is circulated on top of the PV panel to cool down the operating temperature of PV panel. Akbarzadeh et al. [1] investigated a hybrid photovoltaic-thermal (PV/T) system and it is found out that the output power increased by 50%. With the PV/T attached to it, the PV panel's temperature can maintain at lower than 46 °C when exposed for 4 hours. Dubey et al. [2] tested a combined system of PV panel with thermal solar water heater (PV/T) in outdoor condition of New Delhi, India. The authors studied the efficiency of PV panel under three different cases, which are absorbed of fully covered, 50% covered by a PV panel and partially covered by the PV panel. They found out that the efficiency increase from 33% to 64% when the PV panel are fully covered. Meanwhile, Kluth [3] experimented a water cooling system to increase the PV panel efficiency. The surface of the PV panel was sprayed with water by using a fan. Unfortunately, it is not an efficient method because the water was not sprayed entirely over the surface. Hence, there will be some part of PV panel that not cooled and has a high water loss.

From the literature review, it can be concluded that water a cooling mechanism is more effective than air cooling. Thus, the main goal of this paper is to build a PV panel with water cooling system due to solve the PV panel's operating temperature problem. In this paper, the PV panel with water cooling system was studied with 5 different water mass flow rates. These different water mass flow rate was then compared in terms of thermal performance (by using ANSYS CFX) and heat transfer rate.

2. CFD modelling

The geometry of PV panel with water cooling system has been drawn by using CATIA software. The dimension of the PV panel is 120 cm x 54 cm. This dimension and thickness of PV panel were based on the actual dimension. A thin layer of water is added to the surface of PV panel, which was embraced as the water cooling system. The water layer is assumed to flow on the entire surface of PV panel. Usually PV panel consists of 5 different layers, which are glass (covering the surface), Ethylene Vinyl Acetate (EVA1 which act as encapsulation to the silicon), silicon layer (device that convert sunlight into electricity), EVA2 and tedlar (at back surface). Each layer has different thickness with different specific heat capacity, density as well as thermal conductivity as shown in Table 1.

Table 1. Data for thickness of PV panel that used in 3D modelling process[4].

Materials	Thickness (m)	Specific Heat Capacity (J/kg^{O}C)	Density (kg/m^3)	Thermal conductivity (W/m^{O}K)
Water	0.0045	4180	997.0	0.59
Glass	0.003	500	3000	0.98
Eva1	0.003	2090	960	0.23
Silicon	0.004	712	2.329	148
Eva 2	0.003	2090	960	0.23
Tedlar	0.0001	1250	1200	0.36

ANSYS CFX is one of the computational fluid dynamic (CFD) software that has been used for this simulation study. The PV panel with water cooling that has been drawn in CATIA need to be imported into ANSYS CFX software by using .igs format as shown in Figure 1. A hexahedron meshing has been applied to the geometry drawing, with the skewness near 0, which is considered as a good meshing according to Hiren [5]. The PV panel with water cooling system is consists of 6 different domains with different properties, as clarified in Table 1. The water is flowing on top of surface of PV panel, assuming that the flow is laminar and steady flow.

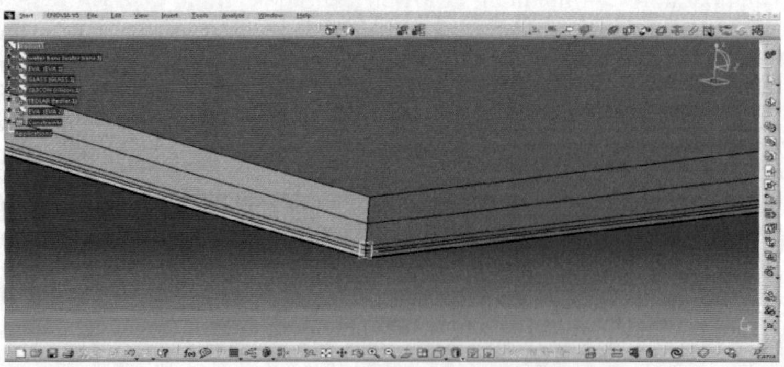

Figure 1. Geometry model of PV panel layer with water cooling system

The initial water temperature is assumed as 25 °C, while the temperature of PV panel follows the ambient temperature. The weather data, such solar irradiance (W/m^2), wind speed (m/s) and ambient temperature (°C) were based on actual data that was taken at Ulu Pauh, Perlis. The data was taken from 8.00 a.m. until 7.00 p.m. The highest ambient temperature and solar irradiance that has been recorded are 36.6 °C and 1037 W/m^2 respectively. The simulation is repeated with 6 different mass flow rates, due to compare its thermal performance and the heat transfer rate. The thermal performance will be the temperature output obtained from ANSYS CFX simulation. While, the heat transfer rate will focused on the convection heat transfer which occurs at the backside of PV panel (tedlar layer). The convection heat transfer is described as the energy transfer between a solid surface and a fluid surrounding of that surfaces [8]. If the surface (tedlar) is at a temperature T_s, and the fluid (surrounding air) has a temperature T_∞ that is lower than the surface temperature, then the equation used to describe such heat transfer is as follow:

$$Q_{conv} = hA_s(T_s - T_\infty) \quad (1)$$

This equation is known as Newton's Law of Cooling because the surface is being cooled by the fluid. If the fluid were hot and heated a cooler surface, then equation is as follow:

$$Q_{conv} = hA_s(T_\infty - T_s) \quad (2)$$

Equation (2) is known as Newton's Law of Heating. In a general form, Newton's Law of Convection can be written as follow:

$$Q_{conv} = hA_s(\Delta T) \quad (3)$$

where ΔT is the temperature difference between solid (tedlar) and the average surrounding temperature of the fluid. The area, A (m^2) is the area of the tedlar and h indicates the convective heat transfer coefficient of air (W/m^2.°C). The convection heat transfer coefficient is not a property of fluid. It is an experimentally determined parameter whose value depends on the variables that influence the convection, such as surface of geometry, nature of fluid motion and properties of fluid. The heat transfer coefficient can be obtained by using correlation below, where V is referred to the wind speed (m/s).

$$h = 2.8 + 3.8V \quad (4)$$

3. Results

The temperature of PV panel with a different mass flow rate is shown in Figure 2. It can be seen that, without a cooling system, the highest PV panel temperature is 70.2 °C at 1.20 p.m. But, with a 0.01 kg/s water mass flow rate being applied, the temperature can be reduced until 48.5 °C. There is almost 20 °C in temperature difference when the water mass flow rate is 0.01 kg/s and 0.05 kg/s. The PV panel temperature started to increase at 8.10 a.m. and slowly drop at 2.00 p.m., following the ambient temperature. It can be concluded that with increasing water mass flow rate, it helps to lower the PV panel's operating temperature.

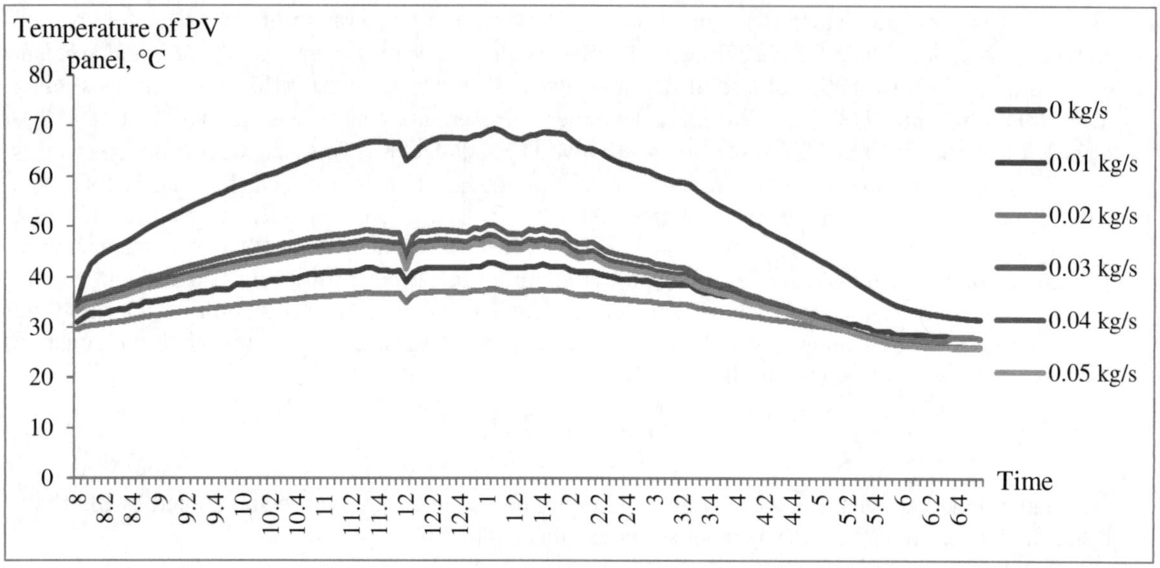

Figure 2. Temperature of PV panel with 5 different mass flow rate of water.

Meanwhile, Table 2 shows the differences in PV panel contour at a water mass flow rate of 0.01 kg/s, 0.02 kg/s, 0.03 kg/s, 0.04 kg/s and 0.05 kg/s. The PV panel contour was taken at 1.20 p.m., whereby the ambient temperature is the highest, which is 36.6 °C. PV panel surface with 0.00 kg/s (without cooling) covered with red spot the most, which indicated that it suffers from a high operating temperature. With a cooling system applied to it, the red spot slowly disappeared. The blue spot on the PV panel surface with 0.05 kg/s indicated that a lower temperature, compared with 0.01 kg/s. It is proven that, with the increase mass flow rate, it helps to reduce the PV panel's temperature slightly.

Table 2. The differences of PV panel contour at different water mass flow rate.

Mass flow rate (kg/s)	PV panel contour at 1.20 p.m.
0.00	
0.01	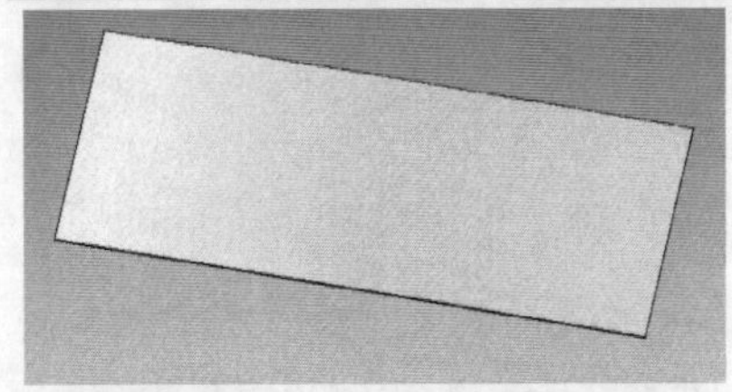
0.02	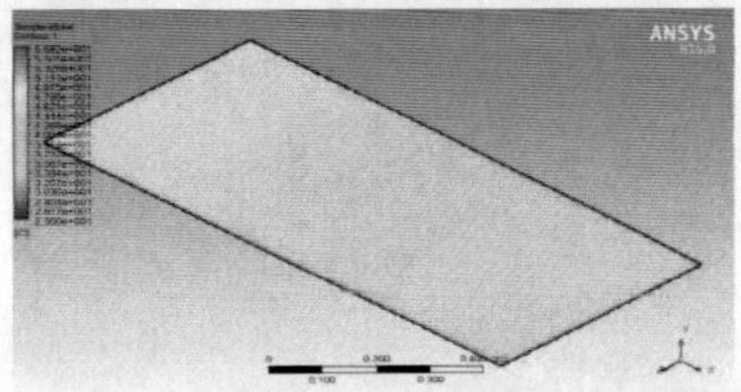
0.03	
0.04	
0.05	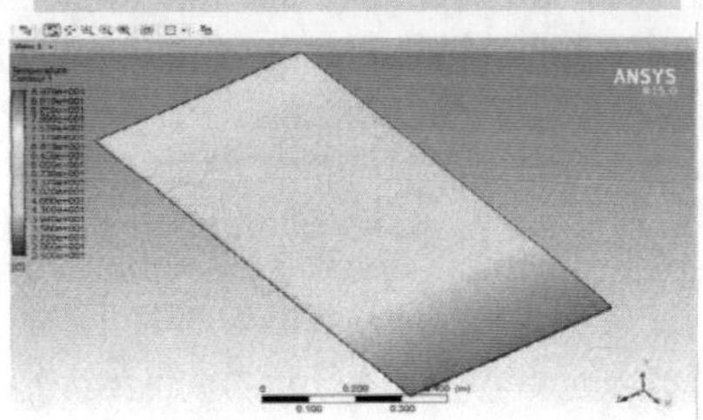

The convection heat transfer rate between tedlar and surrounding air is illustrated as in Figure 3. The heat transfer rate are varied with the increment of water mass flow rate. The heat transfer rate decreases with the increase of water mass flow rate. PV panel without cooling system recorded the highest heat transfer rate of 30 869 W. The average heat transfer rate for PV panel with 0.00 kg/s, 0.01 kg/s, 0.02 kg/s, 0.03 kg/s, 0.04 kg/s and 0.05 kg/s water mass flow rate are 18 804 W, 6 063 W, 3 236 W, 2 918 W, 2 798 W and 235 W respectively. A positive heat transfer rate indicates that the heat transfer is from a high temperature to a lower temperature, which is from surrounding air to tedlar. Meanwhile, a negative heat transfer indicates that the heat is transferred from tedlar to surrounding air, since tedlar has the low temperature. PV panel with water mass flow rate of 0.05 kg/s reach a negative heat transfer rate at 2.35 p.m., since it is the fastest in achieving a lower temperature, compared with others. With both agreement result from thermal analysis and heat transfer rate, PV panel with 0.05 kg/s is considered as the suited mass flow rate in reducing PV panel's temperature.

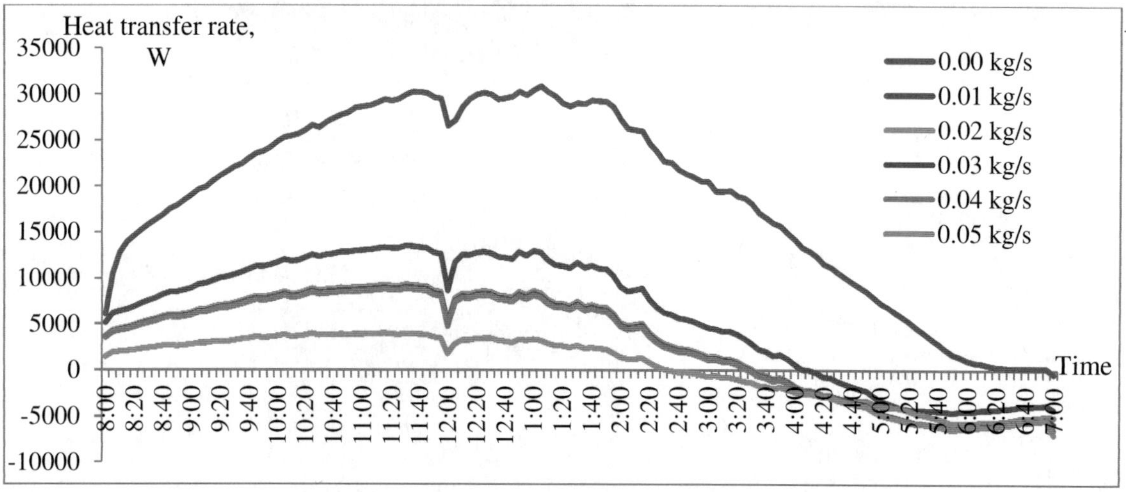

Figure 3. Convection heat transfer rate between tedlar and surrounding air.

4. Conclusion

As the conclusion, PV panel with water cooling system gives a difference in terms of thermal performance and heat transfer rate, compared with PV panel without cooling system. With the help of water cooling system, the PV panel's temperature slightly drops. The impact of water mass flow rate towards thermal performance and heat transfer rate also has been discussed. When the water mass flow rate increase, the PV panel's temperature also decreases. The convection heat transfer rate reduced greatly with the increase of water mass flow rates. PV panel with 0.05 kg/s mass flow rates eventually has the lowest average heat transfer rate of 235 W.

5. Acknowledgement

The authors would like to acknowledge the Ministry of Higher Education of Malaysia and the Universiti Malaysia Perlis, Perlis, Malaysia for the financial support under FRGS (9003-00564).

6. References

[1] Akbarzadeh A, Wadowski T Heat-pipe-based cooling systems for photovoltaic cells under concentrated solar radiation (*Applied Thermal Engineering* vol 16) pp 81-87
[2] Dubey S, Tiwari G N 2008 Thermal modelling of a combined system of photovoltaic thermal (PV/T) solar water heater (*Solar Energy* vol 82) pp 602-612
[3] Kluth A 2008 Using water as a coolant to increase solar panel efficiency (California State science fair, California, USA)
[4] Armstrong S, Hurlery WG 2010 A thermal model for photovoltaic panel under varying atmospheric conditions pp 1488-1495
[5] Hiren DR, Subama M, Ashish M 2014 Computational fluid dynamics analysis and experimental validation of improvement in overall energy efficiency of a solar photovoltaic panel by thermal energy recovery (*Journal of Renewable and Sustainable Energy* vol 6) pp 1-12
[6] Yusuf AC, Michael AB 2011 *Thermodynamic an Engineering Approach* vol 1 (McGraw-Hill) p 54
[7] Heat transfer rate equation 2017 [Online] Avaible : https://image.slidesharecdn.com/htfx-

38130030119042-151025045316-lva1-app6892/95/overal-heat-transfer-coefficient-3-638.jpg?cb=1445748995 Accessed: 20 May 2017.
[8] Kurt CR 2000 *Heat and Mass Transfer* (New Jersey:Prentice-Hall, Inc.) p 205

Thermoelectric as recovery and harvesting of waste heat from portable generator

S N Mustafa[1], N S Kamarrudin[1,2], M S M Hashim[1,2], S A Bakar[1,2], Z M Razlan[1,2], A Harun[2,3], I Ibrahim[1,2], M K Faizi[1,2], M A M Saad[1,2], I Zunaidi[2], W K Wan[1] and H Desa[1]

[1] School of Mechatronic Engineering, Universiti Malaysia Perlis, Pauh Putra Campus, 02600 Arau, Perlis, Malaysia.
[2] Motorsport Technology Research Unit (MoTech), Universiti Malaysia Perlis, 02600, Arau, Perlis, Malaysia
[3] School of Microelectronic Engineering, Universiti Malaysia Perlis, Pauh Putra Campus, 02600 Arau, Perlis, Malaysia.

saifullah@unimap.edu.my

Abstract. Generation of waste heat was ineluctable especially during energy producing process. Waste heat falls into low temperature grade make it complicated to utilize. Thermoelectric generator (TEG) offers opportunity to harvest any temperature grade heat into useful electricity. This project is covered about recovery and utilizing waste heat from portable electric generator by using a TEG which placed at exhaust surface. Temperature difference at both surfaces of TEG was enhanced with supplying cold air from a wind blower. It is found that, even at low air speed, the TEG was successfully produced electricity with aid from DC-DC booster. Results shows possibility to harvest low temperature grade heat and still exist areas for continual improvement.

1. Introduction

Power production devices, regardless of conventional or renewable always face heat loss problems, which decrease the energy efficiency. The heat loss also includes of heat energy that is rejected to the environment with lower temperature, also known as waste heat. Recently, instead of dumping waste heat into the environment, many efforts being done to reuse the lower grade energy into useful energy [1],[2]. Internal combustion engine (ICE) as a power produce device also produce waste heat as by-products, and the fact is the contribution of waste heat from ICE are not negligible. Severe environment pollution, entropy rise and greenhouse effect are conveyed when almost 60 – 70% of energy from internal combustion is released as waste heat. Organic Rankine Cycle (ORC) and Thermo-Electric Generator (TEG) are therefore the examples of a system that is applied to vehicles to increase the power efficiency[3]. Due to TEGs potential to convert heat directly to electrical energy, thermoelectric is the amongst of technology that promising alternative green technology device. In this paper, authors will deal with the potential of converting waste heat from small ICE engine to generate useful energy. Broadly implementation of this alternative green technology will extensively save a large amount of heat energy to useful electrical power and increase the utilization of energy conversion system and reduce the fuel consumption by 6% depending on the waste heat recovery technology [4].

Heat energy trapped at the exhaust of portable generators was used to simulate waste heat, since the portable engine operates at steady condition with longer running time. The main purpose of this project is to study the feasibility of converting waste heat carried out from the small ICE into useful energy. Although it is reported that at the current state, the conversion efficiency is low, this energy recovery from waste heat has a promising future potential as it is low cost and emission free technology.

2. Working Principle of Thermoelectric Generator (TEG)

Thermoelectric generator which also known as a Seebeck generator is a device that can directly convert heat to the electricity. It works when there is temperature difference existing between the two opposite surfaces of TEG, electrons will move from hot side to the cold side and produce electricity. The higher the temperature difference, the higher electricity can be produced. Conversely, the temperature difference (hot and cold at opposite surface) is created when the voltage is transfer to TEG.

3. Applying TEG to small internal combustion engine (ICE)

Recently Malaysia government encourage industry level until consumer level to explore green technology. Nevertheless, at the current state, the dependency on small ICE as a power source still shows high demand, for e.g. small engine (below 100cc) motorcycle, portable electric generator, water pump for agriculture, etc. The usage of these small cc engine however have fewer thermal efficiency, therefore, a lot of heat energy being dumped as waste heat.

Such small ICE engine wastes lower temperature of heat, especially through the engine block and exhaust pipe. Lower temperature of heat energy unfortunately carries lower grade energy. As previously mentioned, the larger temperature difference, the higher efficiency of TEG. However, dealing with small ICE means the temperature difference between waste heat and environment temperature is not relatively high compared with temperature differences of TEG that applied in the industry. This paper tends to convert even lower grade waste heat into useful energy in term of electricity.

A portable electric generator that used 4-stroke gasoline engine was used as the waste energy source due to this type of ICE is often used especially at morning market or night market in Malaysia. The TEG module was attached to the exhaust pipe of the generator to extract direct waste heat from the engine. An external fan was used to supply cold wind to the other side of TEG to enhance greater temperature differently.

The modeling of thermoelectric is based on building block. The complete thermoelectric building block can be divided into three major parts which are a thermoelectric material as a medium of electric current flow, an aluminum plate to absorb waste heat from exhaust plate, a heat sink to increase cooling effect from external wind. The thermoelectric building block is build using 66mm × 76mm heat sink, heat sink Compound ZP - 360 and a 10 watt 5.6 volt with 44mm × 40mm thermoelectric power generation Peltier module, as shown in figure 1.

Preliminary experiment result shows that temperature of the exhaust surface was around 150℃ and commercial TEGs module usually able to operate below 220°C. Even though the exhaust surface temperature was in the range operation of TEG, as a precaution step an aluminum plate is added to reduce direct contact temperature while preserving the TEGs from malfunction [5]. The 66mm × 76mm aluminum plate was sandwiched between TEG module and exhaust pipe surface. The TEG was inserted between the aluminum plate (hot surface) and heat sink (cold surface). To increase the efficiency of heat transfer, heat sink compound was applied between every contact surface.

IOP Conf. Series: Journal of Physics: Conf. Series **908** (2017) 012084 doi:10.1088/1742-6596/908/1/012084

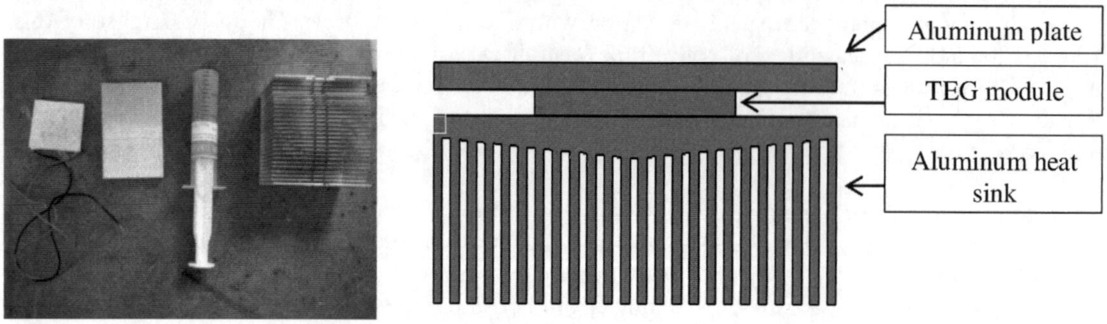

Figure 1. (Left) Actual part and (Right) sub part of the TEG module assembly and thermoelectric building block

4. Jig design

Due to simplified design, almost small ICE face higher vibration problems at the whole body when it operates, which leads to difficulties to ensure the TEG module always intact to the exhaust pipe even though the contact surface area is larger enough. Consequently, a custom TEG module mounting was designed to hold TEG building block tightly on the exhaust contact surface in this module. The jig was design so that able to hold the TEG module without existing of any gap between the TEG hot surface and the aluminum plate surface. The jig was fabricated using 0.25 inch steel plate which joined by welding process.

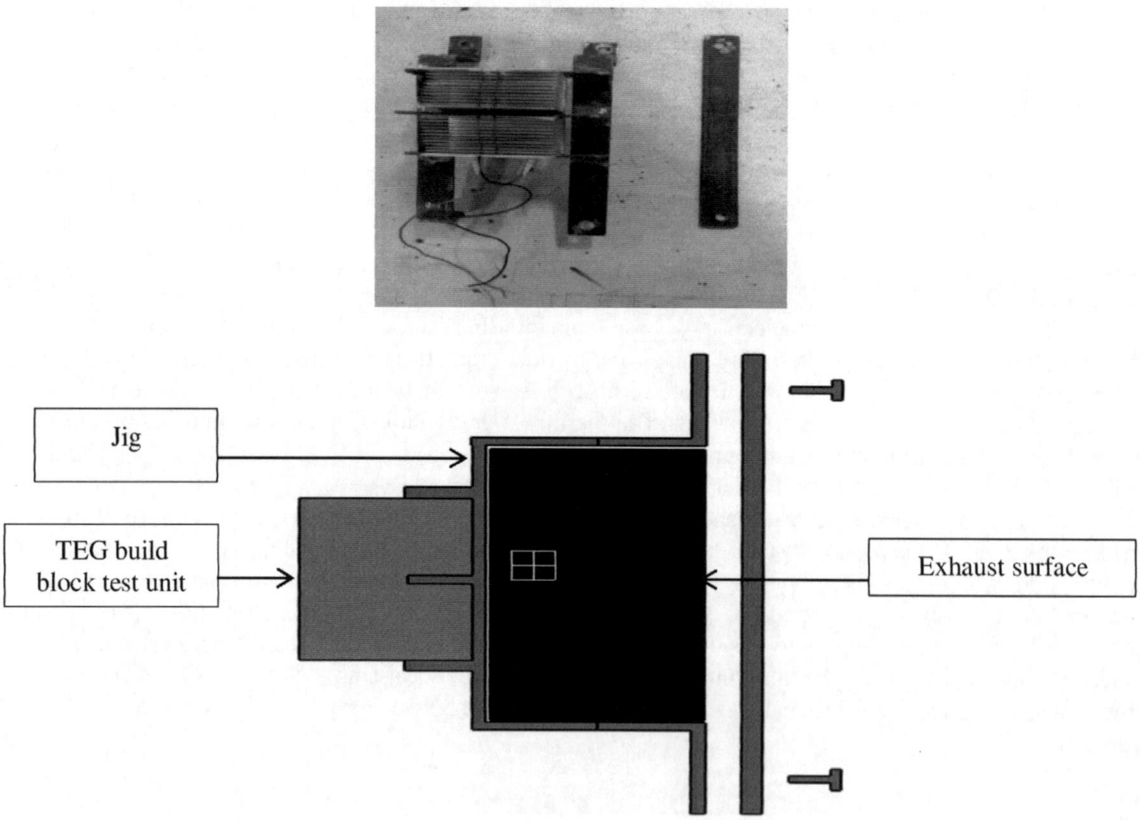

Figure 2. (Top) Actual jig, (Bottom) Jig design that used in the experiment

IOP Conf. Series: Journal of Physics: Conf. Series **908** (2017) 012084 doi:10.1088/1742-6596/908/1/012084

5. Experimental setup – Effect of wind towards TEG

Effect on temperature difference to the harvested energy was investigated in this section. It is important to face the wind direction toward the heat sink fin of the thermoelectric module. Figure 3 shows experiment setup which consists of a portable engine, a small wind tunnel and a wind blower. Multimeters was placed near the engine. The wind tunnel is fabricated using the Galvanic plate which have 508 mm long and has a 635 mm in diameter. The wind tunnel is designed to wrap around the heat sink fins to ensure cooled air supplied from air blower reached the heat sink. The air blower that has three variable control speeds was used as a wind source with ambient temperature. The speed of air was measured using digital anemometer (MASTECH MS6252B). It is found that the air velocity range of wind that breeze the cold area of the heat sink is around 4.44 m/s to 4.69 m/s. For each experiment, the engine was run steadily and data was collected within 15 min. The surface heat temperature data was collected using an infrared thermometer (EXTECH VIR50) and the data of the voltage generated by TEGs and booster was recorded using digital multimeters. For each air velocity, the experiment was repeated three times and data presented in this paper were the average value from collected data. After one experiment had completed, the portable engine will be shut down to allow the temperature of exhaust surface cooled to ambient temperature.

Figure 3. Experiment setup for the effect of wind test

6. Result and Discussion

6.1Exhaust Surface Temperature

The waste heat source in this study was origin from exhaust surface and for experiment purpose, exhaust surface temperature were set to be as the hot temperature (T_h). Exhaust temperature increases up to 153℃ after 570 seconds. As mentioned earlier, the aluminum plate was placed in between exhaust surface and TEG module, therefore it was expected that the highest temperature for the aluminum surface is only at 75℃. This step was necessary to avoid over heat temperature that will cause permanent damage to TEG module. Therefore, for the temperature difference ΔT, temperature of the aluminum plate will be used instead of the temperature of exhaust surface.

6.2 *Effect of wind speed to the temperature different.*

The air velocity was varied from 4.44 m/s to 4.69 m/s. This air was directed to cooling fin which attaches at cold side (T_c) of TEG module. Temperature ambient however not considered in this study. Figure 4 shows the development of surface temperature at the hot side of TEG module. At early of engine running, temperature of aluminium plate was increase at a different rate for each air speed, and reach maximum temperature between 500 s to 600 s. After that period, the temperature at aluminum plate was observed to be maintained throughout the experiment. From the Figure 4, effect of air speed were significant from air speed 1 and air speed 3. It was confirmed that although the difference of air speed 1 and air speed 3 was only about 0.25 m/s, however, the effect of the temperature successfully showed difference of maximum temperature about 20ºC. Overall, even with air speed 1, the temperature different between T_h and T_c was successfully increase which will affect the increment of electricity generated by TEG module. This experiment also shows same behaviour with experiment conducted by Alexander et al [6], which applying TEG on exhaust surface of a motorcycle.

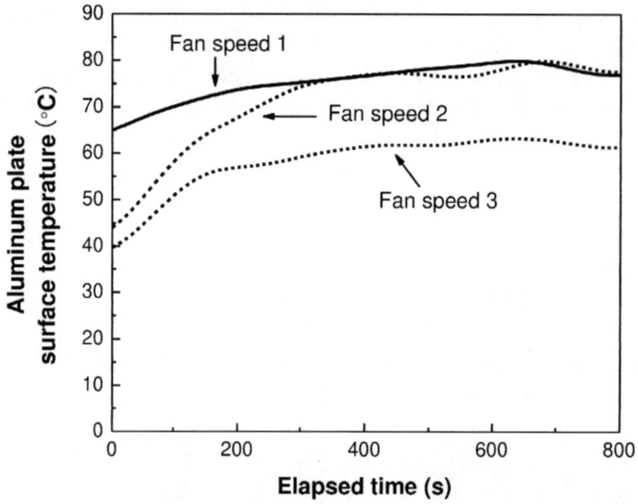

Figure 4. Effect of air velocity to the development of average surface temperature of aluminium plate

Figure 5 shows voltage, current and power that were harvested from experiment based on temperature differences between T_h and T_c. It is obvious that harvested voltage, current and power from TEG module was minimal, therefore it need to boost up in order to get better electrical supply. A DC-DC converter was applied as a booster, and Figure 6 shows the voltage, current and power that being boost up. After the temperature difference increase more than 80°C, the DC-DC converter was able to boost up voltage to the maximum value, 5V. As for this study, it needs three minutes from engine start to achieve temperature different to be 60ºC. The current and power generated for both TEG and booster is calculated using resistance value of 3.07 ohm. It was found that the maximum voltage, current and power generated by TEG is 1.19V, 0.39 amp and 0.46 watt, respectively. For the booster the maximum electric harvest is 5.01V, 1.63 amp and 8.18 watt, respectively.

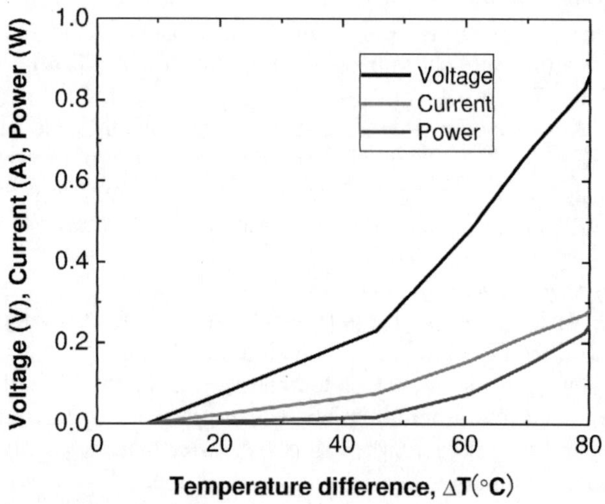

Figure 5. Voltage, current and power harvested from TEG module

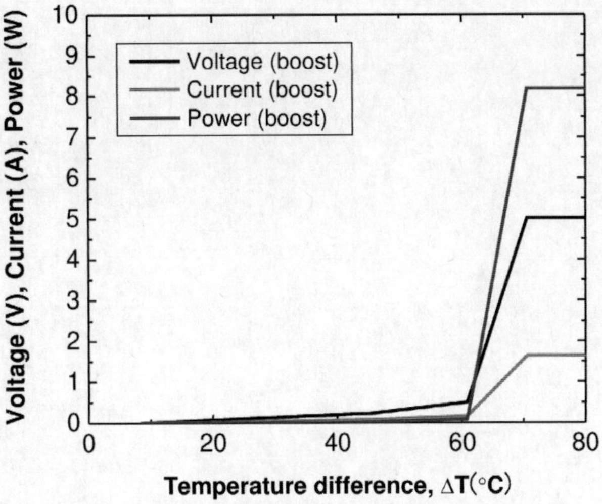

Figure 6: Voltage, current and power output being boost up by a DC-DC converter

7. Conclusion

In this study, we have been successfully developed waste heat recovery power generator by using thermoelectric building block. Waste heat from the exhaust surface of the portable generator had been utilized and recovered and converted to electrical power by using the single thermoelectric generator (TEG) module. Objectives of this study achieved by looking at the highest voltage that harvest by TEGs was 1.19V, the maximum electric current is 0.39 amps and the maximum power output is 0.46 watt. Maximum voltage after boost up by a DC-DC converter was 5V, maximum current was 1.63 amp and maximum power output is 8.18 watt. An aluminum plate was placed between exhaust surface and TEG module which may contribute to the results of lower harvesting current. In next study, TEG may be applied directly to exhaust surface to increase temperature differently and increase possibility of harvesting waste heat. Vibration from small engine makes us difficult to keep TEG stay intact with exhaust surface, therefore need to be taken care in the next study. For future project, fuzzy logic was proposed to enhance DC-Dc converter as presented in paper [7].

8. References

[1] Zorbas K T, Hatzikraniotis E and Paraskevopoulos K M 2007 *In Proc. of 5th European Conference on Thermoelectrics* **8** 292-8
[2] Peter A J D, Balaji D and Gowrishankar D 2013 *IOSR Journal of Engineering* **3** 01-04.
[3] Sharma A and Kaur S G K, 2014 *GJESRM* **1** 14-28
[4] Arnaud L, Ludovic G, Mouad D, Hamid Z, and Vincent L 2014 *J. Energies* **7** 5273-90
[5] Martins J, Brito F P, Goncalves L M and Antunes J 2011 *In SAE 2011 World Congress & Exhibition*, Detroit 1-23
[6] Schlichting A D, Anton S R and Inman D J 2008 *In Proc. of SPIE* **6930** 69300B-8
[7] Raghavendran P S and Asokan R 2016 *AJRSH* **6** 207-218.